AF588433

Foundations of Solid State Physics

Foundations of Solid State Physics

Dimensionality and Symmetry

Siegmar Roth
David Carroll

Authors

Prof. Siegmar Roth
MPI für Festkörperforschung
Heisenbergstr. 1
70569 Stuttgart
Germany

Prof. Dr. David Carroll
Wake Forest University
Department of Physics
100 Olin Physical Laboratory
NC
United States

Library of Congress Card No.:
applied for

British Library Cataloguing-in-Publication Data
A catalogue record for this book is available from the British Library.

Bibliographic information published by the Deutsche Nationalbibliothek
The Deutsche Nationalbibliothek lists this publication in the Deutsche Nationalbibliografie; detailed bibliographic data are available on the Internet at <http://dnb.d-nb.de>.

Print ISBN: 978-3-527-34504-5
ePDF ISBN: 978-3-527-81656-9
ePub ISBN: 978-3-527-81658-3
oBook ISBN: 978-3-527-81659-0

Cover Design Adam-Design, Weinheim, Germany
Typesetting SPi Global, Chennai, India

10 9 8 7 6 5 4 3 2 1

Contents

Preface

There are a great many textbooks on solid-state physics, condensed matter physics, or materials physics. Each has specific points, perspectives, or *foci*, that is, a purpose and an audience. There are, in fact, so many topics and principles that *could* be covered in the broad field of solid-state physics, and it is surely impossible to be comprehensive for a single, accessible text. Thus, each *school of thought* must choose its own emphasis areas for its students from semiconductor physics to soft condensed matter, and that is what we have done here.

In *Foundations of Solid State Physics*, we have presented what is essential for us, the authors, in the field of emerging, exotic, novel materials. The reader will quickly notice *our* passion for *molecular solids* and carbon-based systems and the phenomena associated with them. Conducting polymers, carbon nanotubes, nanowires/nanoparticles, two-dimensional plates of dichalcogenides, perovskites, and organic crystals are systems understood largely through their dimensionality, topological connectedness, and quantum confinements. So studies of these materials expand our most fundamental solid-state models, *and* they offer to us the basic challenge of connecting to deeper physical insights. In our writing, we have tried to embrace that invitation and challenge. We intend our text for the advanced undergraduate or beginning graduate student. But researchers with interests in the areas of dimensionality in solids, organic or molecular electronics, and molecular materials should also find our perspective enjoyable.

We have chosen an unusual presentation style for the text. It is conversational, and throughout the text there are *italicized* words and concepts. We intend these as *focus points* where we want the reader to go outside of the text to supplement their understanding of the concepts. So be ready when we return to these points again and again. There are also a number of graphical components and historical references intended to give discoveries, old and new, the context of their origins. Finally, we expand upon specific topical areas through the use of open-ended exercises. Not surprisingly we have encouraged the reader to look through the references on which the exercises are based and therefore engage with the original authors of the work. We hope our readers find this *engagement approach* mentally stimulating, challenging, and fun. Think of the old adage from Ben Franklin: "Tell me and I may forget, teach me and I may remember, involve me and I learn."

Some may prefer to skip through the mathematical details, problems, and references in a diagonal way to get to the physical models quickly. We believe that the text has been laid out in such a way as to accommodate this style of reading as well. However, as a textbook, the presentation is intended as a two-semester detailed discussion of the world of solid-state physics. Our core premise is that solid-state physics is as fundamental in its nature as any field of physics, with unique models and explanations of reality. Understanding these models and explanations brings us ever closer to understanding the universe in its deepest complexities.

In preparing the text, two desks, one in Munich and one in Winston-Salem, were filled with dozens of reference books. Of these we found that there was a subset we particularly enjoyed, and we used them (coupled with experiences in our labs, our own publications, and journal articles from outside our research groups) to form an outline of our presentation. Some of these texts are getting pretty old by now, and each expresses unique perspectives and passions for the field. But it is always useful to see how others frame things.

1. *Kittel: Introduction to Solid State Physics*, now in its eighth edition. This is the truth as it was revealed at UC Berkeley, wonderful for building a pedagogical understanding at the most fundamental level using elementary models. This book is simply hard to put down. Published by Wiley.
2. *Ashcroft and Mermin: Solid State Physics* first edition. While Kittel may be seen a bit as "Moses on his mountain," this text is the truth as it is known in Ithaca. And it is frequently associated with things far more devilish. With more than 800 pages of electrical and optical properties in solids and one of the first texts to categorize the different models of electron behavior in a crystal, this text provides exquisite detail for every detail you might think of. It is a *must read*. Published by Brooks Cole.
3. *Ibach and Lüth: Solid-State Physics: An Introduction to Principles of Materials Science* now in its fourth edition, a more modern compilation of solid-state

physics with plenty of experimental examples. This laboratory-centered treatment is a favorite in many German universities. It certainly doesn't take long to see why. Published by Springer.

4. *Chaikin and Lubensky*: *Principles of Condensed Matter Physics*, a *tour de force* of thermodynamics in the solid state. These authors make the daring leap of dealing with novel systems in a fundamentally different way and help to define many aspects of modern solid-state physics. From soft condensed systems to liquid crystals and to phase transitions, you will find the foundations here. Published by Cambridge University Press.
5. *Harrison*: *Solid State Theory*, the quantum chemistry of hybridization. The focus of this treatment is on how specific bonding characters arise in crystals. Special emphasis is given to the spatial mapping of bonds within the solid and how they become bands. It is especially important for people studying semiconductors or oxides. Published by Dover.
6. *Marder*: *Condensed Matter Physics second edition*, a graduate-level introduction that has gained rapid acceptance. This text has focused on basic calculation approaches to a wide range of physical phenomena in solid-state physics. Though relatively young, the text is already a classic. Published by Wiley.

Our text started as a fourth edition of the now well-known *One-Dimensional Metals* (ODM) by S. Roth in 1995 by VCH. Over the many years of teaching this material to undergraduates and graduate students at Wake Forest University, we have filled the margins of numerous copies of ODM with ideas, problems, and notes. All of these are penciled in during conversations with each other and with students. So it soon became apparent that ODM was set to evolve into more of a textbook presentation and so came the current text. It has remained important for us both, as authors, to retain the style, humor, and ease of access of that first text. This reflects who we are as scientists and as people. But it is also necessary to recognize the comments and thoughts of students at Wake Forest University and the Max-Planck-Institut für Festkörperforschung in Stuttgart, postdocs, and technical staff at both institutions, as well as our many colleagues that have read through sections of the text. For better or worse, their words and ideas are reflected in its pages as well.

It takes a long time to write a book even when there are two people doing it! This always means there is one group that should receive the most credit for its completion, and that group is our families. Thank you Richard, Jiangling, Lauren, and Melissa for supporting us in this endeavor. Without such families as you, textbooks would rarely be written at all.

Munich and Winston-Salem 2019 *Siegmar Roth and Dave Carroll*

1

Introduction*

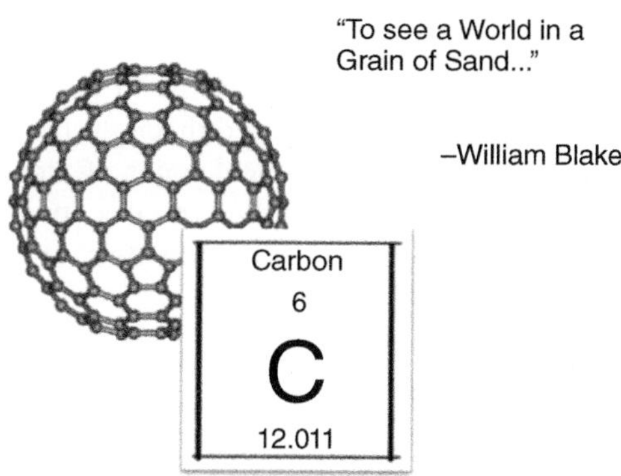

"Dimensionality" and "atomic ordering in finite structures" seem like rather odd principles by which to organize thoughts on solid-state physics. Indeed, this is not a historical approach to understanding solids at all. However, in learning *solid state* today, we must embrace the historical orthodoxy of crystal lattices, phonons, and band structure, as well as a whole *zoo* of emerging exotic materials that range from fullerenes to organic superconductors.

How do we understand two-dimensional (2D) dichalcogenides, atomically layered permanent magnets, perovskites, topological insulators, conducting polymers, quantum dots, graphene, glassy carbon, etc.? And what of the low-dimensional analogues of orthodox collective behavior: charge density waves, excitons, spin waves, and the like? We know these things "live" in/on such low-dimensional structures. An interesting and instructive way to build a framework is to begin with the normative behavior of a special atom, *carbon*, and the *dimensionality* of the structures it makes. Why carbon? Because among the elements it is about the most robust at making compounds and structures. It is extremely flexible in how it chooses to arrange itself. Why dimension? Well, lower-dimensional materials offer new approaches to technology, holding

*Historical Note: some of the hand-drawn images of the text have their origins in the very first edition of *One-Dimensional Metals*. They are an interesting and important reminder of what our state of mind was at a time when *dimensionality* was a new and mysterious science.

Foundations of Solid State Physics: Dimensionality and Symmetry, First Edition.
Siegmar Roth and David Carroll.

the key to everything from quantum computers to new medicines. But most importantly, it introduces the idea of "topology."

Look, the traditional story goes like this. We begin our description of solids with an infinite mathematical construction (the lattice) given by specific point group symmetries. Onto the lattice points we attach some arbitrary set of atoms (generally picking something found in nature). We calculate specified properties based upon idealizations of how free the electron may be at each lattice point or how free the motion of the atom at the lattice point may be. We *adiabatically* add interactions between vibrations, carriers, etc. of the lattice to get more interesting phenomena.

Our story, though, is like a tale of *die Brüder Grimm*[1]: carbon is the central atom of the universe.[2] It forms more compounds in more ways than any other atom. Thus, other atomic systems deviate from carbon by breaking its norms of symmetry. Beginning with large carbon molecules, we form nanometer structures. As we add, subtract, or substitute C atoms in the structure, we design materials with properties that can be examined through the dimensional change we have brought out. It isn't quite a chemical point of view, and it isn't quite solid-state physics in its purest form. It is the type of conversation you hear in working research labs across the world: complementary and an *enjoyable compromise* between the perspectives.

1.1 Dimensionality

The concept of *dimensionality* has been with us for a while, and it is an intellectually appealing concept. Speaking of a dimensionality other than three will surely attract some attention. Some years ago it was fashionable to admire physicists who apparently could "think in four dimensions" in striking contrast to Marcuse's *One-Dimensional Man* (Figure 1.1) [1]. Physicists would then respond with the understatement: "We only think in two dimensions, one of which is always time. The other dimension is the quantity we are interested in, which changes with time. After all, we have to publish our results as two-dimensional figures in journals. Why should we think of something we cannot publish?"

This fictitious dialogue implies more than just sophisticated plays on words. If physics is what physicists do, then in most parts of physics there is a profound difference between the dimension of time and other dimensions, and there is a logical basis for this difference [2]. In general, the quantity that changes with time and in which the physicist is interested is some *intrinsic* property of an object.

1 The Brothers Grimm wrote fairy tales in the southern part of Germany around the early 1800s. In 1812 they published their first collection of folk tales, Kinder- und Hausmärchen (Children's and Household Tales). Their hometown was only a short drive from the author's laboratory at the borders of the Black Forest (where many of their tales were set). They are responsible for almost as many nightmares as organic superconductors!

2 This is clearly a biased and self-indulgent statement, and should only be taken metaphorically. Si-based life forms would certainly have a different opinion. Note that when we say "Si-life" we do not distinguish between that life based on a processor and life based on Si-regulated metabolic mechanisms.

Figure 1.1 Marcuse's man. Simultaneously with Herbert Marcuse's book *One-Dimensional Man*, which widely influenced the youth movement of the 1960s. W.A. Little's paper on "Possibility of Synthesizing an Organic Superconductor" was published, motivating many physicists and chemists to investigate low-dimensional solids.

The object in question is typically imbedded in a three-dimensional (3D) space. Objects themselves, however, may be very flat such as flounders, saucers, or oil films with greater length and width than thickness. In materials such as graphene or MoS_2, thickness can be negligibly small – atomic. Such objects can be regarded as (approximately) 2D. Now, if the intrinsic property that the physicist wishes to study is somehow constrained in behavior, in direct correlation to the dimension of the object, like a boat on the 2D surface of the sea is hopefully constrained to 2D motion, then we say the property is *expressing the dimensionality* of the object. In our everyday experience one-dimensional (1D) and 2D objects and 1D and 2D constraints are more common than you might think. Indeed, low dimensionality should not be particularly spectacular to our expectations. For this reason too, it is reasonable to introduce non-integer, or *fractal*, dimensions [3]. Not much imagination is necessary to assign a dimensionality between one and two to a network of roads and streets – more than a highway and less than a plane. It is a well-known peculiarity that, for example, the coastline of Scotland has the *fractal dimension* of 1.33 and the stars in the universe that of 1.23.

Solid-state physics treats solids both as objects and as the space in which objects of physics exist, e.g. various silicon single crystals can be compared with each other, or they can be considered as the space in which electrons or phonons move. The layers of a crystal, like the *ab*-planes of graphite, can be regarded as 2D objects with interactions between them that extend into the third dimension. But these planes are also the 2D space in which electrons move rather freely. Similar considerations apply to the (quasi) 1D hydrocarbon chains of conducting polymers.

1.2 Approaching Dimensionality from Outside and from Inside

There are two approaches to low-dimensional or *quasi*-low-dimensional systems in solid-state physics: geometrical shaping as an *external approach* and increase of anisotropy as an *internal approach*. These are also sometimes termed *top-down* and *bottom-up* approaches, respectively. For the external approach, let us take a wire and draw it until it gets sufficiently thin to be 1D (Figure 1.2). How thin

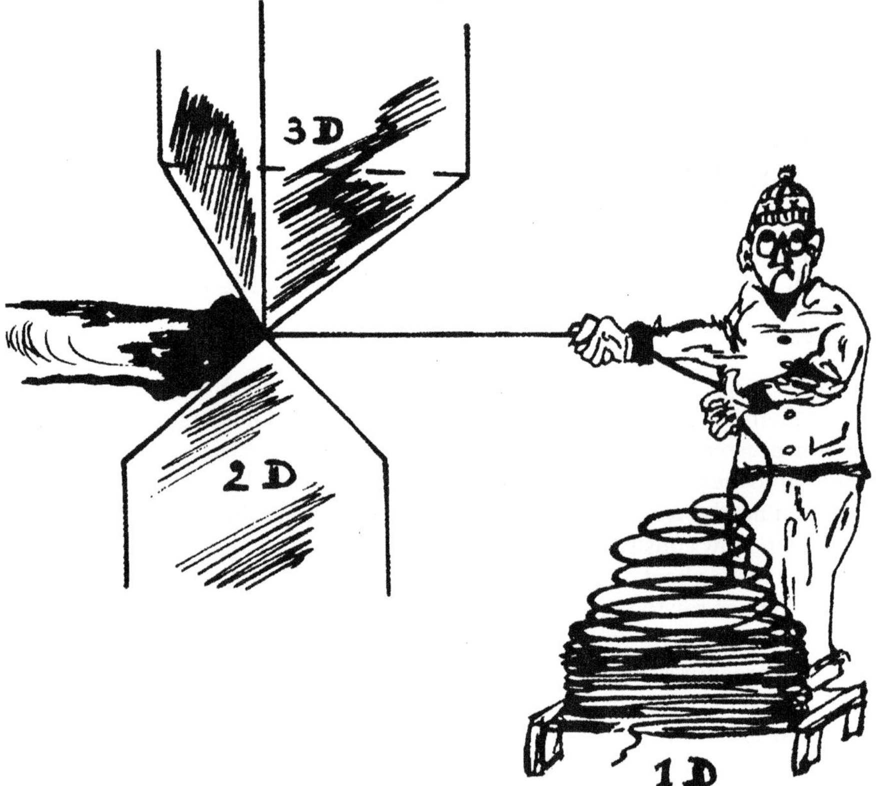

Figure 1.2 Wire puller. An "external approach" to one-dimensionality. A man tries to draw a wire through a mandrel until it is thin enough to be regarded as one-dimensional. Metallic wires can be made as thin as 1 μm in diameter like this, but this is still far away from being one-dimensional. Lithographic processes using focused ion beams and focused electrons can produce some metal and semiconductor structures that are narrow enough to exhibit one-dimensional properties (~nanometers).

will it have to be to be truly 1D? This depends a little on exactly what property of the structure is desired to express low-dimensional behavior. Certainly, thin compared to some microscopic parameter associated with that property. For example, for 1D electrical transport properties, the structure must have length scales such that the mean free path of an electron or the Fermi wavelength is affected by the physical confinement of the structure. We will discuss these concepts further a little later on in the text. But surely the meaning is clear: some fundamental aspect of an internal object responsible for the phenomenon of interest must be dramatically altered by its localization within the structure.

Technology today has made it possible to approach such sizes using methods of *lithography* as well as *chemical assembly*. Lithography is the top-down approach to creating confining structures as it whittles away material until only very small structures remain. Chemical assembly is the "bottom-up" approach, and it forms the structure through chemical reactions. The two approaches offer very different properties to the nanoscale structure created, both in terms of atomic ordering and control over object placement.

To achieve "one-dimensionality" does the wire puller in Figure 1.2 have to draw the wire so extensively that it is finally to become a monatomic chain? Well, the *Fermi wavelength,* a fundamental property of the carrier electron responsible for conductivity, becomes relevant when discussing the eigenstates of all the electrons of the structure. If electrons are confined in a box, quantum mechanics tells us that the electrons can have only discrete values of kinetic energy. The energetic spacing of the eigenvalues depends on the dimensions of the box – the smaller the box the larger the spacing (Figure 1.3):

$$\Delta E_L = h^2/2m(\pi/L)^2 \tag{1.1}$$

where ΔE_L is the spacing, L is the length of the box, m is the mass of the electrons, and h is Planck's constant. For a box containing multiple electrons, the *Fermi level* is the highest occupied energy state (at absolute zero). The wavelength of the electrons at the Fermi level is called the *Fermi wavelength.* At finite temperatures, if the energy difference between levels is much larger than the thermal energy ($\Delta E_L \gg kT$), there are only completely occupied and completely empty levels (not accounting for spin). A thin wire is a small box for electronic motion perpendicular to the wire axis, but it is a very large box for motions along the wire. Hence, in two dimensions (radially), it represents an insulator, and in one dimension (axially), it is a metal! This is simply because the $\Delta E_{\text{radially}} \gg kT$ whereas $\Delta E_{\text{lengthwise}} \ll kT$.

If there are only very few electrons in the box, the Fermi energy is small and the Fermi wavelength fairly large. For *real* materials, these are the electrons that can participate in bonding–antibonding orbitals. This is the case for semiconductors at very low doping concentrations. Wires of such semiconductors are already 1D if their diameter is on the order of hundreds of Ångstroms.

Such thin wires can be fabricated from silicon or from gallium arsenide by lithographic techniques, and effects typical for 1D electronic systems have been observed experimentally [4]. Systems with high electron concentrations have to be considerably thinner if they are to be 1D. It turns out that for a concentration of one conducting electron per atom, we really need a monatomic chain!

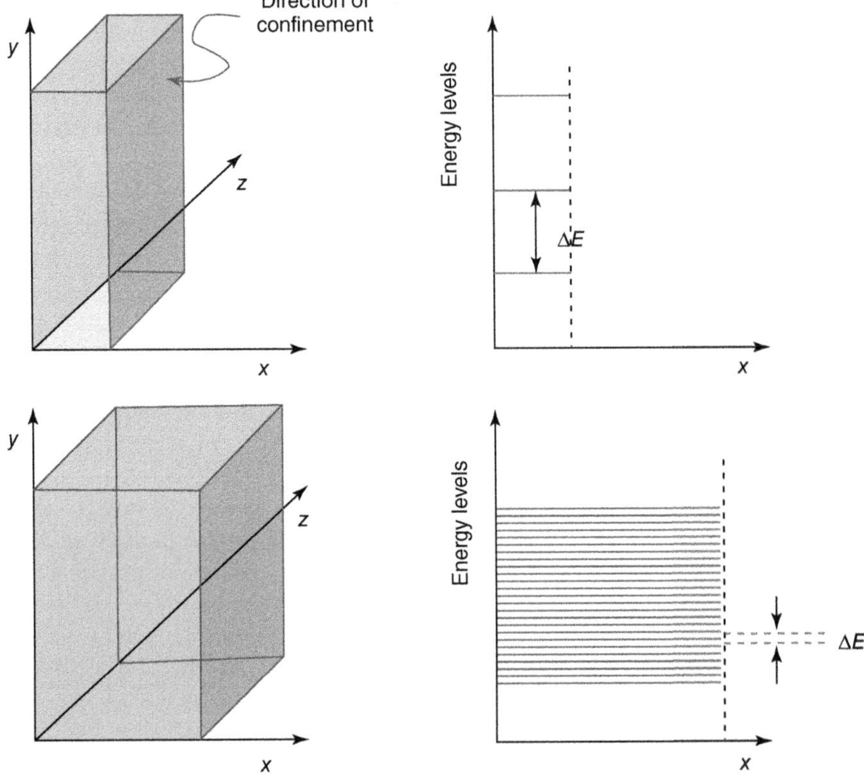

Figure 1.3 Electrons in small and large boxes and energy spacing of the eigenstates. This is an example of dimension based on confinement.

Experiments on single monatomic chains are very difficult to perform, so a bundle of chains is usually used. An example of such a bundle is polyacetylene fiber, consisting of some thousands of polymer chains, closely packed with a typical interchain distance of 3–4 Å. Certainly there are *some* interactions between the chains; however, in the case of small *interchain coupling*, it can be assumed that the net sum of the individual chains determines the properties of the bundle (Figure 1.4). The experiment becomes one of an *ensemble* of 1D chains.

Another method of geometrical shaping employs surfaces or interfaces (Figure 1.5). The surface of a silicon single crystal is an excellent 2D system, and there are various ways of confining charge carriers to a layer near the surface. Actually, the physics of 2D electron gases are an important part of today's semiconductor physics [5], and most of the 2D electron systems are confinements to surfaces or interfaces. The most fashionable effect in a 2D electron gas is the quantized Hall effect or von Klitzing effect [6]. A 1D surface, i.e. the edge of the crystal, is much more difficult to prepare and hardly of any practical use. But one can argue that exposing a sample to a magnetic field would be an excellent example of a 1D electronic system since electrons can be forced into motion along specific paths defined by the crystal and the field. In fact,

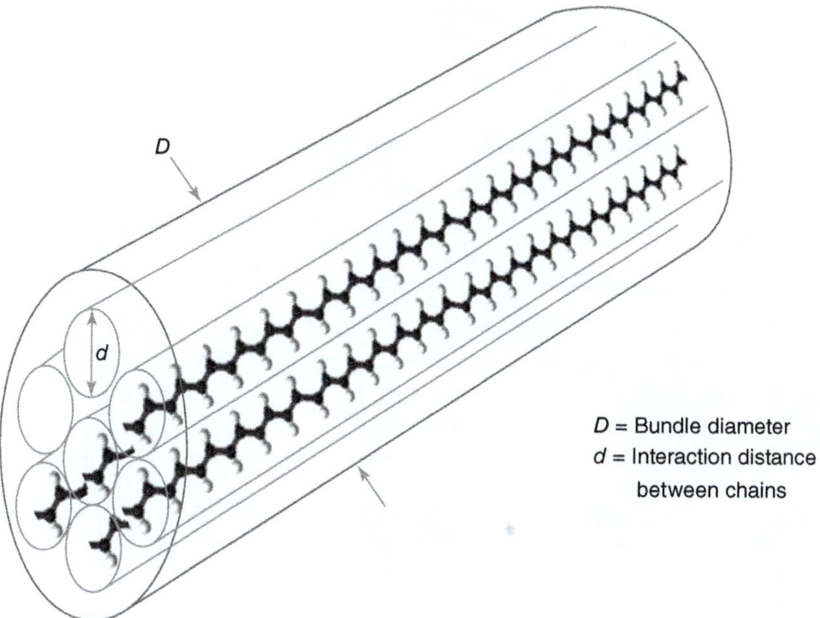

Figure 1.4 Experiments on individual chains are difficult to perform. But bundles of chains are quite common, for example, fibers of polyacetylene.

reducing von Klitzing's sample to "edge channels" is one way of explaining the von Klitzing effect [7].

The *internal approach* to 1D solids comprises the gradual increase of anisotropy. In crystalline solids the electrical conductivity is usually different in different crystallographic directions. If the anisotropy of the conductivity is increased in such a way that the conductivity becomes very large in one direction and almost zero in the two perpendicular directions, a nearly 1D conductor will result. Of course, there is no simple physical way to increase the anisotropy. However, it is possible to look for sufficient anisotropy in already existing solids that could be regarded as (quasi) 1D. Some anisotropic solids are compiled in the next chapter of this book. How large should the anisotropy be to meet one-dimensionality? A possible answer is: "Large enough to lead to an open *Fermi surface.*"

The *Fermi surface* is a surface of constant energy in *reciprocal space* or momentum space. While the Fermi surface and reciprocal space will be discussed in detail later, for the discussion here, it is sufficient to imagine this surface as describing all of the electron states within the solid that are available to take part in electrical transport. For an isotropic solid, the Fermi surface is spherical, meaning that electrons can move in any direction of the solid equally well.

If the electrical conductivity is large in one crystallographic direction and small in the other two, the Fermi surface becomes disklike. The kinetic energy of the electrons can then be written as $E = p^2/2m^*$, resembling the kinetic energy of a free particle (p = momentum, m = mass), with the exception that the mass

Figure 1.5 The crystal cutter. Crystal surfaces are excellent two-dimensional (2D) systems. The cutter here tries to improve the crystal face by mechanical polishing, but the qualities achieved by this method are not sufficient for surface science. Surface scientists cleave their samples under ultrahigh vacuum conditions and use freshly cleaved surfaces for their experiments – leaving large 2D planes of atoms. Another approach is to use highly oriented and polished crystals that are then sputtered with high current ion beams and annealed at high temperatures to reform the surface.

has been replaced by the *effective mass* m^*. The *effective mass* indicates the ease with which an electron can be moved by the electric field. If the electrons are easy to move, the conductivity is high. Easy motion is described by a small effective mass (small inertia), and p must also be small to keep E constant. If it is infinitely difficult to move an electron in a specific direction, its effective mass will become infinitely large in this direction, and the Fermi surface will be infinitely far away. However, the extension of the Fermi surface is restricted: if the Fermi surface becomes too large in any direction, it will merge with the Fermi surface generated by the neighboring chain or plane ("next Brillouin zone" in proper solid-state physics terminology) assuming this hypothetical solid is made up of stacked structures of some sort. This merging "opens" the Fermi surface, similar to a soap bubble linking with another bubble (Figure 1.6).

1.3 Dimensionality of Carbon: Solids

As promised, we now want to put these structures in the context of carbon. But again, why carbon? How will it be different from other atoms? Let's contrast

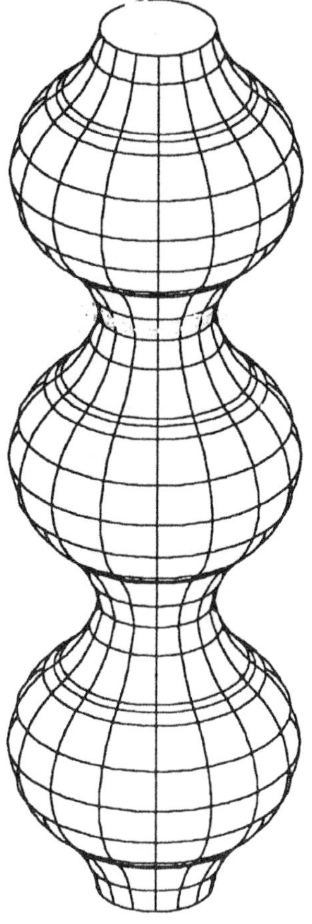

Figure 1.6 Open Fermi surfaces, analogous to merged soap bubbles, as a criterion of low dimensionality. The Fermi surface belongs to a solid that is essentially two-dimensional. The solid will have no electronic states contributing to electrical conductivity along the axial direction but will easily conduct radially, normal to the axis.

it for a moment with a similar element silicon – the basis for much of today's technology. Silicon is unique among solids [8]: it is the most perfect solid producible. That is, there are fewer imperfections in a silicon single crystal than there are gas atoms in ultrahigh vacuum (per unit volume). It is the solid we know most about, and it is the solid that has largely influenced the vocabulary of solid-state physics. Carbon is located directly above silicon in the periodic table of the elements, and just as silicon is outstanding among the solids, carbon is outstanding among the elements. Carbon forms the majority of the known chemical compounds. Much of organic chemistry simply involves arranging carbon atoms (with hydrogen not having any specific properties but just fulfilling the task of saturating dangling bonds). In our context, carbon has the remarkable property of forming 3D, 2D, 1D, and zero-dimensional (0D) solids. This is related to the fact that carbon is able to form single, double, and triple bonds. This ability of carbon to form many types of bonds, at many different bonding angles, sets it aside from silicon in another important way; it leads to biology rather than technology.

1.3.1 Three-Dimensional Carbon: Diamond

Beginning with an example from silicon, diamond appears as the trivial solid form of carbon (Figure 1.7). Diamond has similar semiconducting properties to silicon. Both substances share the same type of crystal lattice. The lattice parameters are different ($a = 5.43$ Å in silicon and 3.56 Å in diamond), and the energy gap between valence and conduction band is larger in diamond, 5.4 eV, compared with 1.17 eV in silicon. Diamond is more difficult to manufacture and more difficult to purify than silicon, but it has better thermal conductivity and can be used at high temperatures. Since the costs for raw material change the final price of electronic equipment only slightly, some people believe that diamond is the semiconductor of the future. Silicon is typically used with added dopants to modify its electronic behavior. Doping diamond has proven to be far more difficult however. Here we mean doping to be a substitution of a lattice atom: in Si it would be Si, and in diamond it would the substitution of a C, with another atom of different valency. The substituted atom adds carriers to the materials, changing its electrical properties. However, the potential dopant atom must fit into the lattice in some way, and this process must be better understood in diamond before the realization of high-quality diamond electronics.

Sometimes, semiconductors and metals are mentioned interchangeably in this book although they are quite different. The reason stems from the idea that a doped semiconductor can be regarded as a metal with low electron concentration. Here, "metal" is essentially used as a synonym for "electrically conductive, solid-state system."

1.3.2 Two-Dimensional Carbon: Graphite and Graphene

In diamond the carbon atoms are tetravalent, that is, each atom is bound to four neighboring atoms by covalent single bonds. Another well-known naturally occurring carbon modification is graphite (Figure 1.8). Here all atoms are trivalent, which means that in a hypothetical first step only, three valence electrons participate in bond formation and the forth valence electron is left over. The trivalent atoms form the planar honeycomb lattice, and the residual electrons

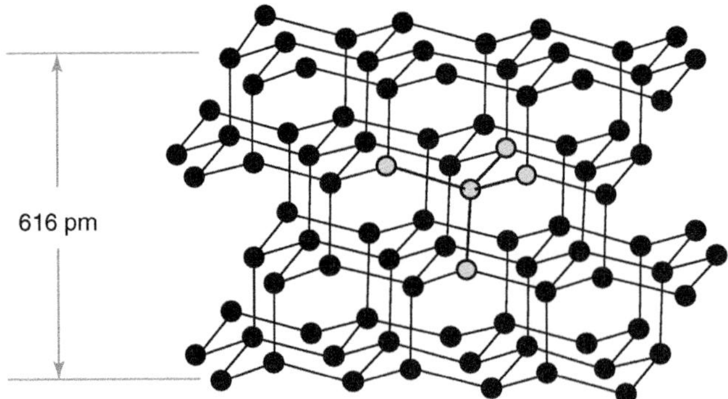

Figure 1.7 The diamond lattice. The diamond lattice can be seen as a "wavy" set of carbon planes connected together by carbon–carbon bonds.

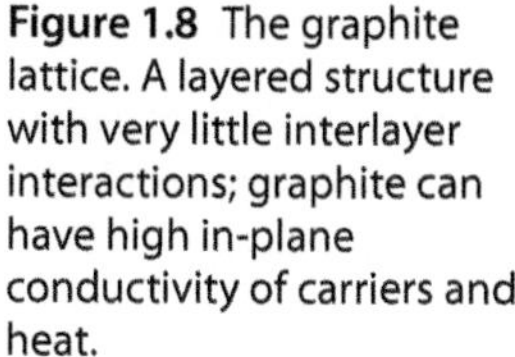

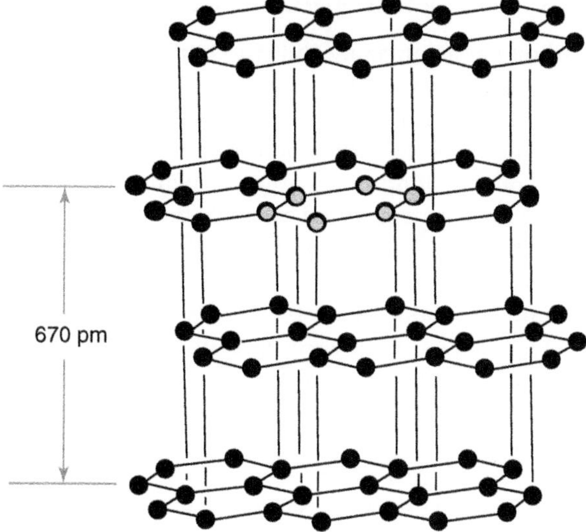

Figure 1.8 The graphite lattice. A layered structure with very little interlayer interactions; graphite can have high in-plane conductivity of carriers and heat.

are shared by all atoms in the plane similarly to the sharing of the conduction electrons by all atoms of a simple metal (e.g. sodium or potassium). The various graphite layers only interact by weak van der Waals forces. In a first approximation graphite is an ensemble of nearly independent metallic sheets. In pure graphite they are about 3.35 Å apart but can be separated further by intercalating various molecules. Charge transfer between the intercalated molecules and the graphitic layers is also possible. Graphite with intercalated SbF_5 shows an anisotropy of about 10^6 in electrical conductivity, conducting a million times better within a layer than between layers.

Diamond is a semiconductor and graphite is a metal (or semimetal). In diamond there are very few mobile electrons; in an undoped perfect diamond single crystal at absolute zero, there are exactly zero mobile electrons; and in graphite there are many, one electron per carbon atom. This difference is not due to dimensionality (three in diamond and two in graphite) but to single and double bonds. Several attempts have been made to build 3D graphite [9]. Theoretically it seems possible [10], but practically it has not yet been achieved.

Of course, since the layers of graphite are very weakly bound together, it is rather easy to separate them mechanically to form graphene – a single sheet of the honeycomb lattice. This lattice is truly 2D, since there is nowhere else for the electrons to go except upon the sheet that essentially defines their "world" for them. Notice though that this 2D sheet "samples" the 3D world in which it lives. If one takes the sheet and bends it in the third dimension while applying a field across it, one can induce phase accumulation in the wavefunction of its electrons – Berry's phase, which comes from the geometrical intersection of the 2D and 3D worlds. Graphene has been studied extensively over the last few years, and transport in graphene led to the 2010 Nobel Prize in Physics [11]. By numbers, the density of graphene is 0.77 mg/m^2. Its breaking strength is 42 N/m, the electrical conductivity is $0.96 \times 10^6\ \Omega^{-1}\ \text{cm}^{-1}$, and thermal conductivity is 10 times greater than copper. We will return to graphene in later chapters.

1.3.3 One-Dimensional Carbon: Cumulene, Polycarbyne, and Polyene

Carbon has an amazing ability to bond to itself in multiple ways. So "1D carbon" comes in several varieties. Using double bonds one can image a monatomic chain as Figure 1.9. (There are no dangling bonds in cumulene and in polycarbyne.) This substance is called *cumulene*; the name refers to the cumulative (meaning consecutive) double bonds. Any organic chemist will tell you that double carbon bonds can be *isolated* (separated by single bonds), *conjugated* (in strict alternation with single bonds), or *cumulated* (placed adjacent to each other) for a wide variety of compounds.

Cumulene has been synthesized for chains 5–10 carbons long [12]. While such long molecules are interesting, they fall a little short of a 1D wire, and polymeric cumulene has not been synthesized. Indeed, quantum chemistry predicts *polycarbyne*, an isomeric structure in which triple bonds alternate with single bonds, is preferred over cumulene. Polycarbyne is shown in Figure 1.10, and it is of particular interest to space scientists since it occurs in interstellar dust, meteorites, and in supernova remnants. It also is seen in trace amounts within natural graphite [13].

If we accept the simplification that in carbon compounds, hydrogen atoms just have the purpose of saturating dangling bonds (making them non-active) and that otherwise they do not contribute to the physical properties of the material, cumulene and polycarbyne are not the only 1D carbon solids. From this point of view, all polymers based on chain-like molecules are 1D.

Let's learn some organic chemistry. On naming conventions, the ending "-yne," as in polycarbyne, is used to indicate triple bonds. The ending "-ene" stands for double bonds and "-ane" for single bonds. A polyane is shown in Figure 1.11. (To add a little confusion to the subject, this substance is typically called polyethylene, ending with "-ene" instead of "-ane." The reason is simply that the names of polymers are often derived from the monomeric starting material, which in this

--- C=C=C=C=C=C ---

Figure 1.9 One-dimensional carbon example one: cumulene.

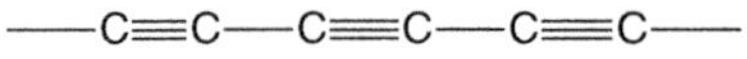

—C≡C—C≡C—C≡C—

Figure 1.10 One-dimensional carbon example two: polycarbyne.

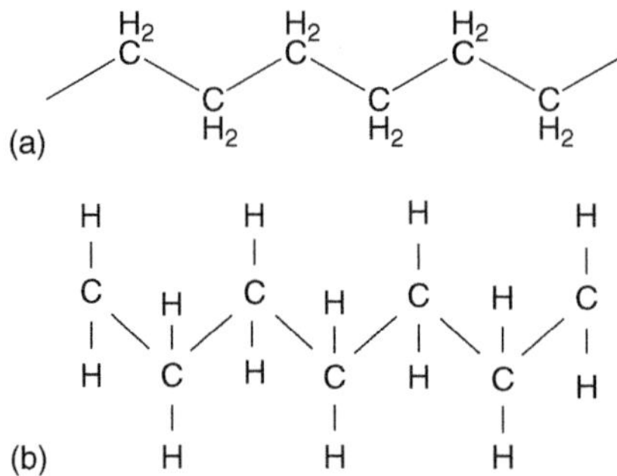

Figure 1.11 Polyethylene, as we might imagine the (a) polymerization of ethylene and (b) arrangement of bonding.

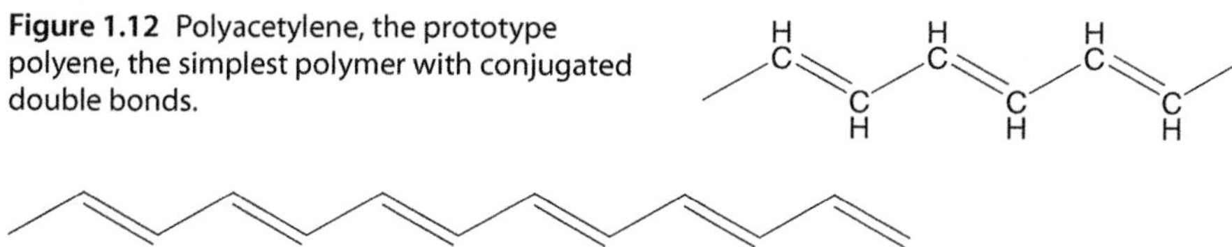

Figure 1.12 Polyacetylene, the prototype polyene, the simplest polymer with conjugated double bonds.

Figure 1.13 Polyacetylene using a simplified notation.

case is ethylene, $H_2C{=}CH_2$. Here the monomer contains a double bond, but during polymerization the double bond breaks to link the neighboring molecules.) Polyanes are insulators and of less interest in the context of this book. (Insulators are large bandgap semiconductors. Because of the large bandgap, it is difficult to lift electrons into the conduction band, and therefore the number of mobile electrons is negligible.)

Figure 1.12 shows polyacetylene, the prototype polyene, the simplest polymer with conjugated double bonds. The structure shown in Figure 1.12 is often simplified to the one in Figure 1.13, since by convention carbon atoms do not have to be drawn explicitly at the ends of the bonds and protons are neglected.

1.3.4 Zero-Dimensional Carbon: Fullerene

If we work our way down in dimensionality from *volume-diamond* to *plane-graphite and graphene* to *lines-polymers*, we will finally end up at the *point* as a 0D object. Do 0D solids exist outside of the obvious (the atom)? In semiconductor physics the "quantum dot" is well known [14]. Historically, this is a small disk cut out of a 2D electron gas. It is small compared to the Fermi wavelength, so that the electrons are restricted in all three dimensions of space (the 1D analogue to a quantum dot is often called "quantum wire"). Following the discussion in Section 1.2, a quantum dot is a 0D object. The present state of the art is to fabricate quantum dots containing more than 1 but less than 10 electrons. Because of the low electron concentration in semiconductors, such quantum dots can exhibit quite large diameters, up to several hundred Ångstroms. More recently, quantum dots have been fabricated as chemically assembled nanoparticles, wherein the structure defines the confinement. Metal nanoparticles of Au, Ag, Cu, etc. have been created using a variety of chemical synthesis routes, and confinement of the electrons occurs at particle diameters of only a few nanometers. Likewise, quantum dots made from semiconductor materials such as Si, Ge, and compounds such as CdS, CdSe, PdS, etc. have been created. Following the rules we have already discussed, these nanoparticles can be many nanometers in diameter and still exhibit confinement because there are fewer electrons in the "box." The ΔE between these electron states can be quite large, leading to some fascinating optical properties that are quite different from their bulk counterparts.

Carbon can form quantum dots in a number of ways – nanodiamonds, nanoplatelets of graphene, and others – as would be expected from carbon's ability to bond in different ways. However, the most famous of these quantum dots of carbon in solid-state physics are the fullerenes [15]. The 1996 Nobel Prize

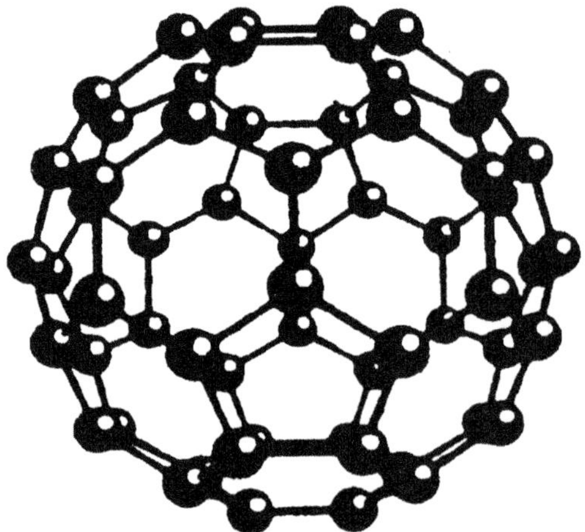

Figure 1.14 A fullerene molecule. This is an example of a C_{60}, but much larger cages can be made.

in Chemistry was given to R.F. Curl, H.J. Kroto, and R.E. Smalley for their role in the discovery of this class of molecules. Under certain conditions, carbon forms regular, cage-like clusters of 60, 70, 84, etc. atoms. A C_{60} cluster is composed of 20 hexagons and 12 pentagons and resembles a soccer ball (Figure 1.14), all bonded together as in graphene. The diameter of a C_{60} ball is about 10 Å and thus is considerably smaller than that of a semiconductor quantum dot. However, in these carbon compounds, the electron concentration is higher than in inorganic semiconductors: in a system of conjugated double bonds, there is one π-electron per carbon atom! (More on π-electrons later.) In other words, there are 60 π-electrons in a fullerene ball of 10 Å diameter, compared with some five electrons on a 100 Å GaAs quantum dot. In quantum chemistry and solid-state physics, 60 is already a quite large number (we are used to counting: "one – two – many"). In fact, a 60-particle system is already a mini-solid, and a fullerene ball plays a dual role in solid-state physics: it is a mini-solid. It can also be a constituent of a macro-solid – fullerite.

We can study electronic excitations in the mini-solid and their mobility and interaction with lattice vibrations. At the same time it is possible to examine unexpected transport properties of the macro-solid, like superconductivity [16], photoconductivity, and electroluminescence [17]. Figure 1.14 shows the graphic representation of a fullerene mini-solid. Figure 1.15 schematically indicates the fullerene macro-solid.

1.4 Something in Between: Topology

Conceptually, we might conceive of a solid that is a combination of dimensions. Imagine, for instance, a single graphene sheet described in the section on graphite. Roll this conductive sheet into a seamless tube in which each atom is threefold coordinated as in the sheet. When the diameter of such a tube is between 14 and 200 Å, we refer to the object as a *carbon nanotube*. For such an object, the electron wavefunction is confined to boxlike states around the

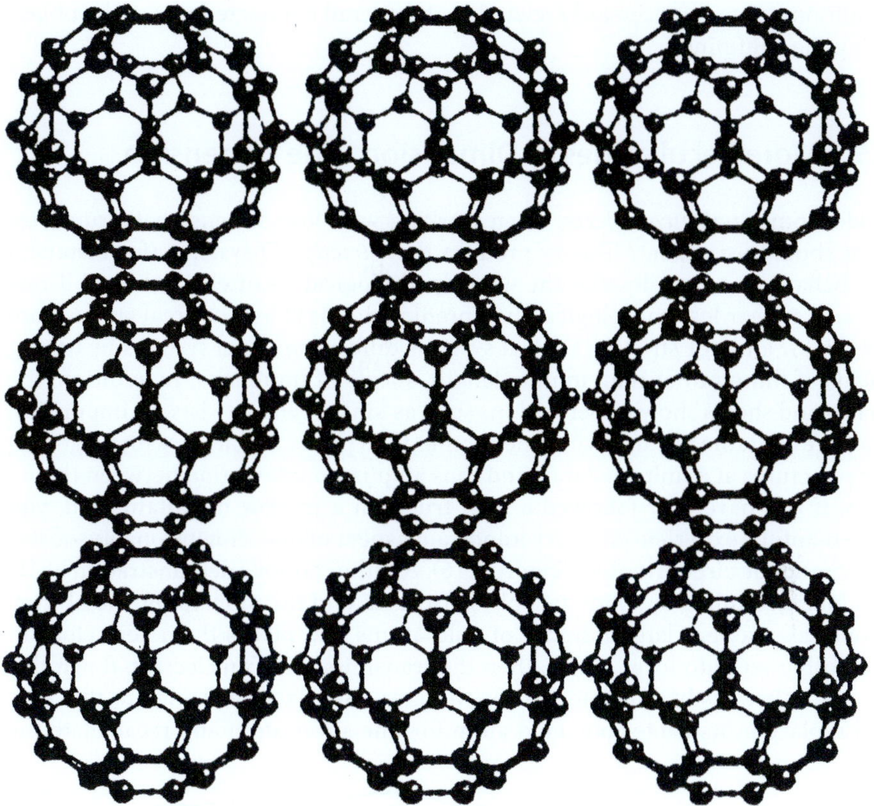

Figure 1.15 The fullerene crystal lattice: "fullerite." These compounds have a rich chemistry. They can be doped by placing atoms between the balls, inside the balls, etc.

circumference. Along the axis of the tube, the electrons move in essentially a 1D system. Normally, this would appear to be similar to the semiconductor wires mentioned earlier. However, this circumference (or rolled-up) dimension allows for a set of spiral-like classical trajectories of the electron as it moves down the tube. In this way, if a 3D field (like a magnetic field) should penetrate the tube, the phase of the electronic wavefunction would be altered, resulting in Aharonov–Bohm effects. Thus, while the tube certainly has the character of a 1D system, it also has a "little more." It is clearly not quite 2D however. For such systems, there is a *topology* that must be considered. That is to say, the object is connected together in such a way as to introduce an additional "dimensional" aspect. Here we mean a physical topology associated with a sheet of atoms rolled into the third dimension from a 2D starting point. However, the topological aspects of low-dimensional systems in general – or the way in which their electronic states might be connected together to form closed manifolds in space – will be a recurring theme. This is actually quite a natural outcome of the whole idea of working with low-dimensional materials. By restricting spatial dimension and confining the electronic wavefunction, we introduce boundary conditions that are necessarily related to the overall connectedness of the object doing the confining.

1.5 More Peculiarities of Dimension: One Dimension

Aside from *topological effects*, when working with low-dimensional structures, what should we expect? Theory predicts that *strictly* 1D systems (for instance) will behave so unusually that the word "pathological" is often used. And if real systems appear less pathological than predicted, this is because real systems are only *quasi* and not strictly 1D. Real systems differ from ideal systems by having chains of finite rather than infinite length, sheets of finite area. In addition, the chains and sheets show imperfections such as kinks, bends, twists, or impurities. They are contained in an environment other than perfect vacuum, with neighboring structures at a finite distance and thus a nonzero interaction between them.

So, if you have ever followed a slow truck on a narrow mountain road, you have painfully experienced a very important aspect of one-dimensionality: obstacles cannot be circumvented! (Figure 1.16). A rather famous demonstration of 1D conduction studied by solid-state physicists is that of the monatomic metal wire. If one takes a very large number of gold atoms and places them very close to each other so as to form a wire, then the transmission of an electron down this wire is rather easily calculated. Now, we offer a very subtle change to this wire and replace in its center one gold atom for one silver atom and recalculate the

Figure 1.16 The road to Kirchberg. A very important aspect of one-dimensionality is that obstacles cannot be circumvented.

Figure 1.17 Bond percolation demonstration on a two-dimensional grid, where bonds are successively cut in a random way. Source: After Zallen 1983 [19].

transmission probability of the electron traveling its length. What is found is that even for very small variations in the periodic atomic potential, reflections of the electron wave on the wire become large [18].

Another more sophisticated conceptualization of *dimensional restriction* can be made in terms of *percolation*. *Percolation* means macroscopic paths from one side of the sample to the other and the threshold for bond percolation in one dimension is 100%! Such macroscopic paths are necessary, for example, for electrical conduction. The concept of bond percolation is quite different in two dimensions as demonstrated by a grid (Figure 1.17) [19] where bonds are cut at random. In this 2D square lattice, a few cuts yield little change in sample's conduction properties. In particular, the conductivity drops only slightly due to the appearing holes. When 50% of the bonds are cut randomly, no path is left that connects one side of the sample to the other, and the conductivity must be zero. The percentage of intact bonds necessary to establish macroscopic paths is the *percolation threshold*. The higher the dimensionality of the sample, the lower the percolation threshold. For a 1D system the threshold – quite simply – is 100%: if we cut one bond, the sample consists of two disconnected pieces.

Another trivial aspect of 1D systems is the low connectivity. Each atom is connected to two other atoms only: one to the left-hand side and one to the right-hand side. In 3D solids there are connections to neighbors in the back and front as well as to neighbors above and below. Connectivity is a topological concept. Chemists usually speak of the coordination number, the number of nearest neighbors. In a 1D chain the coordination number is 2.

A consequence of the low connectivity of 1D systems is the strong *electron–lattice coupling*. If bonds are completely broken, a 1D system separates into two pieces. Usually complete breaking of bonds does not happen, however. Often bonds are only partially cleaved; for example, only one component of the double bond in the system as in Figure 1.9 or Figure 1.13 is broken. In chemical terms, this means that a bonding state is excited to form an antibonding state. In semiconductor physics it would be described as an electron being lifted from the valence band into the conduction band. Such manipulation of valence electrons is quite common in semiconductors, and it is the first step for photoconductivity and photoluminescence. In a 3D semiconductor like silicon, the transfer of an electron from the valence to the conduction band creates mobile charge carriers (the electron in the conduction band and the "hole" left behind in the valence band), but it does not change the arrangement of the atoms in the crystal. This is due to the high connectivity of the silicon lattice, where breaking or weakening *one* bond has not much effect. In low-connectivity 1D systems, where each atom is held in place by two neighbors only, each change in bond strength leads to a large distortion of the lattice. In conjugated polymers, the lattice distortion shows a change in bond length when a double bond is partially broken to yield a single bond.

With strong electron–lattice coupling, the electrons moving in the solid creates a large distortion that polarizes the lattice. If the effect is distinct enough, the electrons receive a new name: *polarons* – that is, the charge plus the distortion. Depending on the strength of the coupling and the symmetry of the lattice, there is a variety of quasiparticles resulting from electron–lattice coupling, the most famous of which (and very typical for 1D systems) is the "soliton" [20]. We will have a closer look at solitons and polarons a little later.

An important peculiarity of 1D systems for our discussions is band edge singularities in the *electronic density of states*. As we will see in later chapters, in a solid, electrons cannot have *any* energy they wish (as they could have in vacuum). There are only allowed energy regions (*energy bands*) separated by forbidden gaps – energies they may not take on. The long-range ordering within the system determines these forbidden and allowed energy bands for the carriers. The *density of states* within a band of allowed energies is simply how closely packed together the allowed states are in energy or the number of states per unit energy interval. This density is not constant in a solid. The form of the density of states function depends on the crystal structure, and surprisingly, near the band edge, it reflects the dimensionality of the system. This is shown in Figure 1.18: in three dimensions the density of state function $N(E)$ is parabolic, in two dimensions it

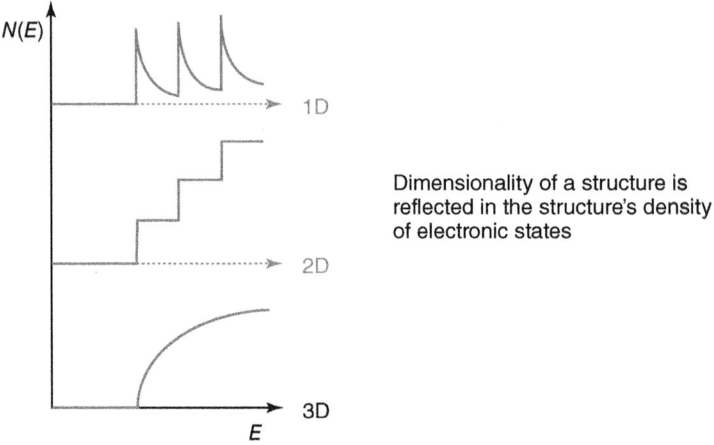

Figure 1.18 Density of states function at the band edge in three-, two-, and one-dimensional electronic systems. Note the singularity that occurs in the one-dimensional case.

is steplike, and in one dimension there is a square root singularity to infinity! In real systems $N(E)$ never reaches infinity, of course, but at least there is a very high density of states.

One-dimensionality also differs from two- and three- dimensionalities in random walk problems. In a higher dimension it is very unlikely that a random walker will return to the place he/she started, whereas in one dimension this happens quite often. Whether or not the random walker comes back to the point of departure is important for discussing the recombination of photogenerated charge carriers and thus for the time constants of transient photoconductivity and of luminescence. Luminescent devices might turn out to be the most important practical applications of 1D metals!

One-dimensional solids are particularly interesting in the context of fundamental studies on phase transitions. In fact, one motivation in the field of 1D conductors arises from the hope of finding the key to high-temperature superconductivity. However, there is a famous theorem of Landau that suggests phase transitions are impossible in 1D systems [21]: long-range order is unstable with respect to the creation of domain walls, because the entropy term in the free enthalpy will always overcompensate the energy needed to form new walls. Whereas phase transitions are impossible, 1D systems might be "close" to a phase transition even at fairly high temperatures. Fluctuations might "anticipate" the phase transition and have already prompted speculation toward some technologically useful properties such as low-resistance charge transport. Perhaps we could allow for "just a little bit" of three-dimensionality and thus obtain a high-temperature superconductor? Organic superconductors are known, but they are closer to two-dimensionality than to one-dimensionality. Their superconducting transition temperatures reach 12 or 13 K (for fullerene even up to 33 K), still far below the recently discovered inorganic oxide superconductors with transition temperatures of 100 K and above [22].

1.6 Summary

In summary, *dimension*, *connectivity*, and *symmetry* show up in many different ways for the solid-state scientist. We note in this chapter that the unusual character of *carbon* with the many structures it is able to make allows us to capture a remarkable number of ways in which a solid can behave in reduced dimension. Basically, the flexibility allows this atom to make a solid in any dimension we might want and with plenty of variants. All of these different allotropes have radically different properties – all based entirely on the organization of the atoms.

As can now be seen, our discussions here will be aimed at introducing both basic and advanced models of solid-state physics in the context of standard chemistry, physics, and materials science. However, *dimensionality* and *topology* will continue to be a unifying language for the materials systems discussed with carbon-based solids as our inspiration. Remember, if you didn't get the full meaning of everything in *italics* the first time around, don't worry; we will come back to it again and again.

Synthetic Metals
T_C πσ 1d
Kyoto thank we

Figure 1.19 Haiku from the ICSM 1986 closing ceremony session in Kyoto [23].

To complete the present chapter however, we reprint in Figure 1.19 a "haiku" that was used during the closing ceremony of the International Conference on Science and Technology of Synthetic Metals in Kyoto [23]. This forum has traditionally focused on the field of conducting polymers, conducting small molecules, low-dimensional organic structures, and similar topics. From it, an international group of scientists formed a community that continues today.

Exploring Concepts

1 *Carbon*: The original identification of *carbon*, known at the discovery of metalworking, has been lost to history. Its electronic ground state configuration is [He] $2s^2 2p^2$, and so the outer shell's four electrons have s and p characters. Its melting temperature is 3550 °C (6420 °F) and boiling temperature is 4827 °C (8721 °F). And carbon is the world's primary fuel source (energy storage medium). More specifically, the CH_x unit is the basis for most of the *energy-dense fuels* that our planet uses. From gasoline to coal to animal fats, mankind has recognized the extraordinary utility of this sub-compound of carbon and exploited it. Take a little time and compare the energy density (J/kg) of animal fat, oil, coal, a Li-ion battery, and TNT. Remember that the carbon compounds must be oxidized, so when computing the energy released by mass, you must also include the weight of the oxygen to get the true energy density. Many references fail to do this: so don't just go to Wiki.

2 *The Euclidean dimension of an object*: Dimension is informally thought of, in physics, as the minimum number of coordinates that are needed to describe any point on or within the object. Likewise, *n-dimensional spaces* extend this idea to include all of the possible coordinate values needed to describe any *n*-dimensional object within the space. However, for our purposes, this isn't quite complete. For example, imagine a ball. It sits in a 3D space, but if confined to the surface of that ball, we are decidedly 2D, and indeed it takes only two coordinates (with reference to some axis set attached to the ball's surface) to describe every point on the ball's surface (longitude and latitude, or θ, φ). If something from a third dimension were to intersect our ball, let's say it is a 2D plane, as seen here, then notice that using the ball's coordinates I could describe only the points of intersection and not anywhere else on the plane (Figure EC1.1). This is similar if I took the point of view of the plane. Moreover, notice that the intersection is a 1D object generally (or 0D in the case of a point). But the set of points that describe the intersection have symmetries reflected in both the flatness of the plane and the spherical nature of the ball. So something else is required to describe the symmetry of the intersection line, and that *something* is related to higher-dimensional objects.
This notion is made quantitative in the *topology* and *geometry* of *manifolds*. So this exercise takes you outside the text a bit. First read up on *topology* and *geometry* and describe how topology is a subdiscipline of geometry.

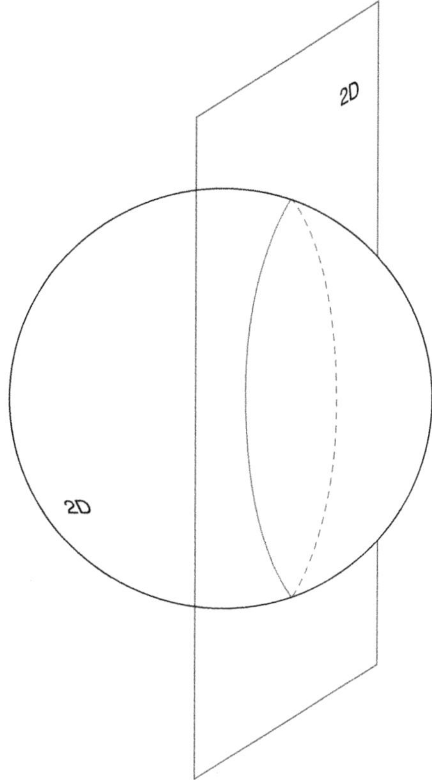

Figure EC1.1 The apparent lower dimensional intersection between two objects of higher dimension.

How does the *topology* of an object pertain to our discussions here? What is meant by the concepts of *connectedness* and *compactness* in topology? Imagine that we could take a single layer of graphene and connect it to itself along one edge such that it formed a Möbius strip. What do you think this would mean for the electrons in/on the strip?

3 *Si and C both form "hybridized" bonding orbitals*: You will learn more about these in the coming chapters. For now however, you can just think of them as orbitals that have very specific directions associated with them. But C is able to allow its bonds to take on a number of different bonding angles (it bends), whereas this is more difficult for Si. Why would you think this might be? What are the ramifications of this for the formation of compounds and crystals? (You may need to journey outside of the text for this one as well.)

4 *Euler's rule*: In simply connected, volumetric polyhedron structures (as in Figure EC1.2), there is a simple rule that must be obeyed before the structure can be constructed and closed using regular polygons (remember *regular polygons* means the 2D structures have many sides but the sides all have the same length).
Now typically, a polyhedron is just one piece. It can't be made up of two (or more) separate parts stuck together at an edge or a vertex or something. It is

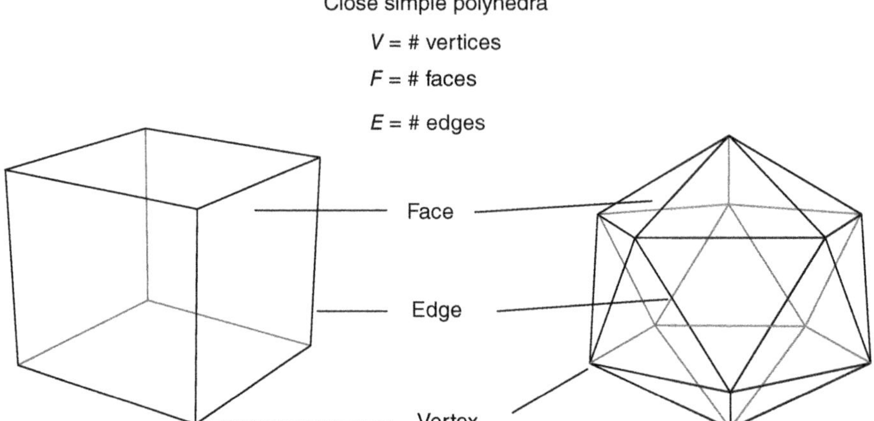

Figure EC1.2 Simply connected polyhedra are simply closed structures with no holes through them.

a "box" and its faces will be made of regularly shaped polygons. But here is the rub. To get that box to close properly without bending the polygons etc., then

$$V - E + F = 2; \text{(known as the } \textit{Euler rule}\text{)}$$

(a) Using this rule, determine how may pentagons plus hexagons of carbon it will take to construct a C_{60} molecule.
(b) Is it possible to make a C_{70}? If so, how many pentagons and hexagons would this require? Draw out what you think this might look like.

5 *Fractals and dimension*: Imagine that for any given *Euclidean dimension, D,* we reduce the overall unit of measure by the factor $1/R$. For $R \in I$ we get the schematic shown in Figure EC1.3. The *measure of the object* (that is its length, area, or volume) increases as

$$N = R^D$$
$$D = \mathrm{Log}(N)/\mathrm{Log}(R)$$

This generalized notion of dimension D is known as the *Hausdorff dimension*, and it doesn't need to be an integer as it is in Euclidean geometry. Indeed, in fractal geometries, it is fractional and can be used as an estimate of *roughness*. This idea was eventually applied to the length of coastlines as in Figure EC1.4. In fact, if we assume that the coastline's "roughness" is reproduced at every scale, say, it's *self-similar*, then the processes of halving and then halving again will converge to an estimate of the length of the coastline that is infinite. Thus, such an estimate doesn't make much logical sense. In other words, to describe the coastline, we can't just ask: "how long is it?" The answer to this question doesn't contain complete information. We need something more. L.F. Richardson found a simple way to think of this as seen in Figure EC1.5.
The log–log plot linearity of length estimates in Figure EC1.5 is known as the *Richardson effect*. Mandelbrot used this effectively to define a

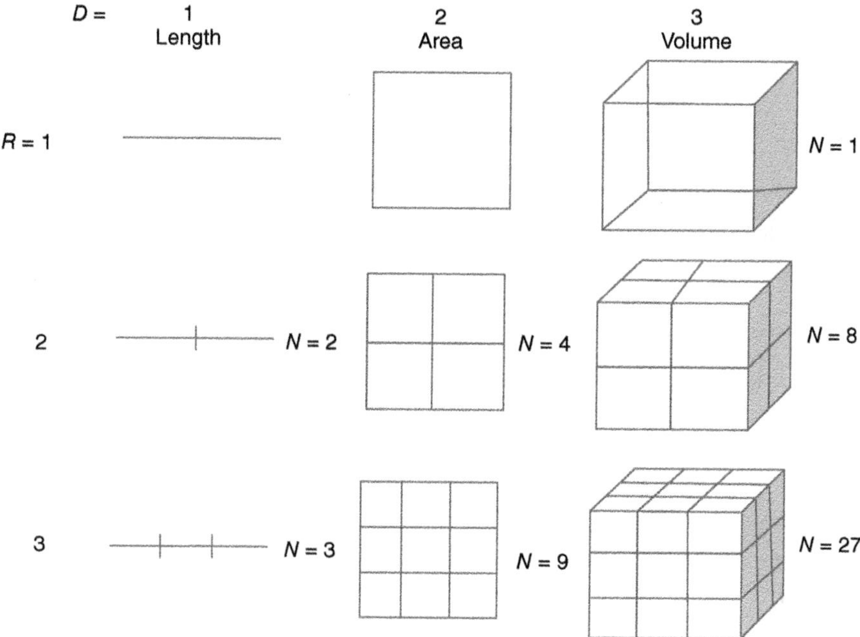

Figure EC1.3 Schematic of the changing unit measure.

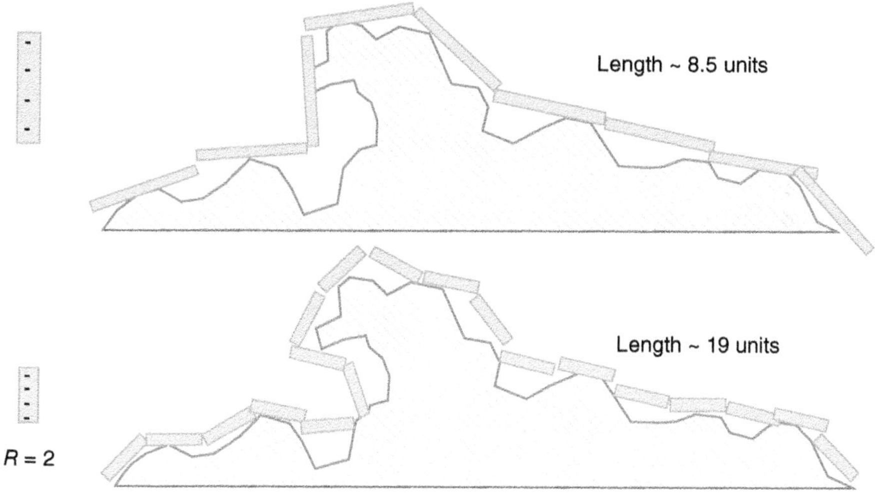

Figure EC1.4 Estimating the length of a coastline. Notice the top estimate gives a length of roughly 8.5 rulers in length. Now we halt the length of the ruler. We get a length estimate of the coastline of roughly 19 rulers, not the 17 rulers we might expect. Imagine halving the ruler's length yet again. It is easy to see that more of the "nooks and crannies" of the coastline will be measured, making the estimate require more than twice the last ruler length again.

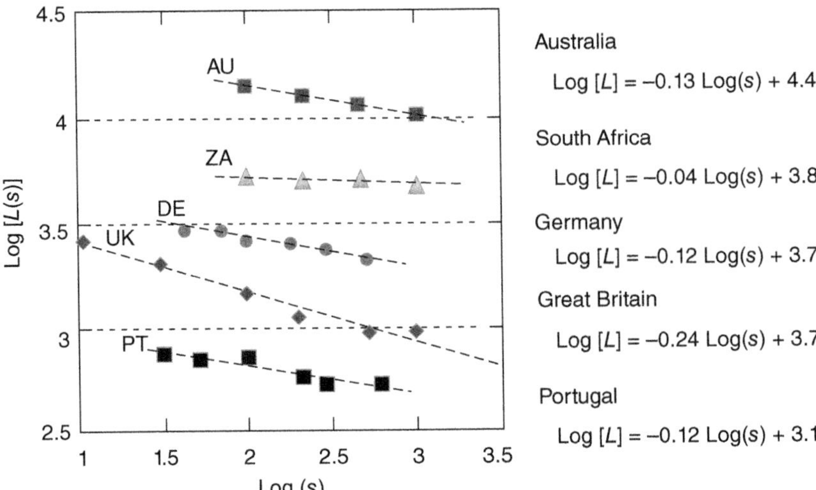

Figure EC1.5 The relationship between the length of coastline estimate and the length of scale used to make that estimate of the coastline is linear on a log–log plot. Indeed, this is true for many naturally occurring structures in the universe, not just coastlines. Of particular interest to us might be polymer lengths, surface areas of rough crystals, and more. Source: Mandelbrot 1983 [3a].

dimensional characteristic to what was being estimated. He assigned the term $(1-D)$ to the slope. This makes the fitted functions look like $\text{Log}[L(s)] = (1-D)\text{Log}(s) + b$ where D is the *fractal dimension*.

Notice in the replotted data above that the UK has a $(1-D) \sim -0.24$. So, $D = 1-(-0.24) = 1.24$, a fractional value. The coastline of ZA is much smoother. So, the slope above is very nearly zero so $D \sim 1$ (i.e. almost a Euclidean object or a line with a dimensionality of one). Generally, the *rougher* the line, the steeper the slope will be. This yields a larger fractal dimension, as though this highly squiggly line is trying to fill space in a nearly 2D way but doesn't quite make it!

In this exercise you will generate line segments of your own coastline. These are a set of line segments that are self-similar over different length scales.

The Example

The **Koch curve** is constructed conceptually by taking a line segment (the *initiator*) and removing the middle third of the line. The gap is then filled with two line segments that are equal in length to the segments on either side of the gap. This is shown in Figure EC1.6. The new structure is called the *generator*. So, starting with these two structures, the rule says to take each line and replace it with four lines, each one-third the length of the original. Notice that as we do this, the "length" of the curve gets greater and greater until it eventually diverges.

(a) As we noted the Koch curve length increases with each iteration, until it diverges. So this means we can only deal with it in a treatment as we show for the coastlines above: Figure EC1.5. Estimate the *lengths* for the next

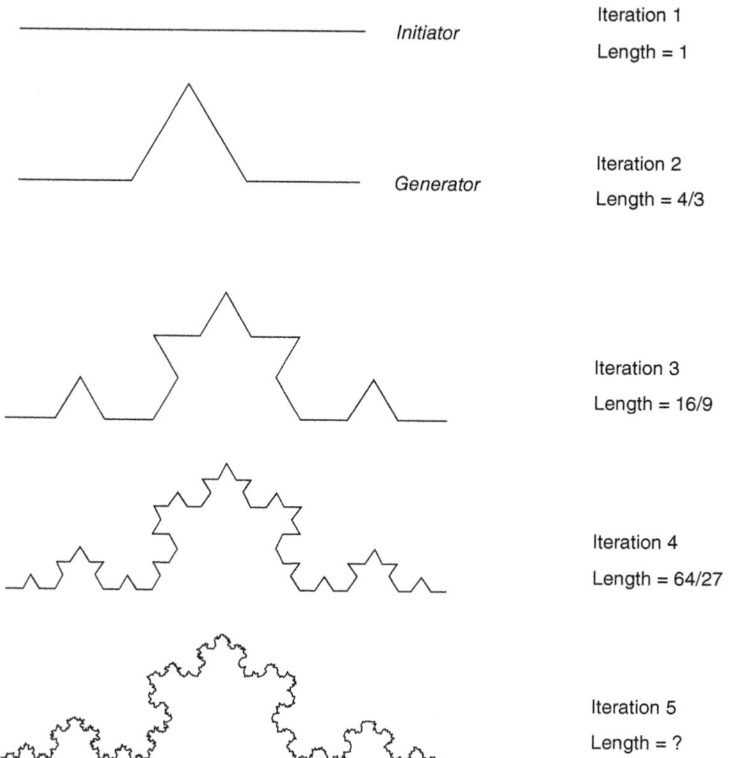

Figure EC1.6 The Koch curve. Starting with the initiator (iteration 1) and the generator (iteration 2), the curve can be continued infinitely.

couple of iterations following the example in Figure EC1.6. Plot them as seen in Figure EC1.5 and then make a determination of the fractal dimension of the Koch curve.

(b) Write a short *Maple, MATLAB,* or *Mathematica* program to produce the Koch curve and estimate lengths for iterations up to 20 or so. How fast is the length diverging? How fast does the fractal dimension converge to a limiting value and how close was your estimate in (a)?

(c) Now let's see how to use this and why we have placed this in a solid-state book. Go to a local atomic force microscope (AFM). Get a sample of fractured glass or metal of appropriate size. Image this sample observing the roughness in the image and in individual line scans (see user's manual for your machine). The software of most AFMs allow for a roughness analysis to be made; you just hit a key and it gives a number. There are a number of different ways to define and calculate this roughness, but generally the machine will determine the mean or average variation from a horizon line defined by the image itself and call this the roughness. However you might suspect that there is some relationship between this number and that of the fractal dimension, and with a few small limitations and caveats, there is. Using MATLAB or one of the other symbolic math programs, write out a code to load one of the line scans into the program and then overlay line segments to determine length in methods similar to that of the above Richardson plot. From this determine the fractal dimension.

Now change the imaging conditions including the size of the scan area, the speed of scanning, etc. (we are assuming you know what an AFM is here of course). Repeat what you have just done. Is the answer different? Yes! Of course it is. But do you know why? How many scales would I need to scan over to ensure that I have a reasonable correlation between what we will call surface roughness and fractal dimension?

Note to the reader about our problem sets: Ever notice how some texts end their chapters with *problem #1* derive equation 2.7... *problem #7* repeat problem #4 for all these different lattice parameters... and on they go. Well, our problems don't work that way. Following the lead of great works like Kittel, we assume that our readers are living and breathing their desire to become true solid-state physicists. So, they are not opposed to reaching outside of the text to understand a concept through a reference or performing an experiment or two to test our conjectures. Occasionally easily accessed references just don't provide enough discussion, and so we walk you through that concept in the problem. In other cases, gems are lying upon the ground waiting to be read and appreciated. We encourage the reader to try *all* of our problems first alone and then within study groups and with their instructors. They are not homework, they are home entertainment, and they are opportunities to go well beyond what we have covered in the text.

References

1 Marcuse, H. (1964). *One-Dimensional Man: Studies in the Ideology of Advanced Industrial Society*. Boston, MA: Beacon Press.

2 von Weizsäcker, C.F. (1986). *Aufbau der Physik*, 2e. München: Carl Hanser Verlag.

3 (a) Mandelbrot, B.B. (1983). *The Fractal Geometry of Nature*. New York: W.H. Freeman & Co. (b) Stewart, I. (1982). *Les Fractals*. Paris: Librairie Classique Eugéne Belin (This is a book of popularized science in cartoon form).

4 (a) Beenakker, C.W.J. and van Houten, H. (1991). Quantum transport in semiconductor nanostructures. In: *Solid-State Physics*, vol. 44 (ed. H. Ehrenreich and D. Turnball). San Diego, CA: Academic Press. (b) van Houten, H., Beenakker, C.W.J., and van Wees, B.J. (1991). Quantum point contacts. In: *Semiconductors and Semimetals*, vol. 35 (ed. M. Reed). San Diego, CA: Academic Press.

5 McCombe, B. and Nurmikko, A. (1994). *Electronic Properties of Two-Dimensional Systems*. Amsterdam: Elsevier.

6 (a) von Klitzing, K., Dorda, G., and Pepper, M. (1980). New method for high-accuracy determination of the fine-structure constant based on quantized Hall resistance. *Phys. Rev. Lett.* 45: 494. (b) von Klitzing, K. (1986). The quantized hall effect. *Rev. Mod. Phys.* 58: 519. (Nobel lecture in physics 1985).

7 Haug, R.J. (1993). Edge-state transport and its experimental consequences in high magnetic fields. *Semicond. Sci. Technol.* 8: 131.

8 Queisser, H. and Krisen, K. (1985). *Mikroelektronik; Wege der Forschung, Kampf und Märkte.* München: Piper.

9 Baughman, R.H. and Cui, C. (1993). Polymers with conjugated chains in three dimensions. *Synth. Metals* 55: 315.

10 Hoffman, R., Hughbanks, T., Kertesz, M., and Bird, P.H. (1983). Hypothetical metallic allotrope of carbon. *J. Am. Chem. Soc.* 105: 4831.

11 Geim, A.K. and Novoselov, K.S. (2007). The rise of graphene. *Nat. Mater.* 6: 183.

12 Januszewski, J.A. and Tykwinski, R.R. (2014). Synthesis and properties of long [n]cumulenes ($n \geq 5$). *Chem. Soc. Rev.* 43: 3184.

13 Bunz, U.H.F. (1994). Polyine–faszinierende Monomere zum Aufbau von Kohlenstoffnetzwerken. *Angew. Chem.* 106: 1127.

14 Geerlings, L.J., Harmans, C.J.P.M., and Kouwenhoven, L.P. (1993). Proceedings of the NATO advanced research workshop on "the physics of few-electron nanostructures". *Phys. B: Condens. Matter* 189: 1.

15 (a) Taliani, C., Ruani, G., and Zamboni, R. (1993). *Fullerenes: Status and Perspectives.* Singapore: World Scientific. (b) Kroto, H.W., Fischer, J.E., and Cox, D.E. (1993). *The Fullerenes.* Oxford: Pergamon Press. (c) Koruga, D., Hameroff, S., Withers, J. et al. (1993). *Fullerene C60: Physics, Nanobiology, Nanotechnology.* Amsterdam: Elsevier. (d) Billups, W.E. and Ciufolini, M.A. (1993). *Buckminsterfullerenes.* Weinheim: VCH.

16 (a) Gärtner, S. (1992). *Festkörperprobleme,* Advances in Solid State Physics, vol. 32, 295. (b) Hebard, A.F. (1992). Superconductivity in doped fullerenes. *Phys. Today* 45: 26. (c) Murphy, D.W., Rosseinsky, M.J., Fleming, R.M. et al. (1992). Synthesis and characterization of alkali metal fullerides A_xC_{60}. *J. Phys. Chem. Solids* 53: 1321. (d) Lüders, K. (1993). Superconducting intercalation compounds of C_{60} fullerene and graphite. In: *Chemical Physics of Intercalation II* (ed. P. Bernier, J.E. Fischer, S. Roth and S.A. Solin), 31. New York: Plenum Press.

17 Werner, A.T., Anders, J., Byrne, H.J. et al. (1993). Broadband electroluminescent emission from fullerene crystals. *Appl. Phys. A* 57: 157.

18 This is given as a computational problem in many solid state courses. To learn more see for example: Datta, S. (1995). *Electronic Transport in Mesoscopic Systems.* Cambridge: Press Syndicate of the University of Cambridge.

19 Zallen, R. (1983). *The Physics of Amorphous Solids.* New York: Wiley.

20 (a) Roth, S. and Bleier, H. (1987). Solitons in polyacetylene. *Adv. Phys.* 36: 385. (b) Heeger, A.J., Kivelson, S., Schrieffer, J.R., and Su, W.P. (1988). Solitons in conducting polymers. *Rev. Mod. Phys.* 60: 781. (c) Lu, Y. (1988). *Solitons and Polarons in Conducting Polymers.* Singapore: World Scientific.

21 Landau, L.D. and Lifshitz, E.M. (1990). *Lehrbuch der Theoretischen Physik*, 7. Auflagee, vol. V/1. Berlin: Akademie-Verlag.

22 Kuzmany, H., Mehring, M., and Fink, J. (1993). *Electronic Properties of High-Tc Superconductors*. Berlin, Heidelberg: Springer-Verlag.

23 Roth, S. (1987). *Synth. Metals* 22: 190. (A haiku is a Japanese poem with 17 syllables distributed in 5-7-5 over three lines using a highly symbolic language.).

2

One-Dimensional Substances

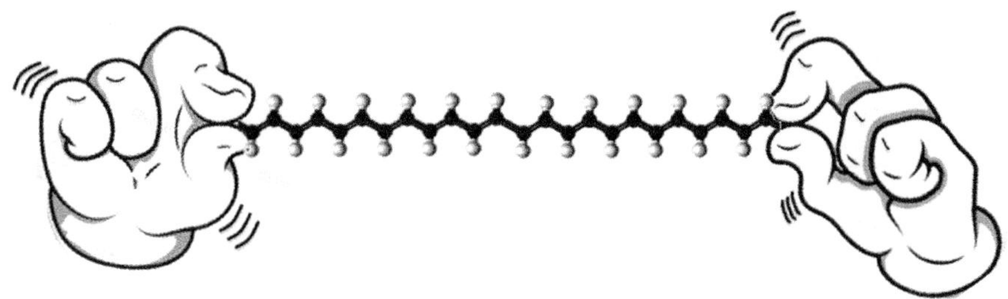

We begin by considering one-dimensional (1D) substances.[1] A more traditional approach might start with lattices in three-dimensional (3D), and we will talk about them later. But here, we want to consider how 1D behavior is observed in a variety of *real solids*, with a combination of elements, and that may or may not include carbon. Interesting material structures also occur in zero-dimensional (0D) or two-dimensional (2D) analogues of our 1D solid. So our discussion translates easily.

The 1D solid is particularly interesting because it is the *least dimensional object* that connects two separate points in space. In other words, it practically defines the field of *transport* in a solid: moving heat, electrons, and ions from one point to another! Physicists, chemists, and material scientists have long been intrigued by the idea of *transport* in a low-dimensional solid ever since the 1964 work on superconductors by Little [1]. Little suggested synthesizing an *organic superconductor* by appropriately functionalizing polyacetylene, a polymer with conjugated double bonds (Figure 1.12). For a simple picture, Little proposed replacing some of the hydrogen atoms in polyacetylene by specifically designed substituents R, as shown in Figure 2.1. In Little's superconductor, electrons are supposed to move

1 This is done in a rather odd way. We will discuss these compounds using the full apparatus of what is to come. So the reader may not have been instructed in all the language or concepts quite yet. This is OK since what we are seeking to do is to set the tone and language that will be important throughout the text. Keep the descriptions and italicized words in mind as you work through the rest of the text, then refer back to these compounds as they become more and more clear to you. In this way, the book is meant to be quite active in its use, like a conversation where you fill in as you learn more.

Foundations of Solid State Physics: Dimensionality and Symmetry, First Edition.
Siegmar Roth and David Carroll.

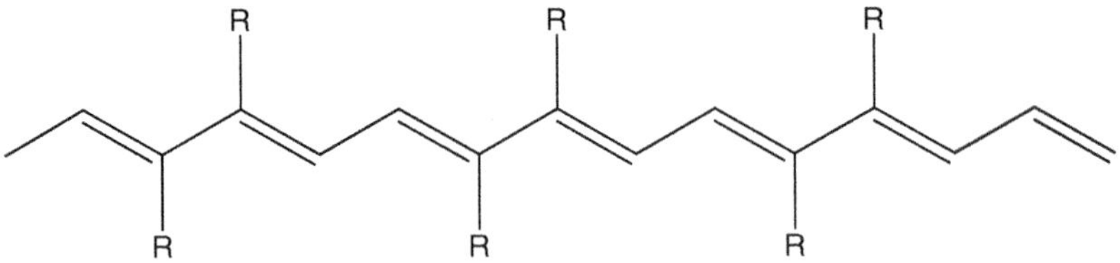

Figure 2.1 Little's superconductor. Specially designed groups are attached to polyacetylene chains so that excitations in the substituent Rs "pair" the electrons moving along the chain.

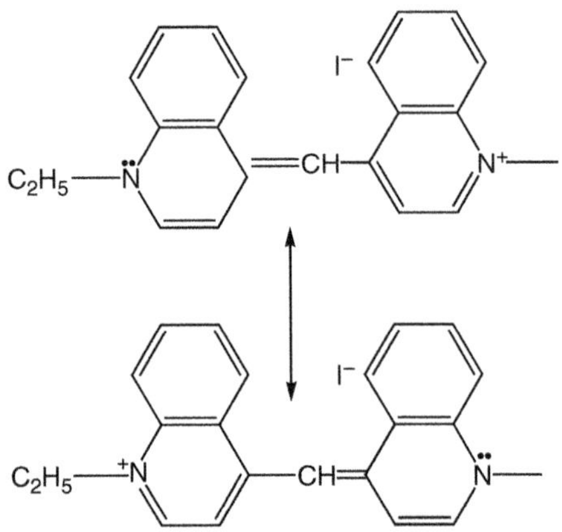

Figure 2.2 Suggestion for a substituent R in Little's superconductor and rearrangement of double bonds upon excitation.

along the conjugated polymer backbone and to excite the R substituents when passing. Conceptually, one electron will deposit an *exciton* (or localized excitation composed of a charge and an atomic distortion) in the substituent R group, and the next electron will reabsorb the exciton. This "exchange" of excitons *couples* with the consecutive electrons in the same way that phonons couple with electrons in the Bardeen, Cooper, and Schrieffer (BCS) theory of superconductivity, forming electron pairs [2]. If the coupling effect is strong enough, high superconducting transition temperatures might be achieved, perhaps even as high as room temperature.

The substituent proposed in Little's paper is shown in Figure 2.2. The arrow indicates the change of the bond arrangement in the substituent during excitation. Clearly, though, any number of species might be proposed for R. The important aspect of the proposed mechanism is that it utilizes the 1D nature of conduction together with the 3D nature (to prevent scattering) of the charge-coupling group.

The hope of *room temperature superconductivity* has not yet been realized. Moreover, no substance has been found in which superconductivity seems to occur according to Little's mechanism, though it might be argued that some Type II, high-temperature superconductors present a similar doping geometry in two dimensions. Little's proposition, however, does suggest that such materials and phenomena might be approached from a dimensional point of view: it changed our perspective.

We note also that *organic superconductors* do exist, and they owe much to Little's concepts. These superconductors have transition temperatures considerably

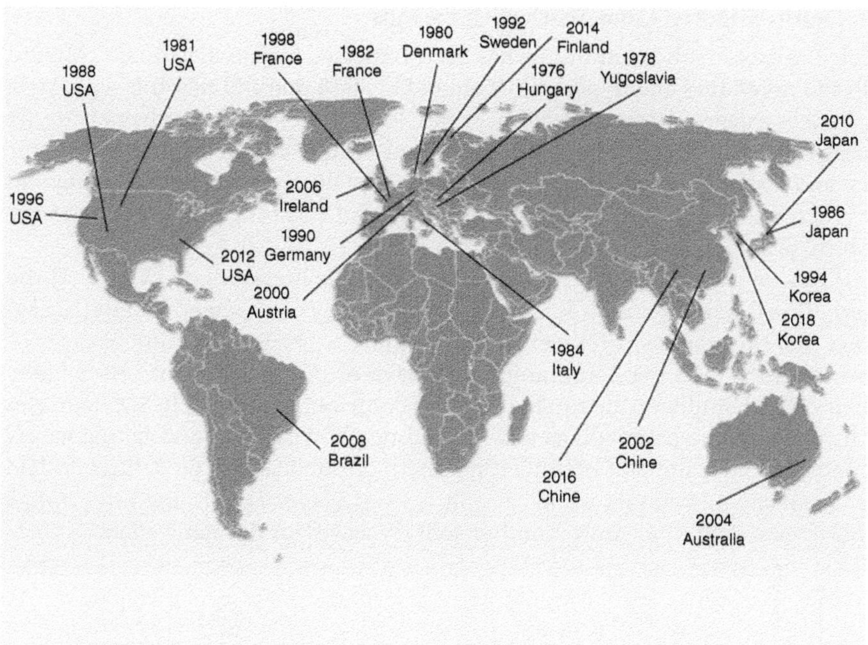

Figure 2.3 Historically, scientists attracted by *exotics* gradually formed a "community of conspiracy," moving like grasshoppers from one fashionable system to the next, a cooperative behavior of their own. Their main forum of discussion, the International Conferences of Science and Technology of Synthetic Metals (ICSM), reflects this topical schizophrenia in its proceedings (as well as a notable desire to visit the whole of the planet). The ICSM meeting proceedings can be found in *Lecture Notes in Physics* (1977, 1979), *Chemica Scripta* (1981), *Molecular Crystal Liquid Crystal* (1982), *J. Physique, Paris, Colloq.* (1983), *Molecular Crystal Liquid Crystal* (1985), and *Synthetic Metals* (1987–today).

below room temperature: the highest of which is $T_c \sim 12$ K [3]. When one includes fullerenes and their compounds, then $T_c \sim 18$ K for K_3C_{60} and 33 K for Cs_2RbC_{60} [4]. These are essentially close-packed fullerene "balls" with alkali metal intercalated between them to donate charge. Here again, notice the delicate interplay of dimension where superconducting pathways are restricted to the C_{60}s of the structure and charge is added from "outside" the structure by placing it interstitially between the balls.

Cooperative or *coupled* behaviors such as superconductivity have been and, in large part, are well understood and well established in more common solids, such as lead, alkali metals, potassium chloride, and silicon. As we will see later, these models are not so hard to follow, and they get mostly the right answers. But in some solids, let's call them *exotics* (for example, *organic and molecular crystals, liquid crystals, amorphous solids, low-dimensional solids*); there is the opportunity to push traditional models of material behavior to the extreme. This is exactly what Little's proposal did. Indeed, the approach has evolved into quite the subfield of physics, chemistry, and materials science as seen in the **inset** below and Figure 2.3.

The Exotic Scientists that Work with Exotics

While the exotic-solid community has been criticized for moving too quickly and with too much fuss and hyperbole, it should be kept in mind that this *work-style* introduces a degree of interdisciplinary intellectual exchange. In other words, by necessity it yields a research community that readily combines physics, chemistry, and materials science, embraces novelty and discovery, and reaches beyond the narrow confines of established disciplines. But this is of course only a philosophical point.

For fun we note that a statistical evaluation of the literature on solitons at the International Conference on Science and Technology of Synthetic Metals (ICSM) shows that the response of the exotic community to the introduction of the soliton moves through the community as a wave of publications that rises like a solitary wave! Similarly, the total number of contributions to the ICSM increases each year with the proceedings now exceeding 5000 pages and weighing nearly 20 pounds! Here, the plot of annual numbers of publications on solitons is laid over Katsushika Hokusai's wood carving *View of Mount Fuji from a wave trough in the open sea off Kanagawa* (another solitary wave) for dramatic effect.

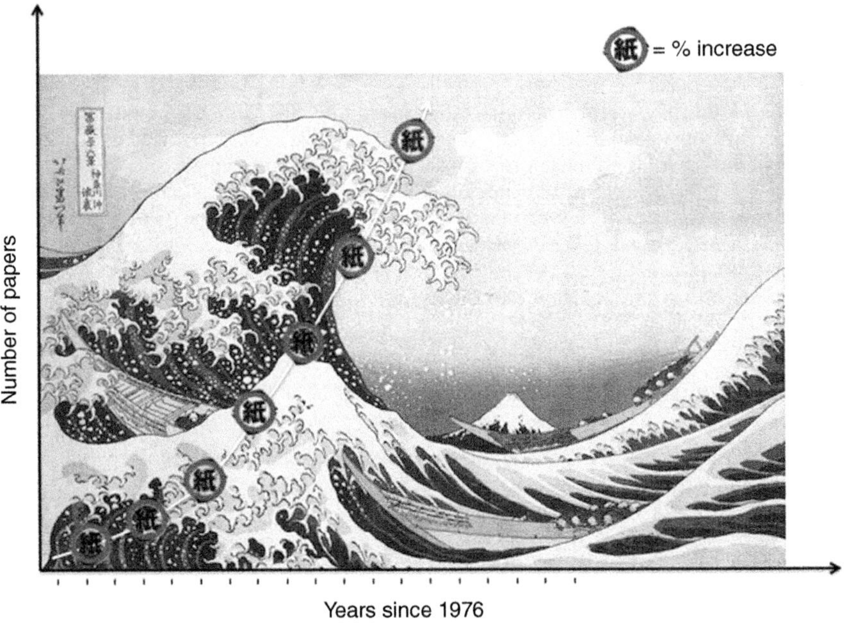

2.1 A15 Compounds

A15 compounds (also known as β-W or Cr_3Si structure types) are *intermetallic compounds* such as V_3Si and Nb_3Sn, where one of the partners is a transition metal. The chemical formula of these compounds is A_3B (where A is the transition metal and B can be any *element*), and the A15 structure is shown in Figure 2.4.

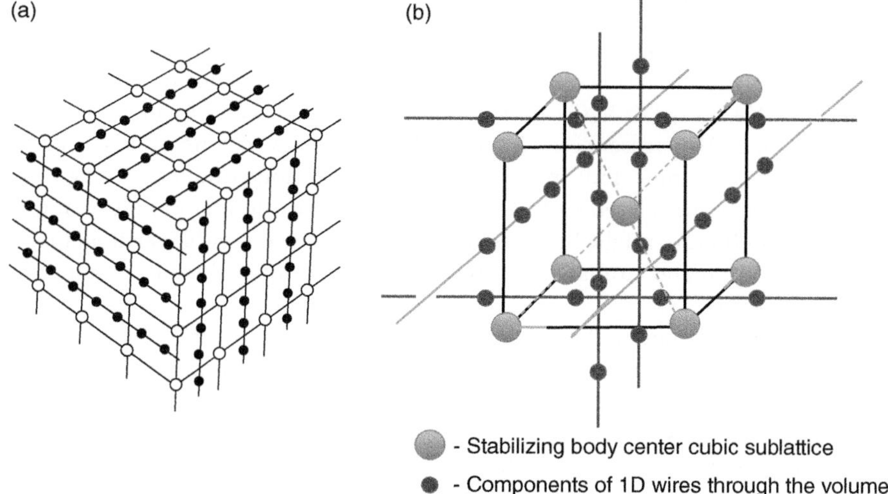

Figure 2.4 Crystal structure of A15 compounds. (a) Filled circles are transition element atoms (vanadium or niobium). They form three sets of mutually perpendicular chains. Open circles symbolize Si, Sn, Ge, or Ga atoms. They mainly serve to support the structure. (b) The unit cell of the structure showing the conduction pathways.

The essential feature of this structure is that the transition metal atoms are arranged in chains. There are three mutually perpendicular sets of chains, pointing to the three directions of 3D space. At first glance the structure seems 3D, and it takes some imagination to recognize the chains. The distance between two vanadium or niobium atoms in the same chain is smaller than between atoms of different chains. In addition the wave functions of the valence electrons of vanadium are oriented in such a way that the in-chain overlap is considerably larger than the chain-to-chain overlap.

The origin of the name *A15 compounds* is quite remarkable; it is the classification of the crystal structure. This can be misleading however. Structures starting with the letter A are designated to structures of elements, not of compounds. A15 would then be the structure of the 15th modification. Since 15 is a fairly large number and numbering probably would start on the simple side, we would expect a complicated structure. The *β-tungsten structure* name for the A15 structure is because for some time people thought that this was the structure of a complicated modification of the element tungsten. Later it was shown that β-tungsten is not elementary tungsten but actually a binary compound, a suboxide of tungsten, W_3O. The wrong assignment in the early days is easy to understand. X-ray diffraction is the most important tool for structural analysis, and the light oxygen atoms with atomic number 8 are difficult to observe in the presence of the heavy tungsten atoms with atomic number 74. However, in the case of 1D metals, the crystallographic misinterpretation does not matter: we are only interested in the sort of atoms that form those metallic chains, and these are the heavy atoms. The light atoms mainly act as spacers.

Why Are A15s Important?

A15 compounds are important as Type II superconductors, and for a long time the highest superconducting transition temperatures were found among the A15s (V_3Si, 17 K; Nb_3Sn, 23 K; V_3Ga, 15 K). These compounds can remain superconducting even in the presence of a magnetic field of up to tens of Tesla. Even today Nb_3Sn is an important material for high field superconducting magnets. Of course, it was speculated early on that the high transition temperature of the A15 superconductors might be somehow related to the chain-like arrangement of the transition metal atoms. From our discussion of dimensionality so far, this might be expected.

In addition to superconductivity there is another interesting phenomenon observed in A15 compounds: a structural phase transition from a cubic crystal lattice to tetragonal lattice occurs. With reference to the crystallography of iron–carbon systems (steel), this transition is often called *martensitic transition*. At the martensitic transition, the chains of one set contract, whereas the chains belonging to the other two sets expand. This transformation is accompanied by a "softening" of lattice vibrations because some of the restoring forces (those which stabilize the cubic crystal) fade at the transition. Perhaps these "soft" lattice vibrations are responsible for the coupling of the conduction electrons to form Cooper pairs and thus lead to superconductivity. In any case, there is an interesting reciprocal action between the superconducting phase transition and the martensitic phase transition: whichever occurs first upon cooling seems to rule out the other. Exposure to a magnetic field modifies both transitions [5]. (A review of A15 superconductors is available in [6]; for general reading on superconductivity, [7] is recommended.)

In Figure 2.5 the soft lattice vibrations are illustrated for a superconducting V_3Si sample, near the martensitic transition and in a magnetic field. The velocity of sound is plotted vs. the applied magnetic field. The polarization and propagation direction of the sound waves are chosen so that the sound velocity is determined by the restoring force that stabilizes the cubic structure. At the martensitic transition this force vanishes and the velocity approaches zero. Close to the martensitic transition, the sample is superconducting. Superconductivity prevents the martensitic transition. As the magnetic field is turned up, superconductivity is gradually suppressed, and another effect of the magnetic field becomes dominant. From there on, the velocity increases, i.e. the sample hardens, because the magnetic field works toward preventing the martensitic transition.

The soft mode behavior of A15 compounds and its modification by superconductivity and by magnetic fields can be traced back to singularities in the electronic structure of these compounds (as we discussed earlier). Figure 2.6 shows the density of states for V_3Si. There is a wide band due to the s-electrons and three narrow d bands. In later chapters we will discuss the electronic density of states of a 1D solid that has square root singularities at the band edges, i.e. the density of states approaches infinity with $(E_0 - E)^{-1/2}$. These singularities are indicated in Figure 2.6a. In V_3Si, there are three mutually perpendicular sets of vanadium chains. In the cubic phase all chains are equivalent, and the singularity occurs at identical energetic positions for all chains. In the tetragonal phase this

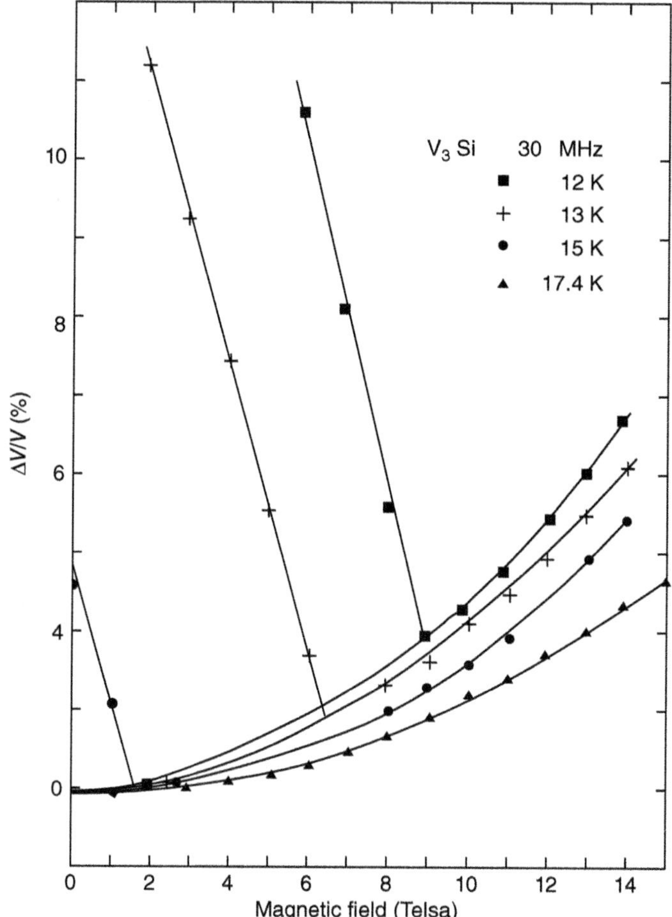

Figure 2.5 Field dependence of the soft mode sound velocity in V_3Si at various temperatures Source: From Cheeke et al. 1973 and Dietrich et al. 1975 [5].

degeneracy is lifted, because in the compressed chains the bandwidth increases and the band edge moves downward; in the expanded chains it moves upward as shown in Figure 2.6b. If the Fermi level is close to the singularity in the density of states, moving the singularity downward causes an overall reduction of electronic energy. Thus the tetragonal phase is energetically favored. At high temperatures there are also occupied states above the Fermi level, and the energy gain by the tetragonal distortion of the lattice vanishes. Therefore, the cubic lattice is stable at high temperatures.

The martensitic transition is a special case of the more general *Jahn–Teller theorem*. It states that an electron shared by degenerate states can always gain energy when the degeneracy is lifted and the electron moves into the lower state, as schematically indicated in Figure 2.7. Another example of energy gain by symmetry breaking is the *Peierls transition*.

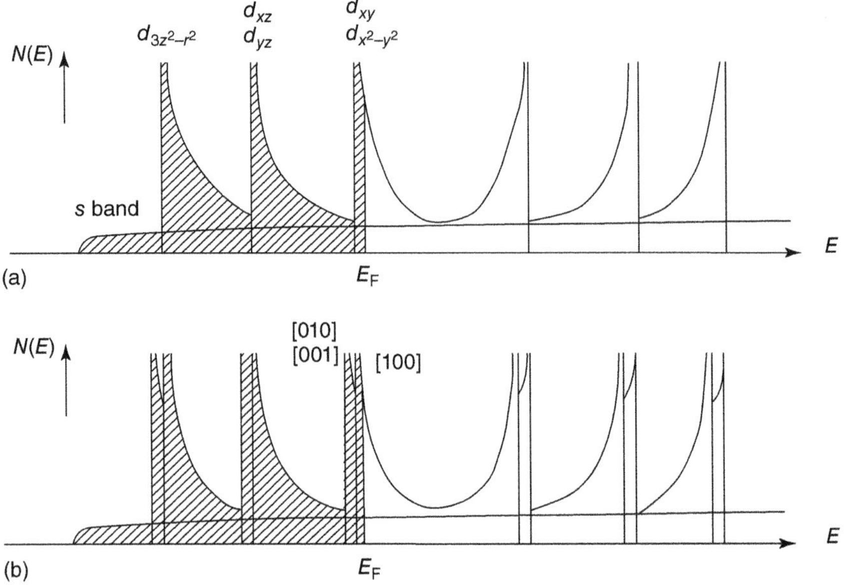

Figure 2.6 Electronic density of states of V_3Si in the cubic phase (a) and the tetragonal phase (b). In the cubic phase the singularities at the band edges are threefold degenerate because all three sets of V chains are identical. In the tetragonal phase this degeneracy is lifted, resulting in a lowering of the Fermi level and, as a consequence, of the overall energy of the electrons. Source: After Labbe 1960 [8].

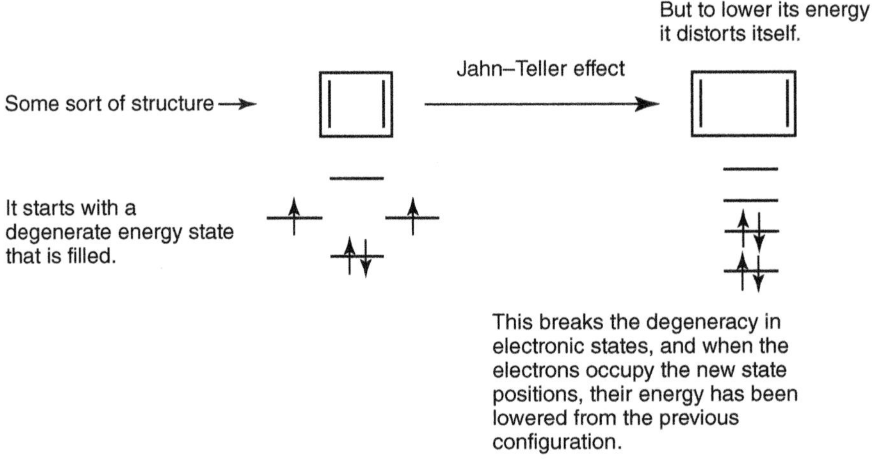

Figure 2.7 Scheme of the Jahn–Teller effect: an electron shared by degenerate states gains energy upon lifting degeneracy and descends to the lower state.

Similarly, the occurrence of superconductivity also stabilizes the cubic phase. Superconductivity creates a gap around the Fermi level; modifications of the band structure within this gap do not affect the energy balance. The same argument can be used to explain the lattice hardening induced by a magnetic field (in the nonsuperconducting state). In a magnetic field Zeeman splitting of the electron energies occurs. This Zeeman splitting alters the singularity in the density of states already, so that a further modification by lattice distortion is less effective.

So the A15 compounds are structures composed of 1D wires running throughout the volume with no interaction between them. The wires are held in place by a sublattice of non-interacting atoms (as far as conductivity is concerned). Yet this simple structure yields several important terms for our structural lexicon of materials: *martensitic*, *Jahn–Teller*, and *Peierls* – all features particularly enhanced by lowering the dimension of a system. Moreover, its electronic structure also reflects its dimensionality.

2.2 Krogmann Salts

As mentioned, A15 compounds contain three perpendicular sets of chains. Therefore the properties of the material exhibit cubic symmetry. If there was only one set of chains, a much more drastic appearance of one-dimensionality would be expected. *Krogmann salts* [9] are a good example for solids with only one set of chains. The prototype of Krogmann salts has the stoichiometric formula $K_2[Pt(CN)_4]Br_{0.3} \times 3H_2O$. This compound is commonly called KCP, an abbreviation of kalium(potassium)tetracyanoplatinate. The essentials of the structure are presented in Figure 2.8 [10]. Figure 2.8a shows a planar complex of platinum surrounded by four cyano groups. Two of the electronic orbitals of platinum are above the plane. In Krogmann salts such complexes are stacked one on top of the other to generate a columnar structure (Figure 2.8b). For the physicist the important part of the Krogmann salt is the metal chain. The function of the cyano groups is to sterically support the structure. In addition to $Pt(CN)_4$ complexes, the crystals also contain alkali metals for electron balance, with the "fine-tuning" achieved by halogen. The (001) and the (010) projections of the elementary cell of KCP are shown in Figure 2.9 [10] (while we have not yet covered this nomenclature for crystals, it is rather easy to imagine that we are talking about rather high symmetry axis of the crystal here).

The one-dimensionality of Krogmann salts becomes apparent when looking at KCP crystals under polarized light. If the light is polarized parallel to the direction of the platinum chains, the crystal appears shiny and lustrous like metals. For perpendicular polarization its appearance is dull as an insulator.

Why Are Krogmann Salts Important?

Some people say that in solid-state physics each effect has its proper substance where the effect is best exhibited and most easily studied. If so – in A15 compounds we discussed the martensitic transition – KCP is the substance belonging

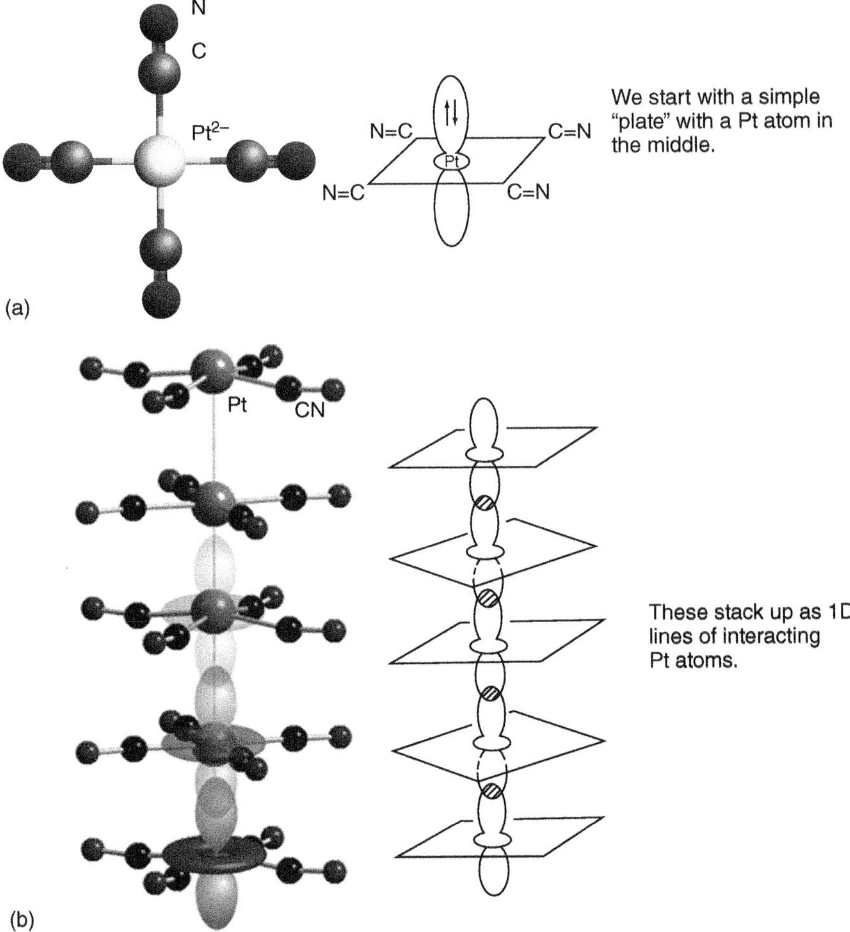

Figure 2.8 Electronic orbitals of platinum above and below the plane of the tetracyanoplatinate complex (a); linear chain arrangement of platinum orbitals (b) Source: From Wagner 1974 and Zeller 1973 [10].

to the *Kohn anomaly*. The Kohn anomaly is the precursor of another structural phase transition, which is likely to occur in 1D solids: the Peierls transition. The Peierls transition consists of pairing of atoms in a metallic chain. This pairing – also an example of symmetry breaking, in this case of a translational symmetry – results in the doubling of the elementary cell, containing one atom before the phase transition and two after. Again there are restoring forces that tend to keep the atoms of the chain equidistant and that become soft when the phase transition is approached.

In A15 compounds the softening is observed for long-wavelength shear waves (i.e. transverse polarization); in KCP it occurs for short-wavelength compressional waves (with longitudinal polarization). The atom pairing – or more general, the rearrangement of the atoms – corresponds to a compressional wave with a wavelength twice the interatomic distance. Figure 2.10 shows the dispersion

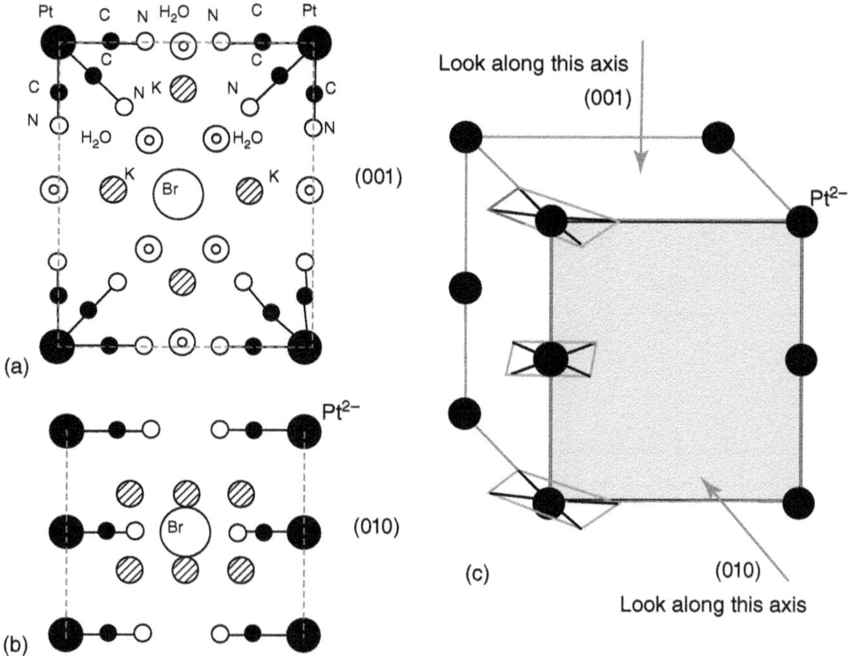

Figure 2.9 The Elementary cell of the Krogmann salt KCP ($K_2[Pt(CN)_4]Br_{0.3} \times 3H_2O$). (a) Projection of the cell along the (001) crystal, (b) Projection along (010). The metallic chains can be seen in both projections. (c) 3D rendition of the cell. Source: From Wagner 1974 and Zeller 1973 [10].

relation for longitudinal vibrations in KCP. The dispersion relation is the dependency of the vibrational frequency on the reciprocal wavelength, and the vibrational frequency is a measure of the restoring force. A pronounced dip in the dispersion relation of KCP can be noticed. A small dip is predicted for the vibration dispersion relation of any metallic solid, and it occurs where the wavelengths of the vibrations match the wavelength of the electrons at the Fermi level. The small dip is called the *general Kohn anomaly*. However in KCP, the Kohn anomaly is *giant*, a dip that actually reaches to zero in frequency, corresponding to the fading of the restoring force, and leads to a structural phase transition.

The velocity of long-wavelength lattice vibrations can be measured by ultrasonic techniques. This was done in the case of V_3Si (Figure 2.5). To measure the frequency of other vibrations and to sample the dispersion relation, neutron scattering has to be employed. The data in Figure 2.10 were obtained in this manner. (The above discussion is actually oversimplified: the giant Kohn anomaly does not occur at a wavelength exactly twice the interatomic distance but at some more general location in the *reciprocal lattice*, the exact position being determined by the Fermi level and ultimately by the electron balance of the crystal due to the alkali and halogen ions.)

So the Krogmann salt, and systems like it, introduces yet another concept into our vocabulary: the *Kohn anomaly* – or a softening of vibrations along the 1D

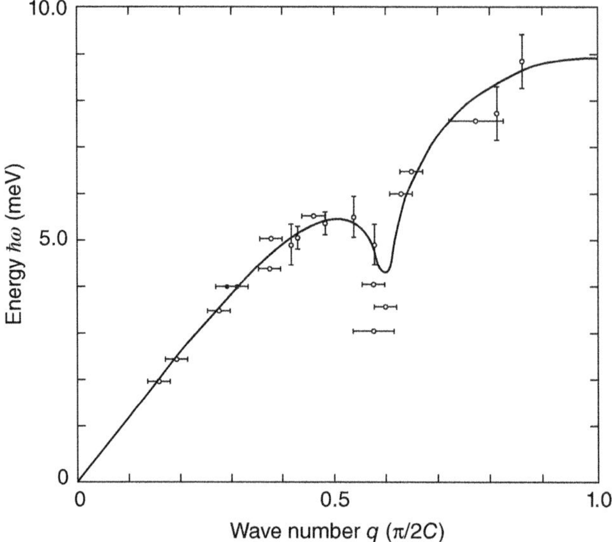

Figure 2.10 Longitudinal acoustic phonon branch in [001] direction of KCP showing the Kohn anomaly as a large dip [11].

chains that compose the solid. We also see a dramatic reduction in isotropy in the system since the chains run in only one direction for this system.

2.3 Alchemists' Gold

A15 compounds contain three perpendicular sets of metallic chains, Krogmann salts one, and – if searched – there will certainly also be a system with two sets. *Alchemists' gold* is such a system. Mercury and gold are direct neighbors in the periodic system of the elements. Perhaps the alchemists thought that exposing mercury to a very aggressive environment would convert it into gold. Arsenic fluoride certainly is aggressive, and mercury can be brought into contact with that substance. Figure 2.11 [12] shows the crystal structure of $Hg_{2.86}AsF_6$. The mercury chains along the crystallographic a direction are clearly seen in this figure. To see both sets of chains, in a and in b directions, Figure 2.12 is more appropriate [13].

Why Is Alchemist Gold Important?

One of the curious properties of this substance is that the mercury in the chains is said to be a "liquid," whereas the AsF_6 matrix is a well-defined solid. How does a 1D liquid differ from a 1D solid? The usual definition given is that in solids and in liquids, the distance between next neighbors is fixed, but in solids the angles are also fixed. In a perfectly straight chain, there is no angular degree of freedom anyhow; therefore the difference between solid and liquid must disappear. (Polymer chains as in Figures 1.11 and 1.12 are not straight but zigzag, and hence they

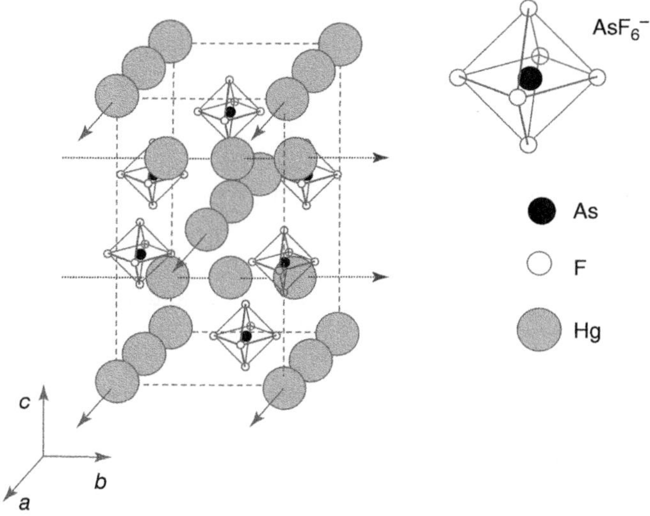

Figure 2.11 Crystal structure of "alchemist's gold," $Hg_{2.86}AsF_6$. The mercury chains follow the crystallographic *a* and *b* axis. Source: After Kaveh and Ehrenfreund 1979 [12].

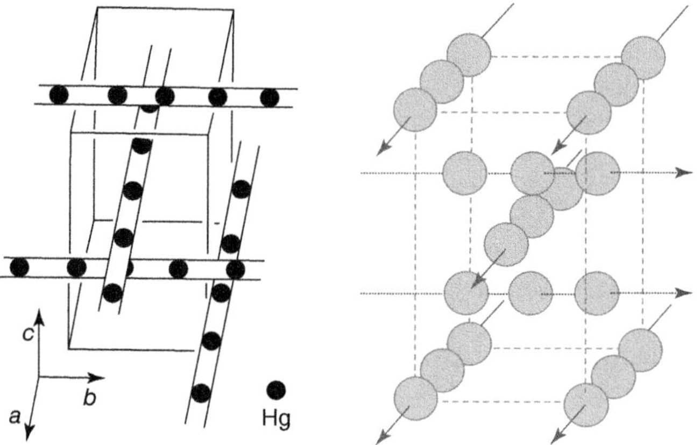

Figure 2.12 Mercury chains in alchemist's gold. Source: From Heilmann et al. 1979 [13].

can melt by rotation around the bonds.) If alchemists' gold is cooled down, the AsF_6 matrix contracts and the crystal "sweats mercury": small droplets of liquid mercury appear on the faces. Upon heating these droplets are reabsorbed into the bulk crystal.

From the fundamental science point of view, the mercury chains are quite interesting. Bulk mercury is a superconductor. Will 1D mercury chains also become superconducting? Is there an anisotropy to be expected, say, in the critical magnetic field? Will the transition temperature of the mercury chains be above or below the T_c of bulk mercury? Or is there no transition to superconductivity because of Landau's interdiction of phase transitions in one dimension? Unfortunately the escaped mercury makes experiments and their interpretation more difficult than expected. There is a transition to superconductivity around 4 K, which coincides with the T_c of bulk mercury and apparently is due to superconductivity in the secreted droplets. There is a further transition around 2 K, where the chains become superconductive.

2.4 Bechgaard Salts and Other Charge-Transfer Compounds

In the Krogmann salts planar units are stacked to form columns, and the molecular orbitals overlap within these columns to form a 1D metal (Figure 2.8; overlap of the d-orbitals of platinum). In a similar way the p-orbitals of *carbon* can be forced to overlap by appropriate stacking of planar organic units.

Figure 2.13 [14] shows the structure of $(TMTSF)_2PF_6$, one example of *Bechgaard salts*. In Bechgaard salts organic donors are combined with inorganic acceptors. The donors are fairly large planar organic molecules, in this case tetramethylenotetraselenofulvalene (TMTSF). For everyday use the names of these compounds are too lengthy and therefore usually abbreviated to some combination of characteristic capital letters, for example, TMTSF. The organic donors contain large numbers of conjugated double bonds. From the point of view of a simpleminded physicist, they are just tiny flakes of graphene [15] (Figure 2.14). The chiplets are stacked like honeycomb planes as in graphite with a certain overlap of the p-electrons along the stacking axis. As a consequence, electronic energy bands are formed. There is a certain amount of charge transfer from the organic donor to the inorganic acceptor. Because of this *charge transfer*, the electronic bands are only partially filled, and the Bechgaard salts are metallic. At low temperatures some of the Bechgaard salts become superconducting; others undergo a *metal-to-insulator transition* via the Peierls mechanism. As in the A15 compounds, superconductivity often competes with other phase transitions.

Figure 2.14 [16] shows another Bechgaard salt: $(fluoranthenyl)_2SbF_6$. This substance is famous for its very narrow electron spin resonance (ESR) lines, which

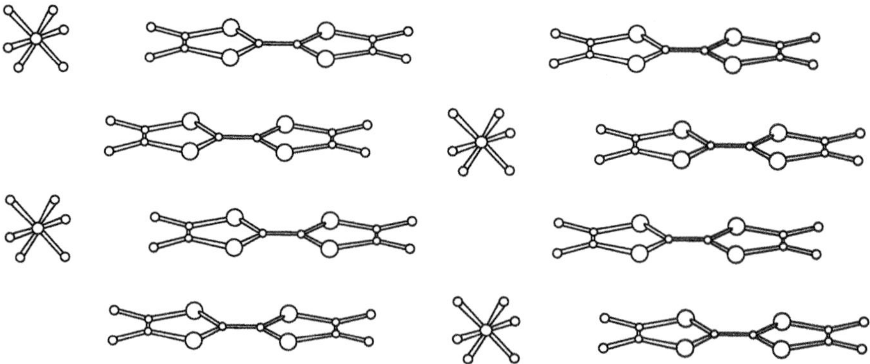

Figure 2.13 Crystal structure of the Bechgaard salt $(TMTSF)_2PF_6$. Source: After Jerome 1981 [14].

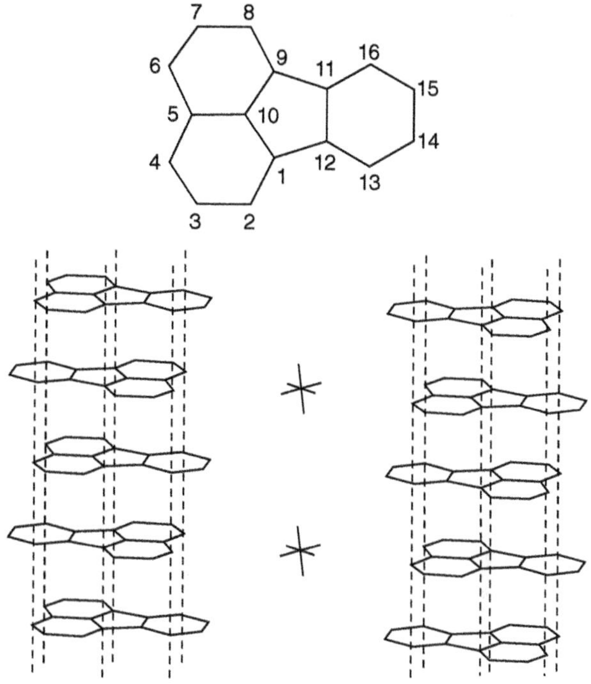

Figure 2.14 Crystal structure of (fluoranthenyl)$_2$SbF$_6$. Source: After Mehring and Spengler 1984 [16].

make it a very sensitive magnetic field probe. The narrow ESR lines are due to the delocalization of the electrons over the fluoranthenyl stacks and thus are typical for good 1D metals. Figure 2.15 [17] summarizes several organic donors, and Figure 2.16 [17] organic acceptors.

In Bechgaard salts organic donors are combined with inorganic acceptors. It is also possible to make all-organic charge-transfer salts by combining an organic donor with an organic acceptor. A very famous example of this approach is TTF–TCNQ. The crystal structure of this substance is shown schematically in Figure 2.17 [18]. The TTF and the TCNQ units are arranged in a herringbone pattern. Because of the tilt, the units are more densely packed with a larger overlap between the molecules in a stack. It is interesting to note that both TTF and TCNQ also crystallize separately, forming insulating solids. Thus, it is the charge transfer from the TTF stacks to the TCNQ stacks that makes TTF–TCNQ metallic.

Why Are Bechgaard Salts Important?

TTF–TCNQ shows many interesting solid-state phenomena, such as high electrical conductivity and Peierls transition from metal to insulator, but it does not become superconducting. In the 1970s superconductivity with a T_c as high as 58 K in some of the TTF–TCNQ samples was reported [19]. Of course, these reports attracted an enormous amount of attention. In fact, the first report is the paper most frequently cited of all papers ever published in *Solid State Communications*. Today, however, many scientists agree that these materials show tricky

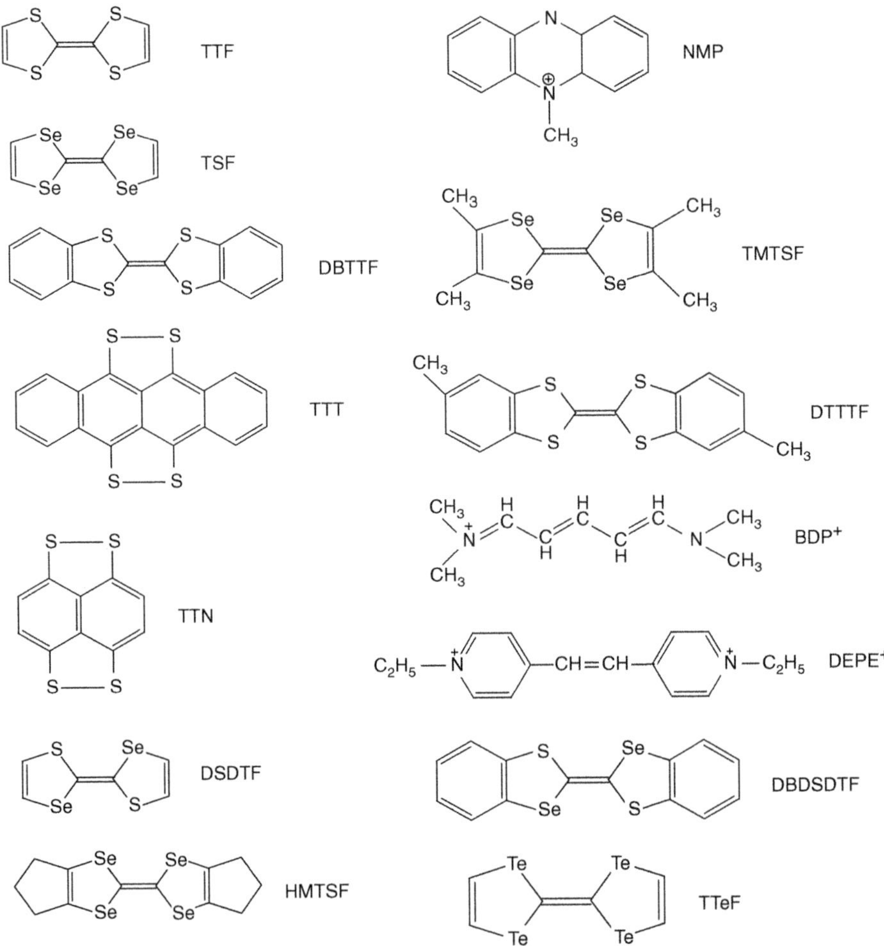

Figure 2.15 Organic donors. Source: After Hamann et al. 1981 [17].

artifacts caused by the high, strongly temperature-dependent anisotropy of the electrical conductivity [20].

The use of heat treatment, various solvents and cosolvents, and other techniques has become a major point of interest in working with these materials. The larger the domains of perfect structure that can be created, without defects or grain boundaries, the larger the *mean free path* of the carriers can become. This has been tied to applications such as organic transistors as we discuss later on. But for our purposes here, the 1D metals formed have been again added to our vocabulary: *charge-transfer complexes* and *metal–insulator transitions* are two ideas that are strongly expressed in such low-dimensional systems.

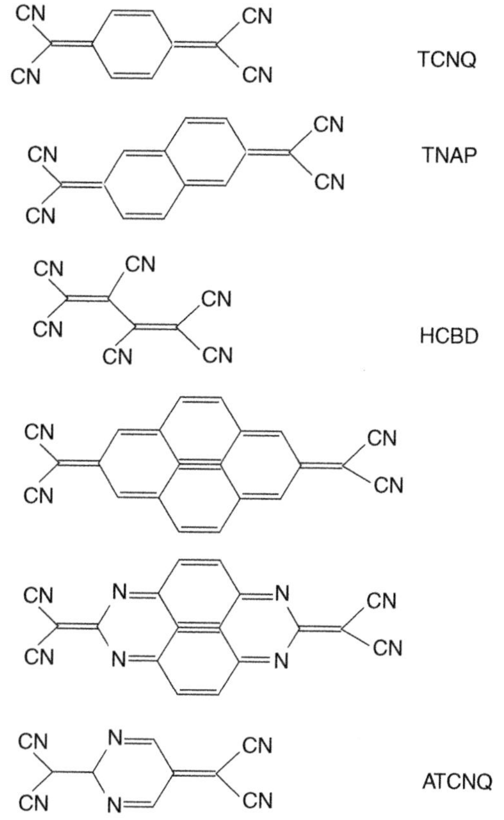

Figure 2.16 Organic acceptors.
Source: After Hamann et al. 1981 [17].

2.5 Polysulfurnitride

For the physicist, it may not be evident that sulfur and nitrogen can be arranged in long chains in which sulfur atoms and nitrogen atoms alternate regularly. But chemists know it can be done; the substance even crystallizes and shows metallic properties. Furthermore, it becomes superconducting, although only at extremely low temperatures ($T_c = 0.26$ K [21]). The structure of polysulfurnitride $(SN)_x$ (also called polythiazyl) is shown in Figure 2.18 [22]. $(SN)_x$ can be doped, for example, with bromine. Doping slightly increases the T_c, possibly due to the Fermi level being moved to a position of a higher density of states, which is more favorable for superconductivity.

Why Is Polysulfurnitride Important?

$(SN)_x$ is usually accredited with being the first conducting inorganic polymer. Probably more important than the doping-induced T_c shift of $(SN)_x$ is that the

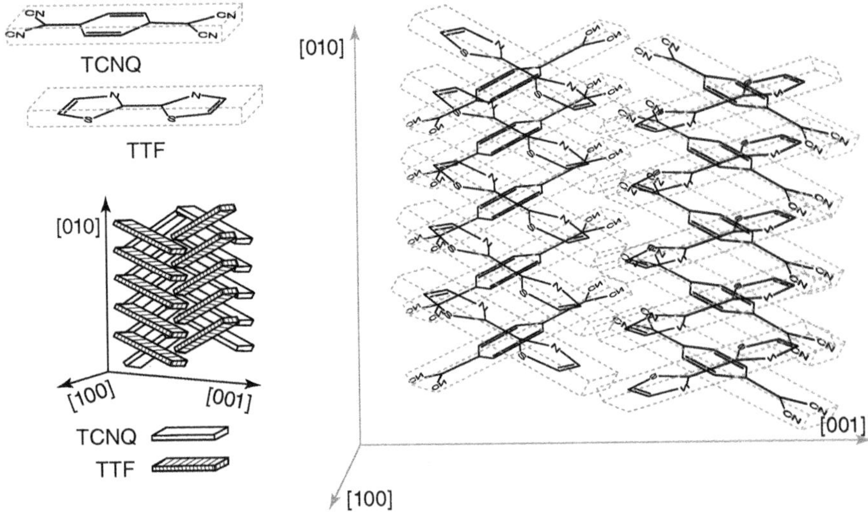

Figure 2.17 Crystal structure of the organic *charge-transfer* salt TTF–TCNQ. Source: After Friend and Jerome 1979 [18].

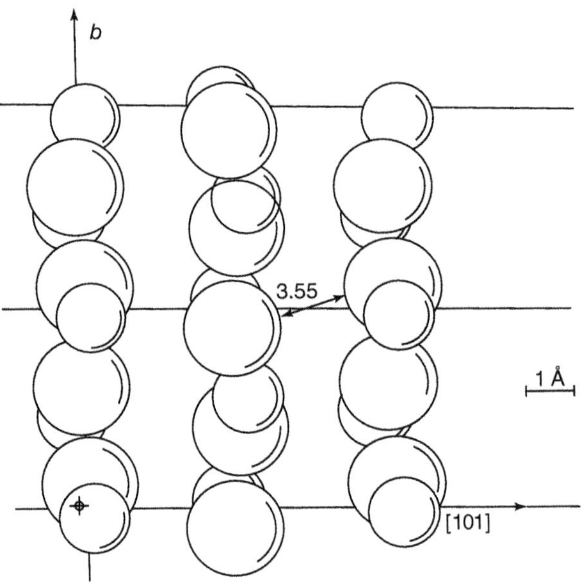

Figure 2.18 Crystal structure of the "inorganic polymer" polysulfurnitride, $(SN)_x$. Source: After Möller 1976 [22].

doping of $(SN)_x$ might have stimulated the idea of also doping polyacetylene, $(CH)_x$, leading to a broadening of the field of conducting polymers generally. Today, this polymer has a wide range of potential uses in industry such as in LEDs, transistors, and batteries.

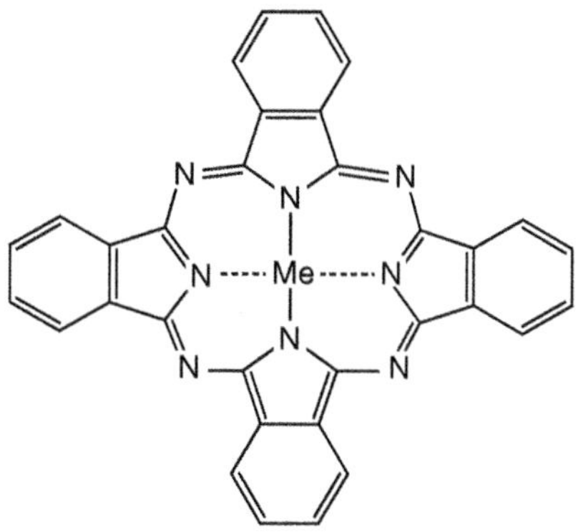

Figure 2.19 Chemical structure of the macrocyclic organic compound phthalocyanine containing a transition metal in the center.

2.6 Phthalocyanines and Other Macrocycles

When discussing the Bechgaard salts, we met with the concept of stacking of graphene-like chiplets. The chemical structure of these macrocycles is shown in Figures 2.19 and 2.20. Both substances contain conjugated systems with 18π-electrons in the "chip." For a physicist, "three is already many and ten is infinite." So these macrocycles are infinitely large plates containing an infinite number of electrons. Properly stacked they become graphite-like.

Why Are Phthalocyanines and Porphyrins Important?

The phthalocyanines and porphyrin both show a feature not present in graphite platelets: there is a void in the center that can hold a transition metal atom. While electrons may be well delocalized *on* the platelet, interactions *between* platelets can proceed through this metal atom. So, the choice of the center metal allows for a "fine-tuning" of the out-of-plane electronic properties of the stacked substance. The real "trick" is to get them to line up just right so charge transport can occur.

A particularly interesting example is lead phthalocyanine. While the other phthalocyanines are planar, in this case with the lead atom being so large, distortion of the molecule occurs. It assumes the shape of a badminton shuttlecock. Shuttlecocks should be easier to stack than flat plates and have the tendency to align the center atom. In addition, a special type of stacking fault might be observed, as indicated in Figure 2.21. The conductivity of a stack largely depends on the presence of such faults, and fault generation could be used for switching [23].

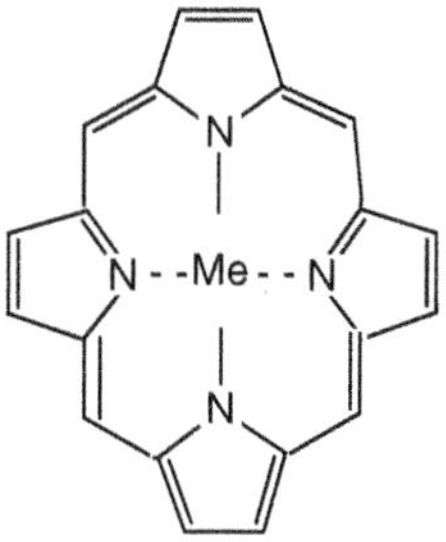

Figure 2.20 Chemical structure of the macrocyclic organic compound porphyrin with a transition metal in the center.

To stack other planar macrocycles, the "shish kebab" method has been developed: covalent connections are placed between the central metals, and the macrocycles are pinned on a polymer backbone like pieces of meat

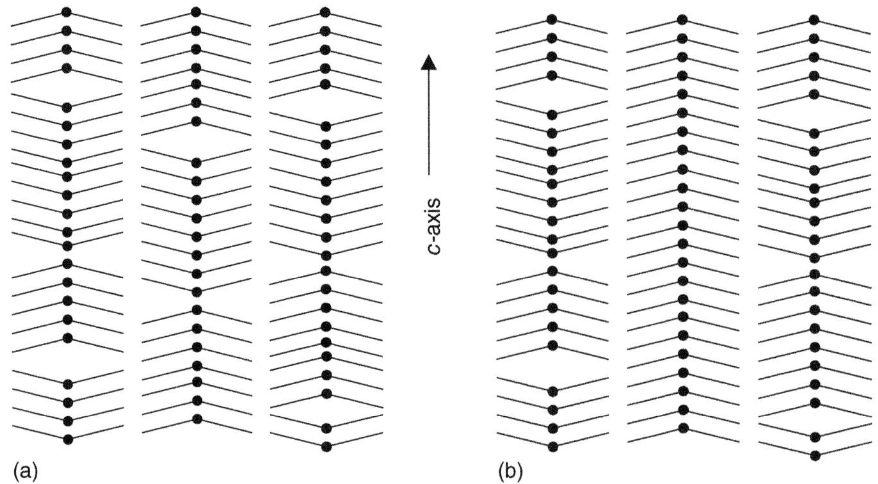

Figure 2.21 Stacks of lead phthalocyanine with stacking faults. Source: After Hamann et al. 1978 [23].

Figure 2.22 Shish kebab polymer of metallophthalocyanine. Source: After Hanack et al. 1981, 1994 [24].

on a barbecue spit (with the spit being the line of metal atoms in the center) (Figure 2.22 [24]).

2.7 Transition Metal Chalcogenides and Halides

There is a large family of transition metal chalcogenides and halides that can be described as condensed atomic clusters [25]. The basic structural unit is a "cluster" consisting of a transition metal M surrounded by group VI or group VII elements X. Figure 2.23 shows a trigonal prism with six selenium atoms at the corners and a niobium atom at the center; Figure 2.24 presents a MoO_6 octahedron.

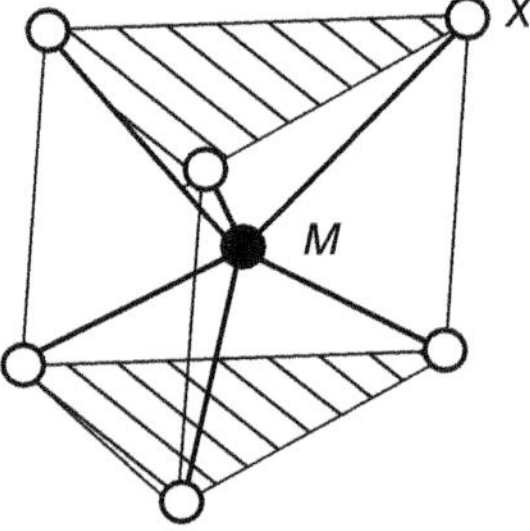

Figure 2.23 Trigonal prism of $NbSe_6$ as basic structural unit of low-dimensional niobium selenide solids.

These atomic clusters "condense" by sharing corners, edges, or faces. The case of prismatic condensation is indicated in Figure 2.25, where edge-sharing leads to the layered structure of $NbSe_2$ (a 2D material) and face-sharing to fibrillar $NbSe_3$ (a 1D material).

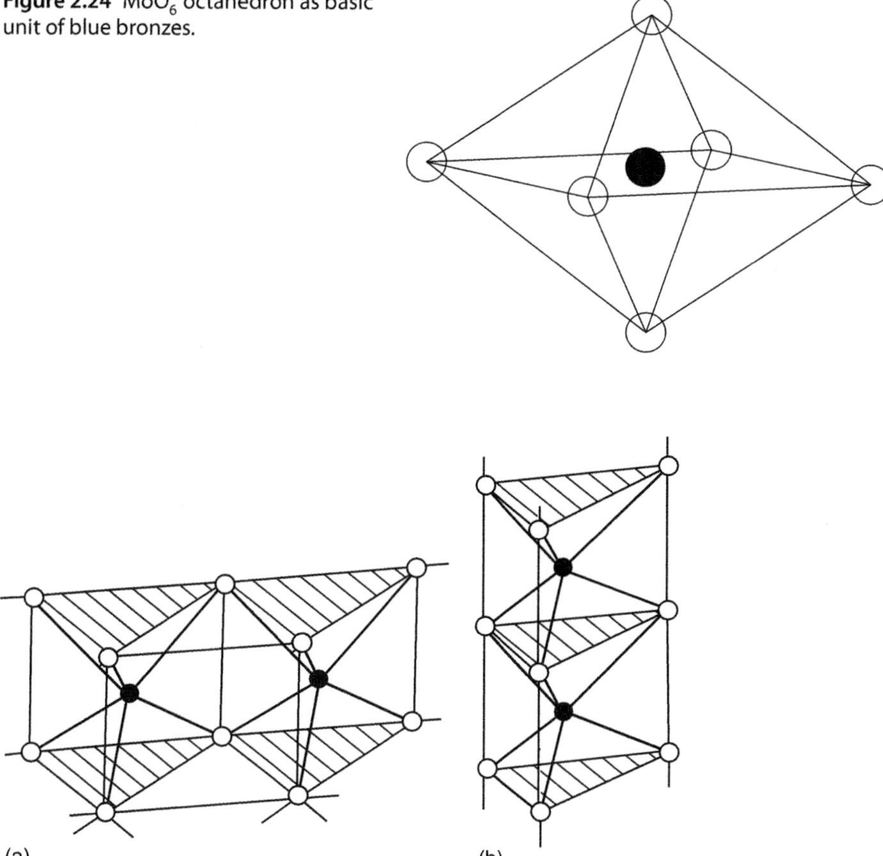

Figure 2.24 MoO_6 octahedron as basic unit of blue bronzes.

Figure 2.25 Condensation of MX_6 prisms to form layers by edge-sharing (a) or fibers by face-sharing (b). Source: After Bullett 1985 [26].

$NbSe_2$ is the representative of a large group of inorganic reduced-dimensional solids, of which the 2D types are thoroughly discussed in a series of monographs [27]. Some of these layered solids are metals and superconductors; others become metallic after "intercalation," i.e. after insertion of organic or inorganic molecules between the layers. But $NbSe_3$ is also one of the most important inorganic quasi-1D solids. It is particularly suited for studying *charge density wave* phenomena.

An example of a corner-sharing cluster condensation (making a 1D wire) is $(MX_4)_nY$, shown in Figure 2.26, where M is a transition metal such as Ta or Nb, X is a chalcogenide such as Se or S, and Y represents halogen ions between the fibers. A more complicated cluster condensate is shown in Figure 2.27, representing $K_{0.3}MoO_3$, one of the blue bronzes. A further member of this cluster condensation family we already encountered is the Krogmann salt KCP in Section 2.3 of this chapter. Here M = Pt and X = CN, the cyano group being a close relative of the halogens.

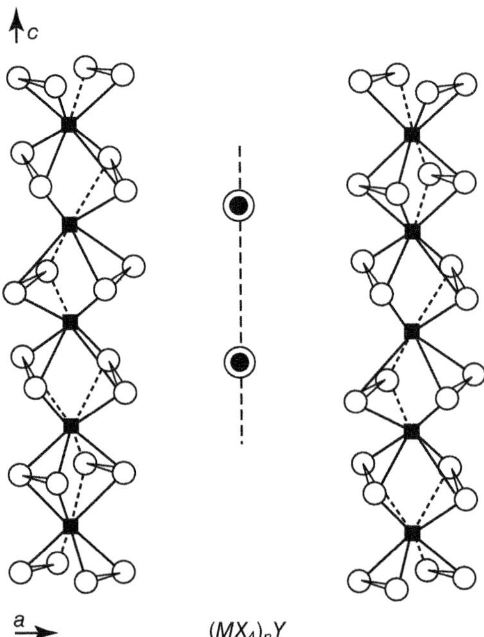

Figure 2.26 Crystal structure of $(MX_4)_nY$. Source: After Gressier et al. 1983 [28].

Why Are Transition Metal Chalcogenides and Halides Important?

Cluster condensation may lead to metals or to semiconductors (insulators), depending on the number of M and Y atoms. When the number of nonmetal atoms in a compound is not sufficient to completely surround the metal cluster, the clusters link up by direct M—M bonds [25], thus forming extended M—M chains or metal planes. The alkali atoms in Figure 2.27 and the halogen atoms in Figure 2.26 help to adjust the electron density and fix the Fermi level. Thus, they determine the details of the electronic properties of these compounds and in particular the charge density wave behavior.

2.8 Halogen-Bridged Mixed-Valence Transition Metal Complexes

Many scientists see halogen-bridged mixed-valence transition metal complexes as an ideal platform for 1D solid-state physics. The long name of these substances has been abbreviated to HMMC; another short colloquial term is *MX chains*, where M stands for transition metal atoms and X for halogen atoms. The essentials of the structure are shown in Figure 2.28: in the chain M and X atoms alternate; here, M = Pt and X = Cl. Ligands are attached to the metals; in some cases there are counterions between the chains. Hence, in a certain respect HMMCs are related to Krogmann salts (cf. Figure 2.8), even though the chains are not exclusively made of metal atoms. An electronic energy band is formed by the d_z-orbitals of the metal atoms and the p_z-orbitals of the halogen atoms.

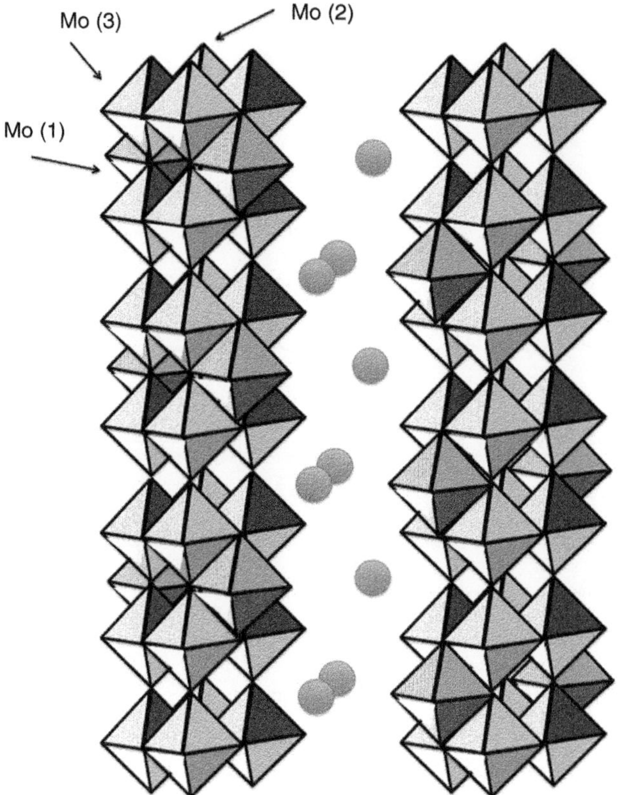

Figure 2.27 Crystal structure of the blue bronze showing the infinite sheets of MoO_6 octahedra, separated by the alkali ions. The sheets contain the infinite chains. Source: After Schlenker et al. 1983 [29].

Depending on structural details, the electrons in this band can have various degrees of localization. The limit on one side is the valencies of the transition metals oscillating along the chains between II and IV (see Figure 2.28). On the other borderline case, all metal atoms have the valence state III.

Why Are MX Chains Important?

The importance of the MX chains arises from the fact that the synthesis of these substances is straightforward and high-quality single crystals can easily be grown – not only for one particular MX compound but for a large variety. Typical metals are M = Pt, Pd, or Ni and the halogens are X = Cl, Br, or I. Common ligands comprise L = halogens, ethylamines, ethylenediamines, or cyclohexanediamine, and as common counterions between the chains Y = halogens or ClO_4^-. One particular way of writing the stoichiometric formula of a HMMC is

$$[M^{\rho-\delta}L_4][M^{\rho+\delta}X_2L_4]Y_4 \tag{2.1}$$

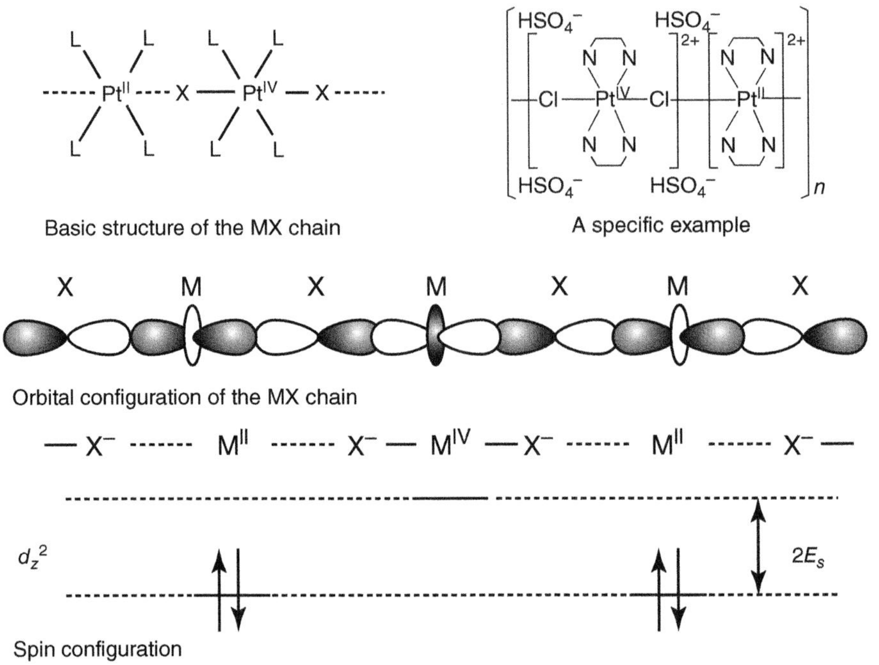

Basic structure of the MX chain

A specific example

Orbital configuration of the MX chain

Spin configuration

Figure 2.28 Halogen-bridged mixed-valence transition metal complex.

Here ρ denotes the average valence of M and δ the deviation from the average. If δ is small, the X atoms are centered between the metal atoms. For large δ the halogens move closer to the metal with the higher valency. In some PtCl chains distortion can be as large as 20%. By changing M, L, X, and Y, it is possible to vary the properties of the substance in a well-controlled way. In particular, the dimensionality can be adjusted (interchain coupling), and the electron–electron and the electron–phonon interactions can be tuned leading to a wide variety of exotic excitations along the chains such as charge density waves. Important articles on MX chains are provided in [30]. Further publications are found in the proceedings of the ICSM starting at 1990 onward (as mentioned above).

2.9 Returning to Carbon

"Let Carbon be your guide." It is now time to return to carbon to compare and contrast how it "handles" dimensionality in 1D, in light of what we have just seen. Indeed, with its ability to flexibly bond, pure carbon does throw us a few curves, with complications not readily seen in the structures we have encountered – though for some cases it isn't too hard to imagine that the tricks pure carbon plays could also be engineered in other materials. So for this section, our discussion centers on the situations in which pure carbon plays the role of the low-dimensional "wire."

2.9.1 Conducting Polymers

No other class of 1D materials has triggered such large numbers of publications or altered the landscape of solid-state physics research, as conducting polymers. In 1977 it was discovered that the conductivity of polyacetylene can be increased by many orders of magnitude, through a process known as "doping" [31]. We use quotation marks here because this type of doping is a little different from what we might know from the more typical substitutional arrangement of impurity atoms in a 3D solid like Si. Indeed, like Little's superconductor and the other many examples we have already encountered, here we mean a transfer of charge over to the low-dimensional structure.

Since 1977, conducting polymers have dominated contributions to the ICSM. Contributions can also be found in the national meetings of a great number of professional societies; the American Physical Society (APS) meetings and the American Chemical Society (ACS) meetings have both begun to incorporate sessions on conducting polymers, for instance. In 1979 the journal *Synthetic Metals* came into life, and most of the articles therein deal with conducting polymers. In 2000, the Nobel Prize in Chemistry was awarded to Alan Heeger, Alan MacDiarmid, and Hideki Shirakawa for their pioneering work on conducting polymers, which basically explained how conductivity arises [32].

The reason for all this popularity is found in three motivations:

1) Conducting polymers have already been shown to have a large number of applications, with more emerging (see later chapters and also [33]).
2) Conducting polymers are derivatives of polyenes, i.e. of compounds with extended systems of conjugated double bonds. Such systems are subject to quantum chemical concepts and calculations [34].
3) There are certain excitations in polyenes, which are related to solitary waves and to solitons. Thus, there is an interdisciplinary link to field theory, hydrodynamics, elementary particle physics, and certain aspects of biology [35].

These three motivations will quickly become the focus of our discussion in later chapters. For now, Figure 2.29 shows the basic chemical structure of some of the most important conducting polymers. The regular array of alternating single and double bonds, characteristic for polyenes, is clearly visible. (In polyaniline the extra electron pair on the trivalent nitrogen atoms participates in band formation, so that this substance is conjugated as well.)

Here we have followed several conventions while making our "stick figure" diagrams. Generally we leave out the carbon labels (C) on the diagram because we understand them to be carbon. If something else is there, we mark it. So the hexagons of the polyphenylene have carbons at each of their vertices. Next we take up the free bonds of the carbons with a hydrogen (saturating the bonds), which we also assume you know, so it's left out. If there is something different there, like in the polyfluorene, we let you know. This shorthand way of writing down the structure allows the "conjugation" or the alternating double-to-single-to-double bonds, to come through rather clearly in the picture, so we can trace the path of carrier transport.

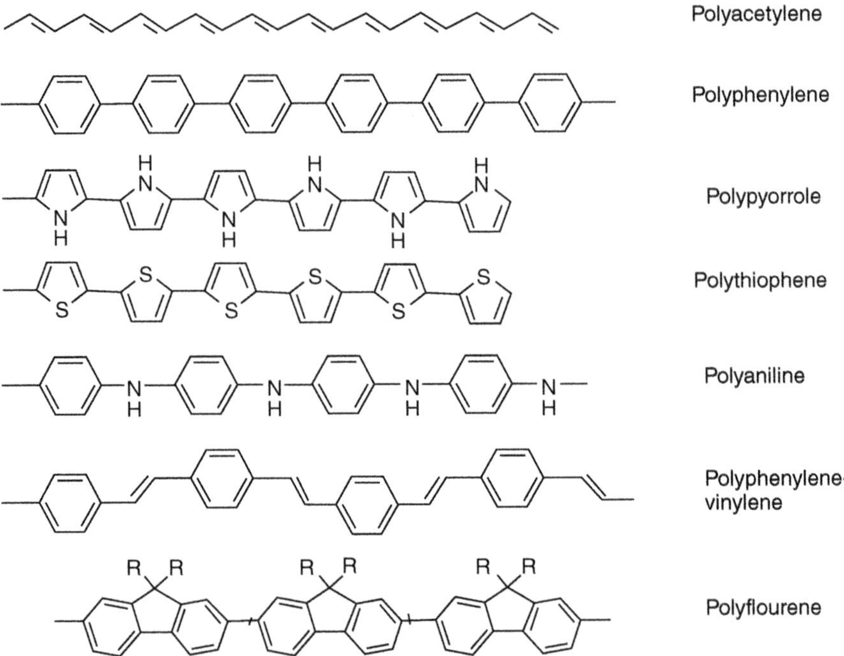

Figure 2.29 Chemical structures of some important conducting polymers.

In Figure 2.30 computer renditions of the polymers above are shown as space-filled models, highlighting the orbitals. A large part of the interesting physics is related not only to the high conductivity after doping but also to the extended system of conjugated double bonds.

Figure 2.31 shows the chemical structure of polydiacetylene. Here the bond sequence is single-double-single-triple-single-double-single-triple. From a certain point of view, it is justified to say that the system is in between polyacetylene and polycarbyne (see Figures 1.10 and 1.12). Polydiacetylene is the only conjugated polymer from which large single crystals can be grown. Actually, these crystals are not grown from the polymer but from monomeric diacetylene. In the crystal, the monomers can be polymerized without disturbing the crystal structure. This is a very rare case of *solid-state polymerization*. It can be achieved because in diacetylene crystals, the monomer molecules are already in the positions and orientations needed for the final polymer. The polymerization is initiated by light or heat. Figure 2.32 shows a scheme of the solid-state polymerization [36]. Polydiacetylene research actually has its own scientific community, and several conference proceedings and monographs have appeared on this topic [37].

It should be noted that there exists a structurally unifying way of visualization for conducting polymers. This involves the use of a single sheet of graphite. As most tunneling microscopists will tell you, to create an excellent, flat surface, on highly oriented pyrolytic graphite, all one needs is a little Scotch tape. By placing the tape on the surface and pulling off again, a few individual layers will be removed. This is due to the extremely weak van der Waals forces that hold the

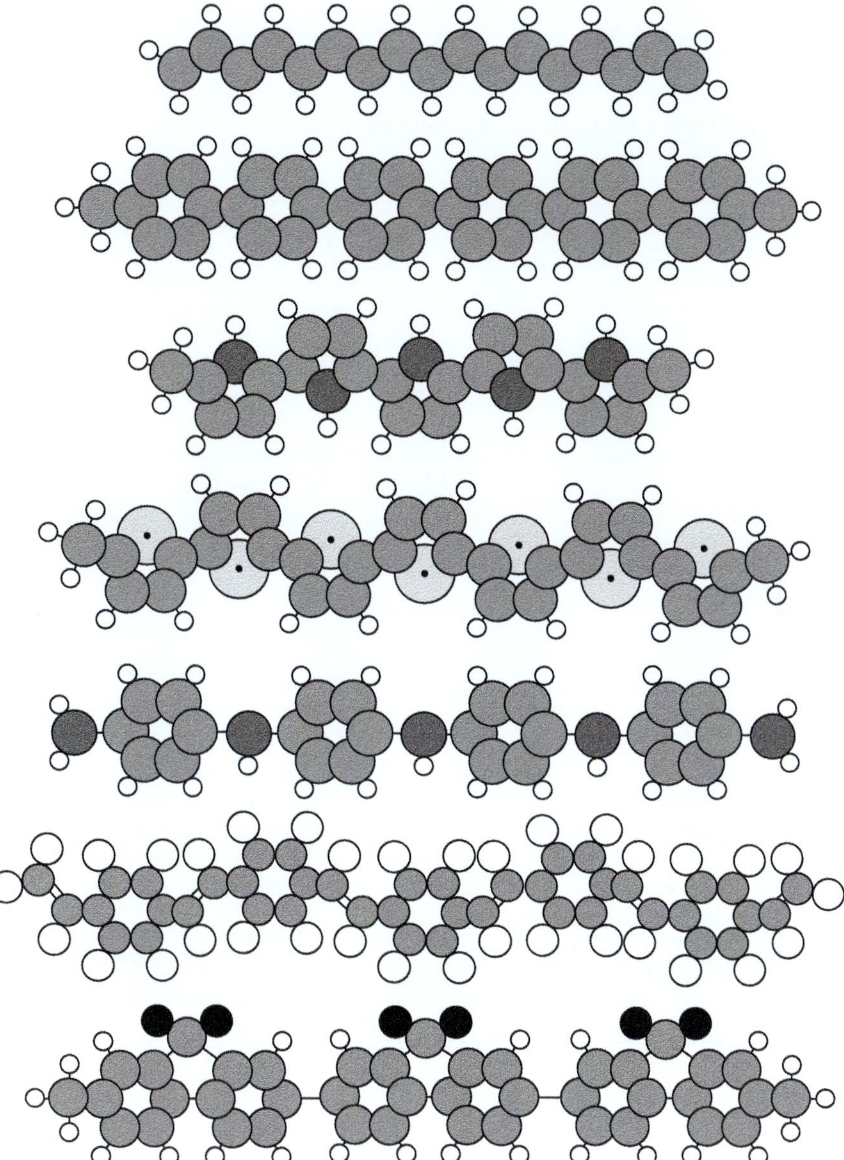

Figure 2.30 Computer-generated space-filling models of the polymers shown in Figure 2.28.

layers together. Imagine now only one, atomically thin, layer: *graphene,* and it can be used as the template for most conducting polymers. To demonstrate this, consider the graphene sheet in Figure 2.33.

2.9.2 Carbon Nanotubes

Carbon nanotubes are unique in the field of 1D structures [38], and we have introduced this odd object already. In Section 1.3.4 a 0D form of carbon was

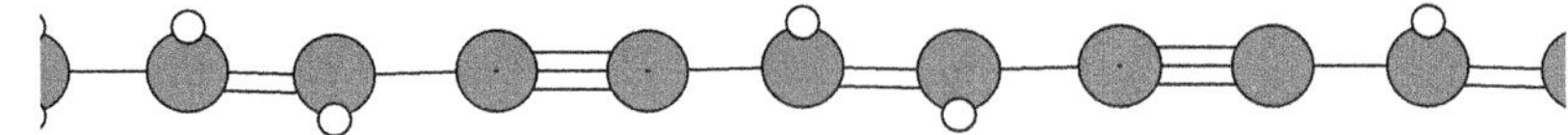

Figure 2.31 Idealized chemical structure of polydiacetylene.

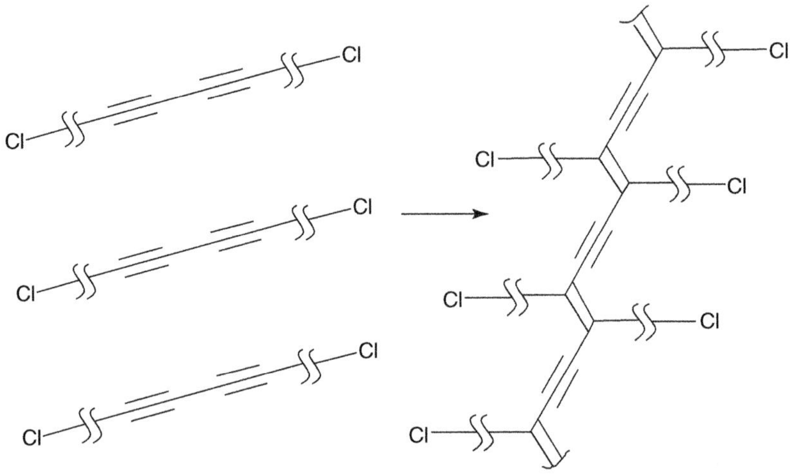

Figure 2.32 Scheme of solid-state polymerization of diacetylenes.

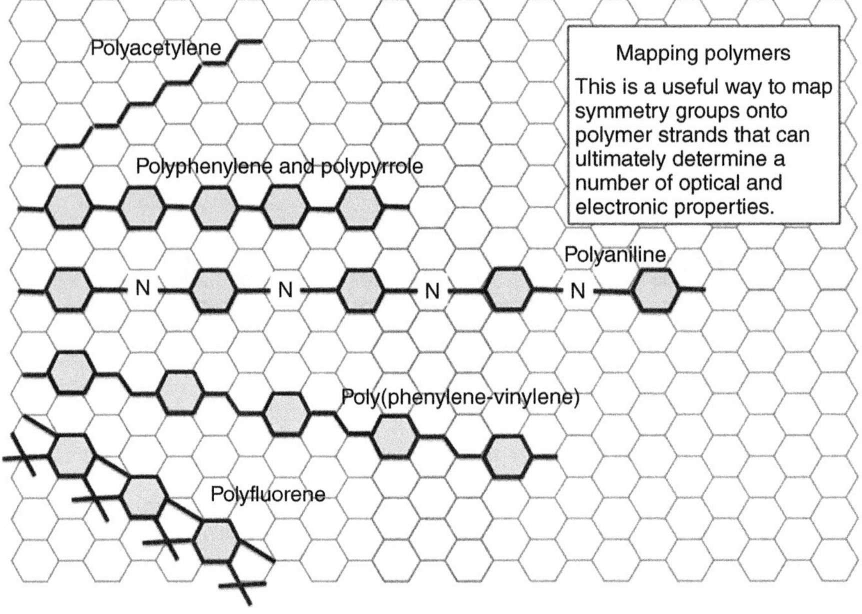

Figure 2.33 The conjugated systems map fortuitously onto graphene. The technique can be used to "map" specific symmetry groups onto polymers, allowing for a deeper understanding of their properties.

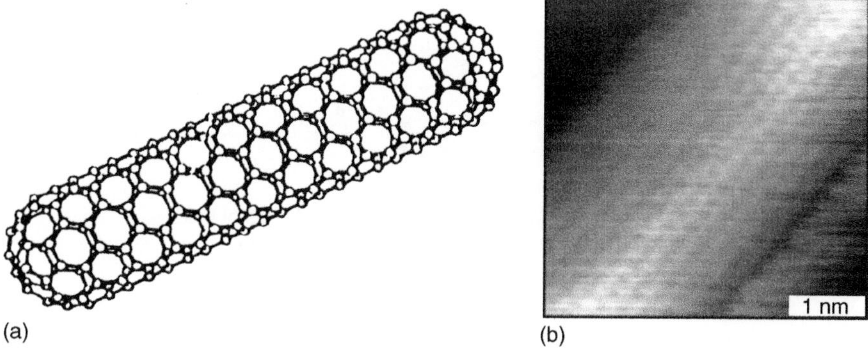

(a) (b)

Figure 2.34 "Nanotube." Carbon can form cylinders with some 10 Å in diameter and several micrometers in length. At the right is an atomic resolution image taken with a tunneling microscope at WFU.

introduced – fullerene. Figure 1.14 shows a football composed of 12 pentagons and 20 hexagons. This structural concept can be extended to molecules with more carbon hexagons, yielding long and narrow carbon tubes: sort of cylindrical graphene with fullerene caps at the ends. In Figure 2.34, one of these tubes served as logo for the "International Winter School on Electronic Properties of Novel Materials: Progress in Fullerene Research" in Kirchberg, Austria, in March 1994 [39].

When discussing the concept of a "mixed-dimensional" or *topological* object, the carbon nanotube came up again. To truly understand what the concept of a 1D topological object has to teach us, we must complete that discussion of the carbon nanotube. Specifically, to construct (conceptually) a carbon nanotube, we begin with a graphene sheet. The sheet is infinite in 2D extent, one atom thick, and composed of threefold coordinated carbon using sp^2 hybrid bonds as in Figure 2.35.

Notice that any number of A–A′ and B–B′ points could be chosen leading to different **R** and **T** vectors as we define them in the figure. So specifying *a specific* way to wrap the sheet up into a tube is done using the indices (n,m). These are derived from the number of unit vectors (upper left) that it takes to specify vector **R**.

Alternatively, one can specify the diameter of the tube and the angle, Θ, that **R** makes with the darkly shaded hexagons running diagonally on the map (we call them the "armchair direction"). This angle is termed the "chiral" angle, but it really adds up to be nothing more than the helicity of the molecule. In the example given above, a (8,5) nanotube will be formed when rolled.

This geometry of the tiling of the surface with hexagons is quite useful for descriptions, but it is also describing another dimensional aspect of the nanotube. For instance, knowing that the above is a (8,5) nanotube tells us that the diameter of the nanotube will have to be

$$\mathrm{DIA} = \mathrm{R}/\pi = (\sqrt{3}a_{\mathrm{c-c}}/\pi)(n^2 + m^2 + nm)^{1/2} \tag{2.2}$$

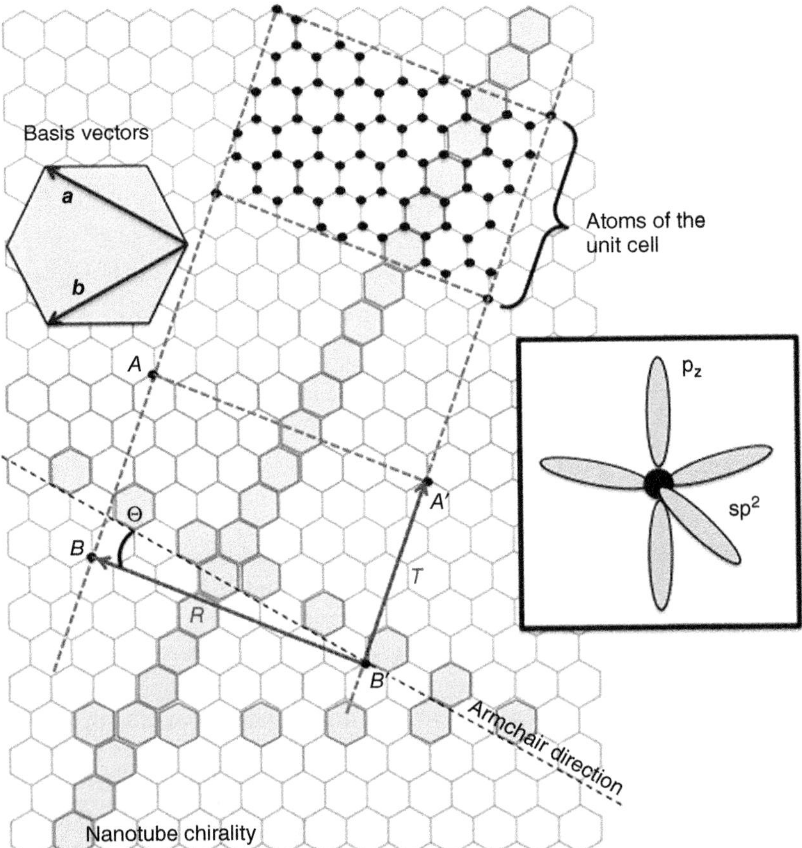

Figure 2.35 The standard map for understanding carbon nanotubes. Cut the graphene along the dotted lines on the right. Roll (along vector "**R**") into a tube connecting A–A′ and B–B′. This gives a tube. Notice that the unit cell of the tube has been marked out on the map with dotted lines across the tube axis. The vector marked "**T**" is the translation vector for this system. At the bottom is what this tube will appear as once it is rolled. Source Courtesy Jannik Meyer, University of Vienna.

where a_{c-c} is the carbon–carbon distance (lattice parameter). Further, we know that the *chiral angle* must be

$$\Theta = \arccos[(2n + m)/2(n^2 + m^2 + nm)^{1/2}] \tag{2.3}$$

or, in our case, ~22.5°.

Putting numbers in Eq. (2.2), we get DIA ~1.3 nm for a (10,10) nanotube ($n = 10$, $m = 10$, and for a_{c-c} we take the bond length in a benzene ring: $a_{c-c} = 0.14$ nm). This essentially describes another dimension of confinement in the system: the width of a box into which the electrons must fit. The vector **T** is also uniquely defined by this and is the distance one must translate along the tube to repeat the atomic order – in common solid-state terms: a unit cell. So it tells us how many atoms are donating electrons to the unit cell and thus how many electronic states must fit into this box. As we will see in coming chapters, since the *chirality* of the tube defines the number of electrons in a unit cell together with lattice parameters of the cell, it allows for both semiconducting and metallic characteristics. So, specifying (*n,m*) also tells you what electronic type of nanotube you have! Such a dramatic difference in properties from something as simple as the "twist" of the molecule is surprising – now you can see why we study carbon.

Carbon nanotubes with diameters $\ll 1$ nm are unstable, because the elastic energy needed to roll the tube (graphene wants to be flat; the sp^2-orbitals are in a plane) is larger than the chemical energy gained from closing the bonds to form a seamless tube. Nanotubes with much larger diameters are also unstable: fat tubes collapse due to the attractive van der Waals forces between opposite segments of the walls.

It is also possible to imagine inserting atoms into the empty space inside the tube – *endohedral chemistry*! Indeed, a number of different species have been added to the interior of nanotubes – from fullerenes to fullerenes with atoms inside and to bimetallic compounds. Practically, the tubes can be as thin as one nanometer in diameter and as long as several microns. A special case of inserting something into a carbon nanotube is to put another, smaller, carbon nanotube into it. This leads us to single-wall, double-wall, and multiwall carbon nanotubes: SWNT, DWNT, and MWNT.

2.10 Perovskites

Perovskite is a mineral of $CaTiO_3$ (calcium titanate). It was first discovered and recorded by Gustav Rose in the Ural Mountains of Russia in 1839 and named in honor of his colleague, the mineralogist, Lev Perovski (1792–1856).[2] In 1926 Victor Goldschmidt described the perovskite structure, and this structure was confirmed in 1945 for $BaTiO_3$ using X-ray diffraction by Helen Dick Magaw. So, unlike many of our examples above, it is quite an old and well-established family of crystals.

The structure itself is quite simple: $({}^{XII}A^{2+\,VI}B^{4+}O^{2-}{}_3)$ or ABO_3 where the A site ion is typically an alkali earth or rare earth element and the B sites are 3d, 4d, and 5d transition metal. The crystal structure is drawn out in Figure 2.36. Historically, these materials have emerged into the scientific eye several times. First it was to mineralogists who realized that substitutions such as (Ca, Ce, Na)(Ti, Fe)O_3 could be used to identify many naturally occurring minerals throughout northern Europe. Then it was realized that materials such as $BaTiO_3$ have astonishing piezoelectric properties. Still later it was shown that many of these materials make excellent substrates for the growth of Type II superconducting thin films. This provided the basis for the technological development of synthesis and growth of such crystals.

2 Sadly, people don't much name things for their colleagues anymore do they? Today, we choose a character from the latest Harry Potter book, or Star Trek movie. But Gustav Rose obviously had a great deal of respect for his colleague Dr. Perovski.

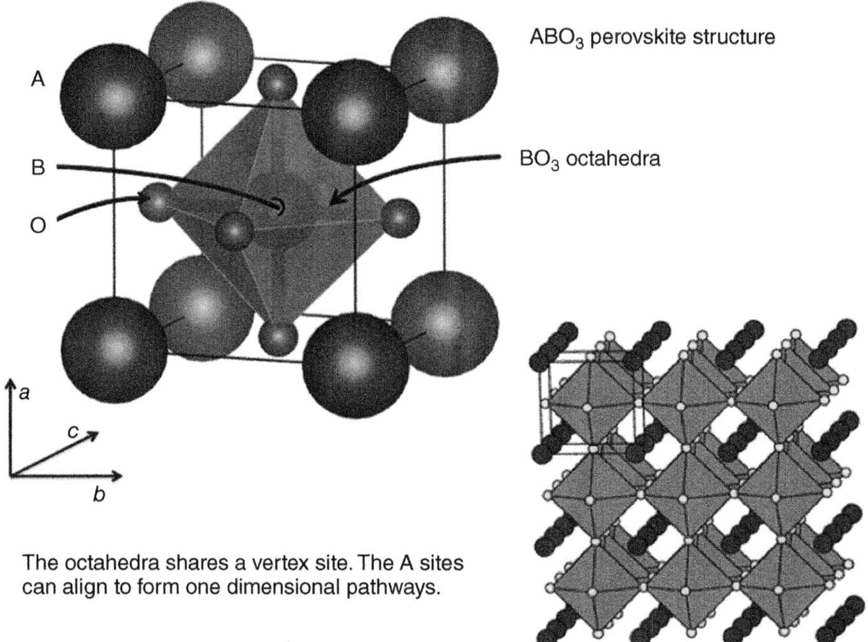

Figure 2.36 The perovskite structure.

However, as the complexities of the ABX_3 structure continued to be explored by the synthesis community, it was quickly realized that A and B didn't need to be a single ion and X didn't need to be oxygen. The structure provided the properties, and those properties could be tuned further by substituting more exotic constituents. Specifically, the community sought to tune properties such as optical absorption bandgaps, absorption lengths (oscillator strength), and carrier mobility.

In 2009 [40], it was demonstrated that such tuning could lead to perovskites that were suitable for photovoltaic applications. Very quickly materials such as the methylammonium lead trihalides $CH_3NH_3PbX_3$ (where X is a halogen atom such as iodine, bromine, or chlorine) were able to achieve conversion efficiencies of greater that 20%, rivaling commercial silicon. In this crystal, the methylammonium cation ($CH_3NH_3^+$) is surrounded by PbX_6 octahedra [41]. Due to concerns about the use of lead, as well as the search for bandgaps neared the ideal for photovoltaics, these compounds were quickly followed by similar constructs: the formamidinum lead trihalide ($H_2NCHNH_2PbX_3$) and the tin-based $CH_3NH_3SnI_3$ [42]. Of course more traditional perovskites have also had some impact in the photovoltaic field, such as $LaVO_3/SrTiO_3$ heterostructures [43]. For the purposes of our discussions, however, we must keep in mind that the properties sought and the ability to vary those properties so widely come from the inherently low-dimensional nature of the perovskite structure.

2.11 Topological States

There are some unusual cases where the shape, size, and connectedness – or *topology* – of a system can be used to create one-dimensionality. Consider the very simple example of the thin conducting bar. If the bar is truly thin, with strong confinement in z as in the above example (Figure 2.37), and a symmetric set of contacts are placed on each end with a potential, then a constant vector field of current will result. The current density along the bar will be the same at all points. Now if a homogeneous magnetic field is applied perpendicularly to the bar and the vector field of current, in the above case out of the page in z, then Landau levels, or circular orbits, will begin to form. Let's consider the case where the $\boldsymbol{B}$-field strength has been raised to the point where the curl of $\boldsymbol{J}$ is a maximum, and therefore all electrons within the plane participate in closed circular orbits: $\boldsymbol{B}_c$ or the critical field as we might call it. At this point something very unusual happens. There is no transport in the center of the bar at all! For most purposes we might refer to the middle part of the bar as an electrical insulator. Here we mean *transport* to be the movement of electrons from one contact to another so that they can be measured (counted). However, according to Stokes theorem, we surely do have a current that runs along the perimeter of the system, defined by the boundaries of the system itself, having some unusual properties. Now we might notice that the net current delivered between the two contacts is zero. However, each edge carries current in only one direction. What if we were to place a scattering impurity upon one of these edges? Could the electrons in this state be scattered backward? In order for the electron to do this, it would have to be transported

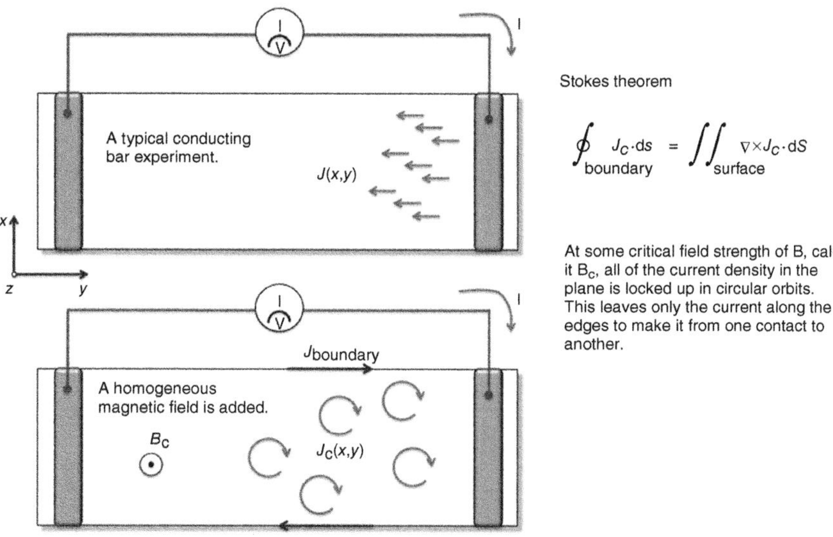

Figure 2.37 A thin conducting bar (two-dimensional) can be made into a one dimensional system easily with a magnetic field.

across the insulating middle of the system and onto the opposite edge, quite an improbable task. Indeed, to handle this scatterer and maintain the Stokes equality, the system must redefine its boundary and allow the current to flow around scatterer! We refer to such states as *topologically protected*. They can be made to occur under a variety of situations, not just applied magnetic fields, and are a result of a fundamental principle in such systems: *time reversal symmetry*. We will discuss this in far more detail in later chapters.

2.12 What Did We Forget?

Let's look at a few more systems in one dimension. In this section we take a look at materials that are a little more rare in the lab.

2.12.1 Poly-deckers

Figure 2.38 shows the *poly-decker* structure: poly(η^5,μ-2,3-dihydro-1,3-diborolyl) nickel [44]. These materials are constructed from five-membered heterocycles that are stacked by sandwiching metal (Ni) atoms in a similar way as Fe is sandwiched in ferrocene. If ferrocene is classified as *double-decker* structure, the compound depicted in Figure 2.38 should be called a *poly-decker*. Early demonstrations using Ni were quickly expanded to include a wide variety of metal atoms: Mn, Fe, Co, Ni, Cu, and Zn [45]. Charge transport is through the

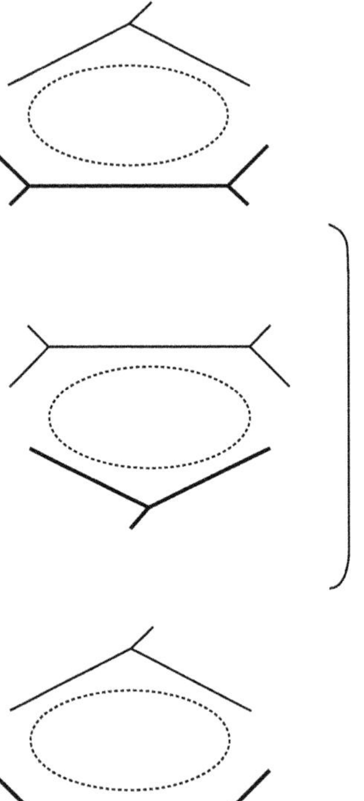

Figure 2.38 Poly(η^5, μ-2,3-dihydro-1,3-diborolyl)nickel as an example of a "poly-decker" structure.

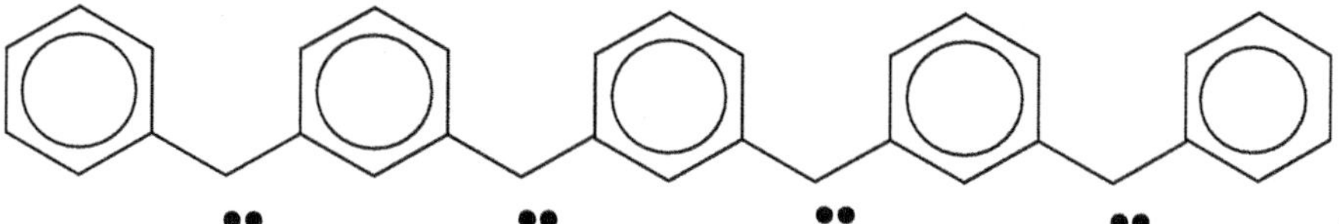

Figure 2.39 Polycarbene as one possible first step toward one-dimensional organic ferromagnets.

metal ions with the rings acting as a scaffold, an *atomic wire*. But of course, this means the system is excellent for the direct study of lattice coupling to the electronic properties of such wires and thus they have been popular for testing predictions on Peierls distortion effects.

2.12.2 Polycarbenes

Carbenes contain divalent carbon atoms with two nonbonding electrons. One way of arranging carbenes in a polymer is shown in Figure 2.39 [46]. If the nonbonding electrons in carbenes are in singlet states, their spins are compensated, and they form a *lone pair* with no net spin.

In the triplet state carbenes carry spin = 1. In the polymer there might be a positive (ferromagnetic) interaction between these spins so that the whole molecule is in a high-spin state. These high-spin carbenes can be interpreted as a first step toward all-organic ferromagnets. *Organic ferromagnet* is also a rapidly emerging field of modern material research [47].

2.12.3 Isolated, Freestanding Nanowires

Many of the materials described so far have utilized asymmetries in bulk conductivity to approximate 1D behavior. The measurement is an *ensemble average* of the behavior of many 1D systems. But, as with carbon nanotubes, an entire cottage industry has developed around the synthesis of individual/isolated nanowires. These synthesis methods use a wide range of atomic building blocks. Indeed, today we know quite a lot about the creation of highly crystalline, high aspect ratio nanorods in monatomic, binary, ternary, quaternary systems, inter-wire, and intra-wire junctions, as well as the chemical manipulation for placement of individual or infinitely many nanowires. And the number of treatises written on the subject grows each year.

There are a few general statements that can be made regarding this vast catalogue of materials. First is that the growth and structure of these, typically solid, nanowires are dominated by heats of formation. Moreover, surface and interface energies dominate faceting, and stoichiometries can be hard to control due to lattice expansion with the "few atom" diameters they present. Second, air sensitivity due to rapid oxidation and fragility (damage) can be a problem in experiments. Mechanical properties are usually dictated by dislocations within the atomic structure, and defects are usually very mobile. Finally, surfactants are typically used to separate and place such nanowires where one wants them. Many times, this must be removed before experiments begin. So as you can see, experiments are a challenge, but they can be done.

A couple of the most common methods of creating these materials are as follows.

2.12.4 Templates and Filled Pores

Many 3D solids contain pores, voids, or channels in which materials can be inserted or intercalated in the liquid or gaseous phase. Examples of such materials are zeolites and clays that provide long and narrow channels ideal for metal insertion (Figure 2.40) [49].

The idea is simple; fill up the channels with the metal of your choice. Then use a chemical bath to etch away the superstructure of the zeolite or clay. This leaves behind the nanowires. Such nanowires can be quite long but usually entangled and not very straight: not too dissimilar from a *Brillo Pad* used to scrub pots and pans, only with nanoscale wires. If one chooses a superstructure that can be removed using supercritical drying techniques, aerogels of such wire meshes can be explored. The optical properties of such metal nanowire aerogels have proven to be quite interesting. This is because the loops and wires within the gel can act as a system of inductors and antennae.

Of course, one can be more systematic about this. Over the past 10 years or so, the technology to create long pores of well-defined diameter in a matrix such as alumina (Al_2O_3) has matured, and these can be made controllably and reproducibly. Indeed the science of template formation has become somewhat of a cottage industry by itself.

The process is quite simple; first aluminum foil is anodized and etched, forming closely packed pores as shown in Figure 2.41. The packing, diameter, and length of the pores depend intricately on the anodizing voltage and time. By suddenly changing the voltage during the process, branching can be achieved (nano-tuning

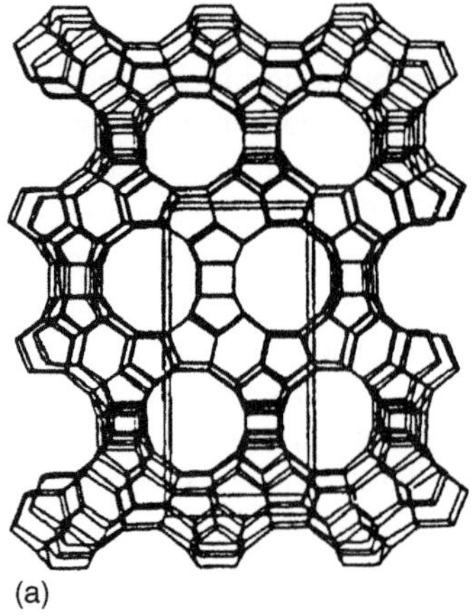

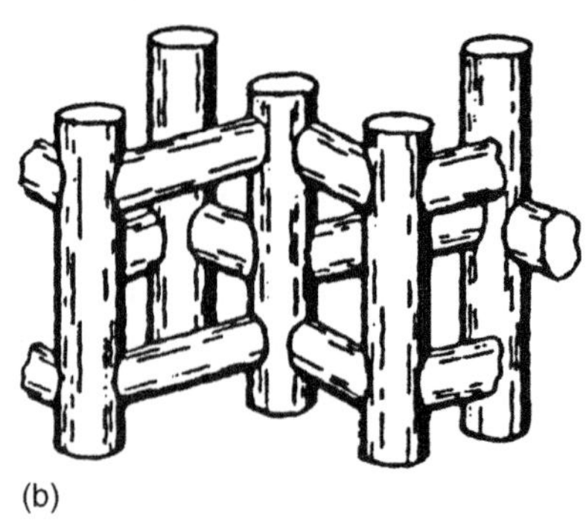

Figure 2.40 (a) Framework structures and (b) tubular representations of the channel systems of zeolites. Source: After Ramamurthy 1991 [48].

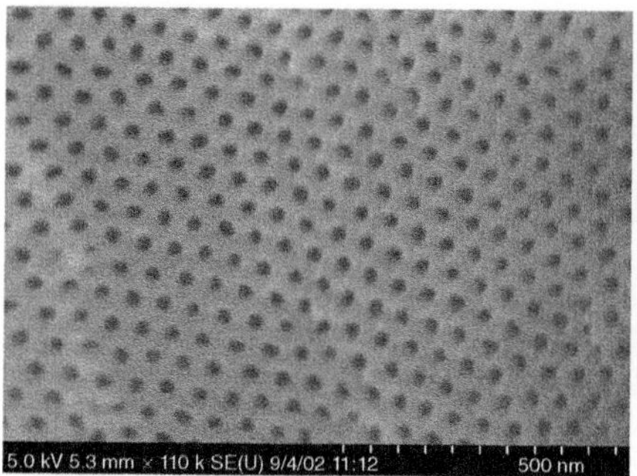

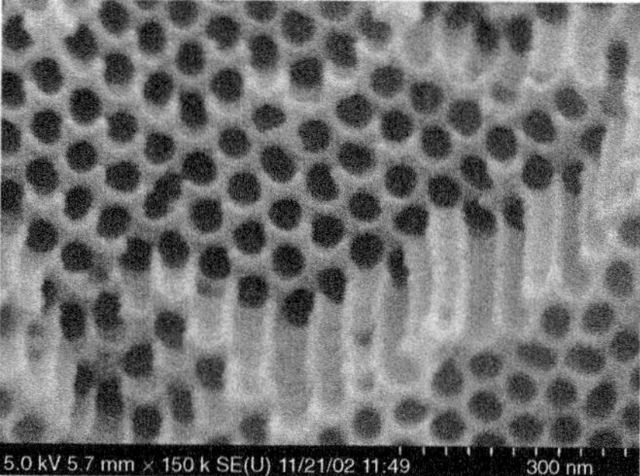

Figure 2.41 Nano-porous templates have become a standard in the fabrication of a widely dissimilar array of nanowires. The pores shown here are approximately 20 nm in diameter and 1000 μm in length.

forks). These pores can then be filled with a substance, monomers for creating *para*-phenylene vinylene (PPV), for instance. Alternatively, metals can be added and even alternated to make heterojunction, metallic nanowires! Because the technique relies on the processibility of the filling material and is not as sensitive to the chemical makeup, the types of wires that have been demonstrated vary dramatically. Once the pores are filled, the Al_2O_3 can be removed using acids, leaving behind only the wires as in the case above.

Of course, as pointed out, nearly any material can be used to create such nanowires, including highly reactive materials. When placed in contact with air, these reactive materials can undergo rapid oxidation, which is generally exothermic. This has been the origin of many lab-based explosions!

2.12.5 Asymmetric Growth Using Catalysts

As templating methods (above) have progressed, so too have direct catalytic growth methods. A wide assortment of nano-width, micron-long wires can now be reproducibly created in rather large quantities, by directly assembling atoms from a gas on a catalyst particle (chemical vapor deposition, CVD) within a high-temperature, atmospherically controlled oven. Typically, small catalyst particles are faceted, and diffusion of adsorbed (or absorbed) gases across (or through) the particle is greater along one crystallographic direction than another. As some atomic arrangement assembles more rapidly on one given surface of the particle than another, an asymmetry in growth direction is introduced resulting in 1D wires. The catalysts can be added either before growth on some substrate placed within the growth oven or can be introduced simultaneously using a vapor coalescence of metal atoms provided by flowing a metal-containing organic with the growth source gas. For 1D wire CVD growth, there are few generalities that must be met. Catalyst particles must be small (typically <20 nm), temperature of growth must be sufficient to allow for breakdown of gas stock, and catalyst-growth material systems must be chosen to allow for some solid solubility of the growth material in the catalyst particle. Examples of such growth (Figure 2.42) now include such semiconducting materials as GaN, once quite difficult to control.

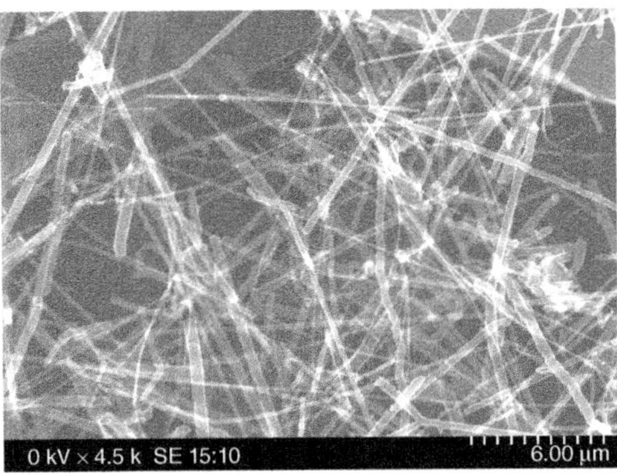

Figure 2.42 A TEM micrograph of GaN nanowires grown using chemical vapor deposition (CVD) techniques shows well-defined, one-dimensional structures. Source: Courtesy of J. Liu, Wake Forest University.

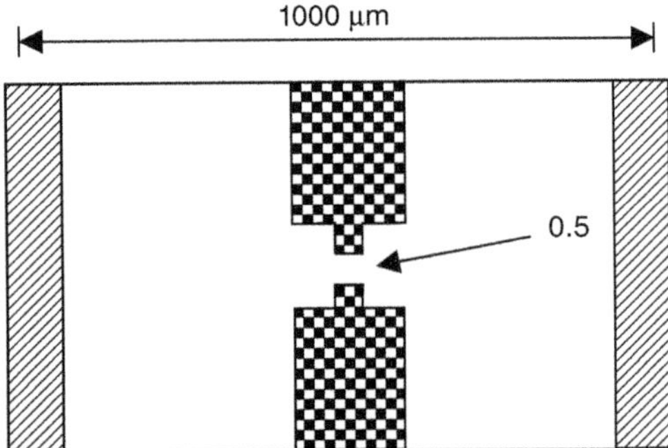

Figure 2.43 Scheme of a semiconductor quantum wire defined by using a split Schottky gate imposed on a AlGaAs–GaAs heterojunction. The white areas between source and drain electrodes indicate the quasi two-dimensional electron gas formed at the AlGaAs–GaAs heterointerfaces.

2.12.6 Gated Semiconductor Quantum Wires

We have already encountered lithographically fabricated silicon and gallium arsenide quantum wires as an example of the *external approach* to one-dimensionality. However, there is also a macroscopic semiconductor device that acts as though it were a quantum wire, shown in Figure 2.43. Since the advent of modern microfabrication technologies and extremely pure substrate growth mechanisms, such "semiconductor quantum wires" have been the subject of a large number of exciting experiments [50]. As substrates, usually AlGaAs–GaAs heterojunctions grown by molecular beam epitaxy are used. At the AlGaAs–GaAs interfaces, a quasi-2D electron layer is formed, which can exhibit extremely high mobilities. The confinement to a 1D wire is then achieved by imposing a microstructured split *Schottky gate* onto the surface. Upon applying a negative bias to the split gate, electrons in the underlying 2D electron system are depleted, thus forming a narrow (or quasi-1D) channel in the gap. The split gates can be fabricated using sophisticated technologies like electron beam lithography making it possible to achieve extremely small dimensions.

2.12.7 Few-Atom Metal Nanowires

Before ending our discussion of 1D systems, we really should mention the fascinating case of nanowires built using the scanning tunneling microscope

(STM). As with a few of our examples, this isn't a material so much as it is a way of coupling a nano-object to an experiment. But these interesting and novel experiments highlight the difficulties that come with measuring phenomena in low-dimensional structures. And, they have also opened up a rich world of phenomena that couple mechanical and electronic properties in metals.

In Chapter 1 we mentioned the "famous" problem of the single line of gold atoms acting as a wire. As we have seen, there are many ways to synthesize compounds that result in ensembles of *atomically thin* metal wires. Of course what we really might want is an isolated, single-strand, atomically thin metal wire with nothing touching or supporting it except perhaps the electrical contacts. After all, to an atomically thin wire, anything that might touch it is obviously a large perturbation of the system! This situation has actually been realized, in a surprising way.

We tend to think of surfaces in terms of their macroscopic thermodynamic energies of formation. However, at the microscopic or atomic scale, they can be quite fluid, and so the dynamics of contacting two materials and then separating them again can be messy – depending on the materials. Using this idea, a number of scientists have sought to study the very thin wire that forms when two macro-objects come into contact and are subsequently removed slowly. Specifically, the measurements involve "crashing" the gold tip of a scanning tunneling microscope into a gold surface and slowly retracting [51]. During retraction, adhesion between the two surfaces allows for creation of a very fine wire between the tip and surface. The further the surfaces are retracted, the thinner the wire becomes. This "simple" experiment has led to a complex variety of behaviors from molecular-scale wires in a "freestanding" state (Figure 2.44).

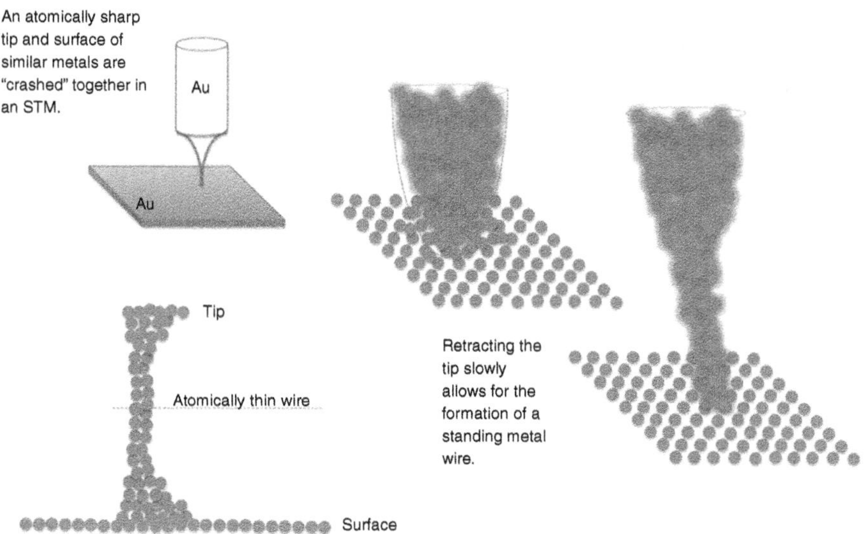

Figure 2.44 The retraction of the tip from the surface after contact leaves an atomically thin wire. Current–voltage (*I*–*V*) curves from the STM are used to determine the electronic properties.

Such wires can exhibit surprising ductility, ordering during "drawdown" and quantum resistance. Moreover, similar experiments can be carried out on "break junctions." In this case a narrow metallic strip on a silicon substrate is broken by bending the substrate. A gap opens in the strip, which might be bridged by a row or individual atoms. Molecules can be squeezed in the gap, and then electrical current flowing through a single molecule can be studied [52].

2.13 A Summary of Our Materials

In this text we are embracing the concept of *exotic material* as the case where *trans-dimensional effects* are strong. What do we mean by trans-dimensional? This is simply a lower-dimensional structure that is influenced by something in a higher dimension. So in 1D nanowires, the electrons are essentially confined to the path of the wire. However dopants and fields can be added from the third dimension by simply bringing a dopant molecule up to the wire and touching it. This intersection from the point of view of the carrier may look only like a simple point, but it puts the wire into contact with the properties, symmetries, and phenomena associated with that 3D dopant object. In other words, the effect traverses the dimensional divide between the two.

This semi-historical tour has provided a number of examples that are instructive to this point. We have focused on 1D materials, but the same kind of chapter could be written for 2D and 0D materials. The point here is only to introduce the many ways in which dimension enters into material structure–property relationships. But it should also be clear by now that in *low-dimensional* materials there can be surprising degrees of complexity.

There is another theme that emerges from this chapter. Regardless of the system under study: the metal "wires" of a Krogmann salt, an MX chain, or the carbon "wire" of the carbon nanotube, the coupling of the *electronic properties* to the *lattice* of the linear arrangement of atoms is exceedingly strong. In most cases *lattice distortions* in low-dimensional systems travel along with the carriers in the wire to yield a rich assortment of transport and optical properties *unique to the dimensionality*.

In Chapter 1 we described how the approach to the organization of thoughts on solid state in this text would be a bit different from other such organizational principles. Now you might see the advantage of such an approach. You might say that the *intrinsic* story of solid-state physics, with its models of lattice and electron dynamics, is not complete without some understanding of the *extrinsic* properties of dimension and symmetry of the solid-state object across many different scales. The examples of this chapter certainly help to reinforce this perspective.

Unlike other chapters in the text, this problem set focuses on *literature searches.* This is done to give a broad-based overview of the different directions that materials engineering and materials sciences have taken. What is out there? What are scientists worrying with?

Exploring Concepts

1 *Chalcogenide nanowires:* Wet chemistry, templating synthesis techniques have allowed the creation of highly crystalline ternary chalcogenide nanowires. Look through literature and describe in detail the crystal structures claimed for two different examples. Examine older attempts at the creation of three component systems that resulted in much wider diameters. Are the crystal structures reported the same as for the extremely thin wires? Discuss why this is the case.

2 *Perovskites:* We mention several hybrid (meaning inorganic/organic) perovskites used for photovoltaic applications. Do a literature search and draw these structures in the same manner we have presented the oxide perovskite. See if you can identify potential pathways that may provide a link to low-dimensional behavior.

3 *Topological insulators:* The first naturally occurring topological insulator was reported by a team at the Max-Planck-Institut für Festkörperforschung in Stuttgart in 2013. The layered *kawazulite* was found to be less defective than its synthetic counterparts. Do a literature search and describe this material in detail, including the crystal structure.

4 *Conducting polymers:* What is the highest conductivity achieved with a conducting polymer? How will you have to look this up in the literature and will you run across a wide variety of candidates? Describe your "champion" polymer and how the measurement was made that showed the high conductivity.

5 *Carbon nanotubes:* It is frequently said that carbon nanotubes are among the strongest materials in nature. In fact people have even proposed to build elevator into space from such materials. Just how strong is such a beast (the single-wall nanotube)? How much force would be required to break *one* nanotube? Compare that to a BN nanotube.

References

Note: In our references we have tried to trace concepts back to primary sources as much as possible. Of course, for many of these topics, there are more contemporary sources and discussions in the literature, but these are rather easy to find today using the Internet and a good library link.

1 Little, W.A. (1964). Possibility of synthesizing an organic superconductor. *Phys. Rev. A* 134: 1416.
2 Bardeen, J., Cooper, L.N., and Schrieffer, J.R. (1957). Theory of superconductivity. *Phys. Rev.* 108: 1175.

3 Schirber, J.E., Wang, H.H., and Williams, J.M. (1990). *Organic Superconductivity* (ed. V.Z. Kresin and W.A. Little), 117. New York: Plenum Press.

4 (a) Rosseinsky, M.J., Ramirez, A.P., Glarum, S.H. et al. (1991). Superconductivity at 28 K in Rb_xC_{60}. *Phys. Rev. Lett.* 66: 2830. (b) Iqbal, Z., Baughman, R.H., Ramakrishna, B.L. et al. (1991). Superconductivity at 45 K in Rb/Tl codoped C_{60} and C_{60}/C_{70} mixtures. *Science* 254: 826.

5 (a) Cheeke, J.D.N., Mallié, H., Roth, S., and Seeber, B. (1973). Direct observation of soft mode stiffening in V_3Si by high magnetic fields. *Solid State Commun.* 13: 1567. (b) Dietrich, W., Roth, S., Mallié, H., and Seeber, B. (1975). *Proceedings of the 14th International Conference on Low Temperature Physics (LT14), Helsinki*, 36.

6 Muller, J. (1980). A15-type superconductors. *Rep. Prog. Phys.* 42: 641.

7 (a) Buckel, W. (1984). *Supraleitung*. Weinheim: Physik-Verlag. (b) Buckel, W. and Kleiner, R. (2004). *Supraleitung*, 6e. Wiley-VCH.

8 Labbe, J. (1960). Structure de bande fine des composés intermétalliques supraconducteurs du type V_3Si et Nb_3Sn. Dissertation. Paris: instabilité crystalline d'origine électronique.

9 Krogmann, K. (1969). Planare Komplexe mit Metall-Metall-Bindungen. *Angew. Chem.* 81: 10.

10 (a) Wagner, H. (1974). PhD dissertation. Universitat Karlsruhe; (b) Zeller, H.R. (1973). Advances in solid state physics. In: *Festkörper Probleme XIII*. Berlin, Heidelberg: Springer-Verlag.

11 Renker, B., Rietschel, H., Pintschovius, L. et al. (1973). Observation of giant Kohn anomaly in $K_2Pt(CN)_4Br_{0.3}{\bullet}3H_2O$. *Phys. Rev. Lett.* 30: 1144.

12 Kaveh, M. and Ehrenfreund, E. (1979). Square Fermi surface model for conduction in linear chain mercury compounds: $Hg_{3\text{-}\delta}ASF_6$. *Solid State Commun.* 31: 709.

13 Heilmann, I.U., Axe, J.D., Hastings, J.M. et al. (1979). Neutron investigation of the dynamical properties of the mercury chain compound $Hg_{3\text{-}\delta}AsF_6$. *Phys. Rev. B* 20: 751.

14 Jerome, D. (1981). Organic superconductivity. *Chem. Scr.* 17: 13.

15 Bechgaard, K., Jacobsen, C.S., Mortensen, K. et al. (1980). The properties of five highly conducting salts: (TMTSF)2X, X = PF6-, AsF6-, SbF6-, BF4- and NO3-, derived from tetramethyltetraselenafulvalene (TMTSF). *Solid State Commun.* 33 (11): 1119–1125.

16 Mehring, M. and Spengler, J. (1984). Locally resolved ^{13}C Knight shifts in the organic conductor $(fluoranthenyl)_2SbF_6$. *Phys. Rev. Lett.* 53: 2441.

17 Hamann, C., Heim, J., and Burghardt, H. (1981). *Organische Leiter, Halbleiter und Photoleiter*. Braunschweig und Berlin: Friedrich Vieweg & Sohn.

18 Friend, R. and Jerome, D. (1979). Periodic lattice distortions and charge density waves in one- and two-dimensional metals. *J. Phys. C: Solid State Phys.* 12: 1441.

19 Coleman, L.B., Cohen, M.J., Sandman, D.J. et al. (1973). Superconducting fluctuations and the Peierls instability in an organic solid. *Solid State Commun.* 12: 1125.

20 Schafer, D.E., Wudl, F., Thomas, G.A. et al. (1974). Apparent giant conductivity peaks in an anisotropic medium: TTF-TCNQ. *Solid State Commun.* 14: 347.

21 Greene, R.L., Street, G.B., and Suter, L.J. (1975). Superconductivity in polysulfur nitride $(SN)_x$. *Phys. Rev. Lett.* 34: 577.

22 Cohen, M.J., Garito, A.F., Heeger, A.J. et al. (1976). Solid state polymerization of S2N2 to (SN)x. *J. Am. Chem. Soc.* 98: 3844–3848. But simultaneous is the work of W. Möller, Universität Karlsruhe (1976) PhD. Thesis.

23 Hamann, C., Höhne, H.J., Kersten, F. et al. (1978). Switching effects on polycrystalline films of lead phthalocyanine (PBPC). *Phys. Status Solidi A* 50: K189.

24 (a) Hanack, M., Mitulla, K., and Schneider, O. (1981). Synthesis of a new kind of low-dimensional conductors. *Chem. Scr.* 17: 139. (b) Hanack, M. and Lang, M. (1994). Conducting stacked metallophthalocyanines and related compounds. *Adv. Mater.* 6: 819.

25 (a) Simon, A. (1981). Kondensierte metall-cluster. *Angew. Chem.* 93: 23. (b) Simon, A. (1981). Condensed metal-clusters. *Agnew. Chem. Int. Ed. Engl.* 20: 1.

26 Bullett, D.W. (1985). *Theoretical Aspects of Band Structures and Electronic Properties of Pseudo-One-Dimensional Solids* (ed. H. Kamimura). Dordrecht: D. Reidel Publ.

27 Moses, E. (1976–1979). *Physics and Chemistry of Materials with Low-Dimensional Structures, Series A: Layered Structures*, 1–6. Dordrecht: Reidel.

28 Gressier, P., Meerschaut, A., Guemas, L. et al. (1983). New one-dimensional Vb transition metal tetrachalcogenides: structural and transport properties. *J. Phys.* C3: 1741.

29 Schlenker, C., Filippini, C., Marcus, J. et al. (1983). Peierls transition in the quasi-one dimensional blue bronzes $K_{0.3}MoO_3$ and $RbO_{0.3}MoO_3$. *J. Phys.* C3: 1757.

30 (a) Gammel, J.T., Saxena, A., Batistic, I. et al. (1992). Two-band model for halogen-bridged mixed-valence transition-metal complexes. I. Ground state and excitation spectrum. *Phys. Rev. B* 45: 6408. (b) Wever-Milbrodt, S.M., Gammel, J.T., Bishop, A.R., and Loh, E.Y. Jr., (1992). Two-band model for halogen-bridged mixed-valence transition-metal complexes. II. Electron-electron correlations and quantum phonons. *Phys. Rev. B* 45: 6435.

31 Chiang, C.K., Fincher, C.R. Jr., Park, Y.W. et al. (1977). Electrical conductivity in doped polyacetylene. *Phys. Rev. Lett.* 39: 1098.

32 (a) Heeger, A.J. (2001). Semiconducting and metallic polymers: the fourth generation of polymer materials. *Rev. Mod. Phys.* 73: 681. (Noble Lecture). (b) Heeger, A.J. (2010). Semiconducting polymers: the third generation. *Chem. Soc. Rev.* 39: 2354.

33 Roth, S. and Graupner, W. (1993). Conductive polymers: evaluation of industrial applications. *Synth. Met.* 57: 3623.

34 (a) Longuet-Higgins, H.C. and Salem, L. (1959). The alternation of bond lengths in long conjugated chain molecules. *Proc. R. Soc. London, Ser. A* 251: 172. (b) Kuhn, H. (1948). Free-electron model for absorption spectra of organic dyes. *J. Chem. Phys.* 16: 840. (c) Kuhn, H. (1949). A quantum-mechanical theory of light absorption of organic dyes and similar compounds. *J. Chem. Phys.* 17: 1198. (d) König, G. and Stollhoff, G. (1990). Why polyacetylene dimerizes: results of ab initio computations. *Phys. Rev. Lett.* 65: 1239.

35 (a) Su, W.P., Schrieffer, J.R., and Heeger, A.J. (1979). Solitons in polyacetylene. *Phys. Rev. Lett.* 42: 1698. (b) Rice, M.J. (1979). Charged π-phase kinks in lightly doped polyacetylene. *Phys. Lett. A* 71: 152. (c) Brazovskii, S.A. (1980). Self-localized excitations in the Peierls-Fröhlich state, Zh. Eksp. Teor. Fiz B 7, 677 (1980). *Sov. Phys. JETP* 51: 342. (d) Lu, Y. (1988). *Solitons and Polarons in Conducting Polymers.* Singapore: World Scientific. (e) Roth, S. and Bleier, H. (1987). Solitons in polyacetylene. *Adv. Phys.* 36: 385. (f) Heeger, A.J., Kivelson, S., Schrieffer, J.R., and Su, W.P. (1988). Solitons in conducting polymers. *Rev. Mod. Phys.* 60: 781.

36 Wegner, G. and Mitt, I. (1969). Polymerisation von Derivaten des 2,4-hexadiin-1,6-diols im kristallinen Zustand. *Z. Naturforsch. B* 24: 824.

37 Cantow, H.J. (1984). *Advances in Polymer Science*, 63. Berlin, Heidelberg: Springer-Verlag.

38 (a) Oberlin, A., Endo, M., and Koyoma, T. (1976). Filamentous growth of carbon through benzene decomposition. *J. Cryst. Growth* 32: 335. (b) Iijima, S. (1991). Helical nanotubes of graphitic carbon. *Nature* 354: 56. (c) Jorio, A., Dresselhaus, G., and Dresselhaus, M.S. (2008). *Carbon Nanotubes: Advanced Topics in the Synthesis, Structure, Properties and Applications.* Berlin, Heidelberg: Springer-Verlag. (d) Hayden, O. and Nielsch, K. (2011). *Molecular- and Nano-Tubes.* Berlin, Heidelberg: Springer-Verlag. (e) Reich, S., Thomsen, C., and Maultzsch, J. (2004). *Carbon Nanotubes: Basic Concepts and Physical Properties.* Wiley-VCH. (f) Hierold, C. (2008). Carbon nanotube devices. In: *Advanced Micro and Nanosystems*, vol. 8. Wiley-VCH.

39 Kuzmany, H., Fing, J., Mehring, M., and Roth, S. (1994). *Progress in Fullerene Research.* Singapore: World Scientific.

40 Kojima, A., Teshima, K., Shirai, Y., and Miyasaka, T. (2009). Organometal halide perovskites as visible-light sensitizers for photovoltaic cells. *J. Am. Chem. Soc.* 131 (17): 6050.

41 Eames, C., Frost, J.M., Barnes, P.R.F. et al. (2015). Ionic transport in hybrid lead iodide perovskite solar cells. *Nat. Commun.* 6: 7497.

42 Noel, N.K., Stranks, S.D., Abate, A. et al. (2014). Lead-free organic–inorganic tin halide perovskites for photovoltaic applications. *Energy Environ. Sci.* 7 (9): 3061.

43 Assmann, E., Blaha, P., Laskowski, R. et al. (2013). Oxide heterostructures for efficient solar cells. *Phys. Rev. Lett.* 110: 078701.

44 Kuhlman, T., Roth, S., Roziere, J., and Siebert, W. (1986). Polymeres ($\eta^5\mu$-2,3-dihydro-1,3diboryl)-nickel, die erste Polydecker Sandwichverbindung. *Angew. Chem.* 98: 87.

45 Böhm, M.C. (1987). *One-Dimensional Organometallic Materials.* Berlin, Heidelberg: Springer-Verlag.

46 Sugawara, T., Bandow, S., Kimura, K. et al. (1986). Magnetic behavior of nonet tetracarbene as a model of one-dimensional organic ferromagnets. *J. Am. Chem. Soc.* 108: 368.

47 (a) Friend, R.H. (1987). Polymers from the Soviet Union. *Nature* 326: 335. (b) Miller, J.S., Calabrese, J.C., McLean, R.S., and Epstein, A.J. (1992). Meso-(tetraphenylporphinato)-manganese(III)-tetracyanoethenide, [Mn^{III}TPP][TCNE], a new structure-type, linear-chain magnet with a T_C of 18 K. *Adv. Mater.* 4: 498.

48 Ramamurthy, V. (1991). *Photochemistry in Organized and Constrained Media*. VCH, Weinheim.

49 Tresos, E. (1993). Contribution a l'Etude des Effets de Taille Quantique de Particules d'Alcalins dans des Cavites de 'Zeolithes. Dissertation. Montpellier: Etudes RPE et RMN.

50 (a) Beenakker, C.W.J. and van Houten, H. (1991). Quantum transport in semiconductor nanostructures. In: *Solid State Physics*, vol. 44 (ed. H. Ehrenreich and D. Turnbull). San Diego, CA: Academic Press. (b) van Houten, H., Beenakker, C.W.J., and van Wees, B.J. (1991). Quantum point contacts. In: *Semiconductors and Semimetals*, vol. 35 (ed. M. Reed). San Diego, CA: Academic Press. (c) Pepper, M. and Wharam, D. (1988). Rewriting the rules of resistance. *Phys. World* 1: 45.

51 Takayanagi, K., Kondo, Y., and Ohnishi, H. (2001). Suspended gold nanowires: ballistic transport of electrons. *JSAP Int.* 3: 3–8.

52 (a) Weber, H.B. and Mayor, M. (2003). Stromfluss durch einzelne Moleküle. *Phys. Unserer Zeit* 34: 272. (b) Weber, H.B., Mayor, M., and Löhneysen, H.V. (2005). Electrical current through single molecules. *Nova Acta Leopold.* 92: 47. (c) Elbing, M., Ochs, R., Koentopp, M. et al. (2005). A single-molecular diode. *Proc. Natl. Acad. Sci. U.S.A.* 102: 8815.

3

Order and Symmetry: The Lattice

Consider these three states of matter: *solid, liquid,* and *gas*. Solids and liquids are collectively called "condensed" matter because they have a large mass density and are *difficult to compress*. Atoms in close proximity have little room for irregularities in interatomic distances whereas, gasses are highly compressible. For liquids and gases, shear forces are negligible. In crystal systems, forces keep the distance between the atoms fixed and lock the relative angular positions of the atoms in place. These materials do not flow. But liquids and gases, together classified as *fluids*, can take the shape of any container. In other words they flow. So this use of properties (compressibility, flow, etc.) is one way to group materials and understand their nature.

Thé Mystéré of thé Cãthédrãl Glãss

There is a wonderful story about the amorphous glasses used in Europe's great cathedrals. It was discovered that the oldest panes of such glass were always thicker at the bottom (as it hung) than at the top. It was also rather quickly suggested that this was because the glass was amorphous and that after one thousand years the glass had slowly run down to the bottom thickening it. Amorphous materials are sort of a cross between a liquid and a solid after all. Unfortunately, the scientists who suggested that such properties of amorphous materials were at work here had never watched a stained glass window maker in action. It turns out that when cutting the glass to fit into the complex window geometries, you first stand it on its end to find the thickest portion and place that at the bottom for better balance when installing it!

Even among solids, *crystalline solid* describes molecules or atoms arranged in a regular way. *Amorphous solid* is when there is a large amount of disorder in a solid's *isotropic ordering*. The existence of shear forces in a solid, due to steric or bonding considerations, implies order. The angular positions of atoms cannot be completely random. For example, a silicon atom is found in the center of a tetrahedron with an angle of 109.5° between the bonds to its nearest neighbors. Therefore, there is *short-range* order even in amorphous solids. In crystalline

Foundations of Solid State Physics: Dimensionality and Symmetry, First Edition.
Siegmar Roth and David Carroll.

solids the order is *long range*. But notice now that we speak of symmetries as opposed to properties as an organizational principle.

If we assume that the ground state of all matter is crystalline, will all amorphous solids crystallize sooner or later? Curiously, as time frames for crystallization approach universal time scales, the answer to this question is "no." If you find a rock formed during the Big Bang, you can't necessarily expect it to be fully crystalline. Nor do you expect it to become so by the time the universe is cold and dark. This is because these systems approach their ultimate ground state fate asymptotically.

3.1 The Correlation Function

Of course, the degree to which something is amorphous or crystalline can be important for some purposes. How defective must a crystalline solid become before it is amorphous? What is *glassy* behavior? To better understand the degree of organization, physicists frequently resort to statistical measures such as the *correlation function*. How does this work? Well starting at any given atom within the solid, the *mass correlation function* is simply the probability of finding another atom at some distance away. Usually the starting atom is chosen so that it is not special in any way (representative). There are other correlation functions as well. They usually link the probability of finding a specific value of some characteristic starting with a value for that same characteristic at the function's origin. *Spin–spin correlation* is an example. Typically, such functions are presented isotropically, meaning you show the probability as you move away from the origin radially as in Figure 3.1. This example presents the mass correlation function of a jar of marbles (hard spheres packed together). Choose any marble and move away, in distance, in any direction. What is the *probability of the mass density* or the *value of the mass density*, if you prefer, in each differential shell as you move out along R? It is $g(R)$, the *correlation function* shown. Notice that for randomly stacked hard spheres, the spheres fit together as tightly as possible. This means that a given sphere is surrounded by a shell of mass centered roughly at $R = 1$ diameter of the spheres used.

For a *spin glass* place a magnetic ion at the center of each marble, and then ask what the probability of finding a specific spin projection is as we move out along R. Again there would be some correlation function $g(R,S_z)$. So this function seems to present us with the geometrical ability of describing the interactions of objects that have been packed together. Some scientists base all discussion of order in condensed matter on this function – as the most basic sense of order. In our discussion, it is important to understand only that a disordered solid may also reflect symmetries and that there are well-defined ways to describe them. The correlation function is one of those ways.

Of course, to be precise we should say that we are describing the *static correlation function*. After all, if we were to describe a system in which the expectation values of some observable could change with changing stimulus, then the correlation function for that observable would also change with the stimulus. This forms the beginning of formal *response theory*, and it is particularly useful at describing

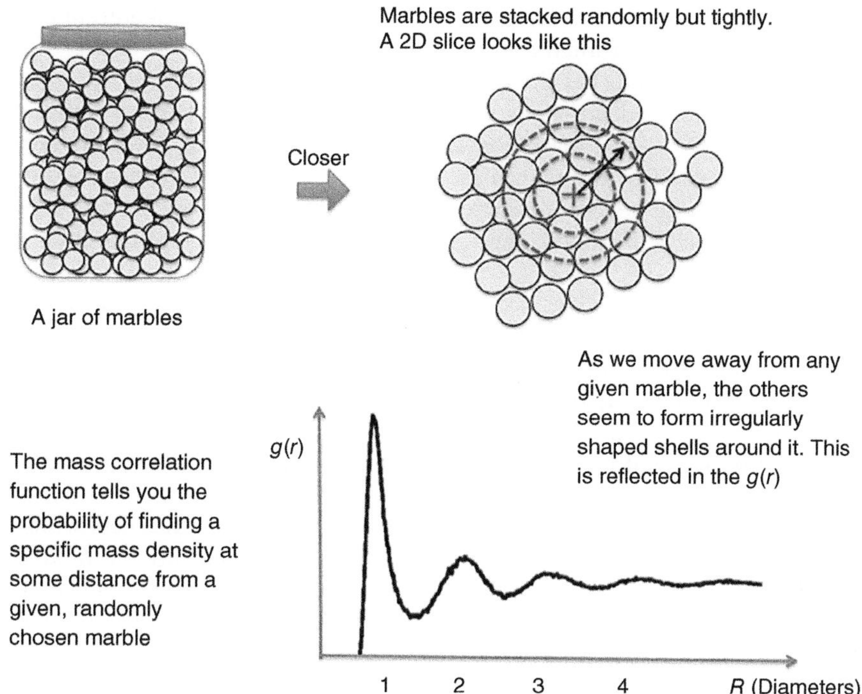

Figure 3.1 Correlation functions can be constructed for any *order parameter*: spin, mass, etc. Here, we give a simple model for how to think of a mass correlation function. Notice that the less ordered the "packing" is, the smaller the maxima in the $g(r)$ curve become. In our example there is very good *local* order, a few marbles out, and they are still in places where you expect them. However when we exceed about 4 or 5 diameters of the marbles, the mass is smeared isotropically in the volume.

phenomena such as phase transitions. The observables become known as *order parameters*, and the distances characteristic of their expectation values become *correlation lengths*. The thermodynamics and statistical mechanics of response theory is especially important because it forms a bridge between critical phenomena in systems of condensed matter and other processes in the universe at a fundamental level. Whether it is the condensation of atoms in a beaker to form a solid or the symmetry breaking of a cooling universe to separate out the fundamental forces, it is all described by the same mathematics more or less. This is exactly why this remains a mystery, but the principle is known as *universality*, and it is reflected in the exponentials that describe temperature dependences of order parameters, near-phase transitions, and the *critical exponents*.

3.2 The Real Space Crystal Lattice and Its *Basis*

What does "long-range" order mean? In *single crystals* the order covers the entire crystal, which is from some microns up to many centimeters. Most solids, however, are *polycrystalline*. They are composed of microcrystalline domains or grains, which stick together to form the bulk solid. The diameter of a grain can be from fractions of a micrometer up to several millimeters.

Interestingly, most solids get their mechanical and electronic properties from the interfaces at these grains: *grain boundaries*. Such boundaries can have atomic registry between the grains or no registry at all. Some can support boundary layers that are an entirely different phase of material than the crystals on either side of the boundary. Phenomena such as the *yield strength* of quenched metal bars or *superplasticity* in some ceramics are directly related to grain boundary properties.

Clearly, long-range order gives rise to crystals and large crystalline grains. This leads naturally to the concept of the *crystal lattice*. In a crystal lattice the regular arrangement of the atoms (or molecules) is periodic in some way; the same pattern is repeated over and over again. For example, the sodium metal lattice is shown in Figure 3.2, the graphene lattice in Figure 3.3, and the sodium chloride lattice in Figure 3.4. Multiple repeating units in each direction are shown. The lattices in Figures 3.2 and 3.4 are *cubic*, while the lattice of Figure 3.3 is *trigonal*. The names of such lattices are pretty obvious; a "cubic" elementary cell, which is a commonly found repeating unit, can be constructed as indicated in bold in the upper right corner of Figure 3.2 (made out of cubes). For the sake of clarity, only the atoms at the surface of the crystal are shown, while the atoms "inside" are omitted.

The example in Figure 3.2 is of a simple cubic point array that has one sodium atom on each point. The sodium atoms are bound together through the sharing of electrons, but the description of the array ignores this for now. Instead we

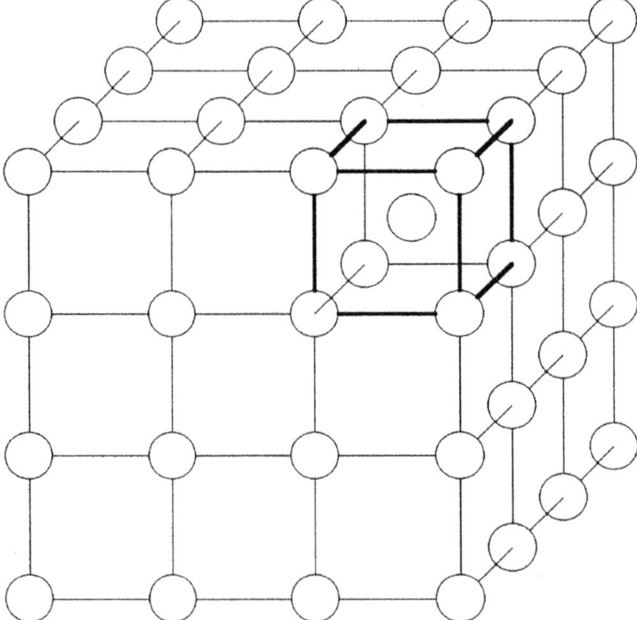

Figure 3.2 Crystal lattice of sodium metal. A cube of atoms is being repeated.

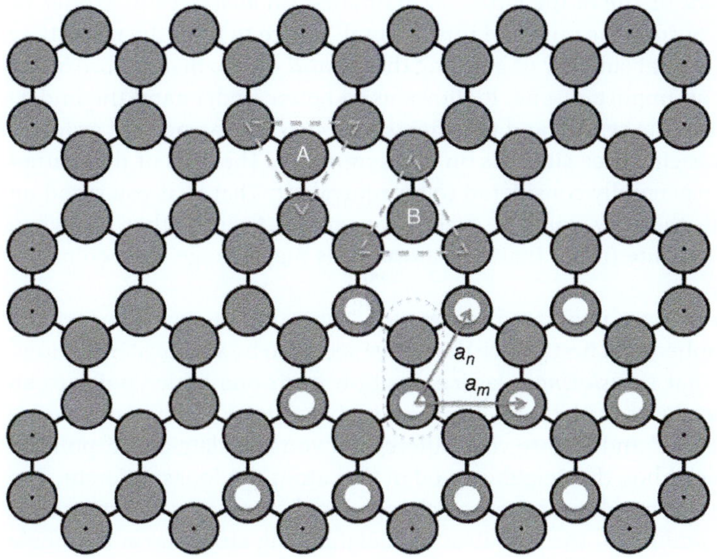

Figure 3.3 The triangular lattice of graphene has two carbon atoms as its basis. The vectors $\mathbf{a}_n$ and $\mathbf{a}_m$ are translation vectors of the lattice. The white dots represent the "lattice," and each site has two carbon atoms.

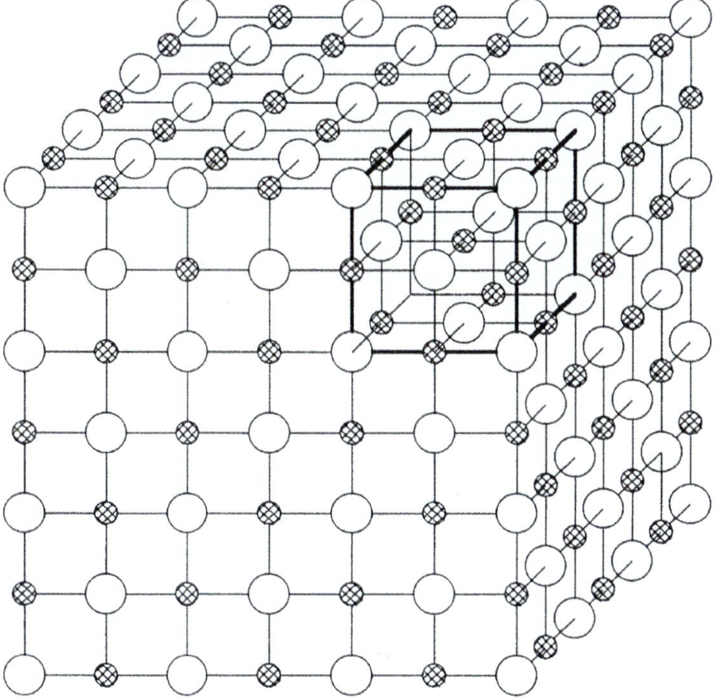

Figure 3.4 Crystal lattice of sodium chloride.

imagine each point or *site* of the lattice in mathematical abstraction. Atoms, or groups of atoms, occupy the points of this "mental" lattice to form the *real* lattice of the solid. The formal subtlety of building the *mental lattice* first, before filling it with atoms, is an important one. It allows us to better understand the underlying symmetries of the system and the effects the *basis* (the associated group of atoms placed on each lattice site) has on this symmetry. The sites of the mathematical lattice are generally considered to be equivalent. That is, if you stood on any particular site, the view would be exactly the same as for any other. But when a *basis* is added to a site (more than one atom), this may change, as seen in our next example.

The use of a *basis set* is illustrated in Figure 3.3, showing the two-dimensional (2D) lattice of graphene. At first glance the lattice seems to be a honeycomb. However, the rule is that *translating* a lattice position from one lattice point to an adjacent one reproduces the lattice – *translational symmetry*. But as can be seen, the points marked "A" and "B" are very different. If you translate an "A" point to its next nearest neighbor, the neighborhood of that atom site looks different. The symmetry of the surrounding atoms has changed with respect to the translated atom. Imagine standing on the "A" sites; the neighboring atoms form a triangle pointing down, whereas on the "B" sites, the triangle of surrounding atoms points up. These sites are clearly not equivalent, and such translations will not reproduce the honeycomb pattern.

Now instead, imagine that the "lattice" is mathematically defined by the sites marked in white (bottom right). Place onto each white site two atoms: the one at the site and another just above it. This is seen in the site where the two-atom *basis* is looped together by the dashed line. Translation of these two atoms to any neighboring site (as defined by the white dots) reproduces the honeycomb pattern exactly. We can choose the translation vectors $\mathbf{a}_n$ and $\mathbf{a}_m$ to find any site in the honeycomb by specifying (*n*,*m*) and atom 1 or 2 of the basis set and then walking n steps along $\mathbf{a}_n$ and m steps along $\mathbf{a}_m$. So in fact the honeycomb of graphene is a triangular lattice with a two-atom basis set (both carbon atoms but with different bond rotations).

In the case of sodium chloride, a two-atom basis, one of Na and one of Cl, is associated with each site of the square lattice (Figure 3.4). The crystal of sodium chloride is built of sodium and chlorine units according the chemical formula NaCl. There are as many sodium atoms as there are chlorine atoms. However, a particular sodium atom does not have only one single chlorine atom as a partner but six equivalent chlorine neighbors. These again are neighbors to other sodium atoms, and there is an equal nearest neighbor spacing among them all. In other words, there is no uniquely defined NaCl unit choice for the crystal. This coincidence does not always happen when there is a basis set.

The distinction between solids with complex basis sets and solids that are elemental (or nearly so) is frequently used to further classify solids. *Elemental crystals* and *molecular crystals* can be distinguished along exactly these lines. Most inorganic materials form *elemental crystals*. Examples include silicon, copper, and carbon (diamond), and we might put the ionic crystal sodium chloride into this category as well as other binary, ternary, and simple combinations thereof. In contrast, organic *compounds* can form molecular crystals, wherein the molecular units are weakly stacked into a crystal structure. The properties of this solid are dominated by the molecules placed on each lattice point. For example, benzene molecules exist in benzene vapor, in liquid benzene (with solvents), and also when incorporated in a molecular crystal. In a benzene crystal there are well-defined C_6H_6 units on each lattice site held there by weak forces. The interactions between the carbon atoms within such a unit (molecule) are much stronger than between the carbon atoms belonging to two different molecules. We have already run into a variety of such molecular crystals in the previous chapters such as the Krogmann salts. (For a monograph on organic crystals, see [1].)

So, we have described the two fundamental components of a real space lattice: (i) an abstract mathematical construction and (ii) a basis set, atoms added to each point in the lattice abstraction. The lattice abstraction and its basis both express a set of symmetries that may differ from each other. These are rotations, translations, reflections, etc. that map the crystal onto itself. Typically, we use these symmetry properties to classify and collect different crystal *types* or *families*. So when discussing a specific crystal, you will usually give the name of the *symmetry groups* that apply to the crystal's family. There are many sources of these crystallographic designations, and Ashcroft and Mermin [2] has long been a standard introduction for crystal symmetries. But it isn't the only one [1, 3]. However, instead of jumping directly into the obscure nomenclature of crystal groups, let's introduce some ideas of lattice coordinates, so we can see how these symmetries come about.

3.2.1 Using a Coordinate System

Symmetry groups are applied to see what rotations, translations, and so forth can be applied to the whole infinite lattice and leave it completely unaltered (that is, it looks the same). There are really two types of symmetries we typically think of when discussing crystals; first is the translational symmetry of points in space. This is almost a definition for how you build a lattice: by repeating some elementary portion of it over and over throughout space. The second is point symmetries such as rotational symmetries. These are based on local relationships among lattice components: what angles and distances are the nearest neighbors placed? So it is then natural to ask: what is the smallest chunk of lattice components that completely express the symmetries that apply to the whole, infinite lattice?

To answer this question we introduce the *elementary cell* – a small chunk of lattice that allows us to make up the whole lattice. An example of an elementary cell is shown for the 2D rectangular lattice in Figure 3.5. A choice of elementary cell is presented as a hatched rectangle in Figure 3.5, and notice we have used it to help in the definition of a local coordinate system. Also notice the choice is not unique; there are other ways to choose. The lattice is built up by moving this cell parallel to the *crystallographic a axis* and the *crystallographic b axis.* (We switched from (a_n, a_m) used by many studying graphene to (a, b) to match more standard conventions for solids.) In this way the whole area (or volume) of the crystal is "tiled" or "filled" without any gaps. Length a and width b of the elementary cell (like the a_n and a_m above) are called *lattice parameters or unit lattice vectors* (the "x's" in Figure 3.5 mark the *lattice points*).

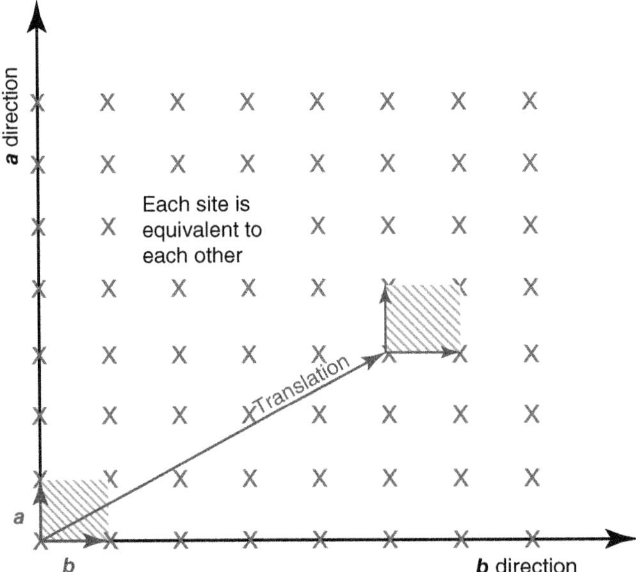

Figure 3.5 A two-dimensional rectangular lattice. There are other potential unit cells that could be chosen. But this one is particularly easy to work with.

Primitive crystal lattices or *Bravais lattices* are mathematically the simplest constructions, or arrangements of points, that reproduce the lattice using translational symmetries. Here, we mean "translational" to apply between any two lattice points: that is, *all lattice points are equivalent* in such a construction. Nonprimitive or *conventional lattice* structures may include basis atoms and additional points to make the lattice easy to visualize. This is like calling the graphene lattice a *honeycomb lattice* instead of a *triangular lattice with a two-atom basis.* The second more accurately describes the symmetries of the system, but the first is easier to "see."

Coordinate systems are generally based on the Bravais lattice, so they will express symmetries of this lattice. That is, reflection, rotation, and translation symmetries typically occur along the principle axis directions. However, the basis may not have the same or even similar symmetries. Yet, we will use this (Bravais-based) coordinate system to describe it.

As always in physics, the choice of coordinate placement is subject to choice. For instance, in Figure 3.5 we have placed the origin on a lattice point. The axis arrows point along crystallographic a and b directions, and lengths a and b form unit lattice vectors **a** and **b**. They are said to "span the elementary cell." Any lattice point can be reached by a linear combination of the unit vectors. In two dimensions and three dimensions,

$$\mathbf{T} = n\mathbf{a} + m\mathbf{b} \tag{3.1a}$$

$$\mathbf{T} = n\mathbf{a} + m\mathbf{b} + g\mathbf{c} \tag{3.1b}$$

T is said to be a *lattice vector* (**T** for translation). We have used the integers n, m, and g as *indices* of the lattice point to which the vector **T** points. A more refined notion of these indices is used to characterize directions and planes in crystals a little later. Since all lattice points are equivalent, the physics at each lattice site is the same, or, in mathematical terms, the physical laws are invariant to the addition of a lattice vector to the coordinates of any lattice point. This is called the *translational invariance* of the crystal lattice and of the laws governing the physics within a crystal.

As with the ambiguities of coordinate choice, the unit cell we use to create this lattice can be chosen in a number of different ways generally. We have said only that it is "basic" in nature (contains only one lattice point) and that it is *space filling*. To be a little more precise, we want to choose a primitive cell that reflects the same symmetries as the overall Bravais lattice (so it will not include the basis). The most common choice of such a cell is the *Wigner–Seitz cell*. The region around the lattice point contained in such a cell is the closest to that lattice point than any other lattice point of the lattice as seen in Figure 3.6. In our discussions of primitive unit cells, we assume that a Wigner–Seitz cell has been chosen.

When considering primitive lattice structures, it is important to realize how many possibilities exist for distinctively different lattices. "Different lattices" in this context has a specific meaning. We enumerate the different lattice types by considering their symmetries. As we have already hinted at above, a *symmetry operation* is doing something to the primitive lattice that leaves it unaltered. For instance, if we take the cubic lattice of above and rotate that lattice by 90°, it looks exactly the same. Thus, a 90° rotation is a symmetry of the cubic lattice. We say that it "maps the lattice onto itself."

As we said, there are two types of symmetry operations: *point symmetries* and *translational symmetries*. However, it is the *ratio* of lengths between the lattice vectors and the *ratio* of angles between lattice vectors that matters in distinguishing lattices. Lattices with the same ratios between angles and lengths are the same lattice, even if the individual *a*'s, *b*'s, *c*'s, etc. are different! This is because the symmetry operations are the same. The lattice point symmetry operations in three dimensions are:

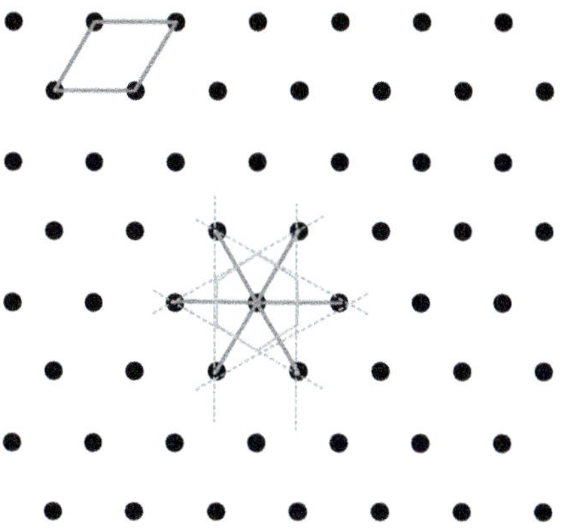

Some choices of unit cells

top – conventional

bottom – construction of the Wigner–Seitz cell. Notice that we take the lines drawn to each of the nearest neighbor sites and use the perpendicular bisectors to form the walls of the cell

Each of the cells has one lattice point in it. This is clear in the W–S cell, but the corner points of the conventional cell count as 1/4th each

Figure 3.6 The construction of the Wigner–Seitz cell in two dimensions. The advantage to using this primitive cell type is that it reflects the full symmetry of the Bravais lattice.

1. Rotations through multiples (or multiples of fractions) of π about the different primary axes of the crystal.
2. Rotations plus reflections.
3. Rotations plus inversions.
4. Pure reflections.
5. Pure inversions.

These represent *point symmetries*, and they form a mathematical group: the *point symmetry group* of the crystal. When we add in translational symmetries, we get a full set of *space group symmetries* that tells us all the symmetries of a given crystal. Remember, combinations of the symmetry operations are also elements of the group.

Since this set of invariant symmetry operations is finite, so too is the number of different types of lattices that can be constructed with them. Specifically, there exist only a finite number of primitive lattice arrangements expressing unique symmetries. Again, Bravais lattices are said to be "equivalent" if they have isomorphic space symmetry groups. Thus, in three dimensions:

1. EVERY Bravais lattice belongs to one of 14 possible symmetry groups.
2. When a basis of arbitrary symmetry is considered, we get 230 different individual space groups. See references such as Ashcroft and Mermin [2].

Though most solid-state physicists likely do not carry an exhaustive catalogue of all the crystal systems around in their head, it is important to know something of the most common ones.

The *difference* between the **Bravais** lattice and the **conventional** lattice is important here. You have seen this already in connection with the hexagonal lattice of graphene. The primitive/Bravais lattice expresses the lattice symmetry, and every point of the lattice is identical. The conventional lattice is easier for us to "see" and includes nonequivalent lattice points. In three dimensions, two common examples are *face-centered cubic lattices* and *body-centered cubic lattices* both shown in Figure 3.7.

One wants to be a little careful however in imposing this appearance-based description. The diffraction and electronic, optical, and other phenomena related to the crystal lattice will reflect the symmetries of the Bravais lattice generally. So while the conventional description helps you to understand the lattice, the Bravais lattice is needed for calculations and the application of group theoretic methods.

The lattice vectors (**a**,**b**,**c**) as marked in Figure 3.8 are a way of describing the positions to each of the points of the lattice. For the square body-centered cubic (BCC), we might typically think of **a**, **b**, and **c** in terms of their components on a standard Cartesian coordinate system with unit vectors (**x**,**y**,**z**). In other words we see $\mathbf{a} = (a_x\mathbf{x} + a_y\mathbf{y} + a_z\mathbf{z})$ and so on. But of course we should describe vectors or directions in terms of the Bravais lattice (**a**,**b**,**c**) coordinate system, where the vectors are non-orthogonal unit vectors of that coordinate system. Because we wish to keep our descriptions confined to the unit cell generally, we use a *reduced intercept* index – the (h,k,l) designation as seen above – to identify points, directions, and planes within a crystal. The brackets tell us what we are identifying: (h,k,l)

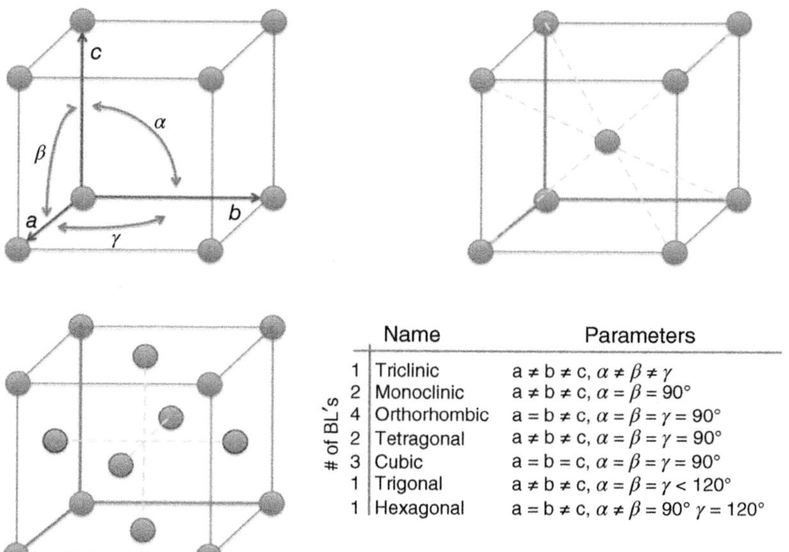

# of BL's	Name	Parameters
1	Triclinic	a ≠ b ≠ c, $\alpha \neq \beta \neq \gamma$
2	Monoclinic	a ≠ b ≠ c, $\alpha = \beta = 90°$
4	Orthorhombic	a = b ≠ c, $\alpha = \beta = \gamma = 90°$
2	Tetragonal	a ≠ b ≠ c, $\alpha = \beta = \gamma = 90°$
3	Cubic	a = b = c, $\alpha = \beta = \gamma = 90°$
1	Trigonal	a ≠ b ≠ c, $\alpha = \beta = \gamma < 120°$
1	Hexagonal	a = b ≠ c, $\alpha \neq \beta = 90°$ $\gamma = 120°$

In the cubic system, three special cases are shown. The simple cubic, the body-centered cubic, and the face-centered cubic systems all have equal a,b,c lengths and equal 90° angles. The table shows the number of Bravais lattices for each lattice class

Figure 3.7 Ratios of lengths and angles define the possible number and type of point symmetry operations that a crystal can allow. This lets us categorize the crystal into families.

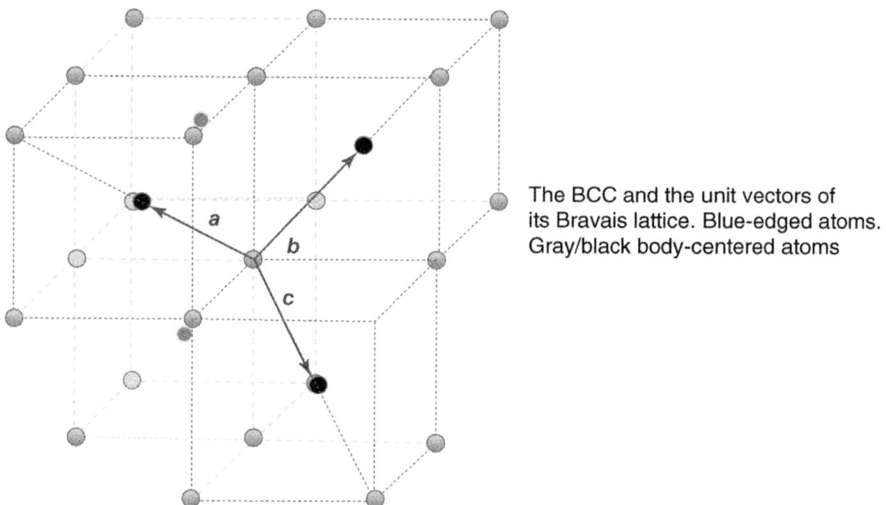

Figure 3.8 A comparison of what we call the body-centered cubic lattice and its primitive lattice can be surprising because they look nothing alike! The conventional lattice is like a pneumonic – it is intended to help one remember where the atoms are.

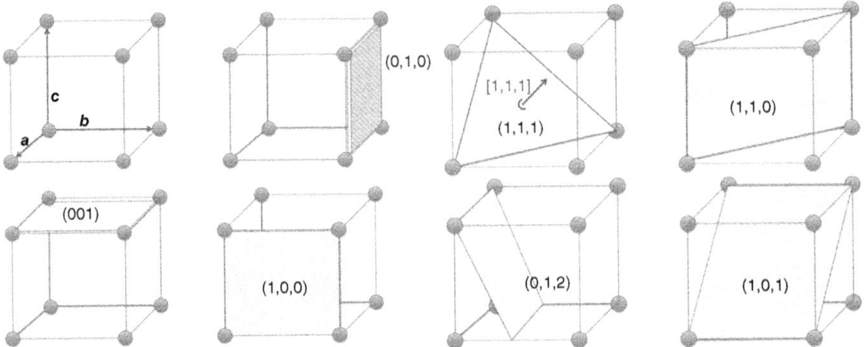

Figure 3.9 Example: identifying the planes in a cubic crystal. Notice that the plane associated with the ½ intersection along the *b* axis gets designated with a "2."

means the plane of atoms associated with the [*h*,*k*,*l*] axis direction. {*h*,*k*,*l*} means the set of equivalent planes to (*h*,*k*,*l*). These are known as *Miller indices.*

To *index* a set of planes or a direction in the crystal, as in Figure 3.9, the procedure is relatively straightforward:

1. Find the intercepts of the plane with **a, b, c** axes. You should use primitive but can use nonprimitive coordinates. This is where it can become confusing: one must know how the crystal is being described. If there is not an intersection with a specific axis, this axis index will be designated with a "0."
2. Take the reciprocals of these numbers, and reduce to three integers having the same ratio as the fractions. For example, the intercepts of plane shown in the second from the right (bottom) are 0, ½, 1. The reciprocals are 0, 2, 1. To get the same ratio between the numbers, we multiply by ½: 0, 1, 2. So the plane is the (0,1,2) plane. It is associated with the [0,1,2] axis vector.
3. If a negative direction is expressed, you do that using a bar over the index $[\bar{h}\ \bar{k}\ \bar{l}]$.

Pay attention to the word "index." It is also used to associate diffraction spots with the set of planes that caused them.

So, Figure 3.10 summarizes how to build a three-dimensional (3D) lattice:

1. Identify the Bravais or primitive lattice.
2. Overlay a set of coordinates onto this lattice.
3. Add a basis set of *real* atoms.

The primitive unit vectors can be used to describe where everything is in the unit cell.

3.2.2 Surprises in Two-Dimensional Lattices

While, in 3D space, there are 14 possible Bravais space symmetry groups, in lower dimensions this situation changes. In 2D, there are only five such space symmetry groups allowed. In 1D, there is only one type of Bravais or primitive lattice.

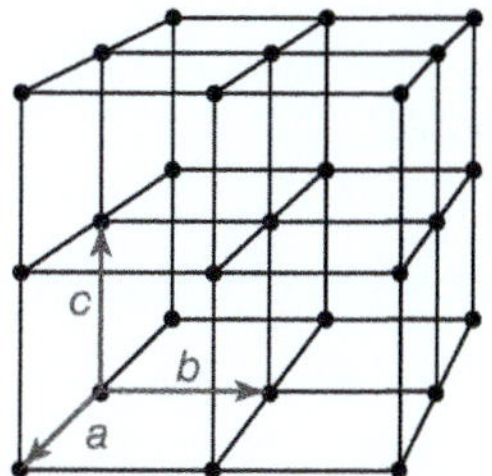

To build a real crystal, we begin with a mathematical abstraction of points with a specific set of symmetries. Here is a square lattice. It is symmetric under translations in any of the three principle directions, rotations of multiples of 90°, inversions of all sorts. It is highly symmetric, and all of the symmetries taken together form a point symmetry group

Next we add a system of coordinates. These are the *lattice vectors*, and they allow us to know what direction in the lattice we are going **(a,b,c)**. Because the symmetry operations form a group, the labels chosen for the unit lattice vectors is pretty arbitrary

Once we have the lattice in mind, we add to each lattice position some chosen arrangement of atoms. This is known as the *basis* set. Notice that the arrangement of the basis set can have a different symmetry than that of the original lattice. Thus the symmetry operations of the real lattice is a smaller group than that of our mathematical construct

The mathematical construct is known as the *Bravais lattice* of the system. The real lattice will have symmetry operations that are some subset of those expressed by the Bravais lattice

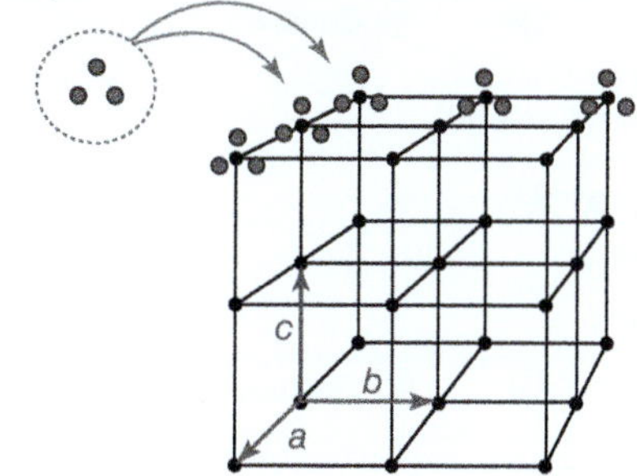

Figure 3.10 A summary of lattice construction starting with the mathematical construct.

This seems simple enough. In fact, one might suspect that the lower dimensions are easier to deal with than the higher ones. However, we can be deceived. Why? It is because the lower-dimensional structures have the added freedom to twist and turn into the higher dimensions, and this can introduce complex "connectivity" between different volumes, areas, or lengths of the material. In other words, we might construct a simply connected closed structure that expresses the symmetries of the lattice locally (that is at the unit cell level), but globally it might break translational and rotational symmetries. Hum, that's odd: you mean that we can translate the unit cell locally and get the lattice reproduced, but if we do this across the entire material, somehow we do not? YES! Using nanoscale self-assembly, such structures are now regularly achieved as we will see.

Let's begin with a look at how the simple 2D lattice might work. We start by enumerating the Bravais lattices of the second dimension.

These (Figure 3.11) represent the symmetries with which atoms can be arranged in a plate, ribbon, or strip of a 2D plane. Notice that the translational and point symmetry groups that describe the arrangement of atoms are all perfectly well described. Curiously though, we can play a game of *topological complexity* with such sheets of atoms. That is, in the 2D system, we can bend the sheet into the third dimension and connect it to itself in a variety of ways. In the mathematical sense now one local area of symmetry is connected to another local area of symmetry across the object. In 2D materials this "connectedness" or *global topology* can be simple loop of ribbon or nonsimple like a donut. As the global topology becomes more complicated, properties at each of the lattice sites, such a spin projection, can lose their translational invariance as it once existed on the plane. Think, for instance, of the Möbius strip as shown in Figure 3.12.

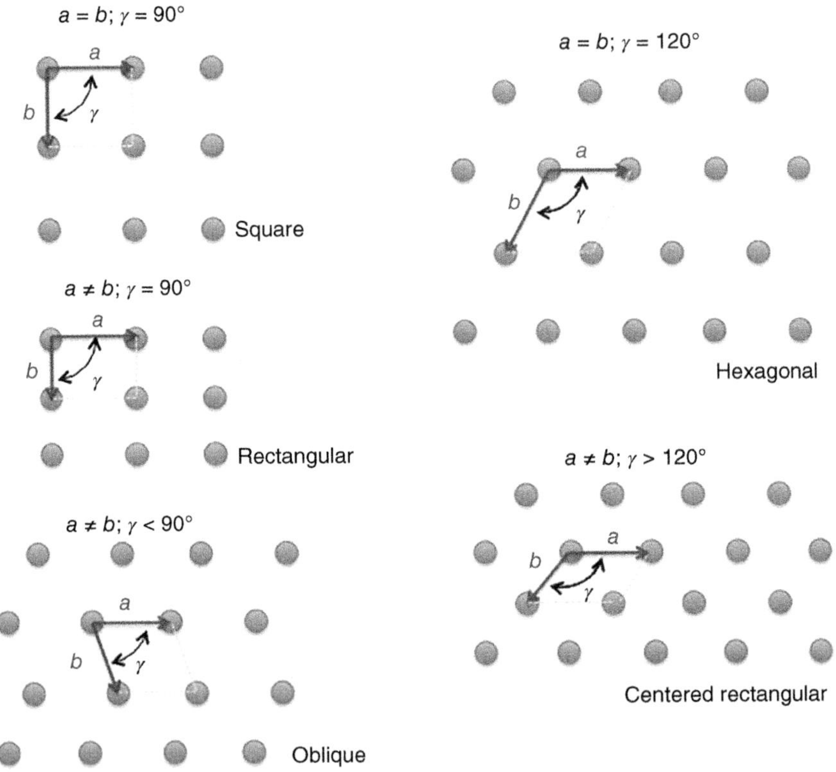

Figure 3.11 The two-dimensional Bravais lattices.

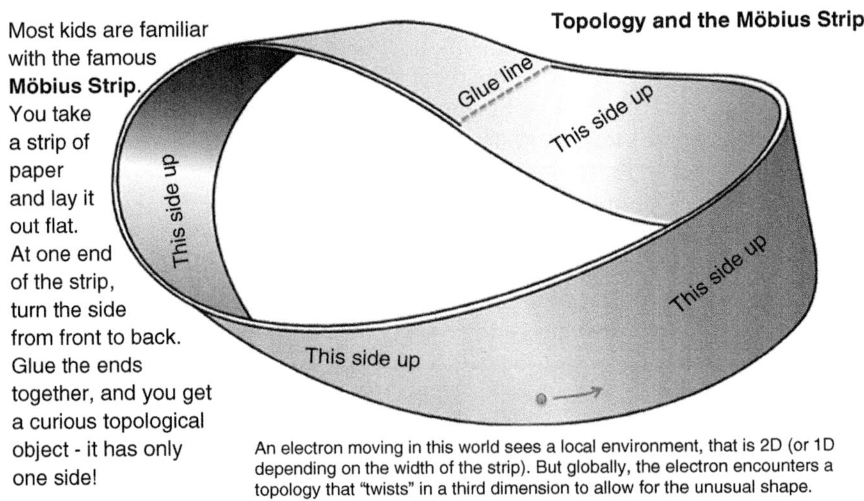

Figure 3.12 A topologically closed surface that can break global symmetries while maintaining local atomic arrangement.

Our physical example of 2D solids is the carbon *allotrope*: graphene, an absolutely extraordinary material. However, there are a large number of 2D materials, other than graphene and its relatives (like BN), that have been demonstrated. In such materials, carriers can move anywhere in a plane, up to the edge of the crystallite of course.

But can such structures as that of Figure 3.12 (or structures like it) occur naturally or be synthesized? Flat ribbons and planes of graphene can, and will, have defects in them. However, the carbon–carbon bond is extremely flexible, willing to take on a wide range of bonding angles. This means that most defects can be accommodated by this "flexibility" of bonding (depending on the type of defect), resulting in little overall strain in the lattice. So the bonding *hybridization* stays intact and the structure remains relatively flat. But this is not the case with atoms that have more rigid bonding requirements. For such flat materials, atomic imperfections lead to significant structural strain. Since the structure consists only of a 2D set of atoms, there are no third-dimensional linkages to take up this strain. This, in turn, affects the overall topology of the system by introducing curvature locally. To see how this works, let's consider the example of Sb_2Te_3.

In the Sb_2Te_3 system [4], when it is grown from a solution of antimony trichloride ($SbCl_3$) and tellurium dioxide (TeO_2), if the temperature is high and the growth is slow, the atoms have the time to add to the edges of the plate perfectly, forming regular crystallites of two dimensions. Details of the growth and further microscopy are shown in the references. One of these plates, a single unit cell thick and many unit cells across, is shown in Figure 3.13.

However when the temperature and growth speeds (governed by the molarity of the reactants) is just right, a defect occurs at the kernel where the plate is first

Figure 3.13 A single, defect-free plate of Sb_2Te_3. The inset shows the atomic ordering using high resolution electron microscopy. (Courtesy of ChaoChao Dun, Wake Forest University.)

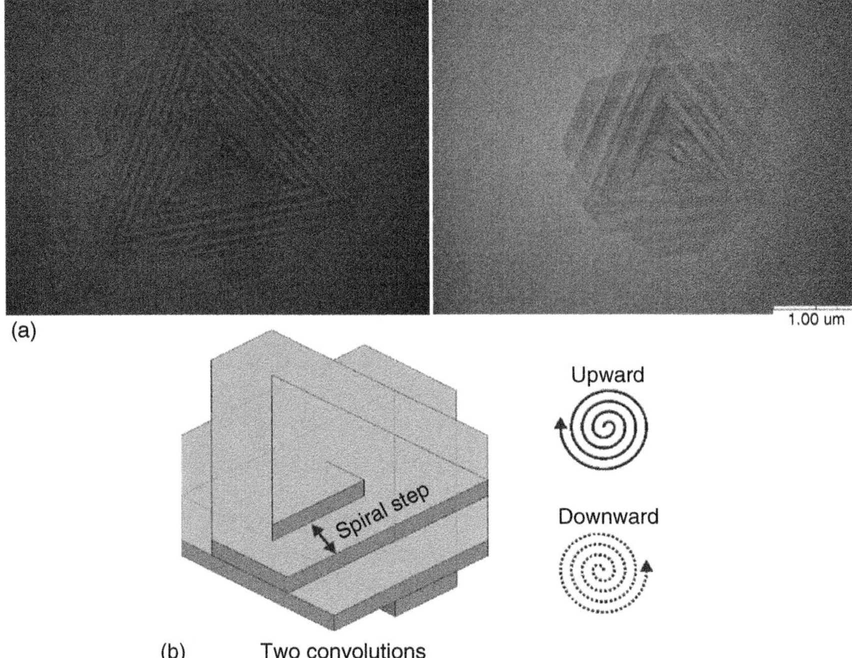

Figure 3.14 The "spiral" growth of Sb_2Te_3. Two examples are shown in panel (a) while a model is presented in panel (b). Note that the upper and lower halves spiral in different directions. (Courtesy ChaoChao Dun Wake Forest University.)

formed, leaving behind a localized strain. This turns into a dislocation of atoms along the out-of-plane axis of the plate. To reduce the strain, the plate growth proceeds along a spiral. Two of these spiraled plates are shown in Figure 3.14a, the top with fewer "steps" and the center with more. The bottom of the Figure 3.14b shows how the steps are arranging themselves during growth.

In systems that utilize atoms with rigid bonds, these spiral structures are relatively common. Again, this can be understood as the introduction of local strain due to defects. The spiraling planes emanating from the center of the plate are simple extensions of an embedded *screw dislocation* at the core of the structure. This is an ordered type of defect. However, the Sb_2Te_3 system is a little unusual in that the screw reverses itself between top and bottom. That is, if we take the screw to be right-handed on the top, then the bottom screw will be left-handed! These type of complexities are far less common in materials such as metals, but not unheard of.

So it is clear that these single unit cell thick plates can become large *chiral* molecular objects with extraordinary topological complexity. And it is hypothesized that electrons on such surfaces will acquire phase shifts associated with the translation through 3D space. This global topologically induced phase factor in the electronic wavefunction is the so-called *Berry's phase,* and it is unique to such objects as far as we know.

3.2.3 The One-Dimensional Lattice

The simple lattice of the one-dimensional row, or line, of equidistant points is shown in Figure 3.15. In a strictly one-dimensional lattice, there is no need for considering the angular positions of the atoms. However, many "real" one-dimensional substances have the atoms arranged in a zigzag line, as indicated in Figure 3.16. Such a zigzag chain can "melt" by generating defects as illustrated in Figure 3.17, where the strict alternation is disturbed. These defects allow for rotation around a bond between. Thus, the one-dimensional object can also "sample" a third dimension without being 3D itself, in the sense of its electronic properties. This ability to "fold" into additional dimensions while constraining carriers to a lower-dimensional space is this object's *topology*. We might ask if any complexities come along with this topology.

An electron moving along a lattice such as in Figure 3.15 would "see" equally spaced potentials and be "unaware" of the third dimension outside of the chain. If a field of some kind were to intersect with this chain, it would appear to the electron as a localized potential. It might be reflected back along the chain or tunnel through the potential barrier. It may also correlate with other electrons along the chain to form excitations. This is, in fact, a fairly famous *random walk* problem set before students of computational methods.

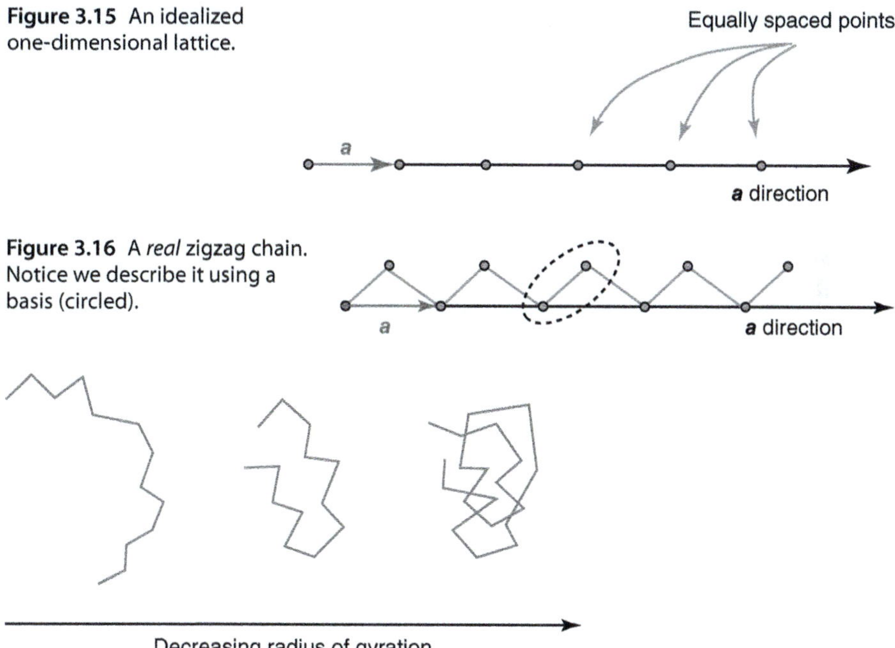

Figure 3.15 An idealized one-dimensional lattice.

Figure 3.16 A *real* zigzag chain. Notice we describe it using a basis (circled).

Figure 3.17 A zigzag chain with defects, twists, and turns. We must turn to other quality factors to describe this. For instance, the *radius of gyration* lets us know the degree of coiling. The radius of gyration is influenced heavily by the solvent used with the polymer. A *good solvent* yields a large radius of gyration, while a *poor solvent* yields a small radius or tightly bound coil.

3.2.4 Polymers as One-Dimensional Lattices

In many compounds, like conducting polymers, atoms have the tendency to pair up and form a zigzag as in Figure 3.16. We actually saw this in previous chapters and will return to it in more detail later. For now it is the structure itself we are interested in, and an electron in transit down such a chain will "see" a one-dimensional path just as above. But the repeating potential and length scales will be very different. Typically the lattice in this case is thought to be composed of pairs of atoms, with each pair separated by the lattice constant. As seen in the previous chapter, the lattice need not be just a pair of atoms, but can be whole collections of atoms, as long as the conjugation (single bond–double bond alternation) remains.

While it is intellectually interesting to consider these polymers as perfect, one-dimensional chains, Figure 3.17 shows what actually happens when a solvent is used to create or process a polymer. The twists and gyrations of the chain fill space as though it were a ball. An electron moving along this chain might well "forget" where it began. That means it would lose the *phase information* of its wavefunction. The interaction of the polymer with proximate solvent molecules determines just how coiled the polymer ends up being. The degree of coiling is known as the *radius of gyration*. This radius is of primary interest to scientists studying the processing of polymers.

As noted in Chapters 1 and 2, *the example* of a conducting polymer system with alternating double and single carbon bonds is that of polyacetylene. The structure is given in Figure 1.12 but is reproduced here in its schematic form in Figure 3.18.

For this one-dimensional system however, this simple schematic is not the end of the story. As discussed, rotations about a bond are allowed in many systems, including in polyacetylene. In fact, such bond rotations can occur in an ordered way, leading to alternative "crystal structures" for the same material. Thermodynamic internal energies dominate the formation of these equilibrium configurations. Experimentally, these equilibrium configurations in a given polymer sample can be identified rather easily through the exotherms and endotherms of a differential scanning calorimetric (DSC) curve (heat vs. temperature). In the case of polyacetylene, there are two well-known forms the structure *could* take. They are shown in Figure 3.19. The question for the experimentalist

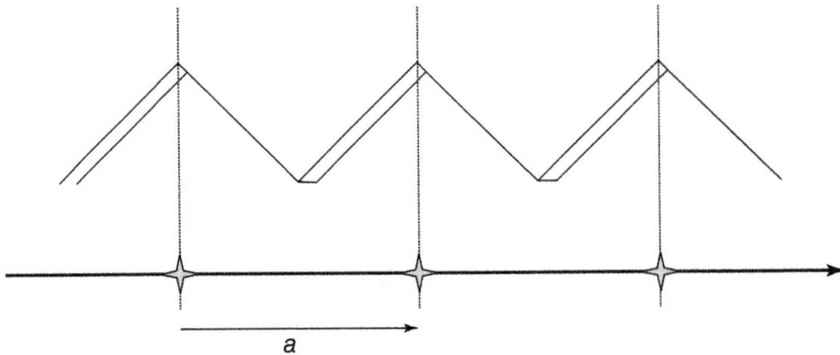

Figure 3.18 The schematic structure of polyacetylene along with its lattice. Notice the lattice points do not fall on every carbon. Rather, there is a two-carbon (along with some hydrogens) basis.

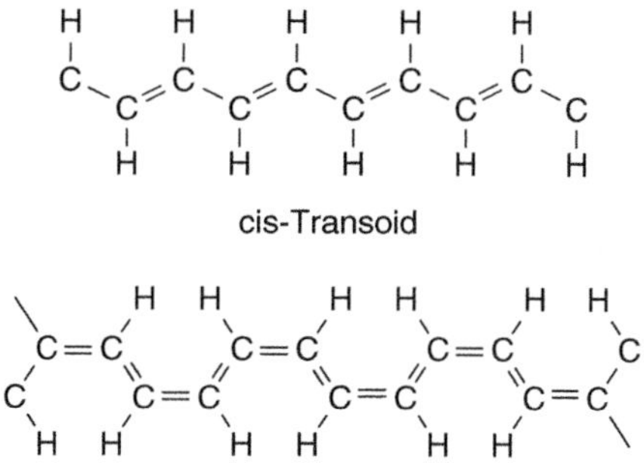

Figure 3.19 The difference between *cis* and *trans* polyacetylene is the rotation of a few bonds. This gives different crystal structures in one dimension, though the Bravais lattice stays the same.

is whether or not the different phases of the polymer structure will alter the local environment a conducting electron might see.

As noted in Chapter 2, the one-dimensional nature of such systems as polyacetylene means that the properties of the material, especially the electronic transport characteristics, are extremely sensitive to defects (or mistakes) within the lattice. Thermodynamics tells us that there will *always* be some mistakes. Consider only an occasional bond rotation within the lattice as seen in Figure 3.19. It makes sense then that our expectation of the electronic properties of such a system of randomly distributed defects is that they are dominated by the electronics of the individual segments as though they are isolated. In this way the prediction of phenomena in one dimension can easily be as complex as that of a 3D system.

3.2.5 Carbon Nanotubes as One-Dimensional Lattices

Carbon nanotubes were introduced in the last chapter. But in the context of the current discussion, it is important to take another look at its dimensionality. Recall that the nanotube is constructed by rolling the honeycomb lattice of graphene seamlessly into a tube. Specific points of the lattice roll onto themselves symmetrically ($A \rightarrow A'$ and $B \rightarrow B'$) as seen in Figure 3.20. The lattice vector connecting A and A′ or B and B′, respectively, is called the *roll-up vector* $\mathbf{R}$; its components (n,m) are written in terms of the unit vectors and are called the *indices* of the nanotube. The orientation of this roll-up vector is given by the *chiral angle* (see Eq. (2.3)).

The "unit cell" of the nanotube is the part that reproduces itself along the translation from points A to B. But notice that this depends on the roll-up vector $\mathbf{R}$ and its angle rather drastically. The (2,2) nanotube shown at the left has relatively few basis set atoms in its unit cell (given by counting up the atoms in the shaded area), whereas the (3,1) nanotube in the center has quite a large number of basis atoms. The lattice parameter (the distance along the nanotube that one must travel to repeat the unit cell) has also changed. Notice too that the (3,3) nanotube on the right has a greater diameter. This also makes the basis set larger. However, each of these nanotubes has a one-dimensional lattice. They have one Bravais lattice with one symmetry group representing the set of operations that take the primitive lattice onto itself.

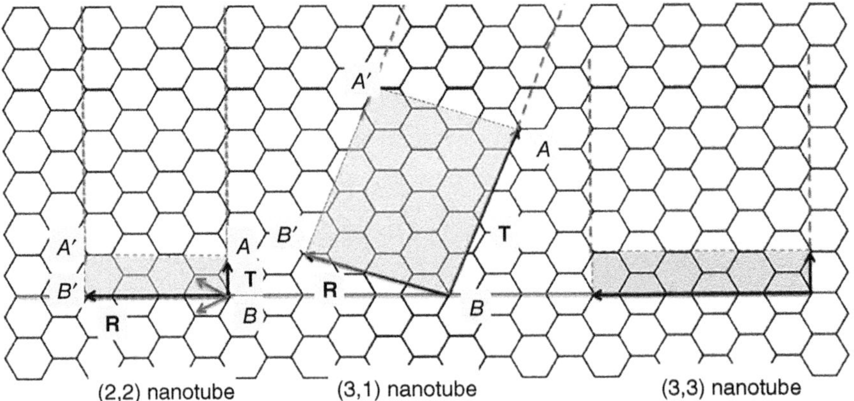

Figure 3.20 Nanotubes are formed by rolling equivalent lattice points onto each other along the vector **R**. The "unit cell" for hypothetical (2,2), (3,1), and (3,3) nanotubes is shown.

Nanotubes can also have kinks, twists, missing or substituted atoms, and other defects that will effect electronic and thermal transport properties, just as in the case of polymers. As it turns out they also have strong coupling between charge carriers and the lattice – a fact that was missed or ignored by the scientific community early on. So in many ways the nanotube lattice has the aspects of a polymer but with this extra "mini-dimensional" aspect of chiral angle.

3.3 Bonding and Binding

We note that we have not said much about how the real atoms "talk" to each other, that is, how they share electrons and how they interact. That is because when speaking of organizational symmetries, we really don't care. Lattices are essentially mass correlation functions of a sort. They tell us where the atoms are. However, it is through these interactions that dimensional phenomena will arise, so we will have to specify: *what makes bonds rigid or flexible*? *What gives the solid and low-dimensional structures their shapes? What tells the atoms where to go?* These questions are afforded a little more attention in the study of low-dimensional materials than in 3D systems because only a few nanoscale, low-dimensional materials are found in nature. Since most are made artificially in a lab, the low-dimensional scientist is rarely working with the most stable form of the element or elements.

So, in lower dimensions, the material system's topology is sensitive to defects, bond rigidity, and local minima in formation energies. How do we know that, given certain conditions during synthesis, we will end up with a carbon nanotube and not a lump of coal? The answer lies in our first-year chemistry course.

Bonding is generally thought of in terms of "bonding character": (i) *weak or van der Waals*, (ii) *ionic*, (iii) *covalent*, (iv) *metallic*, and (v) *hydrogen bonded*. There are other ways to classify as well, but these are quite useful. They represent the most common ways of modeling assemblies of atoms in a solid. Introducing the physical picture that we "see" in our minds for each of these is not nearly so hard

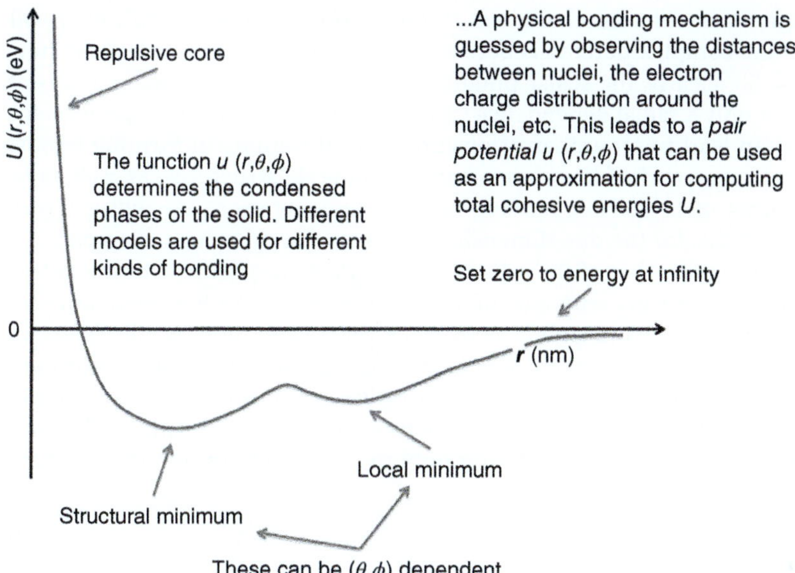

Figure 3.21 A schematic potential well between two atoms that can vary with radial r, as well as azimuthal and polar angles: θ, ϕ.

as trying to come up with mathematical models that accurately capture observed behavior. What we would like is a way to determine the overall equilibrium structure from the lattice to the crystallite. The route to this is through the total internal energy of the system, U, as seen in Figure 3.21. Presumably, this extrinsic parameter will include the surfaces, grain boundaries, inclusions, and defects as well as the arrangements of all atoms. The absolute minimum of this U, at a given temperature and pressure, will predict for us the equilibrium structure of a system.

However, the *devil is in the details* as they say. To get U we have to start with the potential between individual atoms, u. But of course this is the *bonding* we discussed above. We approach this generally by taking the sum of all the potentials over all atoms, adding up the energies of the bonds, and then including special sites such as the surface sites, defect sites, edge sites, etc. For macroscopically large materials, we can ignore the contributions these might make to the *lattice sum*. However in the nanoscale, they count in a major way. This is because there may be as many surface sites as bulk sites in a nanoparticle.

To build a "picture" in our classification scheme, we start with **covalent bonding** as found in carbon compounds, Si, and other semiconductors. The associations atoms choose to make between themselves using these mechanisms typically have to do with valence electron shells and how they are filled. The shells are filled or emptied by charge sharing or charge transfer between neighboring atoms – this means an overlap in the electronic wavefunction somehow. Since the atomic orbitals doing this "sharing" begin (before condensation) as quantum eigenstates of the atomic potentials associated with individual nuclei, the electrons involved with have specific angular momentum and energy characters: s, p, d orbitals. When atoms are brought together, the individual characters and the willingness of different characters to mix (*hybridization* sp^2, sp^3, etc.) will determine the complex potentials between the atomic nuclei that yield the new eigenstates. So, clearly, the scalar potential field between neighbors in a crystal can have not only a radial dependence (distance) but also an angular component to it. This can lead to a number of positional local minima in forming bonds, giving rise to a myriad of low-dimensional, nanoscale structures, as well as to different forms of crystals depending on the elements used. We will get very detailed about this for the one-dimensional carbon system in later chapters.

Ionic bonding like that of the salts NaCl or KBr is simply the extreme in this electron transfer process wherein the electron of one atom has been so completely transferred to its neighbor as to make them both ions. While the exact nature of interatomic potentials might be difficult to guess for nonionic covalent bonding, the ionic bond is, in fact, quite easy to guess since the potential will be dominated by the electrostatic interactions of the ions. Madelung worked out a nice approach to this by considering only the Coulomb attraction and some exponential repulsive core potential:

$$U_{ij} = \lambda e^{-r_{ij}/\rho} \pm q^2/r_{ij} \tag{3.2}$$

r_{ij} is the jth nearest neighbor distance from the ith ion in the lattice, sometimes written as Rp_{ij} with R being the lattice constant. q is the charge of the ions; ρ and λ are constants to be determined (fitting parameters). The lattice sum becomes

$$U = \frac{1}{2}\sum_{i,j} U_{ij} = N\sum_{j}(\lambda e^{-r_{ij}/\rho} \pm q^2/r_{ij}) \tag{3.3}$$

which represents the total lattice energy. This is frequently written as

$$U = N(z\lambda e^{-R/\rho} - \alpha q^2/R) \tag{3.4}$$

$$\alpha = \sum_{j\neq i} \frac{\pm 1}{p_{ij}} \tag{3.5}$$

α is referred to as the *Madelung constant.*

Curiously some noble gasses and large neutral molecules can also form crystals under the right conditions of pressure and temperature. Layered crystals such as graphite should be included in this list, though in plane they are covalently or ionically bonded usually. We call these **van der Waals bonded** solids (a special case of *Casimir forces*), and they are held together by weak interatomic forces. Generally we can break the potential contributions down as shown in Figure 3.22.

The *dispersive* part consists of spontaneous dipole induction due to quantum fluctuations. Fritz London [5] showed that these potentials typically follow a $\sim C/r^6$ form as well. Thus the potential in such systems is typically modeled as a *Lennard-Jones potential:*

$$V_{vdW}(r) = -A/r^6 + B/r^{12} \tag{3.6}$$

where the $1/r^{12}$ is derived from Pauli exclusion repulsion at very close distances. The lattice sum then becomes

$$V_{total} = \left(\frac{1}{2}N\right)\left[\sum_j \left(\frac{B}{p_{ij}R}\right)^{12} - \sum_j \left(\frac{A}{p_{ij}R}\right)^{6}\right] \tag{3.7}$$

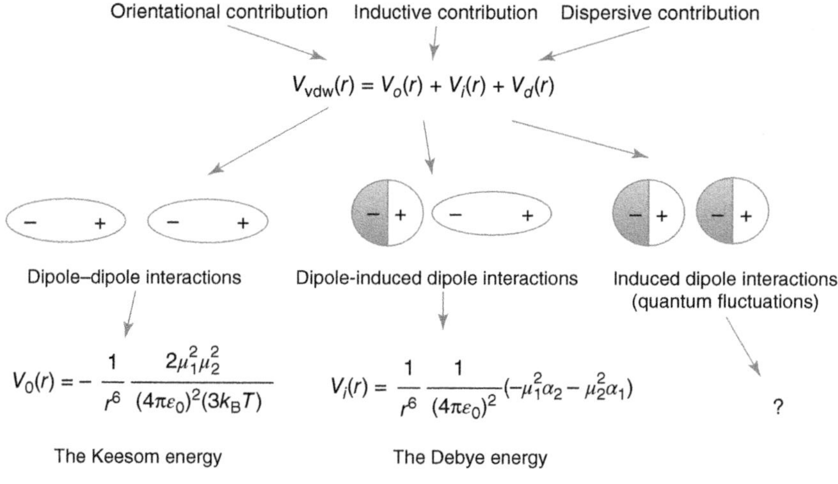

Figure 3.22 The van der Waals interaction can be thought of in terms of dipole–dipole interactions plus interactions between dipoles that have been induced between the atoms of a layer, either through a permanent dipole interacting with a polarizable species or through quantum fluctuations of the two species.

where $p_{ij}R$ is the distance to the jth nearest neighbor from the ith atom. This distance is written in terms of the lattice parameter R. N is the number of atoms and so this gives the total extrinsic energy of the system. The trick is now to find the A's and B's and perform the sum.

In our examples here, the energy per bond is determined according to a model that is guessed based on how electrons might be shared or interact. A potential minimum is calculated, and this is summed across the whole of the crystal to achieve the total internal energy. In large 3D crystals, this task can become cumbersome, and there are issues of summation convergence, long-range interactions, and multiple bonding types that come into play. However, at the nanoscale, with countably large numbers of atoms in a particle or object, the problem can be far more manageable.

To see how this might work, let's consider a simple but effective example: the one-dimensional salt crystal. In this example we will consider alternating ions of Na and Cl, interacting with each other along a one-dimensional strand. This is shown in Figure 3.23. Since this crystal is ionic in nature, we will take the attractive interaction between atoms to be that of the Coulomb field $\sim 1/R$. We will take as the repulsive part of the potential (the part that keeps two ions from occupying the same space on the strand) to look like e^R: (as in Eq. 3.2). We also anticipate that the total force repelling the ions will be related to the Pauli exclusion of electrons, so the number of electrons participating must also enter into it; to do this we add an atomic weight term Z. We might also argue that this includes the repulsion of the ionic nuclei. Notice too that we have used λ and ρ as "strength constants."

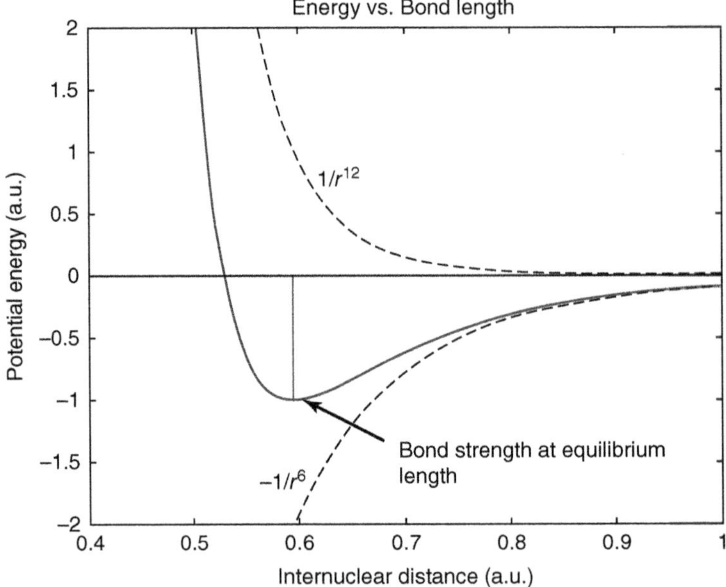

Figure 3.23 A graphical representation of the Lennard-Jones potential.

ρ represents roughly the radius of the ion since the strong repulsion starts some distance out from $R = 0$.

Next, we must find the equilibrium distance where $dU/dR = 0$. We call this value of R; R_0.

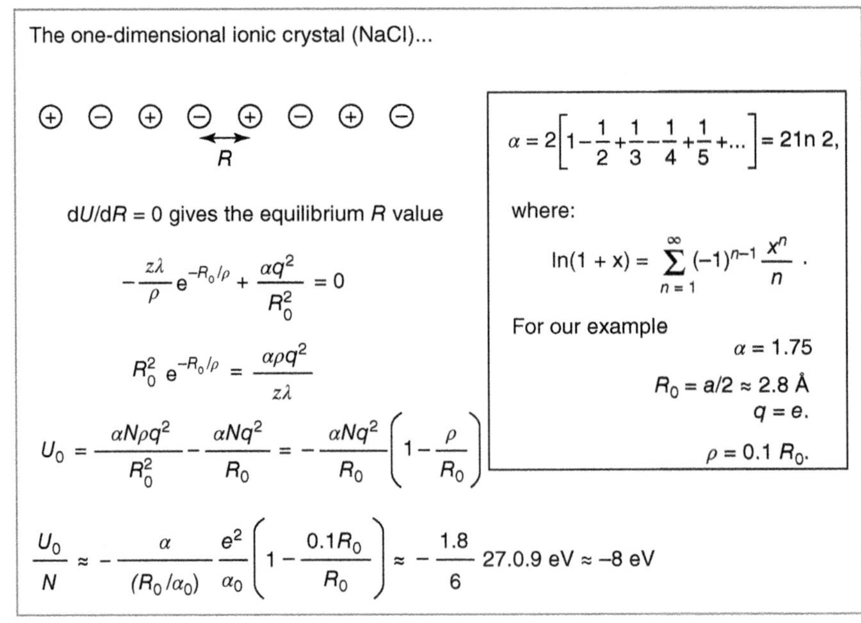

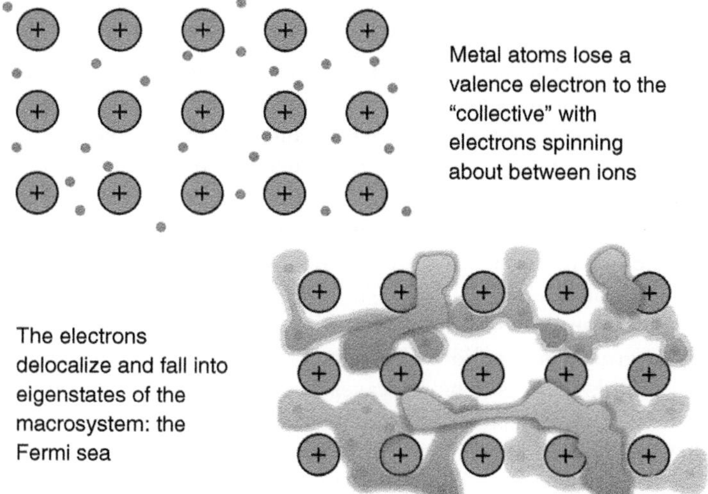

Figure 3.24 Metallic bonding using background electron clouds.

Metallic bonding is rather more subtle than the lattice sums above. In the case of metals, the atom is relatively willing to give up outer shell electrons as in Figure 3.24. This might occur for different reasons as, for instance, between the examples of Ag or Au and alkali metals like Cs or K. Nevertheless, this willingness to release an electron results in a large collection of *correlated* electrons contained within the boundaries of the solid: the so-called *Fermi sea*. Now we have used the word *correlated* here, and by that we mean to imply that this entire sea of electrons is a single eigenstate of the macrosystem wherein the wavefunctions of individual particles have become so delocalized that to speak of an "individual electron" is a little misleading. The electrons are participating in a many-body wavefunction that lowers their imagined individual energies. Thus, this "gas" or "sea" of electrons does not disperse under their own electrostatic interactions. Rather the balance of ions in the volume plus their collective behavior prevents this.

However, the ion background doesn't delocalize as easily as these very low mass electrons. They are attracted to the volume of negative charge, that is, the Fermi sea, but they pack into this charge as localized ions. This appears as a force between the ions, holding them together, and we refer to this form of "bonding" as *metallic bonding*. As one might imagine a reasonable attempt at computing self-energies (U) perhaps should involve the energies of a many-body quantum state, and there are different approaches to this. But it is the picture of the bonding mechanism that interests us here.

The **hydrogen bond** is really a modification of the van der Waals and the covalent mechanisms that is made possible when hydrogen is participating. It comes from the electrostatic attraction between two polar groups: one that occurs when a hydrogen (H) atom is covalently bound to a highly electronegative atom such as nitrogen (N), oxygen (O), or fluorine (F). The other is any highly electronegative atom nearby. The two polar groups experience each other's electrostatic fields, and there is a resulting force. This bonding mechanism may not really seem that different from what we have already discussed – again electrostatic energies brought about by the coordination of electronic positions. However, its importance is to suggest that materials can be held together by more than one mechanism and that the blending of bonding types can lead to additional internal forces, as seen when polarization occurs between the constituents of the crystal basis. In the many different materials introduced in Chapter 2, this has already been seen, though not emphasized.

But how do we know when we have it correct? The approach we have taken relates the bonding energies in a solid to the internal energies that form the ground state of the solid. This is true in cases such as 3D polycrystalline systems, when the ordered lattice sites are far more numerous than the special sites such as grain boundaries and surfaces. In such cases, these bonds – the attractions between the electrons and nuclei of the collection of atoms – are responsible for the cohesion of the solid itself. Generally, crystals (solids) are formed when the total energy of the collection of atoms (kinetic + potential) at infinite separation is greater than the total energy when the atoms are condensed together. The difference between the two (free atom energy) and (condensed system energy) is >0 and is referred to as the *cohesive energy of the structure*. Here we have implicitly made the connection between the energy of cohesion and the modulus of the crystal or structure. We can connect the two through a measurement of the *bulk modulus*. The *bulk modulus* is defined as

$$B = -V\, \mathrm{d}P/\mathrm{d}V \tag{3.8}$$

where V is the volume of the crystal and P is the pressure. The so-called *compressibility* is the reciprocal of the bulk modulus. At absolute zero (or near it), the entropy is constant, and the thermodynamic identity $\mathrm{d}U = P\,\mathrm{d}V$ is the change of internal energy of the crystal corresponding to a change in volume. Thus we get $\mathrm{d}P/\mathrm{d}V = \mathrm{d}^2U/\mathrm{d}V^2$ and

$$B = V\, \mathrm{d}^2P/\mathrm{d}V^2 \tag{3.9}$$

The higher the bulk modulus is, the stiffer the material. Here we are making the approximate identification of this thermodynamic internal energy U with the overall crystal cohesive energy.

So, to get some idea of what this U is, we must first come up with some sort of pairwise potential function $u(r,\theta,\phi)$ with which we could approximate the bonding. Once the atoms are relaxed into the minimum pairwise potential position, we could then sum that energy over the whole crystal. And a measurement of the bulk modulus could tell us if we are right!

It is important to note that this does not apply to *yield strength* of a material since that typically is dominated by defects and grain boundaries. This notion is particularly nuanced as one approaches the nanoscale where there might be very few if any defects in the structure being tested. Thus, objects like a carbon nanotube can appear to be very strong because its failure relies on the direct rupture of carbon–carbon bonds.

3.4 Spatial Symmetries Are Not Enough: *Time Crystals*

This chapter has presented a few details of how the observed symmetries of a general array of crystal lattice points could pertain to properties in real-world materials. These spatial arrangements of points and the symmetry operations that repeat them define crystals, as well as ordered nanoscale objects, rather completely. Or do they?

There is, surprisingly, another way to see arrangements of points, that is, arrangements in time. Frank Wilczek of MIT first proposed the idea, and it imagines a type of matter that exhibits a sort of fundamental oscillation over time. This means that some property of the material goes through a repeating cycle. Moreover, it represents a "crystal" by analogy to space crystals that have regular repeating units, or periodic cycles, of atoms in space. Molecular patterns repeat over and over along their lattices. *Time crystals* repeat some internal state of the system with constant separations in time [6].

Wilczek's model used charged particles in a superconducting ring to break continuous time translation symmetry. This symmetry breaking simply means that the system will look different on a global level from one instant to the next. This is quite different from a normal matter in thermal equilibrium, which has only random internal motion. These are the random motions of the atoms on their lattices from place to place. Notice though that in thermal equilibrium, from one instant to the next, random motion stays the same (random). This means simply that in regular matter, in equilibrium, statistical properties stay the same over time. Wilczek's system breaks this symmetry because there are global statistical differences from one instant to the next. That is, nonrandom patterns in time emerge for statistical properties.

Now, Wilczek's proposal was for a *substance* whose ground state was in perpetual motion while in equilibrium. And this would indeed break time translational symmetry. But it was quickly recognized by Watanabe and Oshikawa [7] that, based on fundamental considerations, time translational symmetry cannot be broken by a quantum system in thermal equilibrium. So, Yao et al. [8] decided to investigate what might happen when the thermal equilibrium constraint is relaxed. To do this they proposed an external input of energy (thus, not in thermal equilibrium) to force the oscillating states. Their calculation aligned a set of ions with specified spin states: a one-dimensional spatial lattice of ions. In such a lattice, spins next to each other like to align or anti-align due to the interactions of the local magnetic fields. In fact, the aligned and anti-aligned states are lower energy than random alignment (see Chapter 11). The next step is to cause the spins to flip back and forth with the input of a collinear laser beam. This spin-flipping occurs because of the sinusoidal changing magnetic field that passes by each spin as the wave moves by. Thus, the spin-flip oscillation will be determined entirely by the period of the laser (it will actually be an integral multiple of the driving frequency). This is the energy that Yao et al. proposed to pump into the system to take it out of equilibrium.

However once the spin wave is established using the laser (prepared it in an excited state), Yao et al. proposed that it should continue and be sustained internally. Specifically, it should resist any change in the input frequency of the laser. Another way to say this is if the input E&M field is randomized, the spin

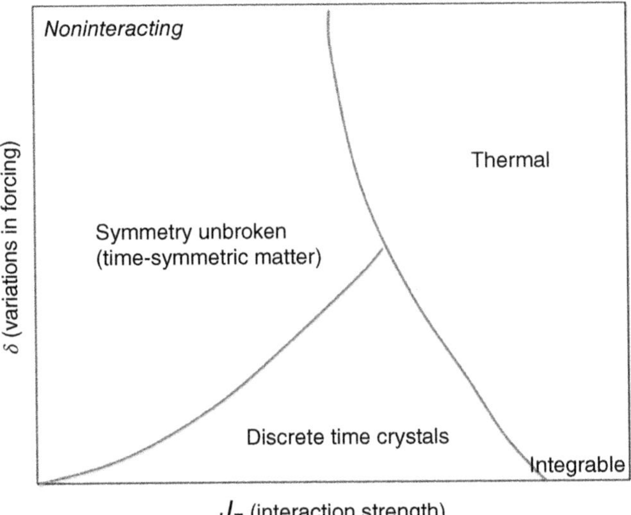

Figure 3.25 A time crystal phase diagram as imagined by Yao et al.

wave should "ring down" for at least a little while. Using this approach Yao et al. were able to construct a "phase diagram" of the system that plots the interaction strength between atoms with imperfections in the spin-flip driving signal (in terms of frequency). An idealized drawing of such a phase diagram is presented in Figure 3.25 (after Yao et al.).

The diagram of Figure 3.25 is easy to read. On the right-hand side (thermal), thermal motions dominate, and stochastic motion is always the answer. On the upper left, the chain of spins follows the laser's frequency no matter what, perfectly time symmetric. However in the lower triangle of the graph, the ion spin does maintain its own rhythm against the changing frequencies of the laser. Thus, it has an internal time oscillation of the spin variables even when the driving force that first prepared this state is randomized.

So, Wilczek proposed it, and Yao showed us how to build one. In 2017/2018 *time crystals* were first realized in the lab by several different teams. Their approaches were quite divergent. The demonstrations ranged from Yao's original idea of linking a row of atoms, from ytterbium held in an ion trap [9] to mono-ammonium phosphate, crystals found in children's chemistry sets, and diamond [10] with dilute magnetic defects. These experiments have just opened the door to these *nonequilibrium solids* with possible exciting applications in areas such as quantum computing by allowing for stable quantum memory elements. But to be sure, crystals can, in some cases, be considered ordered arrays in four dimensions, not just three.

3.5 Summary

The standard approach to categorize crystalline materials is to use point symmetry groups. This science is called *crystallography*. There are a finite number of *point symmetry groups*, and these categorize all known Bravais crystal systems for any given dimension. When a basis is added to the Bravais lattice, the symmetry group is further differentiated.

This scheme fails to easily capture all that we need to know about the symmetry and structure of low-dimensional material structures. Polymers with their one-dimensional lattice can twist and turn randomly into the third dimension. Nanotubes have an ordered "mini" dimension that gives rise to chirality, and 2D crystals can express the topological complexities of *connectivity*. Each of these examples requires additional information about the object to understand how its atomic arrangement yields a given global topology. That doesn't mean that the mathematical apparatus of (a,b,c)'s cannot be used to describe the objects, just that it must be modified so that both local symmetries and global symmetries are captured.

Highly ordered arrangements of atoms into a crystal are based upon bonding interactions. These interactions between components seek the minimization of internal energy for the crystal. This internal energy can be calculated in terms of lattice sums, which are useful in a variety of thermodynamic entities. However, low-dimensional objects frequently involve the bonding of atoms that are not fully coordinated, leading to local minima in the crystal energy. These complexities can stabilize a wide assortment of nanoscale structures: from fullerenes to graphene, to nanotubes, and to diamonds.

Exploring Concepts

1 *The body-centered cubic (BCC) lattice*: Also known as the cubic I structure (I is for the German: *Innenzentriert*), this is a cubic lattice with an additional lattice point at the center of the cell (Figure EC3.1).
As is often the case, the presentation of this lattice is one that is not primitive. And we would like to understand the lattice from its basic symmetry properties, meaning the primitive or Bravais lattice (Figure EC3.2).

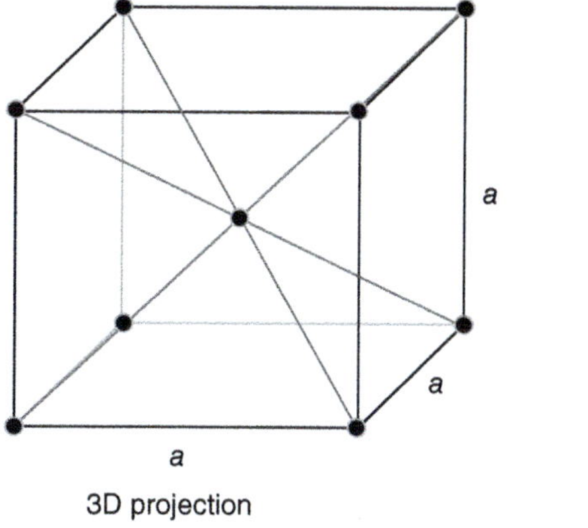

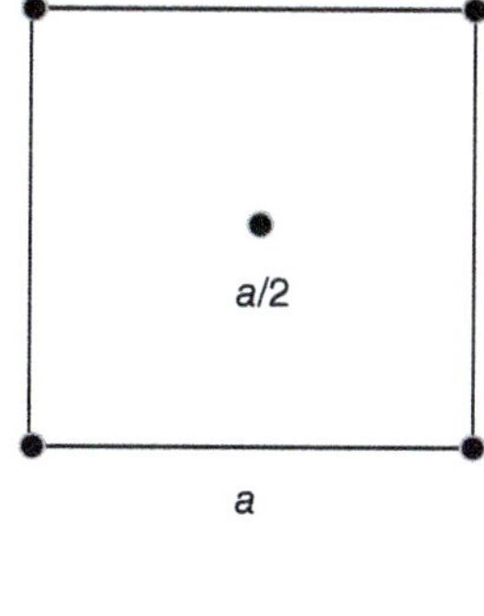

Figure EC3.1 A standard projection image of the cubic I symmetry. This is the "conventional unit cell" as described in the text.

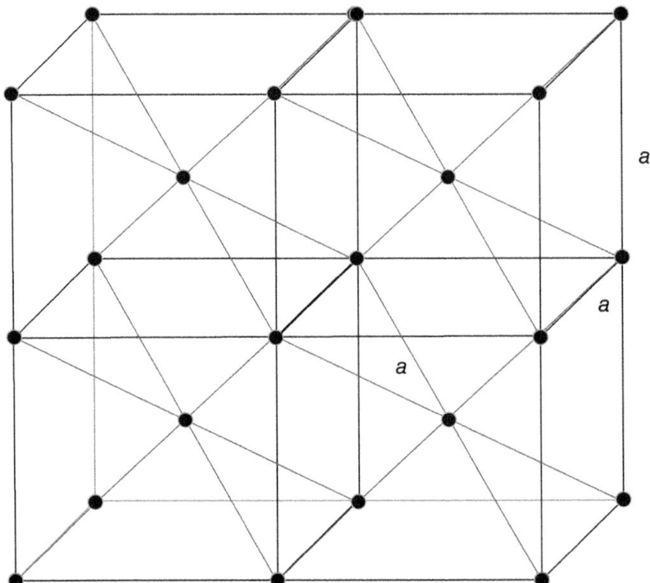

Figure EC3.2 Imagine stacking these unit cells together. Here we have place one on top of another, but there are cells in front and back as well.

Now remember that our lattice vectors can be written simply as

$$[u, v, w] = u\boldsymbol{a}_1 + v\boldsymbol{a}_2 + w\boldsymbol{a}_3$$

For the set of cubic crystals, the choice of the $\mathbf{a}_i$'s is particularly straightforward. Let's use Cartesian: $[x,y,z]$. And to simplify further, each coordinate is some multiple or integral fraction of a single parameter a, since the crystal is of cubic symmetry. This means we write $[x,y,z] = [n_1, n_2, n_3]$ where the n's are merely the number of steps taken in the $\mathbf{a}_1$, $\mathbf{a}_2$, $\mathbf{a}_3$ directions.

(a) Now, for this simple system, notice that we have a similar situation to what we faces with hexagonal graphene. That is, we have two different and nonequivalent lattice positions. The local environments of the corners and the central positions look different to a local observer. Show, by drawing out explicitly, that we can treat this as two interpenetrating simple cubic sublattices with

$$R_{\text{corners}} = [n_1, n_2, n_3]$$
$$R_{\text{center}} = [n_1, n_2, n_3] + (1/2)\,[1, 1, 1]$$

We now move from the conventional cell to a more primitive lattice. Show that the primitive lattice vectors can be written as

$$a_1 = [1, 0, 0]$$
$$a_2 = [0, 1, 0]$$
$$a_3 = (1/2)\,[1, 1, 1]$$

Check to see that any combination

$$\mathbf{R} = n_1\mathbf{a}_1 + n_2\mathbf{a}_2 + n_3\mathbf{a}_3$$

with the n's as integers gives a point within our BCC lattice.

(b) Draw the Wigner–Seitz cell for this lattice, and give the coordination number (the number of nearest neighbors for any point).

2 *The face-centered cubic (FCC) lattice*: The *face-centered cubic* system is another cubic lattice with an additional lattice point added to the faces of cube. A 3D projection of the conventional cell is given in Figure EC3.3. Following the Exploring Concepts (Exercise 1): for the BCC, we might see this lattice in its conventional form as a cubic lattice and some number of basis atoms.

(a) Write down and draw this interpenetrating sublattice system using multiple unit cells as in the case of Figure EC3.2. Using Cartesian coordinates in analogy to the BCC case, write down the coordinates to equivalent sublattice components: R_{corner}, $R_{\text{face-}xy}$, etc.

(b) Now we move to the primitive lattice. What are the primitive lattice vectors in terms of these coordinates? Show that these work.

(c) Finally, determine and illustrate the Wigner–Seitz cell, and calculate the coordination number.

3 *The Zincblende structure*: ZnS typifies the zincblende structure, but the structure is found in many more compounds. Shown in Figure EC3.4 is a top view of such a structure.

Unlike the BCC and FCC examples in the previous two problems, this structure is associated with two different atomic species. If we consider only a single species of atomic constitutive, the structure is known as *diamond* and is exemplified by Si and C.

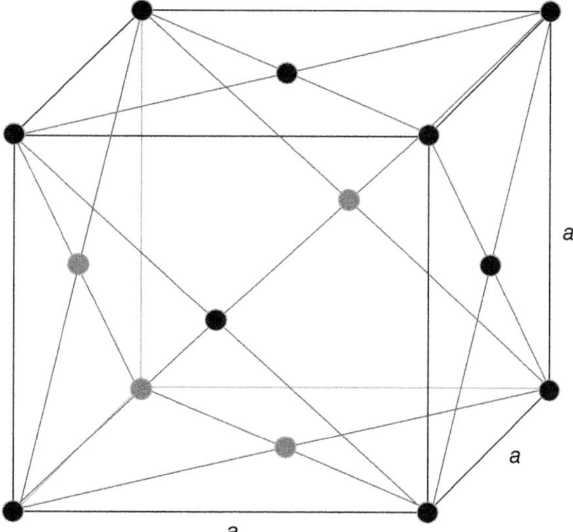

Figure EC3.3 The FCC conventional cell. Notice that the additional atom to the cubic lies along the faces.

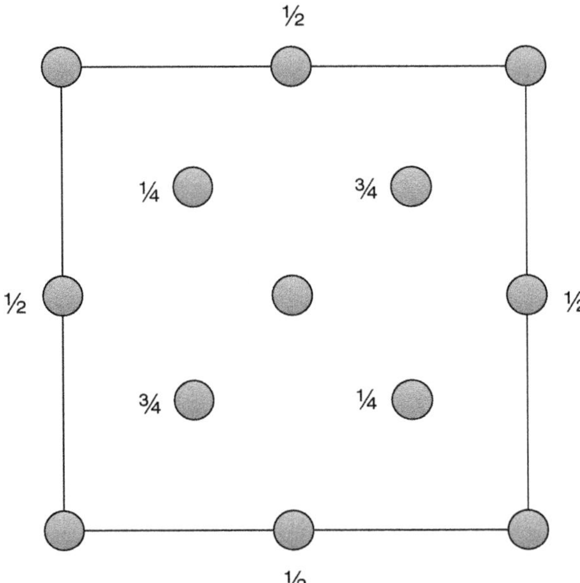

Figure EC3.4 The zincblende structure. The z values of the structure are shown on the lattice sites with those unmarked at $z = 0$ or $z = a$.

Using the methods of the previous exercises, we are going to examine the structure.

(a) From the coordinates we have given, draw out a 3D projection of the zincblende structure, and provide the coordination number for the inequivalent sites.
(b) Draw several such conventional cells stacked together as we have done above, and then show how to construct a primitive lattice for this system. Determine and draw the Wigner–Seitz cell.
(c) Write out the form of the coordinates as they appear in the translation vectors (again following our lead above), and demonstrate the correspondence between the unit vectors for the primitive lattice and the conventional lattice. What is the natural choice of sublattice or basis to simplify the description of the conventional lattice?

4 *The hexagonal close-packed structure* (*HCP*): A curious but common enough structure found in nature is the hexagonal or HCP structure.
As can be seen in Figure EC3.5, it is constructed of alternating layers usually marked A, B, and C or as here 1, 2, and 3. We have shown the ABABAB layering.

(a) Using this conventional cell, construct unit vectors, and provide coordinates for atomic positions as we have done above.
(b) Now imagine and draw how you might rotate layer 3 to form a C fitting where the intercellular triangle of atoms is pointing upward to get an ABCABCABCABC stacking. These are both possible stackings of this cell.
(c) For both the ABAB and ABCABC stacking of the unit cells, determine the primitive lattices and the Wigner–Seitz cells. Draw these in both the projected and top views.

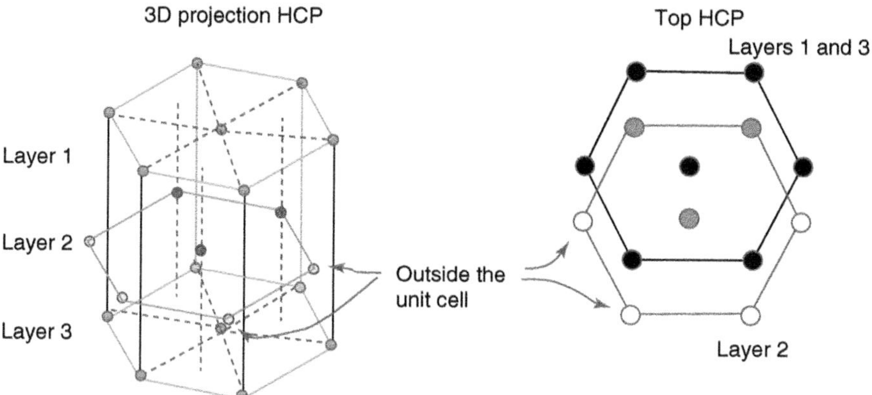

Figure EC3.5 The HCP, ABABAB stacking of atoms.

5 *Crystal planes*: In three dimensions, we use a system of coordinates (*hkl*) to specify a lattice plane. However, if we use different brackets {*hkl*} or [*hkl*], we can also specify a set of equivalent planes of the crystal or direction in the crystal, respectively. Consider the (001) and (100) planes of an FCC lattice (making reference specifically to the conventional lattice). What are the indices of the same planes using the primitive axes?

6 *Penrose tiling*: Look up a two-dimensional Penrose tiling (quasicrystal). Can a Wigner–Seitz cell be constructed for such a lattice? Why or why not?

7 *Molecular solid*: Sucrose, as shown in Figure EC3.6, is a large, sweet molecule. Surprisingly this large nonsymmetric molecule makes a crystal! It does this by fitting together with van der Waals forces. This is an example of what we mean by a molecular solid, as opposed to NaCl, which is of course also composed of the NaCl molecule. In a large laboratory beaker, create a supersaturated solution of pure cane sugar in water. Insert a clean stick or string, and allow the solution to settle out in a warm place over several days, forming sugar crystals on the insert. Select one of the better-formed and larger crystals. Measure the hardness using an Instron® indenter. See why they call it *rock candy*? Now shine a light beam from a laser through it. What happens?

8 *Cohesive energy*: The *Lennard-Jones potential* is a well-known interatomic interaction model used in describing the cohesive bonding of lattice structures broadly. As we have noted already, it consists of two parts: (i) a long-range attractive interaction based on modified or shielded Coulombic forces and (ii) a short-range repulsive interaction that keeps the atoms apart and is based in Pauli exclusion. This phenomenological expression of the potential is surprisingly useful and is generally stated as

$$U(R) = (B/R)^{12} - (A/R)^{6}$$

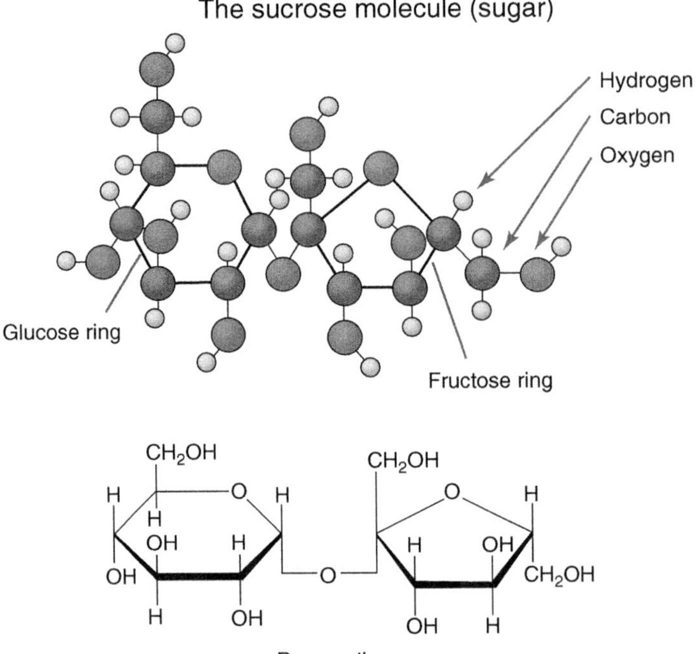

Figure EC3.6 Sugar.

Now typically the A and B are redefined as $B = 4\varepsilon\sigma^{12}$ and $A\ \ 4\varepsilon\sigma^6$. Then if one wants to get a total internal energy of cohesion,

$$U_{\text{total}} = \frac{1}{2}N(4\varepsilon)\sum_j \left[\left(\frac{\sigma}{p_{ij}R}\right)^{12} - \left(\frac{\sigma}{p_{ij}R}\right)^{6}\right]$$

Notice here, to work this integral understanding, $p_{ij}R$ is essential. It is the distance between an ith atom and any other atom, j, expressed in terms of the nearest neighbor distance R. The sum worked over only j, with the $\frac{1}{2}N$ being used to express the sum over all the i's without double counting. This really means that the $(\sigma/R)^6$ and $(\sigma/R)^{12}$ terms are dependent on the kind of atoms used, whereas the $(1/p_{ij})^x$ terms are purely geometrical in nature. They depend on how the balls are stacked together only. Let's see how this might be for a 2D square lattice.

(a) Using purely geometrical arguments, see if you can determine the p terms out to p_{i5}. From this do you expect the sum $\sum[(p_{ij})^{-12} - (p_{ij})^{-6}]$ to converge as $j \to \infty$? Do you need more terms?

(b) In 3D these sums require a great deal of care. For the FCC lattice they have been worked out to be $\sum(p_{ij})^{-12} = 12.13188$ and $\sum(p_{ij})^{-6} = 14.45392$, whereas for the BCC lattice they are $\sum(p_{ij})^{-12} = 9.11418$ and $\sum(p_{ij})^{-6} = 12.2533$. The HCP lattice is given as 12.13229 and 14.45489, respectively (ABABAB stacking). Consider

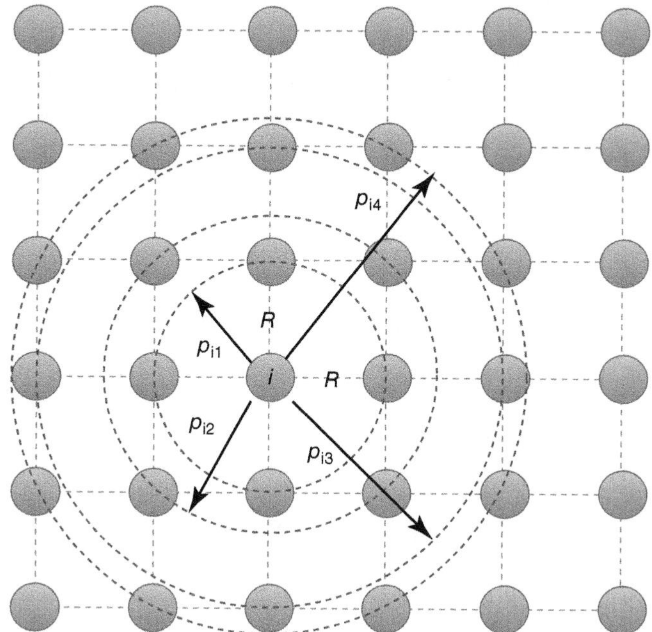

Figure EC3.7 A 2D square lattice highlighting the p_{ij} terms. Notice the similarity of this with the correlation function we mentioned in Chapter 2. Also notice that the summed potential includes the same ($1/pR$) dependence even though there are atoms between the *i*th atom and the *j*th interaction. In fact this is accounted for by the fact that a pure Coulombic potential is not used but rather a screened one.

now a lattice of weakly interacting neon that is well suited to such a potential. It is placed under pressure and low temperatures such that it condenses and forms a lattice. Make an analysis of the lattice sums using the numbers given, and determine the most likely structure that the neon will take.

9 *Ionic crystals*: In ionic crystals (NaCl, KBr, ZnS, etc.), the mathematics of lattice sums behaves in a similar manner to that of Figure EC3.7. However, as we have noted, the interaction strengths have changed. There is also the need to account for positively and negatively charged ions within the lattice. These two complications are the topic of this exercise.
The electrostatic components of the total internal energy are known as the *Madelung energy*, and it dominates the energy landscape of the ionic crystal;

$$U_{ij} = \lambda \exp(-r_{ij}/\rho) \pm q^2/r_{ij}$$

Notice here that we have used CGS units for simplicity and the plus/minus alteration is meant to account for the alternating nature of the ions of the crystal. Introducing the $r_{ij} = p_{ij}R$ as we did before, we get

$$U_{\text{total}} = N[z\lambda \exp(-R/\rho) - (\alpha q^2)/R]$$

Z is the number of nearest neighbors of any ion, and $\alpha = \sum_j{}' (\pm)/p_{ij}$ is the *Madelung constant*. In such a theory λ and ρ are parameters, and α is difficult to compute or determine since it doesn't converge very quickly. For some equilibrium separation of the nearest neighbor ions, R_0, ρ is typically $\sim 0.1R_0$. This is the distance of short-range repulsion, and it holds for very many ionic crystals. Note that we have used $NU_i = U_{\text{total}}$ for N ions of a given charge.

(a) Show that if we use $N\,\mathrm{d}U_i/\mathrm{d}R = 0$ to define R_0 (the equilibrium distance), we can write

$$U_{\text{total}} = -[N\alpha q^2/R_0][1 - \rho/R_0]$$

where $-N\alpha q^2/R_0$ is known as the *Madelung energy*.

(b) For LiF take $R_0 \sim 0.2014$ nm, $z\lambda \sim 0.296 \times 10^{-8}$ erg, and $\rho \sim 0.0291$ nm. Calculate the lattice energy compared to free ions in kcal/mol (answer is 242.2 kcal/mol). How high of a temperature would you have to go to evaporate LiF in a vacuum chamber (10^{-6} Torr) if, say, you were making a thin film organic light emitting device (OLED)?

10 *Solid hydrogen*: Assume spherical H_2 and Lennard-Jones parameters of $\varepsilon = 50 \times 10^{-16}$ erg, $\sigma = 0.296$ nm. Show that the calculated FCC cohesive energy in kJ/mol is much more than the measured 0.751 kJ/mol. Why?

References

1 Silinsh, E.A. (1980). Organic molecular crystals. In: *Springer Series in Solid-State Sciences*, vol. 16. Heidelberg: Springer Verlag.
2 Ashcroft, N.W. and Mermin, N.D. (1976). *Solid State Physics*. New York: Sounders College Publishing.
3 Müller, U. (1982). *Anorganische Strukturchemie*. Stuttgart: B.G. Teubner.
4 Dun, C. et al. (2017). Self-assembled heterostructures: selective growth of metallic nanoparticles on V2–VI3 nanoplates. *Adv. Mater.* 29 (38): 1702968.
5 Heitler, W. and London, F. (1927). Wechselwirkung neutraler Atome und homöopolare Bindung nach der Quantenmechanik. *Zeitschrift für Physik*. 44 (6–7): 455.
6 Wilczek, F. (2012). Quantum time crystals. *Phys. Rev. Lett.* 109: 160401.
7 Watanabe, H. and Oshikawa, M. (2015). Absence of quantum time crystals. *Phys. Rev. Lett.* 114: 251603.
8 Yao, N.Y., Potter, A.C., Potirniche, I.-D., and Vishwanath, A. (2017). Discrete time crystals: rigidity, criticality, and realizations. *Phys. Rev. Lett.* 118: 030401.
9 Zhang, J. et al. (2017). Observation of a discrete time crystal. *Nature* 543: 217.
10 Choi, S. et al. (2017). Observation of discrete time. crystalline order in a disordered dipolar many-body system. *Nature* 543: 221.

4

The Reciprocal Lattice[1]

There are two ways that the crystal lattice is typically used. The first is as a coordinate system for atomic position. In other words, it is a set of spatial coordinates that describes the mass correlation function. You have already seen this; when describing any crystalline compound, we usually provide the positions and relative distances between the atoms as though we could look into the volume and see them there.

A second important use of the lattice is to provide a *frame of reference* for things that move within it. Many times motion in the lattice of a crystal is restricted, such that de Broglie wavelengths correspond to interatomic spacings. That's right; electrons, holes, lattice vibrations, and collective excitations within the lattice can frequently be required to take on *only* discrete values of velocities. The lattice's spacings set the rules for these discrete values (Section 4.1 below). So, to describe *dynamics*, a convenient coordinate system for a given crystal might simply be an array of all the velocity or momentum values that can be realized in that crystal.

As we will see, there is an easy mathematical bridge between the real space and the "velocity points" lattices. They are the *Fourier transforms* of each other. To see why we use the spatial coordinates for mass correlation and the Fourier transform for everything else will take a little work and is what this chapter is all about. But to start with, we will not call the second coordinate system the Fourier transform any further. Instead we will use the word *reciprocal lattices*, and these are constructed in *reciprocal space* just as lattices sit in real space.

4.1 Describing Objects Using Momentum and Energy

The concept of the reciprocal lattice was first developed to describe X-ray diffraction patterns of crystals, and *reciprocal space* is now used to describe all wavelike phenomena in a crystal. It has been mentioned in Chapter 1, (open Fermi surfaces) and in Chapter 2 (the Kohn anomaly). Reciprocal space is not used to describe the position of objects, but to characterize the nature of mobile objects

1 The treatment in this chapter follows the derivations of Kittel, Ashcroft and Mermin, and Simon in its definitions and examples [1]. We have added demonstrations of these systems in low dimensions that are traditionally not included in these texts.

Foundations of Solid State Physics: Dimensionality and Symmetry, First Edition.
Siegmar Roth and David Carroll.

independently of their actual position. For example, consider cars traveling along a highway. To characterize the cars we could specify position at given, equally spaced, times along the path. This is like the real space lattice in that it tells us positions of things relative to some metric (in this case time). However to capture the *dynamics* of the car, which is how it got from point to point, we could have specified a velocity at each position or each time. This too helps us know the manner in which the car moved. Naturally, one can work back and forth between the pictures (Newton's laws), but there are times when the ease of a specific choice of description presents itself. This is exactly the case we find ourselves in trying to determine the behavior of electrons or photons in a crystal lattice. Interactions between electrons or the lattice and an X-ray will conserve momentum generally, regardless of where in the lattice the interaction takes place. So in this case it is best to describe things in terms of velocities (or momentum p). Indeed, a physicist would prefer to use momentum p and kinetic energy E, because this makes some of the physical equations more tractable.

Importantly, using momentum descriptors facilitates the transition from a particle picture to a wave picture of matter. How is this? The equivalence of the momentum and wavelength of a quantum particle link the two pictures. The quantum mechanical wave quantity that is related to the momentum of a particle is the wavenumber $\mathbf{k}$, ($\mathbf{p} = \hbar{\cdot}\mathbf{k}$). Its magnitude is the *reciprocal* of the wavelength of the wave: $\mathbf{k} = 2\pi/\lambda$. The wave quantity related to energy is the frequency, ω ($E = \hbar\omega$). $\hbar$ is the reduced Planck constant, ($\hbar = h/2\pi$), and ω is expressed in oscillations per unit time or dimensions of a *reciprocal* time.

Thus, for a particle moving through a lattice, the wavelength the particle is allowed to have has some definite relationship to the distances between lattice sites. The *Bravais lattice* is a lattice of these distances. So, let's match up the wavelengths of moving objects exactly to the distance values in the B lattice. Remember, its momentum corresponds to a specific wavelength, so this gives *momentum values* that correspond to the natural distances in the lattice, shown schematically in Figure 4.1. The short wavelengths correspond to high energy particles and thus long momentum vectors, whereas the short momentum vectors correspond to the longest wavelengths. Of course, this is a one-dimensional (1D) lattice, but the three-dimensional analogue is easy to imagine.

This argument is a little overly simplified. But it appears that it is quite natural to use a language of describing particle motion in the lattice using terms of natural wavelengths of the structure. A more compact argument might be to simply say that the *reciprocal lattice* is the *Fourier transform* of the Bravais lattice. This reciprocal space language allows some solid-state phenomena to be described more simply. The price for this simplification is to get used to reciprocal space.

4.1.1 Constructing the Reciprocal Lattice

As you might have guessed from the above, there is a simple algebraic procedure used to derive the reciprocal lattice from the crystal's Bravais lattice. The crystal lattice is usually defined by some elementary cell, such as the Wigner–Seitz cell.

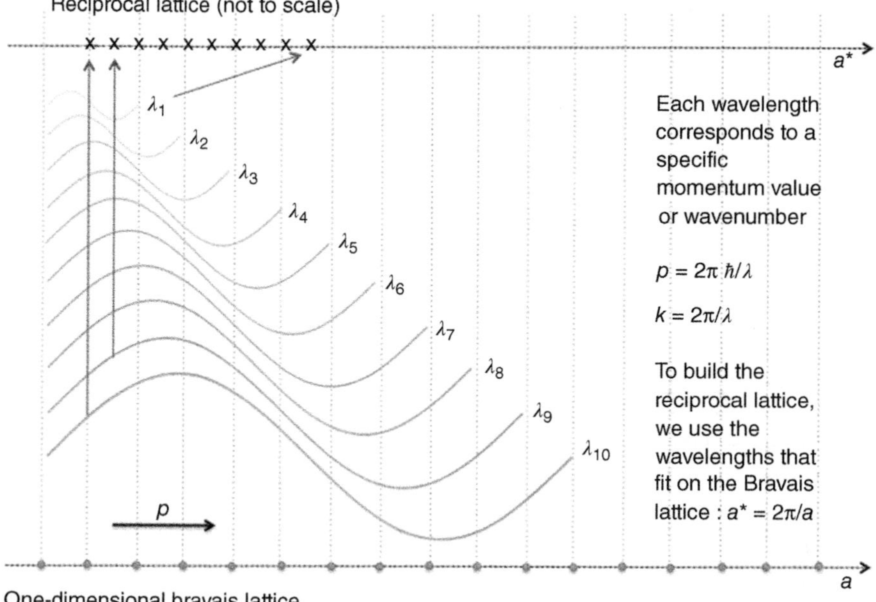

Figure 4.1 This is a diagrammatic way of understanding how the reciprocal lattice is related to the momentum. You might naturally ask: what about shorter wavelengths that match the same lattice points as the longer wavelengths only with more oscillations? For example, λ_5 will also terminate exactly on the λ_{10} lattice point. This represents the same reciprocal lattice point as λ_5.

This, in turn, is laid out by the three unit vectors (**a**, **b**, **c**) that span the elementary cell. The reciprocal lattice is likewise defined by three unit vectors, and they are usually labeled as some derivative of the original lattice vectors such as **a***, **b***, and **c***. In this way **a** and **a*** are linked together.

There is a formulaic way to get from one set of vectors to the other. For a general lattice, the lengths of **a***, **b***, and **c*** are given by Eqs. (4.1a)–(4.1c):

$$\mathbf{a}^* = 2\pi(\mathbf{b} \times \mathbf{c})/(\mathbf{a} \times \mathbf{b} \times \mathbf{c}) \tag{4.1a}$$

$$\mathbf{b}^* = 2\pi(\mathbf{a} \times \mathbf{c})/(\mathbf{a} \times \mathbf{b} \times \mathbf{c}) \tag{4.1b}$$

$$\mathbf{c}^* = 2\pi(\mathbf{a} \times \mathbf{b})/(\mathbf{a} \times \mathbf{b} \times \mathbf{c}) \tag{4.1c}$$

For **a*** this is 2π times the area between **b** and **c** divided by the volume of the elementary cell. For a rectangular lattice, this is 2π **bc**/**abc**. The direction **a*** is perpendicular to the plane defined by **b** and **c**. In a cubic lattice, its length is just 2π times the reciprocal of the lattice parameter a, hence the name *reciprocal lattice.* Numerically, if for a given lattice **a** is 3 Å, the reciprocal lattice, the length of **a*** will be 2π times one-third of a reciprocal angstrom. The factor 2π has been chosen to simplify the equations. The factor 2π is often incorporated into other quantities. Examples are the angular frequency $\omega = 2\pi\nu$ for a wave, used instead of the linear frequency ν, or Planck's constant existing in two versions h and $\hbar = h/2\pi$.

4.1.2 The Unit Cell

The reciprocal lattice is a lattice of points just like the Bravais lattice. So we can treat it as such. In the Bravais lattice we used a procedure above to construct the simplest primitive cell that we called the Wigner–Seitz cell. In the reciprocal lattice there is also a unique choice for the primitive cell. Constructed using the same techniques as for the Bravais lattice, this primitive cell in the reciprocal lattice is called the *Brillouin zone* or *first Brillouin zone.* As with the primitive cell in the Bravais lattice, the first Brillouin zone is repeated to reconstruct the entire reciprocal lattice. This is seen in Figure 4.2. However, the meaning of the reciprocal lattice is different than the Bravais lattice; it represents a dynamic description of the lattice. Thus, the physical dynamics of the first Brillouin zone is repeated for the momentum values in higher zones. We will return to this point in some examples below. For now, it is important only to remember that when trying to understand the relationships between the E and k of specific phenomena within the crystal, it is necessary to plot just the first Brillouin zone. It will contain all the information you need.

Notice that the reciprocal lattice and the Brillouin zone of the simple two-dimensional (2D) example given in Figure 4.2 appear rotated to the real space lattice and its Wigner–Seitz cell. Of course the units of the distances and the distances themselves are all quite different, but the symmetries all seem to be there. In fact, if we had been clever, we might have reasoned that this should be so; the reciprocal lattice should have the same point group symmetries as the original real space lattice. Also of note is that this is an example in two dimensions. Equations (4.1a)–(4.1c) are written for a three-dimensional lattice. The equations used to get the 2D lattice are provided at the top of Figure 4.2,

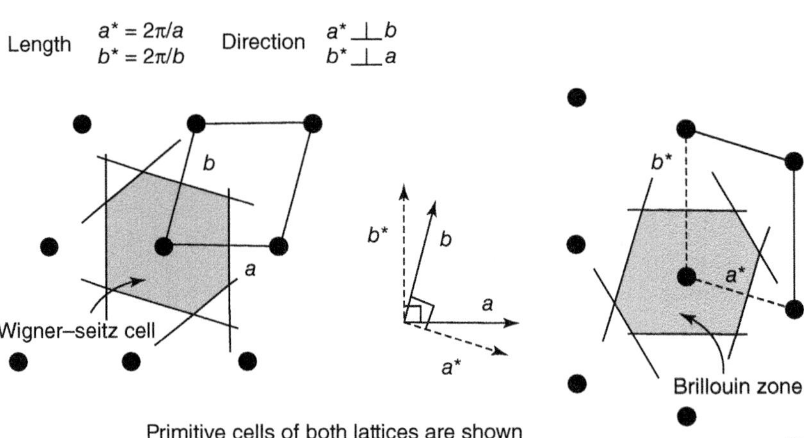

Figure 4.2 Building the reciprocal lattice in two dimensions. In two dimensions the reciprocal lattice appears to be a rotated Bravais lattice with different lattice parameters.

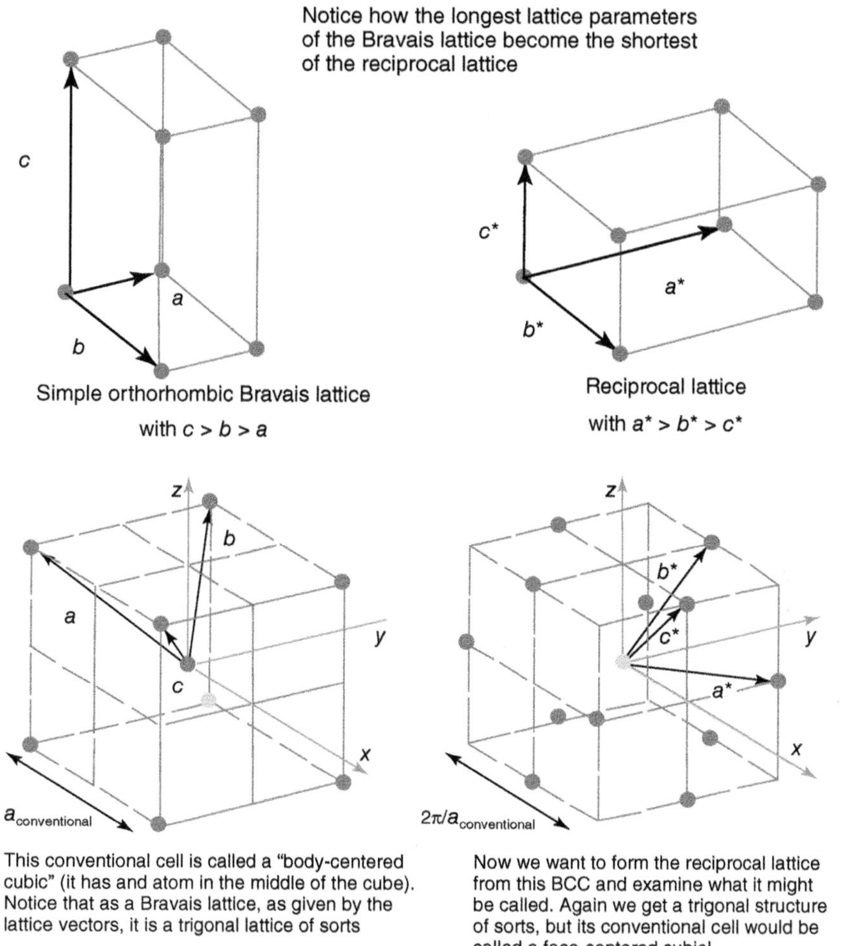

Figure 4.3 (a) The lattice vectors of the simple orthorhombic lattice and its reciprocal lattice and (b) the lattice vectors of the conventional body-centered cubic cell and its reciprocal lattice.

but how did we arrive at them? Can you guess at this point what a 1D reciprocal lattice would look like relative to its real space counterpart?

Now let's examine the relationships between these definitions in three dimensions using the above equations. Examples are given in Figures 4.3 and 4.4.

As you can see, the solid-state physicist or chemist must be adept at switching back and forth between conventional descriptions of lattices, the Bravais lattice and the reciprocal lattice. They must also be aware of the basic and most simple unit cells of these lattices, the Wigner–Seitz cell and the first Brillouin zone. Crystallographers develop an almost encyclopedic knowledge of all of these crystal descriptions and the symmetries that they represent. Indeed, there are whole treatise and texts devoted to the subject.

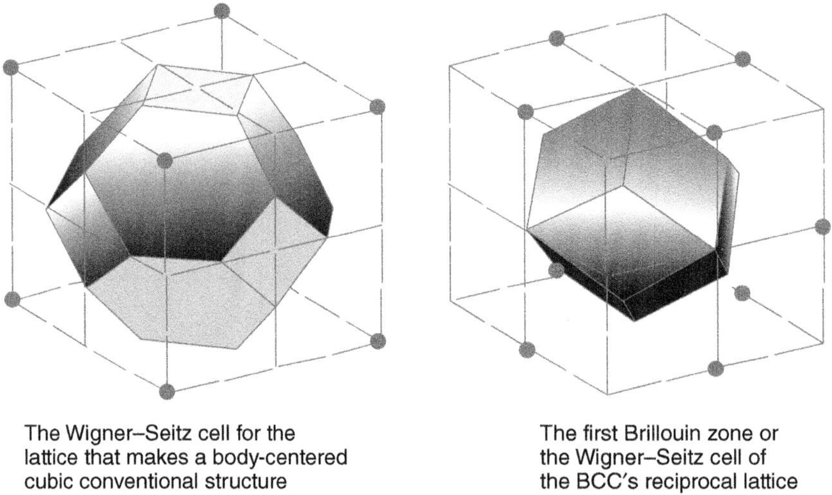

Figure 4.4 The Wigner–Seitz cells of the BCC lattice and its reciprocal lattice.

4.2 The Reciprocal Lattice and Scattering

In practice, the reciprocal lattice and reciprocal space are mainly used for graphical representations. Crystallographers, for example, are interested in predicting the position of X-ray reflections so that they may determine the structure of unknown compounds. They will draw the reciprocal lattice or rather a particular plane of it, choose a certain scaling for it (e.g. centimeter) to represent one reciprocal angstrom, and then proceed according to the description above ($\mathbf{a}^*$ perpendicular to the plane of $\mathbf{b}$ and $\mathbf{c}$). They will then mark the wavevector of the incoming and presumed outgoing X-ray beam (the direction to the detector). The difference between these two vectors is the "momentum transfer" associated with the scattering event.

Naturally, this is just how the use of the R lattice started historically. Reciprocal space has become the *lingua franca* of solid-state physics with nearly every phenomena being expressed in its terms.

4.2.1 General Scattering

Any useful introduction and development of scattering techniques for the purposes of structure determination could fill volumes. And the principles behind different structural characterization techniques are diffractive in nature (scanning probes excepted). For instance, the diffraction patterns of a Laue camera are clearly X-rays diffracting from a crystal as discussed. However, low energy electron diffraction (LEED) patterns and reflection high energy electron diffraction (RHEED) cameras used in semiconductor molecular beam epitaxy (MBE) growth are electrons diffracting from the layers and yielding the atomic structure of crystal surfaces and interfaces. High-resolution transmission electron microscopy

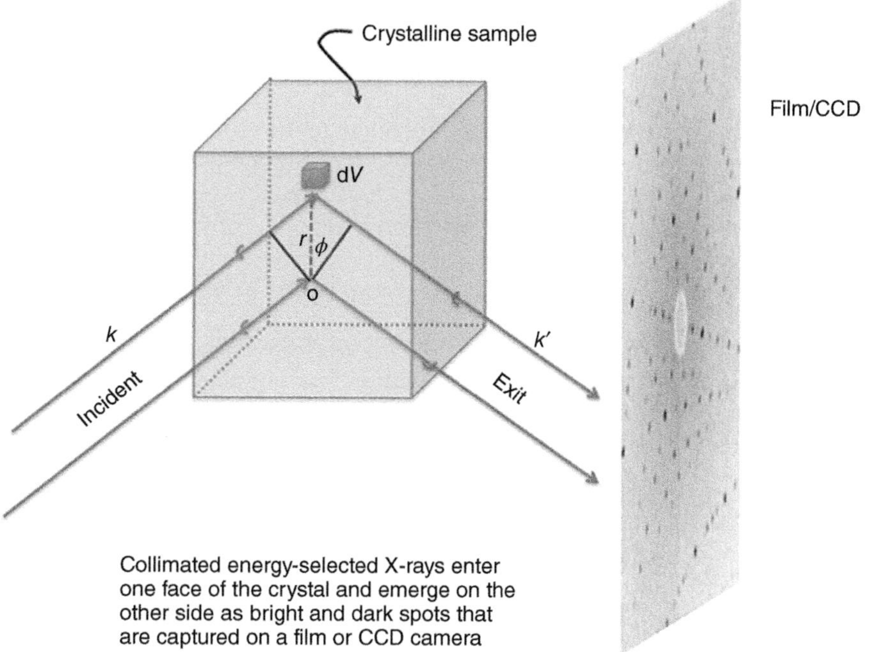

Figure 4.5 The typical X-ray scattering experiment.

(HRTEM) is an electron diffraction-related imaging technique capable of "seeing" the lattice sites directly. Diffraction is central to our study of crystal structure for any dimension. So, our purpose here must be to connect the concept of the reciprocal lattice to specific diffraction conditions, regardless of the nature of the characterization – though X-rays will be an often used example (Figure 4.5).

To start, think of a beam of X-rays, neutrons, or electrons, impingent on a sample with an incoming wavevector: $\mathbf{k}$. On the other side of the crystal where we expect the particles to emerge, they come out at a spectrum of specific angles having a different wavevector: $\mathbf{k}'$. They have been scattered. If we place some sensitized film or a CCD camera there, we will see a series of regularly spaced points. This is the basis of a *Laue camera* for X-rays. What determines the bright and dark spots? Where are the X-rays and why are they not at the exit?

Allow that the incoming beam is fairly well collimated and that the X-rays or electrons, or whatever, are to a large degree in phase with each other. That is, the *correlation length* of the beam is nearly the diameter of the whole spot that hits the surface. To get a hot spot (bright on the CCD, dark on the film) on the other side means that the parts of the exiting "beam" or "beams" must have constructively interfered with each other. Consider two of these individual rays in the beam separated by a distance of r. Let's say some small differential volume $\mathrm{d}V$ at $\mathbf{r}$ scattered one ray and the other was scattered at the origin, both heading away at the angle ϕ. The incoming wave would have looked like $\mathrm{e}^{i\,\mathbf{k}\cdot\mathbf{r}}$, a simple sine wave with wavevector $\mathbf{k}$. The outgoing wave would be similar with $\mathrm{e}^{i\,\mathbf{k}'\cdot\mathbf{r}}$, a new wavevector. If we assume that the magnitudes of these two wavevectors are the same, then the direction has changed. The two rays will still be in phase, but the phase difference between them will have gotten larger by the extra distance one ray had to travel over the other $2r \sin \phi$ or in complex three-dimensional notation $e^{i(\mathbf{k}-\mathbf{k}')\cdot\mathbf{r}}$. Now the intensity of scattering at our differential element for electrons, X-rays, and other scatterers is simply proportional to the electron density at that spot $n(\mathbf{r})$. After all, if there are no electrons there, not much happens. To get the intensity of the diffraction spots due to the whole crystal over all the incoming rays, we must add them all up. First we get

$$F = \int_V dVn(\mathbf{r}) \exp[i(\mathbf{k} - \mathbf{k}') \cdot \mathbf{r}] = \int_V dVn(r) \exp[-i\Delta\mathbf{k} \cdot \mathbf{r}] \tag{4.2}$$

F is called the *scattering amplitude*. This is simply the Bragg diffraction condition spread over a continuous medium with diffracting planes showing up as the modulation of $n(\mathbf{r})$. The spot intensity, I, is proportional to the square of F. Of course we recognize $n(\mathbf{r})$ as being related directly to the crystal lattice, and it can be expressed in terms of its Fourier components (that is, we can write it as a reciprocal lattice):

$$n(\mathbf{r}) = \sum_{\mathbf{G}} n_{\mathbf{G}} \exp[i(\mathbf{G})] \tag{4.3}$$

$n_{\mathbf{G}}$ becomes the Fourier transform. This, of course, has expressed the same function only now in terms of its frequencies instead of its position vectors. Putting Eq. (4.3) into Eq. (4.2), we can rewrite the scattering amplitude as

$$F = \sum_{\mathbf{G}} \int_V dVn_{\mathbf{G}} \exp[i(\mathbf{G} - \Delta\mathbf{k}) \cdot \mathbf{r}] \tag{4.4}$$

When $\mathbf{G} = \Delta\mathbf{k}$, F becomes $Vn_{\mathbf{G}}$ (V is volume) but otherwise it is vanishingly small. This means that we only get "hot spots" at angles where

$$\mathbf{G} = \Delta\mathbf{k} \tag{4.5}$$

or

$$\mathbf{2k} \cdot \mathbf{G} + G^2 = 0 \tag{4.6}$$

Equations (4.5) and (4.6) are the conditions for constructive diffraction, sometimes called the *von Laue diffraction conditions* for Max von Laue. Of course, we might say that they are quite intuitive. Each of these vectors $\mathbf{k}$, $\mathbf{k}'$, and $\mathbf{G}$ represents specific frequencies and "wavelengths." Since we have assumed that no absorption is occurring ($\mathbf{k}$ and $\mathbf{k}'$ have the same magnitude), then the additional path length must amount to a wavelength that fits wholly between diffracting planes $\mathbf{G}$. Naturally it looks a little more complicated since we have expressed it in three-dimensional terms and using exponentials.

At this point we have the model. From here, everything is nuance of application. For instance, (i) the $n(\mathbf{r})$ was treated as though it were the reciprocal lattice, but of course it is the whole of the electron distribution. What if there is a basis set? How would you deal with that? (ii) Notice too that the scattering amplitude as it is written describes something like the *probability* of scattering X-rays into a specific solid angle, but what you actually see is an *intensity* of scattered radiation.

As mentioned, intensities become the square of the scattering amplitude: $I \sim F^2$. But this leaves us with a slight problem because the amplitude goes linearly with volume of the crystal. That means the intensity will become the square of the volume of the crystal. This makes no sense; how much is scattered must obviously be proportional to how much there is to scatter! How do we get around it? (iii) Notice also that the amplitude works exactly the same for (*hkl*) planes pointed in either forward or backward directions (assuming no absorption). This means that the intensity of points for planes (*hkl*) and (*hkl*) is the same (also known as *Friedel's rule*). So all such diffraction patterns will have a natural inversion symmetry even if such a symmetry doesn't exist in the crystal! (iv) Finally, we have expressed scattering as interactions with $n(\mathbf{r})$, but for neutrons the interactions are with ion cores, not the electrons.

The point here is to express that the beautiful, delicate, and subtle science of diffractive techniques goes quite deep. There are many theorems, lemmas, principles, shortcuts, and insights that make up this art – far too many to cover here. Generally, scientists that do electron microscopy become experts in the subtleties of transmitted electron diffraction, crystallographers become experts in X-ray techniques, MBE growers become adept at understanding RHEED cameras, and so on – each with their own special set of rules (Figure 4.6).

An example of one of the more elegant constructs used in the field is the so-called *Ewald sphere*, a sphere with radius equal to the incoming or outgoing momentum vector. It is centered on the origin of the incoming **k**. If the Ewald sphere touches one of the lattice points on the reciprocal lattice, it will show as a peak in the X-ray intensity ("a reflection") at the chosen detector position.

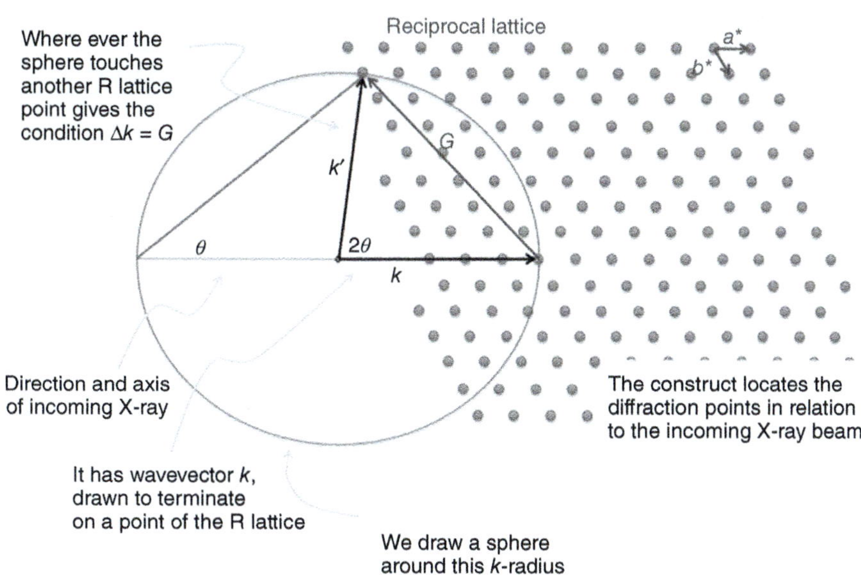

Figure 4.6 The Ewald sphere.

If not, the sample (and with it the reciprocal lattice) will have to be rotated until a reciprocal lattice point will meet the sphere and *the reflection condition is fulfilled.*

4.2.2 Real Systems

Of course in real three-dimensional crystals, many, if not most, will have some sort of a *basis*. And, so, we might return just briefly to the nuance mentioned above. How do we deal with this? What is the effect of a basis on the diffraction spots?

When $\Delta\mathbf{k} = \mathbf{G}$, the diffraction conditions are met, and the scattering amplitude expressed above becomes

$$F_G = \int_{\text{cell}} \mathrm{d}V n(\mathbf{r}) \exp(-i\mathbf{G}\cdot\mathbf{r}) = NS_G \tag{4.7}$$

Now we note here that we have played a fast and loose game with our integrations because we are assuming a perfect crystal. The integral is now over a single unit cell of the crystal, and the scattering amplitude has become a Fourier transform of the electron density over that cell. N is the number of these cells. We call this integral, $\mathbf{S}_\mathbf{G}$, the *structure factor*. To evaluate it when a basis is present, we describe the electron distribution as a superposition of the electron contributions at each atomic site within the unit cell:

$$n(\mathbf{r}) = \sum_{j=1}^{p} n_j(\mathbf{r} - \mathbf{r}_j) \tag{4.8}$$

n_j is the electron distribution of the jth atom in the cell and it is located at $\mathbf{r}_j$. There are p such atoms in the cell. The choice of $n(\mathbf{r})$ is not unique since we must apply models to guess at the actual electron distribution, and that would require we know the exact nature of electron sharing within the basis. But we can get pretty close generally by knowing the *bonding character*. And so we write

$$\begin{aligned} S_\mathbf{G} &= \sum_j \int_{\text{cell}} \mathrm{d}V n_j(\mathbf{r} - \mathbf{r}_j) \exp(-i\mathbf{G}\cdot\mathbf{r}) \\ &= \sum_j \exp(-i\mathbf{G}\cdot\mathbf{r}_j) \int_{\text{cell}} \mathrm{d}V n_j(\boldsymbol{\rho}) \exp(-i\mathbf{G}\cdot\boldsymbol{\rho}) \end{aligned} \tag{4.9}$$

$\boldsymbol{\rho} = \mathbf{r} - \mathbf{r}_j$. Now we take the integral part out. This integral is nonzero over only the jth atomic site, so we can make the bounds of the integral over all space. We get

$$f_j = \int_V \mathrm{d}V n_j(\boldsymbol{\rho}) \exp(-i\mathbf{G}\cdot\boldsymbol{\rho}) \tag{4.10}$$

This is the *atomic form factor* and it contains the bonding information we guessed at above. Combining we get the structure factor for the atomic basis of the crystal:

$$S_G = \sum_j f_j \exp(-i\mathbf{G}\cdot\mathbf{r}_j) \tag{4.11}$$

And from this we get the scattering amplitude, F_G.

Fun Example Here Let's look at a simple example that is covered in most standard texts: that of the alkali metal *sodium*. It has a body-centered cubic (BCC) structure. As we know the BCC is NOT a Bravais lattice; it is a lattice with a basis. We showed above that the basis of the BCC lattice is pretty simple: using a Cartesian coordinate system, we have identical atoms at

$$x_1 = y_1 = z_1 = 0 \tag{4.12a}$$

$$x_2 = y_2 = z_2 = 1/2 \tag{4.12b}$$

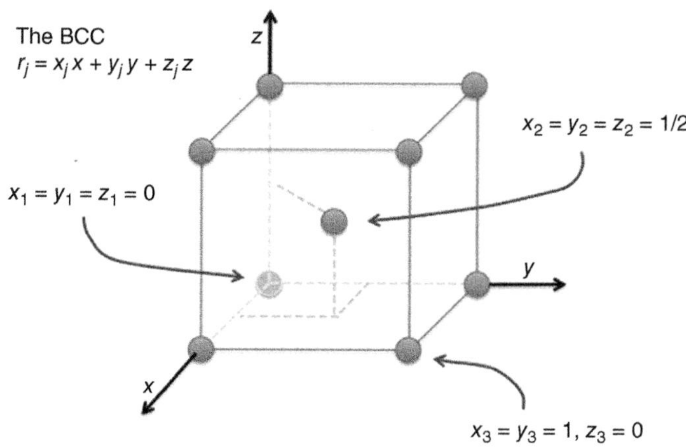

This pair of atoms is placed into a cubic Bravais lattice. The two-atom basis sits on each of the four corners of the original BCC cube, and the second atom of the basis is what gives the "body-centered" appearance. Recall we mentioned earlier that we would work with the Bravais lattice as opposed to the conventional lattice in calculating things like scattering. It happens that this will have some noticeable consequences. To see this we must perform the sums above. To do this, we need to write out the **G**'s and the **r**'s:

$$\mathbf{G} = \nu_x \mathbf{x}^* + \nu_y \mathbf{y}^* + \nu_z \mathbf{z}^* \tag{4.13}$$

x*, **y***, **z*** are the reciprocal lattice unit vectors associated with the **x**, **y**, **z** unit vectors of the real space lattice. In other words we are defining our **G** vectors with reference to the same coordinate system as the real space atomic positions:

$$\mathbf{r}_j = x_j \mathbf{x} + y_j \mathbf{y} + z_j \mathbf{z} \tag{4.14}$$

From the basic definition given above, it is easy to show that

$$\mathbf{x}^* \cdot \mathbf{x} = 2\pi \text{ and } \mathbf{x}^* \cdot \mathbf{y} = \mathbf{x}^* \cdot \mathbf{z} = 0 \tag{4.15a}$$

$$\mathbf{y}^* \cdot \mathbf{y} = 2\pi \text{ and } \mathbf{x} \cdot \mathbf{y}^* = \mathbf{y} \cdot \mathbf{z}^* = 0 \tag{4.15b}$$

$$\mathbf{z}^* \cdot \mathbf{z} = 2\pi \text{ and } \mathbf{y}^* \cdot \mathbf{z} = \mathbf{x} \cdot \mathbf{z}^* = 0 \tag{4.15c}$$

So, we get

$$\mathbf{G} \cdot \mathbf{r}_j = 2\pi (x_j \nu_x + y_j \nu_y + z_j \nu_x) \tag{4.16}$$

and

$$S_G = \sum_j f_j 2\pi(x_j v_x + y_j v_y + z_j v_x) \tag{4.17}$$

Recall that this structure factor goes into the scattering amplitude, and the modulus of that is the intensity of the specific diffraction spot associated with the $(v_{x^*}, v_{y^*}, v_{z^*})$ planes that reflected it. That is, $I \sim S^*S$.

This is interesting because if you will notice, not all selections of $(v_{x^*}, v_{y^*}, v_{z^*})$ give constructive interference. Specifically, in our BCC system, using the coordinates above, we have

$$S(v_{x^*}, v_{y^*}, v_{z^*}) = f\{1 + \exp[-i\pi(v_{x^*} + v_{y^*} + v_{z^*})]\} \tag{4.18}$$

Remember the f is the atomic form factor, and the sum above used for the two-atom basis has the peculiar property of being zero whenever the argument of the exponent is $-i\pi$ (any odd integer). So,

$$S = 0 \text{ when } (v_{x^*} + v_{y^*} + v_{z^*}) \text{ is odd} \tag{4.19a}$$

$$S = 2f \text{ when } (v_{x^*} + v_{y^*} + v_{z^*}) \text{ is even} \tag{4.19b}$$

What this really means is that our X-ray diffraction example of sodium will not have any diffraction spots from planes like (1,0,0), (3,0,0), (1,1,1), etc. But all the even sets of planes will diffract: (2,0,0), (1,2,1), etc. So if we are careful in how we lay out our diffraction film and geometry, we can tell immediately if we have a BCC system.

But what makes this example unique to sodium? Doesn't this hold true for any BCC structure? Yes, so far it does. But the information about the atoms sitting on the lattice points is contained within the f. To see how, remember we must work the integral

$$f_j = \int_V \mathrm{d}V n_j(\mathbf{r}) \exp(-i\mathbf{G} \cdot \mathbf{r}) \tag{4.20}$$

Let's suppose we consider the case where **G** makes an angle α with **r**, so the exponential becomes $-iGr \cos\alpha$. The integral is worked over the individual atomic site. For instance, if we consider the case of the electron density as smooth and symmetric around the site, it will have only an r dependence as we move away from the atomic core:

$$f_j \equiv 2\pi \int_V \mathrm{d}r\, r^2 \mathrm{d}(\cos\,\alpha) n_j(\mathbf{r}) \exp(-iGr\,\cos\,\alpha) \tag{4.21}$$

ϕ part of volume integral.

Electron density in r.

This yields a relatively simple answer for spherically symmetric atoms in unit cells:

$$f_j = 2\pi \int_V \mathrm{d}r\, r^2\, n_j(\mathbf{r}) \cdot \frac{\mathrm{e}^{iGr} - \mathrm{e}^{-iGr}}{iGr} \tag{4.22}$$

$$\text{in the limit}: \frac{e^{iGr} - e^{-iGr}}{iGr} \sim 2 \tag{4.23}$$

$$f_j = 2\pi \int_V \mathrm{d}r\, r^2 n_j(r) = Z \tag{4.24}$$

A curious result: such symmetries reduce to placing all the electrons of the atom on each atomic position symmetrically.

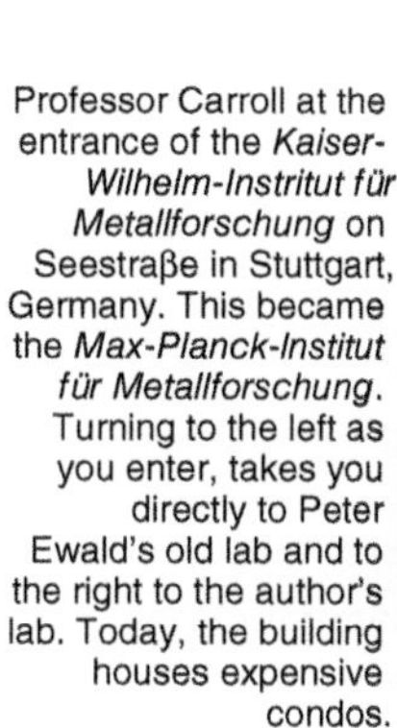

Professor Carroll at the entrance of the *Kaiser-Wilhelm-Instritut für Metallforschung* on Seestraße in Stuttgart, Germany. This became the *Max-Planck-Institut für Metallforschung*. Turning to the left as you enter, takes you directly to Peter Ewald's old lab and to the right to the author's lab. Today, the building houses expensive condos.

4.2.3 Applying This to Real One-Dimensional Systems

Above we asked: what does the reciprocal lattice of a 1D crystal look like? We of course anticipate that the reciprocal lattice will also be a linear arrangement of lattice points, the $\mathbf{a}^*$ axis pointing in the same direction as the a axis and with the length of the unit vector $\mathbf{a}^*$ being 2π times the reciprocal of the length of $\mathbf{a}$.

However the previous discussion described procedures used to design a reciprocal lattice in three dimensions: we absolutely need all unit vectors $\mathbf{a}$, $\mathbf{b}$, $\mathbf{c}$ to find a unit vector of the reciprocal lattice. Yes, we have argued for reciprocal structures in lower dimension, but surely real systems have wavelike phenomena in all three dimensions. With a little reflection, we might try an "anisotropy approach," as in the earlier chapters. To argue more cogently for the reciprocal lattice in one dimension, we start with an anisotropic tetragonal lattice, having three axes at right angles, but with two lattice parameters much larger than the third ($b \gg a$ and $c \gg a$), and then we increase b and c to infinity. The direction of $\mathbf{a}^*$ is perpendicular to the plane defined by $\mathbf{b}$ and $\mathbf{c}$. Since the crystal lattice is tetragonal already, $\mathbf{a}$ is perpendicular to that plane also and hence $\mathbf{a}^* || \mathbf{a}$, pointing in the same direction. What is the length of $\mathbf{a}^*$? 2π times the area $(b{\cdot}c)$ divided by the volume $(a{\cdot}b{\cdot}c)$ gives $2\pi/a$ (a finite value for any value of b and c even those approaching infinity). What about the lengths of $\mathbf{b}^*$ and $\mathbf{c}^*$? $\mathbf{b}^* = 2\pi(\mathbf{a}{\cdot}\mathbf{c})/(\mathbf{a}{\cdot}\mathbf{b}{\cdot}\mathbf{c}) = 2\pi/\mathbf{b}$ and similarly $\mathbf{c}^* = 2\pi/\mathbf{c}$. Both will approach zero as b and c take infinite values (Figure 4.7). So the "infinitely anisotropic" 1D tetragonal lattice indeed consists of a linear arrangement of points as reciprocal lattice.

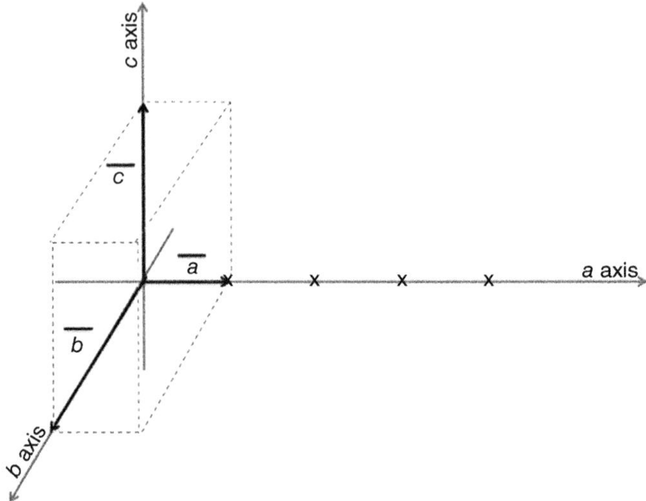

Figure 4.7 In tetragonal lattice, the limits $b \to \infty$ and $c \to \infty$ are the "infinite anisotropy" approach to one dimensionality and illustrate that the reciprocal lattice of a linear chain is again a linear chain.

But if you are working with things like polymers, things aren't so easy. As mentioned in Chapter 2, experiments are usually not carried out on isolated chains, but on *bundles*: arrays of parallel chains. A three-dimensional crystal is a bundle as well, but the term *bundle* is more general. It also covers arrays of chains with random origin and random distances between the chains. In other words, there is no order in the **b** and **c** directions. In this case we still get $\mathbf{a}^* = 2\pi(\mathbf{b}\cdot\mathbf{c})/(\mathbf{a}\cdot\mathbf{b}\cdot\mathbf{c}) = 2\pi/\mathbf{a}$, but $\mathbf{b}^*$ and $\mathbf{c}^*$ are undefined. They are not zero as in the case of the infinitely anisotropic lattice with $\mathbf{b} \to \infty$ and $\mathbf{c} \to \infty$. In a disordered bundle $\mathbf{b}^*$ and $\mathbf{c}^*$ can take *any* value. Consequently, the reciprocal lattice of such objects is not a linear arrangement of equidistant points, but an arrangement of equidistant planes (Figure 4.8). This situation has a very practical background. It actually describes the intensity (usually of X-rays or electrons) observed from samples with high degrees of anisotropy such as polymers, bundles of aligned fibers, etc. This is, of course, assuming that no damage is done to the material by the scattering particle and that is quite easy to do in the case of most isolated 1D materials.

Typically, as we have already pointed out, the X-rays are detected by a blackened spot on a photographic film or brightening of a scintillator screen. Thus, the diffraction from this 1D atomic arrangement creates *lines* on the display, as seen schematically in Figure 4.9: the diffraction pattern of KCP (see Chapter 2). The Peierls distortion (coming a little later in the text in Chapter 9) leads to the reorganization of the platinum atoms within the chains. At room temperature the chains act independently of each other, and, as far as the Peierls distortion is concerned, they form a disordered bundle. The X-ray evidence of a disordered bundle is continuous streaks as shown on the Figure 4.9a of the diffraction diagram. At low temperatures the chains interact, a coherent three-dimensional order is established, and the streaks disintegrate into spots [2].

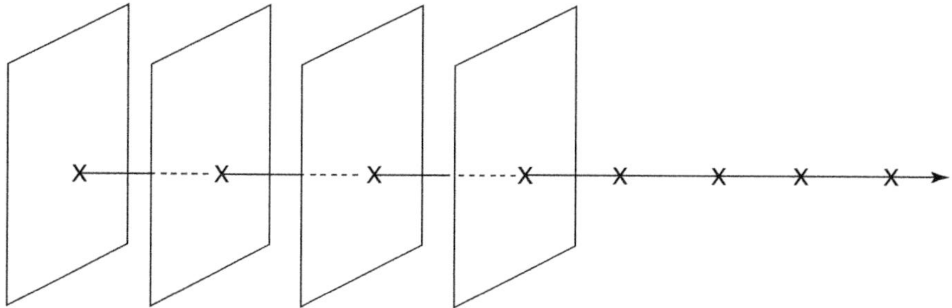

Figure 4.8 Reciprocal lattice of a disordered bundle (order in one dimension only) consisting of equidistant planes. This will apply to noninteracting bundles of chains as well.

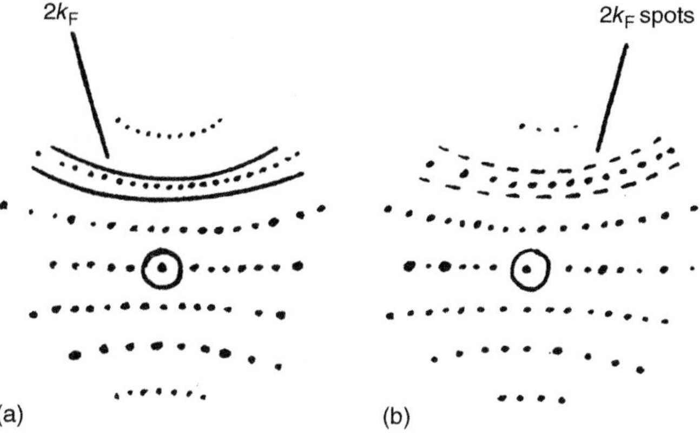

Figure 4.9 X-ray diffraction pattern of KCP. The streak indicates disordered bundle behavior of the Peierls distortion in the platinum chains. Source: From Comes et al. 1973 [[2]].

These representations are of an oriented single crystal of KCP using an apparatus (a *Laue camera*) that aims X-rays into the crystal on one side and places a photographic film (or today a scintillator and CCD) behind the sample. Alternatively the film can be placed in front of the sample, and the X-rays go through a hole in the center. In both cases the transmitted or reflected X-ray pattern is distorted by the geometry of the flat film of the camera and the exiting spherical waves of the scatterers, leading to the curved lines seen. Such a representation is useful for the study of long-range order in single crystals and yields striking signatures of dimensionality. The image is sometimes referred to as a *Wulff net*. To interpret the pattern while taking into account the geometrical distortions, one uses a special graph paper laid over top: *Greninger charts* for reflection and *Leonhardt charts* for transmission.

4.3 A Summary of the Reciprocal Lattice

The tool commonly used to determine structural order and symmetry is diffractive scattering techniques. The language in which this tool is utilized is *reciprocal space* and *reciprocal lattices*. These describe the real space lattice in terms of the natural frequencies of the lattice spacings. Essentially, this reciprocal lattice can be thought of as the Fourier transform of the real space lattice. Due to diffraction conditions the diffraction pattern is a projection of the Fourier transform onto a screen or CCD.

In the real space lattice, the Wigner–Seitz cell is the most basic unit expressing the overall symmetry of the whole lattice. However in the reciprocal lattice, it is the first Brillouin zone that does this job. In scattering experiments with real crystal systems, physical attributes of the crystal such as the basis set or microstructure can modify and modulate the bright and dark spots of the diffraction pattern. A simple set of models to understand these effects is based on point scatterers and is quite effective.

Do not be fooled into thinking that if you don't do X-ray scattering experiments, you will never need to think about reciprocal space though. You see, things within the lattice such as traveling waves on the lattice, or electrons moving from one point to another, interact with the natural spacings of the lattice as easily as objects that enter it from the outside. Reciprocal space, **k**-space, or momentum space is the natural language to describe *all* dynamics within the solid state, so from this chapter forward, it is to be "our friend."

Exploring Concepts

1 *Reciprocal Lattice of the Simple Cubic Structure*: The primitive lattice vectors of the simple cubic lattice are

$$\mathbf{a}_1 = a\hat{\mathbf{x}}; \quad \mathbf{a}_2 = a\hat{\mathbf{y}}; \quad \mathbf{a}_3 = a\hat{\mathbf{z}}$$

$\hat{\mathbf{x}}, \hat{\mathbf{y}}, \hat{\mathbf{z}}$ are the unit vectors in Cartesian coordinates attached to the unit cell, and a is the lattice parameter.

(a) Find an expression of the primitive reciprocal lattice unit vectors in terms of the primitive real space lattice unit vectors.

(b) What is the volume of the primitive lattice unit cell and the reciprocal lattice unit cell? What are the lattice constants?

(c) What are the boundaries of the first Brillouin zone in terms of the two unit vector sets?

2 *The Body-Centered Cubic*: Using the primitive lattice vectors represented in the Exploring Concepts of Chapter 3 for the BCC lattice:

(a) Find an expression of the primitive reciprocal lattice unit vectors in terms of the primitive real space lattice unit vectors.

(b) What is the volume of the primitive lattice unit cell and the reciprocal lattice unit cell? What are the lattice constants?

(c) What are the boundaries of the first Brillouin zone in terms of the two unit vector sets?

3 *The Face-Centered Cubic*: Again we return to Chapter 3 and the FCC lattice. There you also found the FCC primitive lattice vectors. Now:
   (a) Find an expression of the primitive reciprocal lattice unit vectors in terms of the primitive real space lattice unit vectors.
   (b) What is the volume of the primitive lattice unit cell and the reciprocal lattice unit cell? What are the lattice constants?
   (c) What are the boundaries of the first Brillouin zone in terms of the two unit vector sets?

4 *The Hexagonal Lattice*: And finally in the HCP lattice:
   (a) Find an expression of the primitive reciprocal lattice unit vectors in terms of the primitive real space lattice unit vectors.
   (b) What is the volume of the primitive lattice unit cell and the reciprocal lattice unit cell? What are the lattice constants?
   (c) What are the boundaries of the first Brillouin zone in terms of the two unit vector sets?

5 *Structure Factors*: In the text, we have given an example of determining the structure factor and the possible reflections from a BCC lattice. Following this example:
   (a) Repeat the analysis for the FCC lattice.
   (b) Also show this same analysis for the HCP lattice.

6 *2D Lattices*: Construct the reciprocal lattice for graphene. Label directions and lengths. Describe the expected X-ray diffraction pattern.

7 *General Diffraction*: Go get an X-ray source that is collimated and aim it at a crystal of your choice. Make sure it is thin enough for the X-rays to get all the way through, but not so thin as to allow for no interactions. You will need a little trial and error here. Now place a piece of X-ray film on the other side and record the pattern that shows up. Again, a little trial and error is needed to get the right distance. Make sure to stay out of the way of the X-rays; they aren't so good for you! Once you have done this, take your sample; thin it down using cleavage, dimpling, and an ion mill; and then mount it in an electron microscope (say, 120 keV). Go into diffraction mode and record this image. Now compare the two diffraction patterns. Finally, take a sample of this crystal to a synchrotron – you may drive or fly as needed. Mount it in the UHV specimen chamber – ensuring again to use the proper thickness – and observe the diffraction that occurs on a scintillator. Please note that proper cleaning and latex gloves are required for the UHV system – do not contaminate. Compare this diffraction with the other two methods. (This problem is a little harder than others, and it is recommended that instructors allow several days when assigning the problem.)

8 *Atomic Form Factor*: Discuss the physical meaning of the atomic form factor in more detail. What does it indicate in terms of the X-ray–atom interaction? Why is the *electron density* important in the expression?

9 *Rietveld Analysis*: Go outside of the text for this one. Give a detailed overview of the *Rietveld refinement* technique for determining crystal structure.

10 *Neutron Diffraction*: In many cases, neutron diffraction is preferred in structural studies over X-rays. These studies are carried out at nuclear reactors wherein thermal neutrons are "collimated" and large three-axis spectrometers are used to determine the diffraction pattern. Describe in detail the primary differences in these two diffraction techniques and when they are used.

References

1 (a) Kittel, C. (2004). *Introduction to Solid State Physics*, 8e. Wiley. (b) Ashcroft, N.W. and Mermin, N.D. (1976). *Solid State Physics*, 1e. Cengage Learning. (c) Simon, S.H. (2013). *The Oxford Solid State Basics*, 1e. Oxford University Press.

2 Comes, R., Lambert, M., and Zeller, H.R. (1973). A low-temperature phase transition in the one-dimensional conductor $K_2Pt(CN)_4Br_{0.30}{\cdot}H_2O$. *Phys. Status Solidi B* 58: 587.

5

The Dynamic Lattice

From Chapter 4, we have in our minds a crystal as a set of atoms held rigidly in space. But, if we *carefully* examine the diffraction patterns we have discussed, we might notice that the *diffraction peaks* (spots of peak intensity) are attenuated as the temperature of the crystal increases. From this simple observation, we can reason that those atoms must move around! To see why is, we introduce the *Debye–Waller factor*:

$$\mathrm{DWF} = \mathrm{e}^{-2W} = \langle \exp(i\mathbf{G} \cdot \mathbf{u}_i)\rangle^2 \tag{5.1}$$

G is a reciprocal lattice vector and $\mathbf{u}_j$ is the displacement of the ith atom in a Bravais lattice. The brackets stand for a *thermal average.*[1] You may ask how does this help us? Where does such a term *fit* into simple scattering experiments? Well, remember the *Structure Factor* from Chapter 4? S_G, has terms that look like:

$$S_G \sim f_j \exp(-i\mathbf{G} \cdot \mathbf{r}_j) \tag{5.2}$$

f_j is the *form factor* and $\mathbf{r}_j$ the atomic positions in the unit cell. Technically, this would be summed over the basis of the unit cell. **G** is the scattering vector $\mathbf{G} = 2\pi/d$ (d is the plane spacing). But, if the jth atom isn't precisely at $\mathbf{r}_j$, it is, say, oscillating around this point instead. So really,

$$\mathbf{r}_j(t) = \mathbf{r}_j + \mathbf{u}(t) \tag{5.3}$$

$\mathbf{r}_j(t)$ is the instantaneous position of the atom, $\mathbf{r}_j$ is its mean position, and $\mathbf{u}(t)$ is its instantaneous displacement from $\mathbf{r}_j$. The Structure Factor terms will then become:

$$S_G \sim f_j \exp(-i\mathbf{G} \cdot \mathbf{r}_j) \cdot \langle \exp(i\mathbf{G} \cdot \mathbf{u}_i)\rangle \tag{5.4}$$

If we assume the displacements aren't too large, then we could expand the last term:

$$\langle \exp(i\mathbf{G} \cdot \mathbf{u})\rangle = 1 - i\langle \mathbf{G} \cdot \mathbf{u}\rangle - 1/2\, \langle (\mathbf{G} \cdot \mathbf{u})^2\rangle + \dots \tag{5.5}$$

1 For more general references of the above treatment we recommend: (a) Ashcroft, N.W. and Mermin, N.D. (1976). *Solid State Physics.* New York: Sounders College Publishing; (b) Hellwege, K.H. (1976). *Einführung in die Festkörperphysik.* Heidelberg: Springer Verlag; (c) Hunklinger, S. (2007). *Festkörperphysik.* Munchen: Oldenbourg Wissenschaftsverlag. (d) Kittel, C. (1986). *Introduction to Solid State Physics,* 6e. New York: John Wiley & Sons.
Thermal and time averages are used quite a bit in scattering expressions. You should review your statistical mechanics or thermodynamics course notes to recall how to perform them generally.

Foundations of Solid State Physics: Dimensionality and Symmetry, First Edition.
Siegmar Roth and David Carroll.

Now let's make a further assumption that the displacements are *random and uncorrelated*. As we will see, this isn't too good of an assumption, but it will do for now. This means we must take $\langle \mathbf{G} \cdot \mathbf{u} \rangle = 0$. Dropping higher order terms and simplifying the second term:

$$\langle (\mathbf{G} \cdot \mathbf{u})^2 \rangle = G^2 \langle u^2 \rangle \langle \cos^2 \theta \rangle = 1/3 \, \langle u^2 \rangle G^2 \tag{5.6}$$

The one-third comes from the geometric average of the $\cos^2 \theta$ term over a sphere. This yields

$$\exp(-1/6 \, \langle u^2 \rangle G^2) = 1 - 1/6 \, \langle u^2 \rangle G^2 + \ldots \tag{5.7}$$

If we allow for the motion of the atom to be *nearly harmonic*, then we get Structure Factor terms with the form

$$S_G \sim f_j \exp(-i\mathbf{G} \cdot \mathbf{r}_j - 1/6 \langle u^2 \rangle G^2) \tag{5.8}$$

and the intensity of the scattered radiation from the square of this becomes

$$I \sim f_j^2 \exp(-i\mathbf{G} \cdot \mathbf{r}_j + i\mathbf{G} \cdot \mathbf{r}_j - 1/3 \, \langle u^2 \rangle G^2) = I_0 \exp(-1/3 \, \langle u^2 \rangle G^2) \tag{5.9}$$

Here, I_0 is the scattering intensity from a static lattice. So there is an X-ray intensity drop-off. Part of this comes with the increasing angle of **G** and part from the Debye–Waller factor. Of course, for the case of neutron diffraction, the drop-off is entirely due to the Debye–Waller term, as you will understand by the last section of the chapter.

Thus, it would seem that we must include "the motion of things" within our picture of the lattice, even in regimes far from melting. But in this chapter we must do more than this and introduce a powerful new idea for phenomena from diffraction to specific heat, speed of sound, thermal and electrical conductivity, thermoelectricity, and more. What is this idea? The atoms of the lattice move as elastic, *vibrational* waves! In other words the atomic motion within the solid is *correlated*. This fascinating concept must be reasoned through a circuitous route:

1. Introduce a *classical model* to explain the atomic correlation.
2. Examine the *role of dimension* in the dynamics expressions of the modes of vibration.
3. Discuss the modifications of our classical picture that are introduced through *quantum mechanics*.
4. Make the connection between classical elastic vibrational modes and phonon *quasiparticles*.
5. Introduce the *statistics of phonons*.
6. Calculate *thermodynamic entities* on the basis of phonon models.

5.1 Crystal Vibrations and Phonons

The arrangement of atoms or molecules into a crystal is a result of *forces* that keep the atoms in their positions. These forces are due to the chemical bonding between atoms. Now, this is a quantum system so we really should talk about potentials and wavefunctions, but first let's look at the implications of a purely *classical model*. And the most *famous* classical model is to represent each bond as Hooke spring (Figure 5.1).

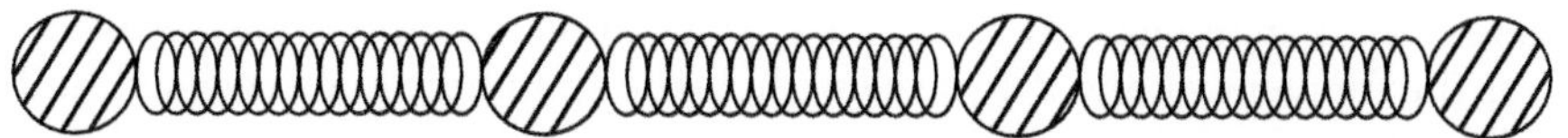

Figure 5.1 The *famous* classical model of a one-dimensional crystal. The lattice forces due to chemical bonding are represented by elastic springs. The problem then becomes a simple undergraduate mechanics problem. Cyclic boundary conditions or fixed boundaries can be used to establish normal modes of vibration.

It is *famous* because of how incredibly well the model works: when an atom is deflected from its position of equilibrium and released, the imaginary spring will pull the atom back. Due to its inertia, the atom will overshoot its equilibrium position and the springs will push it toward the opposite direction. The process repeats, and the atom oscillates. For small displacements, it turns out that this mechanical model of springs, with a linear restoring force, captures much of what we need to know about lattice oscillations.

The strain in consecutive springs causes adjacent atoms to oscillate as well. The atomic oscillations around the lattice points correlate throughout the crystal since they are all linked together through the springs. In fact, according to classical mechanics, a wave will develop. These waves are simply sound waves within the solid.

We can't ignore quantum mechanics forever though. So, the waves are physically allowed to take on only specific energies. Classically we might visualize this as the normal modes of vibration of the chain above in one dimension, where only specific amplitudes of oscillation and multiples thereof are allowed. As we will show, these normal mode vibrations will become associated with discrete packets of energy and momentum, which we can treat as *particles*. The Greek word for sound is *phonos* and the *quasiparticles* corresponding to the sound waves (following the quantum mechanical wave-particle dualism convention) are called *phonons*. The human ear perceives sound up to some tens or hundreds of kilohertz. The phonon frequencies, however, lie in the gigahertz range; thus most of the phonons in solids are *ultrasonic*.

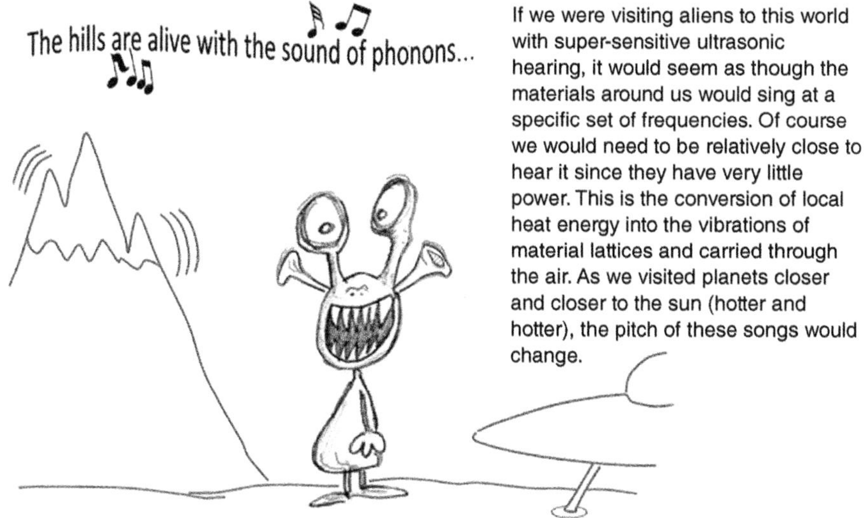

Remember that to characterize a wave, some basic quantities are necessary such as the frequency ν and the wavelength λ. The amplitude is also needed, and we will see later that it corresponds to the number of phonons of a given λ. Recall the frequency and wavelength are related by the *phase velocity*, v_{phase}.

$$\nu = v_{\text{phase}}/\lambda \tag{5.10}$$

The wave *phase velocity* depends on the properties of the medium in which the wave propagates. In our model above, for example, this is related to the force constants of the springs (Figure 5.1). For small values of frequency and amplitude, the force constants are the same for all values of frequency and amplitude; otherwise we would not call them constants. However, this is not true for very high frequencies. In that case, waves with different frequencies move with different velocities and a wave packet (a grouping of different frequencies to form a localized pulse) will *disperse*. That is, it will spread out. Similarly, in optics a prism disperses a polychromatic light beam because the refractive index of the prism material (and hence the light velocity) depends on the light frequency. Because of possible frequency/velocity dependence, Eq. (5.19) is called a *dispersion relation*.

For a complete description of a wave, the *direction* of the wave propagation is also important. The direction of propagation and the reciprocal wavelength are combined into the *wavevector* $\boldsymbol{k}$, which points in the propagation direction and has the length of the reciprocal wavelength multiplied by 2π:

$$|\boldsymbol{k}| = 2\pi/\lambda \tag{5.11}$$

Obviously the wavevector and the reciprocal lattice vectors are tied together. You should ponder this for a few minutes. Using the wavevector for λ and carrying the 2π into ν to get ω (angular frequency in radians per second, which is more convenient to use), then Eq. (5.3) takes on its classic appearance:

$$\omega = k v_{\text{phase}} \tag{5.12}$$

From quantum mechanics and *wave–particle duality*, we know that a wave with frequency ω and wavevector $\boldsymbol{k}$ corresponds to a particle with energy $E = \hbar\omega$ and momentum $\boldsymbol{p} = \hbar\boldsymbol{k}$.[2] The energy travels as a wave but arrives as a particle. So it is also reasonable to write the dispersion relation in terms of a quasiparticle's energy and its k or p dependence, $E(k/p)$ vs. k/p:

$$E = \hbar\omega \searrow \quad \swarrow p = \hbar k$$
$$E = v_{\text{phase}}\, p \tag{5.13}$$

A free particle, for instance, has $\mathbf{p} = m\boldsymbol{v}$, where m is the mass. The phase velocity and group velocity of this particle's quantum wave is equal to its classical velocity. So we get

$$mv^2 = p^2/2m \tag{5.14}$$

However, in a crystal, *phonons* are not free particles. They are bound to the crystal; literally they are a state of the crystal lattice. Therefore $\hbar\mathbf{k}$ does *not* represent

2 Don't forget. When we physicists want to divide a quantity by 2π, we draw a line through the symbol. So $h/2\pi = \hbar$.

a momentum in the "free particle" sense. Instead it is a *quasi-momentum* or a *crystal momentum*. A peculiarity of the *crystal momentum* is that its value cannot exceed that of a reciprocal lattice vector of the crystal. If the momentum of a *quasiparticle* is somehow pushed beyond that of $\hbar x$ (reciprocal lattice vector), then the momentum value $\hbar\mathbf{G}$ will be transferred to the crystal as a whole ($\mathbf{G}$ is a reciprocal lattice vector as in Chapter 4). The recoil energy associated with that momentum, and transferred to the crystal as a whole, does not show up in the energy balance because of $E = p^2/2M$. M is the total mass of the crystal, which is almost infinite compared to the mass of an electron or atom (there are typically 10^{21} atoms in a crystal of 1 cm^3 and an atom is about 10^4 times as heavy as an electron).

Another way to think of this is that waves in the discrete lattice of a crystal differ from waves in a continuum. Because of the discreteness of the lattice, there is a minimum wavelength. Waves with shorter wavelengths than the interatomic distance are not possible, because there is nothing between two atoms that could oscillate. Minimum wavelength means maximum wavevector and consequently maximum momentum. A fancy way of expressing this is to say that the momentum is "only defined modulo $\hbar\mathbf{G}$."

The dispersion relation for phonons in the chain of Figure 5.1 can be calculated using a classical Hooke's law model. And, since our discussions have centered on one-dimensional materials, it is reasonable to take a closer look at phonon behavior in such materials: a monatomic lattice with linear restoring forces on each spring.

5.1.1 A Simple One-Dimensional Model

5.1.1.1 A Model

We begin by placing each atom of mass M at an even spaced position along a chain coupled together with springs. The lattice parameter is a. We index each site by a counting integer s, so that the sth atom is at lattice position a times s. From the static equilibrium state, we can consider two distinctly different types of displacements: a longitudinal displacement – where the sth atom is displaced along the axis of the chain by an amount u_s – and alternatively a perpendicular displacement where the atoms can be displaced perpendicularly to the chain. Both situations are shown in Figure 5.2 with the middle row being longitudinal and the bottom perpendicular (*transverse*) displacements. We note that when we go to solve for the normal modes of vibration in this classically mechanical system, we will have to employ some type of boundary conditions that apply to the atoms at the ends of the chain. We can pin them (no motion), allow them to move independently, or connect the two ends of the line together so one end follows the other. The last choice is known as the Born–von Karman boundary conditions, which we will use. So you could think of this as being a *loop* of atoms. The waves will propagate around the loop.

The fact that three different kinds of displacement are being considered is important. Again, we have: (i) a longitudinal displacement, (ii) an up and down displacement, and (iii) one not shown, into and out of the page. We might refer to each of these as different *polarizations* of the vibration of the atomic chain.

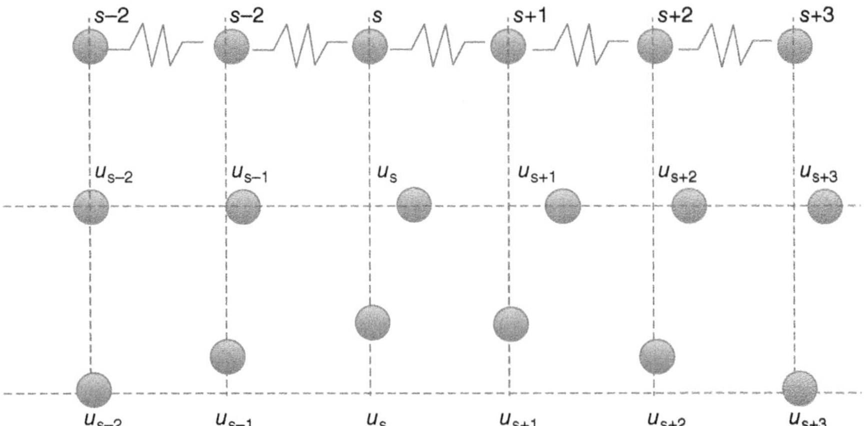

Figure 5.2 The array of atoms indexed by *s*: (middle) longitudinal waves on the chain. (Lower) transverse waves on the chain.

Notice that the restoring forces on each of the do not need to be the same, so it is possible in a two- or three- dimensional system that each polarization has a different vibrational character. Of course, in one dimension this may also be possible, though for now, it isn't necessary for our discussion.

In our development we take the approach of first considering purely elastic, classical waves traveling through the solid. We then add *quantum* considerations since, after all, we are considering atoms and they usually have quantum behaviors. Finally, we must add in quantum statistics. This approach will bring us to one of the first great truths of solid-state physics: nearly every phenomenon in a crystal (except for elastic diffraction) can be described as a box with the volume of the crystal, filled with particles of some sort. These "particles" are not the atoms, but rather the *collective behavior*, or vibrational modes, of the atoms and their electrons.

Now, on to our solution. Here is the derivation from Hooke's law:

$$F = M\,\mathrm{d}^2u_s/\mathrm{d}t^2 = C_1(u_{s+1} - u_s) + C_1(u_{s-1} - u_s) \tag{5.15}$$

We have used C_1 to be an effective "spring constant" of the molecular spring between atoms (the bonding orbital). Further, springs compress and expand according to Hooke's law in this picture. It is traditional to focus on the longitudinal wave solutions at this point. Since we can decompose any general lattice vibration down to a set of amplitudes of normal modes in a Fourier series, we can simply focus on the normal mode solutions. These are given by

$$u_s = u_0\mathrm{e}^{i(kx-\omega t)} \tag{5.16}$$

This harmonic solution has wavenumber k, an amplitude (plus phase factor) u_0, and frequency ω. Since the x values on the chain are discrete, we can substitute x with sa:

$$u_s = u_0\mathrm{e}^{i(ksa-\omega t)} \tag{5.17}$$

and we note

$$u_{s+1} = u_0 e^{i(k(s+1)a-\omega t)} = u_s e^{ika} \tag{5.18a}$$

$$u_{s-1} = u_0 e^{i(k(s-1)a-\omega t)} = u_s e^{-ika} \tag{5.18b}$$

Substituting into the equation above,

$$Md^2u_s/dt^2 = -M\omega^2 u_s \tag{5.19}$$

$$-M\omega^2 = C_1(e^{ika} + e^{-ika} - 2) \tag{5.20}$$

$$-M\omega^2 = C_1(\cos ka + i\sin ka + \cos ka - i\sin ka - 2) \tag{5.21}$$

$$-M\omega^2 = -2C_1(1 - \cos ka) = -4C_1\sin^2(ka/2) \tag{5.22}$$

This yields the dispersion relation for longitudinal waves within a monatomic chain:

$$\omega = 2(C_1/M)^{1/2}|\sin(ka/2)| \tag{5.23}$$

The dispersion relation is shown graphically in Figure 5.3, with the frequency (energy can be used) as ordinate and the wavevector $\boldsymbol{k}$ as abscissa.[3] The dispersion relation is plotted up to the aforementioned maximum value for the wavevector $\boldsymbol{k}$. The graph does not have to end at that point, but from there on, no additional information is obtained in extending it any further, because the energy does not change if integer multiples of a reciprocal lattice vector are added to the wavevector (recall we mentioned that this would happen).

Figure 5.3 is obviously plotted in *reciprocal space*. The wavevector, $\boldsymbol{k}$, has a reciprocal length (the wavenumber) to which any reciprocal lattice vector can be added. Reciprocal space is a convenient tool for discussing dispersion relations, but more than this, it is a generalization of the reciprocal lattice. We know that the reciprocal lattice consists of discrete lattice points, connected by reciprocal lattice vectors with the origin of the reciprocal lattice. Here, the reciprocal

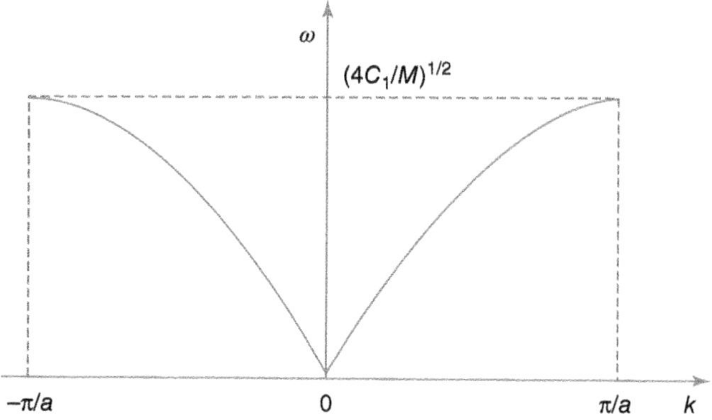

Figure 5.3 The dispersion relation for the model in Figure 5.2, a one-dimensional, monatomic chain.

3 Here is a fine point. This is a one-dimensional problem and the wavevector k is laid out along the direction of the reciprocal lattice – making it a $\boldsymbol{k}$. This convention is adopted even in higher dimensional problems as we will see.

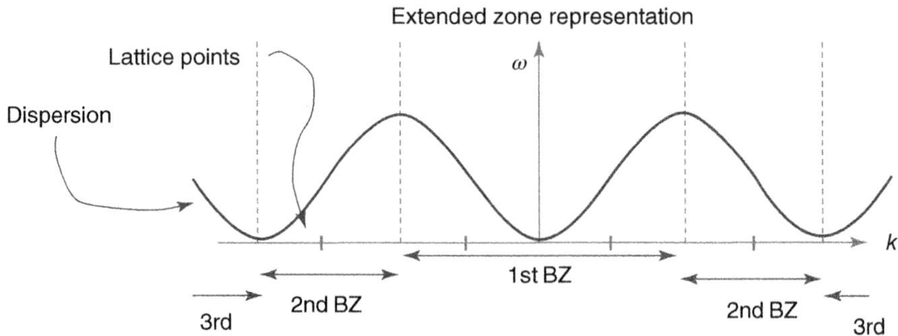

Figure 5.4 Brillouin zones. The extended zone representation is not usually needed. A general dispersion curve is shown.

space also includes the space between the points of the reciprocal lattice. For most purposes only the space between the origin and the reciprocal lattice points nearest to the origin is important: the *first Brillouin zone* (BZ) as introduced in Chapter 4.

Recall the concept of Brillouin zones where Figure 5.4 shows the reciprocal lattice of a linear chain. The arbitrary choice of origin of the reciprocal lattice lies in the center of the figure. The interval between the origin and the first lattice points on both sides is divided into halves; the region "first BZ" marks the first Brillouin zone which is surrounded by the second Brillouin zone and so on. The concept was generalized to two- and three-dimensional space in Chapter 4. For phonons only the first Brillouin zone is relevant: the wavevector and the quasi-momentum was only defined as modulo $[\hbar\tau]$.

5.1.1.2 Long Wavelength Vibrations

Returning to the dispersion relationship of Figure 5.3, for $ka \ll \lambda$ or $\lambda \gg a$, the lattice vibrations are said to be in the *long wavelength limit*. These are the values near zero in the BZ. Here,

$$\sin ka/2 \sim ka/2 \tag{5.24}$$

and

$$\omega \sim ka(C_1/M)^{1/2} \tag{5.25}$$

so ω is *linear* with k. For this case the *phase velocity* is given by

$$v_\mathrm{p} = \omega/k = a(C_1/M)^{1/2} = \text{a constant } (v_0) \tag{5.26}$$

The *group velocity* of a wave packet (the traditional speed of sound in a solid or the velocity with which energy travels through the lattice) is given by

$$v_\mathrm{g} = \mathrm{d}\omega/\mathrm{d}k = a(C_1/M)^{1/2} = v_\mathrm{p} \tag{5.27}$$

So, $v_\mathrm{g} = v_0$ in the long wavelength limit and they are constant with k.

5.1.1.3 Short Wavelength Vibrations

For large k or small λ, a little algebra shows

$$v_{\text{p}} = v_0(\sin ka/2)/(ka/2) \tag{5.28}$$

whereas

$$v_{\text{g}} = v_0|\cos ka/2| \tag{5.29}$$

So at $k = +/- \pi/\text{a}$, $v_{\text{g}} = 0$! No wave is propagated at the BZ boundary. This is simply the Bragg condition for reflection (adjacent atoms vibrate out of phase). And at k near 0, the group velocity is at a maximum.

5.1.1.4 More Atoms in the Basis

The one-dimensional monatomic chain is a rather simple example. What happens when the structure is made more complicated by a basis? Surprisingly, things do not get that much more difficult. As we might expect, there is a set of *in phase motions* of the subsets of atoms and a set of *out of phase motions*. Let us consider briefly the case of a two-atom basis: M_1 and M_2. This is a little like the case of graphene, but we are using this for one-dimensional, more like a conjugated polymer. We assume $M_1 > M_2$. Our labeling system is as shown in Figure 5.5.

Again, our derivation from the basic model, Hooke's law gives

$$M_1\text{d}^2u_s/\text{d}t^2 = C(v_s + v_{s-1} - 2u_s) \tag{5.30a}$$

$$M_1\text{d}^2v_s/\text{d}t^2 = C(u_{s+1} + u_s - 2v_s) \tag{5.30b}$$

The force constants have been assumed to be identical: C. We assume two harmonic solutions:

$$u_s = u\text{e}^{i(ksa-\omega t)} \text{ and } v_s = v\text{e}^{i(ksa-\omega t)} \tag{5.31}$$

Substituting,

$$-\omega^2M_1u = Cv(1 + \text{e}^{-ika}) - 2Cu \tag{5.32a}$$

$$-\omega^2M_2v = Cu(1 + \text{e}^{ika}) - 2Cv \tag{5.32b}$$

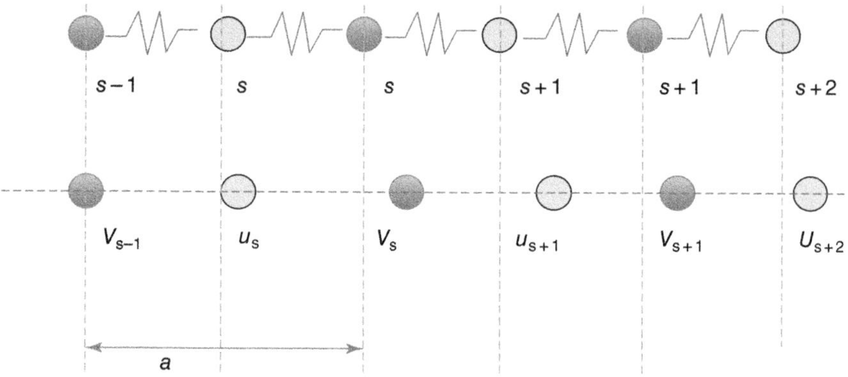

Figure 5.5 An indexing scheme similar to the monatomic model is employed. One may choose either atom for M_1 and M_2 (gray or blue balls).

Manipulating this around we get two simultaneous equations:

$$(2C - M_1\omega^2)u - C(1 + e^{-ika})v = 0 \tag{5.33}$$

$$- C(e^{ika} + 1)u + (2C - M_2\omega^2)v = 0 \tag{5.34}$$

To have a single solution, the determinate of these equations must be zero. This gives

$$M_1M_2\omega^4 - 2C(M_1 + M_2)\omega^2 + 2C^2(1 - \cos ka) = 0 \tag{5.35}$$

From the quadratic equation, we get

$$\omega^2 = \frac{C(M_1 + M_2)}{M_1M_2} \pm C\left\{\left[\frac{M_1 + M_2}{M_1M_2}\right]^2 - \left(\frac{4}{M_1M_2}\right)\sin^2\frac{ka}{2}\right\}^{1/2} \tag{5.36}$$

This is the *dispersion relationship* for the diatomic linear chain.
Notice that at $k = 0$:

$$\omega = 0 \text{ and } \omega = [2C(M_1 + M_2)/(M_1M_2)]^{1/2} \tag{5.37}$$

At the BZ boundary $k = \pm\pi/a$:

$$\omega = (2C/M_1)^{1/2} \textit{ and } \omega = (2C/M_2)^{1/2} \tag{5.38}$$

The dispersion relation is plotted out in Figure 5.6.

The *lower branch* is referred to as the *acoustic* branch and with a little work you can convince yourself that this is composed of coordinated, *in phase* motions of the two sublattices made up of the basis atoms. The *top branch* is referred to as the *optical* branch, and again with a little work it is easy to see that this must be *out of phase* motion of the two sublattices made of the basis atoms.

So the two-atom basis adds another branch to the dispersion curve and a little new physics shows up. However, this model is still rather simple in comparison to some of our real, low-dimensional materials. Consider for instance the carbon nanotube. As we have already shown, there can be a tremendous number of

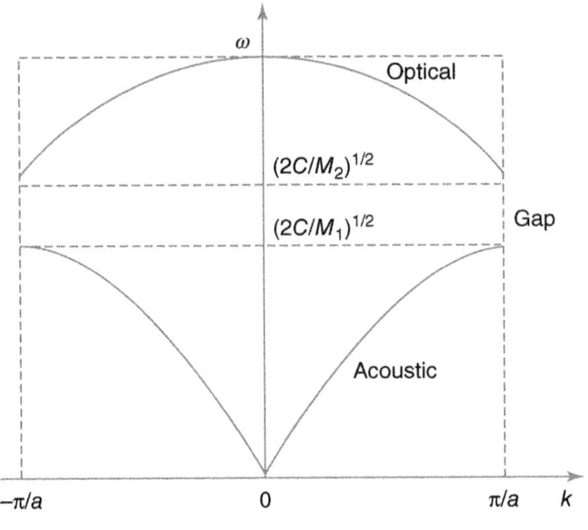

Figure 5.6 Dispersion relations of a diatomic chain.

atoms in the basis set. But there is more than this. Surely the assumption that the displacements and restoring forces are all aligned the same way is rather simplistic. We must recall that the nanotube is "semi-one-dimensional." That is, it is a rolled up graphene sheet and there is a number of different ways to roll it. In this case this means that the ways an atom might move along the tube is more complicated than our model. There are longitudinal modes of course. Then there are also transverse modes: breathing modes as they are called on nanotubes. But then too there are the modes of motion that would have been associated with in-plane vibrations of the graphene lattice before rolling and not necessarily along the axis. These would lead to a vibrational twist of the nanotube – called *twistons*: these are longitudinal modes overlaid onto the topology of a tube. So, the chirality of the object can play an important role in determining the overall phonon dispersion curves, the number of branches, and what each branch stands for in terms of atomic motion. Certainly not difficult to unravel, but beautifully complex for a one-dimensional object!

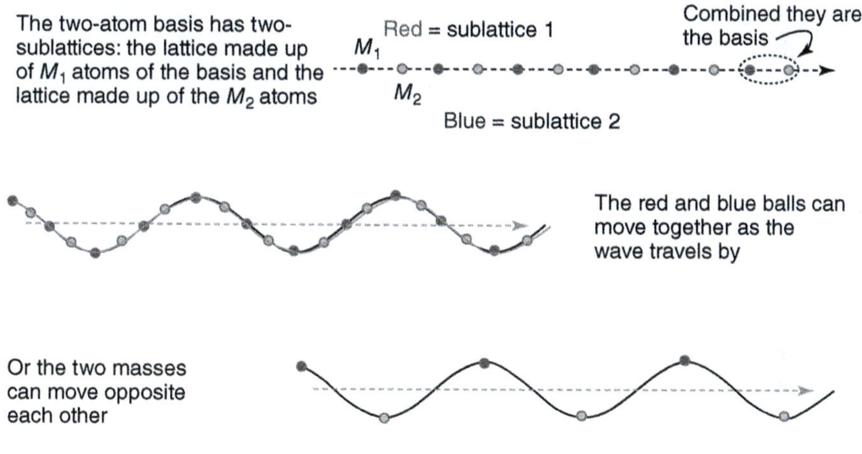

5.1.2 More Dimensions

Extending the above model to higher dimensions can get a little confusing. First there is the "dimension" of polarization. The atoms along a one-dimensional chain can move in three dimensions (forward/back, up/down, side/side). The first is a longitudinal mode and the other two are transverse modes. Of course there is no a priori reason for any of the force constants and restoring forces, to be the same or resulting vibrations and dynamical modes to be degenerate. Naturally they *might be*, but one can easily think of examples where they *would not* be. Thus, *polarization* adds an additional *little dimension*, meaning nonspatial, to our problem.

Like the case of electronic spin, a full description of the phonon must carry along with it the value of this polarization dimension as well. Remember, spatially, the phonon is still one-dimensional in this example as it still moves as a wave, with velocities only forward and backward along the chain. Each polarization adds its own branch to the dispersion curve. If there is a basis, then they too participate in the polarization "dimension."

Obviously, the other dimensional expansion we must consider is spatial. The vibrational modes of a sheet of atoms or of a three-dimensionally array will behave in a one-dimensional manner mainly, along the principle directions of the reciprocal lattice. Now instead of one atom oscillating back and forth, it is a line or a plane of atoms in the array. Let's look to an "old friend" as an example: graphene.

To get some expectation of how vibrations should behave on this trigonal, two-atom basis, carbon lattice, we must first lay down some coordinates and draw out the directions relative to the atoms. In Figure 5.7 we have done just this. The lattice itself is made up of two equivalent sublattices shown in blue and red. This represents the basis set on each primitive lattice point. The unit vectors

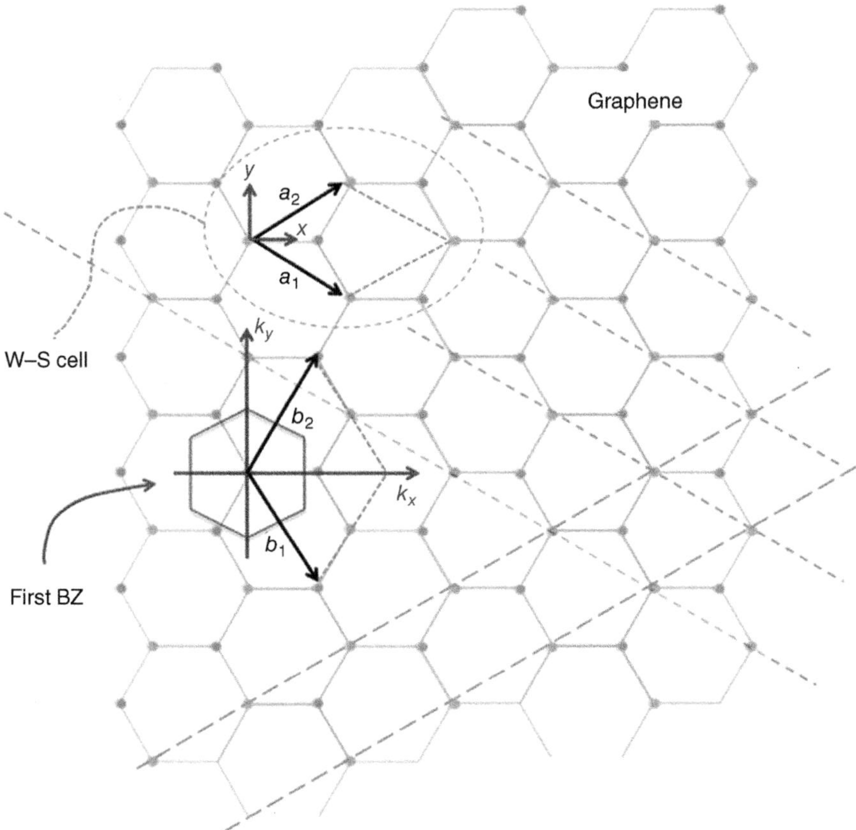

Figure 5.7 The graphene sheet laid out with its primary coordinate systems: real lattice (top) and reciprocal lattice (bottom).

of that primitive lattice are shown as $\mathbf{a}_1$ and $\mathbf{a}_2$ on the top part of the drawing. The Wigner–Seitz (W–S) primitive cell is drawn in grey as a parallelogram and we have oriented it relative to xy coordinates (red). From this primitive cell and the lattice vectors, we can construct the reciprocal lattice vectors $\mathbf{b}_1$ and $\mathbf{b}_2$ as shown at the bottom of the drawing. This follows the recipe in Chapter 4. Again a parallelogram is formed for the first Brillouin zone (not to scale, so that it can be seen), it is rotated relative to the W–S cell, and we have laid out the k_xk_y coordinates parallel to the xy coordinates. Notice that the unit wavevectors lay in the direction of the perpendicular to the dotted green lines, which denote lines of equivalent atoms throughout the real lattice. Thus, the lattice oscillations are described by the primary modes of these lines of atoms oscillating as though they were rigid along the line (the whole line displaces back/forth, up/down, left/right). Notice too that there are two distinct directions of propagation, along $\mathbf{b}_1$ and along $\mathbf{b}_2$. So, let's say we would like to consider the possibility of a line distorting with a wavelike solution. But this can now be described as a superposition of waves in the $\mathbf{b}_1$ and $\mathbf{b}_2$ directions without loss of generality. Thus we may consider a phonon traveling in any direction upon this plane.

How do we describe dispersion in this two-dimensional (2D) world? We would like to use the same type of dispersion graph we have introduced for one dimension. Since our fundamental modes of oscillation are already set up to look one-dimensional, this shouldn't be hard. To do it we must zoom in to that first BZ as in Figure 5.8.

Here in Figure 5.8, we have labeled special points within the reciprocal space: K, M, and Γ. They represent points along the zone boundary and of course the zone center. These happen to be high symmetry points for this structure, and each structure that is encountered will be a bit different. Calculations similar to the simultaneous equations we used above are utilized to derive the expected dispersion curves along the directions marked by these symbols. Since the symbols mark the points of symmetry associated directly with the real lattice, this approach makes it easy for the experimenter to use scattering observations with oriented crystals and determine the accuracy of the calculations. What happens for graphene?

In Figure 5.9 we show the calculated phonon dispersion curves of this 2D system. The dispersion is plotted along the path Γ to K to M and back to Γ with no breaks. Notice that there are a number of our expectations that have been

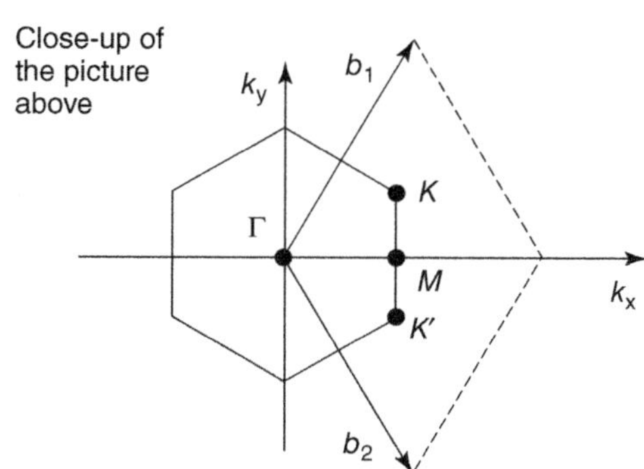

Figure 5.8 A close image of the graphene first BZ with common labeling.

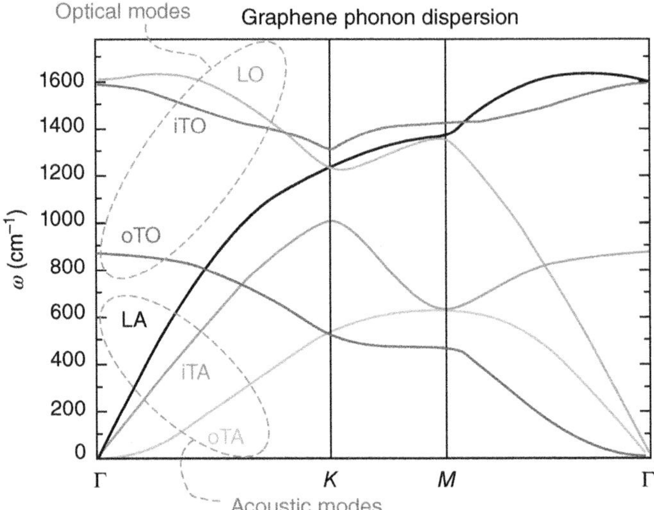

Figure 5.9 The dispersion relations calculated for graphene.

included in the calculation. First of all, this is a two-atom basis system, so there is a set of optical and acoustic modes. Polarization has been taken into account so, as expected, there are longitudinal and transverse modes. Also the spring constants of the atoms are quite different (no degeneracies). Thus we have transverse waves in which the atom vibrates in plane and waves in which it vibrates out of plane. These are marked using a standard notation: iTA means in-plane, longitudinal, acoustic mode, and so on (Figure 5.10). There are, in fact, six different branches

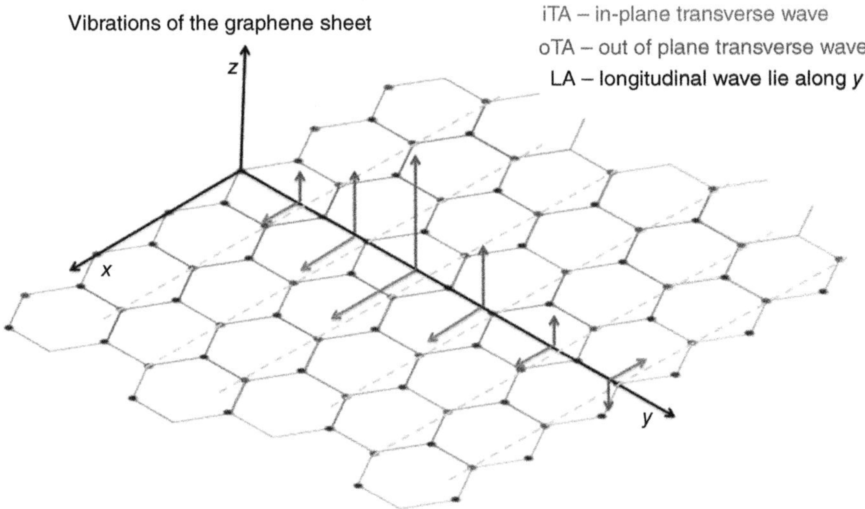

Figure 5.10 The vibrations of a graphene sheet in terms of the directions of atomic motion.

we must keep up with here, two for each degree of freedom of the atoms, and this for a simple sheet of carbon!

Of course an important connection that can be made with these results is the many excellent Raman scattering experiments (inelastic scattering of light in the visible range) on graphene. Knowing such dispersion characteristics can allow the experimenter to identify which type of phonon is doing the scattering. This in turn can give information about the purity or crystallinity of a sample. We introduce Raman in the problem set below.

5.2 Quantum Considerations with Phonons

This simple analysis so far has considered only classical, elastic waves. But, as we have already suggested, the energy of these waves is quantized. In fact, the dispersion curves shown above are not continuous but discrete states. The lines should really be dots, but they would be very close together. The quantized vibration is called a *phonon* – the language we introduced earlier. The energy of a phonon, with angular frequency ω, is given by

$$E = (n + 1/2)\hbar\omega \tag{5.39}$$

where n is an integer and as with other quantum harmonic oscillators in nature these waves have a zero point energy. Here we have used the term "mode" to denote the ω vibrational state, and the n describes the overall amplitude of the wave in terms of discrete steps. This means *the mean squared amplitude* of the vibration is also quantized. From basic mechanics the volumetric kinetic energy of the elastic wave is given by

$$\mathrm{KE} = 1/4\, \rho V \omega^2 u_0^2 \sin^2 \omega t \tag{5.40}$$

where ρ is the volumetric mass density (in one dimension, the linear mass density), V is the volume (or the length in 1D) and the time average of the sine term is ½. And we now say this must occur in steps:

$$1/8\, \rho V \omega^2 u_0^2 = 1/2\,(n + 1/2)\hbar\omega \tag{5.41}$$

$$u_0^2 = 4(n + 1/2)\hbar/\rho V\omega \tag{5.42}$$

This means that the displacement of the atom itself is quantized when participating in such lattice vibrations (the u_0). Each n represents a different number of phonons. Again, we emphasize that this n is an integer and so only specific E's and ω's will occur. This is not surprising since it represents the very meaning of being quantized. An interesting question in this regard is to ask exactly what physical mechanism imposes quantization on the atomic motion. After all a free atom could certainly take on itself any kinetic energy it wished and we have rather artificially stated, "Well atoms are quantum objects, so we guess they must oscillate like a quantum harmonic oscillator."

5.2.1 Conservation of Crystal Momentum

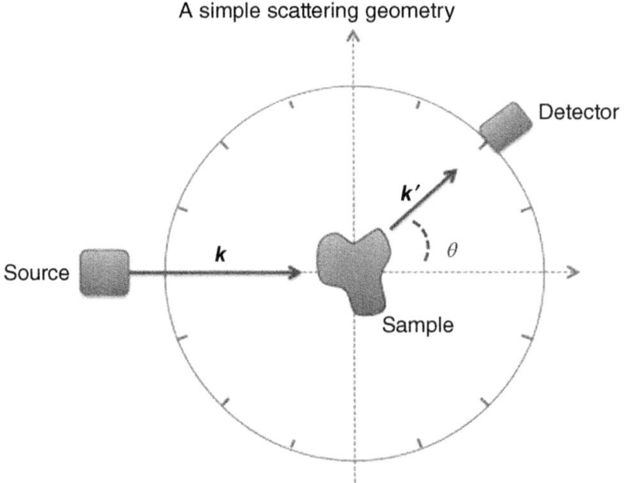

Finally, we restate the above: phonons do not carry physical momentum as we typically think of it. The *crystal momentum* they carry is analogous but it isn't equivalent since there isn't any net mass transport in a crystal vibration. However, like a particle, phonons interact with each other and with electrons, with photons, etc., by obeying the conservation rules of momentum. So, they act like they have momentum when they interact with, for example, a neutron, and this neutron interaction can transfer momentum to start a vibrational mode. But you have to think of the momentum as being transferred to the crystal as a whole. The atoms are not permanently moved from their sites. The amount of momentum that a phonon "thinks" it has is $p = \hbar\, k$. Of course the one exception is the $k = 0$ mode: this represents whole crystal translation. The difference is subtle, and many times it is simpler to treat "momentum as momentum," thereby applying conservation laws broadly. Just keep in mind that the crystal "limits" its momentum transfer to within its first Brillouin zone.

Let's consider what this means to scattering generally. In *elastic scattering* – like diffraction – there is no gain or loss of energy by the interacting particle. There is a selection rule for the momentum vectors involved in this interaction:

$$\boldsymbol{k} + \mathbf{G} = \boldsymbol{k}' \tag{5.43}$$

such that a maximum in the diffracted beam intensity is seen. $\boldsymbol{k}$ is the incoming particle's wavevector and $\boldsymbol{k}'$ is the wavevector of the scattered particle. $\mathbf{G}$ is the set of all reciprocal lattice vectors within the first BZ. In this process the whole crystal will recoil with a momentum of $-\hbar\mathbf{G}$, but we usually ignore this since it is small compared to most anything.

5.2.2 General Scattering

You might have noticed above that we have taken the easiest case for X-ray scattering. After all, we have derived only the resulting patterns created by elastic events in which the magnitude of the momentum of the incoming and outgoing waves are equal as well as the kinetic energy. However, we have now said that momentum and energy can be exchanged with the crystal and that these processes would be conservative. Surely then, with X-rays, or neutrons, or electrons, scattering from the crystal could exchange momentum and energy with the crystal (Figure 5.11).

Let's examine this proposition by considering a more general description of scattering. Remember that our previous idea was that X-rays, electrons, etc., are reflected off from planar mirrors located roughly by $n(\mathbf{r})$ within the solid. Of course that neglects the effects the electromagnetic radiation might actually have on $n(\mathbf{r})$ (for the case of X-rays, for example). Clearly the electronic cloud around an atom, when presented with a time-varying electric and magnetic field, will not just sit there! It seems reasonable to expect that the electrons and consequently the atoms of the lattice will undergo polarization and some small force will be applied as a result of the interaction with the E&M field. That is, there should be a small, time-varying force associated with the interaction (Figure 5.12).

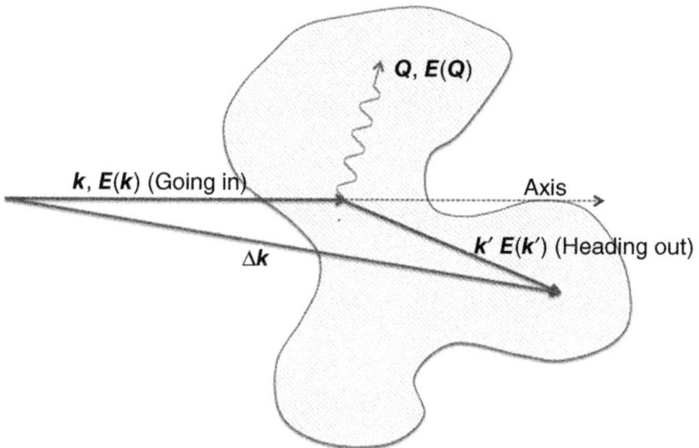

Figure 5.11 When $\Delta \boldsymbol{k} = \boldsymbol{G}$ and $|\boldsymbol{k}| = |\boldsymbol{k}'|$, this problem can be easy, but when there is an excitation absorbed or created, it gets more difficult. Solve. But when an excitation is created or absorbed (destroyed), it can be far more difficult.

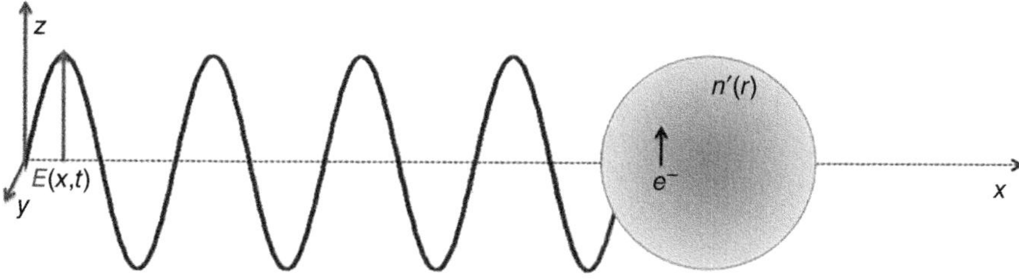

Figure 5.12 We are building a simple model for X-ray interactions in the scattering process. Notice that the interaction here is quite specific, so it would be different for different particles such as neutrons.

That force, (qE), has a time dependence like $\boldsymbol{E}(\boldsymbol{k},\mathbf{r}) \sim E_0 \exp(-i(\boldsymbol{k}\cdot\mathbf{r} - \omega_0 t))$, choosing a simple plane wave as our input wave to be scattered. Thus, the atom will displace, which will lead to wave formation. So we have an idea of how the $u_i(t)$'s from above might behave based on a driving force.

To see how this leads to bright and dark points on a screen outside of the sample, we have two paths: (i) use Fermi's Golden Rule to compute photon transitions from $\boldsymbol{k}$ to $\boldsymbol{k}'$ or (ii) realize that $I \sim |E^2|$ from classical E&M, which means of course that the thing we we're calling the *scattering amplitude* is bound to be proportional to E: $F \sim E$. This is not too different from what we argued before; it keeps the $n(\boldsymbol{r})$ term as it should and we will have to sum or integrate over the unit cell. This time though, we are not ignoring the time dependence of the electric field vector, $\exp(i\omega_0 t)$, like we did last time. (Of course you may have guessed that we did this because we were talking about many such beams over many such interactions and so the outcome in the scattering amplitude would have been a time average over all the oscillatory waves hitting any one particular place on the screen. Thus we ignored it.) Now we analyze each individual scatterer. Either way you will get the scattering amplitude we got before, multiplied by a time-dependent term:

$$F = \int_V n(\boldsymbol{r}) \exp[-i(\Delta\boldsymbol{k}\cdot\boldsymbol{r})]\mathrm{d}\boldsymbol{r}\, \{\exp[i\omega_0 t]\} \tag{5.44}$$

As last time we introduce a time-dependent $n(\boldsymbol{r})$ and consider point scatterers of a simple basis:

$$n(\boldsymbol{r},t) \sim \sum_j \delta(\boldsymbol{r} - \boldsymbol{r}_j(t)) \tag{5.45}$$

$$\boldsymbol{r}_j(t) = \boldsymbol{r}_{j0} + \boldsymbol{u}_j(t) \ :\ \boldsymbol{r}_{j0} \text{ is the static position of the atom} \tag{5.46}$$

$$\boldsymbol{u}_j(t) = \boldsymbol{u}_0 \exp\left[\pm i(\boldsymbol{q}\cdot\boldsymbol{r}_j + \omega(q)t\right] \tag{5.47}$$

and so we get

$$F \sim \sum \exp\left[-i(\Delta\boldsymbol{k}\cdot\boldsymbol{r}_j(t) + \omega_0 t\right] \tag{5.48}$$

$$\sim (1 - i\Delta\boldsymbol{k}\cdot u_j(t))\exp(\pm i\omega t) \text{ for small oscillations} \tag{5.49}$$

So,

$$F \sim \sum \exp[-i\Delta\boldsymbol{k}\cdot\boldsymbol{r}_{j0} - i\Delta\boldsymbol{k}\cdot\boldsymbol{u}_j - i\omega_0 t] \tag{5.50}$$

Putting this all together,

$$F \sim \sum \exp[-i(\Delta\boldsymbol{k}\cdot\mathbf{r}_{j0} - i\omega_0 t)] \times \sum i\Delta\boldsymbol{k}\cdot\boldsymbol{u}_0 \exp[-i\Delta\boldsymbol{k}\cdot\mathbf{r}_{j0} \pm i\boldsymbol{q}\cdot\mathbf{r}_{j0} \mp i\omega(q)t - i\omega_0 t] \tag{5.51}$$

$\mathbf{q}$ is the wavevector of the phonons we introduced above and ω their frequencies. These are associated with normal modes of the lattice as we have already shown. The first sum is simply the elastic scattering component. The second term is far more interesting:

$$F_{\text{inelastic}} \sim \sum i\Delta\boldsymbol{k}\cdot\mathbf{u}_0 \exp[-i(\Delta\boldsymbol{k} + \boldsymbol{q})\cdot\mathbf{r}_{j0} - i(\omega_0 \pm \omega(q))t)] \tag{5.52}$$

The frequency measured at the detector will be

$$\omega = \omega_0 \pm \omega(q) \tag{5.53}$$

and the condition for constructive interference from the inelastic component of the sum will be

$$\Delta \boldsymbol{k} = \boldsymbol{G} \pm \boldsymbol{q} \tag{5.54}$$

In this analysis, the $\boldsymbol{k}$ and $\boldsymbol{k}'$ are the wavevectors of the incoming and outgoing scattered particles, $\boldsymbol{G}$ is the set of all reciprocal lattice vectors that are within the first BZ, and $\mathbf{q}$ is the wavevector of propagation for the interacting phonon. The scatterers can gain or lose momentum consistent with the momentum associated with the reciprocal lattice vectors plus the momentum of the phonons (which is discrete).

Moreover, in *inelastic interactions* generally, energy is gained or lost by the interacting particle (which is also discrete). When a scattering particle is sent into a crystal (an alpha particle, electron, neutron, proton, etc.), it can undergo a change in *energy* given by

$$\hbar^2 k^2/2M = \hbar^2 k'^2/2M \pm \hbar\omega \tag{5.55}$$

where the $\pm$ indicates the absorption of energy from a phonon by the scatterer or the creation of a phonon by the scatterer.

5.3 Phonons Yield Thermal Properties

The "thermodynamics" of material objects is really defined by how heat energy is stored and transported by the solid. How much heat energy, Q, does it take to raise the temperature of the object, how fast does one end heat when the other is heated, when does the object melt, etc. Since lattice vibrations can "store" and "transport" significant amounts of energy within a solid, then certainly these vibrations should be implicated in the macroscopic thermodynamics observed from materials.

To see the extent of this idea, we must first put together everything we have learned about phonons (lattice vibrations) because these characteristics will limit exactly how such vibrational modes are allowed to interact with heat energy. So recall from above, we know…

1. Phonons are *quantized*, meaning they carry energy in discrete chunks. We can picture this classically as discrete vibrational amplitudes at specific frequencies allowed by the lattice. As more quanta of energy are added to a specific vibrational mode, the larger amplitude of vibration gets. We say that we have more *phonons* of this mode ω.
2. Phonons carry no *real momentum*; they carry *crystal momentum*. But we still refer to this momentum as though it were the same as a particle of mass moving at some velocity. It has the restriction of only taking on values within the first Brillouin Zone.
3. Phonons interact with other particles by *obeying conservation laws*. We treat phonons as though they are a particle (called a *quasiparticle*) born of the collective nature of the dynamics of many things within the crystal. These phonon particles obey the laws of being particles for the most part: they conserve

momentum and energy when they interact, for instance. Moreover, like the quantum mechanical *particle in a box*, these phonon quasiparticles have quantum states they occupy, allowed by the lattice and generally orthogonal to each other (available modes of vibration in the lattice). However, we have not adequately described the *filling* of these quantum phonon states (which we enumerated with a K in our derivations above); do we treat these quasiparticles as *fermions* or *bosons*? That is, exactly how many phonons can we pack into any state?[4] We have said that as we add quanta of energy to a given mode, the vibrational amplitude of that mode increases and that this is equivalent to adding phonons. But is there a limit to how many phonons of a specific mode (ω) are allowed? And how does nature choose to fill up these modes for some finite amount of heat energy Q? This we have not yet explored.

4. Finally phonons are *not localized in space*. They extend over the whole crystal. This too is in concordance with the idea of the particle in a box. The nature of this quasiparticle, as should be expected, is to carry a wave and particle characteristic with it at all times. It travels as a wave but arrives as a particle.

With this in mind, can we work out a general scheme for thermodynamic entities such as heat capacity? The answer is "sort of...."

5.3.1 Internal Energy and Phonons

The total internal energy per unit volume of a crystal lattice is composed of several components. The first we have already seen in previous chapters: the configurational energy of the static lattice. This is the binding energy of the atoms in their equilibrium positions and it pertains to *melting*. Added to this energy is the energy associated with the displacement of atoms from equilibrium positions: the *phonons* or vibrational modes. In the context of heat transport and heat capacity, we are concerned with the second component of internal energy and so our models will be constructed toward predictions of solids far from phase transitions generally.[5]

---

4 It is very important to note here that our discussion is taking a strange philosophical path. We are describing a solid volume (the crystal) to have quantum states into which phonons may be added or subtracted. The *creation* of phonons in the crystal populates one or the other of these quantum states. The *annihilation* of phonons removes the phonon from one of these states. The quantum states exist independently from the quasiparticle phonons. Those phonons are excitations of a crystal field. We might imagine this to be consistent with pictures of particle creation in aspects of other field theories, say, positrons and electrons in cosmology.
This is certainly not the historical way of looking at the problem. The phonon population, and indeed phonons themselves, were seen as quantum states of the crystal lattice generally. These quantum states carried momentum and obeyed conservation laws, so they had *particle-like* features but should be counted as vibrational modes of the lattice and not pictured literally as particles. Of course the mathematics is the same between the two pictures and the interpretations largely mean the same thing. So in essence it makes little difference. We do point out that Einstein preferred the second way of picturing things, and while it is a bit tasteless to argue with Einstein, we prefer the first.

5 A more general treatment is necessary for solids approaching phase transitions and this will necessitate the use of the correlation functions introduced earlier.

This *phonon internal energy*, let's call it E_{ph}, can provide us the link to thermodynamic elements associated with lattice vibrations, and as we have already argued, in the temperature ranges we are discussing, this energy should dominate behavior. Let's assume that our volume of solid has some countable number of oscillation modes; we might say some number of phonons. Each vibrational mode or phonon has a frequency of oscillation given by $\omega_{K,p}$. We further assume that our solid is in thermodynamic equilibrium. This means the whole solid sits at the same temperature and no net heat energy Q is entering or leaving. The total energy of the set of oscillators (or oscillations of the lattice, or phonons) is then given by

$$E_{\text{ph}} = \sum_K \sum_p E_{K,p} = \sum_K \sum_p \left[\langle n_{K,p} \rangle + \frac{1}{2} \right] \hbar\omega_{K,p} \tag{5.56}$$

The K is the wavevector index and p is the polarization (there are three of these). Yep, we just add them together. We have also accounted for the fact that there are different polarizations possible and that these may have different modal frequencies. The ½ is a zero point energy. The really important term here is the: $\langle n_{K,p} \rangle$, which is the *thermal equilibrium occupancy* of the state with wavevector K and polarization p. The brackets mean a *thermal average* is to be taken. In thermal equilibrium we equate this to a time average. So we recognize that phonons might *jump* from state to state – though we haven't identified a transition mechanism yet, but what we are asking for is the average *occupancy* (or number of phonons) of a given state $\omega_{K,p}$ over time.

Now notice that at this point, working completely blindly, we have made no assumptions about the nature of $\langle n_{K,p} \rangle$. If we do not make assumptions about the nature of this occupancy rule, then it is clear that the choices we might make to add up to E_{ph} are not unique. However, thermal experiments can be performed with a sense of reproducibility, regardless of the thermal history of sample preparation generally. Thus, we might take the bold step in assuming that the partitioning of energy among these states is not random and that this thermal average can be replaced with a simple distribution function, $f_p(K)$, that counts the number of phonons in each K state, which has a corresponding ω:

$$E_{\text{ph}} = \sum_K \sum_p \left[f_p(K) + \frac{1}{2} \right] \hbar\omega_{K,p} \tag{5.57}$$

We notice here that the term $\omega_{K,p}$ must come from some dispersion relation as we have discussed above, in order to actually work this sum. Further, if we accept that the Ks are relatively close together, this K sum is easier to work as an integral:

$$E_{\text{ph}} = \sum_p \int \left[f_p(K) + \frac{1}{2} \right] \hbar\omega_{K,p} \mathrm{d}K \tag{5.58}$$

We are working the integral over all the possible K values. From # 2 in our list above, we are restricted then to working this over the first BZ. We still need a guess for $f_p(k)$ and a dispersion curve.

5.3.2 Models of Energy Distribution: $f_p(\omega)$ and $\omega_{K,p}$

To work the sum or integral above is to achieve an expression for the component of the internal energy of the system due to physical vibrations of the atomic nuclei. Clearly, given the vast number of circumstances that this might be applied to, the usual approach is to offer approximations that work within specific binding regimes of the crystal and temperature ranges of experiment. It should be specifically noted that generally such approaches are associated with equilibrium situations: so there is no net heat flowing into or out of the system. The system does not change its phase.

5.3.2.1 DuLong and Petit: Equipartition of Energy

Long before a comprehensive, *spectral* theory of phonons was developed, scientists knew that many of the thermodynamic characteristics of solids (internal energy, heat capacity, etc.) had nontrivial dependencies on temperature, T. To address this, *DuLong and Petit* (*D/P*) developed a completely classical approach based on an ensemble of atomic oscillators. The *D/P* approach assumed that atomic vibrations were uncorrelated – just individual harmonic oscillators placed at each of the lattice points. Further, energy was partitioned among the oscillators as would be expected from the classical equipartition theorem. So this means in our sum above we must replace the quantum energy of each mode with its classical statistical mechanics counterpart.The first sum itself is now indexed over the total number of oscillators, not the K states. The second sum is the number of degrees of freedom of the atomic oscillator. Here we have naïvely assumed the oscillator to be an extended body in space, so we have been giving it translational and rotational degrees of freedom (following roughly from what *D/P* did). So the result becomes exceedingly simple. Recall that the heat capacity is defined as

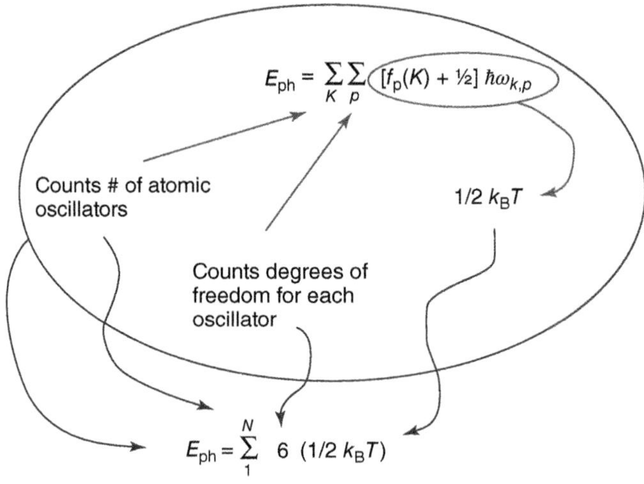

$$C = \Delta Q/\Delta T \tag{5.59}$$

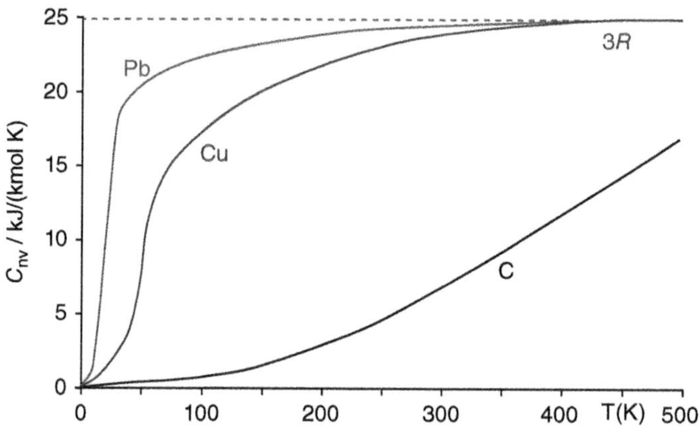

Figure 5.13 The C_v for several elemental solids are shown here. The dashed line is the DuLong–Petit prediction. Notice that some metals do behave very much like the D/P model. But our friend carbon doesn't seem to follow the trend. (If it weren't for carbon, this would all be too easy!)

And through a little manipulation of thermodynamic variables, we can reach the expression we will use to get our heat capacity for constant volume (Figure 5.13):

$$C_v = (\partial E_{\text{total}}/\partial T)_v \tag{5.60}$$

D/P gives $E_{\text{ph}} = 3Nk_BT$ and the $C_v = 3Nk_{B,}$ where N is the number of atoms and this is the vibrational contribution. At high temperatures, this is a pretty good approximation as C_v approaches a constant for many solids as temperatures increase. However, for low temperatures we know from experiment (and experiment is always right) $C_v \sim \gamma T + \beta T^3$ where the first term is due to electronic contributions and is small. So the *D/P* approximation is rather poor for these conditions.

In fact *D/P* is pretty good for light metals that are weakly bound at higher temperatures. Amorphous structures are also pretty well approached using this approximation. Can you see why?

5.3.2.2 Einstein and Quantum Statistics

Clearly this classical model ignores important quantum effects that will be dominant at very low temperatures. So *Einstein* set about to correct for this. Again, beginning with our basic equation (counting integrals):

$$E_{\text{ph}} = \sum_p \int \left[f_p(K) + \frac{1}{2}\right] \hbar\omega_{K,p} \mathrm{d}K \tag{5.61}$$

Or, if we choose to work with ω then the integral can be rewritten as

$$E_{\text{ph}} = \sum_p \int \left[f_p(\omega) + \frac{1}{2}\right] D_p(\omega)\hbar\omega \, \mathrm{d}\omega \tag{5.62}$$

$D(\omega)$ is the density of states at ω for each polarization p.

Einstein's picture also assumed an ensemble of N harmonic oscillators, all with *the same frequency*, ω_E. These N oscillators had 3 degrees of freedom and were located at lattice points as in *D/P*. Each harmonic atomic oscillator could increase or decrease energy by $\hbar\omega_E$ only – thus their oscillation amplitude changed. Now Einstein had to figure out how the heat energy put into the system was distributed among the N oscillators, kind of having N boxes that could be filled with quanta of $\hbar\omega_E$ and this had to total up to some given energy. So how does the energy distribute itself among the boxes (or oscillators?)? Einstein assumed the energy partitioned according to the *Planck distribution* with some small number of the N oscillators having large amplitude vibrations and larger numbers of the N population having smaller vibrational amplitudes.

The Einstein picture itself is, of course, very different from what we have presented above with delocalized phonons, and much closer to that of *D/P*. However, the picture can be made to comport with our own if we replace the idea of *atomic oscillator* with *phonon oscillator*. All phonons have the same frequency ω_E, in this picture, and there are 3N of them that must be filled with different amplitudes. This yields the following expressions for our counting integrals above:

$$f_p(\omega) = [\exp(\hbar\omega/k_B T) - 1]^{-1} \tag{5.63}$$

$$D(\omega) = N\delta(\omega - \omega_E) \tag{5.64}$$

for any solid (a constant dispersion curve). It is straightforward to show that these assumptions lead to the expression:

$$C_v = 3Nk_B x^2 e^x/(e^x - 1)^2 \tag{5.65a}$$

$$x = \hbar\omega_E/k_B T = \theta_E/T \tag{5.65b}$$

where θ_E is the Einstein temperature.

This expression is mathematically equivalent to what Einstein found and approaches 0 functionally as e^{-x} when the temperature approaches 0. This is in accord with the third law of thermodynamics but misses the T^3 rule we observe from experiment. At high temperatures the expression reduces to the *D/P* result.

Einstein's results give us the opportunity to explore one of the many nuances of our counting integrals however. The dispersion curve we have discussed only in terms of a single atom basis. In fact, if we considered a two-atom basis, then there would be two branches of the dispersion curve that must be included in the integral. This is in fact a little easier with the density of states integral than it is with the K-space integral. However, the *Einstein model makes a rather good low temperature extrapolation for the optical modes* of the two-atom basis dispersion relation. It simply represents this branch as a straight, constant line with value ω_E.

5.3.2.3 Debye and the Spectral Analysis

The *Debye approach* includes effects from the different phonon frequencies. Unlike previous models, Debye treats an ensemble of correlated atomic motions, or vibrational modes from the very beginning. Such modes are quantum mechanical harmonic oscillators (phonons) vibrating at different frequencies with their spectrum being described by dispersion relations. A mode with frequency ω can have multiple phonons: n (number of phonons = the amplitude of vibration) and will have an energy of $E_{ph}(n) = (n + ½)\,\hbar\omega$. But this is exactly the scheme we have laid out at the beginning of the chapter – our modern view. What the Debye approach does is to introduce some interpolation schemes or approximations that make this model useful and predictive.

As with the Einstein model, we have to understand how a specific amount of energy is partitioned among the available modal states. Again, we guess that *the Planck function* will do this: $f_p(\omega) = (\exp[\hbar\omega/k_B T]-1)^{-1}$. However, this time our "boxes" to be filled with quanta of energy (available phonon states) each accept slightly different quanta sizes: $\hbar\omega$. That is, the quantum of energy placed into K_1 is different from the quantum of energy placed into K_5. Notice here we presume to index the K's as would be the case in the counting sums and integrals associated with Eq. (5.58). To be specific here, each phonon mode or state can have n phonons occupying it, but it will be $n_{K,p}$ phonons of $\hbar\omega$ energy each (Figure 5.14).

This leaves us with the rather specific task of finding $D(\omega)$ for some general problem (assuming we are working with this form of the counting integral as opposed to the integral written in K). Clearly if we have already some model worked out, then we can easily determine $D(\omega)$. But what if we didn't know precisely the $\omega(K)$ function and so $D(\omega)$ was not *analytically* at our fingertips? How might we think about the problem of finding $D(\omega)$? We can first ask what $D(\omega)$ could possibly be in the most simple cases imaginable: the evenly spaced (square), monatomic, low-dimensional, lattices.

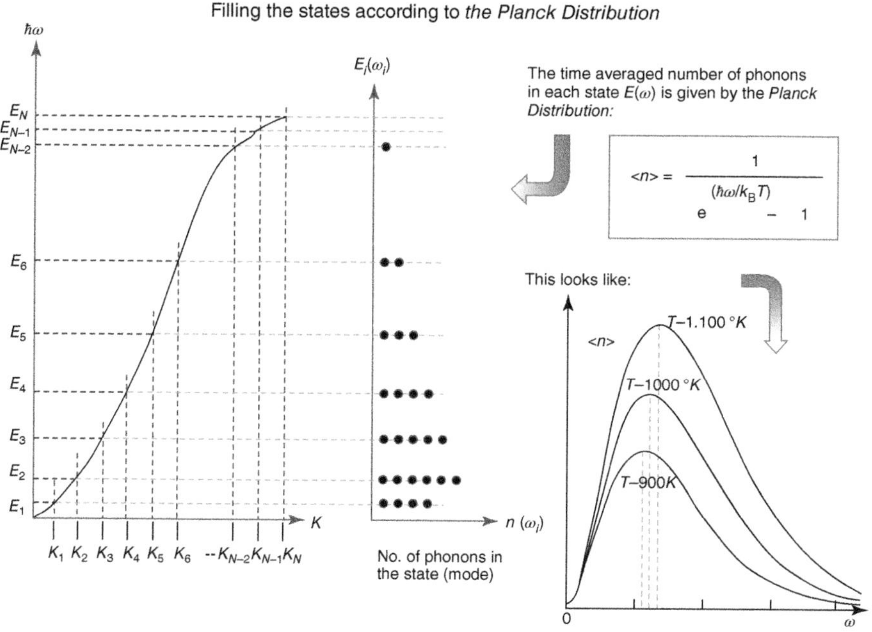

Figure 5.14 The "dispersion" gives a relationship between the E (or ω) and the K state. From this we can enumerate the ω's and determine the occupation of each energy state. Then it is only a matter of coming up with some statistical function that describes the filling of these energy states. The Planck distribution does this nicely.

In One Dimension We consider a line of evenly spaced atoms (spacing a), each with equal mass, as in our spring model above. For ease of counting let's set our boundary conditions to consider the case where our one-dimensional crystal has $N + 1$ actual atoms in it and the end atoms are fixed so that they cannot move. The crystal is a length L, which is long (so $a = N/L$), and we will index the atomic placement with the counting integer s, as seen in Figure 5.2. Now each normal mode of vibration of the string of atoms has a standing wave form that looks like

$$u_s = u(0)\exp[-i\omega_{K,p}t]\sin sKa \tag{5.66}$$

u_s is the displacement of the sth atom at position sa along the length of the crystal. Yes, it is a vibrating standing wave like that of a violin, only there is no mass between the atoms. Of course, K can take on only specific values: $K = \pi/L, 2\pi/L, 3\pi/L, 4\pi/L, \ldots, (N-1)\pi/L$; otherwise the wave would not fit nicely into the L length "box" with $u_s \sim 0$ for $s = 0$ and $s = N$. Moreover, the solution $K = N\pi/L = \pi/a = K_{max}$ and $u_s \sim \sin(s\pi)$ for this K_{max}. For this state each atom is motionless – they all sit at a node of the sine wave. In other words, K_{max} touches the edge of the first Brillouin Zone. This means there are precisely $(N-1)$ independent K values that are allowed for this chain, no more or less. This number is equal to the number of atoms that are allowed to move: add an atom, you add a mode.

Now let's look at a density of modal states. The total length of the line in K-space (that is the length of its reciprocal lattice) is $L^* = (N-1)\pi/L$. As we have argued, the total number of modes is $(N-1)$. So the total number of states per unit *length in* K is $n = L/\pi$. This is of course for $K \leq \pi/a$. Anything where $K \geq \pi/a$ gives $n = 0$, there is a maximum allowed K; K_{max} as we have said. Also, don't forget we have 3 polarizations. Thus, for this *simple one-dimensional* solid,

$$f_p(\omega) = [\exp(\hbar\omega/k_B T) - 1]^{-1} \tag{5.67}$$

$$D(K) = L/\pi \tag{5.68}$$

However, our integrals require the number of modes per *frequency range* (as a function of ω not K). Notice that in the differential length dω, you get the same total number of modes as in the corresponding interval of dK. So,

$$D(\omega)\mathrm{d}\omega = D(K)\mathrm{d}K = (L/\pi)\mathrm{d}K \tag{5.69}$$

$$D(\omega) = \left(\frac{L}{\pi}\right)\mathrm{d}K/\mathrm{d}\omega = L/\pi[1/(\mathrm{d}\omega/\mathrm{d}K)] = L/\pi[1/v_g] \tag{5.70}$$

$$\mathrm{d}\omega/\mathrm{d}K = \text{group velocity of wave } v_g \tag{5.71}$$

$$E_{ph} = \int_0^{\omega_D} 3\{[\exp(\hbar\omega/k_B T) - 1]^{-1} + 1/2\}L/\pi[\mathrm{d}\omega/v_g]\hbar\omega \tag{5.72}$$

From this it seems that we have traded needing to know $D(\omega)$ for needing to know dω/dK, both leading back to knowing the dispersion relation. Notice also that we integrate to K_{max}, which has now become ω_D or the *Debye cutoff frequency*. But we still cannot work this integral until we have a sense of v_g and as it happens this can be estimated using more classical notions of dispersion rather easily. Of course we know exactly what it is from our analysis of this model above, but we will come back to working the integral in the next section where we generalize the approximations to v_g.

A 2D Lattice The classical continuum mechanics analogue of the 2D lattice considered here is the square drumhead.

This 2D atomic net has normal modes of vibration along the K_x and K_y directions. In each of these directions, there are three polarizations: two transverse and one longitudinal. Since we seek what happens in the simplest circumstances, we treat the restoring forces in each direction and for each polarization equally. This is not realistic, but it does give some bounds on how to determine the $D(\omega)$. It is important to note that we must impose boundary conditions on this net to allow for the edges to connect together – the atoms of one edge follow the atoms of the opposite edge. In this way we can ignore modes of vibrations that might be associated with the edges of the crystal. In real crystals we cannot do this and the edge modes play some role in behavior.

The area of the Brillouin Zone, as discussed in the previous chapter, is $A_{BZ} = (2\pi/L)^2 = (2\pi/Na)^2$ In the case of the 1D lattice above, it is relatively easy to visualize why it is that the number of modes is equal to the number of atoms allowed to move, that is, that there is one allowed value of K per reciprocal cell. However, it is much harder to visualize why this would be the case for the 2D example. Nevertheless, it is also true for the 2D system: one K state per reciprocal cell in the net. This means simply that if we wanted to know roughly how many states are encircled by the $\boldsymbol{K}$ marked in red for Figure 5.15, we multiply the area of the K-circle and the areal of K-states per cell: one state per $(L/2\pi)^2$, to get a decent guess: $\pi K^2(L/2\pi)^2 = N(k < K)$, where k is a general k inside the circle. This is for each polarization type.

So we can now get to an expression for density of states in ω:

$$D(\omega) = \mathrm{d}N/\mathrm{d}\omega \tag{5.73}$$

$$\mathrm{d}N/\mathrm{d}\omega = (KL^2/2\pi)\mathrm{d}K/\mathrm{d}\omega = KL^2/2\pi v_g \tag{5.74}$$

where $K(\omega)$ andtheareaoftheatomic net havebeenused, $L \times L$.

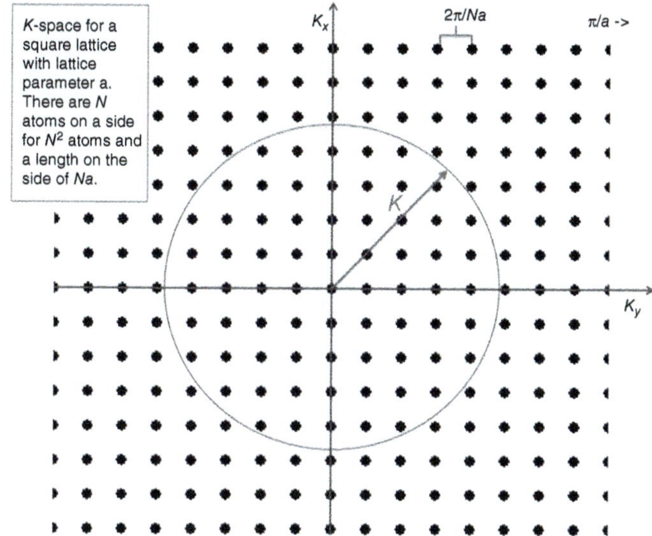

Figure 5.15 The reciprocal lattice for the 2D square lattice.

The Three-Dimensional Lattice This follows along the same example as 2D with few surprises. The circle of Figure 5.15 becomes a sphere in 3D K-space and the radius $\boldsymbol{K}$ is analogous. With this we also argue a single allowed K-state per reciprocal cell, three polarizations for each of the three directions in K-space, and we assume perfect symmetry between restoring forces since (again) we are seeking the most simple understanding of the most simple system we can imagine. So what do we get?

$$D(\omega) = \mathrm{d}N/\mathrm{d}\omega \tag{5.75}$$

$$\mathrm{d}N/\mathrm{d}\omega = (VK^2/2\pi^2)\mathrm{d}K/\mathrm{d}\omega = VK^2/2\pi^2 v_g \tag{5.76}$$

where $K(\omega)$ and V is the volume of the crystal.

5.3.3 The Debye Approximation

In our presentation we have drawn the probably artificial but useful distinction between the *Debye approach* and what we are now going to call the *Debye approximation.* Above is the Debye approach: how to count up modes and state filling using quantum statistics, dimension, symmetry, etc. Actually, our discussion of phonons and oscillatory modes at the very beginning of the chapter *is* this approach also and we now return to it. In other words, our whole intervening presentation (Section 5.3.2) has been a historically circular review of how we got to the more modern interpretations we now hold. However we are still left with the need to know $v_g = \mathrm{d}\omega/\mathrm{d}K$ as well as $K(\omega)$.

For this problem, Debye decided to assume a constant group velocity across the K-states or modes: the *Debye approximation.*

$$\omega = v_g K \tag{5.77}$$

where v_g is the constant.

This obviously makes things a bit easier: for instance, $D_{3D}(\omega) = V\omega^2/2\pi^2 {v_g}^3$. But is it a very reasonable approximation? Well, we have already seen from our classical analysis of the modes of the one-dimensional system near the center of the BZ, that is, dispersion goes linearly. This translates into saying at the lower phonon energies: or when the population of phonons is small (from the Planck distribution). In other words, we can say *at low temperatures.* But this is exactly what we need. After all, the *D/P* and *Einstein* models (with our new interpretations overlaid) are not so bad at predicting thermodynamic behavior at the higher temperatures. Importantly though, it is only a good approximation for the acoustic branch of the dispersion. So it would seem that for a diatomic basis, the *Debye* approximation and *Einstein* approximation might work well together.

Now, let's return to the C_v to see what the results of this approximation are. If there are N atoms in our crystal and N reciprocal cells, then there are N modes in the acoustic branch of our simple monoatomic lattice. Notice that this then gives the largest wavevector K. The total number of wave modes with wavenumbers less than K is given by following the above examples:

$$N = (L/2\pi)^3(4\pi K^3/3) \tag{5.78}$$

So,

$$K_{\max}^3 = 6N\pi^2/V \tag{5.79}$$

and using the Debye approximation,

$$\omega_{\max}^3 = 6N\pi^2 v^3/V \tag{5.80}$$

We usually give these cutoff wavevectors and frequencies the symbols associated with the approximation used: K_D and ω_D, as we introduced above. So these will set the limits of the counting integrals to be used in C_v. Substituting our Debye density of states and distribution functions into the counting integral (Eq. (5.58)),

$$E_{\text{ph}} = \int \{[\exp(\hbar\omega/k_B T) - 1]^{-1} + 1/2\} VK^2/2\pi^2 v_g \hbar\omega \, d\omega \tag{5.81}$$

Then substituting the Debye approximation,

$$E_{\text{ph}} = \int \{[\exp(\hbar\omega/k_B T) - 1]^{-1} + 1/2\} V\omega^2/2\pi^2 v_g^3 \hbar\omega \, d\omega \tag{5.82}$$

We will for simplicity assume again the three polarizations are equivalent – so we multiply the integral by 3 and we make a few notational changes that are common in the field:

$$x = \hbar\omega/k_B T \text{ and } x_D = \hbar\omega_D/k_B T = \theta/T \tag{5.83}$$

θ is the Debye temperature.

Our integral thus becomes

$$E_{\text{ph}} = \frac{3Vk_B^4 T^4}{2\pi^2 v_g^3 \hbar^3} \int_0^{x_D} dx\, x^3/(e^x - 1) + \text{terms with no } T \text{ dependence} \tag{5.84}$$

The terms with no T dependence come from the zero point energy and we ignore them here. Or more commonly,

$$E_{\text{ph}} = \frac{9Nk_B T}{(T/\theta)^3} \int_0^{x_D} dx\, x^3/(e^x - 1) \tag{5.85}$$

from which we get,

$$C_v = dE_{\text{ph}}/dT = \frac{9Nk_B T}{(T/\theta)^3} \int_0^{x_D} dx\, e^x x^4/(e^x - 1)^2 \tag{5.86}$$

It might also be noted that for very low temperatures $T \ll \theta$, the E_{ph} can be simplified significantly.

$$\int_0^\infty dx\, x^3/(e^x - 1) = \int_0^\infty dx\, x^3 \sum_{s=1}^{\infty} e^{-sx} = 6 \sum_{s=1}^{\infty} 1/s^4 = \pi^4/15 \tag{5.87}$$

wherein

$$E_{\text{ph}} \sim 3\pi^4 N k_B T^4/5\theta \tag{5.88}$$

yielding the T^3 dependence in the C_v as observed. Plotted in Figure 5.16 is the calculation for the Debye result.

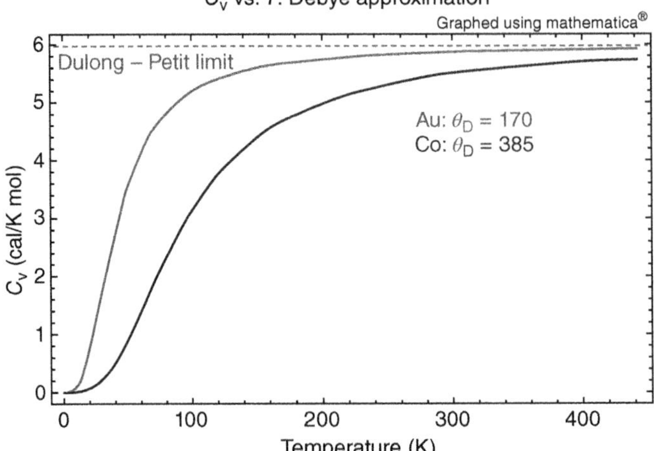

Figure 5.16 Using a standard graphical package like Mathematica®, the Debye approximation can produce fairly acceptable approximations to C_v with only the knowledge of θ. Of course these values must be looked up from experimental work to get accuracy as we have done here. First principles determinations are a little more difficult, considering we have used so many simplifications in the equations presented.

So here we have presented three different approximations to understanding the T dependencies in thermodynamic characteristics of a solid. We have focused on the C_v and E_{ph} calculations, but of course the *picture* of how phonons behave is quite general. While this view was not necessarily held as these approximations were historically introduced, we can see that there is utility in such approximations when we transpose the results to our more modern picture of vibrational modes. What should be most obvious, however, from our discussion of Debye, is that dimension plays a role even here. Notice that the temperature dependencies will be different for one-dimensional and 2D solids than for the three dimensional solid we ended with!

An Example in the Carbon Nanotube: The single-walled carbon nanotube is actually a pretty complex structure as we have already discussed. It can have a huge basis set and has geometrically associated normal modes that twist about the axis. However, in a very real sense, it is simply a one-dimensional object and at low temperatures we might be able to make a simple guess at the behavior of its C_v using the Debye approximation. We gave above the density of states of the one-dimensional system (simplified). With this we can follow the exact same algebraic route of substitutions and simplifications as we did for the three-dimensional system. We will leave this algebra to the reader for this example (you should really do this), but what is obtained is

$$C_v(1D) \sim 3.292[3Lk_B^2T]/\pi\hbar v_g \tag{5.89}$$

Yes, we would naively expect a linear dependence of the nanotube C_v as $T \ll \theta$. In fact, though there is some deviation, this is a pretty good approximation of what has been observed. However, it must be kept in mind that such

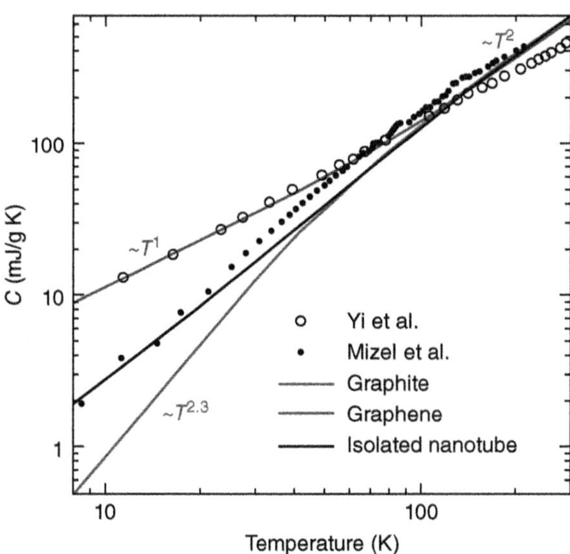

Figure 5.17 Heat capacity measurements show some slight deviation from the purely linear expectations of our most simple Debye approximation. However, it is still astonishing close considering how very complex the nanotube actually is and how hard the measurement is to make. This graphic was taken from Dresselhaus, M.S. and Eklund, P.C.

measurements on individual and isolated single-walled nanotubes are extremely delicate. Figure 5.17 compares C_v of nanotubes with graphene and graphite.

But wait! Didn't Figure 5.13[6] suggest something completely mysterious for carbon? Yes, it did. And now we can see why. We didn't mention there what form of carbon we were talking about. From this we can see that the phase of carbon will effect its θ_{carbon} and the dimension will effect its T dependence at lower temperatures.

In fact, we aren't quite done. The question of dimension can be placed on a more formulaic footing. In fact, using just the tools above it is quite easy now to show that [1]:

$$C_v(1D) \sim \frac{[3D\pi^{D/2}\, Vk_{\text{B}}^{D+1}\, T^D\,]}{[(2\pi)^D(D/2)!\hbar^D v_{\text{g}}^D]} \int_0^\infty \mathrm{d}x\, \mathrm{e}^x x^{D+1}/[\mathrm{e}^x - 1]^2 \tag{5.90}$$

where D is the dimension of the system. Here we have treated all polarizations equivalently again.

5.3.4 Generalizations of the Density of States

The *density of states* we have produced up until now has been of the most simple type, counting each mode by the length, area, or volume taken up by a single reciprocal lattice cell. In reality, there will be numerous branches of vibrational modes to consider along with differences in polarization and directions in K-space. So we really need a more robust understanding of how to do this counting of states. We do this through a generalization of the methods we have already made use of.

We need $D(\omega)$ generally, the number of states unit frequency range. We must begin with our direct tie to the solid in question, the dispersion relation $\omega(\boldsymbol{K})$.

6 Source of graph: https://pawn.physik.uni-wuerzburg.de/video/thermodynamik/g/sg15.html.

The number of allowed states between ω and $\omega + \mathrm{d}\omega$ is given by the integral:

$$D(\omega) = \left(\frac{L}{2\pi}\right)^3 \int_{K-\text{shell}} \mathrm{d}^3K \tag{5.91}$$

The integral is worked over the volume of a shell in K-space bounded on the bottom by the surface of a sphere of constant energy: $\hbar\omega$ and on the top by the surface of $\hbar(\omega + \mathrm{d}\omega)$.

If we define $\mathrm{d}K_{\perp}$ to be the perpendicular component of the K vector at the differential element that connects the two surfaces, then $\mathrm{d}^3K = \mathrm{d}S_{\omega}\,\mathrm{d}K_{\perp}$, a differential volume of the shell at the surface element we have shown in Figure 5.18. This is written in terms of K's and our counting integrals have been cast generally in terms of ω's, so we need to transform these over. Notice that the vector $\nabla_K\omega$ is also perpendicular to the constant surface of $\hbar\omega$. So we can write

$$|\nabla_K\omega|\mathrm{d}K_{\perp} = \mathrm{d}\omega \tag{5.92}$$

which is the differential change in ω between the two surfaces connected by $\mathrm{d}K_{\perp}$. The volume element then becomes

$$\mathrm{d}^3K = \mathrm{d}S_{\omega}\mathrm{d}K_{\perp} = \mathrm{d}S_{\omega}\,\mathrm{d}\omega / \mid \nabla_K\,\omega \mid = \mathrm{d}S_{\omega}\mathrm{d}\omega/v_{\mathrm{g}} \tag{5.93}$$

and we have a more general expression for the density of states:

$$D(\omega)\mathrm{d}\omega = V/(2\pi)^3 \int \mathrm{d}S_{\omega}/v_{\mathrm{g}} \tag{5.94}$$

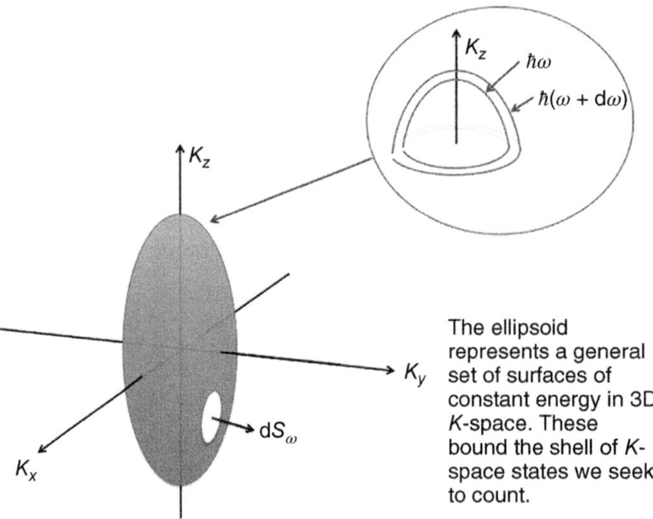

Figure 5.18 Consider a general set of surface in K-space each at a constant energy. The inner surface sits at $\hbar_{\omega}$ and the outer one at $\hbar(\omega + \mathrm{d}\omega)$. The differential element $\mathrm{d}S_{\omega}$ is attached to the inner surface. The ellipsoids shown here are meant to represent the general situation wherein different directions in K-space can result in different dispersions. Because the surfaces are differentially close, we assume they are nearly concentric, but not quite. The task here is to add up the number of allowed k-state that exist between the two surfaces. So, clearly, we must have a detailed knowledge of the solid to do this.

5.3.5 Other Thermal Properties: Thermal Transport

In our discussions we have used the models of phonons together with counting methods to guess at the functional form of thermodynamic entities with temperature. While we may have focused only on E and C_v above, the approach is surely valid for other things we might want to know. Among these are, for example, thermal transport properties. How does heat energy move along a material and what role do phonons play? This provides a nice example of how to extend these methods. Let's see how.

Heat transfer or *thermal conductivity* in carbon nanotubes was originally thought to be around 6000 W/m K or roughly that of diamond at 300 K. To see if this is a reasonable number, we should be able to use our simple models to derive something for low-dimensional materials. In classical heat transfer, the kinetic theory of gases is used as a starting point, like a gas of phonons. Such an approach leads to *Fourier's law* where heat flux, Q, is defined as $Q = k\Delta T$. The k is a thermal conductivity constant: $k = C_{v_g} l/3$. C is the heat capacity, l is the mean free path of the "gas" or the distance a particle goes without scattering, and v_g is the group velocity of the "particles." This is what we are taught in introductory physics.

However, this kinetic theory has a major drawback: it assumes that the system is in local thermodynamic equilibrium. But heat is flowing through it, so how good is such an assumption? For large crystals it isn't too bad. But for smaller crystals, like the nanotube, it can be disastrous. While it is OK to think in terms of our phonons as a gas of particles moving through the structure, it is definitely NOT OK to assume that the filling factors and distribution functions we have discussed will be the same on each end of the structure. They can change differentially as we move along the nanotube, for instance. Instead a more common way to address this is the use of the *Boltzmann transport equation.*

$$\frac{\partial f}{\partial t} + v\nabla f + \frac{F\partial f}{\partial p} = \left(\frac{\partial f}{\partial t}\right)_{\text{scattering}} \tag{5.95}$$

f is the statistical distribution function as introduced above, v is the velocity of the particles, F is any driving force on the particles, p is the particles momentum (notice we use p here instead of K to match with convention), and the last term is the rate of change to the distribution function due to any collisions and scattering in the system. The physical mechanisms of this last term will be discussed more fully later, but for now think of it in terms of real particles interacting. In this way we might assume that the net result of the scattering interactions to the distribution of state filling is to restore or relax a distribution toward some form of equilibrium. This assumption is known as linearization and it allows us to write

$$\left(\frac{\partial f}{\partial t}\right)_{\text{scattering}} \sim (f_0 - f)/\tau \tag{5.96}$$

τ is a time scale for the relaxation of the distribution function from its f condition to its f_0 equilibrium condition. We presume this is related to the mean free path for interaction as $l \sim 1/\tau$.

This allows a solution for $f(p; r, T)$, or at least an approximation. Thermal flux then looks like

$$Q = \int v f \varepsilon D(\varepsilon) \mathrm{d}\varepsilon \tag{5.97}$$

The integral uses the density of state function and integrates over all energies (ω). So as is apparent, the problem reverts back to the same form as we have been addressing above with some augmentation of how f is treated for nonequilibrium purposes and an additional term in the integrand of the counting integral.

The subject of the thermal energy contained in lattice vibrations of condensed matter systems is a rather broad one. In fact, it has become quite specialized and represents a field unto itself at this point. But the foundations we have laid in our phonon models above play the critical role in our modern understanding of material thermodynamics and kinetics. Much of what we can predict about condensed matter systems involves knowing the state filling function and the density of those states and then adding them all up across some index such as momentum K or energy $\hbar\omega$.

5.4 Anharmonic Effects

We have presented a harmonic lattice model. The atomic displacements and vibrations are like a harmonic oscillator in classical mechanics. This clearly has some consequences that we cannot live with observationally.[7]

1. There can be no thermal expansion of a solid in this model – but we know that at high temperatures thermal expansion is observed.
2. In the above section, we mention phonon interaction (scattering), which we do observe. But in a harmonic theory of phonons, the waves cannot interact.
3. Pressure and temperature dependences of the elastic constants will not occur.

To express such phenomena, lattice vibrations must include some *anharmonicity*. By *anharmonic*, we mean higher order terms in the restoring force; so rather than $F_{\text{restoring}} \sim x$, we should really think: $F_{\text{restoring}} \sim kx + k'x^2 + k''x^3 + \dots$ where the k's get small quickly. Typically the k's become more important as the displacement gets larger. From our discussion of crystal binding in the previous chapters, this really isn't so surprising.

It is clear immediately that anharmonicity introduced into a crystal, even without knowing its exact form, will have dramatic effects. Our first example is in thermodynamic systems *far from equilibrium*. Such problems are like the thermal transport discussed above, they characteristically present us with specific relaxation times associated with nonequilibrium forms of the distribution function f. Stated simply, some stimulus prepares f in a nonequilibrium thermodynamic state. The f can be thought of as a descriptor of that specific state. However, because the phonons described by f can interact with each other, they do so,

7 We haven't actually shown that these phenomena would be expected from a harmonic lattice, as the chapter is already quite long. However these references can allow you to derive it yourself.

allowing for a relaxation of f toward a thermodynamically favored state of equilibrium f_0. Thus, the strength of these anharmonic interactions has an almost defining effect on ***nonequilibrium***. But problems with systems at ***thermodynamic equilibrium*** are also effected by the introduction of anharmonicity. The nonlinearity of the restoring force will clearly lead to modifications in $\omega(K)$. This is obvious from a purely classical perspective that the equations of motion have changed.

So how is it that a medium described by nonlinear restoring forces leads to the generation of waves that *interact*? What do we mean by *interact* in this context? A simple way to "see" this might be the basic *Duffing oscillator* [2]. You might recall from the basic mechanics of oscillators: the *hard Duffing oscillator* is a spring that gets stiffer as its displacement from equilibrium becomes large. So at small displacements it is nearly harmonic, but at large displacements the strain within the spring gets large fast – as it would do presumably if it were an atomic bond. The simple form (mass and k normalized to 1) of this oscillator is

$$d^2u/dt^2 + u + u^3 = 0 \tag{5.98}$$

$$du/dt = \dot{v} \tag{5.99}$$

$$\dot{v} = -u - u^3 \tag{5.100}$$

These systems are conservative and so the orbits can be written down as an energy integral:

$$\frac{1}{2}v^2 + \frac{1}{2}u^2 + \frac{1}{4}u^4 = E \tag{5.101}$$

Each such E-orbit is a closed orbit, symmetric about the u and v axis. So the maxima of u and v can be easily related through the energy expression:

$$E = \frac{1}{2}v_m^2 + \frac{1}{2}u_m^2 + \frac{1}{4}u_m^4 \tag{5.102}$$

Thus, for large E, v_m goes like $(E)^{1/2}$, whereas u_m goes as $(E)^{1/4}$ and the ellipse becomes elongated along v as E increases. So generally, we expect the period of oscillation, the time it takes to go around the orbit, to decrease as the amplitude of oscillation increases. This means that the frequency of vibration increases with higher amplitude. This is seen in Figure 5.19.

Now imagine that we have a medium that is governed by such a restoring force and there is a single vibrational mode in this medium. This might correspond to a single phonon quasiparticle. If a second phonon were to come by, and, under the conditions of phase matching, they overlap each other, the amplitudes would add, yielding a modified period of oscillation for that moment of overlap. We might imagine, since these are moving phonons, that their kinetics might be modified a little but ultimately they would slip past each other and go on their merry way. This can happen, but what if the new frequency of the added waves happens to be another eigenstate of the system? And what if that addition also happened to have an amplitude number that was exactly the quantum of energy required for a single phonon in that eigenstate? In other words, what if the new frequency and amplitude matched up with some other point on the dispersion curve of the system? Then one might have a new phonon with a different energy. Such processes

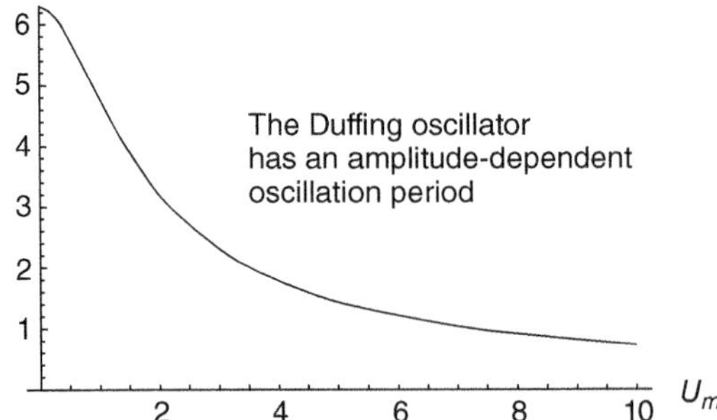

Figure 5.19 The Duffing oscillator has a nonlinear term that yields an amplitude-dependent period and frequency of oscillation. The strength of this term depends on any prefactors that may be placed in front of it, but for our normalized system one can see that the fall-off in period with amplitude is rather rapid (plotted with Mathematica®).

do happen – this is the form of *interaction* we are speaking of above – and this particular form is known as *harmonic* generation of phonons because the frequency of the final phonon is typically a harmonic of the two starting phonons. However, the effect comes about purely due to anharmonic terms in the system's restoring forces.

We note here that this is a very simple model for generating mental pictures and it has a few conditions that must be met. The first is a phase-matching condition analogous to that seen in the harmonic generation of light in nonlinear optical media. The second is of course the conservation of energy and momentum that would accompany quasiparticles traveling in a media. Naturally, there are numerous beginning and ending states that can be imagined and physicists typically use a set of diagrams to keep them all straight. The physics is contained within the vertex of the diagram and is usually handled by some rather complicated perturbation theory based on far deeper models than what we have presented here. However, it is the ending states that help us understand the results of the interactions and what we can expect in experiment. Because we have conservation rules that appear on either side of the vertex, we can more easily talk about the *rates* that specific interactions might occur under specific circumstances. Shown in Figure 5.20 are a few of the many different ways that phonons interact with each other, electrons, and photons. However, they can also interact with other quasiparticles like plasmons, and they can couple particles together such as in superconductivity where phonon exchange mediates electron pairing. All of this from the simple idea that the lattice is anharmonic!

We did mention that there are some simple guidelines that help us with the final states. To make use of these, we break our phonon interactions into two types. Let's focus on phonon–phonon (Table 5.1). We have *Umklapp* processes and *normal* processes (U and N scattering).

Generally, normal scattering acts as though the phonon quasiparticles are real and it conserves momentum as though the particles carry it away as any particle would. However, the Umklapp process recognizes that the *crystal momentum* can also play a role in momentum conservation. How does this work? Recall that for

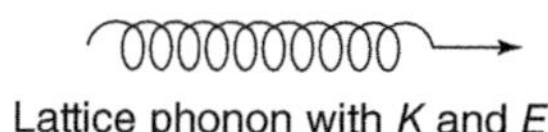

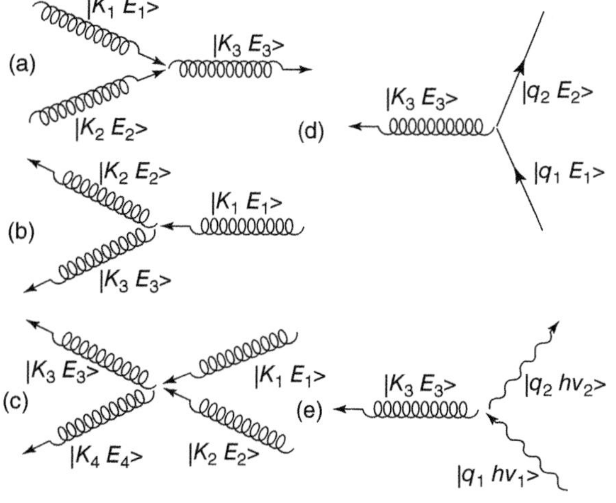

The striking similarity of these diagrams to those of particle physics is no accident. These are the Feynman diagrams the phonon interactions. The vertex represents the physics of interaction and it is usually calculated using quantum perturbation theory.

(a) 2 phonons make one
(b) One phonan decays into 2
(c) A four phonon process
(d) An electron interacts to make a phonon
(e) A photon creating a photon

Figure 5.20 A diagrammatic way of viewing phonon interactions. Hidden in these elegant descriptions are some rather beastly calculations. But the final states are governed by some simple rules. Notice we have tried to stick with the convention on the use of symbols: springs for phonons, wiggles for photons, etc. This is not always done. Straight lines are typically particles, but a squiggly line can also mean a phonon in superfluids, so there can be ambiguity. Hence, it is important to include the labels.

Table 5.1 Umklapp vs. normal interactions among phonons.

Umklapp (U)	Normal (N)
$\boldsymbol{K}_1 + \boldsymbol{K}_2 + \boldsymbol{G} = \boldsymbol{K}_3$	$\boldsymbol{K}_1 + \boldsymbol{K}_2 = \boldsymbol{K}_3$
$E_1 + E_2 = E_3$	$E_1 + E_2 = E_3$

Here we are considering only interactions of type (a) above (Figure 5.20), but a similar table can be constructed for other interactions. $\boldsymbol{G}$ is a reciprocal lattice vector. Notice that normal scattering is a special case of Umklapp wherein $\boldsymbol{G} = 0$.

any phonon with wavevector outside the first Brillouin Zone, it can be translated back into the first BZ by the addition of a reciprocal lattice vector, $\boldsymbol{G}$ without loss of information: the *Nyquist frequency argument*. For phonons scattering off of other phonons, we must think of conservation laws and remember that phonons are bosons and quasiparticles (they can be created and destroyed without much fuss). In the normal scattering processes between phonons, crystal momentum is conserved and all lattice vectors lie within the first BZ: $\boldsymbol{G} = 0$. In Umklapp scattering of phonons on phonons, $\boldsymbol{G} \neq 0$ and some or all of the wavevectors of the participants must be translated back into the first BZ by a reciprocal lattice vector. Thus, the crystal momentum is not conserved. This is shown in Figure 5.21.

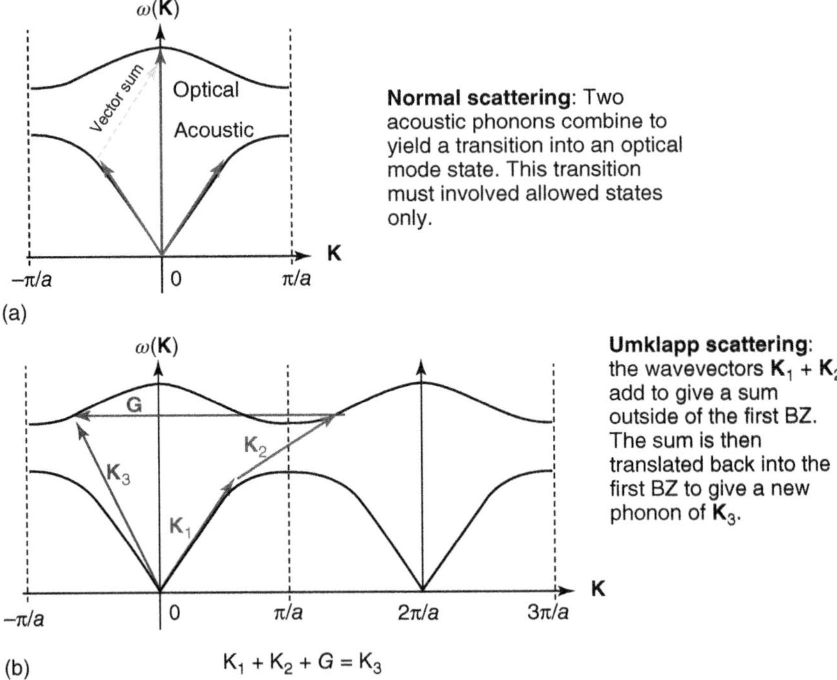

Figure 5.21 The vector sums for normal and Umklapp scattering. (a) In normal scattering the crystal momentum is conserved and the wavevectors of both incoming phonons add to make a transition to the optical branch from the acoustic branch. (b) In Umklapp scattering the resultant vector lies outside of the first BZ and must be translated back into this zone. Thus the resultant vector can be surprising. Energy is conserved in both cases. Shown here are two phonons going into one, but the reverse process is identical.

The first type of scattering (N) provides a relaxation route back to the Planck distribution for the phonon population. However, it doesn't provide for any changes to the thermal conductivity of the system. So, what is the effect that these Umklapp processes have on thermal conductivity? Imagine the following scenario (Figure 5.22). We prepare a sample such that it is heated quickly at one end of a thermally conductive bar. This prepares a nonequilibrium population of phonons at one end that begin to propagate toward the other end. Perhaps this is done with a laser pulse. Most all of the phonons are now moving in unison, parallel to each other toward the opposite end of the bar sample.

Since there is no mechanism for wavevectors to reverse direction in normal scattering, this means that only Umklapp scattering can provide truly diffused scattering and therefore it gives rise to thermal resistance in the bar. We are showing only the v_x components in Figure 5.22. So if we assume that the Umklapp processes dominate at room temperature and above and further that it is proportional to the number of phonons in the system, then we can return to the basic model presented above for an interacting gas of phonon particles and make some guesses about the *mean free path* of scattering. Since the rate of scattering follows the population, then $l \sim 1/T$ since the number of phonons goes up with increasing temperature.

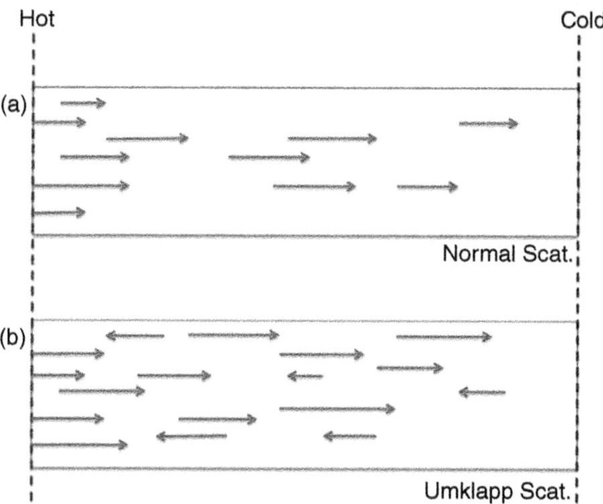

Figure 5.22 Two thermally conductive bars. Each with a distribution of phonons at the hot end prepared to travel along the length. (a) In normal scattering there is no mechanism for the phonon vectors to go any way except straight ahead. So while there may be a redistribution in phonon velocity vectors, there is no net change in direction, meaning that there is no significant resistance to phonon flow. All momentum is conserved. (b) In Umklapp scattering the phonon wavevector can be turned around, providing a *thermal resistance* to the flow of heat energy.

This, however, is not the end of the story with Umklapp processes. As it happens there are nuances and ambiguities we have glossed over. For instance, when Umklapp processes are present, both normal and Umklapp scattering add to the thermal resistance (consider how redistribution of the phonon energies might effect the cross section for Umklapp processes). Moreover, we have discussed Umklapp processes with the idea of a specific choice of Brillouin zone. However, our choice of reciprocal unit cell, though traditional, is not unique nor binding upon us; there are many such choices. In some choices a given process might be Umklapp, whereas that same process might be normal in another choice. The balance of available scattering processes, however, will stay the same, leading to the same statistical spread of outcome states. More about these subtleties and more can be found in the references [3].

Finally, what about anharmonicity and low-dimensional structures? As we have already done several times in this text, we return to our friend carbon. In fact, due to the topology of the single-walled carbon nanotube, no *acoustic only* Umklapp scattering can occur. Thus thermal resistance is derived from three phonon processes only when acoustic–optical scattering is present. In comparison, graphene strongly restricts scattering between out of plane (transverse) phonon modes and so thermal transport is strongly contributed to by the acoustic branch out of plane modes [4]. Thus, for low-dimensional structures, it seems that interactions with the boundaries or topologies of the structure can lead to a restriction (or enhancement) of specific Umklapp scattering events, thereby changing how heat energy is transferred in the structure.

5.5 Summary of Phonons

Our first comprehensive model of phonons is intended to provide some "mental picture" of how energy is partitioned in the lattice of a solid. The picture is based on the classical vibrational modes of a simple, *harmonic*, lattice. But quantum must be used to discretize these vibrations, which we called *phonons*. To understand energy partitioning in the solid, we counted the number of such vibrations with specific energies by applying the statistics of bosons and the *dispersion curve* to get a density of phonon states unique to the solid. This approach yields a reasonable guess of the internal energy of a system, which contains a large contribution from the lattice vibrations, and the heat capacity.

Moreover, the dimension of a material can play a significant role in lattice dynamics. The density of phonon states and symmetries of polarization depend sensitively on whether we are dealing with a carbon nanotube or a sheet of graphene, even though both are covalently bonded carbon. Indeed, we have applied our model for such ideal systems of one and two dimensions. As presented, these ideas of elastic traveling waves can be extended three-dimensional *anisotropic* materials systems, requiring tensor formulations. However, this is really not necessary to understand the underlying meaning of the *phonon*.

Unfortunately, this model or picture isn't quite enough. Experiments from simple thermal expansion to inelastic scattering suggest that the addition of anharmonic terms to the equations of motion is necessary to achieve *interactions* with phonons. As we have stated previously, *interactions* and *correlation* are where the real physics lies. Our physical "picture" becomes a bit more murky here, but the results are clear: phonons are well suited to a *quasiparticle* model of interaction. In experiments, we see the results of such interactions through the effects they have on final state distributions.

Exploring Concepts

1 *Intensity drop-off*: At the beginning of the chapter, we make the astonishing statement that the intensity of scattering drops-off with angle and with the DWF. This is for X-rays of course, not for neutrons. Show that the drop-off of intensity for the X-ray diffraction not associated with the DWF goes as $\sin\theta/\lambda$, where θ is the angle of the radiation detected and λ is its wavelength.

2 *Motion of the two atom basis*: In our analysis of the two-atom basis, linear chain, we make note of the motion of atoms, M_1 and M_2 near the center and the edges of the Brillouin Zone (that is, $k = 0$ and $k = \pm\pi/a$). We make the general hand-waving argument that these extreme values of k represent in phase and out of phase vibrations of the sublattices. Indeed you may have already seen such arguments in a classical mechanics class. Using the solutions presented and the equations of motion for u and v, see if you can prove that in phase and out of phase motion occurs for the k values that we say it does. You might find it easier to start with ratios of these position coordinates.

3 *The conjugated system*: Above, we solved the diatomic system by considering the case where there were two masses: M_1 and M_2. But the interatomic forces were the same: C. However in most conjugated polymer systems, the proposition is precisely the opposite. There is, in fact, an alteration in the force constants C_1 and C_2 that are associated with the double/single bond alteration (Figure EC5.1).

(a) Following through with the analysis we used on $M_1 \neq M_2$, now work out the details and then graph the dispersion curves for $C_1 \neq C_2$ but with equal masses.
(b) Determine the bandgap at the BZ edge.
(c) Using the techniques we discussed, also analyze the motion of the atoms at the special points of $k = \pm\pi/a$ and $k = 0$.
(d) Now imagine that you could grab the two ends of a long chain molecule, holding it away from any supports, and apply a spectroscopic tool to measure the phonon dispersion as you slowly pull the molecule tighter and tighter (like tightening a violin string). Describe what would happen to the dispersion curve in detail according to your model. What does this say about the speed of sound on your molecular string? Compare and contrast this to a violin or guitar string.

4 *Kohn anomaly*: The Kohn anomaly was introduced earlier in the text, but only qualitatively. Here we will consider an interesting model for this effect. Consider the force constant to vary as $Cp = A\ (\sin pk_0a)/pa$. A and k_0 are constants and the term describes the force constant between site s and $s+p$.

(a) Following our examples from before, see if you can plot out the dispersion curve for such a system (assuming equal masses in 1D).
(b) Notice that $d\omega/dk$ becomes interesting around k_0. Find an analytical expression for this and explain its meaning.

5 *Light scattering*: There are two types of visible light scattering used extensively in one-dimensional (1D) materials to characterize phonon interactions. In *Raman scattering* a laser is used to scatter light from the sample and the scattered light is examined for peaks that represent the creation or loss of energy from a phonon scattering event. So a spectrometer is used to

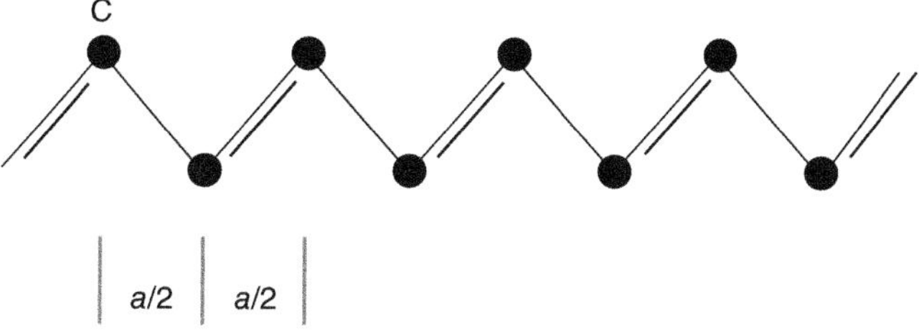

Figure EC5.1 A simple conjugated polymer model for phonon modes. Each carbon atom (C) is spaced at $a/2$ for a unit cell of a.

examine the photon energies off – but near – the primary light energy. In the visible range of light,

$$2k_0 = \frac{4\pi}{\lambda} \sim 2 \times 10^{-3}\text{Å}^{-1}$$

This is about 1/1000 of a reciprocal lattice vector. Thus Raman scattering can probe only the wavevectors near the center of the BZ. A "first-order" Raman effect is the scattering of the photon, which creates or destroys a single optical phonon of $k = 0$. A "second-order" Raman event is the creation or destruction of a pair of phonons (with equal and opposite wavevectors near the BZ center). *Stokes* shift Raman features represent the case of phonon creation and are lower in energy than the primary beam. *Anti-Stokes* shift features represent the situation in which a phonon has been destroyed and are higher in energy than the primary beam. Since we are only examining the optical wavevectors near the BZ center, Raman scattering is not sensitive to angle. *Brillouin scattering* creates or destroys an acoustic phonon. The large difference in the speed of sound and the speed of light means that the energy shift for Brillouin scattering is extremely small. Moreover, such scattering events are strongly dependent on k (see the dispersion curves above). This means that the angle of scattering must be accounted for.

OK, now go to the literature for Raman spectroscopy of carbon nanotubes, graphene, and BN.

(a) Examine the *Raman signatures of single-walled carbon nanotubes*. How does one derive the chirality from this information?

(b) Compare the *Raman signatures of sheets of graphene and BN*. Explore what implications this has for the atomic motions.

6 *C_v for 1D*: Using the dispersion curves derived for 1D systems, derive the C_v. This is usually done in terms of the Debye temperature $\vartheta_D = [\hbar v/k](6\pi^2 N/V)^{1/3}$ but you have to find the 1D equivalent. The answer will look like $C_v = F(T, \vartheta_D) \int G(\hbar\omega/k_B T)\, d(\hbar\omega/k_B T)$.

7 *Umklapp processes*: Peierls showed that for computing thermal resistivity and thermal conductivity in materials, collisions between phonons were important. However, this process was not the expected from $\boldsymbol{K}_1 + \boldsymbol{K}_2 = \boldsymbol{K}_3$ where the $\boldsymbol{K}$'s represent momentum vectors of phonons interacting; instead it was $\boldsymbol{K}_1 + \boldsymbol{K}_2 = \boldsymbol{K}_3 + \boldsymbol{G}$ where $\boldsymbol{G}$ is a reciprocal lattice vector. The idea is actually pretty simple. For two phonons interacting (in a collision, for example), $\boldsymbol{K}_1 + \boldsymbol{K}_2$ must be conserved to yield the resulting momentum vector. However, the addition can result in a $\boldsymbol{K}_3$ that lies outside the first Brillouin zone. This should be translated back into the first BZ by means of adding a reciprocal lattice vector (Figure EC5.2).

Show that the energy required for an Umklapp process to happen is of the order ½ $k_B\, \theta_D$.

8 *The square lattice in 2D*: Finally, when we presented the graphene lattice, we left out quite a few of the details. In this exercise, consider the monatomic square lattice in 2D with lattice parameter a.

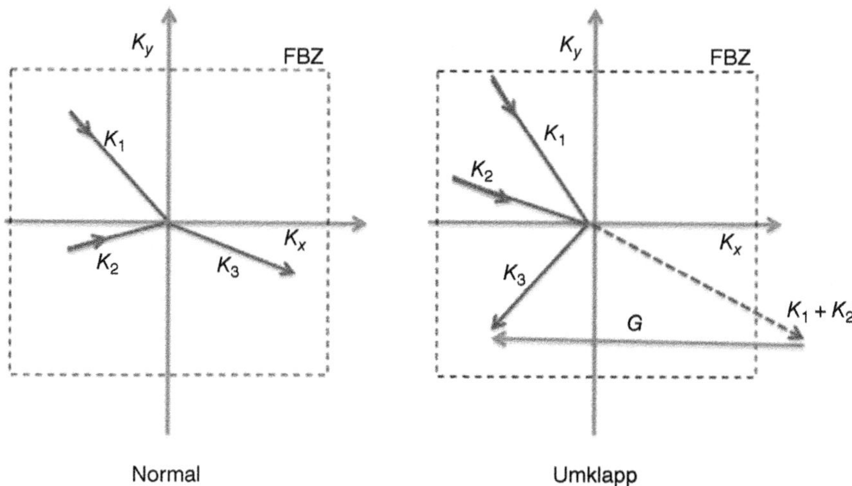

Figure EC5.2 The *k* vectors of a normal vs. an Umklapp scattering event.

(a) Break down the equations of motion into x and y components and set up the linear algebraic expressions to solve.
(b) Now solve and plot these equations in the form of a dispersion curve. Use the X, Γ, M, … presentation that we have used in analogy with graphene.
(c) Explain how this would change if the lattice were rectangular with $a \neq b$. Reason and draw what the dispersion curve might look like.

9 *Ultrasound*: Ultrasonic techniques are used to "look" inside a material and determine faults and voids. This involves placing a transducer on the surface of the material and generating high frequency sound waves that penetrate the material and reflect back to the transducer where they are detected. Is the impulse with which the wave hits the transducer carrying a real momentum or a crystal momentum? Explain your answer.

10 *Melting in 2D*: Imagine that you have a sheet of perfect, unsupported, graphene. You cool it to millikelvin and allow xenon gas to condense on its surface. Even though the xenon is only very weakly interacting with the graphene through van der Waals forces, it is enough to allow for the formation of a 2D "xenon crystal." Introduce a few phonons into the graphene, and describe in detail what should happen to the xenon layer. Base your argument on momentum transfer from the graphene layer to the 2D xenon crystal and remember that the xenon atoms are not interacting with each other very much at all. What would the xenon mass correlation function look like qualitatively? Could this be described as a phase transition? Why or why not? To approach this problem, you will have to decide on a picture of how the xenon atoms are stabilized in their positions to begin with.

References

1 Dresselhaus, M.S. and Eklund, P.C. (2000). Phonons in Carbon Nanotubes. *Adv. Phys.* 49 (6): 705.
2 Zieliński, P., Łodziana, Z., and Srokowski, T. (1999). Harmonics generation and chaos in scattering of phonons from anharmonic surfaces. *Eur. Phys. J. B* 9: 525.
3 Maznev, A.A. and Wright, O.B. (2014). Demystifying umklapp vs normal scattering in lattice thermal conductivity. *Am. J. Phys.* 82 (11): 1062.
4 Lindsay, L.R. (2010). Theory of phonon thermal transport in single-walled carbon nanotubes and graphene. Dissertation. Boston College

6

Electrons in Solids

The Story of the Traveling Electron

When collections of atoms are brought together adiabatically,[1] they interact through Coulombic forces, forming a complicated *potential energy landscape* across the condensate. The electronic state of the solid, which includes the states of all the electrons in the system (the solid's *electronic structure*), is ultimately a single eigensolution of the Schrödinger equation (SE).[2] So solids are many-body

1 Many of the atomic configurations we discuss in solid-state physics are put together this way to ensure thermodynamic stability and lowest energy configuration generally.

2 Technically, we note here that the Schrödinger equation is not consistent with special relativity. While it is adequate for getting a good approximation to electronic structure and viewing the electronic state of the solid as a sum of the individual particle states, it is the Dirac equation that is consistent with Einstein. The Dirac equation also has spin and particle–antiparticle pairs as a natural consequence. Moreover, the Dirac equation leads naturally to second quantization and field theoretic approaches to fundamental particles. But the Dirac approach can be cumbersome. With it,

Foundations of Solid State Physics: Dimensionality and Symmetry, First Edition.
Siegmar Roth and David Carroll.

systems requiring *many-body quantum mechanics* to understand them. The complexity is daunting whether we speak of a macromolecule, a nanoparticle, or a large crystal.

We say *electronic structure* and not *transport* here. There is a subtle distinction. *Electronic transport* is the motion of charge carriers from one place in the solid to another due to some applied outside motivating force. We will get to that later. *Electronic structure* refers to the equilibrium electronic energy states of charge carriers in the solid. For our immediate purposes, these are treated as though they are a bunch of *individual electron states* with no externally applied fields. What?

The full enumeration of all the electronic states in a solid begins with writing down the full Hamiltonian of the system. This includes all electrons (core and valence) and their spins, all ions along with the harmonic and anharmonic motions of the ions, etc. In other words, the full dynamical lattice with all of its parts goes into the Hamiltonian that is then used to make up the SE or Dirac equation if one wants to include relativistic effects. There are two problems with this: (i) it is *really* hard to do, and we are likely to be unsuccessful, and (ii) even if we were successful, the answers would be hard to understand and visualize (obscured by mathematical complexity so that we can't easily see the macro-analogues in our heads).

Thus, we are left to solve the problem by factoring it into smaller problems first. *A priori* we have no reason to suggest that this many-body problem can be separated into components that might be addressed in isolation. But we have done this already with phonons. The electrons there were treated as "classical springs," and the ions moved according to a classical analogue, independent of anything else the electrons might be doing. Sure, we did address this "anything else" when we determined that we needed anharmonic terms to match our observations. This had to be artificially added into our problem because our approximations to the subsystem were too simple. However, the *factorability* of the problem seems to be acceptable. The reason this approach works is not altogether easy to understand, but it also seems to work for electrons.

Evolving Pictures

Taking the perspective above, perhaps the sharing of electrons can be understood as the freedom of *some* electrons to roam the volume of the solid. This is what we might call a *picture.* Of course, *all* electrons are part of a global Hamiltonian. That includes the ions, other electrons, and the collective interactions that they all feel. But we are going to ignore this for a moment and see the mobile electrons as a small set of more-or-less disinterested sojourners that move through the solid

---

it is quite difficult to see basic electronic behavior emerge from the atomic nature of the individual components of the solid. So, for an introductory text, it is usually good practice to stick to the Schrödinger equation and allow it to build our intuitions about the electronic structure in solids. From this starting point more detailed discussions regarding relativistic effects can be had.

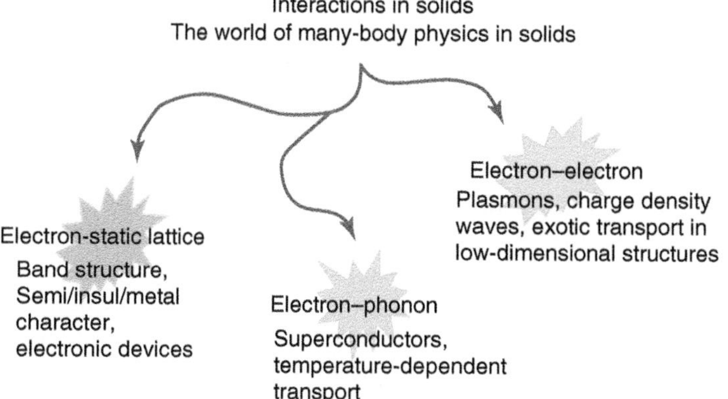

Figure 6.1 Interactions between charge carriers, moving ionic cores, and collective oscillations of polarization, spin, etc. set the stage for the physics of a solid-state system.

and interact weakly with the other things there. How free is the electron to roam? What shall it encounter on its journey? Can we describe the electronic states available to the "quasi-free" electron? These are the questions we must answer if this picture is to be useful. Let's begin with two *expectations* of this picture:

1. We might expect that only the outer shell electrons should be able to roam freely. Inner shell and core electrons are likely to stay put because they are tightly bound to the lattice site. The many-body wavefunction looks like atomic orbitals for these near-core electrons. The outer shell electrons are somehow *screened* from this strong Coulombic interaction with the core. But we remain vague about this for now.
2. Other beasties live in the solid's volume. They include phonons, other electrons, "spin objects," etc., and they will *interact* with free electrons. Such *interactions* can give rise to complex electronic and transport properties. However, to first order, we can think of such interactions as simple scattering events. The electron ball of charge hits another electron ball of charge, or a phonon quasiparticle, or a polariton, or whatever. This is a particle–particle scattering *paradigm* [1]. Such *interactions* are where the interesting physics is to be found (Figure 6.1).

Philosophically, to "evolve" our simple picture above, we will need to artificially add interactions one at a time. This makes the physics easier to picture in our minds. We start simple: let's say with the interaction between the electron and the lattice when the lattice is doing absolutely nothing and in the limit where local lattice potentials are weak. From there we "turn up" the strength of the local atomic potentials, we "add in" electron–electron repulsion and allow the lattice to have some dynamics of its own, etc. We will call this the *adiabatic approach,* and it really consists of a series of refinements to the "individual roaming electrons" picture of the solid. It derives from a basic tendency of physicists to be *reductionist* in the way they see the universe, though reductionism doesn't always serve us well in solid-state systems.

Superconductors

An important example of adding interactions into a picture of a solid is that of lattice vibrations and mobile electrons in *BCS superconductivity* (BCS is a specific type of superconductivity). Typically, in a solid, the solid-state energy states of electrons are dominated by two well-known principles: the *Coulomb force* and the *Pauli exclusion principle*. Oddly enough, however, electrons can *pair up* using lattice distortions caused by the electrons themselves (an interaction). This, in turn, can lead to correlation of behavior between pairs of electrons in the lattice. These pairs are called *Cooper pairs*, and they are allowed to ignore both Coulomb and Pauli! This is because the {electrons + phonon} is its own quasiparticle with its own density of states (DOS) and its own statistics (*bosonic*). What is more interesting is that the phonons associated with this pairing are typically no longer subject to Umklapp-type processes, and so the electrons are "herded" through the lattice with no electrical resistance at all: *superconducting*. A complicated and spectacular phenomenon, superconductivity, arises from a relatively simple interaction. And its explanation can be had in terms of artificially adding interaction terms to our picture that was originally borne of *separability*. Of course! This is not a fully tautological approach to BCS superconductivity, but it does demonstrate the usefulness of pictures in gaining physical intuition about the solid state.

6.1 Properties of Electrons: A Review

Since our approach will treat the electrons in a solid as individual and isolated quantum systems by themselves, it is important to remember or review a few properties of electrons that we know from quantum mechanics.

6.1.1 Electrons Travel as Waves

We normally think of electrons as particles, but according to de Broglie they also have a "wave nature." This wave nature of the electron is particularly important when it is in a crystal lattice since that will define specific, repeating interaction lengths. We also think of phonons as waves, and they too have a (quasi) particle nature, as we have seen. We call the principle *wave–particle duality*, but in a solid it has a nuance. The phonon particle seems to "share" momentum with reciprocal lattice vectors. After all, it is made up of lattice atoms moving about. But what about the electron in the lattice? Does it have a "pure" momentum or one that can be translated by a $\boldsymbol{R}$ lattice vector? It is actually more like the phonon than you might think.

6.1.1.1 Delocalization

In the crystal some free electrons are strongly associated with an atomic site. They are said to be *localized*. Some do not belong to any particular atom in the crystal. These are said to be *delocalized*. The number of localized and delocalized electrons depends on the specifics of the material, but in the *delocalized* case, the electron wavefunction is more like that of a free particle in space, spreading out over the crystal. The delocalized electrons can originate (i) from within the crystal and are associated with valence electrons of some forgotten atom, or (ii) they can be injected from the outside world such as in an electronic device.

The *conduction electrons* of metals or of doped or photoexcited semiconductors are a set of freely roaming electrons. When they are relatively delocalized, they have wavefunctions that can be described most closely as the following:

1. *Plane waves*: In a *simple metal* near room temperature, one assumes a preponderance of free and mobile electrons. After all, really good metals can respond almost perfectly to an applied electric field. So we might imagine many, many electrons that can move readily. But even in a really good metal, there is some *resistance* to electron flow and a *skin depth* of incident radiation. The origins of these phenomena in a metal are well explained by treating the electrons as a *non-self-interacting* gas, moving randomly in a box defined by the crystal boundaries, and having a few phonon particles to "bump into." Such a description goes by several names: *the Fermi gas, the Drude metal, the Sommerfeld model* – all of these are slight variations on a theme. The features of that theme are that the arrangement of the nuclei makes no difference to the electrons. They are free, plane wave particles. When they interact, they enter each collision as a plane wave and emerge as a plane wave.
2. *Modulated waves*: If a quantum mechanical particle moves through a region of space with some repeating field, such as an electron's electrostatic interaction with a lattice, the overall effect of the interaction would be to modulate the electronic wavefunction. The modified wavefunction might now have wavelengths that correspond to the lattice spacings. Indeed, this is not surprising when we remember the quantum problem of a plane wave approaching a potential barrier at an energy higher than that of the barrier. We will later call such waves as *Bloch waves*, but they are really just plane waves "passing over" a series of such low energy potentials. Their form must be written as a Fourier expansion of sines and cosines (Figure 6.2).

If the moving electron is high enough in energy and the repeating potential of the lattice is low enough in strength, then the potential can be ignored. In this case the momentum carried by the electrons is given by $\hbar \boldsymbol{k}$ where $\boldsymbol{k}$ is simply the wavevector of the free-electron wavefunction. However, modulated waves, or Bloch waves, are made up of numerous $\boldsymbol{k}_n$'s, and that combination has been somehow influenced by the spacing of the potential wells of the crystal lattice. Thus, these *interacting* electrons will carry a momentum $\hbar \boldsymbol{K}$, where $\boldsymbol{K}$ is derived from the reciprocal lattice, and we refer to this as *crystal momentum*, in a manner analogous to phonons.

Naturally, both types of solid-state waves will have quantized energy values, as waves confined to a box should. And there is a dispersion relation connecting the energy to the momentum. As we might expect, the shape of the dispersion relation is different from that of phonons (their dispersion was near linear at the

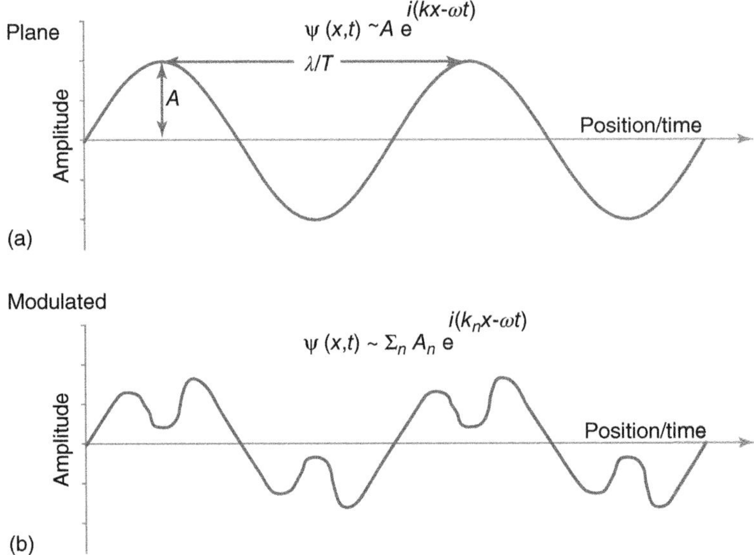

Figure 6.2 (a) A plane wave compared with (b) a modulated wave.

origin of the BZ, resembling a classical wave with constant velocity: $v = \upsilon k/2\pi$). In contrast the electron dispersion relation is roughly quadratic $E = \hbar^2k^2/2m$ near the BZ center – as expected for a classical particle with the kinetic energy of $E = p^2/2m$. For the modulated waves, m is usually not identical with the free-electron mass. It is called the *effective mass* and is designated by m^*. In general, the effective mass is different for $\boldsymbol{K}$ pointing in different directions. To keep the formula $E = \hbar^2k^2/2m^*$, the effective mass is even allowed to *change with* $\boldsymbol{K}$, and this allows us to treat the electrons in a crystal *almost* as if they were free particles. The *effective mass* m^* transforms as a tensor.

6.1.1.2 Localization

If, on the other hand, the electrons of the solid are more localized than delocalized, then they can be better described as "hopping" from place to place. Yes, they will still have some wave properties, and we will address this a little later. But for the first models we treat, we will consider electrons with a relatively high degree of delocalization. This is consistent with our initial picture of free electrons.

6.1.2 Electrons Arrive as Particles: Statistics

Electrons may travel as waves but they arrive as particles. "Arrive" here is a metaphor for being "observed." This fact is no different from the case of phonons. But there is a fundamental difference between electrons and phonons: when they "arrive" **they obey different statistics**! Here, we use the word "statistics" to describe how the particles fill available energy states in the solid. Phonons are bosons, and for bosons there is a natural dictate of spin properties in particles and quasiparticles that says the Pauli exclusion principle *does not* apply. This means it is possible to have many phonons in the same quantum state. And, when we counted up phonons in Chapter 5, we allowed more than one phonon with the same $\boldsymbol{K}$ vector in the same crystal.

Electrons, however, are *fermions*, and they *must* obey Fermi statistics, observing the Pauli exclusion principle. A given quantum state of the crystal system can only be occupied by one such fermion. So for each K value, there are only two electrons possible in a crystal, one with spin-up and one with spin-down. This means the probability of state occupation function, $f_s(\omega)$, we introduced before for phonons will not work for fermions.

This difference is expressed in dramatic fashion as a solid is brought close to absolute zero. All the phonons of the crystal are allowed to transition to the lowest energy state at $K = 0$, and the atoms just vibrate in their zero-point motion (quantum statistical noise). However, at this very low temperature, the electron transition is to the lowest energy allowed by their statistics, meaning to the lowest *unoccupied* state they can find. This means they will fill up the allowed K states of the solid to very high energies, stacking up until you run out of electrons. So unlike the phonons, a lot of energy is still packed in this "sea" of electrons (sometimes called a *Fermi sea*). The energy of the highest occupied state is called the *Fermi energy*, E_F, or the *Fermi level*. The corresponding wavevector is the *Fermi wavevector* $\boldsymbol{k}_F$, and its reciprocal $2\pi/\boldsymbol{k}_F$ is the *Fermi wavelength* λ_F.

In most metals the Fermi energy has very high, several electron volts, much higher than the thermal energy $k_B T$ (k_B = Boltzmann constant, T = temperature), which is only about 30 meV at room temperature. Below the Fermi level, states are all filled with electrons. This means electrons going in one crystal direction are mirrored by states going in the opposite crystal direction, thereby usually having no net influence on measured electronic properties. Only the electrons close to the Fermi energy can be thermally excited into an unoccupied K state above the Fermi level. The same is true for acceleration of electrons by an applied electric field, or absorption of a photon, both of which involve a change of the K state. Consequently, those very few electrons at the Fermi energy determine most of the properties of a Fermi sea of electrons. This is why the Fermi energy, Fermi wavevector, and Fermi wavelength are so important.

What, then, is the *distribution function* for state filling in electron statistics at a finite temperature? In other words, when the electrons are excited above the Fermi level by thermal energy absorbed into the system, how would the electrons distribute themselves among the upper energy states available to them? This is given by the Fermi–Dirac expression

$$f(\varepsilon) = 1/[\mathrm{e}^{(\varepsilon-\mu)/k_B T} + 1] \tag{6.1}$$

$f(\varepsilon)$ is the probability (a number between 0 and 1) of finding the ε energy state occupied. μ is the chemical potential we learned about from electrochemistry (the energy for adding one electron to the system). It is roughly equivalent to the Fermi energy, but technically the Fermi energy changes with temperature since $\mu = F_{n+1} - F_n$ where F is the Helmholtz energy ($U - TS$), where U is the internal energy and S is the entropy.

So, when we add up occupied states to get physical quantities, like total energies (as we did for phonons), we must multiply this statistical probability with the *density of the states* that are to be occupied, $D(\varepsilon)$. The model chosen for the system will determine the DOS, and, as we will see, it makes a big difference! So we will spend a lot of time from system to system determining this quantity.

6.1.3 The Fermi Surface

In three dimensions, the dispersion relation for the electrons is different in different directions, so the set of electronic states to be filled are spaced differently, and in each direction there is a well-defined Fermi vector. The tips of these vectors define a surface in three-dimensional (3D) reciprocal space. This surface is called the *Fermi surface*, and it reflects the symmetry of the crystal being considered. Because of $E_F = \hbar^2 {k_F}^2/2m^*$, the symmetry of the Fermi surface is related to the symmetry of the *effective mass*. For an isotropic solid it will be spherical: the *isotropic approximation*. At one time the construction of 3D Fermi surfaces for metals and alloys was a central theme in solid-state physics because it helped scientists to understand anisotropies in low temperature conductivity.

To construct the Fermi surface of a one-dimensional (1D) metal, let's use the *anisotropy limit* introduced in Chapter 5. A small effective mass means that it is easy to move the electrons. If in the case of a bundle of 1D chains the electrons move easily along the chains but much less easily perpendicular to the chains, the effective mass tensor and the Fermi surface will look like a lens (flat ellipsoid) with a short axis in chain direction and two long axes perpendicular to the chains. In the extreme case the Fermi surface will distort into two planes, as indicated in Figure 6.3.

In Chapter 1 "open" Fermi surfaces were mentioned. The Fermi surface in Figure 6.3 is open, because it consists of two parallel planes. Even if the Fermi surface is slightly curved, it can still remain "open." This effect can be understood with the help of the Brillouin zones. If the Fermi surface has a very flat lens shape, its diameter might be larger than the dimension of the first Brillouin zone. The Fermi surface would close only in the second or even a higher Brillouin zone. However, then it would be allowed to subtract the reciprocal lattice

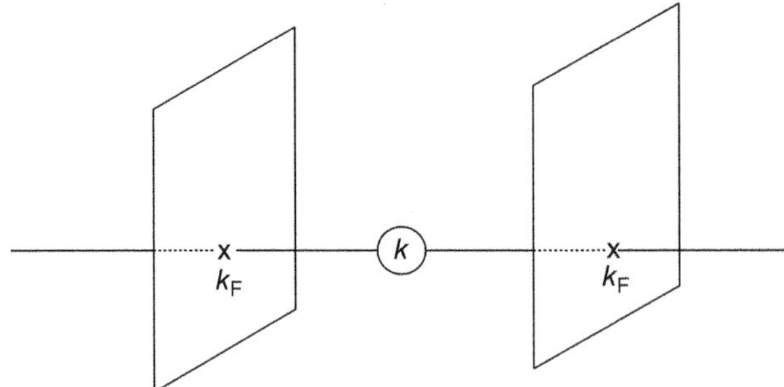

Figure 6.3 An idealized one-dimensional Fermi surface.

vectors from $\boldsymbol{k}_\mathrm{F}$, thus yielding an open Fermi surface, which is a criterion for one dimensionality.

6.2 On to the Models

In metals, the electrons are plane waves as they encounter constant potentials or modulated waves when experiencing locally varying potentials. As particles they obey the Pauli exclusion. We can now introduce models of a solid's electronic structure.

6.2.1 The Free-Electron Model

There are aspects of a metal's response to applied fields that suggest they contain highly mobile electrons in large numbers. Some metals are better than others; for instance, Au, Ag, and Cu are quite good conductors. We think of these "good conductors" as a collection of free electrons that refuse to interact with each other. But how does this happen in real materials? To a lowest-order approximation, it is because the electrons were originally stripped off the lattice atoms, leaving behind an ionic background of the lattice that effectively screens the freely moving electrons from each other more than we might expect. Actually this isn't the best way of thinking of it, but it does provide us a mental picture to get started.

This odd little picture is known as the *free-electron model*. In it the electrons are thrown into a potential "box" defined by the background ions and screening electrons around the ion cores. The electrons behave like a noninteracting gas (thus the term *electron gas* is sometimes used in analogy with *ideal gas*). A central aspect of this model is that the electrons are "noninteracting," meaning they don't even "see" each other's Coulomb repulsion.[3] We will still require them to obey Fermi–Dirac statistics though, so we can also use the term *Fermi gas* for the model.[4]

The confinement of the electron wavefunction, and thus the dimensionality of the system, is defined simply by the dimensions of this box. It is easiest to imagine the box to have an infinite potential at its boundary as in Figure 6.4. Thus the wavefunction goes to zero here. The solutions of the electron waves in this *1D box* look like

$$\psi_n = A\sin(2\pi/\lambda_n x) \text{ where } \lambda_n = 2L/n \tag{6.2}$$

3 This is at odds with our picture from classical electrodynamics. Typically, we *do* think of the free electrons as being influenced by fields internal to the solid – thereby allowing the electrons to respond by moving to the surface of the metal to maintain the condition $E = 0$ inside the metal. Therefore, our picture is a little more complicated. We must think of the electrostatic interactions of all the solid's electrons and ions as balancing, allowing for the electrons to move freely internally until an external field is applied.

4 This is not the only place in physics where the universe seems more concerned with her statistics than her forces.

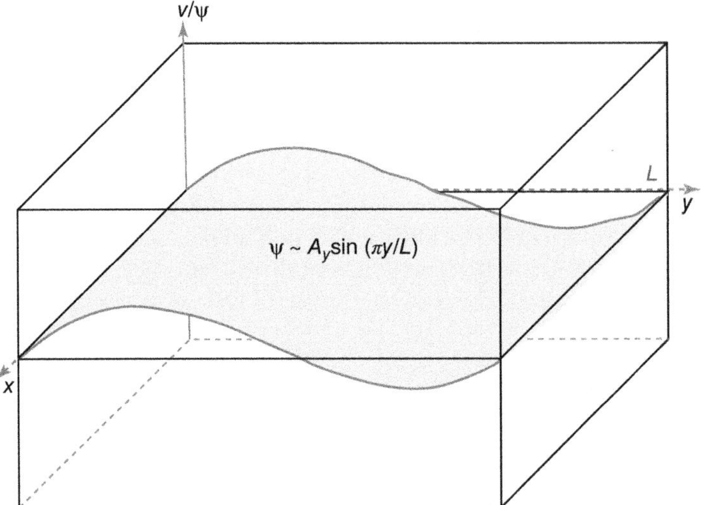

Figure 6.4 An electron in a box. The dimensionality of the problem is defined by the dimensions of the box.

The walls of the box are at $x = 0$ and $x = L$. Plugging this into the 1D SE, we get the energy states of the waves allowed to exist within this box:

$$\varepsilon_n = (\hbar^2/2m)(n\pi/L)^2 \tag{6.3}$$

The energy levels are quantized and are often called "orbitals" to make the chemists happy. If we start tossing more electrons into this box, they will fill up these states one by one according to the Fermi–Dirac statistics (obeying the Pauli exclusion). Each level gets two electrons, a spin-up and spin-down version: $m_s = \pm 1/2$. Of course there may be some symmetry breaking that gives one spin state a slightly higher energy than the other (breaking the degeneracy), but we will ignore this for the moment. The maximum energy level for a given number of electrons is this *Fermi level* we spoke of. We can write n (the state index) in terms of the number of electrons, N, if we want: $n_{\text{Fermi}} = N/2$ at $T = 0\,\text{K}$. Of course we could not do this with bosonic particles like phonons; they can take on any energy they want in any numbers. But with electrons (fermions) knowing how many is knowing what energy.

Since we have begun by discussing the model for the 1D case, let's see how many electronic states we have in a differential element of energy, ε, for the model (remember $\hbar = h/2\pi$):

$$\varepsilon = n^2h^2/8mL^2 \tag{6.4}$$

$$n = \sqrt{8\varepsilon mL^2/h^2} \tag{6.5}$$

$$\mathrm{d}n/\mathrm{d}\varepsilon \sim 1/2\sqrt{8mL^2/\varepsilon h^2} \tag{6.6}$$

$$D(\varepsilon) = \frac{\frac{\mathrm{d}n}{\mathrm{d}\varepsilon}}{\text{unit}}\text{length} \sim \sqrt{2m/\varepsilon h^2} \tag{6.7}$$

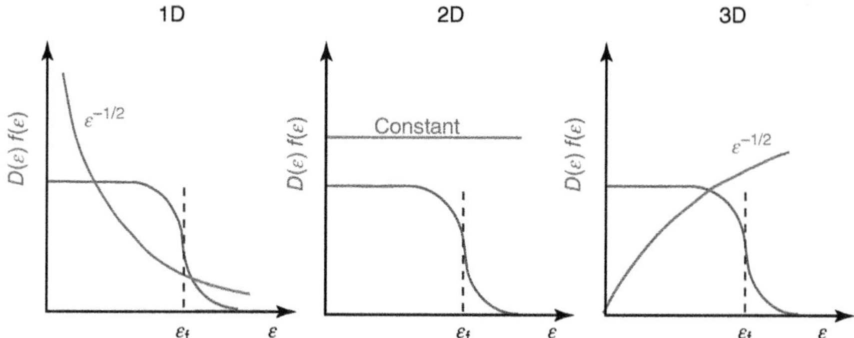

Figure 6.5 The density of electronic states in a 1D, 2D, and 3D free-electron metals ($\varepsilon^{-1/2}$, constant, $\varepsilon^{1/2}$) and the standard Fermi distribution.

Figure 6.5 compares the density of electronic states for a linear chain to other free-electron metals. To get this we have followed the simple road map.

DOS of free-electron systems

Let n be the desired dimension and x_i = x, y and z

Volume of a single state in k space, V_s: $V_s(k) = \prod_i [\pi/x_i] = (\pi/L)^n$

Volume of some set of states in k space, V: $V_n(k) = \dfrac{\pi^{n/2}\,k^n}{\Gamma\,(1 + n/2)}$

Number of filled states in this volume, N: $N = 2\,V_s/V\,(1/2)^n$ (the factors of ½ are for double counting, and the 2 is for spin)

Substitute: $k = (2m\varepsilon/\hbar^2)^{1/2}$

Then take $dN/d\varepsilon$. $D(\varepsilon) = 1/L^D\,(dN/d\varepsilon)$ the density of states per unit volume

What is the internal energy of this collection of electrons in one dimension? What thermal properties might we expect? We count up all the filled electronic states and add the energies together:

$$U = 2\sum f(k_x)\varepsilon(k_x) = 2L\int (\mathrm{d}k_x/2\pi)f(k_x)\varepsilon(k_x) \tag{6.8}$$

U is the internal energy, and the sum is over all k states ($\pm$). As the DOS considered is increased, the sum becomes an integral. Did you notice that we had to put in the density of k states ($L/2\pi$) and the "2" is for spin? This is in terms of k states, and now we convert the integral to the energy variable

$$u = U/L = \int \mathrm{d}\varepsilon\ \varepsilon\ D(\varepsilon)f(\varepsilon - \varepsilon_f) \tag{6.9}$$

u is the energy density and the integral goes from 0 to ∞. As is customary, we have let the chemical potential be approximately equal to the Fermi energy

$$f(\varepsilon - \varepsilon_f) = 1/[\mathrm{e}^{(\varepsilon-\varepsilon_f)/k_BT} + 1] \tag{6.10}$$

At $T = 0\,\text{K}$ this becomes

$$u = 1/3\, n\varepsilon_\text{f} \tag{6.11}$$

an expression for the internal energy associated with the electron distribution of the solid. The C_v at finite temperature is left to the reader as an exercise. Extending this concept to three dimensions is straightforward.

But what if the electrons were just a normal gas? From *classical statistical mechanics*, where Fermi statistics are not used, the electron gas in three dimensions should have had a heat capacity $C_\text{el} \sim 3/2Nk_\text{B}$ where $U \sim 3/2Nk_\text{B}T$ by the equipartition theorem. However, what has been observed experimentally is around 0.01 of this value. So, we see that the electrons contribute to conductivity because they are mobile but are unable to contribute significantly to the heat capacity.

Now we can understand that the reason the classical value fails lies in the Pauli principle and the Fermi statistics. Specifically, when we heat the solid from very low temperatures to higher temperatures, that heat energy is absorbed NOT by all the electrons of the Fermi sea, for where would they go in energy? There is no state to jump to. Instead, only electrons near the Fermi level have states into which they may jump. So only the electrons in states within $k_\text{B}T$ of the Fermi energy absorb energy and make a transition, thereby spreading out the Fermi distribution. This is far, far fewer electrons than the N it takes to make up the solid. Consequently, the energy stored in the distribution of "stimulated" electrons is far less than if all N electrons could have participated.

6.2.2 Nearly Free Electrons, Energy Bands, Energy Gaps, Density of States

The "free-electron model" treats the electrons in a crystal as semiclassical particles with some statistics and confinement added on top. A *completely* free-electron model fails to explain specific heat (and magnetic susceptibility) on its own. So the Pauli exclusion principle had to be taken into account. Thus, we used Fermi statistics to properly count the electrons, yielding the above "free-electron Fermi gas." However, if we treated all materials like this, they would all be metals! To explain why silicon is a semiconductor or sodium is a metal, we have to go one step further and allow for some influence of the crystal lattice on the Fermi gas. This leads to a *nearly free-electron gas model* of Bloch, which differs from the free Fermi gas in two important ways:

1. The free-electron mass m is replaced by the effective mass m^* as we mentioned in the introduction to the chapter. The effective mass can be different for electron motions in different crystallographic directions (see Section 6.1.3 discussion on the Fermi surface for a 1D solid). Consequently, m^* is a tensor, not a scalar (which, when concentrating on one dimension, does not matter, of course). Moreover, m^* will depend on the magnitude of the wavevector $\boldsymbol{k}$, and although the dispersion relation is given as $E = \hbar^2k^2/2m^*$, the functional dependence of E on $\boldsymbol{k}$ will deviate from the parabolic shape. The dispersion relation can even have an inflection point at which m^* will change sign and become negative.

2. The dispersion relation of a nearly free-electron gas is not continuous at the edge of a Brillouin zone but will have jumps. Because of these discontinuities the allowed energy values are confined to *energy bands* separated by *forbidden gaps*.
   Now, let's examine what this means and how it comes about. As with our previous discussions, we begin with the application to one dimension. We should mention here that there are many excellent texts that focus solely on the development of band structure calculations for real materials. But our purpose here is to explore the underlying foundations of lattice–carrier interactions. To do this we begin with a simple model for a periodic potential along a 1D chain. This can be anything – a square well (the Kronig–Penney model), a set of delta functions (the Dirac comb), or a set of atomic potentials. We insist only that the interaction be periodic and relatively *weak* so that the electrons interacting are still basically free to roam.

6.2.2.1 Bloch's Theorem

The model to come next will introduce a repeating potential that somehow represents the crystal lattice. We will think of this as an infinite crystal in the sense that there are many lattice sites and the ends of the lattice are matched boundaries (Born–von Karman boundary conditions). Thus, the differential equation we must set about to solve looks like this:

$$\varepsilon\psi(x) = \left(\frac{1}{2m}p^2 + U(x)\right)\psi(x) = \left(\frac{1}{2m}p^2 + \sum_G U_G e^{iGx}\right)\psi(x) \tag{6.12}$$

Before trying to solve this SE for our simple potentials, we must introduce the *theorem of Bloch*. Bloch proved that SE solutions to any periodic potential must have the form

$$\psi_k(x) = e^{ikx}u_k(x) \tag{6.13}$$

The exponential term is a simple plane wave, and the u term is a modulation of that wave. We predicted above that this would happen. $u_k(x)$ is periodic and must have the periodicity of the potential. Thus in one dimension,

$$u_k(x) = u_k(x + \boldsymbol{T}) \tag{6.14}$$

$\boldsymbol{T}$ is a translation vector of the lattice.

The Bloch theorem is the mathematical statement that the wavefunction will become *modulated*, as we stated above. But it also gives us some specific details as to this modulation. The theorem is among the most consequential in solid-state physics.

6.2.2.2 The Nearly Free 1D Model

This is the Kronig–Penney model. The solution to the SE for a periodic potential such as the one shown in Figure 6.6 is given by plane waves in and out of the wells:

$0 \leq x \leq a \ (U = 0)$

$$\psi = Ae^{ikx} + Be^{ikx} \tag{6.15}$$

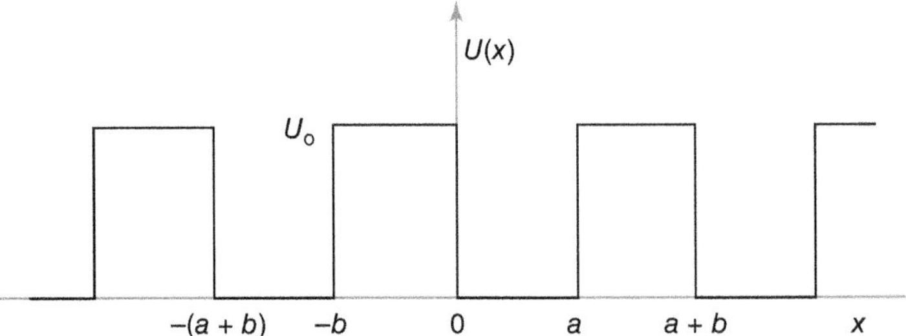

Figure 6.6 The Kronig–Penney potential of repeating square wells. Each well has a width of *b* and is positioned at sites *a* apart along the chain. The translation vector ***T*** is +/− sa where s is an integer.

where

$$\varepsilon = \hbar^2 k^2/2m \tag{6.16}$$

$-b \leq x \leq 0$ $(U = U_0)$

$$\psi = Ce^{iQx} + De^{iQx} \tag{6.17}$$

where

$$U_0 - \varepsilon = \hbar^2 Q^2/2m \tag{6.18}$$

The Bloch theorem states that

$$\psi(a < x < a + b) = \psi(-b < x < 0)e^{ik(a+b)} \tag{6.19}$$

We also introduce conditions at the boundary of the line such that $\psi(a) = \psi(0)$ and $d\psi(a)/dx = d\psi(0)/dx$. This leads to a set of simultaneous equations – four equations for the four unknowns. Setting the determinant of the matrix of these equations to zero, we can find the coefficients

$$\left[\frac{Q^2 - K^2}{2QK}\right] \sin(Ka)\sinh(Qb) + \cos(Ka)\cosh(Qb) = \cos[k(a + b)] \tag{6.20}$$

To get the Dirac comb, as mentioned in the beginning, we take the limit as $b \to 0$ and $U_0 \to \infty$. This gives $Qb \to 0$, $Q^2b \to$ constant, $K^2b \to 0$, $\sin(Kb) \to Kb$, $\cos(Qb) \to 1$. So the above becomes

$$(P/Ka)\sin(Ka) + \cos(Ka) = \cos(ka) \tag{6.21}$$

$P = Q^2ba/2 = a$ constant, $Q \gg K$, and $Qb \ll 1$, the familiar result for this model.

This relationship shows how ε (through K; $\varepsilon = \hbar^2k^2/2m$) is related to the electron's wavevector $\boldsymbol{k}$. Notice that the cosine term on the right can take on terms from −1 to +1 only. This limits the values that k (and thus ε) may have. So for some range of values of the energies, there exists no solution: there is an *energy gap*. These are the so-called *bandgaps* of the system, and they can be shown to exist for any shape of potential, not just the delta function or square potential here. In fact, we can also see that such gaps occur at the Brillouin zone boundaries: here $\cos(ka)$ takes on its maximum and minimum values.

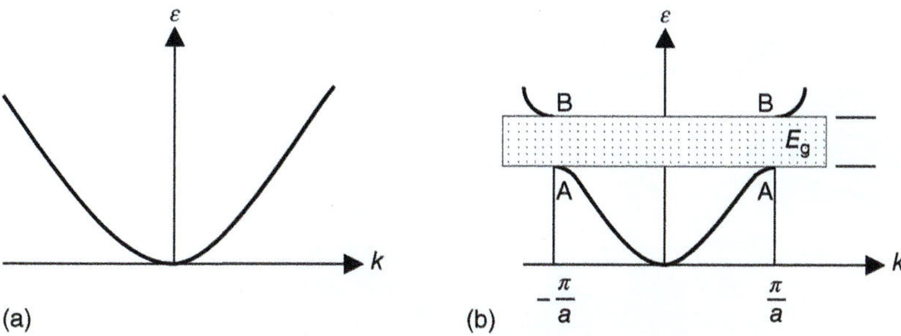

Figure 6.7 Electron dispersion relation for (a) free Fermi gas and (b) nearly free Fermi gas.

In Figure 6.7 the dispersion relation for free electrons and the result above for Bloch's nearly free-electron Fermi gas are compared. In the first case we have a simple parabola, and in the latter case the parabola is distorted and shows jumps where k passes from the first into the second Brillouin zone. At the zone edge ($k = \pm\pi/a$), the dispersion function has two different energy values for the same k value. The energy values are separated by E_g, the width of the forbidden energy gap.

Physically, the energy gap can be explained as an interference phenomenon, which is very simple in a 1D lattice. The electrons (or Bloch waves) are scattered by the electrostatic potential of the positively charged metal ions at the lattice points. The lattice points are separated by the lattice constant a. Waves backscattered from adjacent ions have a path difference of $2a$, and hence waves with this wavelength interfere constructively. A wavelength a implies a wavevector $\boldsymbol{k} = \pi/a$. At this point the discontinuities in the dispersion relations occur. Waves with wavevector $\boldsymbol{k} = \pi/a$ are in geometrical "resonance" with the crystal lattice. They are standing waves, not propagating waves. Two types of standing waves with $\boldsymbol{k} = \pi/a$ can be formed: one type has nodes at the lattice points, and the other has maxima at the lattice points. In one case the electrons are "between" the positive ions, and in the other case they are "at" the ions. Evidently the energy for the two cases is different so that there are two energy values for $\boldsymbol{k} = \pm 2\pi/a$.

So, certain electron energies are not allowed in a crystal for resonance reasons, where the resonance is a consequence of the periodic potential of the positive ions in the solid. Here an analogous example might be helpful. A similar situation exists on certain unpaved roads in the Sahara desert. Because of the camels traveling along, there are periodic bumps in the road. French 2CV cars resonate at 302 km/h. So the driver is forced to either stay below 28 or above 32 km/h. In between there is a forbidden gap (Figure 6.8).

6.2.2.3 Analyzing the 1D Nearly Free Solutions

As we saw in the free-electron model, to get to various physical phenomena, we must know the electronic DOS or the number of electronic states per energy interval. But, unlike the free electron model, interactions with the ion potentials yield a forbidden energy gap where the DOS is apparently zero.

Figure 6.9 shows the electronic dispersion relation between $\boldsymbol{k} = 0$ and $\boldsymbol{k} = \pi/a$ (as explained above, one half of the first Brillouin zone), together with the DOS.

Figure 6.8 The structure of certain Sahara roads allows one only to drive well below or well above 30 km/h. Between 28 and 32 km/h, there is a forbidden gap.

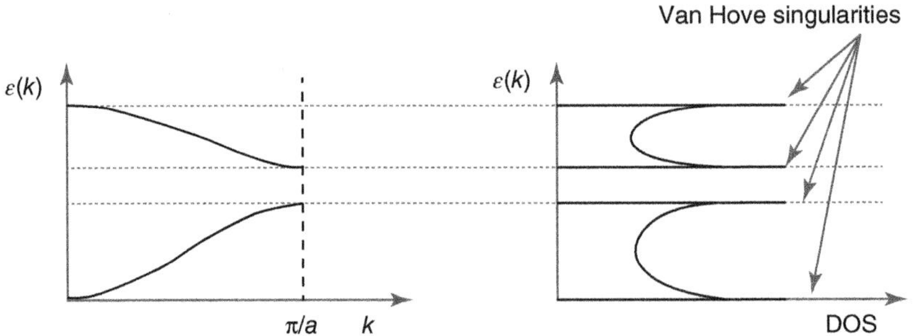

Figure 6.9 1D electron dispersion relation (left) and DOS (right).

The number of allowed k values in a Brillouin zone is large but finite. As we pointed out before, it is discrete because the electronic states are discrete when the electrons are confined in a box. The allowed k values are equally spaced along the x axis. To get the DOS the dispersion relation is projected onto the energy axis. At places where the dispersion relation is flat, the DOS is high and vice versa because $dk/d\varepsilon \rightarrow \infty$. That is, at horizontal parts of the dispersion relation, the DOS is infinite. The points of infinite DOS are called *van Hove singularities* and play an important role in 1D solid-state physics.

Because this is for a rigid 1D solid, Figure 6.9 also shows the van Hove singularities for $\boldsymbol{k} = 0$ and for $\boldsymbol{k} = \pi/a$, i.e. the bottom and the top of the energy band. If the solid is 3D, $\boldsymbol{k}$ vectors point into all directions of space, and the number of $\boldsymbol{k}$ values within a $\boldsymbol{k}$ interval increases with $\boldsymbol{k}$, thus compensating the van Hove singularity at $\boldsymbol{k} = 0$. This $\boldsymbol{k}$ count finally leads to the characteristic dimensionality behavior of DOS as indicated in Figure 1.18: parabolic in shape in three dimensions, step function in two dimensions, and square root singularity in one dimension.

As we now know, many properties of a solid are determined by the DOS at the Fermi energy, $N(E_F)$. We recall that electrons are fermions and obey Fermi statistics, i.e. we can accommodate only one electron in a quantum state (labeled by $\boldsymbol{k}$ and spin-up or spin-down). In simple words, we "pour" the electrons into the crystal, they fill up the dispersion relation, and at places where we run out of electrons, there is the Fermi level, the highest occupied electronic state. If there are as many electrons as there are $\boldsymbol{k}$ values in the first Brillouin zone, the Fermi level is at the top of the band. Adding the next electron causes it to jump to the bottom of the next band. Strictly speaking, this is only true for absolute zero, where the Fermi distribution is sharp and where there are no thermal excitations in the electron system. Otherwise the Fermi level is defined in a somewhat more complicated way, and for a completely filled band, the Fermi level is placed into the center of the gap above the filled band. Technically, as we saw above, this is the electron affinity or chemical potential – not the Fermi level, since it is in the forbidden region.

This is for a *perfectly rigid* 1D material. How would we think of electronic structure in three dimensions? Actually, it really isn't so hard to imagine. Notice that the SE is spatially separable if we use potentials that are well behaved and local: square wells, atomic orbitals, delta functions, etc. To understand the problem of electronic structure in three dimensions, we examine solutions of the SE along specific projections of the crystal lattice, thereby capturing the dynamics of electrons that might move along those directions. Solutions along these directions in space will yield bands, band-state filling and Fermi levels, and every other kind of characteristic that we have seen in the 1D problem. The Fermi points of the 1D problem become Fermi surfaces in three dimensions (we have already alluded to this). And one must adopt a multidimensional plotting scheme to express what the bands are doing along the different directions as is shown in Figure 6.10.

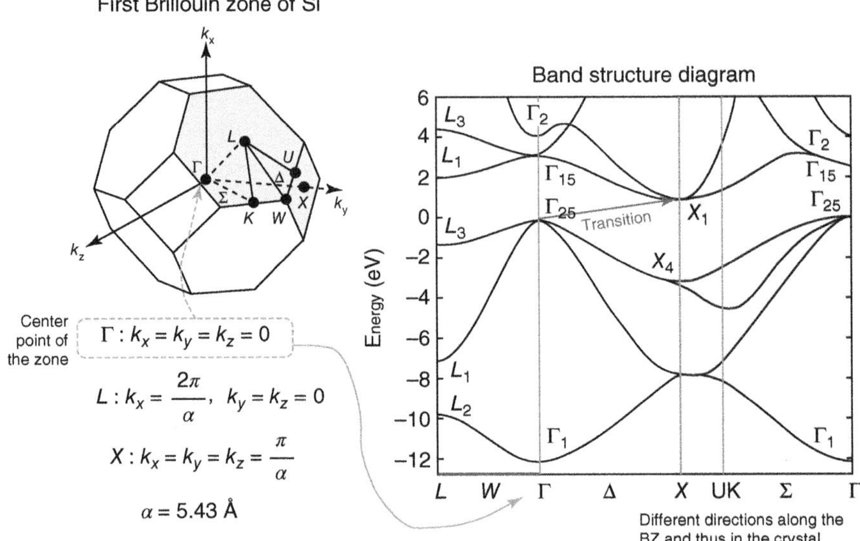

Figure 6.10 The band diagram for silicon.

6.2.2.4 Extending Dispersion Curves to 3D

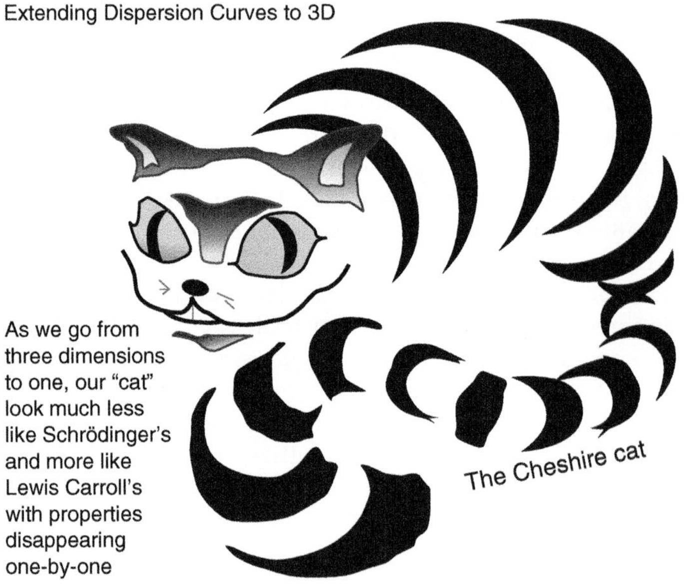

It can be quite entertaining to compare properties of materials in different dimensions. As we shrink a material from three dimensions to one dimension, we see its commonly identified features fade away, like the Cheshire Cat, leaving only a vestige of its fully dimensional self left behind.

Following our 1D example above, in two and three dimensions, we are faced with a really important subtlety. We see this in Figure 6.10: that is, the transition between highest valence band and lowest conduction band states varies across the Brillouin zone. That is, as one looks along different directions of the crystal, the periodicity and local potentials of the atomic arrangement may lead to different band structures and different bandgaps. Moreover, there are many examples where the minimum of the conduction band does not occur at the same $\boldsymbol{k}$ value as the maximum in the valence band. This means for the electron to make such a transition, it must also change the direction of its $\boldsymbol{k}$ vector. Materials for which this is true are called *indirect bandgap* semiconductors. Materials for which the reverse is true, highest occupied molecular orbital (HOMO) and lowest unoccupied molecular orbital (LUMO), occur at the same value of k, and these are called *direct bandgap* semiconductors.

This distinction can be pretty important for specific bandgap transition-related properties such as luminescence. Ever wondered why Si doesn't photoluminesce? Why don't we make light-emitting diodes (LEDs) from this material? Simple, it is an indirect bandgap semiconductor. For a direct bandgap semiconductor, the direct conduction band to valence band transition can yield a photon. However, in an indirect bandgap semiconductor, the electron must go through an intermediate state and transfer momentum to the crystal; a longer process and photons are not emitted. As a general rule transitions across the bandgap are made from the conduction band minimum (LUMO) and the valence band maximum (HOMO). This is because intra-band transitions are much, much faster than bandgap transitions. The intra-band transitions are transitions made among the tightly spaced states of the band itself, and they are on the order of thermal excitations. So the process of an excited electron "settling" to the conduction band minimum quickly is referred to as *thermalization* of the carrier. Importantly, we note that phenomena, like the indirect vs. direct bandgap division in semiconductors, go away as the dimension of the system is reduced. Si nanocrystals do, in fact, luminesce.

Hold on a minute! The bands of Figure 6.10 are not nice and symmetric the way you might think a 3D Kronig–Penney calculation would yield. What gives? 3D Kronig–Penney was not used for this figure, in fact. We might suspect that the Si atom binds its electrons somewhat tighter than our nearly free-electron model would wish. What happens when the electrons are more strongly associated with a given location? Like an atomic orbital? As in Figure 6.10, we apply a new approach known as *linear combination of atomic orbitals* (LCAOs) or *tight binding*. In fact, nearly free-electron approximations are not very good for semiconductors in general. Shown in Figure 6.11 is a comparison between the nearly free-electron bands of germanium and our next approximation, LCAO.

6.2.3 Tight Binding or Linear Combination of Atomic Orbitals

Tight binding or LCAO has two names because it can be thought of in two ways. Depending on if you are a physicist or a chemist, you are likely to see LCAO from very different philosophical perspectives. For instance, a physicist and a chemist are sitting in a bar. A graduate student asks about the deeper mysteries of band structure calculations. Both describe exactly the same approximation:

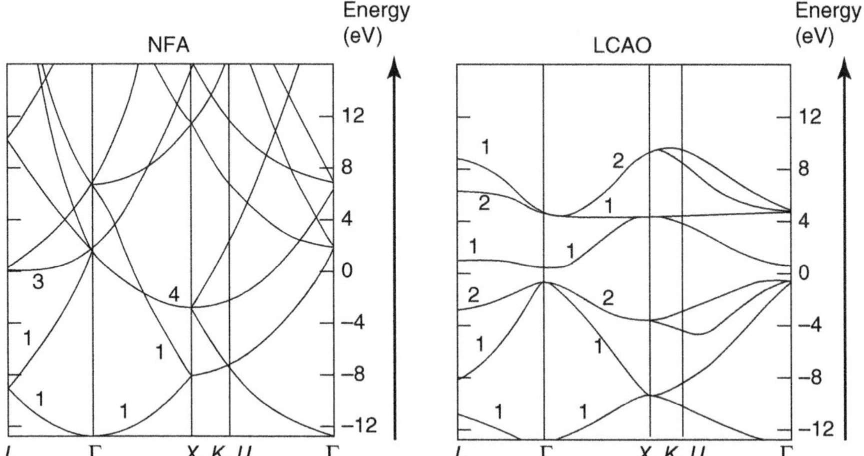

Figure 6.11 Nearly free-electron approximation and the LCAO approximation. The approximation here has been worked for Ge, while the above was for Si. It has been plotted using the 3D representation we introduced in Figure 6.10.

Dr. Physicist: "To "turn up" the electron–static lattice interaction a little more, we start with an infinite array of atomic orbitals as the potential in a Kronig–Penney type of scheme (replacing the artificial square wells). Symmetries and equations stay largely the same as in previous examples, but the shape of the potential changes. This can give great insight into the role this potential plays in band structure symmetries. This is the approach we call *tight binding*. It is simple and elegant. May I have another beer?"

Dr. Chemist: "Based on the assumption that the electrons are localized at atoms or molecules, we begin by using molecular orbitals to build a molecule that is ever increasing in size. For many *organic solids* this is particularly appropriate. As these building blocks come together, their orbitals interact, and the result is a splitting of the energy levels: instead of one electron associated with each building block, each with the same energy, you now get an energy level that is higher and one that is lower from those that you started with, and the two electrons can occupy either of the states. In molecular pairs this is known as *Davydov splitting*. As more building blocks (atoms or molecules) are added to the system, this splitting leads to a grouping of the states (they "bunch" together), and in many-particle systems, like a crystal, the grouping leads to an energy band. The *bandwidth* is directly related to the overlap of the wavefunctions (some million electron volts up to some electron volts). The dispersion relation (band structure) is obtained by passing s-shaped curves through the end points known as the *parabolic approach*. To calculate the actual band structure in this manner, extended expertise in quantum chemistry and fairly large computer power are essential. It is complex and difficult, and I think I shall need something a bit stronger than a beer."

In our discussion here, we shall not take sides, but instead try to introduce the language of both of our hypothetical mentors above.

6.2.3.1 The Formalism[5]

In a crystal, the single particle states are given by

$$H = H_{\text{at}} + \Delta U \tag{6.22}$$

ΔU is the variation of the potential from free space, and H_{at} is the Hamiltonian for a free atom. This ΔU goes to zero at the center of each atomic site. The single particle states are then given by

$$H\psi_{nk}(\boldsymbol{r}) = E_{nk}\psi_{nk}(\boldsymbol{r}) \tag{6.23}$$

where n is the band index and $\boldsymbol{k}$ is the wavevector inside the first BZ. The atomic wavefunctions are simply

$$H_{\text{at}}\phi_i(\boldsymbol{r}) = \varepsilon_i\phi_i(\boldsymbol{r}) \tag{6.24}$$

and ε_i is the ith energy level of the isolated atom. These wavefunctions decay rapidly away from $\boldsymbol{r} = 0$, and so the overlap integral between the wavefunctions of the isolated atoms is small:

$$\int \phi_i^*(\boldsymbol{r})H\phi_i(\boldsymbol{r} + \boldsymbol{R})\text{d}\boldsymbol{r} \tag{6.25}$$

where R describes the positions of the atoms in the crystal. We note that the atomic orbitals we use in this discussion are all orthonormal, so,

$$\int \phi_i^*(\boldsymbol{r})\phi_j(\boldsymbol{r} + \boldsymbol{R})\text{d}\boldsymbol{r} = \begin{cases} 1 & \text{if } i = j \text{ and } \boldsymbol{R} = 0 \\ 0 & \text{otherwise} \end{cases} \tag{6.26}$$

For a simple 1D string of atoms, this would look something like in Figure 6.12.

As before, we insist that our single particle states of the system (ψ_{nk}) obey Bloch's theorem:

$$\psi_{nk}(\boldsymbol{r} + \boldsymbol{R}) = \text{e}^{i\boldsymbol{k}\cdot\boldsymbol{R}}\psi_{nk}(\boldsymbol{r}) \tag{6.27}$$

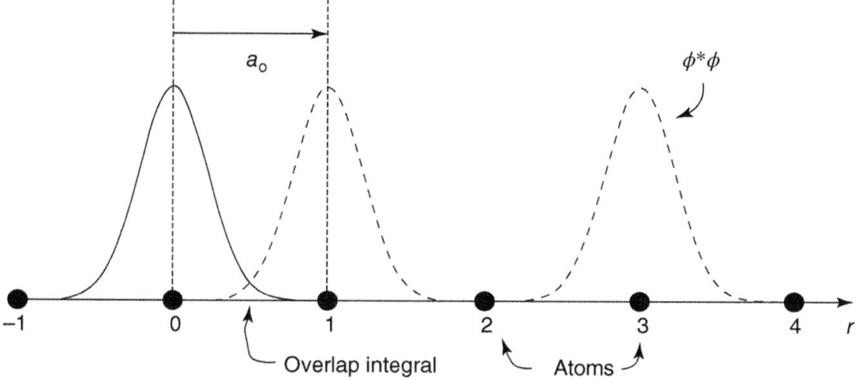

Figure 6.12 There is very little overlap between each of the adjacent sites and no overlap between sites further apart. The translation vector here is $\boldsymbol{R}$ and is modulo (a_0).

5 This treatment was brought to our attention by Professor Mervyn Roy at the University of Leicester in the United Kingdom. We have followed his notes.

where $\boldsymbol{R}$ is the translation vector of the lattice. And we want to build these states from the atomic orbitals (ϕ) since they revert to "atomic like" near the lattice site. Thus, we must construct the state from the atomic orbitals in such a way that it does:

$$\psi_{nk}(\boldsymbol{r}) = \frac{1}{\sqrt{N}} \sum_{\boldsymbol{R}} \mathrm{e}^{i\boldsymbol{k}\cdot\boldsymbol{R}} \phi_n(\boldsymbol{r} - \boldsymbol{R}) \tag{6.28}$$

where there are N lattice sites and the $1/\sqrt{N}$ term ensures normalization.

Believe it or not, this is all we need to start with. From here we must now decide on the atomic orbitals that will interact (that is, overlap) with each other. We have chosen to start with the convention that the atomic orbitals are orthonormal, so the atomic wavefunctions must be of the same type to form a band: thus s orbitals interact with s orbitals, p orbitals interact with p orbitals, d orbitals interact with d orbitals, and so on. Any set of orbitals with significant enough overlap will form a band of single particle states for the system. The generalization to *hybridized bands* such as s interacting with p to form *sp hybridization* is important for semiconductors like Si, but for now, we will stick with the easy case.

6.2.3.2 The s-Band

We begin by considering a set of atoms brought together in which only the s orbitals overlap with each other. Then there will be only one band, and its Bloch state will be given as

$$\psi_{k}(\boldsymbol{r}) = \frac{1}{\sqrt{N}} \sum_{\boldsymbol{R}} \mathrm{e}^{i\boldsymbol{k}\cdot\boldsymbol{R}} \phi_s(\boldsymbol{r} - \boldsymbol{R}) \tag{6.29}$$

and the energy will be given as

$$E(\boldsymbol{k}) = \int \psi_k^*(\boldsymbol{r}) H \psi_k(\boldsymbol{r}) d\boldsymbol{r} \tag{6.30}$$

and the integral of the expectation value is over all space as usual. Substituting the Bloch expression we then get

$$E(\boldsymbol{k}) = \frac{1}{N} \sum_{\boldsymbol{R}} \sum_{\boldsymbol{R}'} \mathrm{e}^{i\boldsymbol{k}\cdot(\boldsymbol{R}'-\boldsymbol{R})} \int \phi_s^*(\boldsymbol{r} - \boldsymbol{R}) H \phi_s(\boldsymbol{r} - \boldsymbol{R}) \mathrm{d}\boldsymbol{r} \tag{6.31}$$

$$E(\boldsymbol{k}) = \frac{1}{N} \sum_{\boldsymbol{R}} \sum_{\boldsymbol{R}'} \mathrm{e}^{i\boldsymbol{k}\cdot(\boldsymbol{R}'-\boldsymbol{R})} \int \phi_s^*(\boldsymbol{x}) H \phi_s(\boldsymbol{x} - (\boldsymbol{R}' - \boldsymbol{R})) \mathrm{d}\boldsymbol{x} \tag{6.32}$$

where in the top integral, we have used the R' to mean a separate lattice sum with $R' \neq R$. In the second integral, we have changed variables with $\boldsymbol{x} = \boldsymbol{r} - \boldsymbol{R}$. Now we notice that the sums are over translation vectors generally. Thus $\boldsymbol{R} - \boldsymbol{R}' = \boldsymbol{R}''$ that is just another translation vector. Since we get the same answer no matter what we call the translation vector,

$$E(\boldsymbol{k}) = \frac{1}{N} \sum_{\boldsymbol{R}} \sum_{\boldsymbol{R}''} \mathrm{e}^{i\boldsymbol{k}\cdot\boldsymbol{R}''} \int \phi_s^*(\boldsymbol{x}) H \phi_s(\boldsymbol{x} - \boldsymbol{R}'') \mathrm{d}\boldsymbol{x} \tag{6.33}$$

but the terms of the two sums are now identical, and so we can write

$$E(\boldsymbol{k}) = \sum_{\boldsymbol{R}''} \mathrm{e}^{i\boldsymbol{k}\cdot\boldsymbol{R}''} \int \phi_s^*(\boldsymbol{x}) H \phi_s(\boldsymbol{x} - \boldsymbol{R}'') \mathrm{d}\boldsymbol{x} \tag{6.34}$$

Notice that the sum we got rid of simply gave us another N.

We now separate out terms in the sums by how much overlap there is between s orbitals: that is, how large the integral is tells us how much each term in the sum counts.

If $\boldsymbol{R}'' = 0$,

$$\varepsilon_s = \int \phi_s^*(\boldsymbol{x})H\phi_s(\boldsymbol{x})\mathrm{d}\boldsymbol{x} = \int \phi_s^*(\boldsymbol{x})\varepsilon_s\phi_s(\boldsymbol{x})\mathrm{d}\boldsymbol{x} \tag{6.35}$$

or just the energy of the atomic orbital itself in a free atom.

If $|\boldsymbol{R}''|$ is large,

$$0 \approx \int \phi_s^*(\boldsymbol{x})H\phi_s(\boldsymbol{x} - \boldsymbol{R}'')\mathrm{d}\boldsymbol{x} \tag{6.36}$$

since we already said the atomic orbitals decay rapidly from the atomic positions.

So generally, the semiempirical tight binding approach includes only the contributions to the sum that are very close. Let's say within one lattice translation away, a_0. We will call this lattice translation τ and we can write

$$E(\boldsymbol{k}) = \varepsilon_s + \sum_{\tau} \mathrm{e}^{i\boldsymbol{k}\cdot\boldsymbol{\tau}} \int \phi_s^*(\boldsymbol{x})H\phi_s(\boldsymbol{x} - \boldsymbol{\tau})\mathrm{d}\boldsymbol{x} \tag{6.37}$$

So, the crux of the matter is that the s band will look like plane waves modulated by the overlap integral – the amount of the wavefunction between the two lattice sites that actually occupies the same space. The energy is likewise shifted as

$$\gamma(|\boldsymbol{\tau}|) = \int \phi_s^*(\boldsymbol{x})H\phi_s(\boldsymbol{x} - \boldsymbol{\tau})\mathrm{d}\boldsymbol{x} \tag{6.38}$$

so,

$$E(\boldsymbol{k}) = \varepsilon_s + \sum_{\tau} \mathrm{e}^{i\boldsymbol{k}\cdot\boldsymbol{\tau}}\gamma(|\boldsymbol{\tau}|) \tag{6.39}$$

At this point it is often common to introduce some sort of empirical relation for the overlap integral as a function of the separation $|\tau|$. This can allow one to examine how applied strain on the crystal, for instance, might alter expected band development.

6.2.3.3 s Bands in One Dimension

The above formulation is really set up for a 1D problem. Let's examine this for a moment. Take the atoms to lie along the x axis. So $\boldsymbol{\tau} = a_0\hat{\boldsymbol{i}}$ using the standard notation for the unit vector in x. Also $\boldsymbol{k} = k\hat{\boldsymbol{i}}$. So we can substitute and write (Figure 6.13)

$$E(\boldsymbol{k}) = \varepsilon_s + \gamma(a_0)(\mathrm{e}^{ika_0} + \mathrm{e}^{-ika_0}) \tag{6.40}$$

$$E(\boldsymbol{k}) = \varepsilon_s + 2\gamma(a_0)\cos(ka_0) \tag{6.41}$$

6.2.3.4 s Bands in Two Dimensions

Similarly, we can set up the problem in two dimensions. Our crystal for this example will be a two-dimensional rectangular net as shown in Figure 6.14.

Working the sum out, we split k into k_x and k_y and τ into a and b to get

$$E(k_x, k_y) = \varepsilon_s + 2\gamma(a)\cos(k_xa) + 2\gamma(b)\cos(k_yb) \tag{6.42}$$

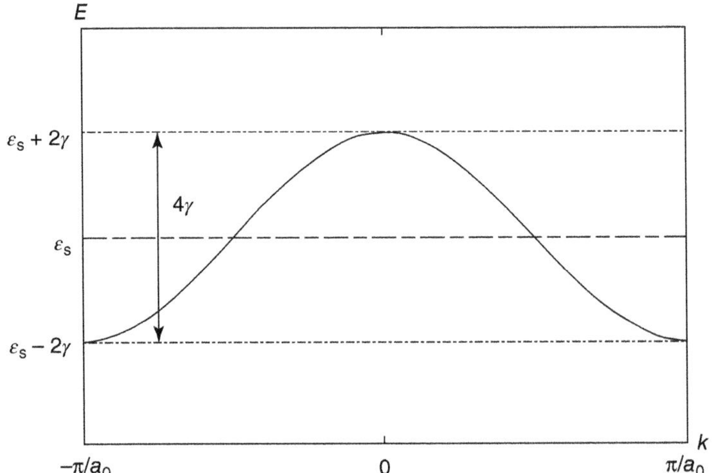

Figure 6.13 The dispersion curve $E(k)$ for a one-dimensional system using tight binding code in Mathematica. Notice however that we have done this in terms of the overlap integral, so such details are not yet included in the calculation, which means absolute energies are not given.

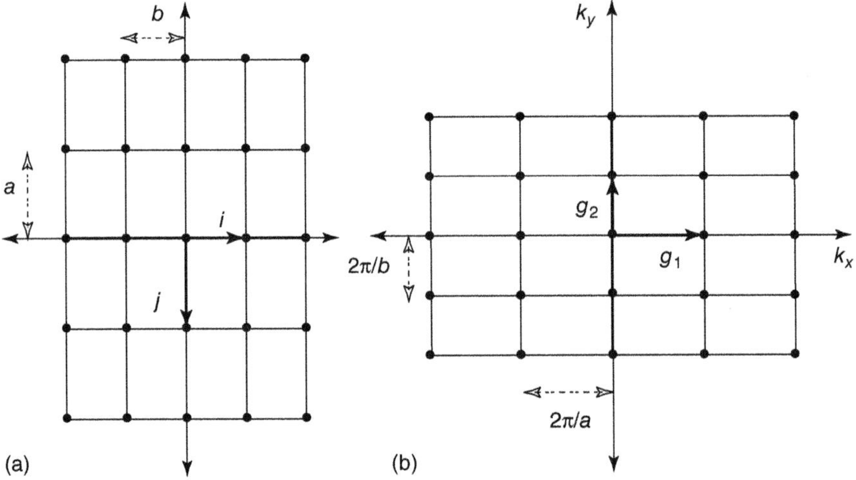

Figure 6.14 (a) The real space lattice and (b) the reciprocal space lattice of our 2D system.

If you have forgotten how to do the dot products, look back at the reciprocal lattice chapter. We can put some numbers onto this: let's take $a = 0.5$ nm and $b = 1.0$ nm and $\gamma(a) = 1$ eV and $\gamma(b) = 0.5$ eV. Finally we take $\varepsilon_s = 2.0$ eV. Then we get Figure 6.15.

6.2.3.5 s Bands in Three Dimensions

Moving to a 3D example is now very simple. We have

$$E(\boldsymbol{k}) = \varepsilon_s + \sum_{\boldsymbol{\tau}} e^{i\boldsymbol{k}\cdot\boldsymbol{\tau}}\gamma(|\boldsymbol{\tau}|) \tag{6.43}$$

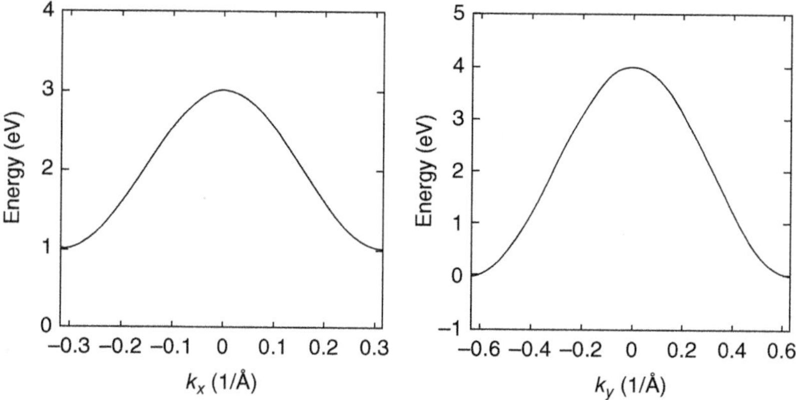

Figure 6.15 Plots of the dispersion curves for both the k_x and k_y directions of the 2D crystal. The choices we made above were really artificial, and the s-like character never really entered into the calculation except through these choices.

$$\boldsymbol{k} = (k_x, k_y, k_z) \tag{6.44}$$

$$\boldsymbol{\tau} = \frac{a}{2}(\pm 1, \pm 1, 0); \frac{a}{2}(\pm 1, 0, \pm 1); \frac{a}{2}(0, \pm 1, \pm 1) \tag{6.45}$$

The translation vectors clearly describe a face-centered cubic (you should graph this out to make sure you see why). There will be 12 nearest neighbors, so a few more terms in the sum than before.

A little algebra and we get

$$E(k_x, k_y, k_z) = \varepsilon_s + 4\gamma(|\boldsymbol{\tau}|)\left(\cos\frac{k_x a}{2}\cos\frac{k_y a}{2} + \cos\frac{k_y a}{2}\cos\frac{k_z a}{2} + \cos\frac{k_z a}{2}\cos\frac{k_x a}{2}\right) \tag{6.46}$$

where $|\tau| = a/\sqrt{2}$.

6.2.4 What About Orbitals Other Than s?

The overlap integral $\gamma(\tau)$ really determines most of the interesting physics here. How it depends on distance is set by the different sorts of orbitals that might be used. s and p, for instance, can extend into space much further than, say, a d orbital. This integral can be fit to data or set equal to some phenomenological model. In doing so, the bandwidth will change as will the curvature overall. The real significance of s vs. p orbitals shows up when more than one orbital is contributing bands. Think, for example, of conjugated polymer systems. They are usually bonded together by the s orbitals, whereas the p orbital-derived bands supply the ability to conduct electrical current. This means that the equilibrium of the s orbitals sets the distances used in the p orbital overlap integrals. We will examine this in more detail in a moment, but there is one more thought to add. The curvature is inversely related to the effective mass of the carrier as we stated above. This means the smaller the overlap integral, the heavier the effective mass, which makes sense.

6.2.4.1 Building Bands in a Polymer

At the beginning of our discussion, we noted the equivalence between the "solid-state physics" picture and a more "macromolecular" or "chemistry." We said this was particularly useful in constructing bands in a polymer system. From one point of view, one dimensionality, as expressed by polymers, is simply the confinement of the Bloch waves in lateral dimensions and the modulation of the traveling wave in the axial dimension of the structure. Alternatively, this can be seen as a very large molecule generally, and charge transport is related to the delocalization associated with specific excited states of the molecule. Now that we have presented the first way of looking at bands, it seems a good place in the analysis to examine this second "macromolecular" way of viewing things [2].

6.2.4.2 Bonding and Antibonding States

To begin we must understand that bonding two atoms together leads to a lowering of the overall electronic energy. Consider the hydrogen atom as depicted graphically in Figure 6.16.

The solutions to the SE for the hydrogen ground state yield a wavefunction that (in space) falls off exponentially as a function of the radius away from the nucleus. The ground state energy of this system is 1 Rydberg (13.6 eV). Now we approach this hydrogen atom with a second hydrogen atom – dihydrogen. As shown in Figure 6.17, this results in an interaction between the electrons such that two possible states can exist – one is a bonding state and the other an antibonding state.

What you get by doing this is obviously an H_2 molecule. The distance between the atoms is set by the balance of attractive and repulsive forces with the attractive force being derived from the lowering of electronic energy. Notice that just as above, the s orbitals are doing the work here, and the overlap integral is fairly easy to calculate. The s orbital is isotropic about the atom and, thus, so is the force of attraction. Thus, of course, it would not be the case if p orbitals were doing the binding. Because the "bonding orbital" is formed using s orbitals from the atom, we use the symbol σ as its designation.

It is important to note that the bonding state yields increased electron density between the atomic nuclei – this is the "bond." In the antibonding state, there is a node in the electron density between the nuclei. This state lies at a much higher energy than the bonding state, and there must be a transition from the lower state, which is filled, to the excited antibonding state, which is empty in the

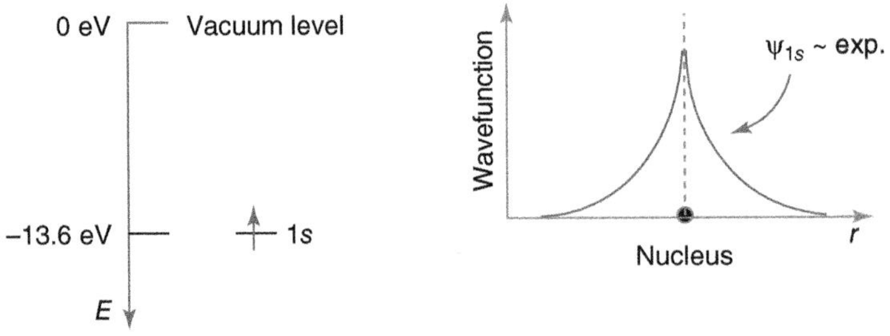

Figure 6.16 The hydrogen atom with its one electron and filled ground state.

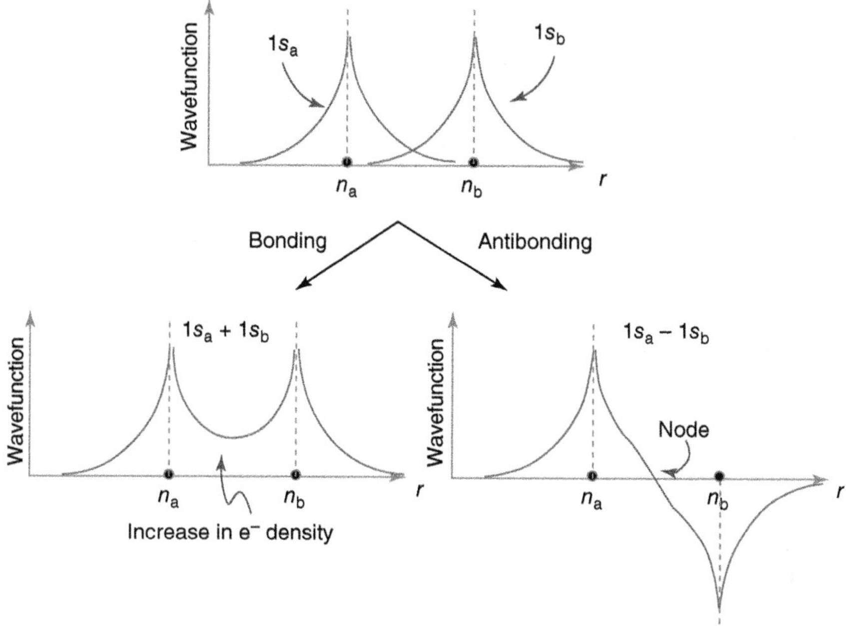

Figure 6.17 The interaction of two hydrogen atoms (a and b) leads to two possible states – bonding and antibonding.

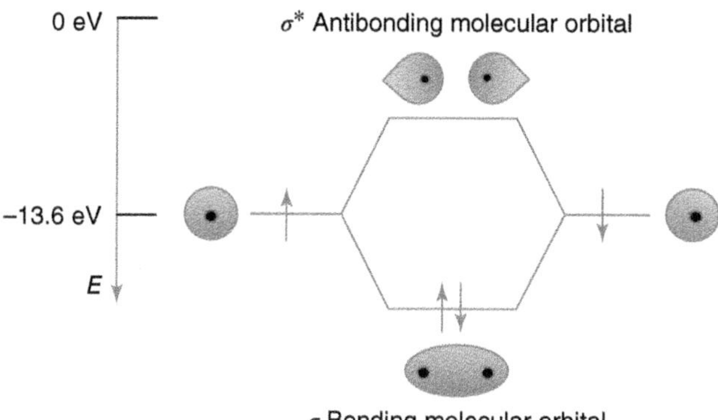

Figure 6.18 The energy states of hydrogen starting with the separate atoms and then combining into the molecule.

ground state of the system. This splitting is generally drawn in an energy state diagram as in Figure 6.18.

6.2.4.3 The Polyenes

Using the same basic reasoning as above, we begin to construct the polymer. But instead of hydrogen atoms, we use methyl radicals. Shown in Figure 6.19 is the methyl radical that will be our basic building block of our first polymer system.

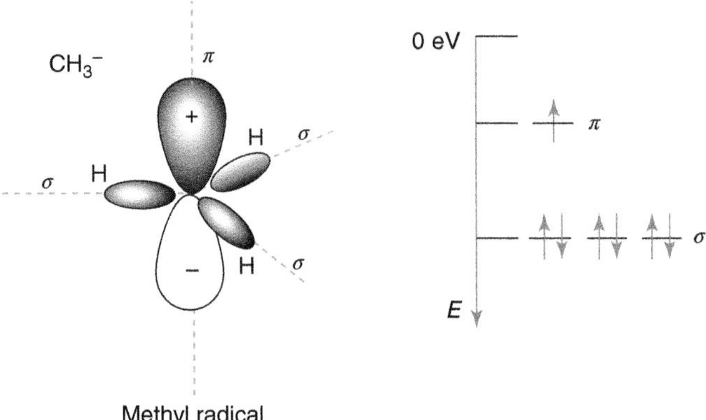

Figure 6.19 The methyl radical that will be used to "build a polymer."

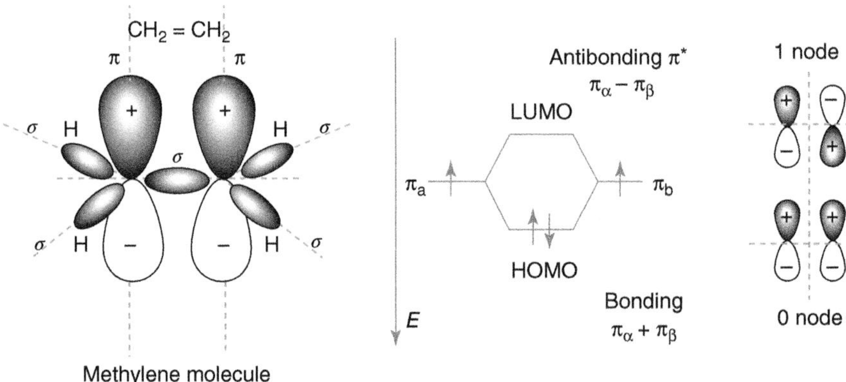

Figure 6.20 The ethylene (=dimethyl) molecule formed by interacting two methyl radicals together.

At the center of the lobes (electron cloud density) in Figure 6.19 is the carbon atom, and it has four electrons in its outer shell to use in the formation of bonds. We have used three of them in plane to make the σ bonds to the hydrogen. These bonds are saturated in the sense of our earlier descriptions, so the energy diagram shows the spins of both the electrons from the C and the electrons from the *H*'s. There is another electron, and it is placed into the p_z orbital to form π bonds. Due to symmetry considerations this electron does not mix with the in-plane σ bonds.

You might have noticed that the σ bonds that are derived from s orbitals no longer have that isotropic character that we would expect from the s orbital. In this figure we are showing the radical with saturating values of *H* attached, and they have some Coulombic repulsion associated with them.

Our next step is to "interact" this methyl group with a second methyl group to form the ethylene molecule. As shown in Figure 6.20, we now have the σ bonds interacting in plane and the π orbitals interacting laterally above and below the plane.

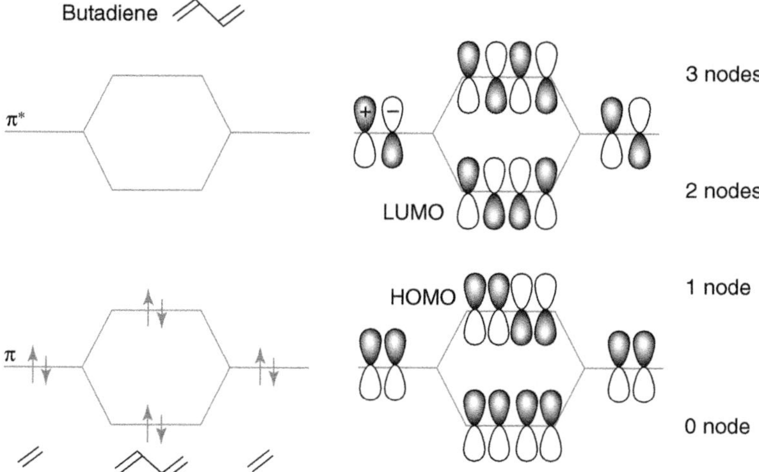

Figure 6.21 Butadiene is essentially two interacting methylene molecules. It now has four π bonds.

Notice on the left that the hydrogens have been removed from the bonds that are now shared between the groups. There are two electrons in this shared σ bond. The π bonds now take on a bonding–antibonding nature as seen on the right. There is a 1 node and a 0 node state just as in the case of the hydrogen earlier. One represents the ground state of the system and it is filled. To excite an electron into the antibonding state takes ~7 eV.

We now add two more methyl groups to form an even longer molecule. As shown in Figure 6.21, this results in four interacting π bonds and is referred to as butadiene. Again, the σ bonds are in place and the π bonds do not mix with the in-plane electrons. Notice that there are many more ways in which the wavefunctions can have nodes and antinodes within the interacting π electrons.

Notice too that we have used the chemical symbolism of double and single bonds to enumerate the wavefunction symmetries and their related state filling as in Figure 6.22.

The highest filled state, the HOMO level, has one node, and the LUMO is with two nodes. All other states have a higher number of nodes and thus are higher in energy. To transition from the HOMO to the LUMO in this system, one must use ~5.4 eV. So the gap has closed slightly between filled and unfilled orbitals.

So far this is easy to reason, and the various states that make up the "bands" of occupied and unoccupied orbitals are simply combinations of interactions between the π electrons. The lowest number of nodes is the lowest energy states and therefore the filled states. If we "interact" three ethylene subunits together, the number of combinations of interactions (bonding and antibonding) becomes even greater as seen in Figure 6.23.

In hexatriene, the six π orbitals can range from having 0 nodes to 5 nodes, and the energy states are beginning to "clump" together into the filled states and the

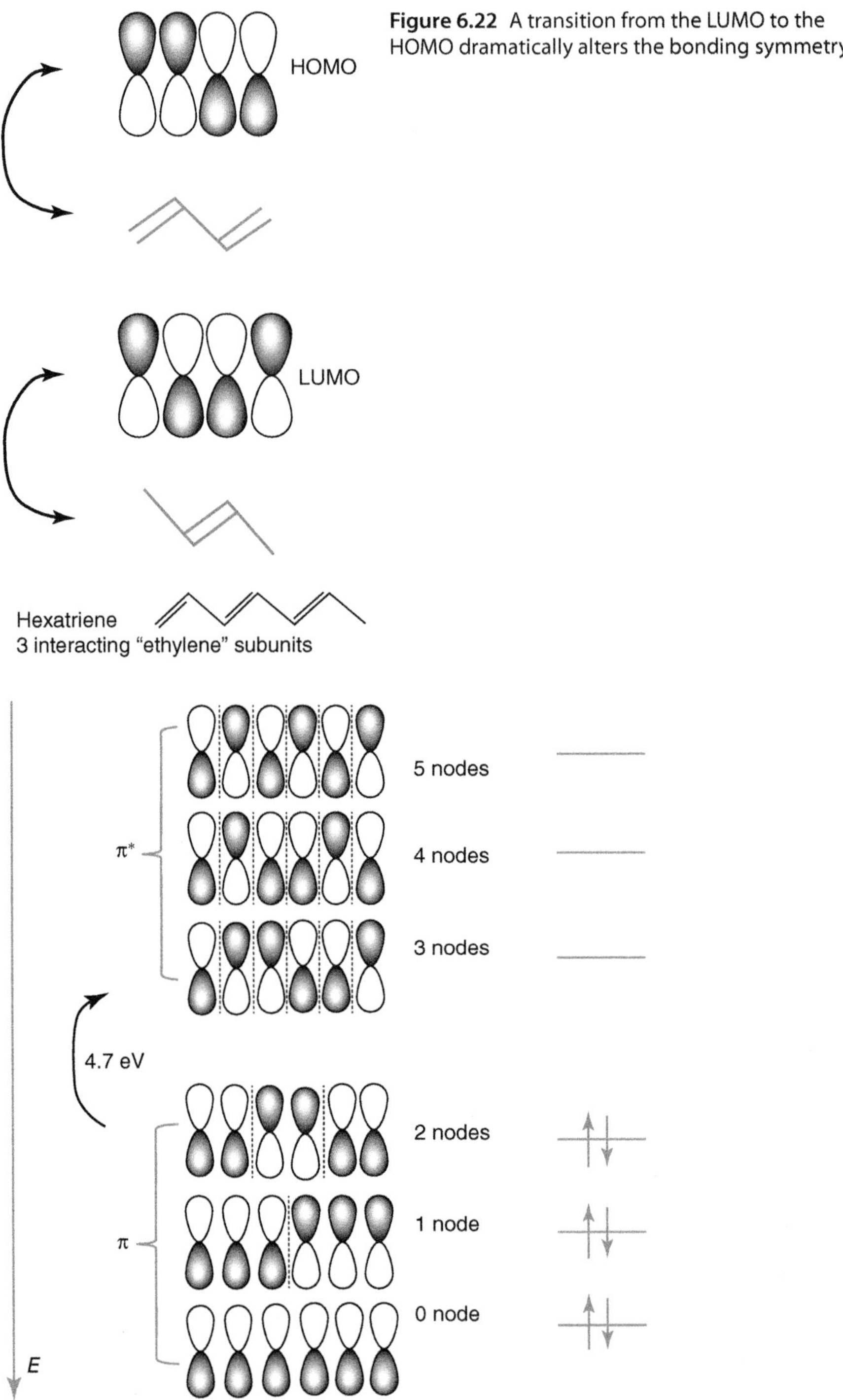

Figure 6.22 A transition from the LUMO to the HOMO dramatically alters the bonding symmetry.

Figure 6.23 Hexatriene with three ethylene subunits yielding six possible interaction sites.

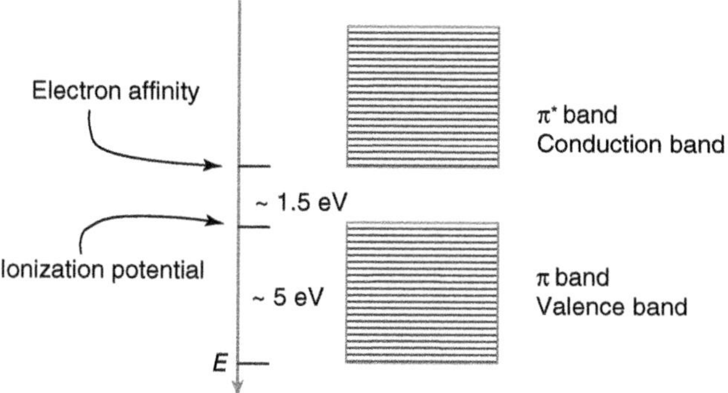

Figure 6.24 The band structure achieved by adding many ethylene subunits together.

unfilled states. Each "band" using solid-state language now has three states in it, defined by the symmetries of the bonding or antibonding wavefunctions. We also notice that the energy gap between the LUMO and HOMO has decreased even further to ~4.7 eV. The symbol we use to describe the hexatriene molecule looks like a truncated conjugated polymer symbol and reflects the LUMO of the macromolecule.

This molecule is already large enough to begin to make some connections with the solid-state language we used above. In the absence of the π electrons – let's say by saturation – we have essentially the alkanes. They have a single bond length of ~1.52 Å. However if this saturation is removed – essentially the same as adding free electrons to this system, these π electrons distribute themselves unevenly by alternating the bonding between single and double bonds of 1.47 and 1.34 Å, respectively. In Chapter 7 we will see that this is the natural result of the competition between bond strain energy and electronic state energy.

If we continue this process, adding more and more interacting ethylene subunits, the energies of the states in the unfilled band and the filled band become closer and closer. The gap in energy between the LUMO and HOMO also continues the trend of getting smaller. In Figure 6.24 we have carried this out for a very large number of ethylene molecules – this is typically done on a computer using a method known as the LCAOs – which is the method we have been following diagrammatically here.

6.2.4.4 Translating to Bloch's Theorem

As we have seen previously, Bloch's theorem uses the translational repeat symmetry of the infinite system to construct a description of the state energies within a band as a function of the "crystal momentum" k. So one can focus only on the repeating unit cell and its interactions with neighbors as in Figure 6.25.

Recall that the theorem states that the electron density at point $\boldsymbol{r}$ in cell j (j = integer) must be equal to the electron density at point $\boldsymbol{r}$ in the origin cell. This is expressed in terms of the modulus of the wavefunction

$$|\Psi(\boldsymbol{r}+j\boldsymbol{a})|^2 = |\Psi(\boldsymbol{r})|^2 \tag{6.47}$$

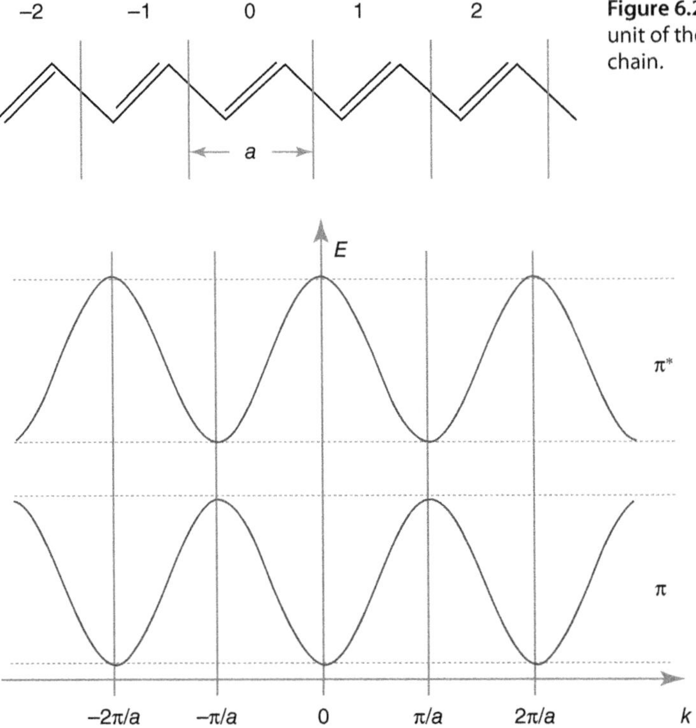

Figure 6.25 The repeating unit of the polyacetylene chain.

Figure 6.26 The extended band structure of the simple Bloch model.

or as we stated it previously,

$$\Psi(\boldsymbol{r} + j\boldsymbol{a}) = e^{i\boldsymbol{k}\cdot j\boldsymbol{a}}\Psi(\boldsymbol{r}) \tag{6.48}$$

and the modulus of this phase factor is simply 1.

For the simple models we presented before, we saw that this gave rise to the band structure diagram of Figure 6.26.

But note that now we have labeled these as π and π^* bands. Let's examine the π and π^* bands a little more closely. Figures 6.27 and 6.28 show the wavefunction symmetries at the band edges (π band and π^* band, respectively).

For polyacetylene the π band is filled and the π^* band is empty. As in Bloch's theory earlier, we expect a semiconductor. But what gap should we expect in energies at $k = \pi/a$? Bloch's approach details this in terms of strength of interacting potentials. In this molecular orbital picture, it becomes clear that the nature (let's say the symmetry or bonding) of the wavefunction at the HOMO and LUMO edges for $k = \pi/a$ must represent different state energies – i.e. nondegenerate. But examine the wavefunction symmetries of the band edges at $k = \pi/a$ closely. The first Brillouin zone is shown in Figure 6.29 along with wavefunction symmetries.

In Figure 6.29 we can see that the interaction symmetries in the A (LUMO) and B (HOMO) wavefunction p orbitals would look the same if we simply shift the definition of the unit cell in B over by one atom. But in doing so in B the

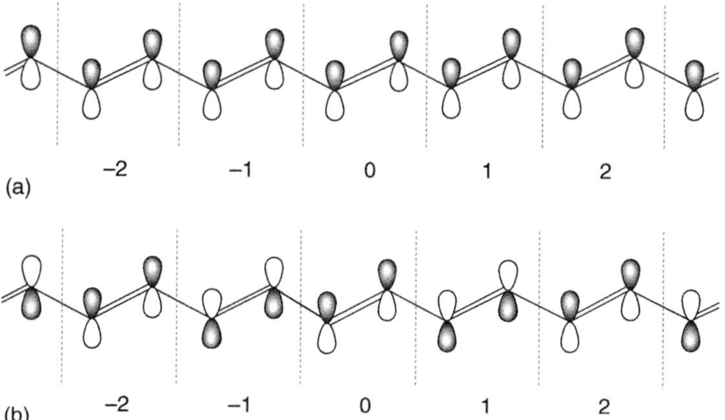

Figure 6.27 (a) Fully bonded wavefunction at $k = 0$: $e^{ikja} = e^0 = 1$. (b) Wavefunction bonding on the double bonds and antibonding on the single bonds at $k = \pi/a$: $e^{ikja} = 1$ for j even and -1 for j odd.

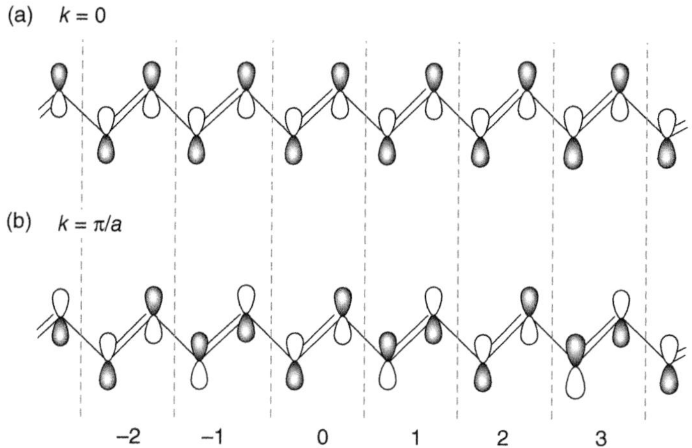

Figure 6.28 The top and bottom of the valence (or π^*) band also goes from one combination of bonding and antibonding states to another. Those combinations now enumerate the k values of the states.

unit cells would be defined as a single-bonded dimer that is double bonded to its nearest neighbors, whereas the A definition would be a double-bonded dimer with a single bond between nearest neighbors. Since this is the HOMO–LUMO of the system, we will refer to this state as the ground state of the system, and if the σ bonds are all of the same length, then clearly this ground state is degenerate. More precisely stated, the π and π^* bands at $k = \pi/a$ would be degenerate. We would expect for there to be *no* bandgap.

As you may have already guessed, polyacetylene is a special example. Specifically, for the *trans*-polyacetylene system, there is no *extrinsic* dimerization, and thus it has no *extrinsic* bandgap. The dimerization occurs entirely due to the addition of the π electrons coupling to the lattice. Thus, the A and B

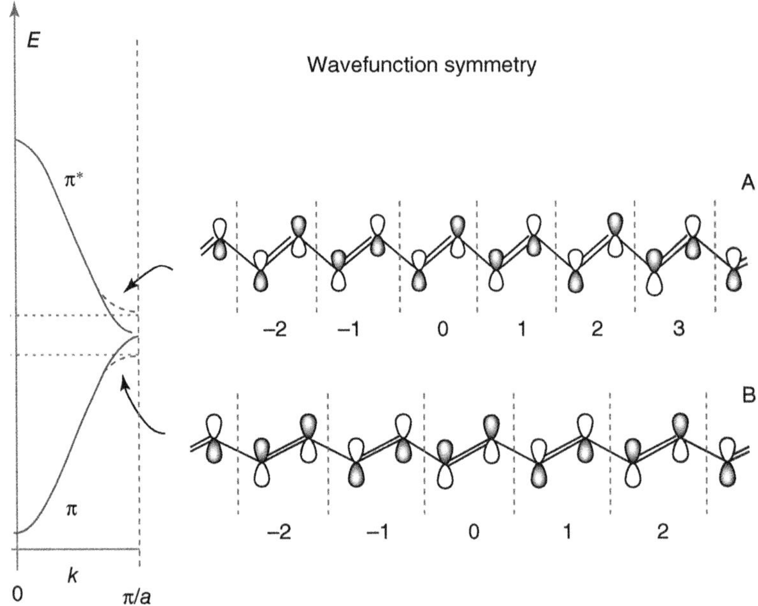

Figure 6.29 The symmetries of the wavefunctions at the $k = \pi/a$ band edge are strikingly similar (A and B above). Notice that one becomes the other if we were to shift the definition of our unit cell over by one atom.

phases are degenerate. This is not true in most other polymers where there is an extrinsic dimerization due to the stereochemistry of the σ bonds. For instance, *cis*-polyacetylene has an extrinsic bandgap due to its σ orbitals. The situation is different for different polymers.

The ground state situation from *trans*-polyacetylene is shown in Figure 6.30. As we will see, if the π electrons are simply added to the system and allowed to delocalize, then the systems energy would sit on the unstable point at $\Delta\pi = 0$. We now know, however, that this does NOT happen. Instead some energy is gained by "strain" in the σ bond "springs," and some is lost in the π bond alternation due to a *Peierls distortion*. The next result is a lowering of system energy. $\Delta\pi = (\pi_{C-C} - \pi_{C=C})$ and the energy gained $E \sim (\Delta\pi^4 - \Delta\pi^2)$. Experimentally, $\Delta\pi_0 \sim 0.08$ Å. This opens up the bandgap seen in Figure 6.29.

6.2.5 Tight Binding with a Basis

Typically, when calculating a band structure for some material, there will be a basis involved: more than one atom in the unit cell. In fact, this is where tight binding approaches can become really helpful, that is, when basis sets get really large.

Actually dealing with this isn't so hard, and we shall use an example in 1D to help us: *trans*-polyacetylene. It looks like this (Figure 6.31).

In our example there are two basis atoms, one at (0,0) and one at $R_{AB} = (a/2, a/2\sqrt{3})$. In a crystal with N_b basis atoms (wherein only one type of orbital is

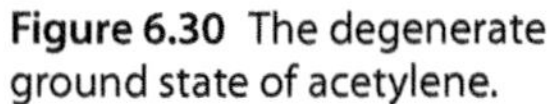

Figure 6.30 The degenerate ground state of acetylene.

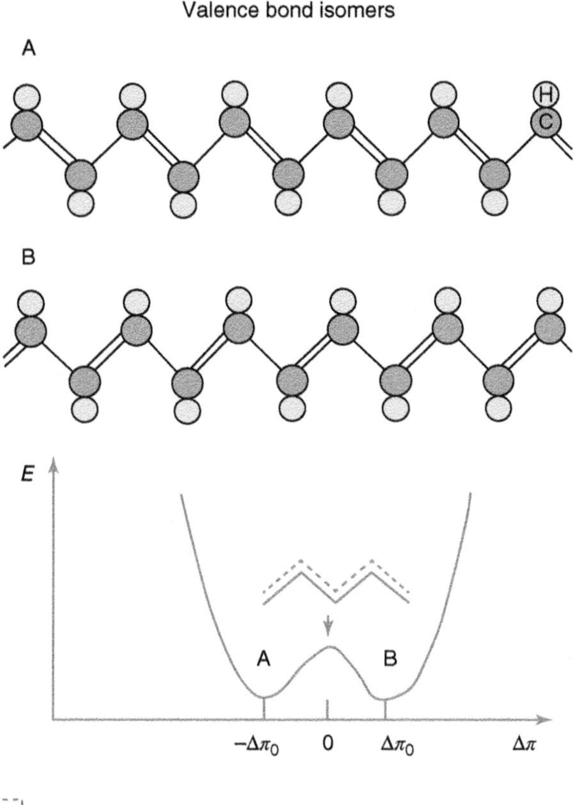

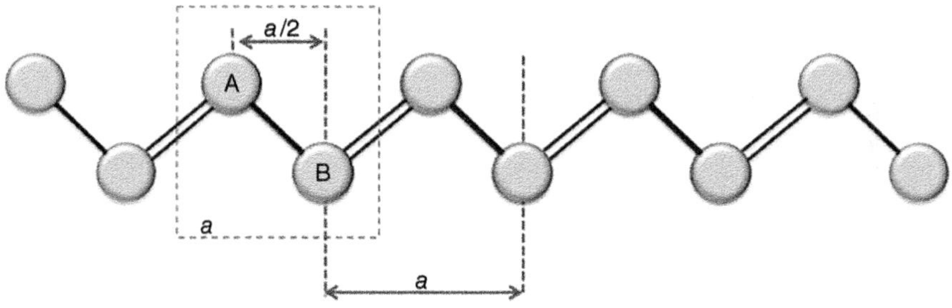

Figure 6.31 Labeling trans-polyacetylene. The translation vector is modulo (*a*), and the basis atoms lie *a*/2 along the *x* axis but are offset. Here we have shown the conjugated bonds: alternating single and double bonds. However, we have not included the slight difference in bond lengths (for simplicity). The two different atoms of the basis are marked A and B.

contributing to the bands), then we can make N_b linear combinations of orbitals that satisfy Bloch's theorem:

$$\Phi_{ik}(\boldsymbol{r}) = \frac{1}{\sqrt{N}} \sum_{\boldsymbol{R}_i} e^{i\boldsymbol{k}\cdot\boldsymbol{R}_i} \phi(\boldsymbol{r} - \boldsymbol{R}_i) \tag{6.49}$$

The index $i = 1, 2, 3, \ldots, N_b$. This labels the different atoms of the basis, and the translation vector R_i is the vector that translates the between atoms of type i. In our simple example i would label atoms A and B. The translation vectors would be $R_A = \pm a\hat{\imath}$, $\pm 2a\hat{\imath}$, $\pm 3a\hat{\imath}$, etc. and $R_B = R_{AB} \pm a\hat{\imath}$, $R_{AB} \pm 2a\hat{\imath}$, etc., where R_{AB} is the vector between the atoms in the basis. $\hat{\imath}$ and $\hat{\jmath}$ are unit vectors in x and y, respectively.

The crystal states can be expanded as

$$\psi_{nk}(\boldsymbol{r}) = \frac{1}{\sqrt{N}} \sum_i c_{ik} \Phi_{ik}(\boldsymbol{r}) \tag{6.50}$$

$$\psi_{nk}(\boldsymbol{r}) = \frac{1}{\sqrt{N}} \sum_i c_{ik} \sum_{\boldsymbol{R}_i} e^{i\boldsymbol{k}\cdot\boldsymbol{R}_i} \phi(\boldsymbol{r} - \boldsymbol{R}_i) \tag{6.51}$$

But how do we find out what the coefficients c_{ik} are? Clearly, the "correct" or physical wavefunction is the one that minimizes the energy in the eigenvalue problem. This means we will have to turn to some *variational principle* to determine the coefficient choices that yield the lowest energy state. Stated in another way, in the above case with only one atom contributing per unit cell, this step was unnecessary. The modulation of the single particle wavefunction was clear because the isolated atoms had only a single spacing and multiples thereof to contribute to the sums. Now that there are multiple atoms contributing at various positions within the unit cell, this translational symmetry is much more complicated, and the balance of how the orbitals add up to yield the band of interest must be determined.

Applying the variational principle to the energy expectation value, we get a set of simultaneous equations:

$$\sum_i (H_{ij} - \delta_{ij} E(\boldsymbol{k})) c_{ik} = 0 \text{ where } H_{ij} = \langle \Phi_{ik} \mid H \mid \Phi_{jk} \rangle \tag{6.52}$$

This means the determinant must be zero:

$$|\boldsymbol{H} - E(\boldsymbol{k})\boldsymbol{I}| = 0 \tag{6.53}$$

where we have written the matrix of the H_{ij} elements as $\boldsymbol{H}$ and $\boldsymbol{I}$ is the identity matrix.

In our two-atom basis example, this means

$$\begin{vmatrix} H_{AA} - E & H_{AB} \\ H_{BA} & H_{BB} - E \end{vmatrix} = 0 \tag{6.54}$$

where we note that obviously $H_{AB} = H_{BA}^{*}$. This simple quadratic equation has two solutions expressed as

$$E(\boldsymbol{k}) = -\frac{1}{2}(H_{AA} + H_{BB}) \pm \sqrt{\frac{1}{4}(H_{AA} - H_{BB})^2 + |H_{AB}|^2} \tag{6.55}$$

This means that we get two solutions for E at every k point or two bands occur! Now we calculate the Hamiltonian matrix elements in exactly the same way we did above for the s band example. For our *trans*-polyacetylene example, each carbon atom is contributing a single p orbital to the *conduction* and *valence* bands. Notice that the s bands are also there, holding things together, but we are interested in the bands responsible for the flow of current. So for the p bands, we have

$$H_{AA} = \frac{1}{N} \sum_{\boldsymbol{R}_A} \sum_{\boldsymbol{R}'_A} e^{i\boldsymbol{k}\cdot(\boldsymbol{R}'_A - \boldsymbol{R}_A)} \int \phi_s^*(\boldsymbol{r} - \boldsymbol{R}_A) H \phi_s(\boldsymbol{r} - \boldsymbol{R}'_A) d\boldsymbol{r} \tag{6.56}$$

$$H_{AA} = \sum_{R''_A} e^{ik\cdot R''_A} \int \phi_s^*(r - R_A) H \phi_s(x - R''_A) dx \tag{6.57}$$

$$H_{AA} = \varepsilon_p + \sum_{m\neq 0} e^{imka} \gamma(|ma|) \tag{6.58}$$

We have used the index m to count over the orbitals along the chain. It can take on positive and negative values. k lies along the x direction. We follow exactly the same for H_{BB}.

We can now make a few assumptions about the nature of the p_z orbital so as to allow for a solution of the integrals. Let's assume that the overlap integral falls off so fast that it can be restricted to having a significant value only for distance $< a$. This means that $\gamma(ma) \sim 0$ for $|m| > 1$. Specifically, this leaves us with

$$H_{AA} = H_{BB} = \varepsilon_p + 2\gamma(a)\cos(ka) \tag{6.59}$$

Now what about H_{AB}? This is the overlap between the bases of a single cell:

$$H_{AB} = \frac{1}{N} \sum_{R_A} \sum_{R_B} e^{ik\cdot(R_A - R_B)} \int \phi_s^*(r - R_A) H \phi_s(r - R_B) dr \tag{6.60}$$

$$H_{AB} = \frac{1}{N} \sum_{R'_A} e^{ik\cdot(R_{AB} + R'_A)} \int \phi_s^*(x) H \phi_s(x - (R_{AB} + R'_A)) dr \tag{6.61}$$

Now we want only the overlap integrals between nearest neighbors to show up in the sums. For the R_A translation, this means $R_a = 0$ and the $R_A = -a\hat{\imath}$. This then leaves only

$$H_{AB} = \sum_{\tau} e^{ik\cdot\tau} \gamma(|\tau|) \tag{6.62}$$

where the τ now must refer to the distance between the basis atoms: $\tau = R_{AB} = (a/2, a/2\sqrt{3})$ and $\tau = R_{AB} - a\hat{\imath} = (-a/2, a/2\sqrt{3})$. But now we must do something pretty peculiar. Remember that k lies only in x: *trans*-polyacetylene is 1D. So when this $k\hat{\imath}$ is dotted into the position vectors, it simplifies the sum into the components along x only:

$$H_{AB} = (e^{ika/2} + e^{-ika/2})\gamma(|\tau|) \tag{6.63}$$

$$H_{AB} = 2\cos(ka/2)\gamma(|\tau|) \tag{6.64}$$

We have, in fact, seen this compression of dimension before.

So we now have H_{AA}, H_{BB}, and H_{AB}. We substitute into the equation above, and we get the energy dispersion curve

$$E(k) = \varepsilon_p + 2\gamma(a)\cos(ka) \pm 2\cos(ka/2)\gamma(|\tau|) \tag{6.65}$$

This is for our special form of *trans*-polyacetylene with all of the simplifying assumptions we have made. Notice the role that the overlap integrals are playing here in determining bandwidth. Plotting this out using a software package such as Mathematica® is straightforward and left to the reader.

6.2.5.1 Hybridization

There is another important case to consider in tight binding. Up until now we have taken the bands to have all of one type of character: s bands, p bands, d bands, etc. Indeed, at the beginning of the section, we explicitly stated that we would take only atomic orbitals that were orthonormal to form our basis of expansion into Bloch states. But in general, bands will contain contributions from more than one type of orbital when they are available. An important example is that of Si and C in the diamond form. In such cases the bands associated with the lowest energies in the valence band and the highest energies in the conduction bands come about due to a mixing of four orbitals: s, p_x, p_y, and p_z. The process is known as *hybridization*, and this particular form of hybridization is known as *sp_3 hybridization*. This is because a mixture of one s and three p orbitals is used to form the band. Similarly, the in-plane bands of graphene are actually sp_2 hybridized. (The p_z orbital makes up another band.)

So how do we expand our treatment to account for this? We begin by indexing the different orbital types as well as the basis atoms. How does this work? Well, consider $N_b = 2$, a two-atom basis. Further imagine each basis atom contributes a s, p_x, p_y, and p_z orbital. As above, we will clearly need to optimize the wavefunction, and this will give us a set of simultaneous equations to solve. In fact, it will be 8 bands (2×4) and an 8×8 matrix eigenvalue problem we must solve if we want the energies for each of the k points. To construct the Hamiltonian matrix elements, we need to enumerate (label) all the orbitals that are at the party. So we have s through p_z on the A atom and the B atom.

Figure 6.32 is a schematic of a possible labeling scheme. The different individual orbitals are added together to get the funny offset lobes at the right of the figure. So, in our example atoms A and B look like these little tetrahedral: the *jacks* that children play with. Indeed, many of the semiconductors do this such as Si. This is an example of sp_3 hybridization as is found in Si and diamond, but one can

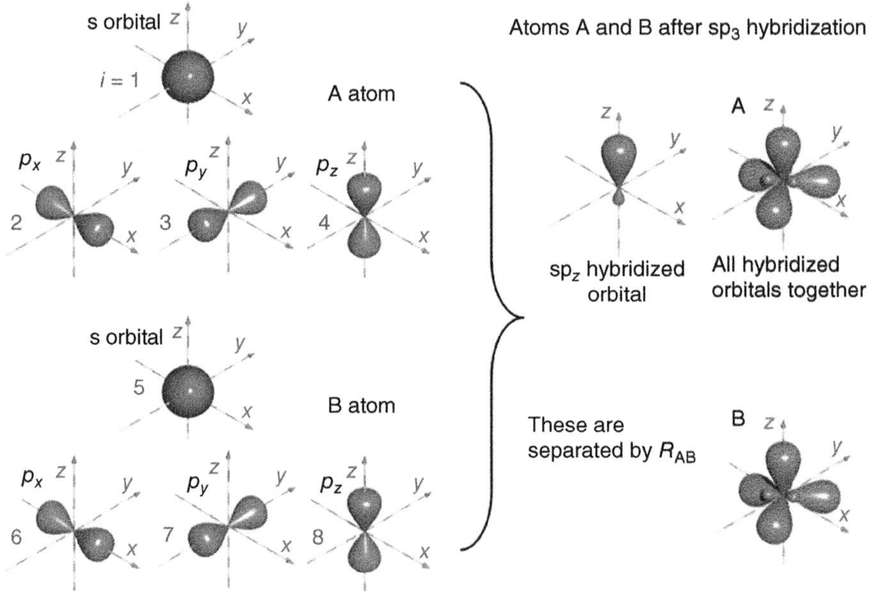

Figure 6.32 Atomic orbitals combine to form a new hybridized orbital that is noncentrosymmetric.

easily imagine the sp_2 hybridization of graphene where the hybridized lobes are in plane and one p_z orbital sits out of plane ready to form its own bands. We note that for the sp_3 case, the orbitals are not planar, but in fact sit at an angle below the x–y plane. The lobes try to get as far apart as they can.

To build the Hamiltonian matrix, we simply run through the indices. Let's take $i = 1$ as the s orbital on atom A and $j = 6$ as the p_x orbital on atom B. Then,

$$H_{ij} = \frac{1}{N}\sum_{R_A}\sum_{R_B} e^{ik\cdot(R_A - R_B)} \int \phi_s^*(r - R_A) H \phi_{p_x}(r - R_B) dr \tag{6.66}$$

And we must now solve

$$|H - E(k)I| = 0 \tag{6.67}$$

to get our dispersion curve.

6.2.5.2 Graphene: A Two-Dimensional Example

Tight binding relies on determining the atomic positions in the lattice, the orbitals taking part in the bands, and the overlap of those orbitals. Once you have the geometry and the orbitals, everything else seems to fall into place. It is all based on the assumption that the electrons are fairly well localized.

Using the examples above, the tight binding bands of graphene are pretty easy to get at. Of course, both the real space vectors and the k vectors lie in x and y. So we have a two-dimensional problem to contend with.

First, the geometry: Previously we argued there that the honeycomb lattice of graphene was actually trigonal with a two-atom basis. That is, the hexagonal lattice of graphene can be thought of as being a superposition of two trigonal carbon sublattices. There are two nonequivalent lattice sites that the electrons might propagate through for any path through the crystal. To compute the band structure using *tight binding*, we consider a hexagonal Brillouin zone.

Next, the orbitals: The orbitals that contribute to conduction in graphene are associated with the p_z orbitals of the carbon atoms at each hexagonal site as in *trans*-polyacetylene. These p_z orbitals can become delocalized over the lattice to allow for conduction. There are also the hybridized sp_2 orbitals in plane. This was also true in our *trans*-polyacetylene example as well as the chemical picture we presented for polymers above. You might say that the work of holding the structure together is done by the hybridized bands, while the p_z's are responsible for conduction. Of course this is a bit simple and not altogether accurate, but it is a good approximation.

We begin with examining the p_z bands. These are the "bunched together states" we showed in the chemical picture above. As we noted, in two dimensions, the work is a little harder since we must designate k_x and k_y directions in reciprocal space and work out the dispersion along these directions. The band structure as derived by simple tight binding is given as

$$\varepsilon(k) = \pm t\sqrt{1 + 4\cos^2\frac{k_y a}{2} + 4\cos\frac{k_y a}{2}\cos\frac{\sqrt{3}k_x a}{2}} \tag{6.68}$$

where t is the energy of hopping between sites (the overlap) and a is the lattice parameter. We note that at six points, which represent the BZ edges, $k = \pm 2\pi/a(0, 2/3)$, $\pm 2\pi/a(-1/\sqrt{3}, 1/3)$, $\pm 2\pi/a(1/\sqrt{3}, 1/3)$, $\varepsilon(k) = 0$.

If we expand the expression above around any of these points, we get

$$\varepsilon(k) = \pm\hbar|v_{\mathrm{f}}| \tag{6.69}$$

$$v_{\mathrm{f}} = \sqrt{3}ta/2\hbar \tag{6.70}$$

Since there are two electrons per unit cell, these bands are filled right up to the BZ boundaries. Thus we have used v_{f} to represent the Fermi velocity, and these are the "Fermi points" or sometimes called the "Dirac points." We note that there is no energy gap in the bands. So graphene is a gapless semiconductor!

What about the rest of the bands? The above is only for the p_z orbitals, which is kind of stand-alone. The other set of orbitals hybridize: s, p_x, and p_y. So while the p_z bands come from solving the two-atom basis matrix equation (a 2×2 matrix), to get two bands, the remaining hybridized orbitals give a 6×6 matrix equation for 6 bands. If we use the numerical values for the overlap integrals suggested by [3], we get the bands shown in Figure 6.33.

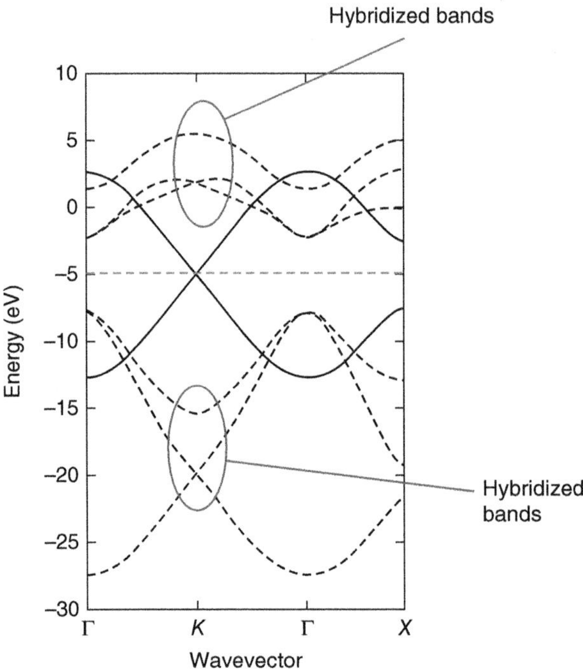

Figure 6.33 The tight binding bands of graphene. Dashed lines stand for the bands that are derived from the hybridized orbitals. The solid lines are the p_z orbital-derived bands. This figure was taken from Professor Roy's notes but can be easily reproduced using a software package such as Mathematica and the literature values of the overlap integrals. Note that the electrons occupy states as high as the light grey dashed line horizontal at −5 eV – a gapless semiconductor. The set of three upper hybridized bands are completely unoccupied. The three lower hybridized bands are completely occupied. It is also important to note the symmetry points of the wavevector chosen for the two-dimensional representation of the bands.

6.2.5.3 Carbon Nanotubes

Carbon nanotubes are a bit of a strange beast. On the one hand, it can be thought of as the simple graphene we have just studied rolled up into a tube and Born–von Karman boundary conditions applied to the circumference. So k_{axis} and $k_{circumference}$ are broken out as the relevant orthogonal directions in k space. This would require that a half integral number of electron waves (wavefunctions of electrons) on the tube's surface must fit around the circumference if we are to employ Born–von Karman boundary conditions. The part of the wavefunction heading down the tube axis, however, looks like a plane wave.

However, k_{axis} and $k_{circumference}$ do not sit collinear with the graphene sheet that was used to make the nanotube. As we have already learned, the tube can be constructed in such a way that the graphene lattice spirals (at a chiral angle) along the axis of the tube. As we have conjectured, the Fermi surface of the graphene looks like six points on the edge of a hexagon as seen in Figure 6.34. When we overlay this hexagon of points with the tube axis, they do not align.

Figure 6.34 shows this hexagon tilted with respect to the axial direction of the tube, and the amount of tilt is given by the chiral angle as shown. However, the states associated with the circumference appear as straight lines along the direction of vector $\boldsymbol{R}$. The points on the vertices of the hexagon (the Fermi surface) are the only electrons allowed to participate in conduction of the object. The states associated with the circumference are the only ones that are allowed so that the wavefunction fits on the tube.

The angular position of the hexagon is determined by the angle that the graphene is oriented relative to the rolling axis. When the hexagon lies at an angle such that the hexagonal vertex intersects one of the circumference states, the tube has electrons that are mobile, and it is a rather good conductor. However, when the hexagon vertex misses the circumference state, the electrons at the vertex energy must be given some energy to make a transition into a conducting state. That is, the tube acts like a semiconductor. So, the nanotube can be either a semiconductor or a metal, depending on its chiral angle. Moreover, the gap

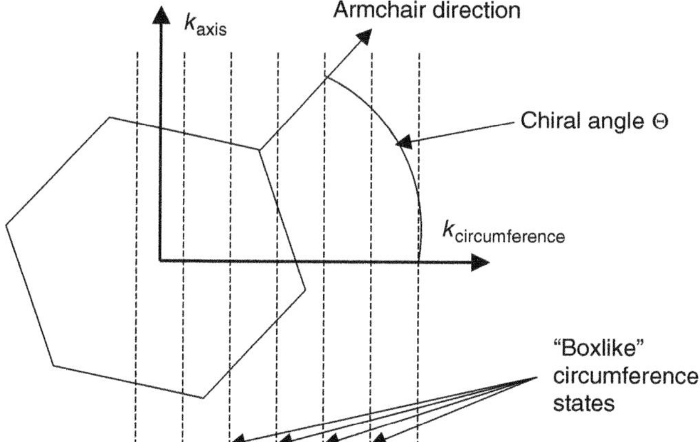

Figure 6.34 The Fermi points of graphene sit at the points of the hexagon shown. However, this hexagon is oriented with respect to the axis of the tube.

depends on the angle and the diameter of the tube! Since we can know the chiral angle in terms of (n,m), we can use a geometric argument to see that if $(n - m)$ is not divisible by 3, then the tube is a semiconductor; otherwise it is a metal. Further, two-thirds of the possible nanotubes will be semiconductors, and only one-third metals. Though we don't show it here, it is clear that as the tube diameter grows, the spacing between the circumference states gets more narrow. Thus, the bandgap of a nanotube is $E_g \sim 1/\text{DIA}$. Lastly, our example above is clearly a semiconductor. This is, of course, only a little hand waving and not a true tight binding argument. But perhaps you can see why people have gained such a fascination for this beautifully symmetric system.

Is this system truly 1D? In what way can we tell? As mentioned earlier, the electronic signature of truly 1D behavior is the occurrence of van Hove singular points in the electronic DOS. Do we, in fact, observe these? There is a way to measure such DOS on nanoscale objects using scanning tunneling microscopy and spectroscopy. In tunneling microscopy, the image is collected by scanning the surface with an atomically sharp tunneling tip. The image contrast is generated through subtle variations in the tunneling current. When one wishes to know the electronic structure at a specific place in the image, the tip is stopped from scanning, feedback systems are disengaged, the voltage is ramped, and current is collected. This tunneling spectrum is then differentiated to yield a differential conductivity. The tunneling differential conductivity is normalized by a term that varies as the tip height above the object, and the result of this normalization is proportional to the density of electronic states in the area where the spectrum was taken. In a stunning set of experiments, two sets of researchers, one at Delft [4] and one at Harvard [5], correlated the electronic structure of single-walled nanotubes with the atomic structure as determined using scanning tunneling microscopy. Since the tunneling microscope can image at atomic resolution, the researchers were able to show that the set of van Hove singularities predicted for a given set of (n,m) were exactly seen in tunneling spectra. These are seen in Figure 6.35.

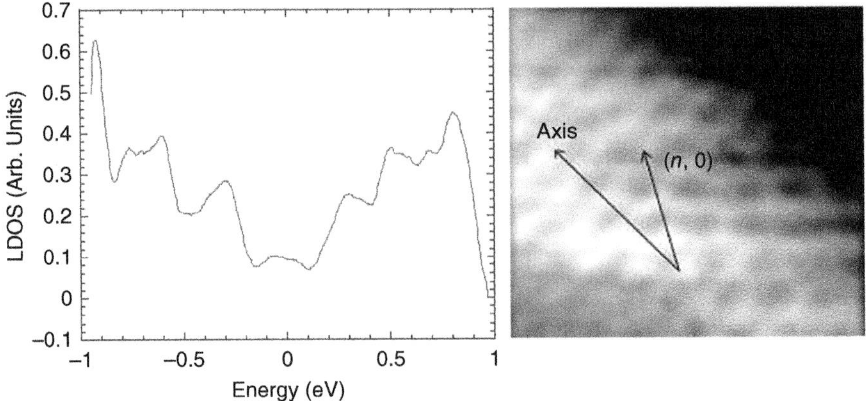

Figure 6.35 Tunneling micrographs and density of electronic states showing the occurrence of van Hove singularities in single-walled carbon nanotubes. Source: Courtesy of R. Czerw Wake Forest University.

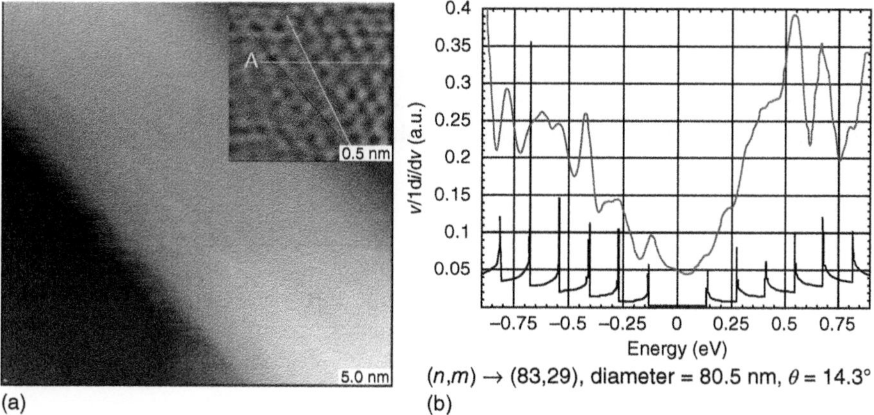

Figure 6.36 (a) The tunneling spectra, converted to electronic density of states, for a multiwalled carbon nanotube. (b) The spectra have been plotted the most simple tight binding prediction for this tube. Source: Courtesy of D. Tekleab, Clemson University.

In fact, van Hove singular points were identified in multiwalled carbon nanotubes earlier. These objects are simply increasingly larger diameter single-walled nanotubes placed into a concentric configuration, giving more than one wall. The interesting point here is that it appears as though there is little cross talk between the shells. In other words, the outer shell of the multiwalled nanotube is a 1D object as well, without regard for what is inside (Figure 6.36).

This leaves very little doubt that carbon nanotubes can reflect a 1D electronic nature. How this 1D nature has manifest itself in thermal, optical, and transport properties of both single-walled and multiwalled carbon nanotubes has been intensely studied over the past 10 years [6]. Further, the role of symmetry breaking in such objects through defects [7], kinks [8], bends [9], and dopants [10] has also been of significant interest to the scientific community. In each case, however, the 1D nature of the nanotube system must be modified with respect to the extra-dimensional degree of freedom around the waist of the tube geometry. This topological modification is especially clear for thermal transport, where the vibrational degrees of freedom must include the so-called "twiston" or the twisting of the tube about the axis as mentioned before [11]. In some sense one might make the claim that a polymer system could exhibit a twisting of the atoms about the axis, and this would be analogous to the twiston. This serves to demonstrate the limited applicability of the concept of "1D" materials.

6.3 Are We Done Yet?

No. There are many approximations and variations on themes we have presented. All are intended to provide better insight into localization and electronic orbital choice. However, such refinements are best left for conversations among specialists in band structure calculations. The *ansatz* is that such refinements do not change the basic meaning of this chapter: single particle states react to their interaction with the static lattice by forming bands.

Nevertheless, there is a final and rather important point that should be made. We have taken a perspective that the only interactions can be found in the forces between the electrons and the ion cores of the lattice. This interaction is mediated by the Coulomb potential represented by a bare ionic core, or some approximate to it, and some electron *screening* placed right at this core's location (this essentially lowers its absolute value).

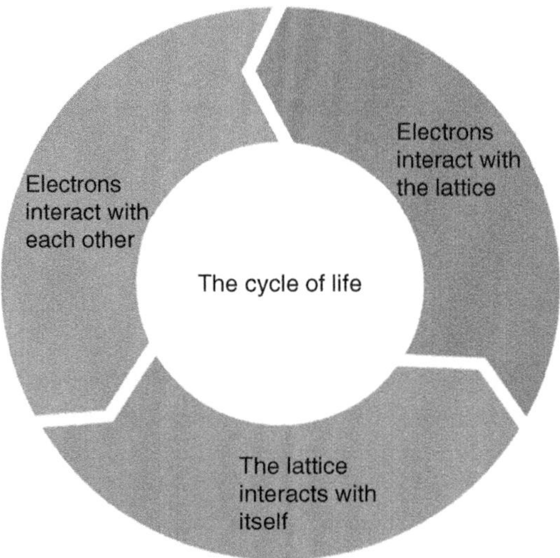

In reality, however, the electrons interact with a dynamic lattice as we saw from Chapter 5 and with each other. These interactions "fill" the excitation spectrum of the many-body state of the system. The lattice, of course, interacts with itself (ion–ion). This allows us to have the dynamics of the lattice that yields phonons. Electron–phonon and phonon–phonon interactions both seem fairly clear, but what about *electron–electron interactions*?

There are, in fact, several different types of electron–electron interactions that we have ignored; or if we didn't, we certainly skipped over their significance. They are the following:

1. *The Coulombic interactions* between the electrons themselves: Typically, we might imagine this as the average electrostatic potential of the electrons and the ions fixed in space, as they push and pull against each other. This simple picture of the electromagnetic environment of the electron in a solid is referred to as the *Hartree approximation*. And the *Hartree potential* is dependent on the electron density.
2. *Pauli exclusion*: This disallows any two electrons to be in the same state. We have seen this before in terms of the statistic of state filling.
3. *Exchange interactions*: These interactions are due to the Pauli principle. So if two electrons have parallel spin, they cannot sit at the same place at the same time. This gives rise to an effective repulsion, and the electrons interact not only via Coulomb but also through their spins.

4. *Correlation interactions*: This is also part of the Pauli principle. In this case there is a correlation of motion between electrons with antiparallel spins due to their Coulomb repulsion.

There are essentially a couple of ways to deal with these interactions. The first is the *Hartree–Fock formalism*. In this approximation the Hartree potential is used in the Hamiltonian, but exchange interactions are forced by insisting on antisymmetric wavefunctions. This has the effect of lowering the total binding energy of the atoms by keeping the parallel spins apart. Its weakness is that it ignores correlated behavior between electrons with antiparallel spins.

The second major approach to this problem is *density functional theory* (DFT). Since the exchange and correlation as well as the Hartree potential are all related to the density of the electronic wavefunction, DFT treat density as a fundamental quantity of the system. So, unlike Hartree–Fock that deals with the electronic wavefunctions, DFT deals primarily with the electron density. The density in the case is a functional – a function of a function – and it depends on space and time. By doing this the degrees of freedom are reduced, and the calculation is much faster than Hartree–Fock. Further, Hohenberg and Kohn showed that the density of any system uniquely determines the ground state properties of that system (the Hohenberg–Kohn theorem). So if we know the electron density functional, we know the total ground state energy of our system.

By focusing on the electron density, it is possible to derive single particle equivalents of the SE. We can then separate out the total energy of the system into individual expressions for the different contributions, all of which will be written in terms of density functionals. The terms will be:

1. Ion–electron potential energy
2. Ion–ion potential energy
3. Electron–electron potential energy
4. Kinetic energy
5. Exchange correlation energy

The derivations and mathematical details are beyond the scope of our discussion here. However, there are some excellent texts on DFT available.

6.4 Summary

In this chapter we have suggested that the electronic structure of solids can be described by single particle wavefunctions and states. The energies of these states are collected together in bands, the shape and curvature of which are described by the specifics of the potentials and the overlaps of the wavefunctions themselves. In metals, a reasonable approximation to these single particle states is plane waves bound in the box of the solid's volume. But in semiconductor materials, the individual lattice sites hold on to their electrons more tightly, and the single particle state's wavefunction becomes strongly modulated by the natural wavelengths of the system's lattice. As we have shown the tools used in physics and chemistry to describe this process are essentially the same, but the language can be quite different. Low dimensions and nanoscale structures really require the scientist to think in both ways.

Exploring Concepts

1 *The Progression of Models*: Consider the simple schematized presentation of the models (Figure EC6.1) we have examined in this chapter.
   (a) Go back through the text, and write down the dispersion curve $E(k)$ for the one-dimensional example of all three models.
   (b) Explain the role of Pauli exclusion in the free-electron model.
   (c) In the nearly free-electron model, what is physically happening near the BZ edge that leads to the formation of an energy gap?
   (d) In the tight binding model, we describe the model in both the physicist language and the chemist language. Explain how these two pictures unite to give an idea of the character of the conduction and valence bands (Figure EC6.1).

2 *Dirac Delta Potentials in One Dimension*: Imagine the one-dimensional lattice, with lattice spacing a, and each atomic potential represented by

Drawing the Pictures of Our Models

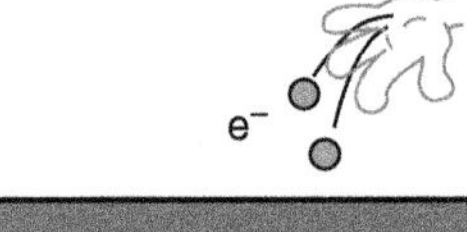

The free-electron model

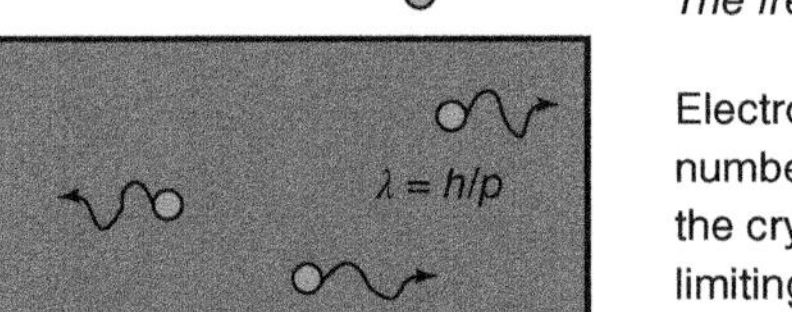

Electron waves must fit into the box with an even number of 1/2 wavelengths. The metaphorical box is the crystal with the facets, surfaces, and boundaries limiting the motion of the electron, but not much else

The nearly free-electron model

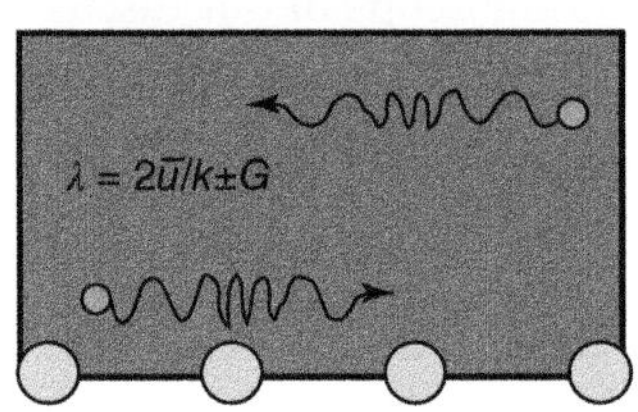

The electrons "feel" the perturbations of the ionic cores but only very weakly. This modulates their wavefunctions when proximal to the core in a way related to the Fourier transform of the local potential. In doing so, some k's and thus some E's are forbidden, giving rise to a bandgap in electronic energy states

Tight binding theory

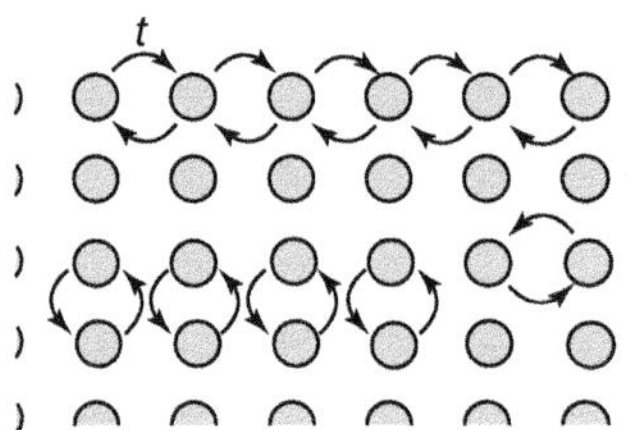

Finally we come to tight binding where the electrons are strongly localized to the cores. Electrons are more imagined to "hop" from site to site as they move through the lattice. The ability to "hop" depends on orbital overlap between nearest neighbors. These electrons are related to bonds

Figure EC6.1 A brief reminder of the models we have examined.

the potential: $V(x) = aV_0\ \delta(x)$. We introduced this in the text as the Dirac comb, but we didn't provide any of the details. Following the example of Kronig–Penney for the *nearly free-electron model*:

(a) Determine E_{gap}, $E(k)$ and the ground state wavefunction.
However, depending on the value of V_0, the energy of the electrons, etc., it can be that the *tight binding approach* is more appropriate. Now, we are asking something rather funny because in the text we have associated tight binding with the use of atomic orbitals (LCAO), and there are no atomic orbitals to be found on delta function potentials. In fact this association isn't quite rigorous, and the principles can be applied to a number of wavefunction choices.

(b) Start with a wavefunction that looks like

$$\psi = A\mathrm{e}^{-k|x|}$$

centered at each point in the lattice. With

$$E = -\frac{\hbar^2 k^2}{2m}$$

and

$$-\frac{\hbar^2}{2m}\left(\left.\frac{\partial\psi}{\partial x}\right|_+ - \left.\frac{\partial\psi}{\partial x}\right|_-\right) + \psi a V_0 = 0$$

find a value for both A and k.

(c) In the tight binding approximation of $ak \gg 1$, show that $E(k)$ is given by

$$E(k) = E_0 - \frac{\beta + 2\gamma\cos ka}{1 + 2\alpha\cos ka}$$

where

$$\alpha = \int \mathrm{d}x\psi^\dagger(x)\psi(x-a)$$

$$\gamma = -aV_0\int \mathrm{d}x\psi^\dagger(x)\partial(x-a)\psi(x-a)$$
$$- aV_0\int \mathrm{d}x\psi^\dagger(x)\partial(x+a)\psi(x-a)$$
$$- aV_0\int \mathrm{d}x\psi^\dagger(x)\partial(x\pm 2a)\psi(x-a)-$$

(d) Show that, under this approximation, this then reduces to

$$E(k) = E_0 - 2\gamma coska$$
$$E(k) = E_0 + 2V_0ake^{-ka}coska$$

(e) Find a general expression for the bandwidth W.

(f) For this potential, show that the electron energy and wavenumber satisfy the relation

$$\cos ka = \frac{\kappa}{K}\sin Ka + \cos Ka$$
$$K^2 = 2mE/\hbar^2$$
$$\kappa = \alpha V_0$$

(g) Calculate the energy gap between the bands or the case where the potentials are weak:

$$V_0 \ll \hbar^2/ma^2$$

Compare this with the nearly free-electron approximation from above.

(h) Now calculate the bandwidth for the strong potential

$$V_0 \gg \hbar^2/ma^2$$

and compare with the result from the tight binding approximation.

3 *The Square Lattice*: In this problem we return to the nearly free-electron model. This time we construct a square lattice with lattice constants: a. It has a general potential of $V(\boldsymbol{r})$, and the Fourier transform of that potential is $V(\boldsymbol{G})$: $\boldsymbol{G}$ is a reciprocal lattice vector. In this case the potential is relatively weak, and so the exact functional form will not be as important. We will write many of our answers in terms of the potential itself.

(a) First we want to investigate band structure around some special points in the reciprocal lattice space, specifically, around the zone edges. First we note that $E(k)$ has a fourfold degeneracy at $(\pi/a, \pi/a)$. Only the $\boldsymbol{G} = (0, 2\pi/a)$ and $\boldsymbol{G} = (2\pi/a, 2\pi/a)$ terms in $V(\boldsymbol{G})$ are important for this nearly free-electron case. Why?

(b) Now let's set $V(0, 2\pi/a) = V_0$ and $V(2\pi/a, 2\pi/a) = 0$. What does this correspond to physically? Find the bandgap with these assignments. How does that bandgap change when $V(0, 2\pi/a) = 0$ and $V(2\pi/a, 2\pi/a) = V_1$?

(c) Sketch the bands along $\Gamma - W$ and $W - X$.

(d) Now let's add the electrons to these states. For one electron per lattice site, sketch out the Fermi surface in (k_x, k_y). Determine if this is an insulator or a metal.

(e) Repeat (d) for two electrons per lattice site.

4 *Tight Binding (LCAO) of the CuO_2 Bands in Superconductors*: Let's talk about superconductors. In type II superconductors (the 2D ceramic kind), CuO_2 planes play a pivotal role in electron–electron coupling as we will discuss later in more detail. For now, let's see if we can calculate the band structure of a 2D structure of this compound using the tight binding approximation.

To begin, we will imagine CuO_2 as a square lattice with three atoms per unit cell (Figure EC6.2).

Interestingly the bonding between the orbitals that we have suggested for the system is actually a mix between ionic and covalent characters.

(a) Following the discussion in the text, show that the tight binding equations for this structure are

$$(E_{\text{Cu}} - E)A_{m,n} + t(R_{m,n} + U_{m,n} + R_{m-1,n} + U_{m,n-1}) = 0$$
$$(E_0 - E)R_{m,n} + t(A_{m+1,n} + A_{m,n}) = 0$$
$$(E_0 - E)U_{m,n} + t(A_{m,n} + A_{m,n+1}) = 0$$

Now if you are following along, you will know what each of these variables and symbols is, so we are not going to tell you other than $A_{m,n}$ is the

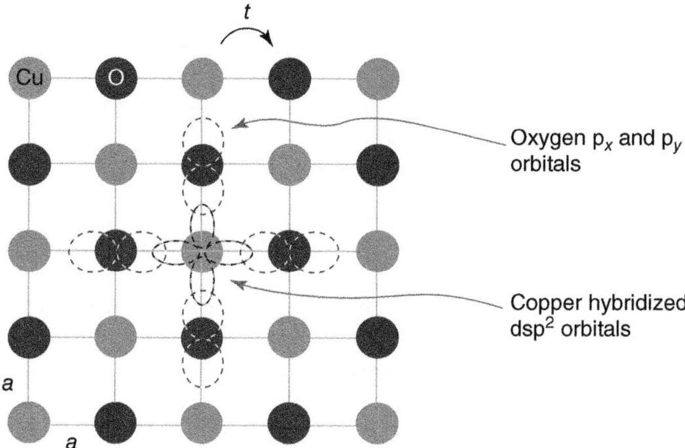

Figure EC6.2 Our model for the CuO_2 sheet. Lattice constants are a, and hopping integral is t. The interaction is actually mediated through the p_x/p_y orbitals of the oxygen and the hybridized dsp^2 orbitals of the copper.

amplitude of the wavefunction on the copper atoms located at (m,n). And remember that (m,n) is the (m,n)th cell.

(b) Next insert expressions for the wavefunction amplitudes:

$$A_{m,n} = A\exp[ia(mk_x + nk_y)]$$
$$R_{m,n} = R\exp[ia(mk_x + nk_y)]$$
$$U_{m,n} = U\exp[ia(mk_x + nk_y)]$$

to get a secular equation with roots as we have done in our examples.

(c) Show that the two solutions or roots to these secular equations are

$$E_{\pm}(k_x, k_y) = \left(\frac{1}{2}\right)(E_{\text{Cu}} + E_0) \pm [(1/4)(E_{\text{Cu}} - E_0)^2 + 4t^2(\cos^2(k_x a)/2 + \cos^2(k_y a)/2)]^{1/2}$$

(d) Now draw the first Brillouin zone and several lines of constant energy so we can get some idea of what the band structure looks like in a two-dimensional projection from the top. Make sure and include the line along $k_x + k_y = \pi/a$. Based on what you have drawn, where would the *free-electron model* be most applicable for this system (in terms of values of k_x and k_y)?

(e) In the tight binding approximation, the bandwidth can be related to the number of nearest neighbors in the system. Specifically, the SE can be written in the form

$$(E_0 - E)A_0 + t\sum_{n=1}^{Z} A_n = 0$$

A_0 is the amplitude on a given site, whereas A_n is the amplitude on the neighboring sites. These are all related by simple phase factors in this approximation, as we just saw. So at some point in the BZ, all the phases are equal to 1. This means the sum is maximized to give $E = E_0 + Zt$. There are also other places within the BZ that give the opposite behavior, and the sum is −1, giving $E = E_0 - Zt$. This gives the bandwidth (B) for the tight binding approximation: $B = 2Zt$.

Notice here that not only are the nearest neighbors important (and the 2D structure is therefore very important in giving very narrow bands), but t also plays a role. Make an estimate of the value of t for this system and then of the bandwidth altogether.

By the way, depending on the specific derivation and the traditions within the field, we have used several symbols for this off-diagonal term we call t. It is the overlap integral between neighboring lattice sites and is given as

$$t = \int \mathrm{d}x\phi^*(m)\boldsymbol{H}\phi(m+1)$$

5 *Shape of Fermi Surfaces*:

(a) The problem above considers the case of the Fermi surface of the 2D square lattice with one or two electrons per lattice site using the nearly free-electron model. If you haven't worked through 3(d) and 3(e), you might want to for this problem. Now examine the hexagonal lattice in the *tight binding model*. Figure EC6.3 sets up the lattice structure schematically, and the bands are discussed briefly in the text when we introduce graphene. First, write down the tight binding bands $E(\boldsymbol{k})$ for this system.

(b) For both cases, one and two electrons per lattice site, draw out the Fermi surfaces, and state whether or not the configuration is a metal or semiconductor.

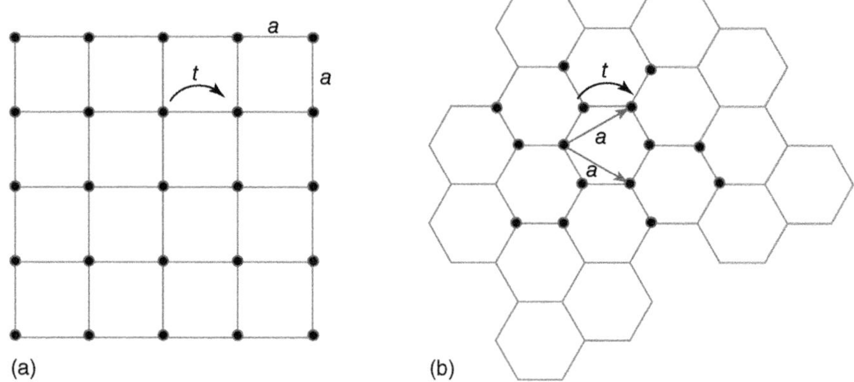

Figure EC6.3 The square (a) and hexagonal (b) lattices. What do their Fermi surfaces look like?

References

1 Physicists special word for *a model* taken fromKuhn, T. (1962). *The Structure of Scientific Revolutions*. University of Chicago Press.
2 Barford, W. *Electronic and Optical Properties of Conjugated Polymers*. Oxford University Press along with lecture notes prepared by Jean-Luc Bedás of Georgia Tech (GA).
3 Popov, V.N. and Henrard, L. (2004). *Phys. Rev. B* 70: 115407. https://doi.org/10.1103/PhysRevB.70.115407.
4 Wildoer, J.W.G., Venema, L.C., Rinzler, A.G. et al. (1998). Electronic structure of atomically resolved carbon nanotubes. *Nature* 391: 59.
5 Odom, T.W., Huang, J.-L., Kim, P., and Lieber, C.M. (1998). Atomic structure and electronic properties of single-walled carbon nanotubes. *Nature* 391: 62.
6 For a detailed and easy to understand introduction to these studies, we recommend Saito, R., Dresselhaus, G., and Dresselhaus, M.S. (1998). *Physical Properties of Carbon Nanotubes*. London: Imperial College Press.
7 Carroll, D.L., Redlich, P., and Ajayan, P.M. (1997). Electronic structure and localized states at carbon nanotube tips. *Phys. Rev. Lett.* 78: 2811.
8 Tekleab, D., Czerw, R., Ajayan, P.M., and Carroll, D.L. (2000). Electronic structure of kinked multiwalled carbon nanotubes. *App. Phys. Lett.* 76: 3594.
9 Tekleab, D., Carroll, D.L., Samsonidze, G.G., and Yakobson, B.I. (2001). Strain-induced electronic property heterogeneity of a carbon nanotube. *Phys. Rev. B* 64: 035419.
10 (a) Carroll, D.L., Curran, S., Redlich, P. et al. (1998). Effects of nanodomain formation on the electronic structure of doped carbon nanotubes. *Phys. Rev. Lett.* 81: 2332; (b) Blase, X. et al. (1999). Boron-mediated growth of long helicity-selected carbon nanotubes. *Phys. Rev. Lett.* 83: 2348; (c) Czerw, R. et al. (2001). Identification of electron donor states in N-doped carbon nanotubes. *Nano Lett.* 1: 457.
11 Kane, C.L., Mele, E.J., and Lee, R.S. (1998). Temperature-dependent resistivity of single-wall carbon nanotubes. *Europhys. Lett.* 41: 683.

7

Electrons in Solids Part II: Spatial Heterogeneity

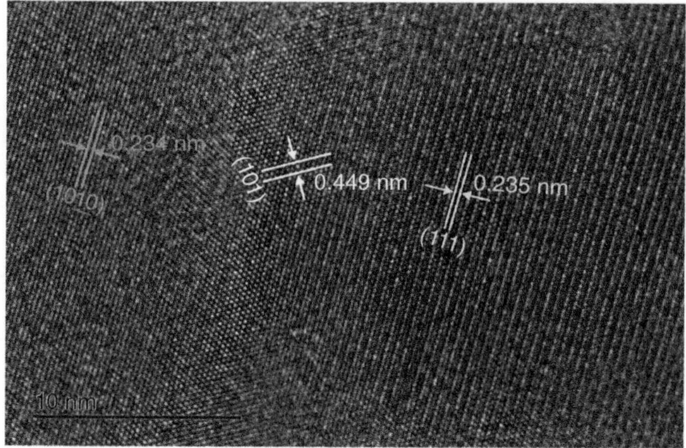

Source: Image courtesy ChaoChao Dun, Wake Forest University and Qike Jiang, Chinese Academy of Sciences, Dalian.

We have, so far, discussed only *homogeneous* systems, perfectly repeating potential landscapes with near-infinite extent. Where *confinement* has occurred, we set up infinite potentials at the material's boundaries so no electron could escape beyond them. A natural point of curiosity then might be the case where dissimilar materials come together forming an interface that allows for electrons to pass or impurity atoms mix randomly in with those of a lattice. These *heterogeneous* situations form the basis for most of solid-state electronics [1].

You might think from Chapter 6 that it could be pretty hard to be quantitative about the electron single particle states in heterogeneous systems. For example, at an interface, the "leaky" terminus to the electronic wavefunction might be thought to add some localized modification to electronic density ($\psi \times \psi$). Additionally, charge might be thought to "pile up" around lattice sites where an impurity has found itself and left uncompensated orbitals. Such local potentials could be hard to guess. In fact, your trepidation would be well warranted. Calculations of band structure near interfaces and defects are notoriously difficult, though they can be done. What is surprising, however, is that a diagrammatic approach to estimating what happens at an interface and with impurities is quite easy to develop

Foundations of Solid State Physics: Dimensionality and Symmetry, First Edition.
Siegmar Roth and David Carroll.

and has tremendous utility. But, as with everything else in our book, dimension can make something simple into something nuanced.

7.1 Heterogeneity: Band-Level Diagrams and the *Contact*

A discontinuous material composition represented by the step function discontinuity is a heterogeneity that we call a *contact*. On one side of "x," we have one material, and on the other side we have another material. Electrons going from one material into another must first exist in an allowed quantum state of material 1 and then enter into an allowed quantum state of material 2. They have some potential landscape to traverse to get from material 1 to material 2, and initial-to-final state transitions must conserve energy. Having said this, it is clear that the alignment of the energy levels on either side of the interface is an essential part of understanding the redistribution of electrons among the states.

To see how we might use this idea, let's start with *metals*: one metal forming an interface with another metal. As shown in Figure 7.1 (right), we introduce the *band-level diagram*. This one describes a simple metal. On the left of Figure 7.1, the relationship between the metal's band structure and band-level diagram can be seen. Such diagrams are intended to provide a quick overview of the energy levels in the bands. Thus, they do not show the bands as a function of k but rather the electron energy levels (relative to some zero) as a function of **x** in the material. Hence, the diagram can capture the spatial heterogeneity.

There are several important things labeled on this band-level diagram. The *chemical potential* or *Fermi level* of the electrons is among them. As we have pointed out before, these are very nearly, though not exactly, the same thing. They are actually derived from different ways of looking at the solid, but their

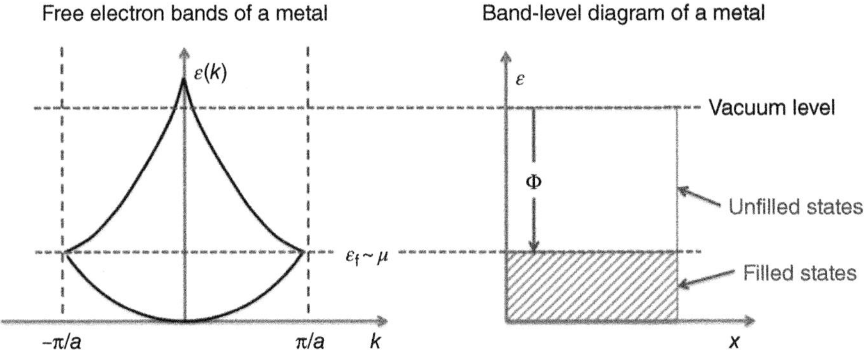

Figure 7.1 Band diagrams in **k** compared with band-level diagrams in **x**. Since the band-level diagram is plotted in x, we can see when electronic energy state changes as we move from place to place in a heterogeneous system. But we have given up **k** information.

equivalence at lower temperatures is frequently taken for granted by physicists. The next of the important labels we find are the *vacuum level* of the system and the *work function*. The vacuum level is lower limit of the energy of the continuum of free particle states in empty space. So, we might imagine it to be the energy of a stationary electron sufficiently removed from the solid so as to set it as a natural zero of potential: electrons trapped within the solid have negative energies relative to vacuum, and electrons free to roam the universe have positive energies. The work function (Φ) is related to this. It is specifically the energy difference at which the electron first overcomes the bounds of the solid. Loosely speaking, the work function is the amount of energy it takes to remove an electron from the topmost occupied state of the system (the Fermi level) and place it at the bottom of the "free" continuum.

These diagrams are particularly useful when it comes to dealing with (trying to visualize) the results of some heterogeneity in the system. This can be seen in the *metal-to-metal* contact shown in Figure 7.2. In the figure, we consider the times both before, during, and after contact is established. At each point, the electronic states in the two materials will follow a few rules. These are reflected by the diagrams.

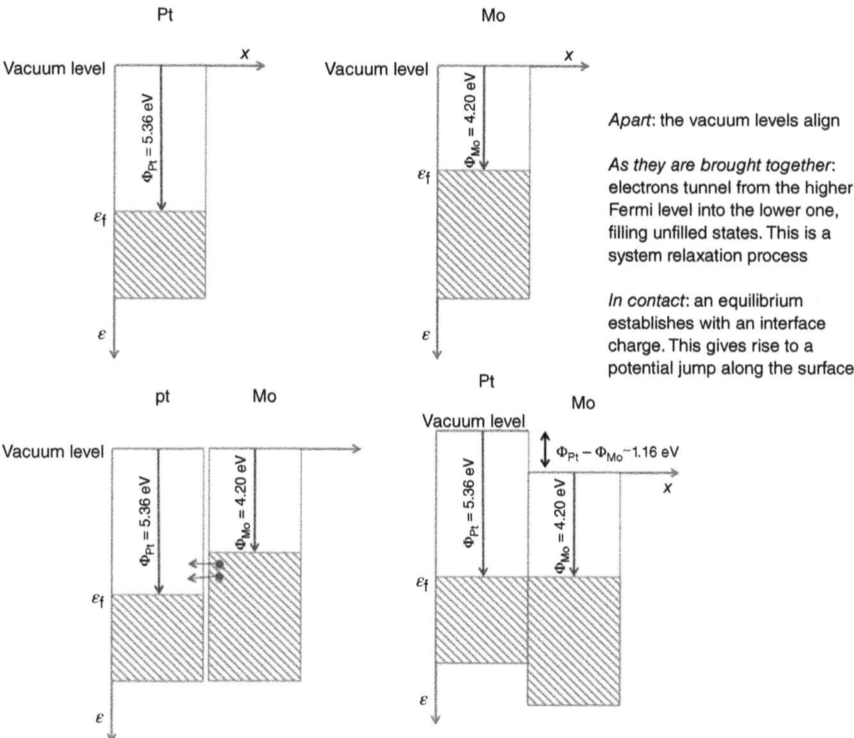

Figure 7.2 An interesting example is contact between Pt and Mo. From far apart to fully formed interface, the process allows electrons to move between the two materials, leaving one material with more electrons than it should have and one with fewer.

Rule #1: "Contact" doesn't always mean atomically well-defined interfaces between materials. Nuances will be discussed a little later, but for now keep in mind that the electrons can tunnel between materials even if the interface is not perfect.

The final diagram in Figure 7.2 (shown bottom right) is the product of a series of imagined events: initially (before contact) all materials share the same vacuum level or zero of potential. Then, when one material begins to contact another:

Rule #2: The Fermi levels will align. In our example, this is because the electrons in Mo have more energy to move than those of Pt. Pt provides those mobile Mo electrons with a place to go: empty electronic states.

In moving from one side of the interface to the other, there is a lowering of the system energy. [*It is pretty straightforward to see this: simply calculate the internal energy of the two systems apart using the tools of the last chapter, and see that it is less than the internal energy of the new system with the new Fermi levels (the range of integration changes). But remember you have to add in the electrostatic energy of the separated charge for both cases.*]

So, as electrons go rushing from one side of the interface to the other and the Fermi level aligns, the vacuum levels above the solids misalign. Zero in potential has not changed of course, but the vacuum level across the interface is now offset by the addition of a voltage. That voltage is $\Phi_{Pt} - \Phi_{Mo}$ in magnitude and comes from a bilayer of charge that is located at the interface. The voltage is known as the *contact potential*. The bilayer of charge is due to the local rearrangement of charge at the interface: on one side, a positive charge relative to the surroundings and on the other a negative. Of course, an extremely interesting situation occurs when one of the solids is of nanoscale. In this case only so many charges can be donated to the interface potential because there are only so many charges in the finite structure. This leaves the nanomaterial side quite charged with tremendous implications for catalytic reactivity.

Of course, in the instant of first contact, there is a lot of dynamics going on. Fields are changing, electrons are rearranging themselves, etc. The precise way in which this works can be interesting, but for now we are only focusing on the final result. That means the system is in some sort of *thermodynamic and static equilibrium*. Under these conditions, this contact potential is only *accessible* by making contacts that will ultimately reverse themselves somewhere else in a circuit, as seen in Figure 7.3. Thus, it can never be used to drive a current. But that doesn't mean it can't be measured. Scanning probe experiments are now certainly sensitive enough to detect it.

Notice that we have not made any statements regarding the correspondence of k values between the states of Mo and Pt that give and receive electrons. But, remembering a little quantum mechanics here, we typically think: (i) electrons will make a transition only into empty states as they move between the materials, and (ii) the most probable transitions should be between states that are similar in the direction of $\boldsymbol{k}$. So, naturally, some mechanism must allow for the $\Delta\mathbf{k}$ as we have discussed before with indirect bandgaps.

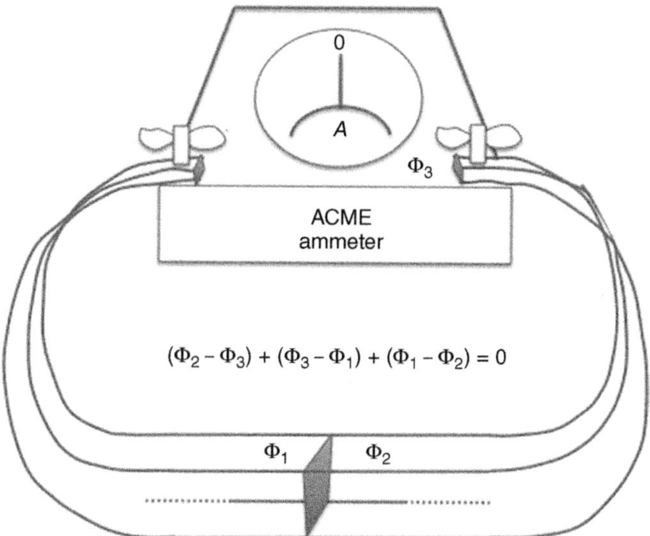

Figure 7.3 No matter how you set it up, closing the circuit means that you subtract out whatever the contact potential was in the junction. Here Φ_3 stands for the contact potential with the ammeter. There is one subtlety however. If you are doing photoemission measurements, you must take into account the work function differences between the sample and the spectrometer. Note, however, this is for thermodynamic equilibrium, meaning that all points of the circuit sit at the same temperature. If, for instance, the ammeter was at a very different temperature than the junction between materials 1 and 2, a current certainly could flow. This is the basis of *thermocouple* thermometry, and it is used all the time in laboratories.

7.2 Heterogeneity in Semiconductors

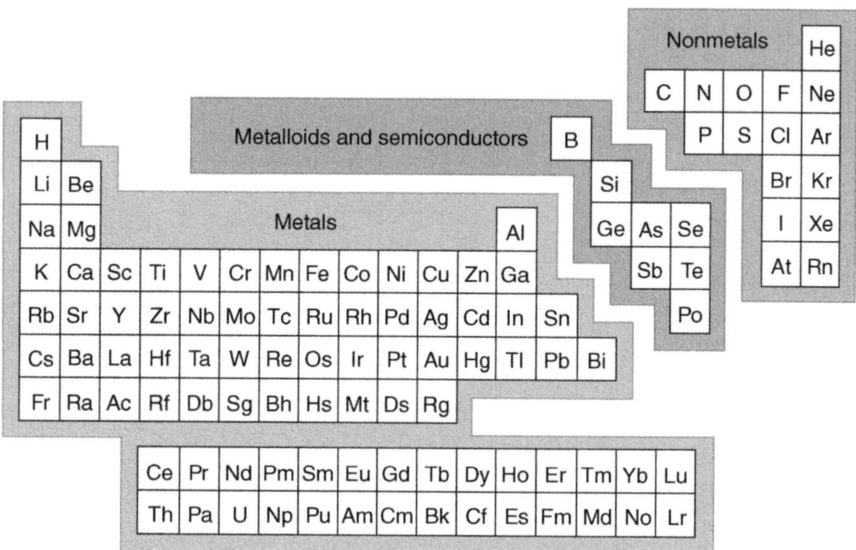

The following is a simple classification scheme for materials:

1. *Insulators and semiconductors*
2. *Semimetals and metals*

Naturally, there is a lot of nuance to material taxonomy that we will not address here. But the distinguishing characteristic between our considerations is how large of a bandgap the material has and what bands are filled. And this is true of other monikers one might choose: inorganic, organic, ceramic, metal, etc. Large, small, and no bandgap materials (insulators, semiconductors, and metals) are all found in our most advanced communications and computational technologies. So we should really explore how all of these materials form junctions/interfaces with all the other forms of materials. Luckily, if we understand the *semiconducting* case, everything else seems to fall into place easily. So let's begin by providing a little more definition and insight into materials with a bandgap (semiconductors).

7.2.1 Semiconductors: Bandgaps and Doping

We have already seen some band structure calculations performed for semiconductors such as Si. And, while we have discussed their properties, it might be best to remember a few of the more interesting points and add one or two new ones:

1. They have a *bandgap* or an energy gap between the HOMO and LUMO. The higher energy unfilled band is called the *conduction band*, and the filled lower energy band is called the *valence band*. There are other gaps between bands in the electronic structure, but they are between filled and filled or empty and empty bands, so they are of less interest.
2. Transitions across this bandgap are necessary to make the semiconductor conductive. For small bandgap materials, this can occur with only thermal stimulation, but for larger bandgap materials, something like the absorption of a photon is needed.
3. Another way of getting electrons into the conduction band without resorting to valence to conduction transitions is by adding electrons that should not have been there. When this is done with impurities, it is known as *doping*, another form of *heterogeneity*.

7.2.1.1 Band-Level Diagrams

The band-level diagram for semiconductors can be a little more challenging than in a metal and allows for more variation as well. In Figure 7.4 we consider an *intrinsic* (no impurities) semiconductor next to metals and insulators.

Now take a closer look at the semiconductor part of Figure 7.4. Figure 7.5 allows us a bit more detailed examination of the components of this diagram. We present the diagrams for three different situations.

7.2.1.2 Doping

As we see in Figure 7.5, the "Fermi level"[1] moves as the semiconductor material is *doped*. But what is this process? As we will see throughout this text, there are

1 Here we use the term Fermi level and *chemical potential*, μ, to be interchangeable, because they almost are. We will see shortly that when we are talking about semiconductors, the more precise term is *chemical potential* for this level.

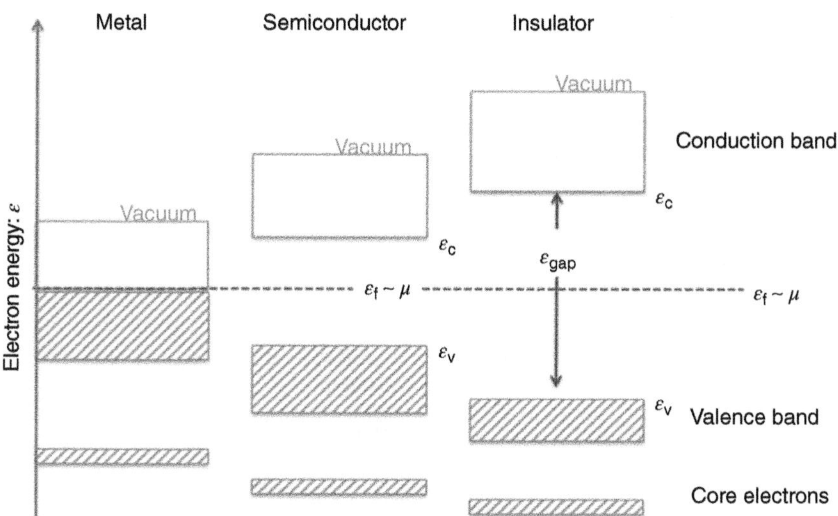

Figure 7.4 A comparison of the band-level diagrams of metals, semiconductors, and insulators. This diagram is shown with the Fermi levels aligned, so the vacuum levels appear offset (don't take this literally yet).

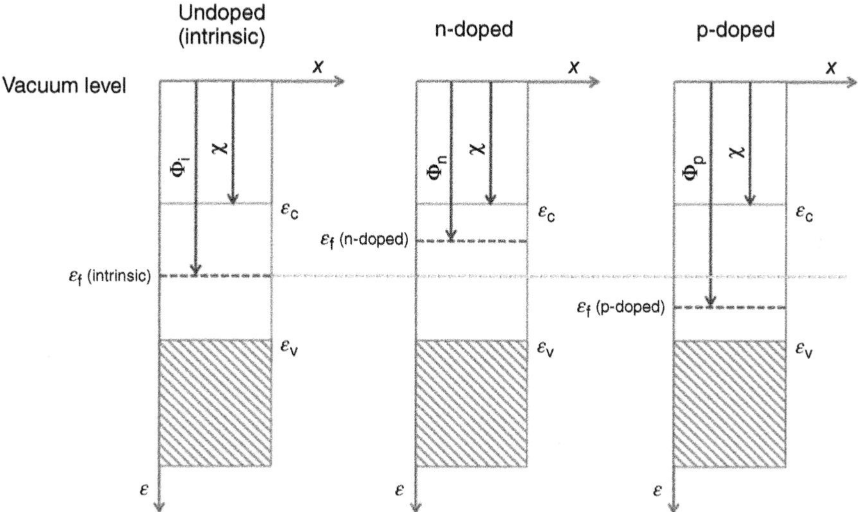

Figure 7.5 The band-level diagrams for a semiconductor in its undoped and n/p-doped states. Here the various quantities we will need to anticipate how junctions will behave have been added. Here Φ still stands for the work function, and the χ represents the *electron affinity* of the material.

many ways to "dope" a material, but generally it can be said that it is the act of adding electrons to, or subtracting electrons from, the existing band structure of a system. We do this by very slightly altering the composition of the material. Here "very slightly" means that the addition of the impurity atoms does not alter the overall shape and function of the bands that have been formed by the existing lattice. So, we are NOT forming a new alloy or new molecular solid; we are "sprinkling" very few foreign atoms in with our crystal's atoms so that our crystal doesn't notice (other than to say "hey, I have too many or too few electrons!"). In actual fact, there will be some additional scattering due to the dopant impurities as well.

To see one way that this is done, let's consider the quintessential semiconductor example: Si. Think of the standard Si structure with the substitution of one of the Si atoms in the lattice for some other atom: known as *substitutional doping* or *extrinsic doping*. What kind of atom should we choose? Recall, as shown in Figure 7.6, that Si will use its four outer shell electrons to form four covalent bonds with neighboring atoms, thereby creating a stable filled outer shell configuration. These are all sp^3 hybrid bonds as we have discussed.

Let's pluck one of those atoms out of its position and replace it with another atom. For this example we will choose some atom that has five valence electrons. Our choices are shown in column V of the periodic table in Figure 7.7. We do have a couple of choices, but notice that not all atoms will fit neatly into the position of the Si atom. It must have about the same atomic radius; otherwise strain will occur and ultimately atomic migration. Let's choose P (shown in green in

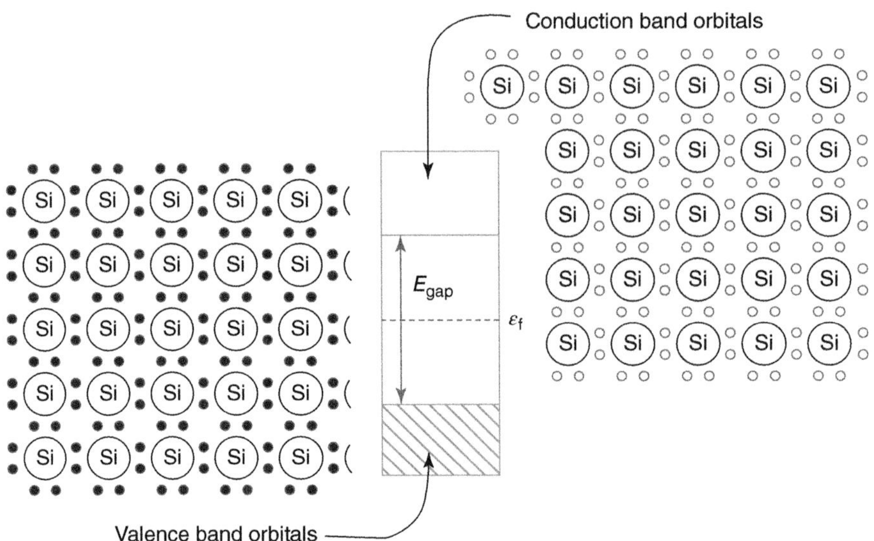

Figure 7.6 The stable bonding configuration of Si. The valence band of bonding orbitals is filled completely. The conduction band of antibonding orbitals is completely empty. Shown here, filled circles mean "filled," and empty circles mean the orbital has no electron occupancy. This is the intrinsic (pure) case, and the Fermi level is in the middle of the bandgap and is shown for 0 K.

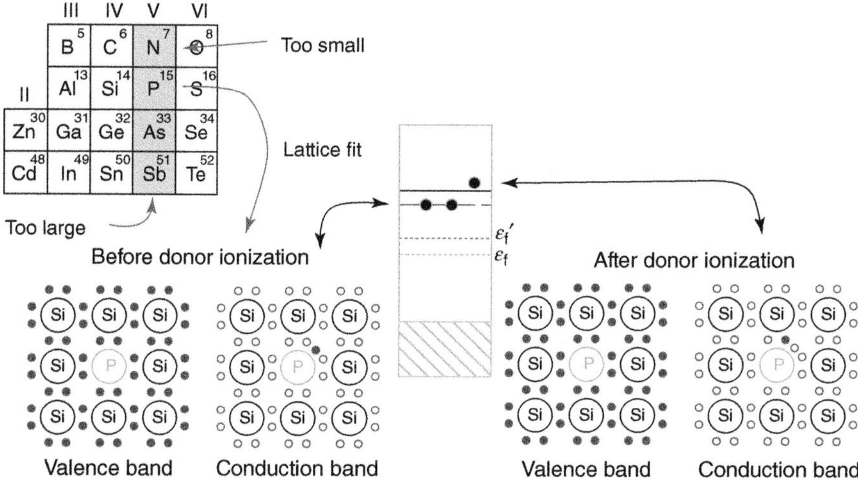

Figure 7.7 The substitution of P into the Si lattice results in the creation of filled states within the bandgap because P has five outer shell electrons and only four are needed for sharing with the surrounding Si atoms. The state that the extra electron sits in is near the conduction band edge, and some small amount of thermal stimulation allows it to ionize, thereby donating the charge to the conduction band of the system.

Figure 7.7). Notice that the surrounding atoms only wish to share four of its electrons, so one filled orbital that is associated with P is unused. Since this electron is not bonded, then we will guess (using our molecular orbital argument from last chapter) that its energy state lies in the bandgap between the bonding and antibonding states of the other four electrons. This is the red line of the band-level diagram in Figure 7.7. In our bond orbital picture, it shows up as an extra circle surrounding the P atom. But notice that at zero temperature this circle is filled, so the electron is bound to P.

What happens when there is a little thermal energy around? If there is enough, then the P-bound extra electron can make a transition to the nearest free state. In other words, the P-bound state becomes ionized by *donating* its electron to the conduction band states of the Si. This type of dopant atom is known as a *donor*. Since the bandgap of Si is roughly 1.12 eV and we know from electrostatic arguments that the donor state must sit somewhere above the middle of the bandgap and near the bottom of the conduction band edge, then it should be clear that it doesn't take very much thermal energy at all to make this state ionized. Another way to say this is that if a volume of Si was loaded up with some number of such substitutions, then at room temperature its conduction band would be populated with free and mobile electrons. These extra electrons occupy Si antibonding states and are quite mobile.

Because these additional electrons are packed onto the top of the Fermi sea of electrons the solid already has, the "Fermi level" of the system must rise as shown in Figure 7.7 as ε_f'. When we do this type of doping to a semiconductor, it is said to be n-type or n-doped. As we will discuss later, it also has *electron majority*

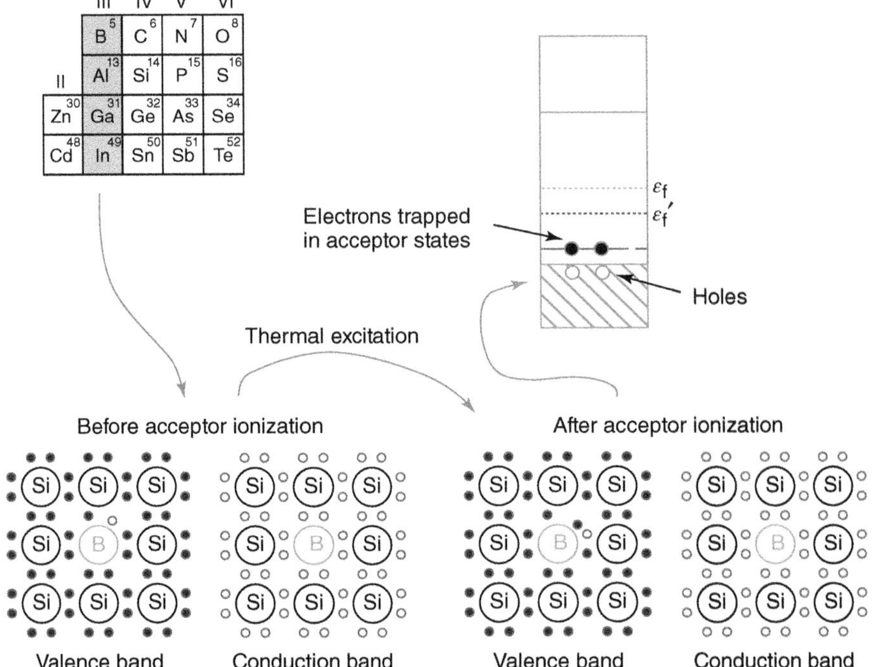

Figure 7.8 P-type doping, or p-doping as it is known, places an atom with too few electrons for the required local bonding of the lattice. In this case we have substituted a B into the lattice site of Si. The result is acceptor states in the bandgap near the top of the valence band. These states can "grab" electrons from the valence band and hold onto them, allowing the sea of electrons in the previously filled valence band to move collectively. We refer to this collective motion as a *hole*, and it behaves as if it is the *antiparticle* of the electron that it replaced. Thus the semiconductor has become a *p-type* conductor. The Fermi level has moved down correspondingly, ε_f'.

carriers, meaning that its conductivity is primarily through free electrons in its conduction band.

Naturally, there is another choice we could have made. Notice, in Figure 7.8, we could choose an atom with only three shell electrons to replace one of the Si atoms of the lattice. Of course the neighboring Si atoms would prefer four electrons in that lattice site so they can have a filled shell. This will leave us with a deficit of one electron.

The example we have chosen for Figure 7.8 is that of B. B doesn't have quite enough electrons for complete shells for all of its neighbors, and so there is an additional trapping state formed in the bandgap. Electrons that are thermally stimulated to transition into these localized B states are trapped there, leaving behind a deficit of one electron each in the valence band. This was the completely filled Si band, but now there are missing electrons into which other electrons from the Fermi sea can transit. Since, with some of the Si electrons being trapped around the *acceptor* B in nonmobile states, there are no longer enough electrons to completely fill the Si valence band, a so-called *hole* is left behind, and it moves about as though it were a positive charge in the system. This is *p-doping* and we have *hole majority carriers*.

What should we know about *holes*? Well.

1. Dispersion relations are typically symmetrical with respect to the bottom and the top of the energy bands. They are usually parabolic at either edge. This type of symmetry is the so-called *electron–hole symmetry* and suggests symmetries in dynamical quantities at the band edges.
2. Holes indicate missing electrons, and if a band is nearly filled, it is more convenient to look at those (few) states that are unoccupied than to look at all the many occupied states. The collective motion of electrons *is* hole motion in a semiconductor pretty much in the same way as empty seats "move" in a concert hall from the expensive front rows to the cheap back rows: people move forward, and the empty seats backward. More formally, holes are not simply a way to look at collective electron behavior; they are instead better thought of as positively charged *quasiparticles*.
3. These quasiparticles are responsible for the conductivity in p-doped ("acceptor-doped" or "oxidized") semiconductors just as excess electrons carry the electrical current in n-doped ("donor-doped" or "reduced") semiconductors.
4. Holes have their own effective mass, they follow Pauli exclusion (they are fermions), and they tend to move in opposite directions to electrons (they have an opposite charge).

7.2.1.3 Carrier Concentrations in Intrinsic and Doped Semiconductors

The picture that we now have in our minds regarding semiconductors should follow our band-level diagrams.

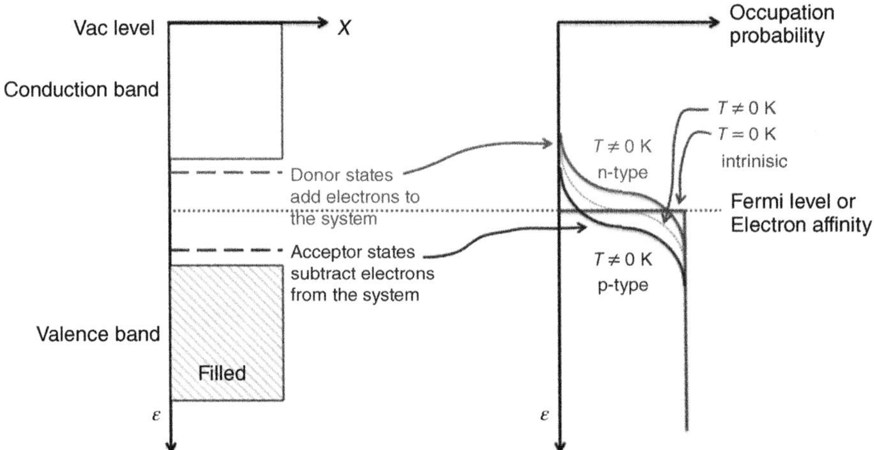

Figure 7.9 The band-level diagram next to the Fermi distribution function showing how the electrons will distribute themselves when the temperature is increased or the electron density is increased/decreased. Note as a reminder that acceptor states "remove" electrons from the system by binding them up in stationary states in the bandgap. This gives us holes that can participate in conduction.

In Figure 7.9 we have placed the Fermi distribution against the band-level diagram and aligned the energies appropriately. There are two entities that we must know to determine how the electrons will fill available states: the temperature that smears out the Fermi "S" curve, allowing for a higher energy tail to place some electrons into the states that would be unfilled at 0 K, and the carrier (electron or hole) density that moves the center of this "S" up or down.

When the system is at cold equilibrium (~0 K), the electron distribution seems rather obvious regardless of the gap states. This is the solid curve in Figure 7.9 (right, marked as $T = 0$ K); electrons of any semiconducting system will have fallen into the lowest energy states allowed to them according to Pauli. From the arguments we have made so far, this means the valence band is completely filled with electrons and the conduction band is empty (the intrinsic case). If there are any acceptor states in the midgap, then they are empty. If there are any donor states in the midgap, they are filled, and the occupation of the valence band and the conduction bands is exactly the same as in the intrinsic case because there is no way for these impurity states to ionize and give carriers to their nearest bands. If there happen to be acceptors and donors at the same time, a situation known as *compensation doping*, then the filled donor states will empty out, thereby filling the acceptor states as would be expected by the Fermi distribution. This is assuming that such a transition is allowed. But the band occupation will stay the same as intrinsic.

When the temperature of the crystal is increased, this heat energy can be transmitted to the electrons through interactions with the phonons. In the case of the *intrinsic semiconductor*, to know how many electrons make it from the valence band into the conduction band, we need to know the bandgap and the total temperature (so we can know how much heat energy there is). If the bandgap of the crystal is large ($E_g \gg k_B T$), then a lot of energy is required for the transition of an electron from the valence band to the conduction band. But if the bandgap energy is more reasonable (say, some factor of $k_B T$), then it is quite easy for electrons to transit to the conduction band – leaving behind holes. So for reasonable bandgaps, the semiconductor becomes more conductive as the temperature is increased because the higher temperatures allow for more electron transitions.

At finite temperature (and in equilibrium), we can know the number of carriers in each band through the *law of mass action.* Let n_c be the number of electrons in the conduction band per unit volume of the solid. p_v is the number of holes in the valence band per unit volume of the solid. $g_c(\varepsilon)/g_v(\varepsilon)$ are the densities of electronic states in the conduction band and valence band, respectively (the limits will actually define which $g(\varepsilon)$ we mean so later we leave these out). We then know

$$n_c = \int_{\varepsilon_c}^{\infty} d\varepsilon \frac{g(\varepsilon)}{1 + e^{(\varepsilon-\mu)/k_B T}} \tag{7.1}$$

$$p_v = \int_{-\infty}^{\varepsilon_v} d\varepsilon\, g(\varepsilon) \left(1 - \frac{1}{1 + e^{(\varepsilon-\mu)/k_B T}}\right) \tag{7.2}$$

$$p_v = \int_{-\infty}^{\varepsilon_v} d\varepsilon \frac{g(\varepsilon)}{1 + e^{(\varepsilon-\mu)/k_B T}} \tag{7.3}$$

Here we have been a little more careful to write the electron affinity μ as opposed to the Fermi level ε_f in the Fermi distribution

$$f(\varepsilon) = 1/[1 + e^{(\varepsilon-\mu)/k_B T}] \tag{7.4}$$

We will come back to this point a little later. For now, it is best to use the μ.

In the case of the intrinsic semiconductor, $n_c = p_v$. So we really only have to figure out one of these (Figure 7.10).

Now let's consider the case where we have n-type or p-type doping. Figure 7.11 shows these situations.

As we note above, the position of μ determines how the dopants will influence the population density of the carriers. We can be a little more precise if we assume

$$\varepsilon_c - \mu \gg kT \tag{7.5a}$$

$$\mu - \varepsilon_v \gg kT \tag{7.5b}$$

where ε_c is the conduction band edge energy and ε_v is the valence band edge energy. This is known as the *nondegenerate approximation,* and with it we can make the following simplification:

$$1/[1 + e^{(\varepsilon-\mu)/k_B T}] \approx e^{-(\varepsilon-\mu)/k_B T} \tag{7.6a}$$

$$1/[1 + e^{(\mu-\varepsilon)/k_B T}] \approx e^{-(\mu-\varepsilon)/k_B T} \tag{7.6b}$$

which yields

$$n_c = \int_{\varepsilon_c}^{\infty} d\varepsilon\, g(\varepsilon) e^{-(\varepsilon-\mu)/k_B T} = e^{-(\varepsilon_c-\mu)/k_B T} \int_{\varepsilon_c}^{\infty} d\varepsilon\, g(\varepsilon) e^{-(\varepsilon-\varepsilon_c)/k_B T} \tag{7.7}$$

$$p_v = \int_{-\infty}^{\varepsilon_v} d\varepsilon\, g(\varepsilon) e^{-(\mu-\varepsilon)/k_B T} = e^{-(\mu-\varepsilon_v)/k_B T} \int_{-\infty}^{\varepsilon_v} d\varepsilon\, g(\varepsilon) e^{-(\varepsilon_v-\varepsilon)/k_B T} \tag{7.8}$$

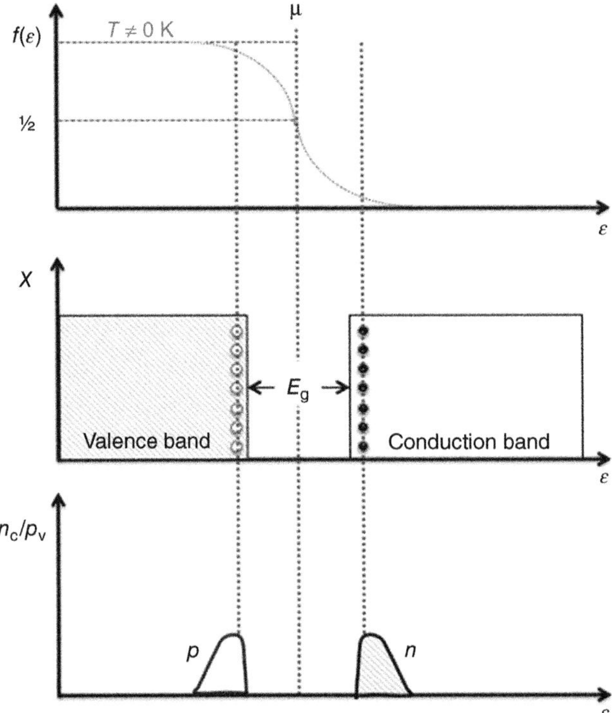

Figure 7.10 The nonzero temperature Fermi distribution spreads the occupation of electrons across the bandgap (assuming high enough temperature). Thus, the electrons in the valence band interact with phonons that give them enough energy to span the gap. Remember, $f(\varepsilon)$ is telling us the probability of finding a filled state, or $1-f(\varepsilon)$ an unfilled state at any given energy ε. At equilibrium electrons are making transitions back and forth at a given rate, leading to a steady population in the bands. At the band edge there is a sharp cutoff of the n/p because no states exist there. If the bandgap were really large as in the case of an insulator, then the temperature needed to raise an electron to the conduction band would approach the melting temperature of the material. We note here that we have assumed electron–hole symmetry, that is, the bands "look" essentially the same right at the energy gap's edge. Thus the distribution in energy of the n/p is mirrored.

This gives us

$$n_c = N_c e^{-(\varepsilon_c-\mu)/k_B T} \tag{7.9}$$

$$p_v = N_v e^{-(\mu-\varepsilon_v)/k_B T} \tag{7.10}$$

so, we still can't get at the actual values without the value of μ. But we can combine these to get the *law of mass action*

$$n_c p_v = N_c N_v e^{-(\varepsilon_c-\varepsilon_v)/k_B T} = N_c N_v e^{-E_g/k_B T} \tag{7.11}$$

where

$$N_c \equiv \int_{\varepsilon_c}^{\infty} d\varepsilon\, g(\varepsilon) e^{-(\varepsilon-\varepsilon_c)/k_B T} \tag{7.12}$$

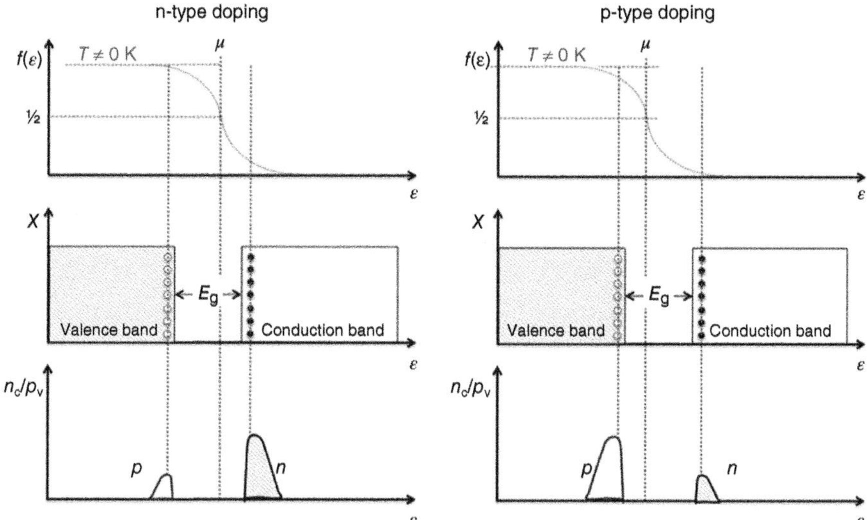

Figure 7.11 On the left, donor gap states (not shown) add electrons to the distribution, thereby moving the μ toward the right. On the right, acceptor gap states remove electrons, thereby moving the μ to the left. The donor and acceptor states thus only influence the n/p by where they place μ.

$$N_{\rm v} \equiv \int_{-\infty}^{\varepsilon_{\rm v}} {\rm d}\varepsilon\, g(\varepsilon) {\rm e}^{-\frac{\varepsilon_{\rm V}-\varepsilon}{k_{\rm B}T}} \tag{7.13}$$

This is a quite general finding and applies to the use of this approximation in doped and undoped semiconductors.

At around room temperatures, we note that the usual purpose of doping a semiconductor is to place the bandgap states rather near the conduction band or valence band edges. And therefore the donor or acceptor states are likely to be *all* ionized, leaving the density of carriers in the band equal to the total density of donor–acceptor states plus the carriers that have transitioned across the gap: $n_{\rm i}$. So, it will simply look like $n_{\rm total} = n_{\rm i} + n/p$. Of course, if we do not make this assumption and instead consider the idea that the donor–acceptor states are only partially ionized (say, for deep donor or acceptor states), it is quite easy to imagine that the donor–acceptor states will contribute to the conduction/valence band using the thermally weighted averages as above.

7.2.1.4 The Fermi Level vs. the Chemical Potential

When the temperature of the solid is NOT zero, the electrons at the Fermi level spread out. Let's just remind ourselves of this fact. Shown in Figure 7.12 is the edge of the Fermi distribution as a function of temperature. As the temperature goes up, that straight Fermi cutoff begins to "smear" and spread. Notice the spread is "S" shape, and for even fairly high temperatures, it is distributed evenly about the point $f(\varepsilon) = ½$. $\varepsilon_{\rm f}$ is chosen as the energy at which the sharp cutoff occurs, between filled states and unfilled states at 0 K. It is independent of T. μ is another matter. It is uniquely defined by the grand canonical ensemble (visit your nearest

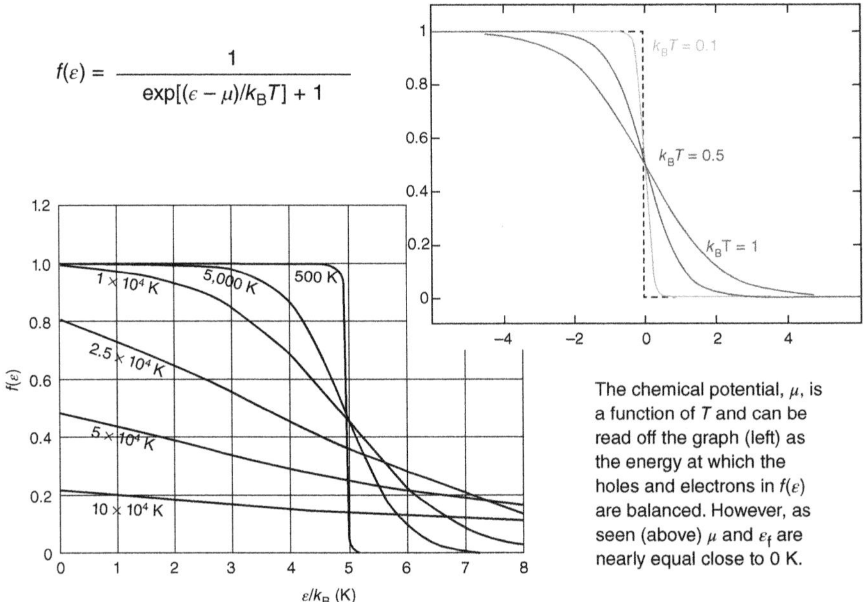

Figure 7.12 The Fermi distribution spreads out as the temperature increases. This means that higher energy electronic states are occupied.

Statistical Mechanics Store) for all finite systems and represents the change in *Helmholtz free energy* (U-TS) of the system with the addition of a single electron [2]. We notice that μ typically (at reasonable temperatures) occurs where the electrons in that "S" curve are exactly balanced against the holes: the inflection point of the "S" curve. This is $\mu \sim dF/dN$, where F is the Helmholtz energy and N the number of electrons according to Baierlein [3]. This value *IS* temperature dependent of course. However, for temperatures under about 200 °C, the two are almost the same value, so it hardly makes any difference, and we can approximate $\mu \sim \varepsilon_f$. But for finite temperature semiconductors, we really should be saying "chemical potential" instead of "Fermi level."

7.2.1.5 Spectroscopy of the Dopant Levels

In the design and development of technologies based upon doped semiconductors, it is necessary to characterize the dopant states, that is, to know how deep in the gap they are and their number. There are numerous ways that physicists have used to do this from temperature-dependent deep-level transient spectroscopy (DLTS) to Hall measurements. These are mostly based in transport measurements.

However, a common method to examine dopants in semiconductor crystals is optical spectroscopy. The experiment is really rather simple. First one cools the sample to the point where the dopant levels are not ionized using a cryostat with an optical window. Then you illuminate the sample: scanning through the different wavelengths of light and measuring the reflected (or transmitted) light to see what wavelengths have been absorbed. The caveat is that you must calibrate this

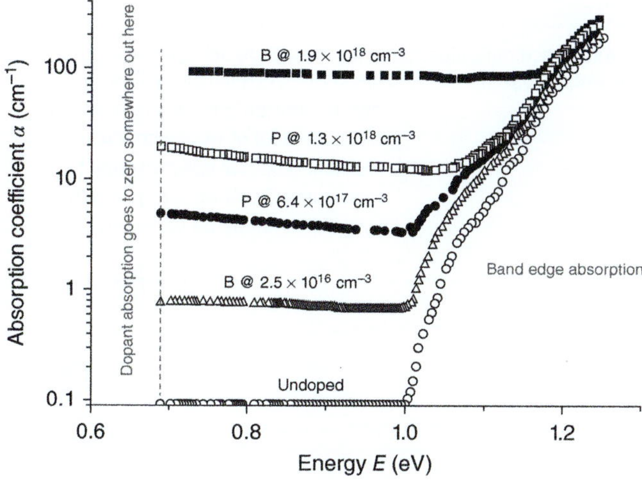

Figure 7.13 Shown is the absorption spectrum for n/p-doped Si. Taken from Ref. [5].

against an undoped sample that is otherwise identical so that you know the effects of absorption in the window and by the surface states, assuming there are any. The result for Si is shown in Figure 7.9. Caution must be used when employing such techniques however. Ionization of the dopant levels is a resonant scattering effect, and it is embedded with scattering among the nearly continuous states of the band as well as phonon scattering. So, the two different kinds of scattering have energies similar to each other and are in proximity to each other. This can lead to a particular kind of interference known as *Fano resonances*, giving sharp features in the spectra [4].

In our illustration of Figure 7.13 [6], we have shown the absorption coefficient (α) as a function of the doping levels (cm^{-3}) for boron (B) and phosphorous (P): p- and n-type dopants, respectively. Recall that the absorption coefficient is the exponential factor in Beer's law: $I \sim I_0 \exp(-\alpha x)$ where x is distance into the sample. This is a good measure of where the strong absorption is occurring. Notice that we cut off the scan for photon energies below about 0.7 eV where the energy states of the dopants are expected to sit. This is due to a limitation of the experiment. However, for the energies shown, it is clear that the dopant levels are being ionized by the incoming photons and end up in the conduction (electrons) and valence (holes) bands.

Notice that for a macroscopic crystal-like Si, the dopant is forced into the specific bonding configuration that the 3D lattice of Si makes for it. That is, the dopant atom must "think" it is Si in a way. But this is not true for nanosystems in which there are simply not enough surrounding atoms to force the dopant atom into a template. Instead, in these cases, the dopant can form complexes that leave states anywhere within the bandgap of the object! Thus, an atom that one thought would surely result in p-doping actually yields n-doping. While this can happen in bulk semiconductors as well, it has been a particularly tricky detail in nanomaterials as with the case of carbon nanotubes.

The Chemist's Cat

Cats have always played a role in scientific illustration such as ***Schrödinger's* cat**. But there has been a persistent story told at North Carolina State University for years about a rather famous chemist who did quite a lot of work with high resolution optical spectroscopy of molecules. Such work required that monochromators be very long

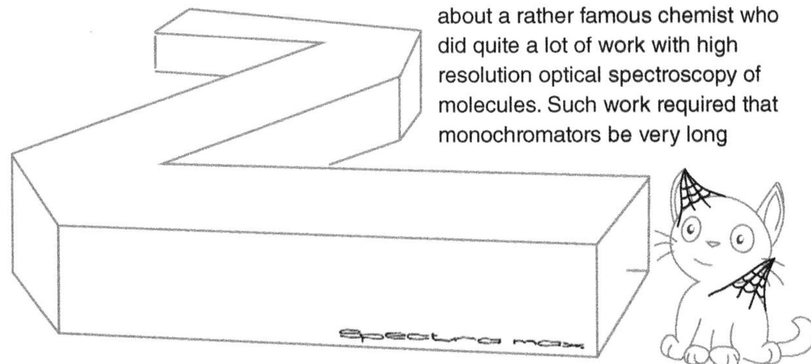

and their gratings must be carefully balanced and synchronized. Every so often this chemist would find that transient fringes and features would begin to occur in his spectra, making interpretation impossible. When this happened he would go home and get his cat, return to the lab, and place the cat into one end of the monochromator. The cat would then emerge from the other end covered in spider webs, and the spectra would return to normal!

7.2.1.6 Carbon Does Not "Dope" Like Si

In Si, extrinsic doping is typically treated as though the crystal sets a template for the insertion of the foreign dopant atom, and the mismatch in the local electronic bonding results in an extra charge (positive or negative as the case may be). This is the archetype and the point at which many discussions leave doping altogether. GaAs and InP both actually follow Si closely. However, there is another class of semiconductors to consider, and they are based on materials like carbon. What happens when it is energetically favorable for the lattice to change its local symmetry as opposed to accepting an impurity atom with a bond dangling in space? Moreover, what happens when the thing we are trying to dope is nanoscale? Such considerations are of particular importance to the field of organic electronics. From diamond doping, to carbon nanotube doping, and to the doping of a conjugated polymer, the rules here can be a bit different from Si and quite unexpected at first.

In the case of *diamond* (3D carbon), we might expect that if we choose N, we would add an electron near the band edge, as we did with Si and P, to the system since it is one column over and on the same row as carbon. In fact, this doesn't happen. The donor state introduced is far too "deep" in the bandgap to help us much in increasing the conductivity of this material. This is, in fact, due to the preferential formation of a distortion wherein the N prefers to form a lone pair of electrons on the N, leaving one of its carbon neighbors with a dangling bond [6]. This means that the extra electron will be localized on a carbon atom somewhere as opposed to the impurity atom. One reason for this to occur can

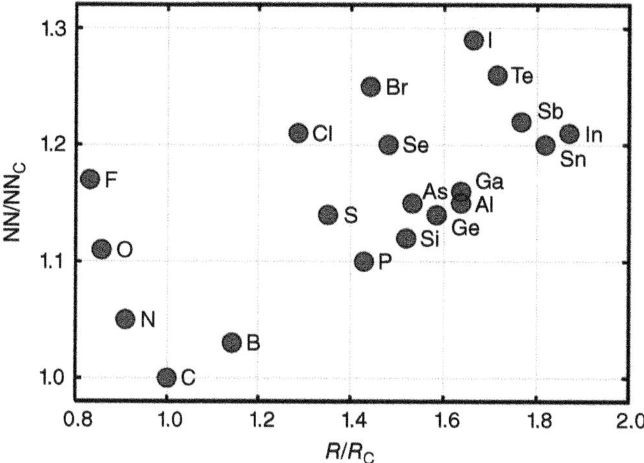

Figure 7.14 Here the nearest neighbor (NN) bond lengths have been calculated. The ratio of nearest neighbor in diamond (NN_c) to these NN distances is plotted as the relative covalent radii (R) for species in diamond (normalized to the radii of carbon R_c). Atoms about the same size as carbon (when in diamond) show up as a 1 on the abscissa. On the ordinate is a number that tells you how close the atom wants to be to its nearest neighbors. Again if it matches carbon, then it is 1. So even though N is a little smaller than its C counterparts, it prefers to be further away when it forms bonds.

be seen in Figure 7.14 taken from Ref. [6]. The range of impurities that will nicely fit into the lattice is limited. Even for species with a smaller covalent radius than carbon, the surrounding lattice can be pushed outward to accommodate bonding angles and lengths. So, as the deformation in the vicinity of the impurity increases, so too does the energy required to form the defect. This means, for the most part, that the equilibrium solubility of dopants in bulk diamond is often very low.

Carbon nanotubes are really no different. The addition of N into the lattice of a single-walled carbon nanotube (SWNT) can be nonobvious. There have been numerous models proposed for the incorporation of N into the SWNT lattice, and one of them is shown in Figure 7.15.

Of course nanotubes, particularly single-walled nanotubes consisting only of a single sheet of atoms, can be "doped" in many ways. Chemical attachment of electron-donating or electron-withdrawing groups, adsorption of species such as oxygen on the outer wall, and introduction of species within the tube itself have all been shown to lead to doping or the exchange of charge. Indeed, this would suggest that at the molecular level there is some ambiguity between *atomic-scale contacts* and *contact doping*. So if we lay a nanotube onto a gold surface and find that it has more electrons now in its conduction band (say, through tunneling spectroscopy), do we say that a low-dimensional contact has been formed like that of the above examples in metals, or do we say that the nanotube has been doped by the gold? As we will see in the next chapters, for nanostructures we can really treat these two things as the same [8].

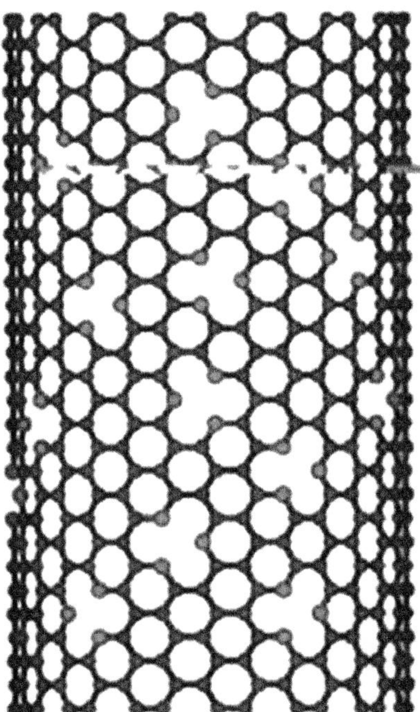

Figure 7.15 This is one of the many models proposed for N incorporation into the SWNT lattice. Such models range from single, isolated N that bonds out of plane into the lattice to these pyridine-like rings. The final result depends on growth of the nanotube and the nitrogen environment used to provide the doping [7].

7.2.2 Junctions with Semiconductors

Now, we move on to consider the combining of the doping inhomogeneity with the inhomogeneity of contacts. Junctions, interfaces, and boundaries between materials of different electrical characteristics can be used to form the basis of modern electronic components. *Metal–semiconductor*, *metal–insulator–semiconductor*, and *p-doped to n-doped semiconductors (p/n junctions)* are all technologically important junctions, forming laser diodes to megabyte chips [9]. Naturally, this is a study area unto itself, and it supports a massive worldwide industrial base. However, using our simple introductions and definitions so far, we can lay down some ground rules of dealing with "electronic" junctions.

We have already established our classification scheme using band-level diagrams as seen in Figure 7.16.

To make this useful, we need to attach numbers to the levels in *different* solids. So we might think of measuring energies relative to the fundamental energy it takes to remove an electron from the solid's surface: the *work function* (usually given the symbol ϕ□ or *W*). Then the energies of the bands and the Fermi level can be measured relative to a *vacuum-level* point as seen in Figure 7.17. This means calling the energy of the "just free" electron the "zero point" on the energy scale. So, for a metal the work function energy is quite simple – it is the energy from the Fermi level (highest occupied orbital) to the vacuum level. This would be the amount of energy it takes to remove the electron. For the room temperature semiconductor, this is a little less obvious: it is the energy to the vacuum level from the chemical potential (loosely; the Fermi level). This is shown in Figure 7.17 on

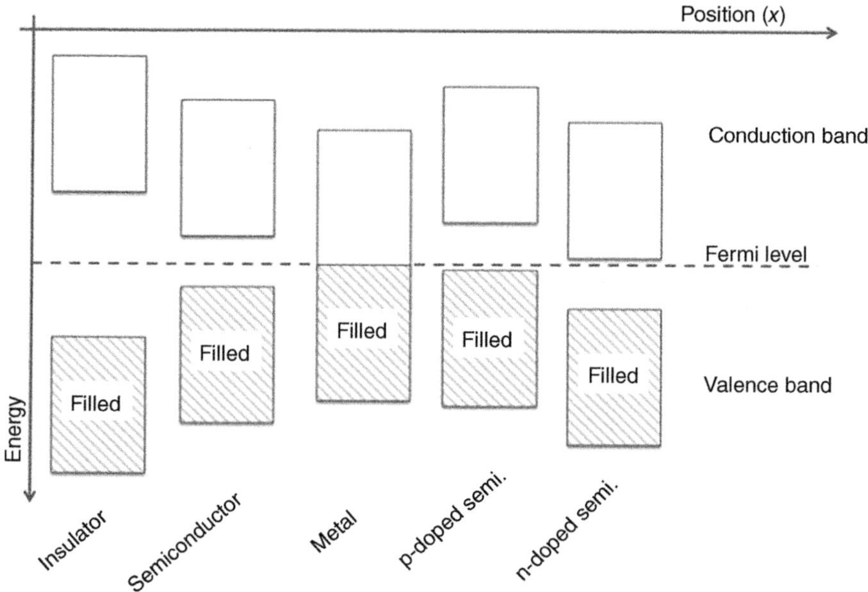

Figure 7.16 Classification of solids according to their conduction properties. The Fermi level or chemical potential is the highest energy level of the electrons, and the different systems are shown normalized to this level.

the left-hand side. χ_e is referred to as the electron affinity, and it is the energy from the vacuum level to the bottom of the conduction band. This is actually a convenient scale choice because it does allow one to compare materials.

To form a junction, the surfaces of two atomically clean half spaces are brought together (or as in reality, a film of one material is grown on a single crystal surface of another). Electronically, the exact response of the combined systems depends sensitively on the details of the surfaces and materials involved. There are many possibilities for order and interphase formation, all of which will influence the electronics of the interface. However, we can form a general set of expectations. As shown in Figure 7.18, as two materials approach each other, nothing happens to first order. *Casmir forces* and other effects are small. When electrons are allowed to transfer from one material to another, contact has been achieved. Charge flows to fill the lowest possible states on either side of the junction. But how does it choose to do this?

Rule #1: Metal–semiconductor junctions follow the *Schottky–Mott rule*. Originally this rule was used to estimate the potential difference (or barrier height) that would occur between a metal and semiconductor in terms of the work function of the metal and the electron affinity of the semiconductor: $\Phi_{\text{barrier}} = \varphi_{\text{metal}} - \chi_{\text{semi}}$. This does work quantitatively sometimes, but other times it does not because, as Bardeen pointed out, it doesn't account for trapping states within the bandgap of the semiconductor that might occur due to defects and reconstructions of that surface. It does however give a general sense of how the bands will bend at the interface.

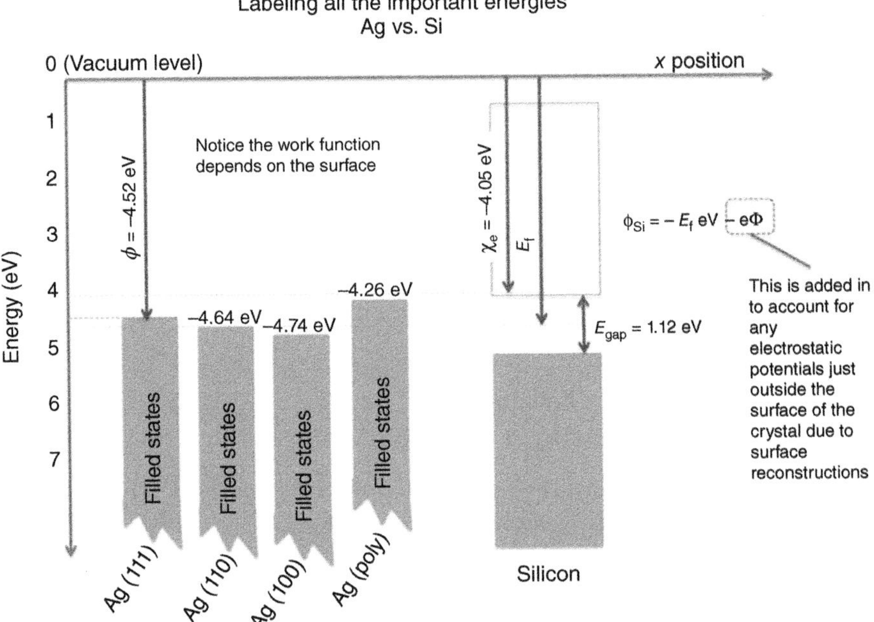

Figure 7.17 Labeling the band diagrams. We note that our definition of the work function of the semiconductor includes surface potentials Φ for completeness. We also note that the actual position of E_f depends on doping levels.

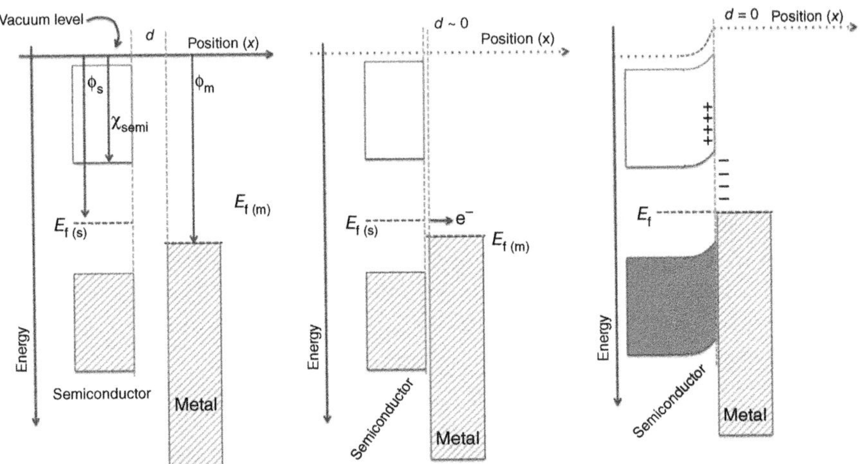

Figure 7.18 The formation of a simple semiconductor–metal interface and its electrostatic barrier. Such barriers are sometimes called "Schottky barriers," and they come about as the material attempts to form a shared Fermi level by exchanging charge across the interface.

In the metal–semiconductor junction, the system will now share one Fermi level as seen in Figure 7.18. Locally, a field is formed due to the unbalanced charge on each side of the boundary necessary to equilibrate the Fermi level. This field holds the charge in place. The states of the energy bands adjust their position in the local field, usually referred to as *band bending*. In some cases this can form a barrier to current flow in one direction, but not the other – *rectification*. The height of the energy barrier, giving rise to the rectification, is *roughly* $\varphi_2 - \chi_1$, where the φ is the work function of material 2 and χ the electron affinity of 1. This scenario, or something similar, holds whether the system is low dimensional or not. Now you can see why we worried about contacts "doping" the system in our discussion above (Figure 7.19).

There are many caveats and nuances to the process of Schottky barrier formation that we have not discussed here, for instance, *Fermi level pinning* where charge becomes trapped at the interface in bandgap states that are due to imperfections or impurities. However, the "flavor" is captured here. The barrier is due to compensation charge at the interface. The field of this charge extends into the semiconductor, creating a zone of depleted carriers: the *depletion zone*. If an electric field is placed across the structure, opposing this field, then current will not flow until it has exceeded this barrier field. However, the current flows freely if

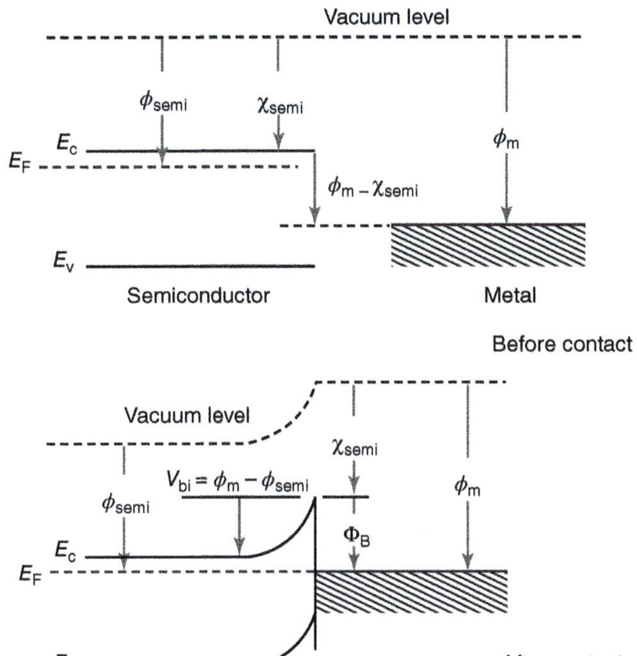

Figure 7.19 If an n-doped semiconductor is used, the amount of band bending is even greater, and quite a large potential barrier, Φ_B, can be accomplished. Here we have used the Schottky–Mott rule, so this represents an *ideal* case. A p-doped semiconductor can bend the bands in the opposite direction.

the applied field is in the direction of the V_{bi}. This non-ohmic nonlinearity (*rectification*) is called a *diode*, and this is an example of a *Schottky diode*.

Of course the bands can bend dynamically with a variable applied voltage as well, making the barrier appear larger: *dynamic band bending*. It should be clear that the type, mobility, and number of available charges in the solids will have a large impact on the kind of barrier formed at the interface. We return to this point when we discuss conducting polymers in more detail.

Rule #2: Semiconductor–semiconductor *heterojunctions* follow the *Anderson rule*. In this case we start with two pieces of the same material – say, Si – one that has been doped n-type and the other that has been doped p-type. So the bandgaps $E_{g1} = E_{g2}$, but the Fermi levels $E_{F1/2}$ are different.

In the diagram of Figure 7.20, we identify the variables that you need to know to "guess" what will happen when current flows through this junction of materials. First notice that the interface doesn't bend at a single position as in the metal–semiconductor example. Here energy levels on both sides of the junction

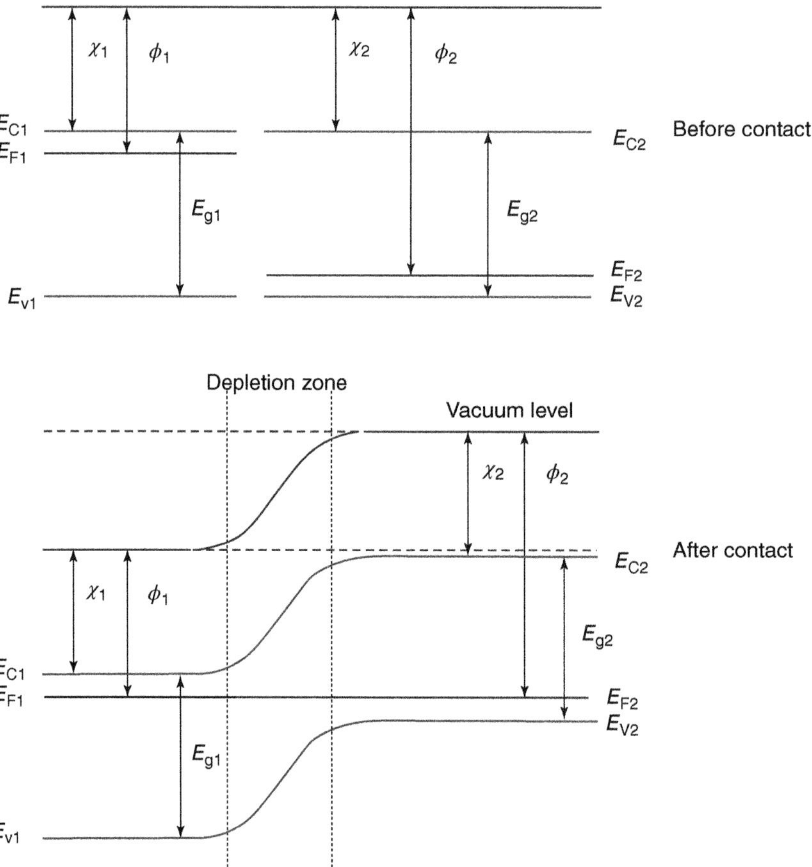

Figure 7.20 A band diagram drawing of the standard p/n junction as is used to form diodes.

bend smoothly away from the position of the physical boundary. This is because the field required to align the Fermi levels penetrates slightly into both materials. It is due to the exchange of charge locally.

7.3 Other Types of Heterogeneity

Heterogeneity can actually take many forms in solids, and we have seen two of them. But as technologies progress, and the use of organics in electronics increases, the types of heterogeneities we encounter also change. Perhaps among the most interesting of these are composite structures. Organically based electronic/photonic composites have been studied for some time now and are just finding their niche in technology, but they represent a fascinating field of physics. The idea is quite simple as shown in Figure 7.21.

At first glance this doesn't seem a soluble problem, and perhaps it might be thought to hold less basic instruction of physical principles for the scientist. In actual fact, however, this problem of embedded systems in low dimensionality actually poses some rather deep questions of how space can be connected in a conducting system.

Important starting points for identifying variables in this problem are the following (let's call them the basic rules of the game):

1. The interphase can interact with the host and vice versa, meaning charge can transfer and cause localized doping, and an antenna-like nanophase can lend

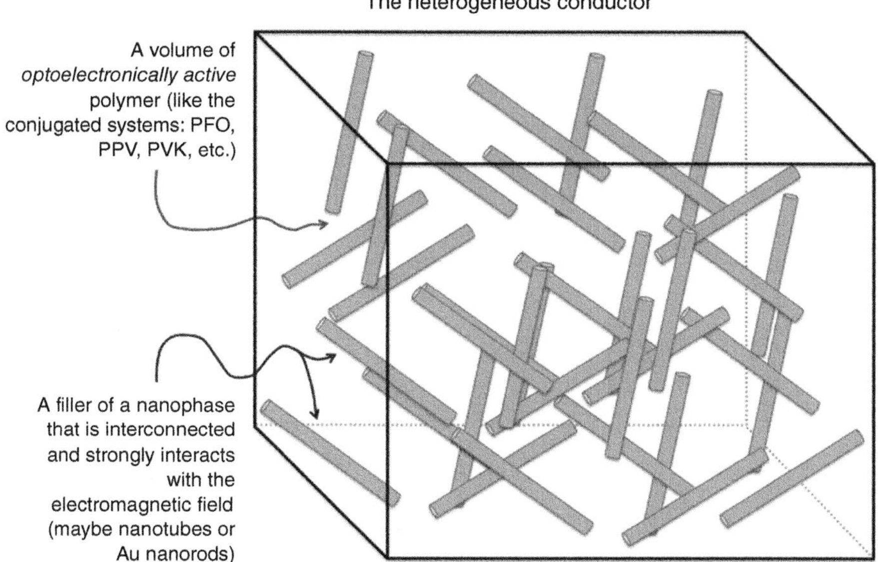

Figure 7.21 The *heterogeneous conductor* is typically thought of in the case of an organic such as a polymer containing a minor conducting phase such as carbon nanotubes. However, there exist liquid crystal systems that also fit this pattern. So the concept is quite general.

oscillator strength to the surrounding matrix modifying its absorption (we will discuss this in more detail later).

2. The fractal dimension of the interphase collection depends on the length of the nanowires together with the loading and relative angle with respect to each other. Controlling this dimension has long been sought by scientists and engineers, who have applied chemical assembly techniques to chaotic mixing.
3. The fractal dimension, along with the oscillator strength of the nanowires, and the conductivity of the host will also determine the interaction of the collective with incoming electromagnetic radiation.
4. The transfer of charge through the network of the interphase depends on the quantum mechanical ability to tunnel, and this is dominated (as we shall see in our chapter on transport) by two different mechanisms: fluctuation-assisted tunneling (FAT) and variable range hopping (VRH). But this in turn depends on an element of nonlocality (since the buildup of charge in one place on the network can affect the tunneling potentials at another) as well as the specific arrangement of the nanoparticles (a *very* local element). We see the second argument in Figure 7.22.

Clearly there can be rather a lot of interesting physics in these so-called *matrix composite* or *nanocomposite* systems. There exists an odd "symbiosis"

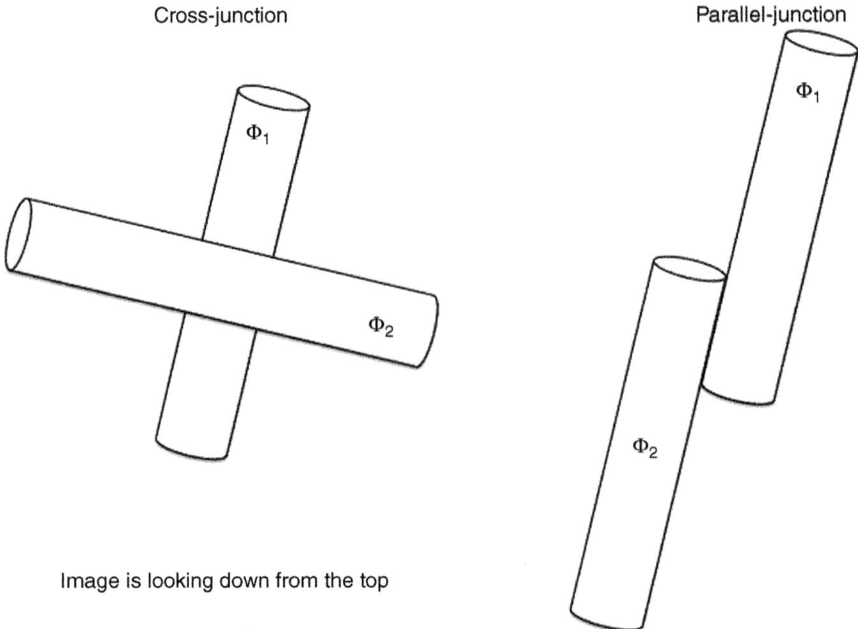

Figure 7.22 Two different types of nanotubes are contacted. However, the contact is geometrically different in the two cases as shown. One nanotube contact is crosswise, and the other is parallel. For different chiralities we naturally expect different work functions and some contact potential. However, tunneling matrix elements will also be sensitive to the *k* values of the states that are transferring the charge. This then also modifies the tunneling probability at every junction within the network.

of dimension, localized properties of modification, and "rule of mixtures" that creates materials characteristics quite different from constituent phases. Moreover, this suggests to many in the field that such systems be seen as a class of materials by themselves. We will return to matrix composites when we introduce transport properties in conjugated systems. For now, however, we want only to introduce the extreme level of heterogeneity that can occur in *dimensionally engineered* materials.

7.4 Summary

We have presented a basic glimpse of *heterogeneity* in the materials we studied in the last chapter. This has included *contacts* between different types of materials as well as (electronic) *doping* that modifies band filling. A simple *band-level diagram* approach toward predicting the behavior of homogeneity has been introduced and used to understand how technologically important systems such as the metal–semiconductor interfaces behave. Finally, what we thought we had just learned, just when we were getting comfortable, such processes in heterogeneity are challenged with the introduction of carbon systems and dimensionality.

Exploring Concepts

1 *Silicon and carbon*: Let's explore the differences and similarities in 3D Si and C (diamond) when it comes to doping. Remember both have this diamond structure, but there is a big difference in the "willingness to accept different bonding angles" of the two.
   (a) From what you can find in the literature, what are the most common substitutional dopants found in each system? Describe their bonding positions and the relative position of the electronic state in the bandgap. So, in these cases we are specifically examining the substitution of one atom for another, such as Si with P, while leaving the rest of the lattice unaltered.
   (b) Just as these substitutions can be carried out without significant alterations to the lattice, there are cases where local alterations do occur. This can mean that the substitution of a single atom, so B into the C diamond structure, may have consequences beyond just an unfilled orbital: different hybridizations between the adjacent atoms. This results in doping levels that are unexpected high or low. Describe and explain which of the two systems (Si/C) you think might be more susceptible to this.
   (c) Again, back to the library. Find a few examples of (b) for each system. Draw them out and explain the what and where of the local bonding. For each example, what happens to the doping state in the bandgap?
   (d) Finally, let's compare and contrast the p/n junction in Si and C (diamond). Of course the bandgaps are quite different. But, given the common dopants in Si and in C (now being studied for technology applications, we will let you look this up), estimate the barrier heights at the p/n junctions and the Schottky barriers for silver contacts to the p- or n-type material.

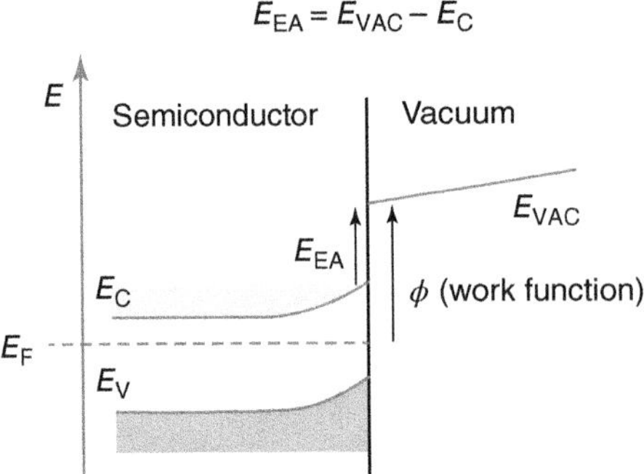

Figure EC7.1 The band diagram for the semiconductor surface. Notice that due to reconstruction and termination of atomic order, band bending can occur here. There is a subtle difference between the electron affinity and the work function as shown here. The level of doping will determine what is actually measured using X-ray photoelectron spectroscopy.

Now, for this exercise, we realize we have asked the reader to search through literature and use numbers found there to get the answers. And, of course, literature will have different numbers reported from different research groups. So, it is important to note how the numbers and how your estimate are generated and the level of confidence you may have in that numbers based on other measurements in the literature of these barrier heights.

2 *Negative electron affinity materials*: In this section, we learned about the *electron affinity*, essentially the work function for the semiconductor.
Specifically, for the semiconductor–vacuum interface, the electron affinity, E_{EA} or χ, is defined as the energy obtained by moving an electron from the vacuum just outside the semiconductor to the bottom of the conduction band just inside the semiconductor as seen in Figure EC7.1:
So, for intrinsic semiconductors near-zero temperature, this is straightforward. Added electrons go to the bottom of the conduction band to which they have been added. For *intrinsic and lightly* doped semiconductors, the *work function* may change with the doping level, but ideally the electron affinity doesn't since the electron is going to the bottom of the conduction band anyway. But at higher temperatures and *heavily* doped semiconductors, an added electron will instead go to the Fermi level on average, which now may be well into the conduction band, so the situation is more complicated.
Like the *work function*, the *electron affinity* depends on the surface termination (crystal face, surface chemistry, etc.) and is strictly a surface property. And, curiously, in some circumstances, the electron affinity may become negative. As it happens, diamond (111) surfaces have an electron affinity very near the vacuum level. If an alkali metal is added to the surface, the electron affinity of the surface becomes negative!
(a) Look up the current best values for the electron affinity of the diamond (111) surface. What alkali metals would bring this surface up to an negative electron affinity state (NEA)? Give estimate numbers.
(b) Draw the band diagram for this system. What is the physical meaning?

3 *Modulation doping: Conductivity* in a material requires charge carrier numbers and charge carrier mobility. Carrier mobility, as we have already seen, is a function of the bandwidths and band shapes of the materials (as well as scattering if there is any). As for carrier number, we can add charge carriers to the conduction band, as has been explored in this chapter, through doping. There is a slight problem though. The addition of large numbers of additional electrons into the conduction band of a material requires that the lattice be riddled with additional dopant atoms. These atoms are "foreign" to the lattice and introduce localized potentials that can extend for some numbers of *lattice parameters* in length away from the dopant site. Consequently, they introduce significant scattering in the system, lowering the carrier mobility. We have lower mobility the higher in dopant levels we go and overall conductivity begins to roll off.

In 1977 at Bell Labs, Horst Störmer, Ray Dingle, and Arthur Gossard thought they had an answer to this. They introduced the idea of *modulation doping*. This clever little technique simply separates the free charge carriers in their conduction channel from the location of the donors. Of course it presumes systems in which the "added" carrier can easily and quickly migrate to regions of interest, quite distant from the atomic donor that placed it into the conduction band in the first place. Now this does eliminate scattering from the donors, and semiconductors that are modulation doped can have very high carrier mobilities and high carrier numbers simultaneously. In fact Störmer and Dan Tsui used a modulation-doped semiconductor to discover the *fractional quantum Hall effect* that same year.

In this problem we are going to take a look at such experiments to understand the setup. This will help us understand modulation doping.

(a) To begin, three layers are grown. The first two are GaAs and $Al_x Ga_{1-x} As$ where the Al content is below 40%. The third layer is a contact of metal. The lattice constants of the semiconducting two layers are less than 1% different, so the lattice structure continues across the interface with few defects (epitaxial). However, the bandgap of the AlGaAs is direct and larger than that of the GaAs. This results in the band offsets shown in Figure EC7.2. For the common values given in literature in undoped GaAs/AlGaAs (we

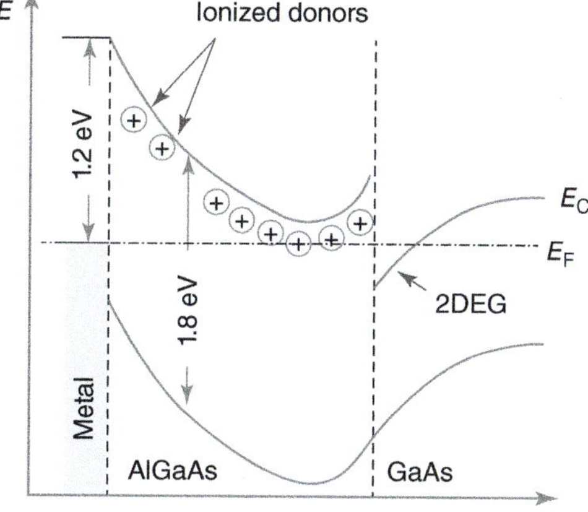

Figure EC7.2 Implementing the modulation doping mechanism in real life is much more complicated than it seems. This is how the Bell Labs team did it. It involves three layers: a metal layer, an n-doped AlGaAs layer, and a layer of non-doped GaAs. Notice in the very thin AlGaAs layer, the bands are never really "flat." They are bent due to the local fields that are established by the exchange of interface charge on either side. This particular structure became the standard for the study of two-dimensional electron gases (2DEGs).

already gave you one), give a rough estimate of the band offset and expected barrier height for the undoped system.

(b) Now, consider the case where the AlGaAs is doped heavily with shallow (near the conduction band edge) donor states. Let's say that we are going to run our experiment at near 1 K; how shallow do we want these states to be if they are to contribute fully to the conduction properties of the system? What happens to the Fermi level in the AlGaAs for this level of doping? What about the Fermi level for the GaAs, where is it? How does this happen?

(c) Notice that in our Figure EC7.2 diagram, the bands bend with a simple curve. We see this curve upward both at the metal/AlGaAs interface and the GaAs interface; explain. We have marked a region in the GaAs near the interface as a 2D electron gas. Why are there electrons pooled there? Assuming there is little interaction with the interface itself by these electrons, why would an electron gas be a good approximation for their behavior?

(d) These electrons have been added to the GaAs system from the AlGaAs system, but they are placed in a position where little scattering can take place due to the dopants themselves. This is the classical definition of *modulation doping*. Now let's use some very simple electrostatics: $E = n_{\_}/\varepsilon$ where $n_{\_}$ is a variable and is defined as the carrier density in the 2D gas, ε the dielectric, and E the field (remember that above is plotted in eV) to write down an expression for the dopant density (as a function of temperature and bandgap position) to get the carrier density $n_{\_}$. This is of course a simplification.

4 *Doping in low dimensions*: Low-dimensional materials such as *carbon nanotubes* and *graphene* can also be doped. For carbon, nitrogen, and boron, this is pretty common. But these raise some rather interesting questions. Now that we know a little more about dimensionality and heterogeneity, we may wonder if these interfaces will behave the same as their 3D counterparts.

(a) Do a literature search, and determine the lattice positions of these dopants and their energies for both CNTs and graphene. How are they alike, and how are they different from each other? Why might you think that they would differ? Could it have anything to do with lattice strain, and so does the size of the nanotube make a difference? Explain.

(b) In the above problem we asked you to make a really big simplification: assume the field looks like that of a plane of charge at the interface. This means $E = \sigma/\varepsilon$ where σ is the trapped charge density, whatever it might be. While this isn't the best assumption, we can see that the bands will bend in a V/d fashion (d is the distance from the interface). Assume now that we have a sheet of graphene. One half of the sheet is doped positively and the other negatively. The "interface" is a line between the two regions. What should the band bending look like in this case using similar simplifications? In this case, graphene is pretty conductive, and so there is significant screening. To start with ignore this. Then try to add it in if you feel up to the challenge. Would your guess for the band bending hold true for carbon nanotubes as well? Under what conditions?

5 *Deep Level Transient Spectroscopy (DLTS)*: Developed in 1974 by D.V. Lang [10], DLTS is an experimental technique for establishing fundamental parameters and concentrations of charge carrier traps, such as donors and acceptors in semiconductors. More specifically, DLTS examines the space charge or depletion region of simple electronic devices such as Schottky diodes or p/n junctions by measuring the dynamical transients of their capacitance.

Quite simply, it works like this: a reverse biased p/n junction or Schottky barrier will "look" like a capacitor for that bias (no current flow and charge will build up on the contact plates). If an AC over voltage (say, ~1 MHz) is placed on the bias, the rf capacitance of the sample depends on the charge state of deep levels in the space charge region (SCR). In the *total depletion approximation*, the rf capacitance of a sample having a homogeneous doping concentration can be written as

$$C_0 = A\sqrt{\frac{\varepsilon\varepsilon_0 e(N_D - N_A)}{2(V_r + V_d)}}$$

A is the area of the "capacitor," $N_D - N_A$ is the total net charge concentration in the SCR, V_r is the reverse bias voltage, and for a p/n junction, let's say, V_d is the built in diffusion voltage of the SCR. V_d is the point where the extrapolated plot of $1/C^2$ vs. V_r intersects with the V_r axis (this is known as a CV curve). $N_D - N_A$ is calculated from the equation above.

If charged trapping levels exist in the SCR, then their charge must be added to $N_D - N_A$. So let's assume a donor-like trap state with a concentration of N_t. Then

$$C_0 = A\sqrt{\frac{\varepsilon\varepsilon_0 e(N_D - N_A)}{2(V_r + V_d)}} - A\sqrt{\frac{\varepsilon\varepsilon_0 e(N_D - N_A + N_t)}{2(V_r + V_d)}}$$

ΔC is the change in capacitance when N_t states are charged and uncharged.

(a) Show that the N_t can be estimated from the charging and discharging of the trapping states as

$$N_t = \frac{2\Delta C}{C_0}(N_D - N_A)$$

(b) In conventional DLTS, the capacitance transients appear as shown in Figure EC7.3. From such plots we can get a rough idea of what is happening in the system even if we know little to nothing to start. For instance, immediately after the pulse, the capacitance changes by ΔC. This ΔC is negative for majority carrier traps and positive for minority carrier traps, so Figure EC7.3 is for majority carriers. In the case of a Schottky diode, or if a p/n junction is pulsed only by a small voltage, $V_{pulse} \leq V_r$, only majority carrier traps are recharged. If during the pulse a p/n junction is forward-biased into injection, minority carriers may also be recharged.

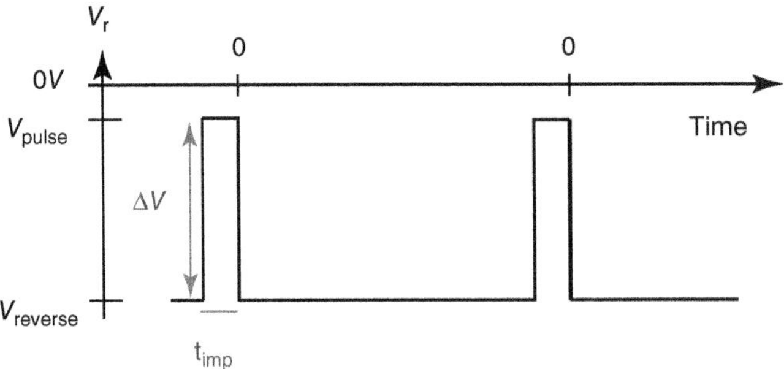

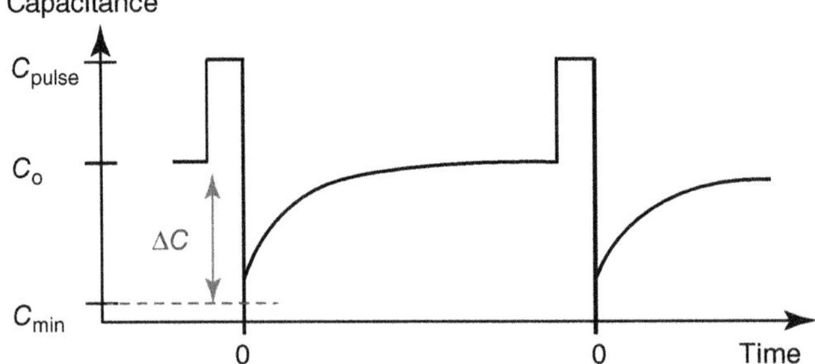

Figure EC7.3 The basic DLTS procedure is shown. At $t = 0$, majority carrier traps are filled by the bias pulse. This yields a change in the capacitance according to the above equation. Thermal relaxation of the point defect traps between two pulses is approx. exponentially in time. The pulses are spaced as close as 1 MHz.

The density of trap filling depends on t_{imp} and on the capture coefficient of the traps $c_{n;p}$. This is usually expressed in terms of the thermal velocity times the capture cross section for the trap. For large pulse widths, all deep-level traps in the SCR should be filled. For narrow pulses, only some of the levels are filled, so the signal gets smaller. The pulse width t_{capt}, for one e-folding (signal height has reduced to $1/e = 0.367$ of its maximum value), lets one measure the capture coefficient c_n (for electrons) or c_p (for holes) by using

$$c_n = \frac{n}{t_{capt}}; \quad c_p = \frac{p}{t_{capt}}$$

n and p are the free carrier concentrations for electrons and holes, respectively.

Now for $\Delta V < V_r$ show that

$$N_t = \frac{2\Delta C V_r}{C_0 \Delta V}(N_D - N_A)$$

This means that for a semiconductor with a homogeneous trap distribution, the DLTS peak height, ΔC, should be proportional to the filling pulse height ΔV. Moreover, N_t can be calculated if $N_D - N_A$ is known.

(c) Finally, we must consider the effects of temperature. Traditionally, the temperature of the experiment is varied from about LN_2 temps to about room temperature or higher. Clearly thermal activation of the traps should be of some importance in determining the trap depth.

The exponential relaxation seen in the capacitance is due to the charged traps emitting their charge once the pulse bias is removed. The time constant τ_e of this thermal emission of charge and the thermal emission rate itself, $e_{n;p}$, both depend on the trap energy E_t and on the temperature T of course. From what we have learned about such statistics already, we can write

$$e_{n,p} = \frac{1}{\tau_e} = \frac{N_{C;V} c_{n,p}}{g} e^{-E_t/k_B T}$$

The $N_{C;V}$ term is the density of trap states in the conduction (C) or valence (V) bands. g is the degeneracy of the level. This is usually 1. Show how this is derived.

How would we use such a curve to determine the trap energy and the capture coefficients? The pulse rate, or multiples thereof, is referred to as the *rate window*. When this rate window is tuned to the emission rate $e_{n;p}$, what should happen when the temperature is scanned slowly? Draw your expectation.

DLTS has a very high sensitivity. In fact it is higher than almost any other semiconductor diagnostic technique. In silicon, for example, it can detect traps at a concentration of better than one part in 10^{12} of the material host atoms.

References

1 For further references see: (a)Ashcroft, N.W. and Mermin, N.D. (1976). *Solid State Physics*. New York: Saunders College Publishing; (b) Hellwege, K.H. (1976). *Einführung in die Festkörperphysik*. Heidelberg: Springer Verlag. (c) Hunklinger, S. (2007). *Festkörperphysik*. München: Oldenbourg Wissenschaftsverlag. (d) Kittel, C. (1986). *Introduction to Solid State Physics*, 6the. New York: John Wiley & Sons.

2 Kaplan, T.A. (2006). The chemical potential. *J Stat. Phys.* 122 (6): 1137.

3 Baierlein, R. (2001). The elusive chemical potential. *Am. J. Phys.* 69: 423.

4 *Fano resonances* are found in a number of fields of physics, not just condensed matter, and are a general property of scattering systems: (a)Fano, U. (1961). Effects of configuration interaction on intensities and phase shifts. *Phys. Rev.* 124: 1866. (b) Gaji, R. et al. (2003). Boron-content dependence of Fano resonances in p-type silicon. *J. Phys. Condens. Matter* 15 (17): 2923.

5 Abroug, S., Saadallah, F., and Yacoubi, N. (2007). Determination of doping effects on Si and GaAs bulk samples properties by photothermal investigations. *Physica B* 400: 163.

6 Goss, J.P., Briddon, P.R., Payson, M.J. et al. (2005). Vacancy-impurity complexes and limitations for implantation doping of diamond. *Phys. Rev. B* 72: 035214.

7 (a) Carroll, D.L., Curran, S., Redlich, P. et al. (1998). Effects of nanodomain formation on the electronic structure of doped carbon nanotubes. *Phys.*

Rev. Lett. 81: 2332. (b) Blase, X. et al. (1999). Boron-mediated growth of long helicity-selected carbon nanotubes. *Phys. Rev. Lett.* 83: 2348. (c) Czerw, R. et al. (2001). Identification of electron donor states in N-doped carbon nanotubes. *Nanoletters* 1: 457.

8 See, for example, the excellent work: Chen, W. (ed.) (2010). *Doped Nanomaterials and Nanodevices*. USA: Journal of Nanoscience and Nanotechnology.

9 (a) Seeger, K. (1973). *Semiconductor Physics*. Wien: Springer Verlag. (b) Böer, K.W. (1993). *Survey of Semiconductor Physics*. New York: Van Nostrand Reinhold. (c) Sze, S.M. (1985). *Semiconductor Devices – Physics and Technology*. New York: John Wiley & Sons.

10 Lang, D.V. (1974). Deep-level transient spectroscopy: A new method to characterize traps in semiconductors. *J. Appl. Phys.* 45: 3023–3032.

8

Electrons Moving in Solids

When we say "electronic transport problems" in solid-state physics, it seems clear that we mean the purposeful movement of electrons from one place to another within the material's volume. So given some driving force and injection of carriers, we would measure the flow of electrons as a current (I) and ask: "how does this current correlate with the applied driving forces and properties of the materials through which the current flowed?" Therein, we have the classic statement of the transport problem.

Historically, we owe this problem largely to Georg Simon Ohm. In purely *ohmic transport*, we have a simple resistance R and an applied potential U[1]:

$$U = IR \tag{8.1}$$

Of course everyone has played a little with those tiny, striped resistors made of carbon and connected batteries and ammeters to them (Figure 8.1).

But we have just introduced several models for the electronic structure of solids. What do these have to say about the mechanisms of this process? That is, where does the resistance come from? What about measurements on a finer scale? If the resistor becomes thinner and thinner, do things remain the same? These questions are what we are going to address here.

8.1 Phenomenology of Electron Dynamics in a Material

The model used to describe the flow of electrons under a driving force obviously depends on just what the exact circumstances are that we want to understand. Like our choices for the models in electronic structure, the material and dimension of a system will play a significant role in our choice of how to view transport in the system. So one model doesn't really fit all situations.

8.1.1 Free-Electron Metals

Let's begin with the easiest example possible: a material that is well described by the free-electron metal model. How would we handle the transport question for such a system (shown in Figure 8.2)?

1 I had a college roommate, majoring in electrical engineering, that once claimed the whole universe could be reduced to this simple equation.

Foundations of Solid State Physics: Dimensionality and Symmetry, First Edition.
Siegmar Roth and David Carroll.

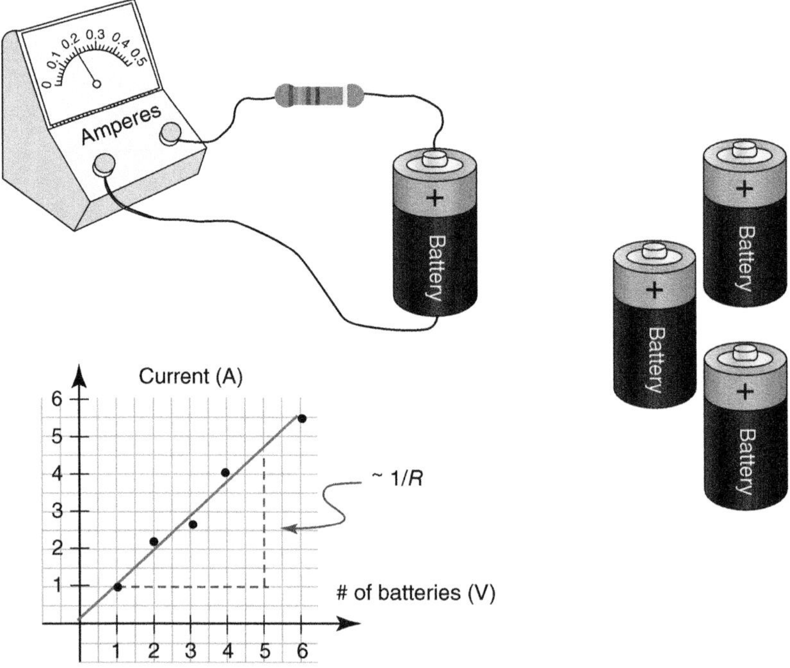

Figure 8.1 Transport is simply the act of measuring how electrons are pushed through a material. Carbon resistors are pretty common, but we assume in this scenario that the wires add nothing to the measurement.

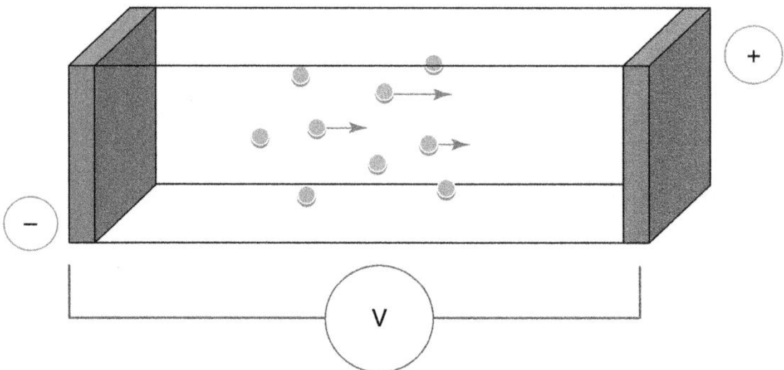

Figure 8.2 Consider a simple slab of Au metal that is very pure. It has dimensions: L_x, L_y, and L_z. A voltage is placed along the L_x direction and current I is allowed to flow. Before the voltage is put in place, we already know what the electrons are doing and their occupation of energy levels. But immediately after the voltage is put in place, a whole new set of eigenvalue states appear. As we will see, this new set of states is directly related to the set of states we began with.

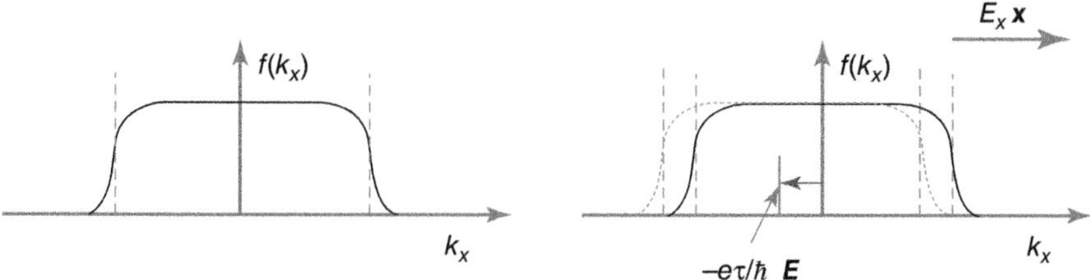

Figure 8.3 The distribution of electron states shifts as an electric field is added, resulting in more electrons with ***k*** values along $-\boldsymbol{k}$ (in this example) than in $+\boldsymbol{k}$. These electrons will compose the measured current.

We place a voltage on this bar of metal and measure the current as it flows through it and out the contacts. We already know that the wavefunctions of the electrons are those of plane waves confined by the solid's boundaries. These *single particle states* are occupied up to the Fermi level ε_f at k_f. Thus below ε_f is a whole sea of electrons moving to and fro with an *ensemble average of the velocity* equal to zero. With the application of the electric field, E_x, the distribution of electrons $f(k)$ in $\boldsymbol{k}$-space shifts, as a whole. The distribution is shifted by the amount of energy that has been added to the system.

If the system is homogeneous, you get a picture as in Figure 8.3. The momentum of each of the free electrons is $mv = \hbar k$. The energy it gains under E_x is $eE_x/v\tau$, where τ is the time between randomizing collisions, and so $v\tau$ is the distance the electron travels under the influence of E_x gathering energy. v is obviously the *average velocity* over the interval.

The expression for I is simply the electronic charge e, times the number of electrons with a given velocity v, added up for all states that are filled. The states that are filled have changed however because of this field:

$$I = -2e \int \left(\frac{\mathrm{d}k_x}{2\pi}\right) f\left(k_x + \frac{e\tau}{\hbar}E_x\right) v(k_x) \tag{8.2}$$

With the application of the electric field, the $\boldsymbol{k}$ values go from k_x to $k_x - e\tau/\hbar$ E_x where τ is related to the time the electron can move on average before some scattering event in the wire. From the expression above we have

$$I = -2e \int \left(\frac{\mathrm{d}k_x}{2\pi}\right) f(k_x) v\left(k_x - \frac{e\tau}{\hbar}E_x\right) \tag{8.3}$$

$$I = -\frac{2e\hbar}{m} \int \left(\frac{\mathrm{d}k_x}{2\pi}\right) f(k_x)\left[k_x - \frac{e\tau}{\hbar}E_x\right] \tag{8.4}$$

$$I = \frac{\tau e^2}{m}\left[2 \int \left(\frac{\mathrm{d}k_x}{2\pi}\right) f(k_x)\right] E_x \tag{8.5}$$

$$I = \frac{n\tau e^2}{m}\boldsymbol{E} \tag{8.6}$$

$$I = \sigma \boldsymbol{E} \tag{8.7}$$

where σ is the conductivity found in the familiar *Drude metal* models, and this result resembles the well-known *Ohm's law*! Writing the law in its more familiar form,

$$I = U/R; \quad R = U/I; \quad U = IR \tag{8.8}$$

where I is the current, U is the voltage, and the proportionality constant R is the resistance. The reciprocal of the resistance is the conductance. Not all conductors obey Ohm's law of course. Gas discharges, vacuum tubes, and semiconductors often deviate from Ohm's law, as do most of the one-dimensional (1D) conductors we have discussed. When resistivity or conductivity values are quoted, the current or voltage range has to be specified where these values were obtained.

In an *ohmic material* the resistance is proportional to the length l of the sample and inversely proportional to the sample cross section A:

$$R = \rho l/A \tag{8.9}$$

ρ is the *resistivity*, measured in (Ω cm). Its inverse is the *conductivity* ($\rho^{-1} = \sigma$). The unit of the conductance is siemens (S). Siemens is the reciprocal of ohm and sometimes is written as Ω^{-1} or mho (ohm backward). The unit of the conductivity is S/cm (in SI units [m] should be used rather than [cm] but [cm] is rather common).

In many solids σ depends on the crystallographic direction and hence is not a scalar quantity but a tensor. Such solids are said to have an *anisotropic conductivity*. The combination of high conductivity in one direction and zero conductivity in the two perpendicular directions leads to 1D solids: the anisotropy approach we have already explored in Chapter 2. In KCP, for instance, the conductivity parallel to the platinum chains is about 200 times larger than the conductivity in the perpendicular direction [1]; in SbF_5-intercalated graphite the conductivity within the *ab*-plane can be up to 1×10^6 times that of the *c* direction [2]. The anisotropy of "highly conducting stretch aligned polyacetylene" has been determined as $\sigma_{\text{perpendicular}}/\sigma_{\text{parallel}} = 25$ [3].

8.1.2 The Free-Electron Metal as a Fluid

Most of our concepts of electric current in the Drude model are derived from the theory of fluids. This is evident in the terminology: current, cross section, source, and drain. Ohm's law is an immediate consequence thereof. Indeed, this *free-electron gas* as a *fluid* picture was first introduced by P. Drude [4] himself only three years after J.J. Thomson's discovery of the electron. Drude applied the kinetic theory of gases to a metal, which he considered as a gas of electrons. As we just saw, in his model the conductivity is given by

$$\sigma = n\tau e^2/m \tag{8.11}$$

where n is the electron density, e is the charge of an electron, m is its mass, and τ is an effective collision time (relaxation time). Introducing the mobility as

$$\mu = \tau e/m \tag{8.12}$$

a typical expression for conductivity is then

$$\sigma = n\mu e \tag{8.13}$$

Replacing the free-electron mass m by the effective mass m^* leads to a transport expression for the *nearly free-electron* model that we introduced some chapters back. The terminology of fluids is retained, and most of solid-state physics (in particular the collision time τ) is put into the asterisk at m^*. The mobility is related to the diffusion constant D via the Einstein relation $\mu = eD/k_B T$.

Why do moving electrons in a solid behave like a fluid? And what about electron waves? The answer lies in Bloch's theorem, wave packets, and the net balance of ionic and electronic charge of course. A more cumbersome detail, however, is an explanation of where the electrical resistance comes from. Assuming partially filled bands, the resistance is a result of deviations from the perfect crystal periodicity (defects), interactions with quasiparticles like phonons and others, and interactions between the electrons themselves (correlation effects). Of course, when a conducting wire is of very small dimensions, scattering is no longer a *statistical outcome* of the flowing current but individual scattering *events*. In such cases, Drude transport is modified such that the electrons are not "moving balls" but rather plane waves entering the wire. Randomly dispersed impurity scatterers then explain repeatable fluctuations in the current–voltage characteristics of the system. This pattern of *mesoscopic universal conductance fluctuations*, as seen in Figure 8.4, has a direct analogue in nuclear scattering [5].

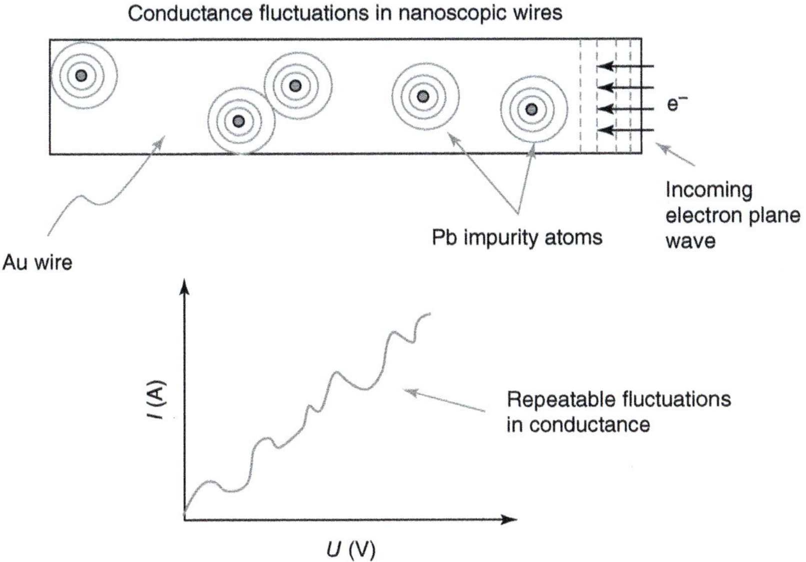

Figure 8.4 Mesoscopic universal conductance fluctuations appear in very low-temperature conductance curves of very narrow conducting metal channels with impurities. If the voltage is scanned up and down, the fluctuations repeated themselves exactly, but if the conducting wire is heated and then cooled again, the pattern changes. They are due to the interference between the incoming plane wave of electrons introduced into the wire and the spherical waves scattering off of the individual impurities scattering that wave. So they are an interference pattern of sorts. When the wire is heated, the impurities migrate, changing their positions and thus the pattern.

8.1.3 Temperature and Conductivity

Often solids are classified by their conductivity at room temperature: for typical conductors the conductivity is greater than several thousand S/cm, for typical insulators it is less than some 10^{-12} S/cm, and for semiconductors it is in between. Probably more significant is the classification by temperature coefficients of resistance. In this case the resistance is plotted vs. the temperature: a positive slope indicates "metallic" materials, and a negative value implies a semiconductor (or an insulator). However, this is also only a very rough classification and does not necessarily agree with the band structure point of view, where metals are characterized by partially filled conduction bands, while semiconductors and insulators have only completely filled and completely empty bands.

To learn more about the mechanism of electrical conductivity, the functional dependence of the conductivity with temperature must be determined. Generally speaking, this is the first quantitative determination an experimentalist would use when faced with a novel conductor, and, indeed, historically this is what was used in conducting polymers, for instance. For systems in which it is expected that a small energy barrier exists to allow charge carriers to flow (or be generated), it is reasonable to "guess" a functional dependence expressed by some power law such as

$$\sigma \sim T^n \tag{8.14}$$

In fact, as we will see later, this is theoretically expected for *intersoliton hopping* in slightly doped polyacetylene. Alternatively, in some systems, such as crystalline semiconductors, where a well-defined charge reservoir is present, an exponential function of activated behavior may be appropriate:

$$\sigma \sim \sigma_0 \mathrm{e}^{-\Delta E/k_\mathrm{B}T} \tag{8.15}$$

Finally, there are occasions to consider a "soft" exponential functional form for the temperature dependence such as

$$\sigma \sim \sigma_0 \mathrm{e}^{(-\mathrm{T}_0/T)^\gamma} \tag{8.16}$$

with $\gamma = 1/2$, $1/3$, or $1/4$. This is particularly applicable in amorphous semiconductors, moderately doped polymers, and other systems with hopping conductivity and charge localization (forms of electron correlation). A given system may actually express all three of these functional forms, and so careful microscopic analysis must generally be coupled with transport determinations to decide on the most applicable model.

A note to budding experimentalists out there is that to decide on the functional dependence, it is not sufficient to change the temperature by several percent (e.g. to warm up the sample from room temperature to 100 °C). Temperature changes of several orders of magnitude are needed. Large temperature changes are more easily obtained by cooling than by heating. Cooling from room temperature to the temperature of liquid helium is a change of nearly 2 orders of magnitude (from 300 to 4.2 K), heating from room temperature to 1000 K is only a factor of 3, and at 1000 K many solids, in particular most organic solids, decompose. As a consequence, electrical transport measurements require the use of liquid nitrogen and in most cases of liquid helium.

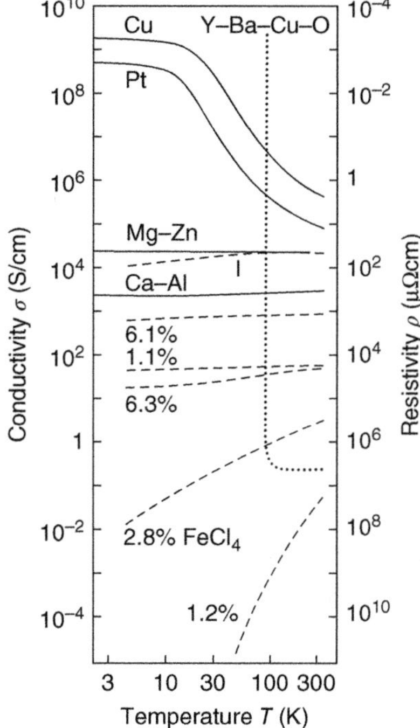

Figure 8.5 Comparison of the conductivities of metals (solid lines) and various doped polymers (dashed lines). Source: After Kaiser and Müller [6, 7].

As an example for the temperature dependence of the electrical conductivity of various materials, Figure 8.5 shows a data compilation after A.B. Kaiser [6]. The format of the figure demonstrates an important fact about the electrical conductivity: the conductivity is the parameter with the largest variability range in solid-state physics. Although it is a logarithmic plot and the insulators are not included, the graph has to be much greater in height than in width to accommodate the conductivity values. (The ratio between the room temperature conductivity of insulators and that of copper is of the same order as the diameter of the universe in kilometers!)

The solid lines in Figure 8.5 represent "metals." Copper and platinum are well-behaved metals. Their conductivity increases upon cooling up to a saturation point at about 10 K. This increase is a result of freezing out lattice vibrations. The cold lattice is more perfect than the warm lattice, and consequently there is less resistance. Below about 10 K collisions of electrons with impurity atoms are more important than collisions with phonons. It is not possible to freeze out the impurity atoms. This temperature-independent resistance is called the *residual resistance.* The resistance of platinum above some 20 K is fairly linear, and in this range platinum can be used as a thermometer.

Mg–Zn and Ca–Al are amorphous (glassy) metals. They have no crystal periodicity, and their thermal properties (lattice vibrations) tend to be dominated by higher energy phonons. Thus, there is little variation in conductivity/resistivity upon cooling. However, as the temperature is raised significantly, there can be effects of the higher energy phonons.

The dashed lines in Figure 8.5 refer to conducting polymers at various doping concentrations. Note that highly conducting polymers behave very similar to glassy metals. Polymers with lower conductivity show a negative temperature coefficient in resistivity (positive in conductivity), as do semiconductors. However, the decrease in conductivity upon cooling is much slower than in the case of crystalline semiconductors. Data for silicon, for example, would be outside the range in Figure 8.5. In crystalline semiconductors phonons are frozen out in the same way as in metals, but the freezing out of phonons (approximately linear with temperature) is by far overcompensated by the freezing out of electrons (exponential). Conducting polymers behave more like *amorphous* semiconductors, where the electrons are not moving in bands but are located at specific states in the gap. They hop between these localized states. *Hopping* is an abbreviation for *phonon-assisted tunneling* [8]. As in crystalline semiconductors, the assisting effect of the phonons is more important than their destructive effect, but in conducting polymers the temperature dependence is smoother, because the phonons do not have to excite the electrons across the gap. They act between localized states within the gap. The excitation energies are smaller, and, in addition, the distribution of excitation energies is continuous. Hopping conductivity will be discussed in more detail later.

At room temperature there is not much difference between amorphous metals and amorphous semiconductors. Actually the temperature coefficient is not the right criterion for distinction. More relevant is the asymptotic behavior as the temperature approaches zero: in a metal the conductivity stays finite at $T \to 0$, and in a semiconductor or an insulator $\sigma \to 0$ as $T \to 0$. As an example the conductivity of moderately doped and highly doped polyacetylene is presented in Figure 8.6. In the moderately doped sample, the conductivity vanishes at absolute

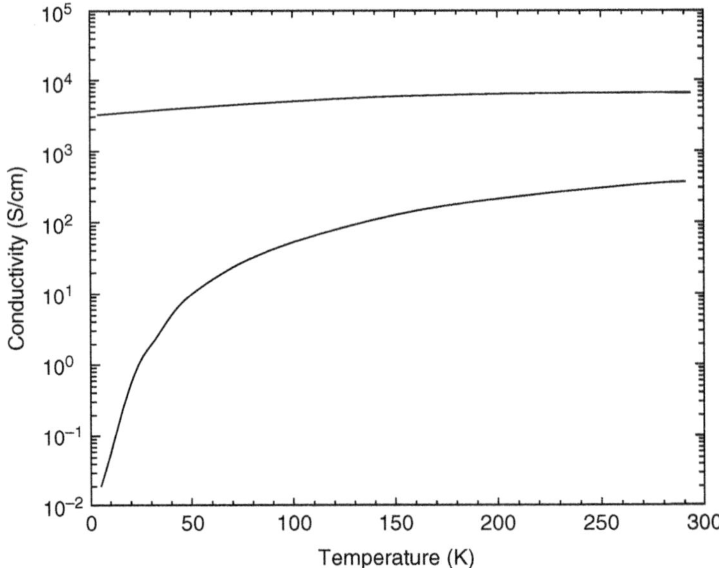

Figure 8.6 Temperature dependence of the conductivity for a highly doped and a moderately doped polyacetylene sample. Note the different asymptotic behavior as the temperature approaches zero.

zero, and the sample becomes insulating. The conductivity of the highly doped sample stays finite, and the sample remains "metallic." Figure 8.5 also shows the behavior of a superconductor (Y–Ba–Cu–O, dotted line). The resistance disappears at $T_c \sim 100\,\text{K}$. (In a superconductor, theory requires the conductivity to be infinite. Of course, infinity cannot be measured, but $\sigma > 10^{24}\,\text{S/cm}$ has been observed.)

8.2 The Semiclassical Approach: The Boltzmann Equation

While the *Drude model* gets us to *Ohm's law* easily, it fails to account for the many subtle attributes of the conductance in different materials at different temperatures. Moreover, it seems only comfortable with static and uniform $\boldsymbol{E}$ and $\boldsymbol{B}$ fields. Let's now examine electronic transport in a more general way using the *semiclassical approach.*

In this approach we will allow for the presence of $\boldsymbol{E}$, $\boldsymbol{B}$, and ΔT, and they can all be functions of positions and time ($\boldsymbol{r}$ and t). However, we will restrict this treatment to:

1. The *independent electron* approximation (same as we have been using).
2. Semiclassical motion between collisions.
3. No interband transitions: conservation of band index, n.
4. *No spin change*: conservation of spin state.

Naturally, everything except the multiparticle wavefunctions can be added back in (with a lot of trouble) if we wanted. But this is an instructive place to start.

8.2.1 The Sources of Electron Scattering

First we must name all of the ways we can think of that electrons could scatter. In the independent particle approximation, we have the following:

Lattice defects: These can be point defects such as missing or substituted atoms (as in the case of dopants), interface defects such as grain boundaries, and other dislocations. Such spatial inhomogeneities within the lattice have associated with them some Coulomb-derived potential.

Thermal effects: This is scattering with phonons (as we have discussed before). Small vibrations of atoms and their electrons about equilibrium positions have amplitudes that depend on T. This is an important scattering source in DC transport and is primarily responsible for the T dependence of conductivity around room temperatures. Again the primary scattering force is derived from Coulomb fields (local dipoles).

So, as $T = {>}0\,\text{K}$, defects will dominate scattering. However, at higher temperatures the role of phonons becomes greater and greater. Again, we emphasize this is for the *independent particle picture.* If we allow for correlation in a multiparticle wavefunction, electron–electron interactions become important, leading to temperature-dependent phonon scattering at very high T and impurity effects at low T. But we aren't there yet.

8.2.2 The Nonequilibrium Distribution Function

For our *more general* approach to transport, we start with consideration of our "old friend" the distribution function

$$g_0(\boldsymbol{k}) = f(\varepsilon(\boldsymbol{k})) = 1/[\mathrm{e}^{(\varepsilon(\boldsymbol{k})-\mu)/k_\mathrm{B}T} + 1] \tag{8.17}$$

For a system of electron states in thermodynamic equilibrium, the distribution function is expressed as above. However in the presence of applied fields or temperature gradients, a nonequilibrium distribution function $g(\boldsymbol{r};\, \boldsymbol{k};\, t)$ occurs in phase space, as the external forces act to drive the system and its distribution function away from equilibrium.

From here we make a pretty important assumption about the nature of entropy flow in the solid. We assume that the main outcome of scattering events in the solid is to effectively relax the system to equilibrium: $g(\boldsymbol{r};\, \boldsymbol{k};\, t) = > g_0(\boldsymbol{k})$. Therefore, if we were to make some simplifications such as semiclassical equations of motion between collisions and simple treatments of the collisions themselves, then a closed form expression for the time development of the system in transport might be possible.

8.2.3 The Relaxation Time τ

To derive a time-dependent relaxation theory of transport, we begin with the introduction of the relaxation time constant (τ). It is the average amount of time between scattering events. Therefore an electron experiences a collision in a time interval $\mathrm{d}t$ with probability $\mathrm{d}t/\tau$, and τ is in general τ $(\boldsymbol{r},\boldsymbol{k})$. This all seems reasonable, since collisions are not entirely random and uncorrelated. After all, they do depend on the occupation of all $\boldsymbol{k}$ states.

8.2.4 The Differential Equation for $g(r;\, k;\, t)$

How does the $g(\boldsymbol{r};\, \boldsymbol{k};\, t)$ evolve in time as it goes to t from $t' = t - \mathrm{d}t$ when *NO* collisions are present during the $\mathrm{d}t$ interval but there is applied electromagnetic driving force $\boldsymbol{F}(\boldsymbol{r};\, \boldsymbol{k})$? Well, the equations of motion will develop classically according to

$$\dot{\boldsymbol{r}} = \boldsymbol{v}(\boldsymbol{k}) \tag{8.18}$$

$$\hbar\dot{\boldsymbol{k}} = -\mathrm{e}\left(\boldsymbol{E} + \frac{1}{c}\boldsymbol{v} \times \boldsymbol{H}\right) = \boldsymbol{F}(\boldsymbol{r}, \boldsymbol{k}) \tag{8.19}$$

Explicitly, the evolution of the system as we go from t' to t (in the first order of $\mathrm{d}t$) looks like

$$t' = t - \mathrm{d}t \rightarrow t \tag{8.20}$$

$$\boldsymbol{r}' = \boldsymbol{r} - \boldsymbol{v}(\boldsymbol{k})\mathrm{d}t \rightarrow \boldsymbol{r} \tag{8.21}$$

$$\boldsymbol{k}' = \boldsymbol{k} - \frac{\boldsymbol{F}}{\hbar}\mathrm{d}t \rightarrow \boldsymbol{k} \tag{8.22}$$

The number of electrons that occupy the phase space volume of $\Delta\boldsymbol{r}\Delta\boldsymbol{k}$ centered on $\boldsymbol{r}$ and $\boldsymbol{k}$ at time t is just

$$\frac{\Delta\mathbf{r}\Delta\mathbf{k}}{8\pi^3} g(\boldsymbol{r}, \boldsymbol{k}, t) = \boldsymbol{v}(\boldsymbol{k}) \tag{8.23}$$

We are ignoring spin for the moment. Similarly, the number of electrons occupying phase space volume $\Delta \boldsymbol{r}' \Delta \boldsymbol{k}'$ centered on $\boldsymbol{r}'$ and $\boldsymbol{k}'$ at time t' is

$$\frac{\Delta \boldsymbol{r}' \Delta \boldsymbol{k}'}{8\pi^3} g(\boldsymbol{r}', \boldsymbol{k}', t') = \frac{\Delta \boldsymbol{r}' \Delta \boldsymbol{k}'}{8\pi^3} g\left(\boldsymbol{r} - v(\boldsymbol{k})\mathrm{d}t, \boldsymbol{k} - \frac{\boldsymbol{F}}{\hbar}\mathrm{d}t, t - \mathrm{d}t\right) \tag{8.24}$$

If there are NO collisions in $\mathrm{d}t$, then all the electrons that are in $\Delta \boldsymbol{r}' \Delta \boldsymbol{k}'$ centered on $\boldsymbol{r}'$ and $\boldsymbol{k}'$ at time t' end up in $\Delta \boldsymbol{r} \Delta \boldsymbol{k}$ centered on $\boldsymbol{r}$ and $\boldsymbol{k}$ at time t. Therefore,

$$\frac{\Delta \mathbf{r} \Delta \mathbf{k}}{8\pi^3} g(\boldsymbol{r}, \boldsymbol{k}, t) = \frac{\Delta \mathbf{r}' \Delta \mathbf{k}'}{8\pi^3} g\left(\boldsymbol{r} - v(\boldsymbol{k})\mathrm{d}t, \boldsymbol{k} - \frac{\boldsymbol{F}}{\hbar}\mathrm{d}t, t - \mathrm{d}t\right) \tag{8.25}$$

From the Liouville theorem,

$$\frac{\Delta \boldsymbol{r} \Delta \boldsymbol{k}}{8\pi^3} = \frac{\Delta \boldsymbol{r}' \Delta \boldsymbol{k}'}{8\pi^3} \tag{8.26}$$

and

$$g(\boldsymbol{r}, \boldsymbol{k}, t) - g\left(\boldsymbol{r} - v(\boldsymbol{k})\mathrm{d}t, \boldsymbol{k} - \frac{\boldsymbol{F}}{\hbar}\mathrm{d}t, t - \mathrm{d}t\right) = 0 \tag{8.27}$$

Now we expand the second term in this expression around $g(\boldsymbol{r}; \boldsymbol{k}; t)$ to the first order in $\mathrm{d}t$:

$$\begin{aligned} g\left(\boldsymbol{r} - v(\boldsymbol{k})\mathrm{d}t, \boldsymbol{k} - \frac{\boldsymbol{F}}{\hbar}\mathrm{d}t, t - \mathrm{d}t\right) &= g(\boldsymbol{r}, \boldsymbol{k}, t) - \frac{\partial(g \cdot v(\boldsymbol{k}))}{\partial \boldsymbol{r}}\mathrm{d}t \\ &\quad + \frac{\partial(g \cdot \boldsymbol{F}/\hbar)}{\partial \boldsymbol{k}}\mathrm{d}t + \frac{\partial g}{\partial t}\mathrm{d}t \end{aligned} \tag{8.28}$$

yielding

$$-\frac{\partial(g \cdot v(\boldsymbol{k}))}{\partial \boldsymbol{r}} + \frac{\partial(g \cdot \boldsymbol{F}/\hbar)}{\partial \boldsymbol{k}} + \frac{\partial g}{\partial t} = 0 \tag{8.29}$$

when there is NO scattering present.

8.2.5 Introducing Collisions

Now, as we introduce collisions, the right-hand side of the above expression is no longer zero.

$$\underbrace{-\frac{\partial(g \cdot v(\boldsymbol{k}))}{\partial \boldsymbol{r}} + \frac{\partial(g \cdot \boldsymbol{F}/\hbar)}{\partial \boldsymbol{k}} + \frac{\partial g}{\partial t}}_{\text{Drift term}} = \underbrace{\left(\frac{\partial g}{\partial t}\right)_{\text{coll}}}_{\text{Collision term}} \tag{8.30}$$

The right-hand side of the equation is known as the *collision term*, and the left-hand side is the *drift term*, and this equation has become known as the *Boltzmann equation.*

If the forces and the specific effects of the collision terms are specified, then this becomes a simple initial value problem to determine $g(\boldsymbol{r}; \boldsymbol{k}; t)$.

We have already supposed that the interactions or collisions are local in nature, derived from Coulombic forces that are likely to be dipole-like or shorter range. Thus, for collisions that happen to electrons near $\boldsymbol{r}$ at time t, they are determined only by properties of the system at $\boldsymbol{r}$ and t and nowhere or time else. Therefore, we can simplify our notation by dropping explicit $\boldsymbol{r}$ and t dependencies in $g(\boldsymbol{r}; \boldsymbol{k}; t) => g(\boldsymbol{k})$.

So, we consider collisions that instantaneously change the crystal momentum from $\Delta\boldsymbol{k}'$ centered on $\boldsymbol{k}'$ to $\Delta\boldsymbol{k}$ centered on $\boldsymbol{k}$. In doing this we must distinguish those collisions that increase and decrease the occupancy of states at $\boldsymbol{k}$:

$$\left(\frac{\partial g}{\partial t}\right)_{\text{coll}}^{\text{IN}} > 0; \left(\frac{\partial g}{\partial t}\right)_{\text{coll}}^{\text{OUT}} < 0 \tag{8.31}$$

OUT: The electrons scattered out of the volume $\Delta\boldsymbol{k}$ centered at $\boldsymbol{k}$ will go somewhere else in k-space. This will presumably be $\Delta\boldsymbol{k}'$ centered at $\boldsymbol{k}'$ where $\boldsymbol{k}'$ is any other vector not already occupied and available by conservation of energy. We can write the contribution of these scattering events to $g(\boldsymbol{k})$ as

$$\left(\frac{\partial g}{\partial t}\right)_{\text{coll}}^{\text{OUT}} = -g(\boldsymbol{k}) \int \frac{\Delta\boldsymbol{k}'}{8\pi^3} W_{\boldsymbol{kk}'}(1 - g(\boldsymbol{k}')) \tag{8.32}$$

It is negative because it is a net reduction of occupancy in $\boldsymbol{k}$, and the $g(\boldsymbol{k})$ weighting factor is necessary because the states must be filled to scatter of course.

IN: These are the electrons that are scattered into the $\Delta\boldsymbol{k}$ volume from elsewhere. Again presumably they come from some other region of k-space: $\Delta\boldsymbol{k}'$ centered at $\boldsymbol{k}'$. Using the same reasoning as above, we have

$$\left(\frac{\partial g}{\partial t}\right)_{\text{coll}}^{\text{IN}} = (1 - g(\boldsymbol{k})) \int \frac{\Delta\boldsymbol{k}'}{8\pi^3} W_{\boldsymbol{k}'\boldsymbol{k}} g(\boldsymbol{k}') \tag{8.33}$$

Notice now that we have introduced a term that points to specific models for the interaction: the matrix element, $W_{\boldsymbol{k}'\boldsymbol{k}}$. This is naturally the quantum mechanical probability of making the $\boldsymbol{k}'$ to $\boldsymbol{k}$ transition and will look something like these matrix elements often do: $\langle k\prime|$ Operator $|k\rangle$. Operator here simply means the specific model for the interaction: like dipole (see Fermi's golden rule).

The *detailed balance* of these events now adds up to give

$$\left(\frac{\partial g}{\partial t}\right)_{\text{coll}} = -\int \frac{\Delta\boldsymbol{k}'}{8\pi^3} \{W_{\boldsymbol{kk}'} g(\boldsymbol{k})[1 - g(\boldsymbol{k}')] - W_{\boldsymbol{k}'\boldsymbol{k}} g(\boldsymbol{k}')[1 - g(\boldsymbol{k})]\} \tag{8.34}$$

8.2.6 The Relaxation Time Approximation

The relaxation time approximation is an attempt to get around direct calculation of the $W_{\boldsymbol{k}'\boldsymbol{k}}$ matrix element. In it we assume locality such that OUT events depend only on $g(\boldsymbol{k})$ and IN events depend only on $g_0(\boldsymbol{k})$, the local equilibrium distribution prior to collisions. So we have

$$\left(\frac{\partial g}{\partial t}\right)_{\text{coll}}^{\text{IN}} = \frac{g_0(\boldsymbol{k})}{\tau(\boldsymbol{k})}; \left(\frac{\partial g}{\partial t}\right)_{\text{coll}}^{\text{OUT}} = -\frac{g(\boldsymbol{k})}{\tau(\boldsymbol{k})} \tag{8.35}$$

and

$$\left(\frac{\partial g}{\partial t}\right)_{\text{coll}} \approx -\frac{g(\boldsymbol{k}) - g_0(\boldsymbol{k})}{\tau(\boldsymbol{k})} \tag{8.36}$$

giving a Boltzmann equation as

$$-\frac{\partial(g\cdot \boldsymbol{v}(\boldsymbol{k}))}{\partial \boldsymbol{r}}+\frac{\partial(g\cdot \boldsymbol{F}/\hbar)}{\partial \boldsymbol{k}}+\frac{\partial g}{\partial t}=-\frac{\delta g}{\tau(\boldsymbol{k})} \tag{8.37}$$

This form of the equation is just what we expected that the collision term serves to restore equilibrium by balancing the drift terms.

8.2.7 Isotropic Scattering from Stationary States

If we consider the case where scattering is to be isotropic and from stationary states, then $g_0(\boldsymbol{k})$ is not a function of $\boldsymbol{r}$ and t at all. So writing the difference between distribution functions as we did above

$$\delta g(\boldsymbol{k})=g(\boldsymbol{k})-g_0(\boldsymbol{k}) \tag{8.38}$$

we can approximate

$$\frac{\partial(\delta g)}{\partial \boldsymbol{r}}=\frac{\partial g}{\partial \boldsymbol{r}};\ \frac{\partial(\delta g)}{\partial \boldsymbol{k}}=\frac{\partial(g-g_0)}{\partial \boldsymbol{k}};\ \frac{\partial(\delta g)}{\partial t}=\frac{\partial g}{\partial t} \tag{8.39}$$

for the system. Then the Boltzmann equation reduces to

$$-\frac{\partial(\delta g\cdot \boldsymbol{v}(\boldsymbol{k}))}{\partial \boldsymbol{r}}+\frac{\partial((g_0+\delta g)\cdot \boldsymbol{F}/\hbar)}{\partial \boldsymbol{k}}+\frac{\partial(\delta g)}{\partial t}=-\frac{\delta g}{\tau(\boldsymbol{k})} \tag{8.40}$$

So, in the case of purely isotropic perturbations from stationary states ($\nabla_{\mathrm{r}}T=0$), this last approximation is what we use. However in the case of non-isotropic perturbations ($\nabla_{\mathrm{r}}T\neq 0$ and $\mu=\mu(\boldsymbol{r})$), then we must return to the full Boltzmann equation above.

8.2.8 A Simple Example: Ohm's Law

In case of isotropic scattering with stationary states and with a static and *small* uniform applied $\boldsymbol{E}$, we can write

$$\frac{\partial(\delta g)}{\partial t}=0\ \text{(stationary state)} \tag{8.41}$$

and

$$\frac{\partial(\delta g)}{\partial \boldsymbol{r}}=0 \tag{8.42}$$

(deviations from equilibrium cannot depend on **r**). Also

$$\frac{\partial(\delta g\cdot \boldsymbol{E})}{\partial \boldsymbol{k}} \tag{8.43}$$

since it will be of the second order in $\boldsymbol{E}$. Then we have

$$\frac{\partial g_0}{\partial \boldsymbol{k}}=\nabla_{\mathrm{k}}\varepsilon(\boldsymbol{k})\frac{\partial g_0}{\partial \varepsilon}=\hbar \boldsymbol{v}(\boldsymbol{k})\frac{\partial g_0}{\partial \varepsilon} \tag{8.44}$$

$$\delta g=e\boldsymbol{E}\cdot \boldsymbol{v}(\boldsymbol{k})\tau(\boldsymbol{k})\frac{\partial g_0}{\partial \varepsilon} \tag{8.45}$$

Returning to the equations used to relate the velocity and the current in the Drude model above, we now have

$$j = -e \sum_{k} \delta g(k) v(k) = -e^2 \int \frac{dk}{(2\pi)^3} [E \cdot v(k)] v(k) \tau(k) \frac{\partial g_0}{\partial \varepsilon} \tag{8.46}$$

We have included now the tensor properties of the conductivity unlike before:

$$j_i = \sum_j \sigma_{ij} E_j \tag{8.47}$$

with

$$\sigma_{ij} = e^2 \int \frac{dk}{(2\pi)^3} \tau(k) v_i(k) v_j(k) \left[\frac{-\partial g_0}{\partial \varepsilon(k)} \right] \tag{8.48}$$

So what is different?

1. *Anisotropy*: Typically j is not parallel with E. Of course it would be in cubic materials but not generally.
2. *Filled bands*: Only the deviations from completely filled bands are important.
3. *Importance of the Fermi surface*: Notice that the integral is nonzero only in an interval around ε_f since the term in the square brackets is zero everywhere else. So conductivity is determined only by the conduction band in an interval of $k_b T$ around ε_f.

8.2.9 Parabolic Bands

If we take

$$\frac{-\partial g_0}{\partial \varepsilon(k)} = \delta(\varepsilon - \varepsilon_F) \tag{8.49}$$

and the average velocity of the electrons as

$$\langle v_i v_j \rangle = \langle v_i^2 \rangle \delta_{ij} = \frac{1}{3} v^2 \delta_{ij} \tag{8.50}$$

as well as counting 2 for the spin degeneracy in the system, then

$$\sigma = 2e^2 \int \frac{dk}{(2\pi)^3} \tau(k) \frac{1}{3} v^2(k) \delta(\varepsilon - \varepsilon_F) \tag{8.51}$$

$$\sigma = \frac{e^2}{12\pi^3} \int_{\text{Fermi surf}} \tau(k)\, v^2(k) \frac{dS}{|\nabla_k \varepsilon(k)|} \tag{8.52}$$

$$\sigma = \frac{e^2}{12\pi^3} \int_{\text{Fermi surf}} \tau(k)\, v(k) \frac{dS}{\hbar} \tag{8.53}$$

Now consider the case of isotropic and parabolic bands. The Fermi surface is a Fermi sphere, and the integral above gives only the surface area of that sphere. This gives us

$$\sigma = \frac{e^2}{12\pi^3} \tau_F v_F \frac{1}{\hbar} 4\pi k_F^2 = \frac{ne^2 \tau_F}{m^*} \tag{8.54}$$

This is exactly the expression we said we would get above using the Drude model, with the mass of the electron being replaced by the effective mass m^*.

8.2.10 Another Simple Example: AC Conductivity and Linear Response

Again we begin with the linearized Boltzmann equation for δg, but now we introduce a time-dependent $\boldsymbol{E}$. Of course this means that unlike before we must keep all $\mathrm{d}/\mathrm{d}t$ terms in our equation. However, like last time, $\frac{\partial}{\partial k}\partial g\cdot\mathbf{E}$ terms will be ignored. This gives us

$$-\frac{\partial(\delta g\cdot v(\boldsymbol{k}))}{\partial \boldsymbol{r}}-\frac{e\boldsymbol{E}}{\hbar}\cdot\frac{\partial g_0}{\partial \boldsymbol{k}}+\frac{\partial(\delta g)}{\partial t}=-\frac{\delta g}{\tau(\boldsymbol{k})} \tag{8.55}$$

Consider now

$$\boldsymbol{E}(\boldsymbol{r},t)=\boldsymbol{E}_0\mathrm{e}^{i(\boldsymbol{q}\cdot\boldsymbol{r}-\omega t)} \tag{8.56}$$

In the linear response approximation,

$$\delta g(\boldsymbol{r},\boldsymbol{k},t)=\Phi(\boldsymbol{k})\mathrm{e}^{i(\boldsymbol{q}\cdot\boldsymbol{r}-\omega t)} \tag{8.57}$$

So we have

$$-i\omega\Phi+\mathbf{v}\cdot\mathbf{i}\mathbf{q}\Phi-\frac{e}{\hbar}\boldsymbol{E}_0\frac{\partial g_0}{\partial \boldsymbol{k}}=-\frac{\Phi}{\tau} \tag{8.58}$$

and

$$\Phi(\boldsymbol{k})=\frac{e\tau\boldsymbol{E}_0\cdot\boldsymbol{v}}{1-i\tau(\omega-\boldsymbol{q}\cdot\boldsymbol{v})}\frac{\partial g_0}{\partial\varepsilon} \tag{8.59}$$

Recalling the expression for the current in terms of electron velocity,

$$\boldsymbol{j}=-e\int\frac{\mathrm{d}\boldsymbol{k}}{(2\pi)^3}\delta g(\boldsymbol{k})\boldsymbol{v}(\boldsymbol{k}) \tag{8.60}$$

$$\boldsymbol{j}=-e\int\frac{\mathrm{d}\boldsymbol{k}}{(2\pi)^3}\Phi(\boldsymbol{k})\mathrm{e}^{i(\boldsymbol{q}\cdot\boldsymbol{r}-\omega t)}\boldsymbol{v}(\boldsymbol{k}) \tag{8.61}$$

$$\boldsymbol{j}=-e\int\frac{\mathrm{d}\boldsymbol{k}}{(2\pi)^3}\frac{e\tau\boldsymbol{E}\cdot\boldsymbol{v}}{1-i\tau(\omega-\boldsymbol{q}\cdot\boldsymbol{v})}\frac{\partial g_0}{\partial\varepsilon}\boldsymbol{v}(\boldsymbol{k}) \tag{8.62}$$

and the conductivity

$$\sigma_{ij}=e^2\int\frac{\mathrm{d}\boldsymbol{k}}{(2\pi)^3}\frac{\tau(\boldsymbol{k})v_i(\boldsymbol{k})v_j(\boldsymbol{k})}{1-i\tau(\boldsymbol{k})(\omega-\boldsymbol{q}\cdot\boldsymbol{v}(\boldsymbol{k}))}\left[\frac{-\partial g_0}{\partial\varepsilon(\boldsymbol{k})}\right] \tag{8.63}$$

which reduces to the previous case when $\boldsymbol{q}=0$ and $\omega=0$.

8.2.11 An Example with Anisotropy: $\mu=\mu(r)$ and $\nabla_r T\neq 0$

In this case g_0 depends on $\boldsymbol{r}$, so we have to use the full Boltzmann expression from which we get

$$\frac{\partial g}{\partial \boldsymbol{r}}=\frac{\partial g}{\partial\mu}\cdot\nabla\mu+\frac{\partial g}{\partial T}\cdot\nabla T \tag{8.64}$$

If we take the linear expansion and equate perturbative terms,

$$\frac{\partial g}{\partial\mu}\approx\frac{\partial g_0}{\partial\mu}=\frac{g_0}{k_\mathrm{B}T}\mathrm{e}^{(\varepsilon-\mu)/k_\mathrm{B}T} \tag{8.65}$$

$$\frac{\partial g}{\partial T} \approx \frac{\partial g_0}{\partial T} = \frac{(\varepsilon - \mu) g_0^2 k_B}{(k_B T)^2} e^{(\varepsilon-\mu)/k_B T} \tag{8.66}$$

And, since

$$\frac{\partial g}{\partial \varepsilon} \approx \frac{\partial g_0}{\partial \varepsilon} = -\frac{g_0^2}{k_B T} e^{(\varepsilon-\mu)/k_B T} \tag{8.67}$$

we can express the right-hand side of the above equation in terms of

$$\frac{\partial g_0}{\partial \varepsilon} \tag{8.68}$$

So

$$\frac{\partial g}{\partial \mu} \approx -\frac{\partial g_0}{\partial \varepsilon} \tag{8.69}$$

$$\frac{\partial g}{\partial T} \approx -\frac{\varepsilon - \mu}{T} \frac{\partial g_0}{\partial \varepsilon} \tag{8.70}$$

Likewise $\frac{\partial g}{\partial k}$ can be expressed in terms of $\frac{\partial g_0}{\partial \varepsilon}$. So (8.71)

$$\frac{\partial g}{\partial \boldsymbol{k}} = \frac{\partial g}{\partial \varepsilon} \frac{\partial \varepsilon}{\partial \boldsymbol{k}} \approx \frac{\partial g_0}{\partial \varepsilon} \frac{\partial \varepsilon}{\partial \boldsymbol{k}} = \frac{\partial g_0}{\partial \varepsilon} \hbar \boldsymbol{v}(\boldsymbol{k}) \tag{8.72}$$

Substituting this into the above and taking only the stationary regime

$$\frac{\partial g}{\partial t} = 0 \tag{8.73}$$

we get

$$\tau(\varepsilon(\boldsymbol{k}))\boldsymbol{v}(\boldsymbol{k}) \cdot \left(\nabla_r \mu + \frac{\varepsilon - \mu}{T} \nabla_r T + e\boldsymbol{E} \right) \left(-\frac{\partial g_0}{\partial \varepsilon} \right) = g_0(\boldsymbol{k}) - g(\boldsymbol{k}) \tag{8.74}$$

ignoring higher-order terms in $\boldsymbol{E}$ and $\nabla_r T$.

8.2.12 The Seebeck Effect and Thermopower

This example highlights the fact that an electrical potential isn't the only way to push electrons around a material. As we have seen, the electronic density and mobilities of a material are temperature dependent. Thus, if there is any thermal asymmetry in a material such that, for instance, one part is hot and the other cold, then electrons will move quite differently in the parts. This naturally implies a direct buildup of charge somewhere. So, if one end of a metal bar is hot and the other cold, the hot electrons, moving much faster, will find their way to the cold end long before any cold electrons make it to the hot end. The electron imbalance yields an internal field that can be used to drive current so long as the asymmetry is maintained. We call this a *thermoelectric voltage*, and it is part of a larger mechanism called the *Seebeck effect*. This is exactly what is being expressed in the above equation. (Note in the derived expression we also allow for an applied $\boldsymbol{E}$.)

The experimental condensed matter scientist frequently uses *thermoelectric power* measurements to understand how heat energy is coupled into a materials system (Fermi level spreading, phonon scattering, etc.). The setup can be visualized in a simple diagram as in Figure 8.7.

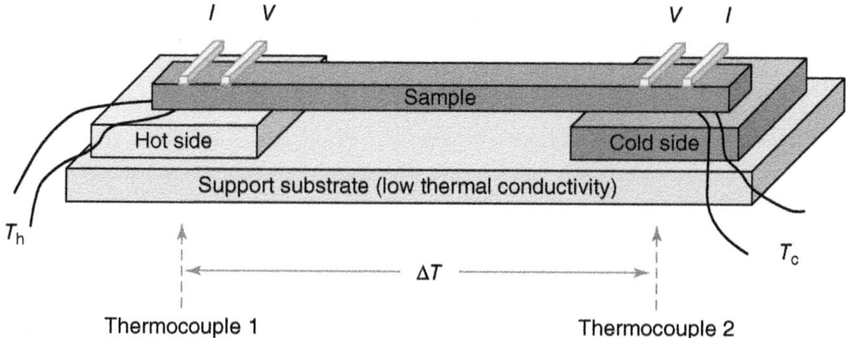

Figure 8.7 A simple thermoelectric power measurement is shown schematically. A thermal gradient is established and the current and voltage generated is measured. This is usually done as a function of different background temperatures.

In our drawing, essentially a temperature gradient is established between the current leads of a four-probe measurement. The charge carriers at the warmer end of the sample have more kinetic energy than those of the cooler end and are allowed to diffuse as described by the Boltzmann expression above. This results in a thermoelectric potential and a measured current when the leads are closed. The current supplied at a given thermovoltage is then the thermopower in simplest terms. It is easy to see why such a measurement could be of use. First, the sign of the carrier is the sign of the thermopower. So it is easy to determine if the sample is a hole majority carrier or an electron majority carrier. Secondly, the thermopower is the largest where there is a small population of phonons and a high carrier mobility and density. In fact, it is quite sensitive to electronic features at the Fermi level of the sample and their coupling to the phonon modes of the system as should be evident from the above expressions.

There are many *quality factors* associated with the thermopower of a material at a given ΔT. The most common starting place is the *Seebeck coefficient* "*S*," which gives the voltage ([volts]/degree of temperature difference [kelvin]). The expression for this Seebeck coefficient is

$$S = -\Delta V/\Delta T \text{ (the minus sign is convention)} \tag{8.75}$$

And it is temperature dependent. That is, it has different values for different background temperatures.

To understand the *thermoelectric effect* from our expression above, we must use it to construct a picture. Consider the microscopic diffusion of carriers (electrons, for example) under a thermal gradient. In the hot volume of the sample, the carriers occupy a larger variation in energy states (as discussed previously) as compared to the cooler part of the sample. That is, the high energy states have a higher carrier occupation per state in the hot part of the sample. Of course this means the hot volume of sample also has a *lower* occupation per state at the lower energy states. This can be visualized as a spreading of the Fermi function in the hot side of the sample. The high energy carriers, with their relatively higher velocities, diffuse away from the hot side of the sample. As they diffuse, they interact with the lattice and carry entropy as they approach the cold side of the sample.

Simultaneously, low energy carriers slowly drift back toward the hot end of the sample to fill the low energy states. This "backflow" of current competes with the current from the hot side, and both processes carry some entropy. A *net* current occurs when one of these diffusive drifts is stronger than the other, and the net current is given by $J = \sigma S \Delta T$, where σ is simply the bulk conductivity at the overall temperature of the system. To be sure, the diffusion rates can depend on velocities, density of states, and rates of scattering, so it is not a forgone conclusion that the hot electrons will dominate. The Seebeck coefficient and thermoelectric power can be either positive or negative, representing hole dominance or electron dominance, respectively. Another way to say this is that the holes or electrons may have the higher conductivity in the system. This language is most useful when describing doped semiconducting materials.

We can use this picture to describe the thermoelectric effect in a material for which the carriers are relatively noninteracting (an electron gas). In this situation of low correlation, we can combine the expressions above to write the conductivity as a simple integral (left as an exercise):

$$\sigma = \int \sigma'(\varepsilon)\left(-\frac{\partial f}{\partial \varepsilon}\right) \mathrm{d}\varepsilon \tag{8.76}$$

where $\sigma'(\varepsilon)$ is the conductivity as a function of energy and f is the familiar Fermi function. The integral is worked over the whole energy range. In the linear response limit, we can then write the thermoelectric coefficient

$$\sigma S = -\frac{k_\mathrm{B}}{e}\int \left(\frac{\varepsilon-\mu}{k_\mathrm{B}T}\right)\sigma'(\varepsilon)\left(-\frac{\partial f}{\partial \varepsilon}\right) \mathrm{d}\varepsilon \tag{8.77}$$

where k_B is Boltzmann's constant, μ is the chemical potential (Fermi level), e is the electronic charge, and σ is the total conductivity. Since σ is frequently an independently measured entity, we can write this as

$$S = -\frac{1}{eT\sigma}\int (\varepsilon-\mu)\sigma'(\varepsilon)\left(-\frac{\partial f}{\partial \varepsilon}\right) \mathrm{d}\varepsilon \tag{8.78}$$

where we now have to figure out how to get $\sigma'(\varepsilon)$. These expressions are known as the *Mott relations* for thermoelectrics. They require more input specific to a given system under study. For example, we can write a simple expression:

$$\sigma'(\varepsilon) = e^2 g(\varepsilon) D(\varepsilon) \tag{8.79}$$

where $g(\varepsilon)$ is the density of electronic states and $D(\varepsilon)$ is the diffusion constant. Again, this is following our "noninteracting" model. To apply the above relations to semiconductors or metals, specific considerations and simplifications are used to write $\sigma'(\varepsilon)$ and to work the integral.

The Mott expression has been used in conjunction with analytical models to predict thermoelectric behavior in simple "noninteracting" systems and generally fails for systems in which there are strong correlations. However, it has been found to be surprisingly useful for mats of conducting fibers (polymers, nanotubes, etc.). Referred to as *heterogeneous conduction,* the Mott approach has become quite powerful in determining the effects of dopants on transport in 1D systems [9]. Typically, the Mott expressions are modified to allow for a metallic, a

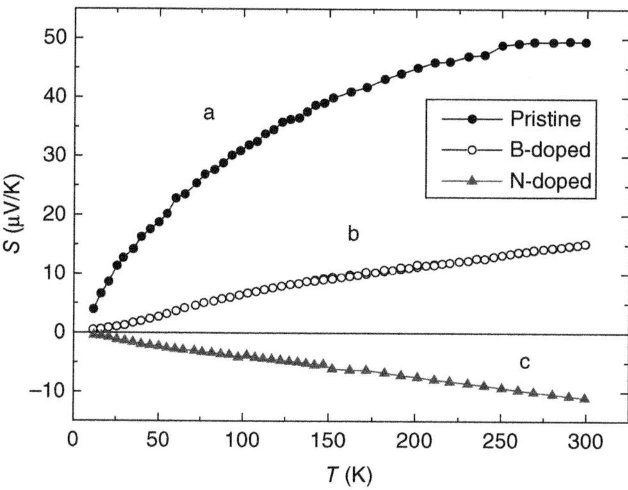

Figure 8.8 Thermopower measurements express majority carrier and dopant levels within fibrillar transport. In the case shown here, different dopants are compared: oxygen (a), boron (b), and nitrogen (c). Courtesy: R. Czerw, Wake Forest University.

semiconducting, and a variable range hopping contribution to the thermopower, and they are each dominant at different temperatures. The metallic term is generally linear, a quadratic term is seen for the semiconducting component, and an exponential term expresses the variable range hopping between fibers in a dense mat. These are added together as

$$S = \beta T + qT_{\mathrm{p}}(e\sigma T^2)^{-1}e^{T_{\mathrm{p}}/T}[e^{T_{\mathrm{p}}/T} + 1]^{-2} \tag{8.80}$$

$$T_{\mathrm{p}} = (\varepsilon - \mu)/k_{\mathrm{B}} \tag{8.81}$$

where β can be used as a fitting parameter, q is the charge and takes on positive and negative values, and the other constants are defined above. As an example of this, consider mats of multiwalled carbon nanotubes. One might wish to know the effects of "doping" impurities in such materials, and thermopower or Seebeck coefficient is a particularly good way to examine these phenomena in detail. Shown in Figure 8.8 is the thermopower as a function of dopant type. Line a is for multiwalled nanotubes exposed to oxygen (thought to weakly dope the tube positively), and Line b for tubes that are heavily doped with boron. This thermopower curve is positive and linear, indicating a hole majority carrier and nearly metallic behavior of the fibers. Finally, Line c shows a mat of nitrogen-doped nanotubes where the nitrogen is introducing donor states, yielding an electron majority carrier material (negative curve) and again linear (metallic) behavior. This follows the expression above very well.

8.2.13 A Final Example: Static *E* and *B* Applied but $\mu \neq \mu(r)$ and $\nabla rT = 0$

In this final case a static $\boldsymbol{E}$ and $\boldsymbol{B}$ are applied to the material. The linearized Boltzmann equation with the Lorentz force has now become

$$\frac{\partial g}{\partial t} - ev \cdot \varepsilon\frac{\partial g_0}{\partial \varepsilon} - \frac{e}{\hbar c}v \times \boldsymbol{B} \cdot \frac{\partial g}{\partial \boldsymbol{k}} = \frac{\delta g}{\tau} \tag{8.82}$$

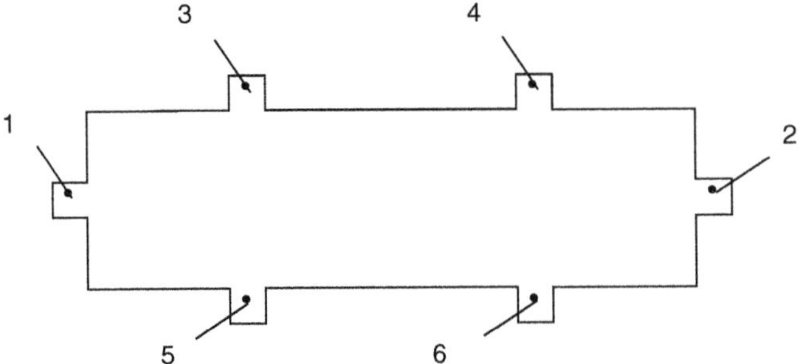

Figure 8.9 The "Hall bar" for magnetotransport measurements. E is applied along the 1–2 direction, whereas the B field is perpendicular, coming out of the page.

$$-e\boldsymbol{v}\cdot\boldsymbol{\varepsilon}\frac{\partial g_0}{\partial \varepsilon} - \frac{e}{\hbar c}\boldsymbol{v}\times\boldsymbol{B}\cdot\frac{\partial g}{\partial \boldsymbol{k}} = 0 \tag{8.83}$$

$$g = \frac{\hbar c\varepsilon k_x}{B}\frac{\partial g_0}{\partial \varepsilon} \tag{8.84}$$

We are going to consider the *high field limit* here. Therefore, we will ignore the collision terms. The last equation comes from a specific choice of field directions: $\boldsymbol{E} = E\boldsymbol{y}$ and $\boldsymbol{B} = B\boldsymbol{z}$. This is done to correspond to the typical *Hall effect* measurement geometries found in most research labs and shown in Figure 8.9. Notice also that we are again expressing the drift of electrons in this equation.

We note here that k_x is not a smooth, single-valued function over the Brillouin zone due to the Bloch periodicity. So our equation only really makes sense if the derivative $\partial g_0/\partial\varepsilon$ confines $\boldsymbol{k}$ to the FBZ. So

$$j_x = \frac{2ec\varepsilon}{B}\int_\Omega \frac{d^3k}{(2\pi)^3}k_x\frac{\partial \varepsilon}{\partial k_x}\frac{\partial g_0}{\partial \varepsilon} \tag{8.85}$$

$$j_x = \frac{2ec\varepsilon}{B}\int_\Omega \frac{d^3k}{(2\pi)^3}k_x\frac{\partial g_0}{\partial k_x} \tag{8.86}$$

Integrating by parts and assuming g_0 vanishes at the zone edge,

$$j_x = \frac{2ec\varepsilon}{B}\int_\Omega \frac{d^3k}{(2\pi)^3}g_0 = -\frac{nec\varepsilon}{B} \tag{8.87}$$

From this we can conclude

$$\sigma_{xy} = -\sigma_{yx} = -\frac{nec}{B} \tag{8.88}$$

and this does not depend on any details of the band structure at all.

For a hole majority material (p-type), the problem is worked the same way though we must replace $g_0{}' = (1 - g_0)$, so

$$j_x = -\frac{2ec\varepsilon}{B}\int_\Omega \frac{d^3k}{(2\pi)^3}g_0' = \frac{nec\varepsilon}{B} \tag{8.89}$$

and

$$\sigma_{xy} = nec/B \tag{8.90}$$

where n is the hole density.

In such measurements we typically will define the following parameters:

$$R_{\mathrm{H}} = -\rho_{xy}/B;\ \text{Hall coefficient} \tag{8.91}$$

$$z_{\mathrm{H}} = -1/n_{\mathrm{ion}}ecR_{\mathrm{H}};\ \text{Hall number} \tag{8.92}$$

where n_{ion} is the ion density in the material. We note that for high field values, the off-diagonal terms of the ρ (ρ_{ij}) and σ (σ_{ij}) matrices can be ignored. Thus $R_{\mathrm{H}} \sim \pm 1/nec$ and $z_{\mathrm{H}} = \pm n/n_{\mathrm{ion}}$.

8.2.14 The Hall Effect and Magnetotransport

So what does this mean physically? If electrical carriers moving in a current are subjected to an applied magnetic field, they will "feel" a Lorentz force. This has the effect of bending the path of the carrier as it moves through the material. There are two consequences of this: firstly, the resistance will increase. The resistance change is called *magnetoresistance*. Secondly, for a finite-sized sample, the carriers will "accumulate" along the edges of the sample for a $\boldsymbol{B}$ field applied perpendicular to their original flight path – assuming little scattering. This accumulation of charge is responsible for the observed *Hall voltage*. The movement of these carriers in different directions can be very anisotropic. So it is no surprise that the determination of the *Hall voltage* as a function of current and $\boldsymbol{B}$-field strength can lead to some important insights into the materials properties. For instance, in a very rudimentary way,

$$V_{\mathrm{Hall}} = -IB/net \tag{8.93}$$

where I is the current, B is the field strength (applied perpendicularly), t is the sample thickness, e is the elementary charge, and n is the carrier density of the material. Thus the Hall voltage gives the number density of carriers as well as the sign of the primary current carrier.

For measurements in a magnetic field, often "Hall bars" are used. An example is shown in Figure 8.9. The current passes through the contacts 1 and 2; a voltage drop proportional to the resistance or magnetoresistance is picked up between probes 3 and 4 and 5 and 6. Between 3 and 5 or 4 and 6, the Hall voltage can be determined. An example of this is seen in Figure 8.10. In this example the resistivity, marked as ρ_{xx}, and ρ_{xy}, represents the flow of current in the different directions. The curves shown are taken at different temperatures. So, the higher the temperature, the more washed out any fine structure in the curve becomes due to thermal agitation of the carriers. Now as we mentioned above, the Lorentz force gives rise to close circle orbits for the electrons that are injected (*Landau levels*). Along the edges there is a half circle "hopping" orbit. When integral numbers of such orbits correlate with the sample dimensions, resonances can occur, and this is seen in the figure.

Of course there are many other variations on this theme such as the quantum Hall effect, the fractional quantum Hall effect, the spin Hall effect, the quantum spin Hall effect, and the anomalous Hall effect. These all take into account the various ways electron orbits will be modified in the magnetic field and look for fine structure within the Hall voltage and resistivity. This, of course, means experiments are carried out at low temperatures.

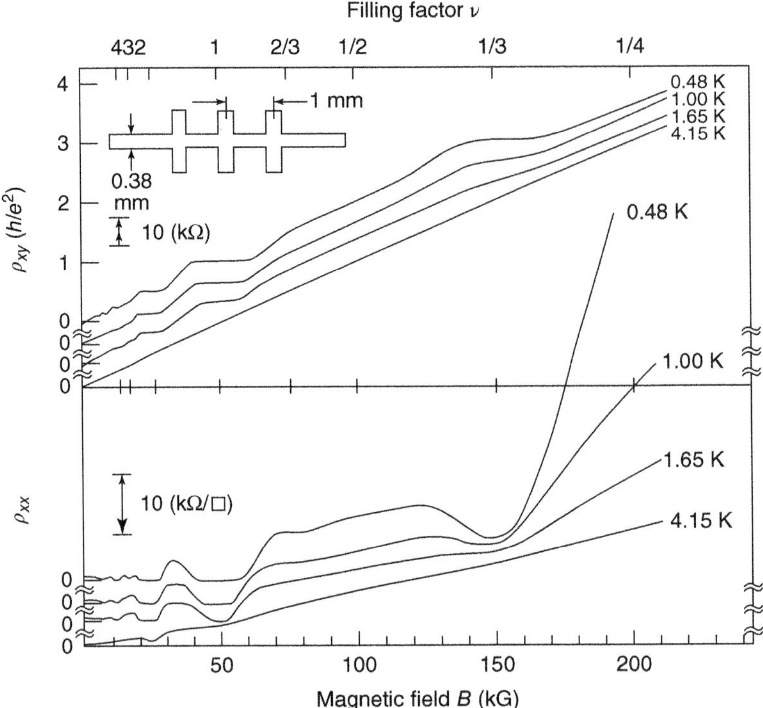

Figure 8.10 Shown here is the magnetotransport data that first demonstrated this effect in a clean, two-dimensional electron gas at a GaAs–AlGaAs inversion layer. ρ_{xy} and ρ_{xx} are shown vs. magnetic field for a set of four temperatures. The Landau level filling factor is $\nu = nhc/eB$. At $T = 4:2$ K, the Hall resistivity obeys $\rho_{xy} = B/nec$ ($n = -1.3 \times 10^{11}$ cm^2). At lower temperatures, quantized plateaus appear in $\rho_{xy}(B)$ in units of h/e^2. Source: Taken from Tsui et al. 1982 [10]. Reproduced with permission of American Physical Society. The work eventually won the Nobel Prize in Physics.

8.2.15 The Curious Case of Al

So if we carry this measurement out on most metals, we should get a positive R_H, reflecting the electron majority nature of the metal. However, we are surprised by Al. In this case at high fields, we get $R_\mathrm{H} = -1$. Why is this negative? Well, as it happens Al has both electron and hole bands. Its valence is 3: two electrons go into a filled band, and one electron is left to split between unfilled electron and hole bands. So $n = 3n_\mathrm{ion}$ and the Hall conductivity is

$$\sigma_{xy} = (n_h - n_e)ec/B \tag{8.94}$$

We can argue a value for $(n_h - n_e)$ in this expression as the following. The electron density in the hole band is given by

$$n'_e = 2n_\mathrm{ion} - n_h \tag{8.95}$$

with the total density of states in the band with two states per unit cell minus the number of empty levels, which are the holes, so

$$n_h - n_e = 2n_\mathrm{ion} - (n_e + n'_e) = n_\mathrm{ion} \tag{8.96}$$

where we have used

$$n_e + n'_e = n_{\text{ion}} \tag{8.97}$$

since one electron from each ion is shared between two partially filled bands. Thus

$$\sigma_{xy} = n_{ion}ec/B = nec/3B \tag{8.98}$$

$$z_{\text{H}} = -1 \tag{8.99}$$

At lower fields we have

$$z_{\text{H}} = +3 \tag{8.100}$$

which is of course the free-electron gas result with suppressed interband scattering.

8.3 Coherent Transport: The Landauer–Büttiker Approach

This semiclassical treatment of Boltzmann's drift and scattering that relaxes the system to g_0 is compelling. It works well for a large number of experimental cases as we have seen. However, when dimensions are reduced in the structures that are carrying the current, the approach can still be a little simplistic or misleading. To see how, consider one of the isolated 1D systems we have already discussed. In any such system such as templated dichalcogenides, to metal atoms that are aligned along the step edges of an insulating crystal's surface, transport is really only related to the transitions between $\boldsymbol{k}$'s going forward and backward along the wire. But how do we place these electrons onto that wire without creating some sort of contact that will introduce a heterogeneous potential at the interface? Or alternatively think of our carbon 1D system: conjugated polymers. In this case we have already suggested that the carrier and then lattice are strongly coupled. Thus the electron–phonon interaction isn't necessarily a scattering interaction with limited and finite time scales (allowing for our "relaxation time" approximation). Instead these interactions can "lock" the carrier and lattice distortion together as we shall see in the next chapter. But this surely cannot be thought of as a simple relaxation of the system to g_0.

To move beyond simple drift equations for these low-dimensional structures, we will need two important observations. The *first observation* is that carrier confinement in the structure will reduce the number of electrons allowed down the wire. The *box modes* that arise due to confinement must obey Pauli's principle. So, each state gets only two electrons (spin-up and spin-down). Thus, realistically we must think of the conducting channel in such mesoscopic systems as waveguides for these modes.

We have also erroneously assumed above that the contacts make no difference. But as we have pointed out already, they are made from metals and will have a different Fermi energy than the channel we are calculating. This can add or subtract electrons into or from the channel. This *second observation* takes us into experimental details that we will not address here.

These two observations for transport in mesoscale structures were dealt with rather elegantly by Landauer in 1957: known as the Landauer–Büttiker (LB) formalism [11–13]. In this approach we view the problem more formally as a transmission problem. The current integral above is modified only slightly as

$$I = \frac{2e}{h}\int M(\varepsilon)g(\varepsilon)T(\varepsilon)\mathrm{d}\varepsilon \tag{8.101}$$

where M represents the number of propagating modes in the channel, g is the deviation from the equilibrium distribution function as we saw before (expressed ion as the variable ε here), and T is the transmission probability function. The integral is worked over only the difference in energies between the metal contacts, that is, the applied voltage plus the differences in Fermi energies. The M and g functions are again only counting functions, whereas the T is related to impurities, excitations, or anything else that might scatter the electron.

We note here that the important subtlety is enumerating and counting the modes because this can depend on the width of the channel. As the channel gets smaller and smaller, the number of electrons allowed through at a time goes down. But think of how appropriate this is in view of our earliest section describing how current flow is like liquid flow. And here we see some correspondence again to that same idea, as the conductance of a fluid-carrying tube is directly related to its diameter. And just as there is a smallest fluid conductance (one atom at a time is allowed down the tube), we can imagine the same with electrical conductivity. If we take the accepted definition of electrical conductance $G = I/V$, then in the limit of a small number of modes, this integral reduces to

$$G = 2e^2MT/h \tag{8.102}$$

Assuming $T = 1$ for ballistic transport, then

$$G_n = 2e^2n/h \tag{8.103}$$

where n is the number of transmission modes. In other words, the conductance becomes quantized, even for ballistic transport (no scattering). Moreover, LB formalism holds generally for electrons that are coherent – they maintain phase information along the length of the wire. So even for perfectly ballistic systems, there is a finite smallest resistance of 12.9 kΩ *per conduction channel.* This is an important number to keep in your mind.

The quantity $G_0 = 2e^2/h = 19.37\ \mu\mathrm{S} = 1/(12.9\ \mathrm{k\Omega})$ is known as the *conductance quantum*. It is interesting to note that the conductance quantum is related to the famous *fine-structure constant* $\alpha = 2\pi e^2/hc$. *G_0 is a combination of general physical constants only, and there is no material-specific parameter entering in the conductivity of 1D metals. In other words it is a constant of the universe: a truly surprising result!*

LB formalism lends itself rather well to the addition of interaction physics within the wire. There are numerous generalizations that include time-dependent fluctuations and dissipation [11]. In systems such as carbon nanotubes where the mean free path for scattering is of the order of 1 μm, LB formalism can be used to understand what excitations couple most strongly to electrons traveling through the lattice – in this case it happens to be optical phonons [14].

8.4 General Remarks on Measurements[2]

Let's make some connection with the lab here. There are many techniques for actually measuring current flow in a wire, even for some very small and very difficult samples. However, what we might refer to as *conductance* in a thin film and a nanowire can be very different things measured in different ways (or at least with really different units). So here we have listed the most common measurements for reference.

8.4.1 Simple Conductivity

The conductivity measurement seems to be quite simple: take an ohmmeter from the shelf and connect it with two alligator clips to the sample. In many cases this will do, but in many others there will be difficulties. The major problems are as follows:

1. The conductivity can vary over a large range, either from sample to sample or within a sample if the temperature, the pressure, or the doping levels are changed.
2. The contacts could influence the measurement.
3. Homogeneity and anisotropy of the sample complicates geometries.

Actually, measurements are only simple when the resistance of the sample is in the range of some kilo-ohms (kΩ) to some mega-ohms (MΩ). In this case a current of some milli- or microamperes is passed through the sample, and the corresponding voltage is around one to 10 V. Large currents can easily cause the sample to warm up by *Joule heating* (changing the resistivity), and small voltage signals are often difficult to separate from the thermopower and contact voltages at interconnects.

With very low resistance of the sample, an ohmmeter would just determine the resistance of the leads. With very low conductivity only leakage currents through the cable insulation would be measured rather than currents through the sample. To avoid lead and contact resistances for low-resistance samples, the "four-lead method" is favorable, where current leads are separated from voltage leads. An example is shown in Figure 8.11. Four thin gold wires are glued to the sample with silver paste. The current I is applied through the outer leads, and the voltage V is picked up at the inner leads (voltage probes).

The resistance is calculated using Ohm's law, and the resistivity is obtained from above, where l is now the distance between the voltage probes. In Figure 8.11 an equivalent circuit to the one in Figure 8.12 is shown. The current

2 This is included to give the reader some idea of how to examine historical literature. While it is clear that experimental techniques in the field are today quite complex, it wasn't always so. Thus, when looking at older result, it is a good idea to ask how trustworthy the results actually are.

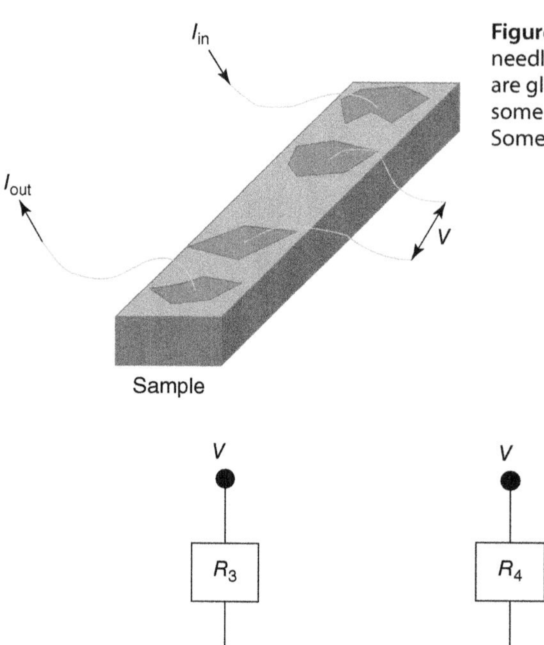

Figure 8.11 Four-lead method on a needlelike sample. Four thin gold wires are glued to the sample by silver paste or some other metallic/conducting glue. Sometimes just "alligator clips" will do.

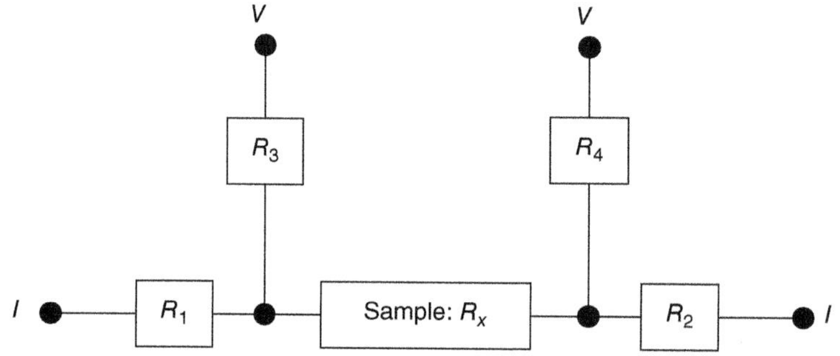

Figure 8.12 Equivalent circuit for four-lead method.

I passes through the sample and creates a voltage drop R_xI. The equivalent resistors R_1 and R_2 represent the resistance of the leads and the contact resistance from lead to sample, respectively (which can be much higher than any other resistance in the circuit). There are voltage drops R_1I and R_2I; the current source sees the sum $R_1I + R_xI + R_2I$. The voltmeter, however, measures the voltage drops $R_3I' + R_x(I + I') + R_4I'$, where I' is the current in the voltmeter circuit and $I' \ll I$, so that it actually determines R_x even if $R_3, R_4 \geq R_x$.

Naturally, nothing is as easy as it seems. For the case of an organic crystal sample, the contact glue has to be carefully chosen in order not to dissolve the sample. If the sample contains iodine, as in iodine-doped polyacetylene, silver paste should be replaced by gold paste. Silver reacts with iodine to form highly insulating silver iodide, resulting in extremely high values for R_1 and R_2. Thus, not enough current can be passed through the sample to yield sufficiently large voltage signals. Organic crystals are often very brittle and break on cooling due to elastic tension. These tensions can be avoided by using very thin gold wires and thermally annealing the wires before pasting them to the sample. In the case of conducting polymers, brittleness is less severe, and it is often sufficient to just press the sample against four equidistant platinum wires (Figure 8.13).

A fairly large error, from the geometrical factor l/A, affects the determination of resistivity for several reasons: crystals are often odd shaped, the thickness of thin films is difficult to measure, and the distance between two blots of silver

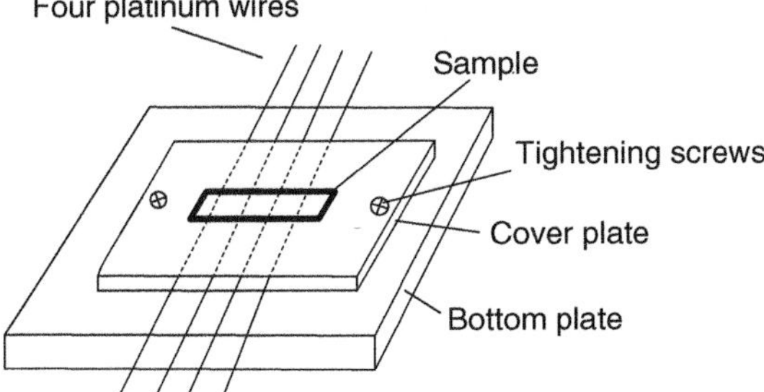

Figure 8.13 Simple sample holder for measuring the conductivity of conducting polymers by the four-lead method. Typical sample size 2 mm × 8 mm and 100 μm thick. The platinum wires are about 0.2 mm in diameter and squeezed into groves of the bottom plate. For room temperature measurements, top and bottom plates can be made of acrylic glass, and at low temperatures ceramic or sapphire plates are advisable.

Figure 8.14 Four-point probe method for conductivity measurements.

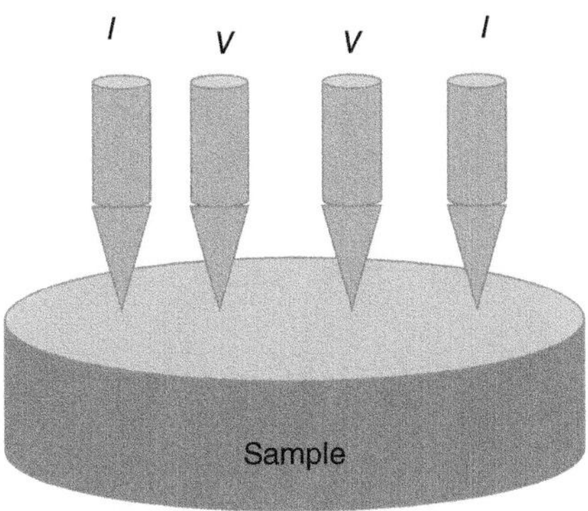

paste is ill-defined. Semiconductor physicists have developed more accurate methods – the *four-point method* and the *van der Pauw method* [15]. The four-point method is indicated in Figure 8.14. It is a special case of the four-lead method discussed above. Here the contacts must be point contacts. Four-point contacts are equidistant and in a straight line. They are pressed against a plane surface of the sample. The sample must be much larger than the contact arrangement, and it must be "infinitely thick." The geometrical factor can then be obtained by integrating over all current paths, and the resistivity is

$$\rho = 2\pi \mathrm{d}V/I \tag{8.104}$$

where I is the current through the outer contacts, V the voltage drop at the inner contacts, and d the distance between two contacts.

For the van der Pauw method, a thin sample of any shape is used. Here the only requirement is that it should be *singly connected*, meaning it should not have any holes. Four contacts are applied close to the edge, as indicated in Figure 8.15. The resistivity is then calculated using the equation

$$\rho = \frac{\pi d}{\ln 2} \frac{R_{12,34} + R_{23,41}}{2} f\left(\frac{R_{12,34}}{R_{23,41}}\right) \tag{8.105}$$

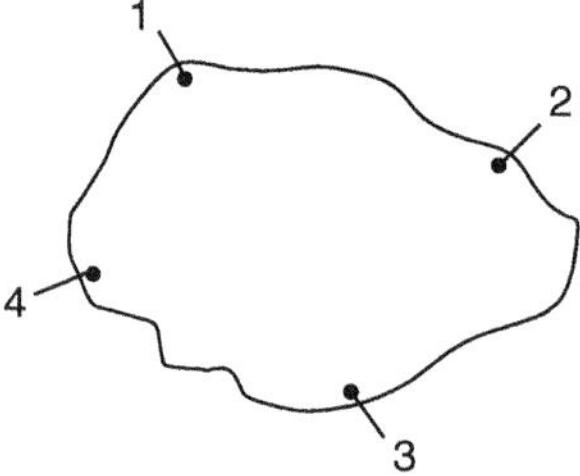

Figure 8.15 Van der Pauw method for conductivity measurements. Notice that this holds for a randomly shaped conducting film.

where d is the thickness of the sample and $R_{12,34}$ is the voltage between contacts 1 and 2 divided by the current through contacts 3 and 4. $R_{23,41}$ is obtained by cyclic interchange of the contacts. f is a function of the ratio of the measured resistances, which varies between 1 and 0.2 when $R_{12,34}/R_{23,41}$ is between 1 and 10^3. Tabulated data for f are found in van der Pauw's original paper [16] as well as in several handbooks [17].

A special case of the van der Pauw method is the use of a square sample with contacts at the corners (Figure 8.15). Using symmetry arguments, we can rewrite the van der Pauw relation above for the symmetric square:

$$\rho = \frac{\pi d}{\ln 2} R_{12,34} \tag{8.106}$$

The anisotropy of quasi-1D materials poses special problems to transport measurements. All the above methods have been developed for isotropic materials, and the influence of anisotropy has to be carefully evaluated in each case. The most straightforward procedure to assess the anisotropy of the conductivity involves cutting out two thin long strips of a stretched polymer film, one parallel and one perpendicular to the stretching direction. These strips are then placed in a sample holder like the one shown in Figure 8.16. A small misalignment will not significantly influence the determination of $\sigma_{\text{perpendicular}}$. However, in σ_{parallel}, we will mainly record the contribution of misaligned $\sigma_{\text{perpendicular}}$ regions. H.C. Montgomery [18] published a procedure to deal with anisotropic samples more accurately. (In *Chapter 2* a notorious anisotropy pitfall was already mentioned in context with the quest for high-temperature superconductivity in TTF–TCNQ.)

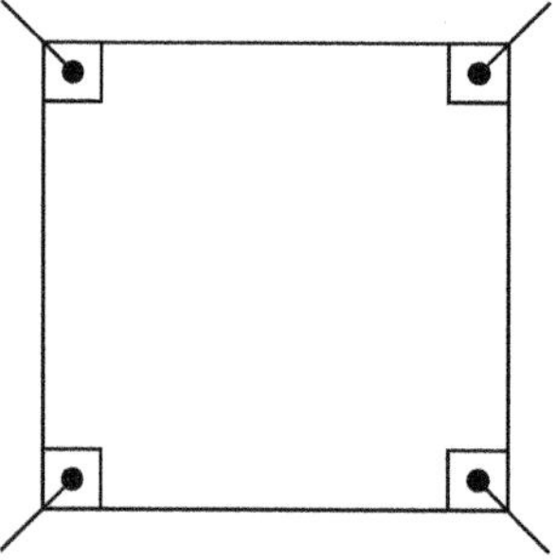

Figure 8.16 Square sample method to measure the conductivity.

8.4.2 Conductivity of Small Particles

The physics of 1D materials often require measurements to be carried out on microstructure contact arrangements. In some cases the samples might just be of some micrometers in size (growing larger crystals was not successful), in others the samples are inhomogeneous, and conductivity microprobes should be applied at different parts of the sample. For these purposes a pattern of microcontacts can be arranged on top of a substrate by lithographic processes, and the sample is then laid over the contacts. For a thin film of a conducting polymer, it is often sufficient to wet it in an organic solvent and to lay it onto the structured substrate. After drying it sticks tightly to the contacts. Figure 8.17 shows a microcontact arrangement that was used to measure transport through a thin organic heterolayer of a phthalocyanine and a perylene derivative [19].

In recent years, lithographic techniques, for the formation of contacts directly onto nanoscale systems, have come into vogue. Electron beam lithography and focused ion beam (FIB) writing have both been extensively used in the creation of nanoscale versions of Figure 8.17. In FIB, an ionized beam of heavy atoms (such as Gd) is created in an acceleration column and focused down to a tight (30 nm) spot. A background metal–organic gas (a gas containing a metal such as Pt) is introduced into the chamber, and the ions at the beam's focus break down the gas near the surface. This results in the deposition of the meal onto the substrate in a very well-defined way. Unfortunately, it also frequently leads to damage to the sample, since it is being bombarded with heavy ions. It also may contaminate the sample, leaving some metals and neutral heavy atoms behind on the substrate. Figure 8.18 is an example of how a single-walled carbon nanotube can be made into a circuit element ("field-effect transistor (FET) configuration")

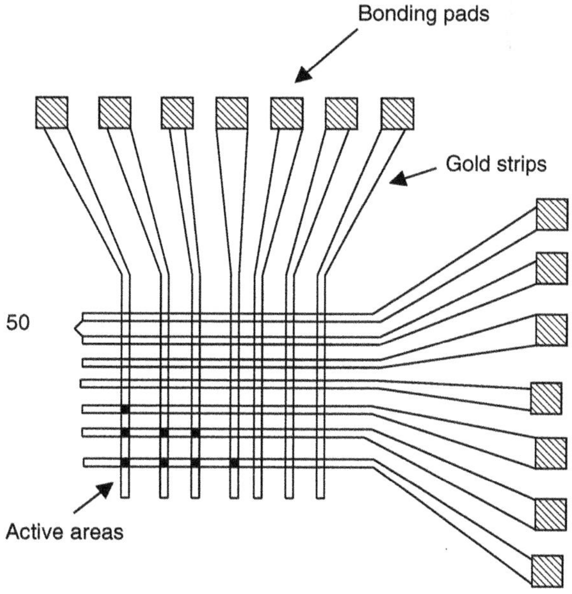

Figure 8.17 Microcontact arrangement for local conductivity measurements. The active area can be as small as a few micrometers squared.

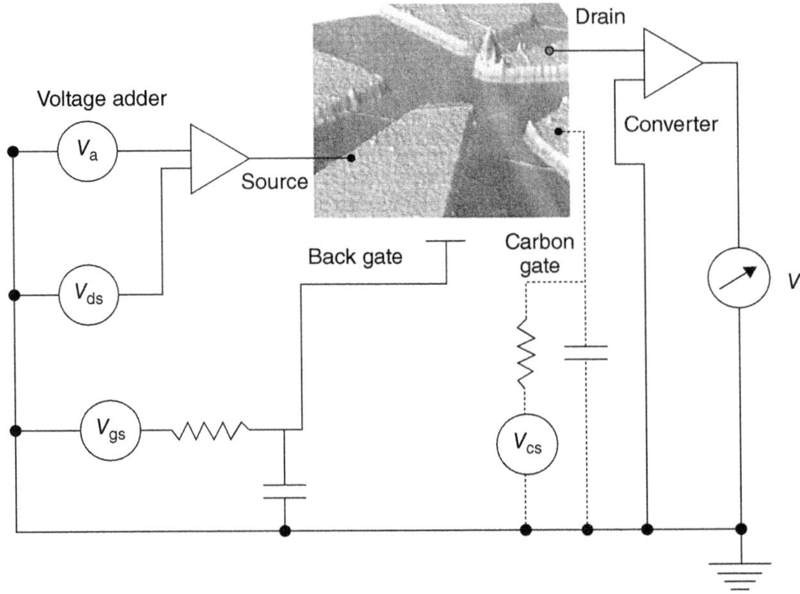

Figure 8.18 A schematic diagram of the setup used for electrical transport measurements. Sze 1985 [20]. Reproduced with permission of John Wiley and Sons.

Electron beam lithographic techniques have the benefit of not damaging the sample as severely. Unfortunately, in this technique, liftoff procedures require that the position of the object of study be well known throughout several procedures of writing and deposition of metals. A clever way of doing this has been worked out, however, that combines atomic force microscopy to determine the positions of the samples on the substrate after they have been laid down. This position is determined relative to larger "markers," which can be easily imaged in the scanning electron microscope used to do the writing [21]. Curiously, this leads to another complication. What happens when the leads and contacts themselves become as small as the sample? We have discussed contact resistance above; however, this is far more subtle. As the leads from the outside world are "whittled" down to create very small (almost atomic point) contacts with the sample, localization, confinement, and back reflection of carriers might well be expressed in the leads. Such circumstances must be analyzed in a far more formal way as worked out by Landauer and Büttiker, where the leads are included as part of the measured system [12]. The actual application of this analysis has been discussed extensively in several excellent texts [13].

8.4.3 Conductivity of High Impedance Samples

In high resistivity samples it could happen that leaks in the insulation of the cables are measured rather than the resistance of the sample. It is also possible that the current passes along a humidity layer on the sample surface instead of through the inside of the sample. This source of error can be avoided by using a guard ring

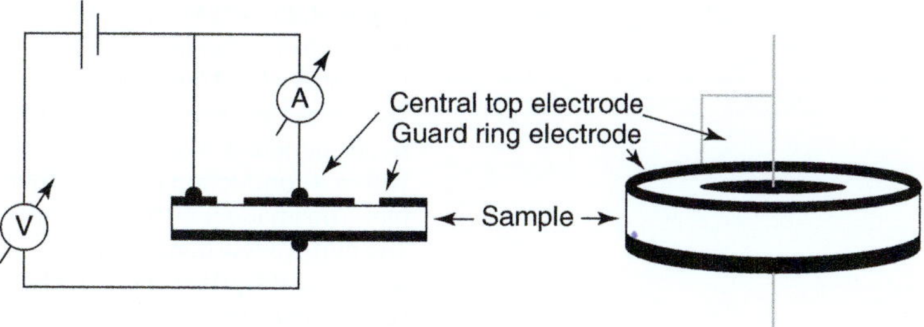

Figure 8.19 Guard ring method for high resistivity samples.

electrode as indicated in Figure 8.19. The central electrode and the guard ring are at the same potential so that no current flows between these two leads and the current between the guard ring and the back electrode does not pass through the ammeter. Many commercial electrometers are supplied with triaxial cable so that the guard can extend into the cable. Complications may arise from the time constant. If the capacitance of the cable is 10 pF and the sample resistance is $10^{12}\ \Omega$, the RC time constant is already 10 seconds, i.e. it takes 10 seconds to charge the cable capacitance. This excludes the application of a lock-in technique. During simple DC measurements on high-resistance samples, on the other hand, space charges build up easily from slow electron trapping and detrapping processes, so that the interpretation of the experiment is by no means straightforward.

8.4.4 Conductivity Measurements Without Contacts

A very serious problem arises from the fact that many materials can only be obtained as powders. The synthetic chemist normally presents a flask with a black powder at the bottom to his physicist friend to find out whether the substance is "interesting." Only for interesting substances the efforts of growing single crystals are worthwhile. Powders have to be compressed to pellets, and transport in pellets is usually not dominated by the bulk properties but by the grain walls, which can either act as low conductivity blockades or as high conductivity carrier accumulation regions.

Many polymers can be formed into *films*. In fact, physicists did not become aware of polyacetylene when it was first synthesized as a powder by Natta in 1958 [22]; it only became famous 16 years later, after Shirakawa had prepared films [23] – as history tells, by a postdoc's mistake in the directions for the catalyst concentration. Although conducting polymer films look smooth and shiny to the eye, their morphology is far from being a physicist's pleasure, as we see in Figure 8.20, a typical thin conducting film used in organic solar cells.

For reasons of contact and powder problems, methods not using contacts have been proposed to determine the conductivity. Most important in this respect is the resonance perturbation of a microwave cavity [24]. The sample is put into a small glass tube that is inserted into a microwave cavity. Microwave absorption in the sample changes the Q-value of the cavity and the resonance frequency. From these two parameters, the conductivity of the sample can be evaluated. Of

Figure 8.20 A TEM image of lamella formation in a highly conductive P_3HT/PCBM blend. This shows the complexities and heterogeneities of the pathways within a conducting polymer thin film. The image is 10 nm × 10 nm. The film is a few nanometers thick. Courtesy: Wanyi Nie, Wake Forest University.

course, the conductivity at microwave frequency (e.g. at 1010 Hz) is not necessarily identical to the DC conductivity. Indeed, if conductivity takes place by a hopping mechanism, large differences have been observed as would be expected [25]. At even higher frequencies the conductivity can be determined from optical absorption [26] ("optical conductivity"):

$$\sigma = \omega\varepsilon_2/4\pi \tag{8.107}$$

where ε_2 is the imaginary part of the dielectric function.

8.5 Complications: Localization, Hopping, and General Bad Behavior

The resistivity is a material constant that can be calculated from the resistance when the sample geometry is known. In very thin wires at low temperatures, a new feature occurs: the resistivity depends on the sample geometry, in particular on the length of the sample. We might have guessed this from our simple introduction to LB theory. Chaudhari and Habermeier [27] prepared very thin (50 Å) and narrow (700–5000 Å) strips of the amorphous alloy W–Re, some 100 μm long. Current leads at the ends and six voltage probes along the wire were applied, as shown in Figure 8.21. The experiment demonstrated that the total resistance of the sample is larger than the sum of resistances between the various leads. In Figure 8.22 the excess resistance is plotted vs. the sample length. The excess resistance represents the amount by which the total resistance exceeds the sum of the partial resistances. The sample length is measured by the total resistance, which reaches up to some 3 MΩ. An excess resistance of 3 kΩ consequently corresponds to an effect in the order of 10^{-3}.

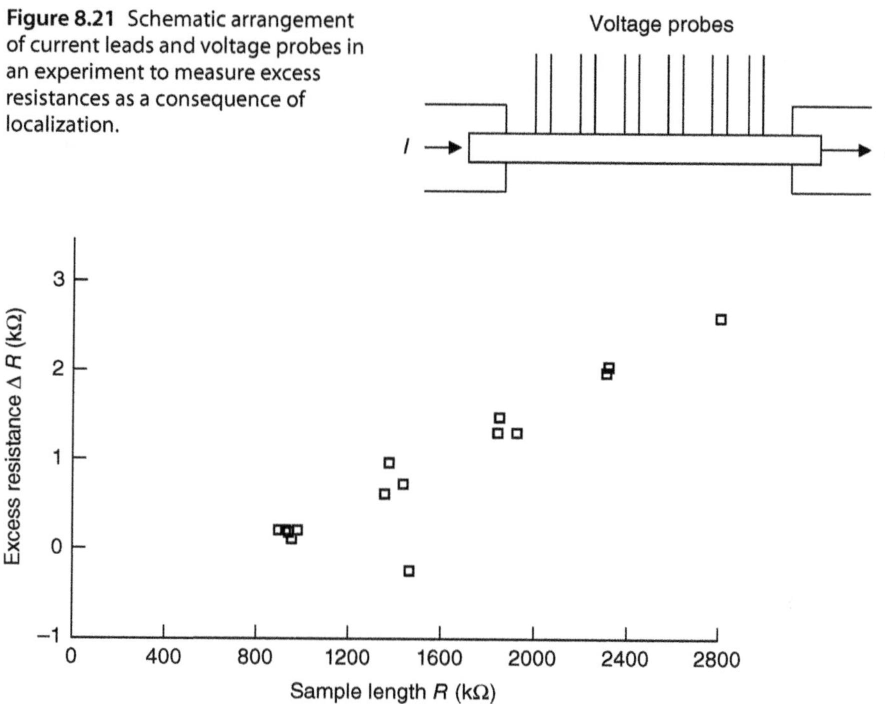

Figure 8.21 Schematic arrangement of current leads and voltage probes in an experiment to measure excess resistances as a consequence of localization.

Figure 8.22 Length dependence of resistance of high-resistance samples measured at 4.2 K. Source: After Chaudhari and Habermeier [27].

The observed excess resistance is interpreted in terms of localization [28]. As already pointed out, the periodicity of the crystal potential is a necessary condition for the existence of Bloch waves. Bloch waves are extended states, delocalized all over the crystal. Occasional defects in a three-dimensional lattice can be treated as scattering centers, giving rise to an ohmic resistance. With many defects, however, the entire concept of Bloch waves fails, and all states in the solid become localized [29]. Between the localized states the electrons move by phonon-assisted tunneling (hopping). At low temperatures when the phonons are frozen, the materials act as insulators. We can quantify the disorder in a crystal by introducing the parameter Δ as average fluctuation of the crystal potential. If, in a three-dimensional crystal, Δ becomes comparable to the bandwidth, all states become localized: the *Anderson–Mott transition.*

The lower the dimensionality, the less disorder is needed to localize the electronic states. In strictly 1D systems, any finite Δ will localize all the states. Since there is always some disorder present, all 1D conductors should become insulators at absolute zero. The localization experiment as shown in Figures 8.21 and 8.22 was carried out at a finite temperature (4.2 K) and on a sample of finite cross section. The observed excess resistance is interpreted as a precursor to the metal-to-insulator transition.

We now give a hand-waving argument for the excess resistance: excess resistance means that the resistance increases faster than linearly with increasing sample length. Theory predicts that it should increase exponentially. If we think of defects as semitransparent mirrors rather than colliding spheres, the exponential decrease of an incoming intensity becomes evident. How could we even have thought that the resistance would increase linearly! In higher dimensionality the electrons can bypass obstacles, and their mirror properties are less pronounced.

How thick must a wire be so that excess resistance can be observed? Thouless [30] has introduced the concept of "maximum metallic resistance in thin wires." There is a maximum resistance a metallic wire can achieve. Whenever the wire is sufficiently long or thin or disordered so that the resistance exceeds a critical value, the wire will not be metallic anymore – it will be insulating in the sense that at absolute zero the resistance will be infinite.

How large is the critical resistance? Since it is independent of the material and of its geometry, it can only be a combination of natural constants. It is not difficult to guess that the quantum of electric charge e and the mechanical quantum h will be involved. The combination $\hbar/e^2$ has the dimension of the electrical resistance and a value of about 4 kΩ. Thus, whenever the resistance of a wire is greater than some 4 or 10 kΩ, the wire is insulating at $T \rightarrow 0$! (Note that here we are content with an order-of-magnitude consideration. However, $\hbar/e^2$ is the famous von Klitzing constant, which in the quantum Hall effect can be determined up to eight significant figures. Multiplied with the velocity of light, $\hbar/e^2$ is the reciprocal fine-structure constant $1/\alpha = 137.036$, the keystone of quantum electrodynamics.) We saw this previously in association with basic transport theory.

Let us now take a closer look at polyacetylene. The conductivity of highly conductive polyacetylene is about 10^5 S/cm. For a single polymer chain, we assume the cross section to be 1 Å^2. Using the equation above a resistance of 10^4 Ω for a polyacetylene chain of 10 Å length is obtained! Thus individual, noninteracting molecular chains will be insulators, not conductors. In the next chapter it will be demonstrated that polyacetylene samples are neither single crystals nor well-separated individual chains but spaghetti-like bundles of chains (fibrils). Figure 8.23 shows such a spaghetti network schematically. A fibril consists of about 100–1000 chains, and its resistance will be some 10^7–10^8 Ω/cm (if highly doped). A fibril of 1 μm length will have a resistance of

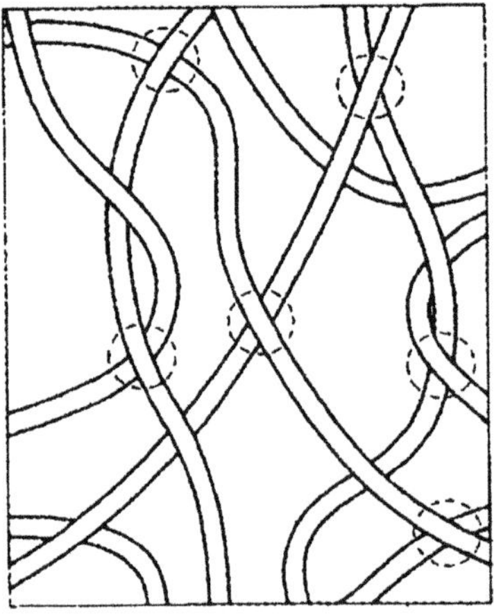

Figure 8.23 Schematic view of the fibrillar structure of a polymer. The rings indicate the interfibrillar cross-links. Source: After Prigodin and Efetov [31].

1–10 kΩ and so will be close to the metal-to-insulator transition. The meshes in the network (Figure 8.23) are about 1 μm wide. The interfibrillar contacts will favor the metallic behavior. Prigodin and Efetov and others [31] have studied the metal-to-insulator phase diagram of a fibrillar network by varying the resistance in the fibers and the interfibrillar contacts.

8.6 Summary

We have presented several detailed methods of understanding electrical transport in well-behaved systems. From the extremely simple ideas of Drude or Boltzmann, many more complex phenomena may be derived. Of course, an important message of this section is that ideas of resistance or conductance cannot be scaled down to small dimensions. A very thin wire behaves differently than a thick wire, and a monomolecular chain is not a molecular wire. This has to be kept in mind when speaking of polymer chains connecting molecular devices in a way as macroscopic wires connect transistors.

Exploring Concepts

1 *Surface Resistance for Drude Models*: Consider the square sheet of conductor $L \times L \times d$. It has a resistivity ρ. The *surface or sheet resistance* is measured between two opposite edges of the sheet, ignoring the effects of the contact: $R_\square = \rho L/Ld = \rho/d$. This is expressed in terms of *ohms per square* (thus the little subscript square), and it is independent of L^2 (Figure EC8.1).
   (a) Using the expression for the conductivity in the Drude metal, show that $R_\square = m/nde^2\,\tau$.
   (b) Now assume that the minimum time between scattering events is just the time it takes the carrier to travel between the upper and lower faces of the slab. That is, consider the case where the surfaces provide the primary source of scattering. This means that $\tau \sim d/v_F$, where v_F is the Fermi velocity. Write down the maximum surface resistivity $R_\square$ and explain why this is.
   (c) Finally calculate this for a sheet of metal atoms, one atom thick ($\sim$4 k$\Omega/_\square$).

2 *Phonons and Resistance*: In real metals, measured resistance is strongly affected by electron–phonon scattering. Clearly, the more phonons one has, the more flowing electrons will scatter into different directions other than the original path, and thus the more apparent resistance one has: it takes longer for the electrons to make it from electrode to electrode since they have a longer path to travel.

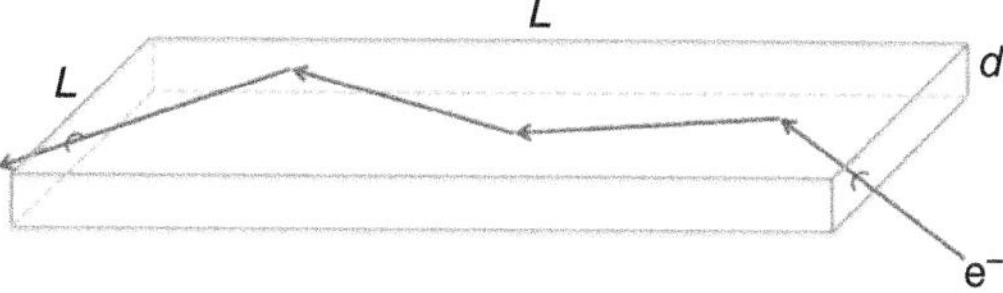

Figure EC8.1 A thin film of metal. The electron flows through the solid with occasional scattering.

Table EC8.1 A 100 Ω Pt resistor.

T (K)	*R* (Ω)
14.0	1.797
20.0	2.147
30.0	3.508
40.0	5.938
50.0	9.228
70.0	17.128
100.0	29.987
150.0	50.815
200.0	71.073
300.0	110.45
400.0	148.62
1000.0	353.402

(a) Previously we made some simplifications in the relaxation rate of scattering wherein we considered this to be a constant. Now let's make an equally simple assumption that it is not a constant but instead depends on the number of scattering phonons present in a given sample and some temperature T. In this case show that it is reasonable to expect that the scattering relaxation time should go linearly with the number of phonons, $N_{\rm ph}$.

(b) However, Landolt and Börnstein [32] worked this problem in more detail, taking into account all the angles into which the electrons could be scattered. They showed that the resistivity could be written as

$$\rho \sim \frac{T}{\Theta_{\rm D}} \int_0^{T/\Theta_{\rm D}} \frac{z^5 {\rm d}z}{(e^z - 1)(1 - e^{-z})}$$

where $\Theta_{\rm D}$ is the Debye temperature. This is known as the Bloch–Grüneisen formula. Using this fact, Pt is often used as a low-temperature thermometer. For a 100 Ω Pt resistor (with a Debye temperature of around 230 K), the following table is given for the resistance in Table EC8.1.
Show that the B–G equation above provides an excellent fit to this data. At what temperature would you estimate the usefulness of this type of thermometer might cut off in temperature?

(c) Give a quick estimate of how the dimensionality in a carbon nanotube might change this result.

3 *Two-Component Models*: As you may already know from the *two-fluid model* of superfluid He flow, *a multicomponent picture* of a system with complex behavior can be a very handy description. Generally, systems described in this way must be in thermodynamic regimes where there is some intimate mixing of a material's phases taking place such as in phase transitions and not at the extremes of stability.

The utility of such descriptions in these regimes is quite simple: different properties can be ascribed to the different individual components of the system. When the system's properties are measured, we see the combined thermal average of the different components. This allows us to "mix" properties that are seemingly incompatible. As in the case of the superfluid where it can mainly act as a fluid with no viscosity but still share some interaction behaviors with some experiments, it is a mixture of the superfluid and a normal fluid.

Now let's take this *model idea* and apply it to conducting sets of charge carriers. Consider a conducting solid that has two distinctly populations of charge carriers. One population is electrons with $q = e^-$, $m = m_e$, and an overall scattering relaxation time of τ_e. The second population is holes with $q = e^+$, $m = m_h$, and relaxation time τ_h.

(a) Using the Drude approximation (i.e. no interaction between the carriers other than hard sphere scattering), calculate the magnetoresistance and the Hall coefficient for applied field $\boldsymbol{H}$. You will have to do this in terms of the concentrations of the charge populations: n_e and n_h.
(b) In an undoped semiconductor we use $n = n_0 \exp[-\Delta/kT]$ to describe the temperature dependence of the carrier concentration. In the model of part (a), describe the temperature dependence of the magnetoresistance and the Hall coefficient.
(c) Now as a little more of a challenge, what would you suspect the result might be, and how would it differ from the Drude two-component model, if we were to allow a small amount of carrier to carrier electrostatic interaction? Think in terms of statistical fluctuations of density.

4 *Thermionic Emission*: Consider a thin conducting channel oriented such that the axis of current flow is in the positive z direction. If we place this within the context of the ballistic transport model (LB-like), then we can write for the total current

$$J = -\frac{q}{4\pi^3} \int_{k_z>0} \mathrm{d}^3k \frac{\hbar k_z}{m^*} f(k)[T_\mathrm{L}(k) - T_\mathrm{R}(k)]$$

k_z is in the direction of current flow (z) and $T_{\mathrm{L/R}}$ express the transmission probability as the carrier is heading to the left or right, respectively. Typically, $T_\mathrm{L} = T_\mathrm{R}$, but certainly not always. Applied fields, heterogeneity, etc. can make the two terms quite different.

(a) Now consider the case of a device in this ballistic regime such that the below is true (Figure EC8.2):

$$T_\mathrm{L}(k) = \begin{cases} 0 & \text{if } E_\mathrm{CL} + \dfrac{\hbar^2 k_B^2}{2m^*} < E_0 \\ 1 & \text{if } E_\mathrm{CL} + \dfrac{\hbar^2 k_B^2}{2m^*} > E_0 \end{cases}$$

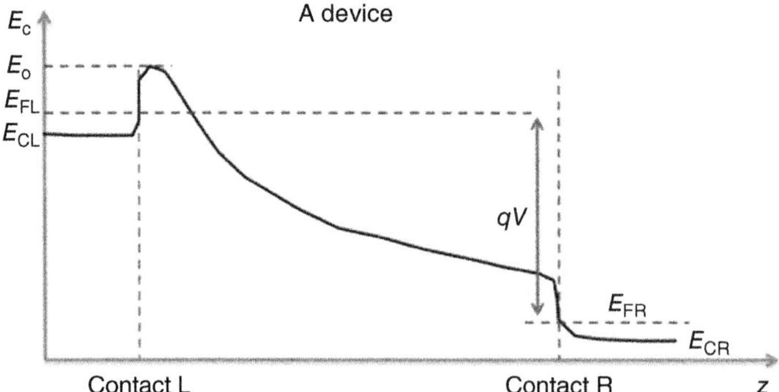

Figure EC8.2 A conducting channel with metal contacts on either side. At the contact region there is a contact barrier.

$$T_R(k) = \begin{cases} 0 & \text{if } E_{CR} + \dfrac{\hbar^2 k_B^2}{2m^*} < E_0 \\ 1 & \text{if } E_{CR} + \dfrac{\hbar^2 k_B^2}{2m^*} > E_0 \end{cases}$$

Show that

$$J = -\frac{qm^* k_B T}{4\pi^2 \hbar^3} \int_{E_0}^{\infty} dE \ln \frac{1 + e^{-(E-E_{FL})/k_B T}}{1 + e^{-(E-E_{FR})/k_B T}}$$

(b) For large barrier heights, show that this reduces to the famous Richardson equation

$$J = J_0(e^{qV/k_B T} - 1)$$

$$J_0 = -\frac{qm^*(k_B T)^2}{2\pi^2 \hbar^3} e^{-(E-E_{FR})/k_B T}$$

(c) What are the physical meanings of these equations?

5 *Conductivity in Tight Binding*: Consider the two-dimensional system with a square lattice (lattice parameter a) and a conduction band given by the tight binding approximation

$$E = E_0 + E_1(2 - \cos k_x a - \cos k_y a)$$

Now consider the case where the relaxation time constant τ is a constant. This means it does not depend on electron momentum or energy. Moreover, the bands are half filled, so this is a metal.

(a) Using the solutions to the Boltzmann equation discussed above, compute the conductivity tensor of this band.

(b) Now compare this result with the Drude model with the same electron density. Discuss why these differences occur.

References

1 Krogman, H. (1969). Planare Komplexe mit Metall-Metall-Bindungen. *Angew. Chem. Int. Ed. Engl.* 81: 10.

2 Vogel, F.L. (1980). Some potential applications for intercalation compounds of graphite with high electrical conductivity. *Synth. Met.* 1: 279.

3 (a) Schimmel, T., Rieβ, W., Gmeiner, J. et al. (1988). DC-conductivity on a new type of highly conducting polyacetylene, N-$(CH)_x$. *Solid State Commun.* 65: 1331. (b) Kim, G.T., Burghard, M., Suh, D.S. et al. (1999). Conductivity and magnetoresistance of polyacetylene fiber network. *Synth. Met.* 105: 207.

4 (a) Drude, P. (1900). Zur Elektronentheorie der Metalle. *Ann. Phys.* 1: 566. (b) Drude, P. (1900). Zur Elektronentheorie der Metalle: II Teil Galvanomagnetische und thermomagnetische Effekte. *Ann. Phys.* 3: 369.

5 (a) Altshuler, B.L., Kravtsov, V.E., and Lerner, I.V. (1991). *Mesoscopic Phenomena in Solids* (ed. B.L. Altshuler, P.A. Lee and R.A. Webb). Amsterdam: Elsevier. (b) Beenakker, C.W.J. (2011). Applications of random matrix theory to condensed matter and optical physics. In: *The Oxford Handbook of Random Matrix Theory* (ed. G. Akemann, J. Baik and P. Di Francesco). Oxford: Oxford University Press. (c) Stone, A.D., Mello, P.A., Muttalib, K.A., and Pichard, J.L. (1991). *Mesoscopic Phenomena in Solids* (ed. B.L. Altshuler, P.A. Lee and R.A. Webb). Amsterdam: Elsevier.

6 Kaiser, A.B. (1987). *Electronic Properties of Conjugated Polymers*, Springer Series in Solid-State Sciences 76 (ed. H. Kuzmany, M. Mehring and S. Roth). Heidelberg: Springer Verlag.

7 Müller, U. (1982). *Anorganische Strukturchemie.* Stuttgart: B.G. Teubner.

8 Zallen, R. (1983). *The Physics of Amorphous Solids.* New York: Wiley.

9 Kaiser, A.B., Park, Y.W., Kim, G.T. et al. (1999). Electronic transport in carbon nanotube ropes and mats. *Synthetic. Metals* 103: 2547.

10 Tsui, D.C., Störmer, H.L., and Gossard, A.C. (1982). *Phys. Rev. Lett.* 48: 1559.

11 (a) Pastawski, H.M. (1991). Classical and quantum transport from generalized Landauer-Büttiker equations. *Phys. Rev. B* 44: 6329. (b) Pastawski, H.M. (1992). Classical and quantum transport from generalized Landauer-Büttiker equations. II Time-dependent resonant tunneling. *Phys. Rev. B* 46: 4053.

12 Imry, Y. (1997). *Introduction to Mesoscopic Physics.* Oxford: Oxford University Press.

13 Ferry, D.K., Goodnick, S.M., and Bird, J. (eds.) (2009). *Transport in Nanostructures*, 2e. New York: Cambridge University Press.

14 Koch, M., Ample, F., Joachim, C., and Grill, L. (2012). Voltage-dependent conductance of a single graphene nanoribbon. *Nature Nanotechnol.* 7: 713.

15 See for example:Lautz, G. and Taubert, R. (eds.) (1968). *Kohlrausch – Praktische Physik Band 2*, 305. Stuttgart: B.B. Teubner Verlag.

16 (a) van der Pauw, L.J. (1958). A method of measuring specific resistivity and hall effect of discs of arbitrary shape. *Philips Res. Rep.* 13: 1. (b) van der

Pauw, L.J. (1958). A method of measuring the resistivity and hall coefficient of lamellae of arbitrary shape. *Philips Tech. Rev.* 20 (8): 220. /69.

17 Nalwa, H.S. (ed.) (2001). *Handbook of Thin Films: Deposition and Processing*. San Diego Academic Press.

18 Montgomery, H.C. (1971). Method for measuring electrical resistivity in anisotropic materials. *J. Appl. Phys.* 42: 2971.

19 Fischer, C.M., Burghard, M., Roth, S., and von Klitzing, K. (1994). Organic quantum wells: molecular rectification and single-electron tunneling. *Europhys. Lett.* 28: 129.

20 Sze, S.M. (1985). *Semiconductor Devices – Physics and Technology*. New York: Wiley.

21 (a) Chiu, P.W. (2003). Towards carbon nanotube based molecular electronics. PhD thesis. Universität München. (b) Liu, K., Roth, S., Wagenhals, M. et al. (1999). Transport properties of single-walled carbon nanotubes. *Synth. Met.* 103: 2513. (c) Kim, G.-T., Waizmann, U., and Roth, S. (2001). Simple efficient coordinate markers for investigating synthetic nanofibers. *Appl. Phys. Lett.* 79 (3479).

22 Natta, G., Mazzanti, G., and Atti, P. (1958). Stereospecific polymerization of acetylene. *Acad. Nazl. Lincei Rend. Cl. Sci. Fis. Mat. Nat.* 25: 3.

23 Ito, T., Shirakawa, H., and Ikeda, S. (1974). Simultaneous polymerization and formation of polyacetylene film on the surface of a concentrated soluble Ziegler-type catalyst solution. *J. Polym. Sci.: Polym. Chem. Ed.* 12: 11.

24 (a) Buranov, L.I. and Shchegolev, I.F. (1971). Method for measuring the conductivity of small crystals at 1010 Hz. *Prib. Tekh. Eksp.* 2: 171. (b) Joo, J. and Epstein, A.J. (1994). Complex dielectric constant measurement of samples of various cross sections using a microwave impedance bridge. *Rev. Sci. Instrum.* 65: 2653. (c) Rabenau, T., Simon, A., Kremer, R.K., and Sohmen, E. (1993). The energy gaps of fullerene C_{60} and C_{70} determined from the temperature dependent microwave conductivity. *Z. Phys. B* 90: 69. (d) Schaumburg, G. and Helberg, H.W. (1994). Application of a cavity perturbation method to the measurement of the complex microwave impedance of thin super- and normal conducting films. *J. Phys. III France* 4: 917. (e) Peligrad, D.N., Nebendahl, B., Mehring, A. et al. (2001). General solution for the complex frequency shift in microwave measurements of thin films. *Phys. Rev. B* 64 (22): 224504.

25 (a) Ehinger, K. and Roth, S. (1986). Non-solitonic conductivity in polyacetylene. *Philos. Mag. B* 53: 301. (b) Kaiser, A.B. and Park, Y.W. (2002). Conduction mechanisms in polyacetylene nanofibers. *Curr. Appl Phys.* 2: 33. (c) Kaiser, A.B., Flanagan, G.U., Stewart, D.M., and Beaglehole, D. (2001). Heterogeneous model for conduction in conducting polymers and carbon nanotubes. *Synth. Met.* 117: 67. (d) Misurkin, I.A. (1996). Electric conductivity of polyacetylene in a three-dimensional model of conducting polymers. *Zh. Fiz. Khim.* 70: 923. (e) Yoon, C.O. et al. (1995). Hopping transport in doped conducting polymers in the insulating regime near the metal-insulator boundary: polypyrrole, polyaniline and polyalkylthiophenes. *Synth. Met.* 75: 229.

26 Genzel, L., Kremer, F., Poglitsch, A. et al. (1984). Millimeter-wave and far-infrared conductivity of iodine- and AsF_5-doped polyacetylene. *Phys. Rev. B* 29: 4595.

27 Chaudhari, P. and Habermeier, H.U. (1980). Quantum localization in amorphous W–Re alloys. *Phys. Rev. Lett.* 44: 40.

28 (a) Krishnan, R. and Srivastava, V. (2000). Localization and interactions in quasi-one-dimension. *Microelectron. Eng.* 51 (2): 317. (b) Krishnan, R. and Srivastava, V. (1999). Resistance of quasi-one-dimensional wires. *Phys. Rev. B* 59 (20): R12747. (c) Kramer, B., Bergmann, G., and Bruynseraede, Y. (1985). *Localization, Interaction, and Transport Phenomena*, Springer Series in Solid-State Sciences 61. Heidelberg: Springer Verlag. (d) Khavin, Y.B., Gershenson, M.E., and Bogdanov, A.L. (1998). Strong localization of electrons in quasi-one-dimensional conductors. *Phys. Rev. B* 58: 8009.

29 Anderson, P.W. (1958). Absence of diffusion in certain random lattices. *Phys. Rev.* 109: 1492.

30 Thouless, D.J. (1977). Maximum metallic resistance in thin wires. *Phys. Rev. Lett.* 39: 1167.

31 (a) Prigodin, V.N. and Efetov, K.B. (1993). Localization transition in a random network of metallic wires: a model for highly conducting polymers. *Phys. Rev. Lett.* 70: 2932. (b) Epstein, A.J., Lee, W.P., and Prigodin, V.N. (2001). Low-dimensional variable range hopping in conducting polymers. *Synth. Met.* 117: 9. (c) Prigodin, V.N. and Epstein, A.J. (2001). Nature of insulator-metal transition and novel mechanism of charge transport in the metallic state of highly doped electronic polymers. *Synth. Met.* 125: 43. (d) Mitra, T. and Thakur, P.K. (1998). Electronic transmittance and localization in quasi-1D systems of coupled identical n-mer chains. *Int. J. Mod. Phys. B* 12: 1773. (e) Dupuis, N. (1997). Metal-insulator transition in highly conducting oriented polymers. *Phys. Rev. B* 56: 3086. (f) Prigodin, V.N. (1997). One-dimensional localization effects in three-dimensional structures of highly-conducting polymers. *Synth. Met.* 84: 705.

32 Landolt, H. and Börnstein, R. (1982). *New Series*, vol. 15 (ed. K.H. Hellwege), 287. New York: Springer.

9
Polarons, Solitons, Excitons, and Conducting Polymers

Part of a letter from Wolfgang Pauli to Rudolf Peierls: a copy of which can be found at the coffee corner in the Max-Planck Institute for Solid-State Research in Stuttgart. The relevant line reads: Der Restwiderstand ist ein Dreckeffekt und in Dreck soll man nicht wühlen. (Residual resistance is a "filth effect" with filth being something one should not rummage in.)

In Chapter 8, we introduced the notions of basic transport. We described how these notions were limited when the size of the structures began to approach that of the wavelengths of the carriers flowing in them. *Interactions* in this picture of transport were described phenomenologically as scattering events that led to the general relaxation of the system. However, there is a second type of interaction that should be considered: *correlation* and *coupling*. These are interactions that do not act over short times relative to the movement of the carrier as in the *relaxation time approximation*, but instead, we can imagine them to take place over the entire path of transport. There is a large catalogue of examples, from *superconductors*, to *mott insulators*, to *charge density waves* (CDWs), and more. In this chapter we consider an example that is of growing technological importance: the *conducting polymer*.

We originally introduced conducting polymers as an example of one-dimensional conducting behavior. However, in the polymer's case, the carrier and the lattice are strongly *coupled* in a way not covered by our previous *scattering theory* discussions. Roughly, what this means is that the charges and phonons of the system move about together. Now that we know a little more about basic electronic transport, we are in a better position to understand what this coupling in the polymer system might mean.

Foundations of Solid State Physics: Dimensionality and Symmetry, First Edition.
Siegmar Roth and David Carroll.

Tradition says that Confucius was once asked to name the duties of a good king or ruler. "First of all, put terminology in order", the master replied.

First we should make sure the language of our conducting polymer discussion and that on crystalline systems is the same (or at least correlated):

1. Most conducting polymers are not good electrical conductors when chemically pure. They are insulators or, at best, semiconductors. Only after treatment with oxidizing or reducing agents do they become conductors. This procedure is called *doping*, and this is in direct analogy to the term introduced previously in semiconductor physics.
2. In semiconductor physics, undoped semiconductors are *intrinsically* conducting and doped semiconductors *extrinsically* conducting. In contrast, doped polymers are often referred to as *intrinsically conducting polymers*. This is to distinguish them from polymers that acquire conductivity by loading with conducting particles, such as carbon black, metal flakes, or fibers of stainless steel or carbon nanotubes. These *intrinsically conducting* materials are also called *conjugated polymers* or *synthetic metals*.
3. Polymers that conduct because they are blended with a percolating network of micro- or nanowires are called *matrix composites* or sometimes *nanocomposites* as we saw before. They are said to be *loaded* with the nanophase, not *doped*.
4. Finally, as we will learn from this chapter, polymers do not conduct using "bare" electrons flowing about; rather they have charge bound to local lattice distortions. Based upon properties such as spin symmetry and charge, these will have a variety of names – soliton, exciton, polaron, bipolaron, etc. – usually some sort of "-on." The names are all meant to emphasize the single-particle nature of the coupled excitation.

As you might have guessed, a detailed treatment of this field is a little beyond an introductory presentation such as ours. And there is already enormous amount of specialist literature on conducting polymers (together with their close cousins the conducting organic small molecule). For instance, a standard in the field is *Electronic and Optical Properties of Conjugated Polymers* by William Barford (Oxford University Press). Skotheim's *Handbook of Conducting*

Polymers [1], the Springer publication [2] edited by Kiess, and Przyluski's [3] and Chien's [4] monographs also provide excellent insight and historical context. As we have already discussed, the matrix composite approach falls between a one-dimensional and three-dimensional system somewhere, and there is a book devoted to *nanoparticle-loaded* conducting polymers[1] published by Mair and Roth [5]. Finally, it can be instructive to examine the series of International Conference on Science and Technology of Synthetic Metals (ICSM) proceedings, the Kirchberg International Winterschool on Electronic Properties of Novel Materials (IWEPNM) volumes [6], the proceedings of the Polish Autumn Schools [7], the NATO Advanced Study Institute in Burlington [8], and the Lofthus meeting [9].

Wow! And with all of these, it is a bit surprising that most presentations and texts fail to give a good *visual* model for exactly how the polymer conducts charge. So let's see if we can make a little headway on that in this chapter – *that will be our goal*. We will begin with the famous *Peierls transition*, a lattice instability. From there we introduce an electron–lattice coupling mechanism based on such instabilities leading to domain walls, solitons, polarons, and excitons. Finally, we touch briefly upon the effects of electron correlation for these systems. This type of introduction borrows heavily from review articles published by Roth and Bleier [10] and by Heeger et al. [11], as well as the monograph published by Lu [12].

9.1 Distortions, Instabilities, and Transitions in One Dimension

In our previous examples, electrons in a solid interacted with moving atomic positions. So the periodicity of the potential experienced by the electrons is that of the lattice *modulated* by the waves of the atoms upon those sites (the Bloch waves of the phonons). But this problem is different in three dimensions than in lower dimensions. In 3D, atoms are highly coordinated and held in position by bonding forces exerted from the surrounding neighbors. In one-dimensional solids, atoms have only two neighbors. Consequently in one dimension, lattices become very *soft* compared with 3D counterparts.

9.1.1 Coupling Charge with the Lattice

In Figure 9.1 we show the simplified schematic structure of polyacetylene (extended chain instead of zigzagged, with a bond angle of 180° instead of 120°). In polyacetylene single and double bonds alternate, and these bonds keep the carbon atoms in their position. Double bonds are stronger than single bonds (although not twice as strong), and consequently the bond length of a double bond is shorter than that of a single bond. Splitting a single bond means breaking

1 As an interesting side note, when the nanophase or interphase of a matrix approach interacts with the host polymer through doping, it can create a two-carrier system. And just as in the case of the two fluid model for superfluids, such electroactive matrix composites provide an interesting proving ground for fundamental physics.

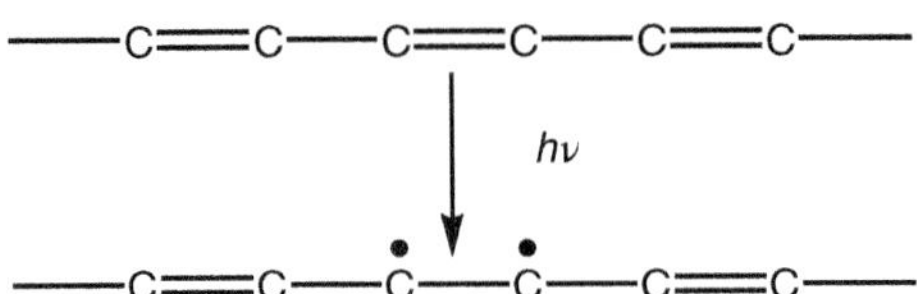

Figure 9.1 *Bond scission* in polyacetylene demonstrates the strong electron–lattice coupling in one-dimensional solids. Raising the electron of a double bond into an excited antibonding state causes strain within the polymer's one-dimensional lattice.

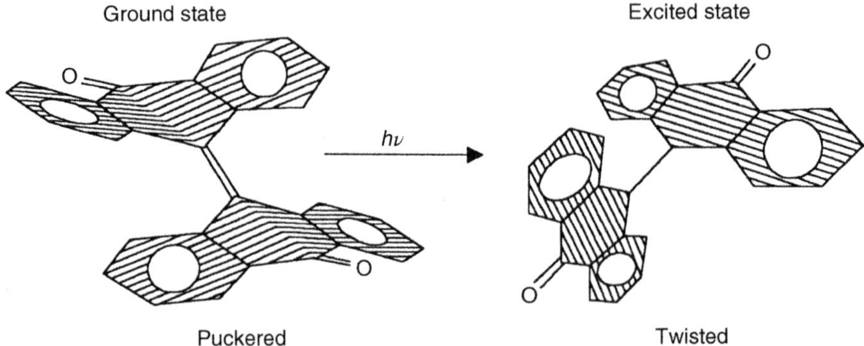

Figure 9.2 Here the two configurations of bianthrone demonstrate the complicated example of electron–lattice coupling. Absorption of light can create an electronic transition that effects local bonding introducing strain. Such mechanisms have been proposed for molecular-scale memory systems.

the chain. If one part of a double bond is cleaved, the double bond will be weakened by the transformation into a single bond. As a result the lattice will be *distorted* due to the lengthening of the respective interatomic distance.

A chemist or a molecular physicist would not be surprised by the large lattice distortion. If there is any surprise at all, it is because of the solid-state terminology. Cleaving a double bond – for example, by the absorption of light – means raising the molecule into an excited state (note that the chemists speak of molecular states, whereas in all previous chapters we used the term electronic states, which the chemist would call orbitals). The shape of the molecule in the excited state is different from its configuration in the ground state. For solids, this *configuration* corresponds to the crystal lattice.

An example of this *excited state – lattice coupling* – is illustrated in Figure 9.2. Here a *bianthrone molecule* is shown, which twists and folds when changing state [13]. In the field of *molecular electronics* (the use of organic materials to beat silicon in device miniaturization), this principle comes into play. Such coupling might help to better localize excitations; thus smaller devices should be possible. In this respect, electron–lattice interactions in organic components are often referred to as *coupling to conformational degrees of freedom*. Excitations would be confined to a molecule, which will extend over some 10 or 20 Å (the oxygen–oxygen distance in bianthrone is 11 Å). We will discuss *solitons* as an example of such localized excitations.

9.1.2 Peierls Instability

An important case of electron–lattice coupling is the *Peierls distortion* (or *instability*). This lattice distortion belongs to a more general class of symmetry breaking or degeneracy lifting transitions, for which the *Jahn–Teller effect* stands as the prototype. In 1955 Peierls [14] showed that the specific example of a chain of equally spaced metal atoms is unstable and will spontaneously undergo a *metal-to-insulator transition* at low temperatures.

As an example, consider a linear equidistant arrangement of sodium atoms as in Figure 9.3. In this case the atoms readily *contribute* one outer shell electron to delocalization and dynamics while effectively screening the core potential with the rest of its electrons. This is true in a number of atomic species we might choose. The Gedankenexperiment, of course, cannot be realized since sodium atoms will naturally not arrange in chains. They tend to form clusters when tossed together. To keep them in line, like beads in a necklace, some sort of a string to thread them would be necessary. That is not so easy and requires a lot of chemistry. Actually all the chemistry in Chapter 2 is just an attempt to approach this idea.

So, although we cannot synthesize an isolated chain of equidistant monatomic sodium, it *is* possible to calculate the electronic dispersion relation and the electronic density of states (DOS) as we have already seen. Because of the extreme delocalization expected of such a system, we expect a close approximation to the nearly free electron model. Thus we have an "S"-shaped dispersion relation approaching a parabola on both sides near the Brillouin zone (BZ) center and flattening out at the BZ edges as in Figure 9.5. The DOS has square root singularities at the top and at the bottom of the band, which are of no significance in this context. The band is half-filled because each sodium atom contributes one delocalized electron to the solid, and two electrons – spin up and down – can be accommodated in each state. The Fermi wavevector $\boldsymbol{k}_F$ is halfway between 0 and

Figure 9.3 Alkali atoms do not arrange in chains but prefer to form clusters. One might argue that we could use templates, say, the step edges of a Si surface, perhaps, to align the atoms against. But again the stabilizing forces due to surface–chain interactions might interfere with electron flow so we wouldn't have a truly 1D conductor. Of course, $CH^{\bullet}$ radicals can be lined up on the σ-bonded backbone of a polymer, like beads on the string of a necklace.

π/a, and the Fermi energy E_F is at the band center. There is a finite DOS at the Fermi level, $N(E_F) \neq 0$. To calculate the total energy of the electrons, we have to sum the overall electron states up to E_F.

Now we call for a *Maxwell's demon* that arranges the atoms in pairs. It displaces every second atom by the amount δ, so that the atoms are no longer equidistant; the short distances $a - \delta$ and long distances $a + \delta$ alternate. Again the arrangement is periodic but with repeat distance $2a$ instead of a. Consequently, the reciprocal lattice changes from $a^*_{old} = 2\pi/a$ to $a^*_{new} = \pi/a$. In the former lattice the border of the first BZ was at π/a; now it is at half that distance, just at the Fermi wavevector $\boldsymbol{k}_F$. The former lattice caused the electrons to have a gap at $\pi/2a$. The demon has transformed the system from a metal with no gap at the Fermi level into a semiconductor with a gap. All states below the gap are filled at absolute zero and all above the gap are empty.

If the demon then leaves, will the system stay semiconducting (or insulating depending on the gap), or will it relax and return to the undistorted lattice? It is easy to see that in the distorted case the electronic energy is lower, because the states in the gap have been accommodated above and below the gap (states cannot disappear; there are as many states as atoms in the lattice, because the states are formed from atomic orbitals). Consequently, when summing the electronic energies from zero to E_F, it shows that the creation of the gap had reduced the electronic energy. But to achieve this, the demon had to exert work: there are springs between the atoms; the demon has alternately compressed and expanded them.

Let's examine this in more detail. We begin by noticing that in the region of the Fermi energy, the dispersion curve looks like the free electron system $\varepsilon = \hbar^2 k^2/2m^*$. At this point the electrons behave as though they have an "effective mass" and a uniform background of ionic potential. As our demon acts, it introduces a strain to the bonding "springs" between atoms. We use a very simple expression for this (though more complicated expressions could be chosen):

$$E_{mech} = s\delta/2 = Y2\delta^2/2 = Ye^2 \tag{9.1}$$

where s is the stress in the spring, Y is the Young's modulus, and δ is the overall strain the demon introduced. The introduction of this δ results in a periodic modulation of the background potential – which at this point on the dispersion curve we are treating as a constant $U_0(x) = 0$. We don't know exactly how much of a change in the potential will result since electrons will redistribute themselves slightly, but a simple approximation might be given by

$$U(x) = 2A \cos \pi x/a = 2A \cos 2\boldsymbol{k}_F x \tag{9.2}$$

This is about the simplest formula we could use. Notice that A depends a bit on how much the system will rearrange its electrons around the atomic site. It is sometimes called the *coupling constant* and describes the coupling between electrons and phonons (lattice distortions). The resulting band structure is given above in Figure 9.5b. The bandgap that occurs is easy to compute:

$$E_{\text{gap}} = 2A\delta \tag{9.3}$$

Clearly, as we have already argued the energy of the electronic distribution is lowered (the topmost energy went down by $2A\delta$), but this came at the cost of E_{mech}. To see if it was "worth it":

$$E(\delta)_{\text{total}} = E(\delta)_{\text{mech}} + E(\delta)_{\text{electronic}} \tag{9.4}$$

The minimum occurs:

$$\frac{\mathrm{d}E_{\text{total}}}{\mathrm{d}\delta} = \frac{\mathrm{d}}{\mathrm{d}\delta}\left[\int E(k,\delta,A)\mathrm{d}E + Y\delta^2\right] = 0 \tag{9.5}$$

This is referred to as a *master equation* and the integral is worked from 0 to $\boldsymbol{k}_F$. This equation either has a solution for a choice of δ or it does not. If it does, this means the *Peierls distortion* of the lattice will occur and be stable because it is energetically favorable. The gap may be arbitrarily small, but there will be one. For our simple model of distorted potentials, it so happens that there *is* a solution:

$$\delta = \frac{\hbar^2 \boldsymbol{k}_F^2}{Am^*}\left[\sin h - \frac{\hbar^2 \boldsymbol{k}_F \pi Y}{4m^* A^2}\right]^{-1} \tag{9.6}$$

Manipulating this around a little, making some guesses about the range of A, and so on, we can estimate that the argument of the $\sin h$ term is $\gg 1$. Thus Eq. (9.6) can be estimated by

$$\delta \gg \frac{\hbar^2 \boldsymbol{k}_F^2}{Am^*} \exp \frac{\hbar^2 \boldsymbol{k}_F \pi Y}{4m^* A^2} \tag{9.7}$$

This is a rather interesting equation since it comes up in a number of physical problems. Particularly it appears in *superconductivity*, which is a peculiar form of *Peierls transition*.

The analysis shows that the electronic energy is approximately linear in δ, whereas the elastic energy depends quadratically on δ. For sufficiently small displacements, the gain in electronic energy dominates the elastic term. Consequently, under Peierls assumptions, there is always a gap at absolute zero. How large it is and at what temperature enough electrons are excited to cross the gap so that it becomes ineffective depend on the parameters of the system (spring constants, bandwidth, etc.). As a result the demon is allowed to leave. We don't even need the demon to start the Peierls distortion. At low enough temperatures the system will know (by fluctuations) that it gains energy upon distortion and thus it will distort. Electron–lattice coupling will drive the one-dimensional metal into an insulator.

This is exactly the situation we have for conjugated carbon, such as a polymer. The Peierls instability can have rather large effects. Typically such systems can have strain-related bandgaps $\gg$1.6 eV. It is important to realize that this comes from a modification of bonding orbitals. Here, polymers (that have an orbital to delocalize) become conjugated – alternating single and double bonds.

9.1.3 Results of Peierls in Real Systems

9.1.3.1 Phonon Softening and the Kohn Anomaly

The monatomic metal chain of Figure 9.4 exists only in a Gedankenexperiment. Real one-dimensional metals are more complicated, as we have seen in Chapter 2. The Peierls transition, however, has been observed in several systems (for a review article, see [15]). What would evidence for this look like? In Figure 9.5, the effects of the Peierls distortion are shown: the giant Kohn anomaly in KCP.

The restoring force stabilizing the equidistant lattice decreases, and the respective phonons become softer. These are phonons with wavevector $2k_F$. In KCP the conduction band is not half filled as it would be for ideal alkali chain. There is charge transfer from the platinum chain to the bromine ions and band filling is more complicated. Therefore in Figure 9.5 the wavevector $2k_F$ does not correspond to the border of the first BZ as it would in the alkali chain. Figure 9.5 shows what to expect from an inelastic neutron scattering experiment on the alkali chain: the zone-end phonons soften and finally condense into a new lattice point at π/a. Therefore the BZ have to be altered. From the segment between $\pi/2a$ and π/a, we can subtract a lattice vector of the new reciprocal lattice and transfer the whole segment to the left-hand side of the origin.

The soft phonon at $2k_F$ also corresponds to the requirement of energy and momentum conservation in electron–phonon scattering. This pronounced electron–phonon scattering is the high-temperature precursor of the *Peierls transition.* Energy conservation requires that only electrons at the Fermi surface are involved: the Fermi energy is much larger than the phonon energies and so only quasi-elastic scattering processes can occur. The Fermi surface of a

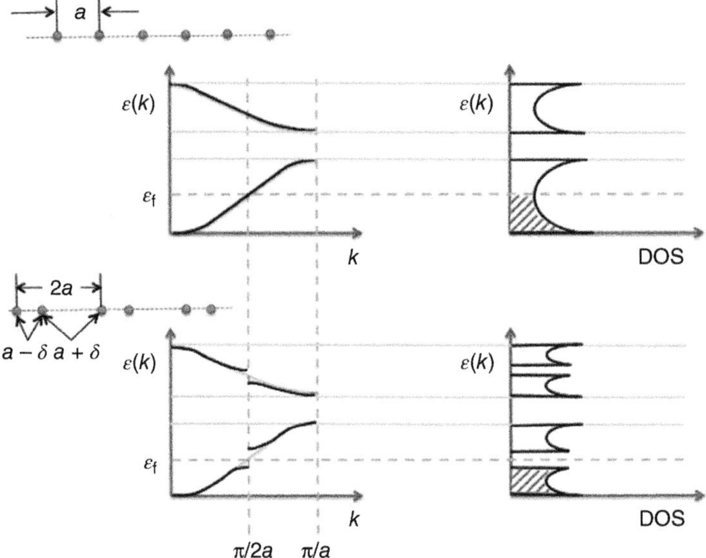

Figure 9.4 Peierls distortion of an isolated chain of equidistant monatomic sodium. The density of states and the band structure for an idealized and perfectly spaced chain is shown above. Below shows those same things only for the chain with an alternating distortion.

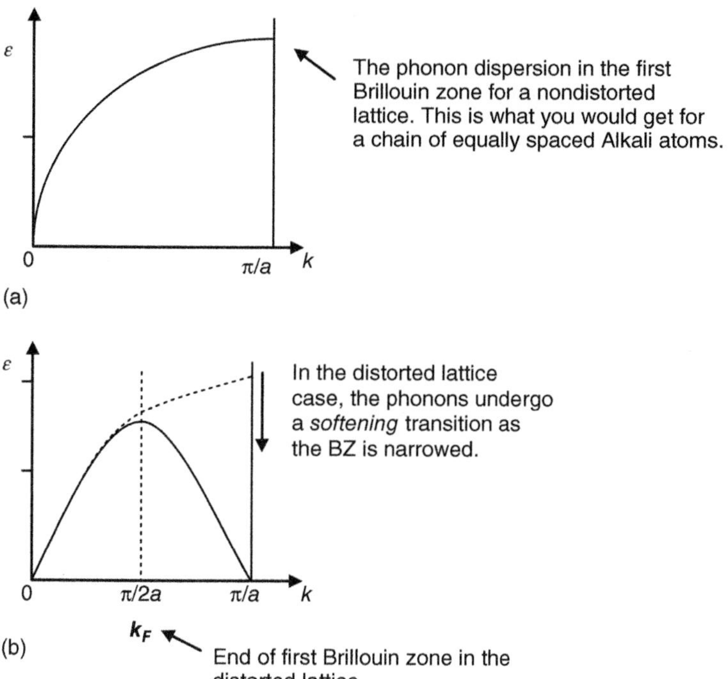

Figure 9.5 Phonon dispersion and Kohn anomaly in a hypothetical inelastic neutron scattering experiment on an ideal monatomic alkali chain: (a) before distortion and (b) after distortion.

one-dimensional metal consists of two planes perpendicular to the metal chain. The phonons scatter the electrons from one plane to the other; this means they just reverse the electron momentum, from $+\boldsymbol{k}_F$ to $-\boldsymbol{k}_F$ and vice versa. Such momentum flipping is called an *umklapp process,* as we have seen. If a phonon wants to change the electron momentum from $+\hbar\boldsymbol{k}_F$ to $-\hbar\boldsymbol{k}_F$, it must have the wavevector $2\hbar\boldsymbol{k}_F$.

Electronic umklapp processes play an important role in the theoretical limit of the conductivity of polyacetylene. In an ideal polyacetylene crystal, electron–phonon scattering is again umklapp scattering, and this scattering is the only contribution to electrical resistivity in an *ideal* one-dimensional crystal. In the alkali metal chain, the $2\boldsymbol{k}_F$ phonons become soft and drive the system from metal to insulator. If the softening is blocked by some other interaction, the $2\boldsymbol{k}_F$ phonons can be frozen out at low temperatures, and there is no contribution to resistivity, thus turning the metal into an ideal conductor at $T = 0$ (but not into a superconductor, which would require a collective state and a finite transition temperature $T > 0$, not simply vanishing resistance at zero temperature).

9.1.3.2 Fermi Surface Warping

In an ideal one-dimensional metal, the Fermi surface consists of two parallel planes. If it is less ideal and some interaction to adjacent chains occurs, the Fermi surface "tries" to become spherical or cylindrical – it bends. Figure 9.6 shows the

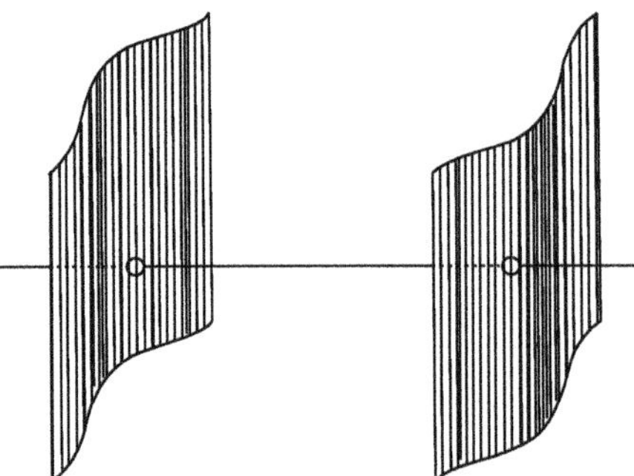

Figure 9.6 Warped Fermi surface of a one-dimensional metal with slight two-dimensional character.

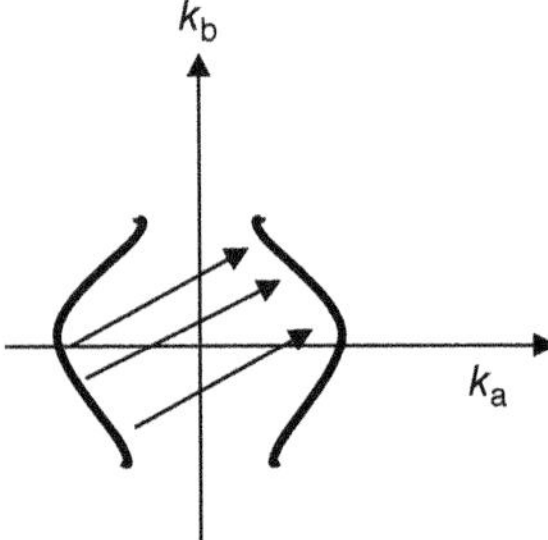

Figure 9.7 Projection of warped Fermi surface and "nesting."

warped Fermi surface of a system that has a slight two-dimensional admixture. In Figure 9.7 a projection of the warped Fermi surface is given. This allows us to demonstrate the concept of "nesting," which makes quasi one-dimensional metals look more one-dimensional than expected. Umklapp processes scatter electrons from one sheet of the open Fermi surface to the other. With parallel sheets all umklapp processes involve parallel vectors of momentum transfer. In the case of a spherical Fermi surface, there are no parallel transfer vectors. If the surface is appropriately warped, however, there are some or even many parallel transfer vectors, as indicated by the arrows in Figure 9.7. In that case phonons moving along the chain can flip the electron momentum by conserving the projection of the momentum onto the chain. If there are "enough" such parallel transfer vectors, the Peierls transition can occur even if the Fermi surface is not planar. Peierls transition and nesting play an important role in CDW phenomena as we shall see.

9.2 Conjugation and the Double Bond

From organic chemistry we know several types of double bonds: isolated, cumulated, and conjugated. *Conjugated* means double bonds separated by single bonds. Looking at Figure 2.28 we notice strict alternation of double and single bonds in all polymers shown (with the exception of polyaniline). But in polyaniline there is an extra electron pair on the nitrogen atoms and "conjugation passes through these extra pairs." Conjugated double bonds behave quite differently from isolated double bonds. As the word implies, conjugated double bonds act collectively, "knowing" that the next nearest bond also is a double bond. To stress this, a colleague once quoted the Chinese proverb: "You pull a hair and you excite the whole body!" [16] (Figure 9.8).

The conjugation of bonds is a defining characteristic of conducting polymers. To discuss the physics of conjugated double bonds, we look at the conjugated polymer with the highest symmetry: polyacetylene. In the Gedankenexperiment

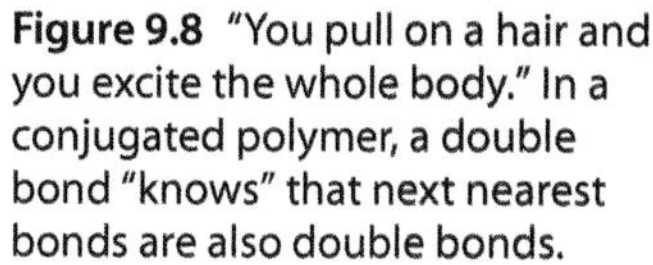

Figure 9.8 "You pull on a hair and you excite the whole body." In a conjugated polymer, a double bond "knows" that next nearest bonds are also double bonds.

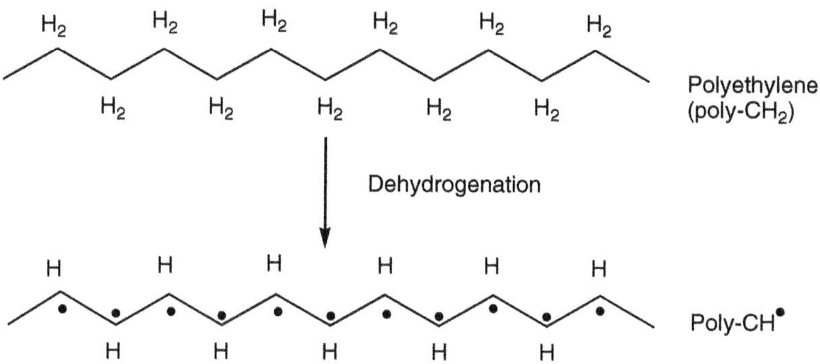

Figure 9.9 Gedankenexperiment, producing polyacetylene by dehydrogenation of polyethylene. In the first step a chain of CH• radicals is obtained analogous to a chain of alkali atoms.

of Figure 9.9, polyacetylene can be prepared from polyethylene, which is the simplest of the *saturated* linear chain polymers (*saturated* organic compounds are those without double bonds). Polyethylene is a zigzagging chain of carbon atoms with two hydrogen atoms per carbon atom. Removing one hydrogen from each carbon would leave us, in a first step, with unbonded electrons everywhere. We would obtain a chain of CH· radicals. A CH· radical has some similarity with an alkali atom: both have an extra electron. The poly-CH· chain in Figure 9.9

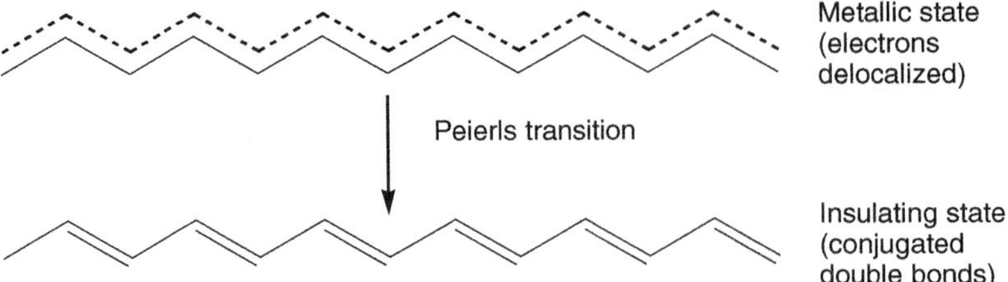

Figure 9.10 As in alkali metals the electrons of the CH• radicals will delocalize all over the solid. The Peierls distortion will then lead to bond alternation.

Figure 9.11 Electron delocalization in a benzene ring to illustrate that "metallic behavior" is not unexpected in organic chemistry.

thus resembles the sodium chain. (In alkali metals the extra electrons are *s* electrons. In a CH· radical, the extra charge is a pelectron. For the discussion in this chapter, the difference between s and p electrons does not matter.) At first glance a metal could be expected, and it is not surprising that polyacetylene is "the synthetic metal par excellence." However from the previous discussion we know that electron–lattice coupling will cause a Peierls transition and drive the metal into an insulator, at least at low temperatures. (Later on, we will see that doping suppresses the Peierls transition.)

In Figure 9.10 the "metallic state" of polyacetylene (top) and the transition to the insulating state (bottom) are illustrated in a slightly different way. In the metallic state the electrons are depicted as delocalized over the entire chain. The dashed line should allude to the Bloch wave behavior of the electrons.

This kind of delocalization is not unfamiliar to the organic chemist and is found in aromatic rings. Often delocalization is symbolized by "resonance structures" (Figure 9.11). We note here that there are a number of polymer structures introduced in Chapter 2 that seem to have a series of such rings connected by a bonding structure of some sort.

So, unlike benzene, a sufficiently long, extended chain is not aromatic and a metallic state will not actually be realized. In fact, an alternation of short and long bonds will occur, as indicated at the bottom of Figure 9.10 (where short bonds are drawn as double bonds and long bonds are single bonds). Due to this alternation, there is a gap in the electronic DOS. All states below the gap are occupied and form the "valence band"; the states above the gap are empty and form the "conduction band." Chemists call these bands the π and π^* bands. As we have already seen in Chapter 6 on electronic structure, chemists might prefer an approach via molecular orbitals rather than solid-state physics. Recall, the chemist might first link two CH· radicals to a $(CH)_2$ pair with a double bond between the two CH groups that consists of a σ and a π bond. (In a polymer chain, s electrons form σ bonds, and p electrons form π bonds.) There will be a bonding π and an antibonding π^* orbital. Forming a macromolecular chain with these $(CH)_2$ pairs, the π and the π^* orbitals will split to give bands. What solid-state physicists call the "top of

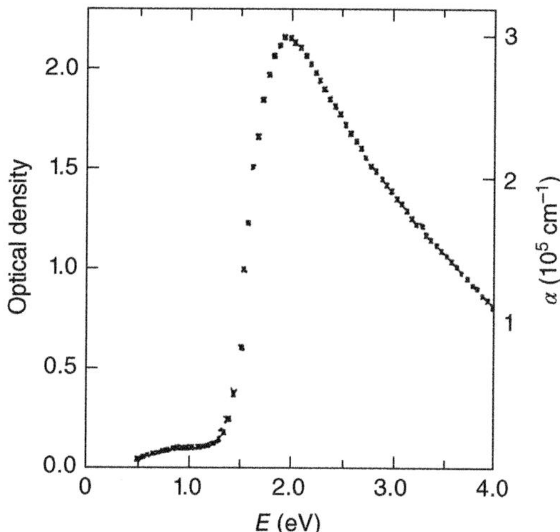

Figure 9.12 Optical absorption in polyacetylene. Like a classical semiconductor the sample is transparent for light with quantum energy smaller than the bandgap. From the onset of the absorption, the bandgap can be determined.

the valance band" translates in chemical language to "highest occupied molecular orbital (HOMO)." The "bottom of the conduction band" is then called the "lowest unoccupied molecular orbital (LUMO)." The solid-state physicist starts with a band, the width of which is given by the average interaction between *single carbon atoms*, and then bond alternation will open a gap; the chemist starts with the gap, followed by broadening the π and π^* orbitals through interaction between *carbon pairs*.

The π–π^* gap in polyacetylene is about 1.7 eV. This is well within the region of known semiconductors such as diamond, 5.4 eV; GaAs, 1.43 eV; Si, 1.14 eV; and Ge, 0.67 eV. The gap can be determined by optical absorption, and this is shown in Figure 9.12.

The polyacetylene bandgap is much larger than the Peierls gap in KCP or in charge-transfer salts (see Chapter 2). The Peierls transition occurs at a temperature T_p that is in the same order of magnitude as the gap energy E_g (just as with superconductivity $E_g \sim 4k_B T_p$), because above that temperature many electrons are thermally excited across the gap and the solid does not "notice" the gap anymore. Since 1 eV corresponds to a temperature of about 10 000 K, at room temperature polyacetylene is far below the Peierls transition. We cannot heat polyacetylene into the metallic phase, because the polymer decomposes at some hundred degrees Celsius.

9.3 Conjugational Defects

In polyacetylene the strict bond alternation is trivial to most chemists, since dimers (pairs) were polymerized. This molecular picture, though it may use a slightly different language, translates well into the picture that physicists use. For a physicist, *dimerization* behaves like a phase transition from a metallic to a semiconducting state, though strictly speaking phase transitions are not allowed in one-dimensional systems according to Landau and Lifshitz [17].

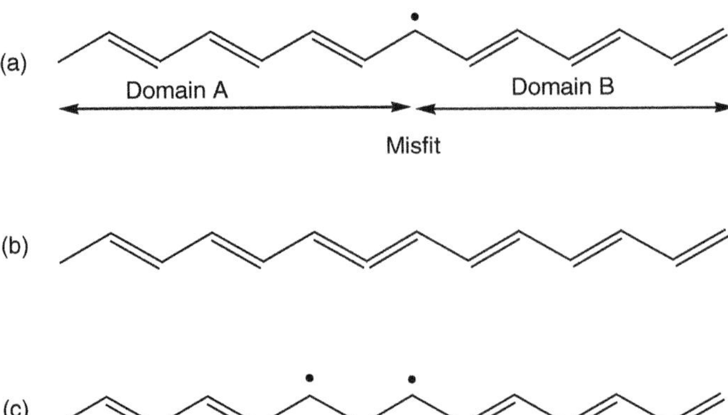

Figure 9.13 Misfits or "domain boundaries" in a conjugated chain: (a) two adjacent single bonds plus a dangling bond, (b) two adjacent double bonds, leading formally to a pentavalent carbon atom, and (c) three dangling bonds on a misfit.

But *dimerization* is a somewhat special case, and this hypothetical phase transition ($T_p \sim 10\,000$ K) can be imagined to nucleate at multiple points along the chain. Domains of the different phases would grow around these nucleation centers in the typical way of phase transitions. "Misfits" will be created at the boundaries where domains touch, as shown in Figure 9.13.

These *misfits*, or *domain boundaries*, are very interesting from the point of view that they break the translational symmetry of the system. As seen in Figure 9.13 there is more than one way the boundary between two phase domains might be imagined. In Figure 9.13a the bond is interrupted by two adjacent single bonds. To keep the carbon atom tetravalent in the *domain boundary* or the *domain wall* (at the defect), a dangling bond must exist (two bonds join neighboring carbons, one the hydrogen atom; the fourth bond has no partner). Dangling bonds are not unusual in semiconductor physics, for example, in amorphous silicon they are quite common. In polymer chemistry they are also known. The defect shown in Figure 9.13a is a radical on a polyene chain, and there is an energy cost to change the phase state on either side to "repair" this radical. The radical can be detected by electron spin resonance (ESR) because of its unpaired spin.

Are domain walls with two double bonds touching possible, as in Figure 9.13b? Technically this would lead to a carbon atom with five bonds, and therefore the defect would preferably be shown as in Figure 9.13c. But is there really a difference between Figure 9.13a,c? In Figure 9.13c it just remains unresolved as to which of the three electrons form a pair and which one is left alone. We will see later that misfits can be neutral or positively or negatively charged, depending on the electron occupation. Therefore the reader should not protest if he/she sees two electron pairs in Figure 9.13b and three dangling bonds in Figure 9.13c.

In reality, these misfits are not as well localized as shown in Figure 9.13. The defect extends over some 10 bonds, modifying the bond alternation gradually.

$$p = \frac{\text{left bond} - \text{right bond}}{\text{average bond}} \tag{9.8}$$

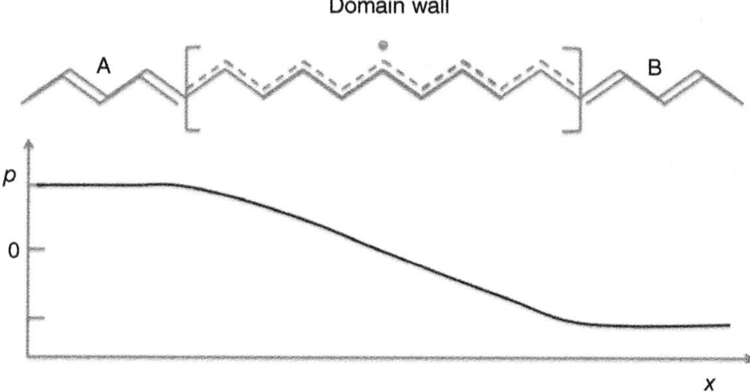

Figure 9.14 Bond alternation parameter changing sign at a misfit on a conjugated chain.

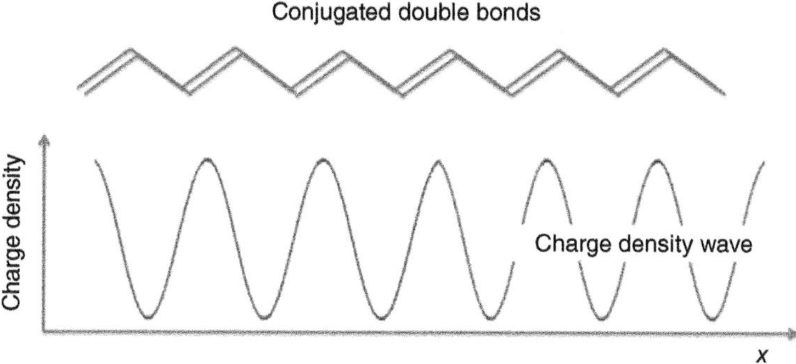

Figure 9.15 Conjugated double bonds as charge density wave in a π electron system.

We define a bond length alternation parameter as the parameter p. It will change sign at the defect (Figure 9.14).

We might naturally ask about where within the band structure the energy state of such a domain wall or misfit might fall. Clearly such nonbonding "localized" states will occur within the bandgap of the material.

From our discussion above regarding the nodes of the wavefunctions between bonding and antibonding π electrons, we can more properly see the Peierls transition as associated with a CDW along the chain. Figure 9.15 illustrates the CDW for polyacetylene. If we interpret the bonding dashes as electron pairs then the CDW is trivial. It is a periodic modulation of the π electrons that are located at the double bonds. Due to the bond alternation there is a periodic modulation of the π electron density. At the domain wall, or misfit, the CDW has a 180° phase slip. Here it should again be stressed that the Peierls transition in polyacetylene is oversimplified in our discussion. In addition to the electron–lattice coupling, there are further equally important interactions, often referred to as correlation effects. To take these into account, the parameter p can be generalized as *bond order parameter* and the CDW as *bond order wave* [2].

Figure 9.16 The soliton thesaurus in conjugated polymers.

Conjugated double bonds have been studied for more than half a century, not because of their importance in conducting polymers but because they are essential for dyestuffs and interesting for fundamental considerations in quantum chemistry. The pioneering work of Kuhn [18] and Longuet-Higgins and Salem [19] should be reviewed. As seen in Figure 9.16, many names have been used to describe these defect or misfit states. Today *soliton* is used frequently. We will see why in Section 9.4.

Can conjugational defects be observed experimentally? The ESR signal of unpaired electrons has already been mentioned. In *trans*-polyacetylene this signal is rather strong, corresponding to about 1 spin per 3000 carbon atoms. The g-factor is 2.0026 and this is associated with the π-conjugated system. Dynamic nuclear polarization (DNP) experiments further suggest that these spins are very mobile. For *cis*-polyacetylene the ESR signal is exceptionally weak, and DNP shows the electronic spins to be immobile [20].

Of course, we have already mentioned that there can exist charged and uncharged defects (just as there are charged and neutral lattice defects in an inorganic semiconductor). In polyacetylene there is a peculiarity that the charged defects have no spin. Consequently charged defects are invisible in ESR experiments – but charged and neutral defects can be seen in optical absorption experiments as states *in the gap* (*midgap states*). Following our discussion above, if the gap is explained as the separation of bonding and antibonding states, a dangling bond, which is neither bonding nor antibonding but *nonbonding*, must be *in the gap*. A similar line or reasoning might be that if the gap is a Peierls gap originating from bond length alternation, the alternation parameter will be zero at the misfit; there is a chain section without Peierls distortion, and there must be some part of the chain without a gap. These states will fall within the overall gap of the polymer. This suggests a transition exists between the localized gap state and the LUMO of the polymer that might be observed through optical methods. Typically such effects in a pristine polymer would be small (there may not be many states), but they can be induced through doping as we shall see. The doping-induced emergence of the midgap absorption is shown in Figure 9.17.

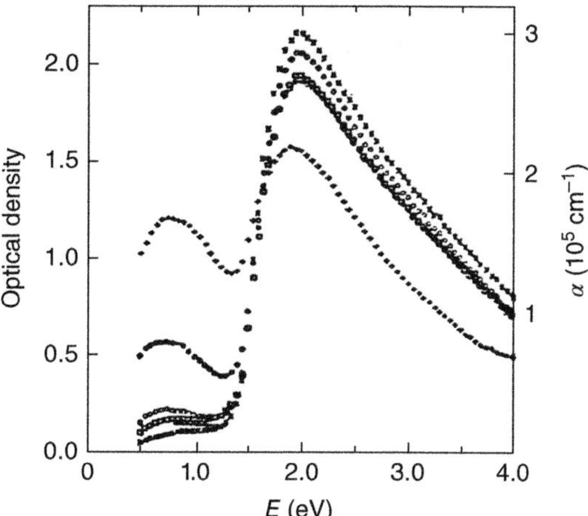

Figure 9.17 Optical absorption of polyacetylene showing the emergence of a midgap state upon doping. Note that the absorption curves cross at an isosbestic point, because the midgap peak grows at the expense of the band states.

9.4 The Soliton

In 1962 Pople and Walmsley [21] investigated the conjugational misfit and suggested that these defects could be mobile and carry charge – giving rise to conductivity. In Figure 9.16 several synonyms for this defect are arranged, each one in a different context and stressing different aspects. The term "soliton" is listed as one of the names for the conjugational defect in *trans*-polyacetylene. This is largely due to the seminal works of Su, Schrieffer, and Heeger as well as M.J. Rice [22]. Referred to as the *soliton model* today, this work equated the domain wall with a soliton as a refinement of the Pople–Walmsley model for defects using nonlinear phenomena.

A *soliton* is a self-reinforcing solitary wave that moves without dispersion (maintains its shape) at a constant speed. First indications of similarity between polyacetylene defects and solitary waves can be obtained from Figure 9.14, showing the bond alternation parameter resembling a tidal wave. This wave concept in physics has been originally developed from water waves. Water waves are generally very complicated; only at small amplitudes are they harmonic waves, looking like a sine or a cosine function.

A harmonic wave is the solution of the wave equation:

$$c^2\phi_{xx} - \phi_{tt} = 0 \tag{9.9}$$

This is a linear differential equation of second order. It states that the second derivative with respect to time ϕ_{tt} must be proportional to the second spatial derivative ϕ_{xx} of the amplitude function $\phi(x,t)$. This is fulfilled with the trigonometric functions sine and cosine:

$$\phi(x,\ t) = \sin(\lambda x - \omega t) \tag{9.10}$$

In water, the wave equation (5.1) is only a small amplitude approximation. For large amplitudes, nonlinear differential equations must be used, for example, the Korteweg–de Vries equation:

$$\phi_t + 6\phi\phi_{xxx} = 0 \tag{9.11}$$

This equation is nonlinear, because it contains the product of ϕ and its derivative ϕ_{xxx}. Other nonlinear differential equations are the Sine–Gordon equation

$$c^2\phi_{xx} - \phi_{tt} = \omega^2 \sin\phi \tag{9.12}$$

and the equation of ϕ^4 potentials:

$$c^2\phi_{xx} - \phi_{tt} = \omega^2(\phi^3 - \phi) \tag{9.13}$$

The wave equation (9.9) applies to motions in a parabolic potential. A chain of balls coupled by springs, as used in our phonon model, would be a good mechanical example. If the potential is non-parabolic, other equations have to be used, e.g. the Sine–Gordon equation (9.12) for a sinusoidal potential or Eq. (9.13) for double-well potentials. The alternation of short and long bonds in polyacetylene can be imaged by double-well potentials, as indicated in Figure 9.18. This figure also shows a defect at the center of the chain.

The wave equation (9.9) has harmonic waves, i.e. a sine or cosine function as solution. The nonlinear equations (9.11)–(9.13) also have analytical solutions, however, in this case, not periodic waves but pulse- or steplike functions. For example, the solution for the double-well equation (9.13) is

$$\phi(x, t) = \phi(x - vt) \tag{9.14}$$

with

$$\phi(x) = \tan h\, x/\xi \tag{9.15}$$

and

$$\xi^2 = (c^2 - v^2)/\omega^2 \tag{9.16}$$

Again a trigonometric function is found as solution, not the periodic sine or cosine but the steplike tan h function.

Since we are used to harmonic waves, we would like to construct steps and pulses as wave packages. Any shape can be Fourier synthesized by appropriate superposition of harmonic components. However, wave packages will *disperse*, if the velocity of a wave depends on the frequency. For this reason Eq. (9.16) is called the *dispersion relation*. The particular steps and pulses we are discussing here, however, *do not disperse*. Therefore they are called *solitary waves*.

In hydrodynamics solitary waves have been studied for more than one and a half centuries. They were observed even earlier, and they have been feared because of their nondispersive properties! In fact, Hokusai's woodcutting is related to tsunamis – disastrous tidal, solitary waves generated by earthquakes under the Pacific Ocean. In Europe solitary spring tides are known to propagate

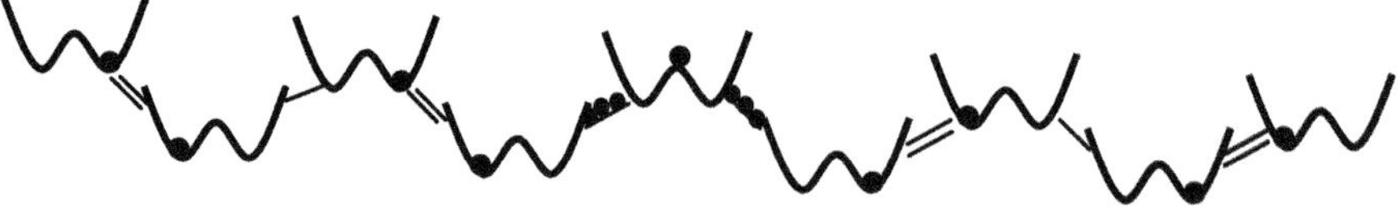

Figure 9.18 Coupled double-well potentials to illustrate conjugational defects in polyacetylene.

far inland into river estuaries at the Atlantic; for example, before the river Seine was dredged, they even went up to Paris!

When we encounter an entity that looks like a pulse or a step, it is not easy to decide whether it is a wave package or a solitary wave. Experimentally, it would have to be examined in motion to see whether or not it will disperse. But it might well turn out that the ocean is too small, so that the entity does not move far enough for us to see any difference between dispersion and nondispersion. Analogously, polymer chains might be too short, and in addition it would be difficult to observe the motion (although, in principle, with magnetic resonance techniques, it could be realized). If the properties of the medium are known the amplitude of the entity could be measured. It could be discussed whether an excitation with such a large amplitude is still compatible with the validity of the harmonic approximation. However, usually our knowledge of the medium is not sufficient.

9.4.1 Doping

In 1977 it was discovered that polyacetylene can be doped and that doped polyacetylene has a very high electrical conductivity [23]. Soon after this observation, M.J. Rice [19, 22]; Su, Schrieffer, and Heeger ("SSH") [19, 24]; and S.A. Brazovskii [24] noticed that nonlinear differential equations with solitary wave solutions can be formulated for polyacetylene and that there is a relation between conjugational defects and solitary waves. This observation triggered a long and vivid discussion on whether or not solitons exist in polyacetylene ("La chasse aux solitons" Figure 9.19) [25]. It was also speculated whether these solitons might be responsible for the high electrical conductivity. It is accepted today that such high doping levels in polyacetylene are related to soliton formation.

Today, if we search for the term "soliton" in literature, many papers on solitons in telecommunications will turn up. If we codify a message in pulses and send these pulses along a cable, it is desirable that the pulses do not disperse

Figure 9.19 *La chasse aux solitons*, reminiscence of the butterfly hunt in a popular song by George Brassens; however, what is hunted there does not seem to be butterflies.

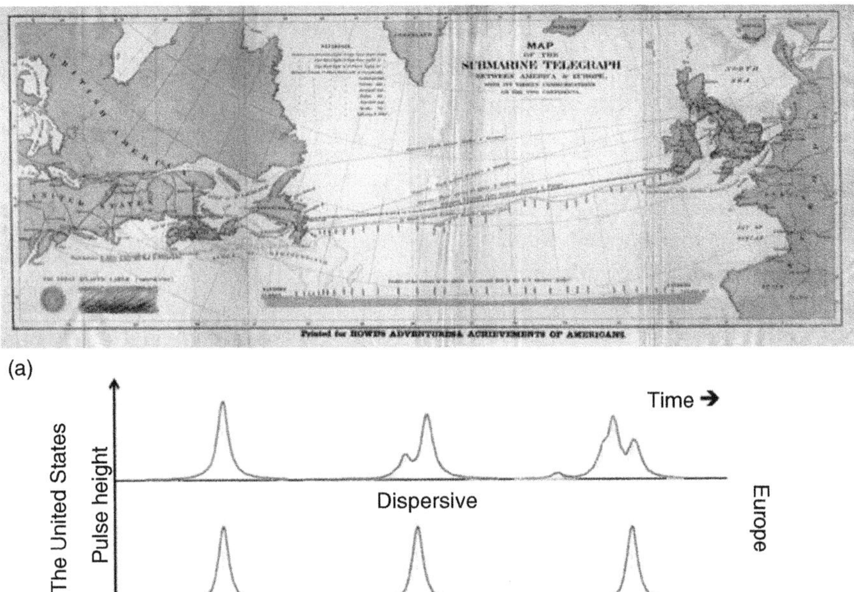

Figure 9.20 (a) A map of the transatlantic cables made by Siemens in 1901 and laid across the ocean floor. (b) The advantage of non-dispersing solitons in telecommunications is that signals can be packed together very tightly without loss of time resolution or signal-to-noise ratio. Today telecommunications solitons are mainly optical. https://www.cbsnews.com/pictures/the-underwater-engineering-feat-of-the-19th-century-the-transatlantic-cable/.

(Figure 9.20). In conventional cables, pulse shaping devices are inserted every couple of kilometers, for example, nonlinear amplifiers that amplify the high pulse center more than the low tails. In glass fibers, dispersion (frequency dependence of the velocity) is compensated by nonlinear optical effects (intensity dependence of the refractive index). It is not surprising that glass fibers are described in a "continuous" way, by differential equations and material parameters (dielectric constant) entering these equations. For cables with amplifiers we would tend to describe in a discontinuous way. However with the cable long enough, it would be possible to average over the amplifiers and to incorporate the mean value into the material parameters.

There is a solid-state analogue to the continuous cable-plus-amplifier description: the jellium model. In jellium the positive ions of a solid are not localized at the lattice sites but homogeneously spread over the volume. Maki and coworkers [26] have studied solitons in polyacetylene in the continuum model. Moreover, an excellent summary of solitons in polymers is given by Bredas and Street [27].

9.4.2 Quasiparticles

Solitons are quasiparticles corresponding to solitary waves, similar to phonons corresponding to sound waves and photons to electromagnetic waves (light waves). There are many properties of solitons that remind us of elementary

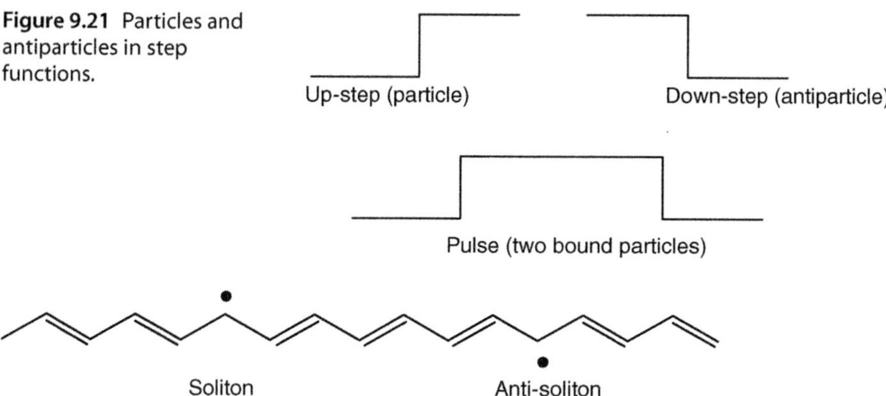

Figure 9.21 Particles and antiparticles in step functions.

Figure 9.22 Soliton and anti-soliton in polyacetylene.

particles, for example, the existence of solitons and anti-solitons. Speaking in terms of step functions, particles and antiparticles are easily identified as up-steps and down-steps, as illustrated in Figure 9.21. In CDWs they correspond to phase-slip centers of +180° and −180°. Figure 9.22 shows a soliton and an anti-soliton for polyacetylene chains. Note that it is acceptable to call the first noticed conjugational defect a soliton or an anti-soliton, but once we have made our choice all further particles on the same chain are defined either as solitons or as anti-solitons.

There will always be some solitons on a polyacetylene chain as a consequence of synthesis (as we noted, typical ESR results are 400 radicals per 10^6 carbon atoms). Additional solitons can only be created in pairs, as soliton/anti-soliton pairs ("conservation of particle number"). From a chemical point of view, pair creation is evident, since it involves the breaking of a bond, which leaves two dangling bonds (or the splitting of an electron pair that leaves two radicals). Particle–antiparticle annihilation corresponds to closing the bond again. Figure 9.23 shows solitons migration. A soliton moves by pairing to an adjacent electron and leaving its previous partner unbound. Note that in this way the soliton always occupies odd-numbered sites; the even-numbered sites are reserved for anti-solitons.

When talking about solitons and anti-solitons, we can introduce the concept of "pseudo-spin": for electrons we have the spin quantum number and we can say there are two kinds of electrons – those with spin up and those with spin down. Similarly we can say there are two kinds of solitons: those moving on sites with odd numbers in Figure 9.23 and those moving on sites with even numbers. The first kind we label "pseudo-spin up" and the latter "pseudo-spin down." Since there are spin conservation rules, there are also pseudo-spin conservation rules. To conserve pseudo-spin we have to create solitons in pseudo-spin up and pseudo-spin down pairs.

The concept of pseudo-spin is also important for electrons and holes moving in graphene. In Figure 3.3, we have shown that the honeycomb lattice of graphene consists of two trigonal lattices, one with A atoms and the other with B atoms. Like solitons in polyacetylene, electrons and holes in graphene also move

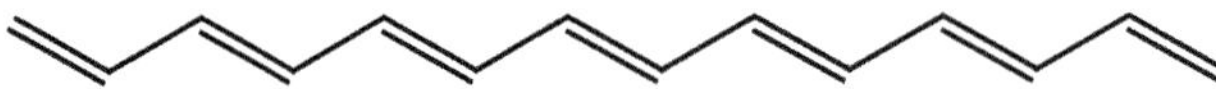

Creation of soliton and anti-soliton

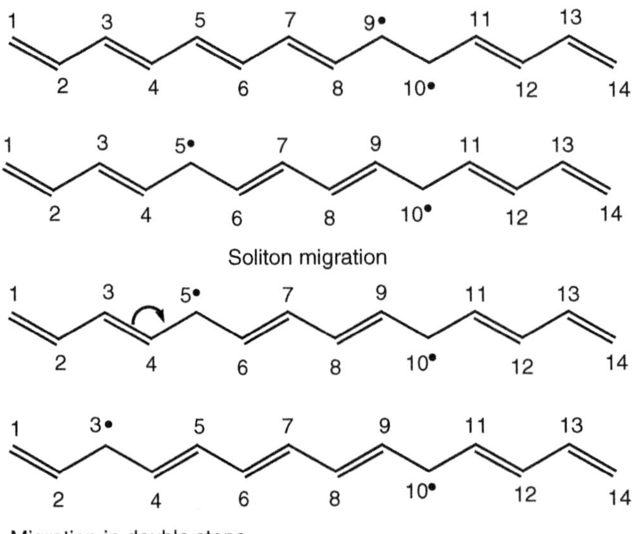

Figure 9.23 Solitons and anti-solitons. The presence of a soliton on a polyene chain allows the classification of the lattice sites as even and odd and of all further conjugational defects as solitons and anti-solitons.

in double steps (graphene also has conjugated double bonds!); one kind of electron manifests itself on the A sites of the graphene lattice, while the other on the B sites.

Figure 9.24 shows the results of a quantum chemical calculation on a polyacetylene chain with 61 carbon atoms [28]. Since polyacetylene synthesis starts with acetylene with the carbon atoms already paired, it is unlikely that polymer chains with an odd number of carbon atoms occur; however calculations are certainly possible with these chains. Evidently, an odd-numbered chain must have an unpaired electron, i.e. there must be a soliton in the ground state. Figure 9.24 shows the spatial distribution of the electron density at the midgap state as well as the density in the valence (π) and the conduction (π^*) band. Note that the density at the midgap state is built up at the expense of the π and π^* densities. Furthermore we see that the soliton extends over about 20 lattice sites (this is the "thickness" of the domain wall); in addition, the midgap density accumulates only on every second lattice site, in this case on the even sites, leaving the odd sites for anti-solitons.

Several attempts have been made to get experimental information on spatial extension of the soliton wavefunction. The most reliable results are obtained from

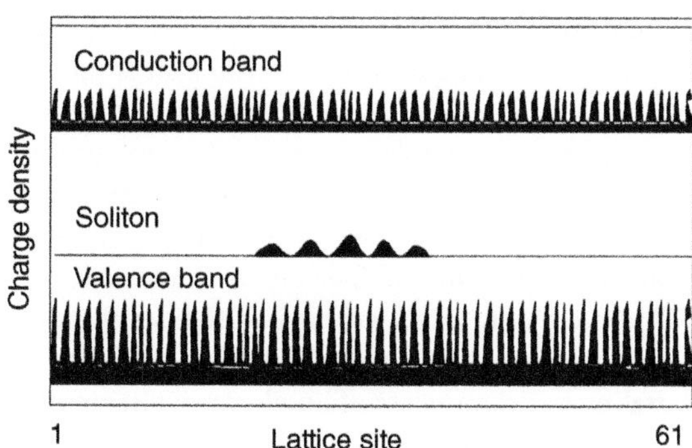

Figure 9.24 Charge distribution of a conjugational defect on a polyacetylene chain on 61 carbon atoms [28, 32].

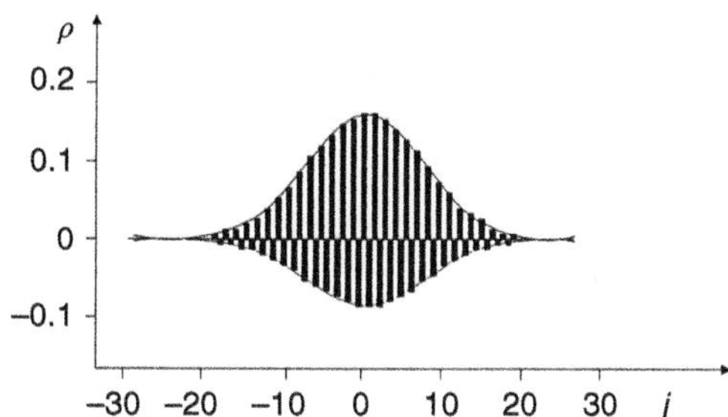

Figure 9.25 Distribution of the spin density of a soliton defect on a polyene chain with $l = 11$ and $u/g = 0.43$ used for simulations of ENDOR spectra [30, 34].

electron nuclear double resonance (ENDOR) and from pulsed triple resonance experiments. Grupp et al. [29] have fit their data to the following spin distribution function:

$$\rho_i = (1/l)\sec h^2(j/l)[g\cos^2(j\pi/2) - u\sin^2(j\pi/2)] \tag{9.17}$$

where j is the lattice site index, $2l$ is the full width at half maximum of the soliton wavefunction, and g and u represent the population of the even- and odd-numbered lattice sites, respectively. The fit yields $l = 11$ and $u/g = 0.43$. The corresponding wavefunction is shown in Figure 9.25. If there is electron–lattice coupling only, u/g indicates that the soliton wavefunction is nonzero also at the sites reserved for anti-solitons. The observed anti-solitonic admixture can be taken as experimental evidence of electron correlations.

A soliton is free to move, because the total energy of the system does not depend on the position of the soliton if the chain is long enough. In short chains, end effects will push the soliton to the center. A more detailed analysis shows that the soliton has to overcome an energy barrier when moving from one site to the next (more exactly, to the next but one, because it moves in double steps). The existence of this kind of barrier leads to "self-trapping" of the soliton on the chain. Free soliton motion is possible only in polyacetylene. Only here the energy of the system does not depend on the position of the soliton; in the other polymers shown in Figure 2.28, motion of a soliton changes the energy. This is because polyacetylene has a "degenerate ground state" as we have already seen. Conjugational defects in "nondegenerate ground-state polymers" will be discussed below.

One can estimate the effective mass of a soliton. This is not the free electron mass, because it is not the electron that is moving. The defect is moving, and this means that the electrons change their partners, not their position. Of course, this requires a slight displacement of the ions to invert short and long bond lengths. These ionic displacements determine the effective mass of the soliton. Within this model Su et al. [22a] estimated the effective mass of a polyacetylene soliton to amount to about six free electron masses.

There are neutral as well as positively and negatively charged solitons. So far we have encountered only neutral solitons. See, for example, Figure 9.23 where the dangling bond consists of an electron sitting on a lattice site that is also occupied by a positively charged ion (carbon nucleus plus closed electron shell). Electronic and ionic charge compensate at the defect as they do in the undisturbed part of the chain – leaving it neutral. However in redox reactions the defect is more sensitive than the rest of the chain. The first electron to be removed during oxidation is that of the dangling bond, and the first additional electron in a reduction reaction will also go to the dangling bond. Figure 9.26 shows the π and π^* band (hatched areas), and the soliton midgap state. This state can be occupied by up to two electrons (as with any orbital because of spin-up and spin-down degeneracy). No matter how many electrons occupy the midgap state, there is always a lattice distortion at the defect interrupting the band alternation. The ionic charge is compensated only when there is one electron at the defect. If there is no electron the ion is left uncompensated and the soliton is positively charged; two electrons overcompensate the ion and the soliton becomes negative.

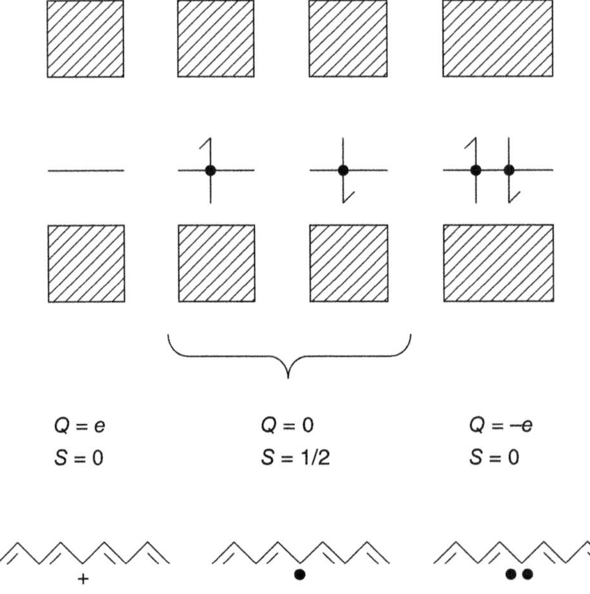

Figure 9.26 Spin–charge inversion of a conjugational defect. Charged solitons are spinless; neutral solitons carry a magnetic moment.

Chemists refer to the neutral soliton as a *radical*, the positive soliton a *carbocation* and the negative soliton a *carbanion* (note, however, that in our case radical, carbocation, and carbanion are not isolated but incorporated into a conjugated chain, which leads to a modification of the adjacent chain segments, often called *relaxation*).

In Figure 9.26 the spin of the solitons is also marked. The neutral soliton is an unpaired electron with spin either +1/2 or −1/2. On the positive defect there is no electron and hence no spin, and the negative defect has two electrons with opposite spins, so that the net spin is zero again. Notice the famous *spin–charge inversion*: whenever the soliton bears charge it has no spin and vice versa! This is trivial considering the chemistry, but it is very surprising when thinking of solitons as particles.

It has repeatedly been emphasized that the approximations in this chapter are an oversimplification; the most exciting result, the soliton, is an oversimplification as well. But simple models bear their justification. Solitons play a similar key role in nonlinear dynamics as the harmonic oscillator does in linear dynamics.

9.5 Generation of Solitons

It has already been discussed that solitons can only be generated in pairs, as solitons and anti-solitons, provided they are not already present from synthesis. During polyacetylene synthesis only very few solitons are created. There should be a soliton on every chain with an odd number of carbon atoms, but there are far fewer odd-numbered chains than even numbered because the synthesis starts from carbon pairs (acetylene).

There are three methods to generate additional solitons: (i) chemical doping, (ii) photogeneration, and (iii) charge injection. Doping of conjugated polymers is a chemical redox reaction. As mentioned before, some people dislike the term "doping" in this connection because of apparent difference to doping of silicon or germanium. One difference is the doping level. Polymers are doped up to several percent, whereas typical doping concentrations of classical semiconductors are in the ppm range. Saturation doping of a polymer certainly leads to a new material with another chemical formula (for example, $[(CH)_7I_3]_n$ for iodine-doped polyacetylene instead of $(CH)_n$ for undoped polyacetylene) as well as to a new crystallographic structure. In the case of doped silicon, we do not describe it as a new material. But in our opinion there are more similarities than differences. Apart from the high doping level, iodine doping of polyacetylene is very similar to lithium doping of silicon: the dopant goes to interstitial sites of the crystalline lattice of the host; charge is transferred from the dopant to the host, thus moving the Fermi level of the host. The main discussion is concerned with the question of whether doping is a continuous or a stoichiometric reaction and whether there is phase segregation between doped and undoped regions or whether the doped material "dissolves" in the host. However, we note that the field has come to use the term "doping" rather widely at this point.

In Figure 9.27 the process of polyacetylene doping is shown schematically. In the first step a double bond is broken followed by the transfer of an electron

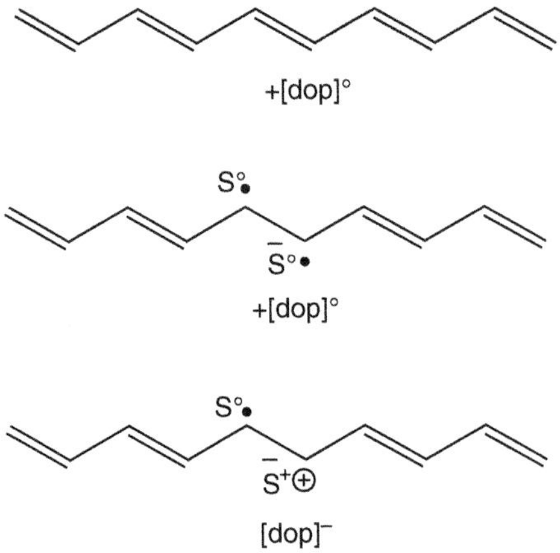

+[dop]°

S°•

$\bar{S}$°•

+[dop]°

S°•

$\bar{S}^{+}$⊕

[dop]⁻

Figure 9.27 Creation of solitons by chemical doping (oxidation) of a polyene chain.

from the polymer chain to the dopant. Thus two solitons are created, one neutral and the other positively charged ("p-doping"). The next dopant will then react with the neutral soliton, because in most p-doping reactions, stoichiometry requires the transfer of two electrons, for example:

$$[PA]^{\circ} + 3I_2 \rightarrow [PA]^{++} + 2I_3^- \tag{9.18}$$

In n-doping only one electron will be transferred:

$$[PA]^{\circ} + K^{\circ} \rightarrow [PA] + k^+ \tag{9.19}$$

But nevertheless only very few neutral solitons will survive, because they will diffuse along the chain and annihilate with anti-solitons from other doping events.

Chemical doping changes the conductivity of conjugated polymers by many orders of magnitude. In fact, there is a doping-induced insulator-to-metal transition. This transition is easy to understand: doping generates solitons. Solitons have midgap states that interact and finally the gap will disappear. If a soliton is a local suppression of the Peierls distortion, doping will just reverse the Peierls distortion. From the spatial extension of the soliton wavefunction over about seven lattice sites (see Figure 9.25), the stoichiometry of 7 : 1 for the ratio of CH to I_3 in saturation doped polyacetylene seems quite plausible.

Photogeneration of solitons is illustrated in Figure 9.28. The figure is drawn in a way to stress the similarity of electron and hole generation in semiconductors. In a first step an electron is lifted from the valence (π) band to the conduction (π*) band. Thus an electron–hole pair is created. Now the lattice relaxes around the electron and the hole, i.e. the bond lengths readjust, and a negative soliton S^- and a positive anti-soliton $\overline{S}^+$ are formed. The solitons S^- and $\overline{S}^+$ can move in an electric field and thus give rise to photoconductivity, or they can recombine. If they recombine radiatively, photoluminescence will be generated. Alternatively to Figure 9.28 it could have been assumed that in the first step the double bond is cleaved by absorption of light, leading to two dangling bonds. However in this case it has to be demonstrated that not only the bond is split but charges are also

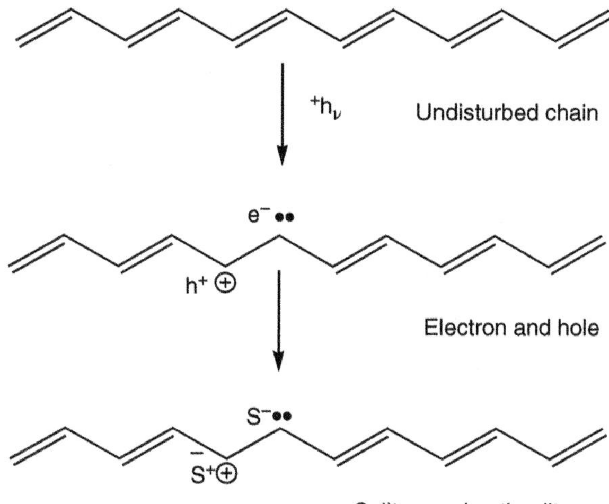

Figure 9.28 Photogeneration of solitons.

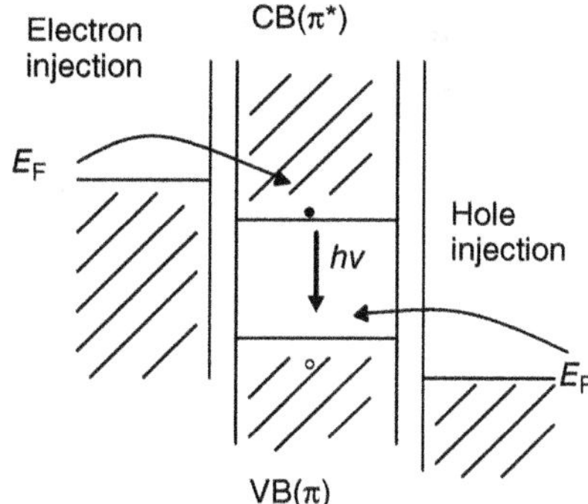

Figure 9.29 Soliton generation by charge injection. In a first step electrons and holes are injected from the electrodes. These then relax to form solitons and anti-solitons.

separated, so that the final products are not two neutral solitons $S°$ and $\overline{S}°$, but two charged solitons S^- and $\overline{S}^+$ [30].

Charge injection is also known from semiconductor physics. A semiconductor is sandwiched between two metal electrodes, as shown in Figure 9.29. By choosing metals with appropriate work functions and additionally applying an external voltage, the Fermi levels of the metals are adjusted in such a way that one electrode injects electrons into the π^* band and the other extracts electrons from the π band (which is the same as injecting holes into this band). The lattice relaxes around the injected carriers and solitons are formed. Finally the solitons recombine, some of them by emission of light. Provided that the quantum efficiency for radiative recombination is high enough, an organic light-emitting diode (OLED) is obtained. We should note one very important difference in the band diagram for charge injection with organics as shown: that is, the absence of so-called "band bending" in the system. Because itinerate, or free, charge doesn't exist within the polymer before it is injected, bands at interfaces are generally represented in a "flat band" condition. However, defects at the interface or doping in the polymer can drastically alter this situation. Charge injection dynamics and soliton formation across the metal–polymer interface has been studied extensively within the last decade and is still hotly debated. Most importantly though, it has been shown by Friend et al. in optical absorption experiments that solitons are formed before the injected carriers recombine [31].

9.6 Nondegenerate Ground-State Polymers: Polarons

It has already been mentioned that there are degenerate and nondegenerate ground-state polyenes. Polyacetylene represents the first group, and all other polymers presented in Figure 2.28 belong to the second group. Degenerate means that the energy does not change when single and double bonds are interchanged. The situation is illustrated in Figure 9.30. For polyacetylene it does not matter whether the double bond is on the left-hand or the right-hand slopes. In polyphenylene, however, interchange of single and double bonds leads from aromatic state A (three double bonds within the ring) to the quinoidal state B with only two double bonds within the ring, but the rings linked by double bonds instead of single bonds. The quinoidal structure has a much higher energy than the aromatic state.

In Figure 9.31 a soliton in polyacetylene is compared with a hypothetical soliton in polyphenylene. Because of the degeneracy in polyacetylene, the position of

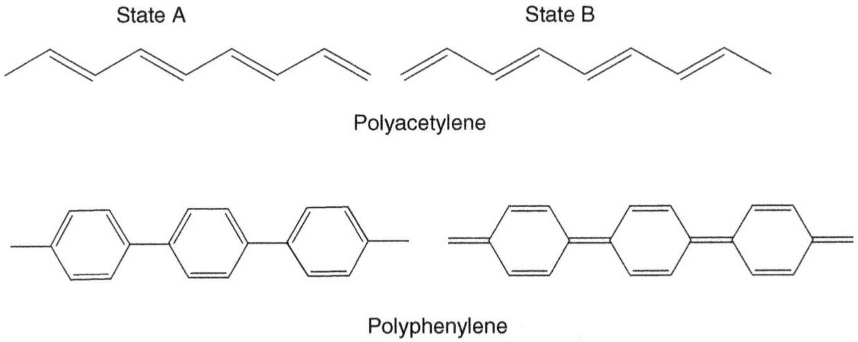

Figure 9.30 Degenerate and nondegenerate ground-state polymers.

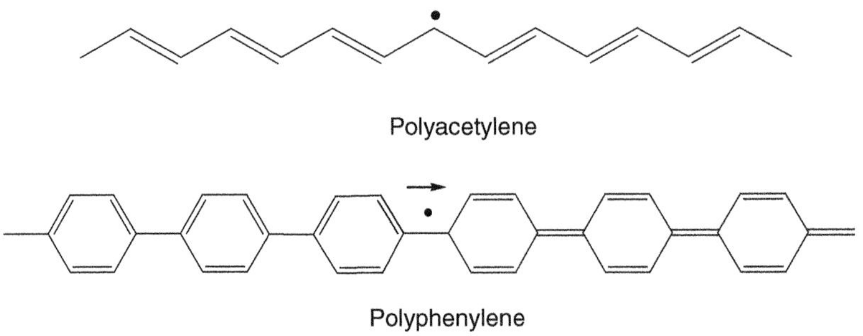

Figure 9.31 A soliton is free to move in polyacetylene, whereas in polyphenylene it is pushed to the chain end by lattice forces.

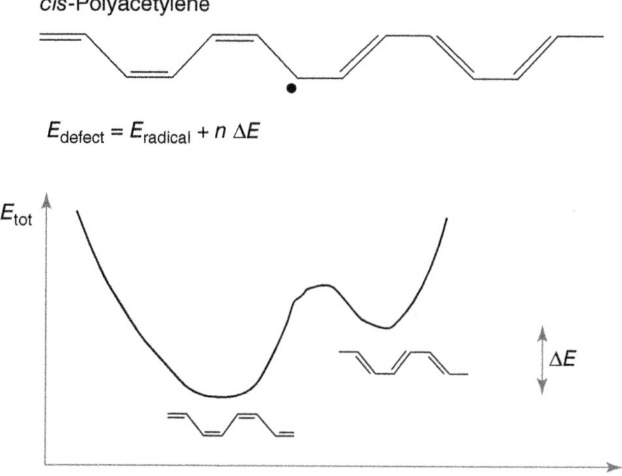

Figure 9.32 The nondegenerate ground state of *cis*-polyacetylene.

the soliton does not matter energetically. In polyphenylene the soliton separates a low energy region from a high energy region. The soliton will, of course, be driven to the chain end, changing the high energy quinoidal rings into low energy aromatic rings as it moves. The situation is comparable to magnetic domain walls in an external magnetic field or to dislocations in a crystal under elastic strain: the magnetic or the elastic field will drive the defects out of the solid. So individual, mobile solitons do not form in a nondegenerate ground-state polymer (though we loosely refer to domain boundaries in these systems as solitons).

Of course we have been discussing *trans*-polyacetylene. *cis*-Polyacetylene also has a nondegenerate ground state as seen in Figure 9.32. Shown in Figure 9.33 are several other polymers we have already met. Notice that these too have nondegenerate ground states.

To stabilize conjugational defects in a nondegenerate ground-state polymer, we have to create bound double defects. Such double defects are called "polarons." An example is shown in Figure 9.34. There is a neutral and a positive soliton. These defects are pushed toward each other by the lattice in order to minimize the length of the quinoidal part of the chain. However these solitons and anti-solitons cannot recombine because one single electron cannot form a bond. A defect composed of two positive solitons is shown in Figure 9.35. This species usually is called a "bipolaron," since in case two polarons meet, the two neutral solitons can form the bond, and only the two charged solitons are left over.

Figure 9.36 presents the whole particle *zoo* for polyacetylene. In the particular case of polyacetylene, there is no evident binding force for polarons (unless it is assumed that interactions between neighboring chains in a polyacetylene crystal exist). Both solid state and the chemical terms of the defects are presented in the figure, for use as a "physical–chemical dictionary."

As demonstrated above, the optical signature of a soliton is the midgap state (because of correlation effects it will not be exactly at the midgap position). A polaron is characterized by two states in the gap. The emergence of two states can

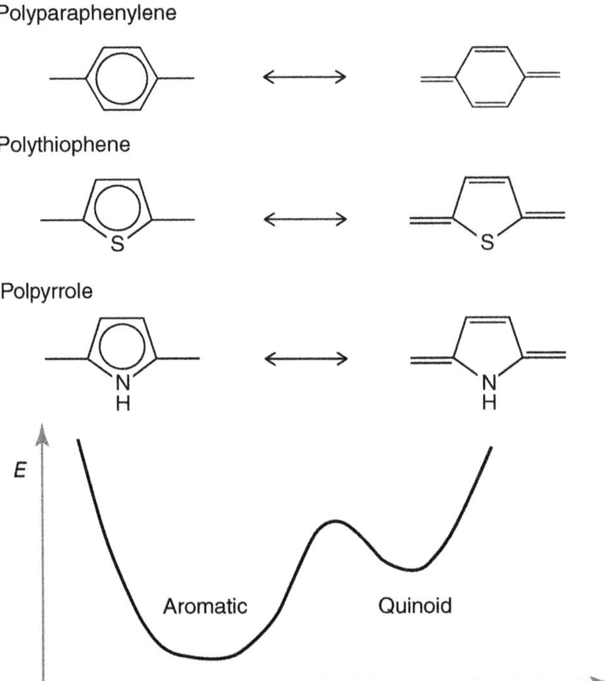

Figure 9.33 The broken degeneracy in a few polymer examples.

Figure 9.34 Polaron in polyphenylene.

Figure 9.35 Bipolaron in polyphenylene.

be rationalized to occur through interaction between the midgap states belonging to the soliton components of the polaron (Figure 9.37).

The two gap states of the polaron can be occupied by zero, one, or two electrons each, as indicated in Figure 9.38. Depending on the occupancy, the defect is either neutral or charged and does or does not carry a magnetic moment. The two possibilities (low spin and high spin) of accommodating two electrons on a polaron are reminiscent of excitons in molecules and in inorganic semiconductors. Therefore, they are called *polaron excitons* or sometimes just *excitons*. Incidentally, polarons are also known in semiconductor physics: an electron moves through the lattice by polarizing its environment, thus becoming a "dressed" electron. It is a lattice distortion that is small compared to the polaron defect in conjugated polymers.

The fact that solitons cannot move freely in a nondegenerate polymer is often called *soliton confinement*. The soliton confinement has an appealing analogy to quark confinement in elementary particle physics. If we tried to separate the two parts of a polaron by pulling them apart (Figure 9.39), we would create a

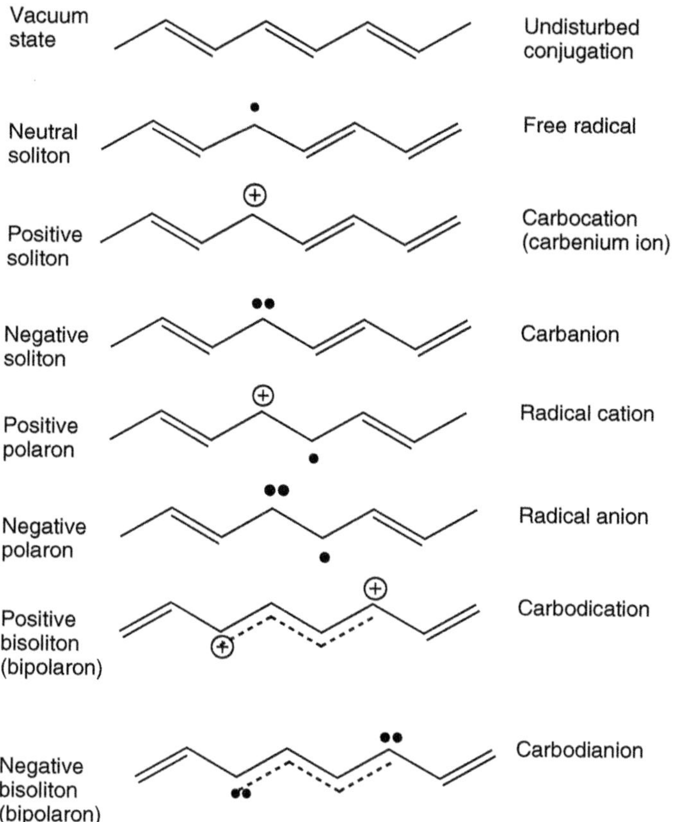

Figure 9.36 Complex conjugational defects constructed from solitons for polyacetylene: a "physical–chemical dictionary."

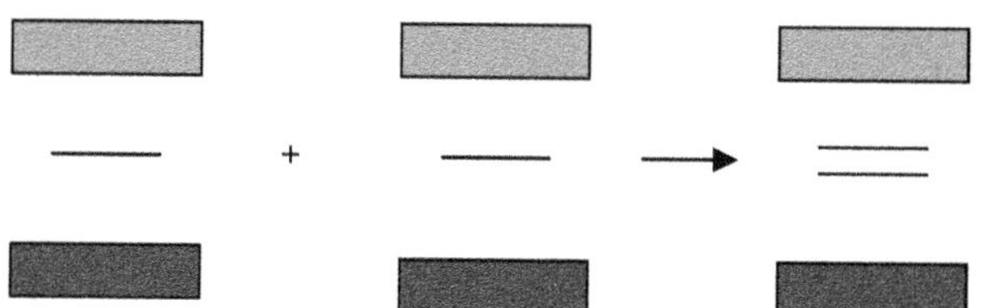

Figure 9.37 Two gap states of a polaron, resulting from splitting the solitonic midgap state by the interaction between the two components.

large chain section with high energy. If the high energy section is sufficiently long (a few quinoidal rings will do), enough energy will be generated to split a double bond and create two new solitons, which then will form new and smaller polarons with the previous solitons, so that the major part of the chain can again relax into the low energy form. This will happen far earlier than the possible cutting of the polymer chain into pieces and isolation of a soliton in one of the fragments. Similarly, quarks cannot be pulled out of elementary particles.

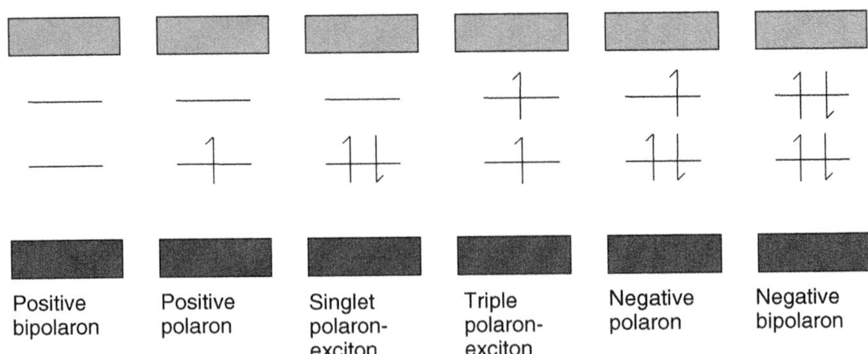

Figure 9.38 The polaron "menagerie."

Figure 9.39 Soliton confinement is a nondegenerate polymer. If the solitons forming a polaron are pulled apart, a high energy section of the chain will be created between the solitons. Soon the energy stored will be large enough to break a double bond and create two new solitons.

9.7 Fractional Charges

Another relation to elementary particles is the concept of fractional charges. Free particles carry integer charges, usually $+e$ or $-e$, where e is the electronic charge. Quarks can have charges like $2/3e$. If we put electrons into a solid, they will of course keep their charge, but they will use it to compensate the ionic charges so that the solid will become neutral. If Bloch waves and wave packages out of Bloch waves are constructed, is it trivial that a wave package must have an integer charge? And what about solutions of nonlinear differential equations? Must they carry integer charges?

Figure 9.40 shows again polyacetylene in the extended version. In the next step two negative solitons are added to the chain. These solitons are marked by arrows. The solitons are 180° phase-slip centers and they amount to the insertion of an additional single bond. To keep the chain length constant, two bonds are removed at the end, a single and a double bond. Counting the electrons we notice that the newly formed chain has two electrons less than the former chain. Since the former chain was neutral, the new chain bears the charge of $+2e$. If this amount of charge is equally distributed over the two defects, each defect will carry the charge of $+e$. For polyacetylene the charge at a defect is evident and such a complicated electron counting procedure is not necessary.

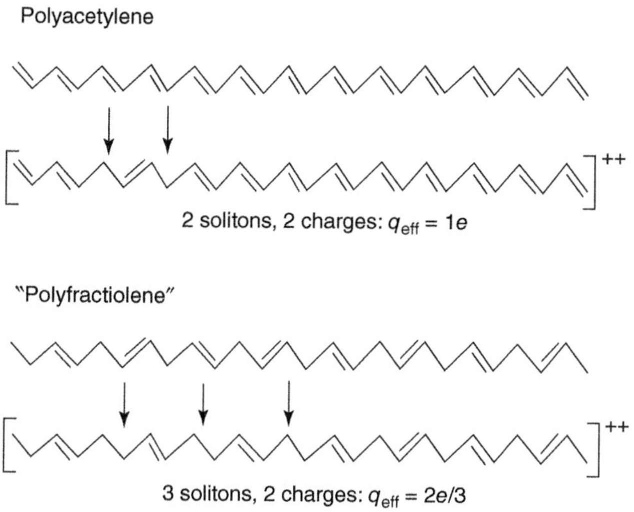

Figure 9.40 Electron counting to demonstrate fractional charges on one-dimensional chains.

The lower part of the figure shows the hypothetical *polyfractiolene.* This is a hypothetical polymer with double bonds separated by two single bonds. Polyacetylene without a Peierls distortion would have a half-filled band. The band in polyfractiolene would be filled to 1/3 (less electrons than in polyacetylene)! Insertion of an extra single bond will again be a phase-slip center, in this case, of 120°. Counting the electrons of polyfractiolene in the same way as in polyacetylene will show that two charges are to be shared by three defects! It is not quite clear whether it is allowed to look at individual solitons. If all solitons created simultaneously must be viewed together, one might argue that fractional charges are just a bookkeeper's trick. On the other hand, fractional charges on solitons are a nice analogy to fractionally charged quarks. Further relationships between charge fractionalization in one-dimensional metals and relativistic field theory are discussed by Jackino and Schrieffer [32]. The similarity between fractionally charged solitons and the fractional quantum Hall effect has been pointed out by Schrieffer [33].

Polyfractiolene cannot be synthesized. The bond sequence shown in Figure 9.40 can only be stabilized by inserting protons onto the chain, and the proton arrangement would break the required symmetry. But in the crystalline one-dimensional metals mentioned in Chapter 2, band filling of ⅓ or ¼ exists, and in these systems fractional charges might occur. Bak and Jensen [34] has even proposed an experiment to put their existence into evidence: via the current flowing through a conductor, the number of charges is measured and the number of carriers is inferred from analysis of the noise spectrum of the current. Divide the number of charges by the number of carriers and so on. The proposal was made in 1982. Today one might go a step further and propose to use the tip of a scanning tunnel microscope as a local probe.

Figure 9.41 *Polyspinolene*, a linear arrangement of magnetic dipoles (a) and (b) forms of the degenerate ground state. (c) One dipole reverse (like a soliton, two domain walls). (d) Domain walls (magnetic charges = poles) separated.

Polyacetylene also stimulates a Gedankenexperiment on magnetic monopoles [35]. Figure 9.40 shows *polyfractiolene*. In Figure 9.41 we show a further hypothetical substance, *polyspinolene*.

Polyspinolene is just a linear arrangement of magnetic dipoles. Here, we are less inspired by real polymers, nor by polycarbene (Figure 2.35). Our intuition comes from stacks of phthalocyanines (Figure 2.21) where the central metal could bear a magnetic moment, or from KCP in Figure 2.8, or from MX chains in Figure 2.33. In any case, we assume that there are some ligands keeping the dipoles in line. Some people will speak of a crystalline field acting on the dipoles. Traces (a) and (b) in Figure 9.41 show that the two degenerate ground states of polyspinolene, while in Trace (c) we have created a soliton–anti-soliton pair (two domain walls). One can easily convince oneself that the domain walls correspond to magnetic poles (at head–tail junctions of the arrows, the poles compensate; at head–head or tail–tail junctions, there are *magnetic charges*). In Trace (d) the domain walls are separated (by "soliton motion"), and if we can cut the chain in between, we have magnetic monopoles!

Polyspinolene of Figure 9.41 exists only in our Gedankenexperiment. But a more curly and interwoven version does exist in nature. These are the so-called *spin ice* compounds. An example is $(Dy,Ho)_2Ti_2O_7$ [35d]. Magnetic order, magnetic domains, and magnetic domain walls can be investigated by neutron scattering. The correlation function of spin ice is different from that of conventional ferromagnetic domains. Neutron scattering experiments have shown evidence [35e] of spin ice with entangled strings carrying magnetic monopoles at their ends.

9.8 Soliton Lifetime

Most measurements of conducting polymers are fairly simple: optical spectroscopy, electrical conductivity, and magnetic susceptibility. But the theoretical concepts are so exciting that many scientists engage in very sophisticated experiments. Critics then might ask, "Why would you waste such a beautiful experiment on such ill-defined samples?" (And compared to perfect silicon single crystals, conjugated polymers certainly are ill-defined.) An example of this type of experiment is picosecond photoconductivity in polyacetylene. However, since samples have improved as has our understanding of statistical methods to treat order, these experiments have become far more common in characterizing new materials (one of the authors, SR, was involved in developing these techniques in polymers).

The idea of the experiment is simple: take *stretch-aligned* polyacetylene, i.e. polyacetylene in which the chains have been more or less aligned by stretching (certainly a large single crystal would be better, but we don't have that). Electrodes are evaporated on the sample and an electric field is applied. Short light pulses are affected on the sample that photogenerate solitons. Charged solitons will be displaced by the electric field, a photocurrent will flow, and the decay of the photocurrent with time can be measured. The photocurrent will flow as long as the charge carriers move. If the experiment is carried out on a silicon single crystal, the carriers move until they hit the electrodes at the crystal surface. From the decay of the photocurrent, the time the carriers need to cross the crystal, *their time of flight*, is obtained. From the electrode separation we know the distance the carriers migrate, and their velocity can be calculated. Division by the electrical field yields the mobility. If the crystal has many imperfections that act as traps, the carriers will get trapped before they reach the electrodes. If we know the average distance between the traps, we can again calculate the mobility from the decay of the photocurrent.

In a very imperfect semiconductor, the carrier lifetime is only a few picoseconds. Such a fast decay of the photocurrent can be difficult to measure directly; standard preamplifiers typically have rise times of about 100 ps. Therefore, picosecond photoconductivity experiments often employ optical correlation methods. The sample is regarded as a fast optical switch; two such switches are put in series. They are illuminated by light pulses, one delayed with respect to the other (Figure 9.42). If both pulses arrive simultaneously, both switches are on, and an oscilloscope records a signal (of course, distorted by the slow response of the scope, but that does not matter). If the delay is so large that the first switch is off already and the second is not yet on, there will be no signal, but if the opening time of the switches overlaps, there is some signal. By changing the delay between the pulses, fast events can be probed (*pump and probe technique*). The two pulses are usually generated by sending the laser beam through a beam splitter followed by change of the optical path length in one of the branches.

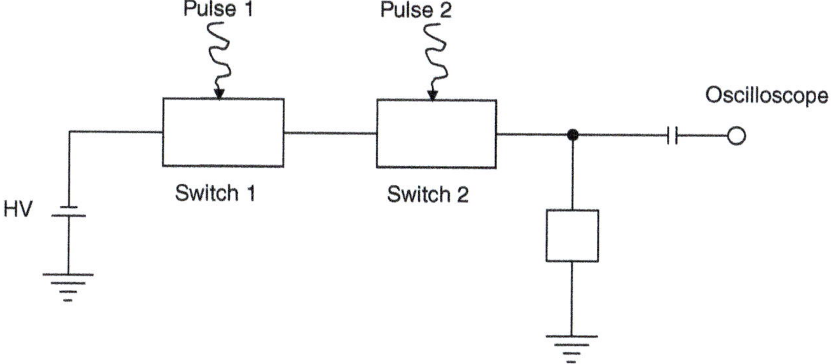

Figure 9.42 Optical correlation experiment for fast photoconductivity decay.

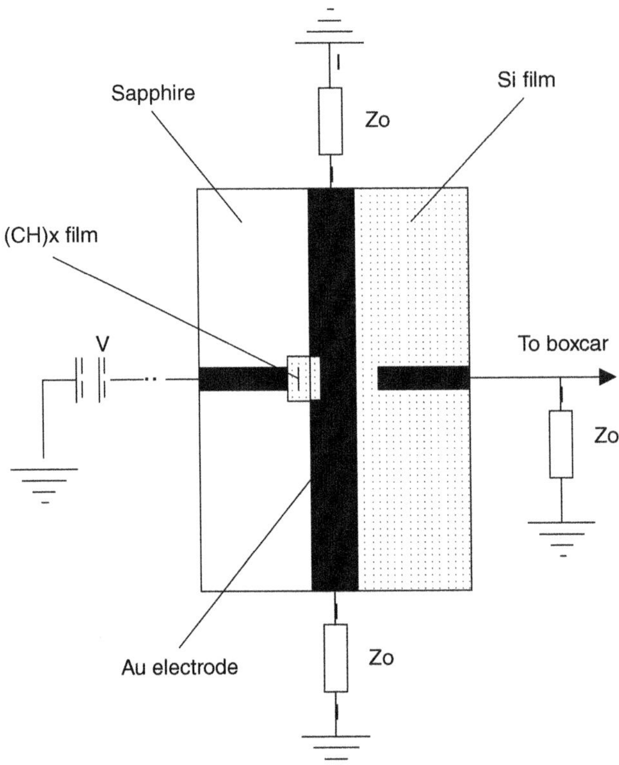

Figure 9.43 Two fast optical switches built into 50 Ω striplines for a picosecond correlation measurement to study the fast decay of the photocurrent in polyacetylene (*soliton lifetime*).

Picoseconds circuitry is somewhat more sophisticated than it would seem in Figure 9.42, mainly because a 50 Ω wave resistance has to be kept throughout the circuit to avoid pulse distortions. Therefore the switches have been constructed as part of 50 Ω *stripline*. The stripline is shown in Figure 9.43. One of the gaps in the copper strips is covered by ion-bombarded silicon, and the other by a polyacetylene flake. The width of the gaps is about 20 μm.

The experimental result is shown in Figure 9.44 [36]. The signal intensity is plotted over the pulse delay. The signal rises within 2.5 ps and decays within 6 ps. The analysis shows that the rise time is mainly determined by the silicon switch, the decay by the polyacetylene sample. From this experiment we know that the photoconductive decay time is about 6 ps. From other optical experiments we estimate the *schubweg*, i.e. the distance the solitons migrate before they are trapped; we obtain a mobility of about 1 $cm^2/V\,s$ for charge carriers in polyacetylene. This is in remarkable agreement with theoretical calculations for solitons (and polarons) in polyacetylene [37].

Just for the sake of interest, we note that the short lifetime of solitons in polyacetylene was at one time proposed to be used as the basis to build a fast holographic computer. Photogeneration of solitons makes polyacetylene a nonlinear optical material: the refractive index depends on the number of

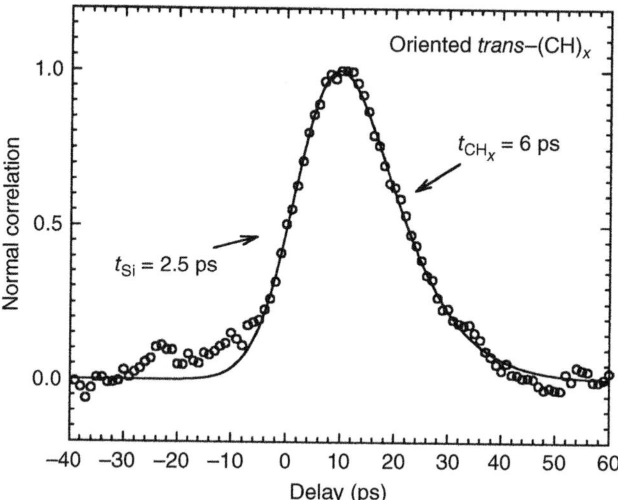

Figure 9.44 Optical correlation measurement of picosecond photoconductive decay in polyacetylene.

solitons. Consequently holograms can be stored in polyacetylene. These holograms fade within a few picoseconds, but this time is long enough to compare two holograms (*holographic pattern recognition*). After a few picoseconds the polyacetylene is ready to analyze the next hologram. This way a computer with the world record in data throughput per time can be constructed [38].

9.9 Conductivity and Solitons

We have discussed conjugational defects. We have seen that these defects are quasiparticles, that they can be neutral or positively or negatively charged, and that they can move along a polymer chain with an effective mass only slightly larger than the free electron mass, $m^* \sim 6m_e$. Therefore it is tempting to associate the conductivity of polyacetylene with the motion of charged solitons.

To estimate how important charge transport by solitons might be, we look at the Drude model. What is the *collision time* for solitons? Jeyadev and Conwell [39] analyzed the scattering of solitons on one-dimensional longitudinal acoustic phonons that are compressional waves along the chain. The soliton energy of the compressed section differs from that in the decompressed section. As with electron–phonon scattering in a semiconductor, soliton–phonon scattering can be calculated by means of the *deformation potential method* [40]. Jeyadev and Conwell obtained values in the order of $1\,\text{cm}^2/\text{V s}$ for the mobility of solitons in polyacetylene at room temperature and about $100\,\text{cm}^2/\text{V s}$ at 50 K. These calculations neglect impurity scattering, which will reduce the mobility particularly at low temperatures. Below 20 K the solitons will be *self-trapped* due to the discreteness of the lattice, so that in the low-temperature limit, a simple polyacetylene chain will be an insulator, both from the electron localization point of

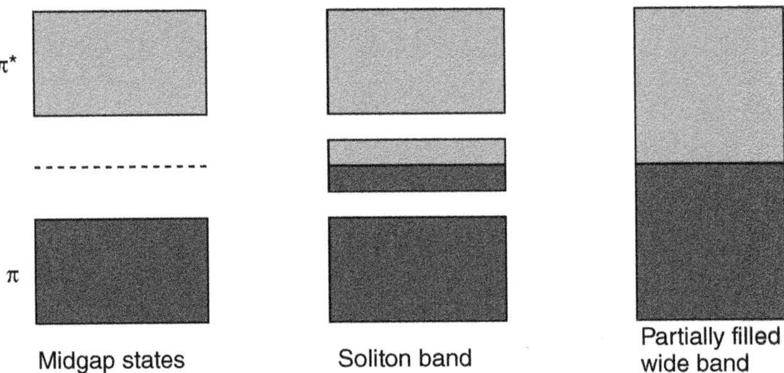

Figure 9.45 Development of a soliton band from midgap states and final suppression of the π–π* gap upon the increase of soliton concentration.

view and from soliton transport considerations. Picosecond photoconductivity experiments allow for experimentally checking the calculated mobility [41]. The agreement is surprisingly good.

If the solitons are not photogenerated but created by doping, the electrostatic interaction between the charge of the solitons and that of the counterions has to be taken into account. For example, by iodine doping two positive solitons and two negative I_3^- ions are created. The solitons will be trapped by the Coulomb field of the I_3^- ions; the electrostatic energy will be much larger than the thermal energy $k_B T$ or the energy generated by the applied electric field. Hence, in doped polyacetylene, charged solitons will not be mobile and will not contribute to conductivity.

In heavily doped polyacetylene, the midgap states associated with the solitons will interact and form a soliton band. The soliton band will be partially filled and band conductivity might occur. If the soliton band is wide enough to overlap with the π and π* bands, little of the exciting soliton particles is left. In Figure 9.45 the development of a soliton band from midgap states is schematically shown as well as the final vanishing of the π–π* gap upon increasing the soliton density. Band conductivity in polyacetylene will be further discussed below.

Another conductivity mechanism related to solitons is *intersoliton hopping* [42]. In this model, charged and neutral solitons are present in (slightly) doped polyacetylene. The charged solitons are trapped by the dopant ions, but the neutral solitons are free to move. Whenever a neutral soliton passes close by a charged soliton, an electron can hop between the midgap states belonging to the solitons (Figure 9.46).

As seen previously, there is spin–charge inversion for solitons in polyacetylene: charged solitons carry no spin, while spin-carrying solitons have no charge. There was much discussion on whether or not experimental evidence of spin–charge inversion could be obtained, for example, by relating data of electrical conductivity to those of magnetic susceptibility. In Figure 9.47 the paramagnetic *Curie susceptibility* and the electrical conductivity of polyacetylene are plotted vs. the doping concentration [43]. The susceptibility decreases, while the conductivity increases. The experimental data are consistent with the assumption that – at low doping concentrations – doping first converts neutral spin-carrying solitons into

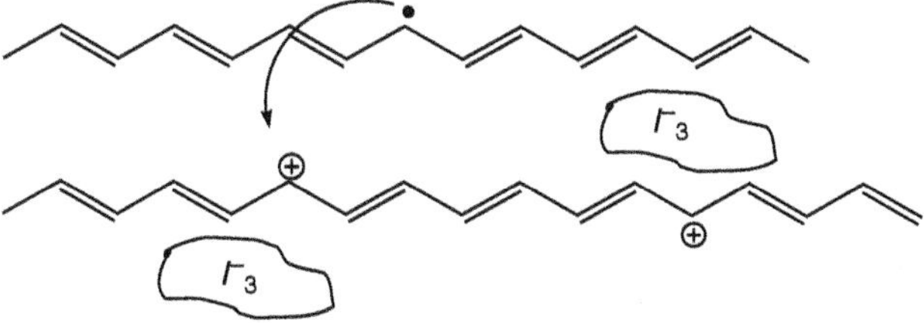

Figure 9.46 Intersoliton hopping: charged solitons are trapped by the dopant counterions, but neutral solitons are free to move. When a neutral soliton is close to a charged soliton, the electron hops from one defect to the other.

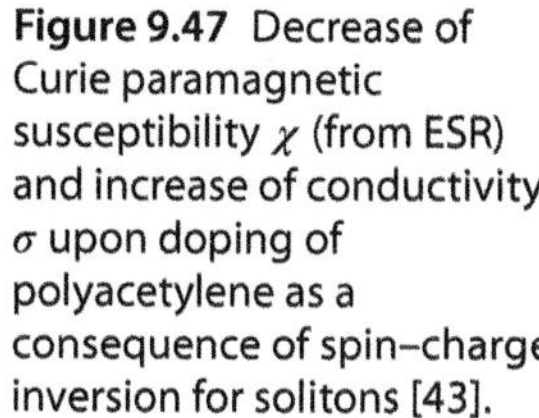

Figure 9.47 Decrease of Curie paramagnetic susceptibility χ (from ESR) and increase of conductivity σ upon doping of polyacetylene as a consequence of spin–charge inversion for solitons [43].

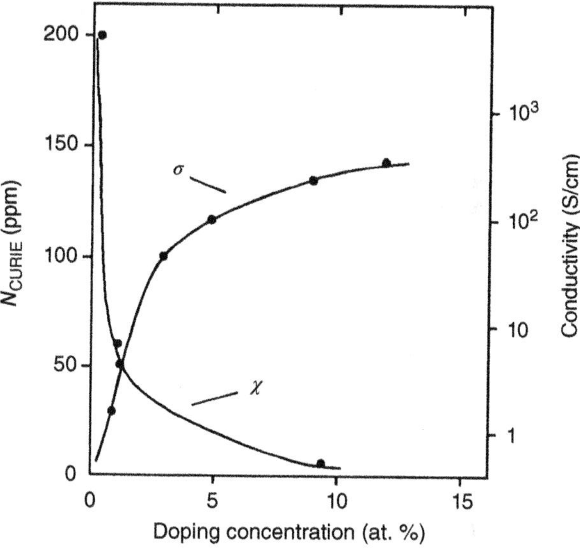

charged spinless solitons. There are some 200 ppm of neutral solitons in undoped polyacetylene, and the first electrons to be oxidized upon doping are those clinging to the solitons.

Simultaneous to charging of existing solitons, doping creates new solitons, as shown. At high soliton concentrations the solitons interact, and a soliton band is formed, which widens until the π–π^* gap is finally suppressed (Figure 9.45). Noninteracting spin-carrying solitons fully contribute to the Curie susceptibility, which, decreasing with $1/T$ (Curie law), is the susceptibility of isolated particles fighting individually against thermal disorder. Interacting particles tend to compensate their spins, until for particles in a band, *Pauli susceptibility* finally replaces Curie susceptibility. To obtain the Pauli susceptibility, the sample temperature T has to be replaced by the Fermi temperature $T_F = E_F/k$. As a result, the *Pauli susceptibility* is temperature independent. Because of $T_F \gg T$ the Pauli susceptibility is much smaller than the Curie susceptibility. Consequently, the decrease in susceptibility as shown in Figure 9.47 could also be due to the gradual transition from Curie susceptibility to Pauli susceptibility.

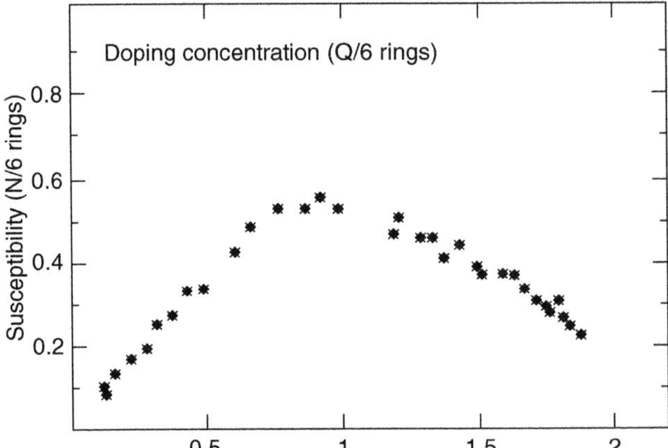

Figure 9.48 Susceptibility vs. doping concentration in electrochemical doping polypyrrole. The susceptibility is measured in spins per six pyrrole rings, and the doping concentration in electronic charges transferred per six pyrrole rings [44].

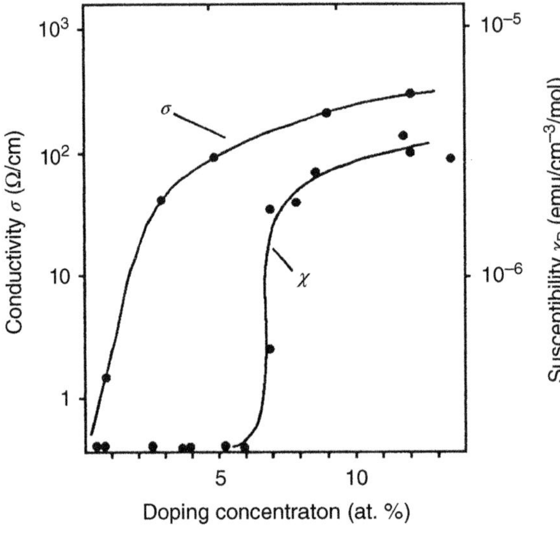

Figure 9.49 Pauli susceptibility χ_P and electrical conductivity σ plotted vs. the doping concentration in polyacetylene. Source: After [45].

In nondegenerate ground-state conjugated polymers like polyparaphenylene and polypyrrole, solitons are unstable, and there are well-defined compound particles like polarons. Polarons carry a spin, and bipolarons are spin compensated. Upon doping, at first susceptibility increases, then, from a certain doping level onwards, susceptibility decreases due to the formation of bipolarons. An example for electrochemical doping of polypyrrole is shown in Figure 9.48 [44]. It is obvious that the susceptibility reaches a maximum when one electronic charge has been transferred to about every sixth pyrrole ring, but this maximum corresponds to only half as many spins as there are charges.

Experimentally, the Pauli susceptibility can be separated from the Curie susceptibility by temperature dependence. In Figure 9.49 the Pauli susceptibility of polyacetylene is plotted vs. the doping concentration [45], and the conductivity is taken from Figure 9.47. There is a pronounced step in the Pauli susceptibility at a doping concentration of 6%. This step is not consistent with a gradual conversion of isolated solitons into a soliton band. It rather suggests a phase transition. A similar step has been observed in sodium-doped polyacetylene [46]. Kivelson and Heeger [47] introduced a theory for a first-order phase transition driven by soliton interactions. An alternative driving force for this phase transition might be a rearrangement in chain packing due to the accommodation of dopant ions [48]. (It is also possible that both soliton interaction and intercalation of counterions lead to separate phase transitions, and these two-phase transitions then interlock.) Despite the origin of the phase transition, it is remarkable that between 2% and 6% doping, high conductivity is observed but no detectable magnetic susceptibility, neither Curie nor Pauli.

9.10 Fibril Conduction[2]

Polyacetylene has been our "model" system for these discussions. Except for materials with very high conductivity ($\sigma > 1000\,\text{S/cm}$) similar data are obtained for all of the conducting polymers. But a peculiar problem arises for the interpretation of data that is complicated by the fibril morphology of conjugated polymers.

Figure 9.50 shows an electron micrograph of a polyacetylene film. Note the *spaghetti structure.* An idealization of this structure is presented in Figure 9.51, stressing the interfibrillar contacts. Small crystalline domains are discernible

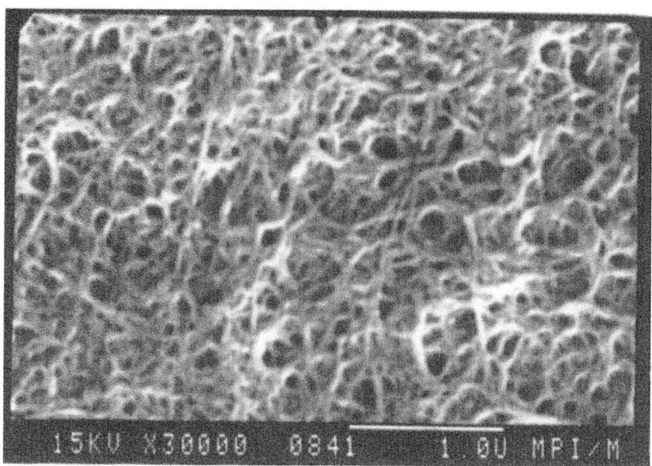

Figure 9.50 Scanning electron micrograph of polyacetylene film showing the "spaghetti" or fleece-like morphology. The scale marked on the rim corresponds to 1 μm.

2 This section goes beyond the results for correlation in transport that is the overall focus of this chapter. It is included to give a view as to the difficulties in interpreting such transport measurements specifically.

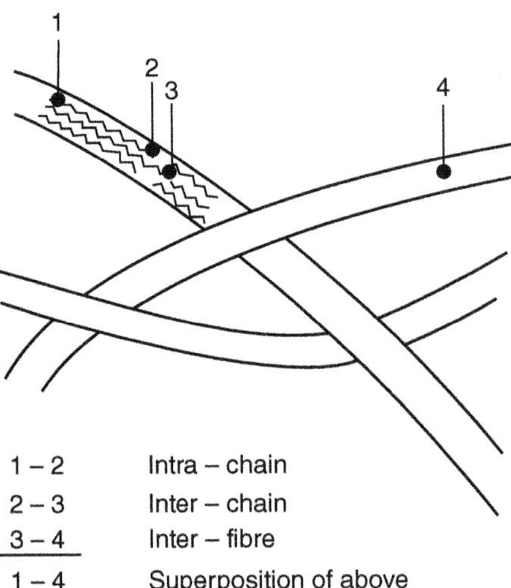

Figure 9.51 Schematic view of the fibrillar structure of polyacetylene. The fibers are bundles of chains, containing some 100 or 1000 polymer chains. They intersect in fractions of 1 μm.

within the fibers in which the chains are well ordered. These areas can be analyzed by X-ray or neutron diffraction to give the structure of the elementary cell. Amorphous regions are situated between crystalline domains. A crystalline domain, however, is not perfectly ordered; there are point defects such as chain ends, cross-links, etc. The overall electrical conductivity will be a superposition of *intrachain, interchain,* and *interfiber* charge transport mechanisms. If the polyacetylene film is stretched the fibers are aligned and the crystallinity within the fibers is improved.

As far as the conductivity is concerned, at present we distinguish between two types of polyacetylene samples: (i) standard or Shirakawa polyacetylene ("Type S") – when saturation doped with iodine, these samples achieve conductivity values of some hundred S/cm – and (ii) new, or Naarmann, or highly oriented polyacetylene [49] ("Type N") in which the conductivity of doped samples exceeds 10^3 S/cm and may even reach 10^6 S/cm. The highly doped polyacetylene, which stays metallic at low temperature belongs to Type N, whereas the Type S, becomes insulating as $T \to 0$. The conductivity in Type S is dominated by hopping, and the conductivity of Type N is not. It is generally assumed that Type N samples exist only for polyacetylene, whereas samples of all other conjugated polymers are of Type S. However, it is possible that material perfection will also lead to Type N samples in polyaniline, polyphenylene vinylene (PPV), and others.

Figure 9.52 presents a conductivity chart that compares the room temperature conductivity of doped conjugated polymers with that of other materials. Note that the polymer conductivity values extend over the whole region from insulating such as diamond to highly conducting such as copper. The top part of the polymer arrow represents Type N polyacetylene.

In Figure 9.53, the conductivity change upon doping is shown for Type S polyacetylene. A sharp initial increase is followed by saturation at some 5 or 6 mol%. These high doping levels contrast with the doping concentrations we used to form conventional semiconductors like silicon or germanium, which are in the ppm range. (The high doping level is one of the reasons why many scientists

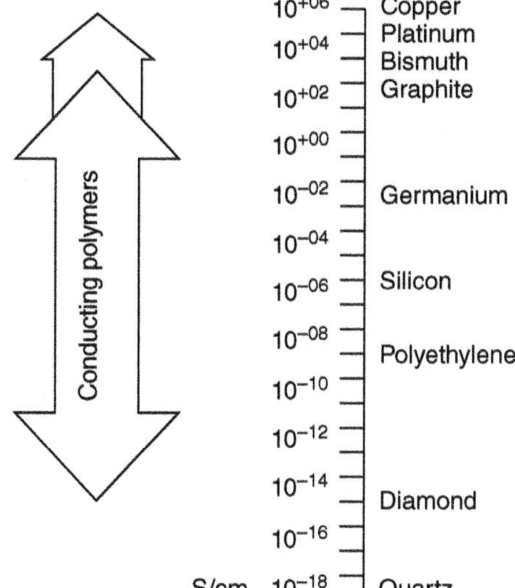

Figure 9.52 Comparison of the room temperature conductivity values of conducting polymers with conductivities of other materials.

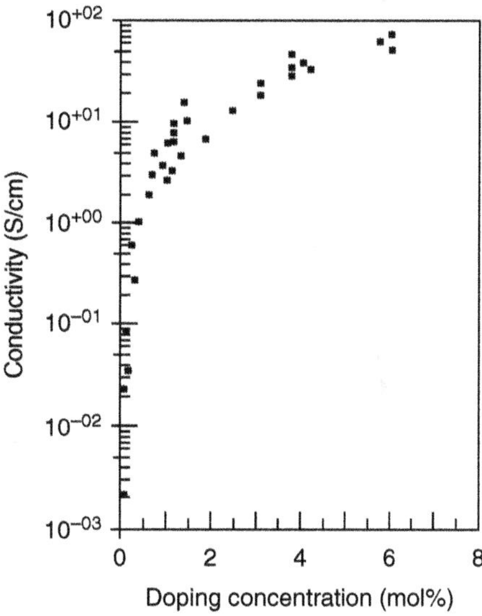

Figure 9.53 Conductivity of polyacetylene as a function of doping concentration (type S polyacetylene, room temperature, iodine doping).

dislike the term doping in connection with conducting polymers.) The initial rise is even more pronounced, when accidental doping by atmospheric oxygen or other impurities is compensated by treating the sample in ammonia. In that case it is possible to observe conductivity changes of more than 10 orders of magnitude.

Figure 9.54 shows the temperature dependency of the conductivity of iodine-doped (Type S) polyacetylene at various doping levels [50]. At all levels of iodine concentration, the conductivity decreases upon cooling of the sample, with the decrease being much smaller for highly doped than for slightly doped samples. The set of curves is consistent with variable range hopping (VRH) that

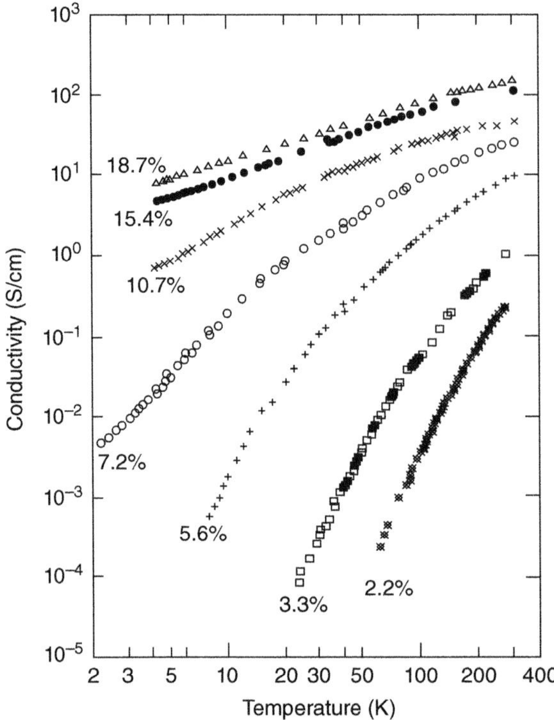

Figure 9.54 Temperature dependency of polyacetylene conductivity at various concentrations of iodine doping (Type S polyacetylene) [50].

will be discussed in Section 9.11. At very low doping levels, power laws (with powers larger than 10) fit the data equally well. Figure 9.55 shows Epstein's data on undoped (accidentally doped) polyacetylene [51]. The straight line corresponds to $\sigma \sim T^{14}$.

Highly doped Type N samples do not behave according to VRH. They are metallic in the sense that the conductivity does not vanish as $T \rightarrow 0$. In some cases the conductivity is found to even *increase* upon cooling. An example for such a behavior is plotted in Figure 9.56. Note the maximum at 250 K; between room temperature and 250 K, the slope is negative [52].

The highest conductivity values published so far are close to 10^6 S/cm. One of the highest is shown in Figure 9.57 [49d, e]. A Type N polyacetylene sample is exposed to iodine, and the conductivity climbs up to 500 000 S/cm, which is almost as high as the room temperature conductivity of copper.

With a few exceptions it turned out that it is very difficult to synthesize samples with conductivities above 10^5 S/cm. On the other hand, it is easy to decrease the conductivity and to convert Type N samples into Type S. One method is the deliberate introduction of defects. The chemical structure formula in Figure 9.58 shows an O=C—CH_2 group within the polyacetylene chain. Such groups cut the conjugated system into shorter segments, and the conductivity can be studied as a function of the average segment length. The result of such an investigation is shown in Figure 9.58. There are two sets of samples. The samples have been segmented first and then doped to 3.5% and 15% iodine. We see that the conductivity decreases exponentially with the defect concentration, or, in other words, it

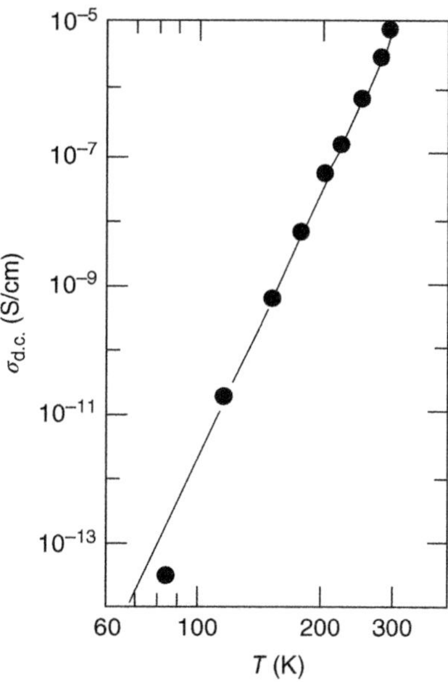

Figure 9.55 Temperature dependency of the conductivity of undoped polyacetylene according to Epstein et al. [51]. The solid line corresponds to the power law $\sigma \sim T^{14}$.

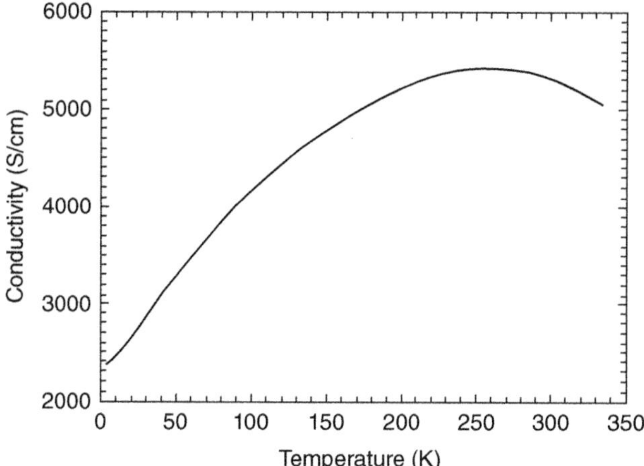

Figure 9.56 Negative (metallike) temperature coefficient of the conductivity of highly conducting iodine-doped Type N polyacetylene [52].

increases exponentially with the segment length (in the range of lengths studied, i.e. from about 20 to 100 Å) [53].

9.11 Hopping Conductivity: Variable Range Hopping vs. Fluctuation-Assisted Tunneling

As already stated the experimental data on the conductivity of conjugated polymers can be explained by *hopping mechanisms*. This is with the exception

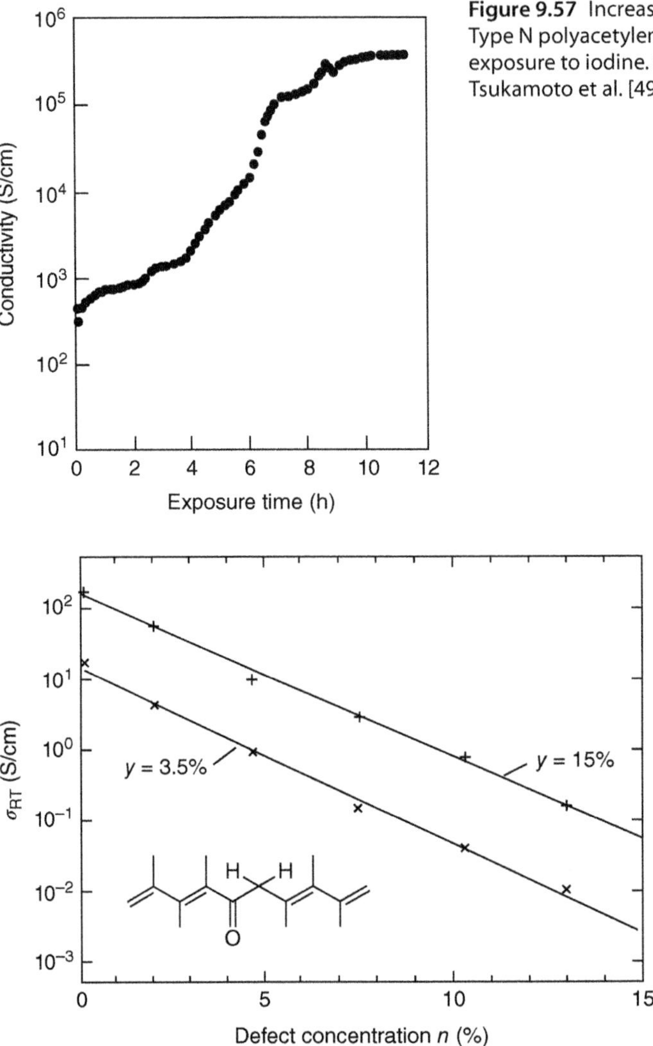

Figure 9.57 Increase in conductivity for a Type N polyacetylene sample during exposure to iodine. Source: After Tsukamoto et al. [49d,e].

Figure 9.58 Conductivity of doped polyacetylene as a function of the defect concentration. Defects interrupt the conjugated system of double bonds ("segmented polyacetylene") [53].

of highly conducting, highly oriented "Type N" polyacetylene, where metallic (band) conductivity has to be assumed. We also know that "hopping" is an abbreviation for *phonon-assisted quantum mechanical tunneling* and that the temperature dependency of hopping conductivity is described by *soft exponential equations*. Hopping, of course, is an anthropomorphic term and reminds us of a man crossing a river by jumping from stone to stone (Figure 9.59). The stones are spread out at random. It is quite obvious: the more stones, the higher the conductivity. Also, the alternating current (AC) conductivity will be greater than the DC conductivity. Coming to a place where the next stone is far away, the hopper has to rest for a long time to sufficiently recover before he/she can do the big jump. While still waiting the field will reverse and he/she could do an easy

Figure 9.59 Hopping transport: man trying to cross a river by jumping from stone to stone.

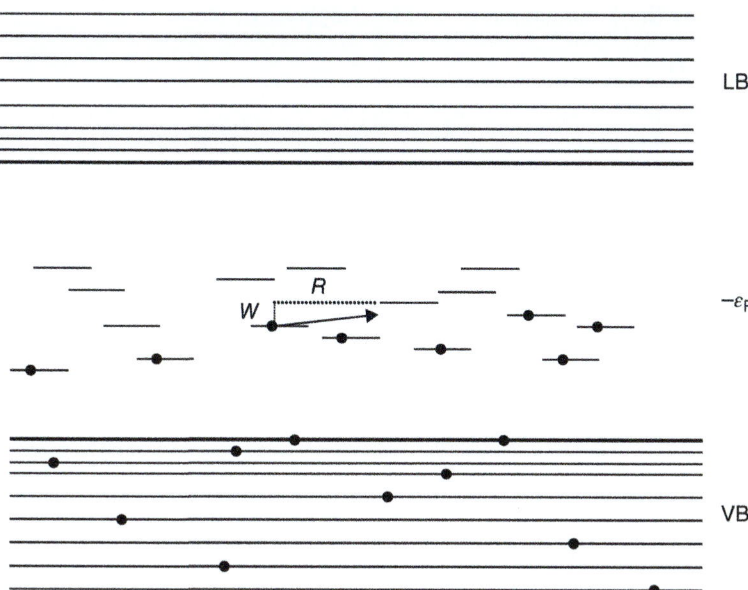

Figure 9.60 Electronic-level scheme of a disordered solid to demonstrate the hopping conductivity (CB, conduction band; VC, valence band; E_F, Fermi energy; W, energetic distance between states; and R, local distance between states).

hop backward. Whenever the field changes with high frequency, the hopper will just head back and forth between near stones. This is the pair approximation to hopping conductivity [54].

To discuss the temperature dependency of hopping conductivity, we have to assume that in addition to the random special distribution of the stones in the river, there is also a random distribution in height. Furthermore, on a warm day it is easier to hop up onto a high stone. At this point the anthropomorphic model begins to fail, and it is more reasonable to look at the band structure representation in Figure 9.60. There are localized states in the gap, randomly distributed in space as well as in energy. The Fermi energy level is about at gap center, the states below E_F are occupied, and those above are empty (except for thermal excitations). Electrons will hop (tunnel) from occupied to empty states. Most of the hops will have to be upward in energy. At high temperatures there are many phonons available that can assist in upward hopping. As these phonons freeze out, the electron has to look further and further to find an energetically accessible state. Consequently the average hopping distance will increase as the temperature decreases, hence the name *variable range hopping*. Since the tunneling probability decreases exponentially with the distance, the conductivity also decreases. But, as already mentioned, the decrease is smoother than in a semiconductor with a well-defined gap, because there is a continuous distribution of activation energies (energetic distances between states).

The calculation leads to a "soft exponential equation" (cf. Eq. 9.20):

$$\sigma = \sigma_0 \exp[-(T_0/T)^{\gamma}] \tag{9.20}$$

where γ depends on the dimensionality d of the hopping process:

$$\gamma = 1/(1 + d) \tag{9.21}$$

For three-dimensional VRH, $\gamma = 4$, and we get Mott's famous "$T^{-1/4}$ law" T (ln σ proportional to $T^{-1/4}$) [55]

$$\sigma = \sigma_0 \exp[-(T_0/T)^{1/4}] \tag{9.22}$$

with

$$\sigma_0 = e^2 N(E_F) R \nu_{ph} \tag{9.23}$$

where R is the average hopping distance,

$$R = [(8/9)\pi\alpha N(E_F) k_B T]^{-\gamma} \tag{9.24}$$

and

$$T_0 = [8\alpha^3/9\pi N(E_F) k_B] \tag{9.25}$$

with $N(E_F)$ as electronic DOS at the Fermi energy, α as inverse localization length (spatial extension of localized wavefunction), and ν_{ph} as typical phonon frequency.

As already stated, Eq. (9.22) is consistent with the temperature dependence of the conductivity shown in Figure 9.56. It turns out that any γ value between ½ and ¼ fits the experimental data, so that it is difficult to distinguish between one-, two-, or three-dimensional VRH.

In Eq. (9.22) there are two general parameters: σ_0 and T_0. Both depend on the microscopic parameters α and $N(E_F)$ (Eqs. (9.23)–(9.25)) of which the former is the inverse spatial extension of the localized electron wavefunction and the latter the electron DOS at the Fermi level. If doping creates defects and defects give rise to states in the gap, $N(E_F)$ will be roughly proportional to the doping concentration y and we obtain:

$$\ln \sigma \sim y^{-\gamma} \tag{9.26}$$

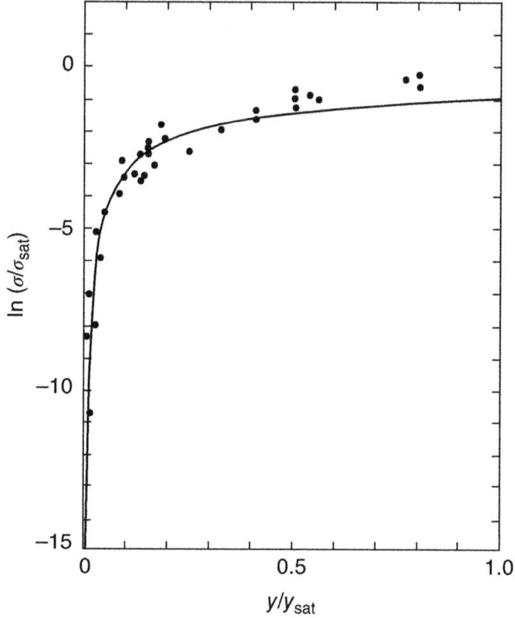

Figure 9.61 Conductivity of iodine-doped polyacetylene vs. doping concentration. The solid curve corresponds to a fit of the variable range hopping model.

Figure 9.61 shows a fit of Eq. (9.22) to the data displayed above. It is clear that the model of VRH explains the doping dependency of the conductivity quite well. The dependency on the localization length can be studied in segmented polyacetylene. We assume that the localization length is proportional to the segment length [56], leading to an approximately exponential decay for the conductivity when the segments are shortened.

A large electronic DOS implies both high conductivity at room temperature (large σ_0 in Eq. (9.23)) and small temperature dependency of the conductivity (small T_0 in Eq. (9.25)). This correlation is evident from the "universal curve" in Figure 9.62, where T_0 is plotted vs. the room temperature value of the conductivity for various conducting polymers [57].

It has already been mentioned in connection with the man trying to cross a river by jumping from stone to stone that in hopping transport the AC conductivity is higher than the DC conductivity. Figure 9.63 shows the frequency dependency of undoped polyacetylene (i.e. only accidentally doped by oxygen and other atmospheric contaminations). The curves refer to different temperatures. At low frequencies the conductivity is constant; from a certain point on it increases more or less linearly with the frequency. The higher the conductivity at zero frequency, the higher the onset frequency. This relation is quite general, regardless of whether we change the low frequency conductivity by heating, or doping, or using a different sample. The solid lines in the graph correspond to calculations by Ehinger et al. [58], who combined VRH and pair approximation to form the "extend pair approximation model" [59].

The data in Figure 9.63 were taken from undoped polyacetylene. In doped polyacetylene the conductivity is so high that the onset frequency is not in the hertz or kilohertz regime, but in the gigahertz region. Therefore the frequency

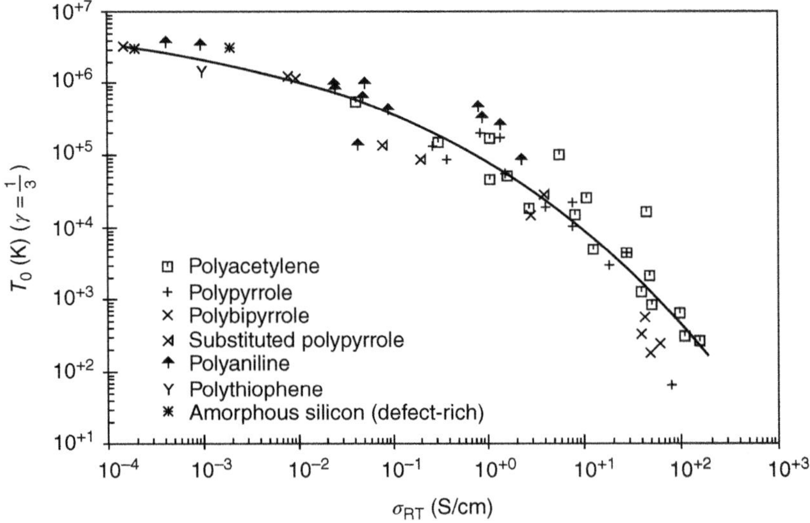

Figure 9.62 "Universal curve" relating the room temperature value of the conductivity to the parameter T_0, which is a measure of the temperature dependency of the conductivity. Data are taken from various conducting polymers [56].

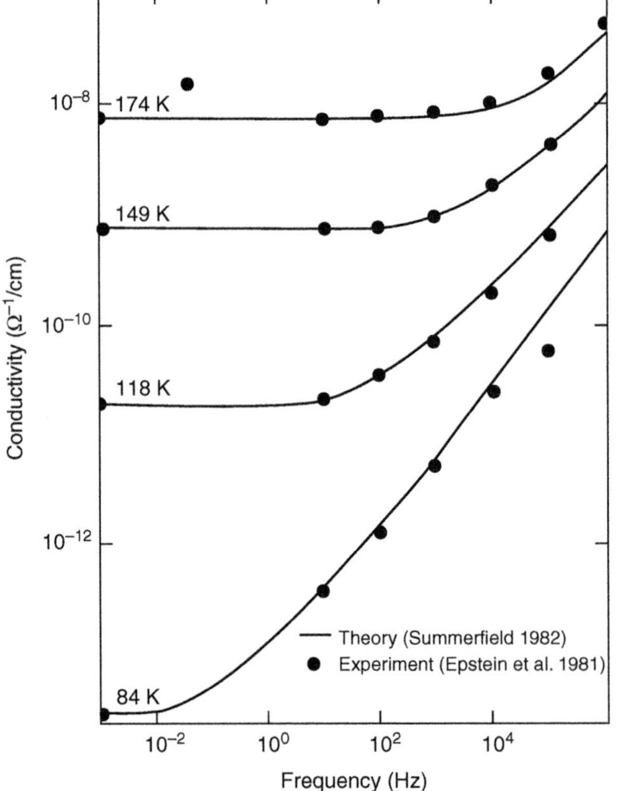

Figure 9.63 Frequency as a function of the conductivity of undoped polyacetylene. Data acquisition by Epstein et al. [51]; calculations by Ehinger et al. [58].

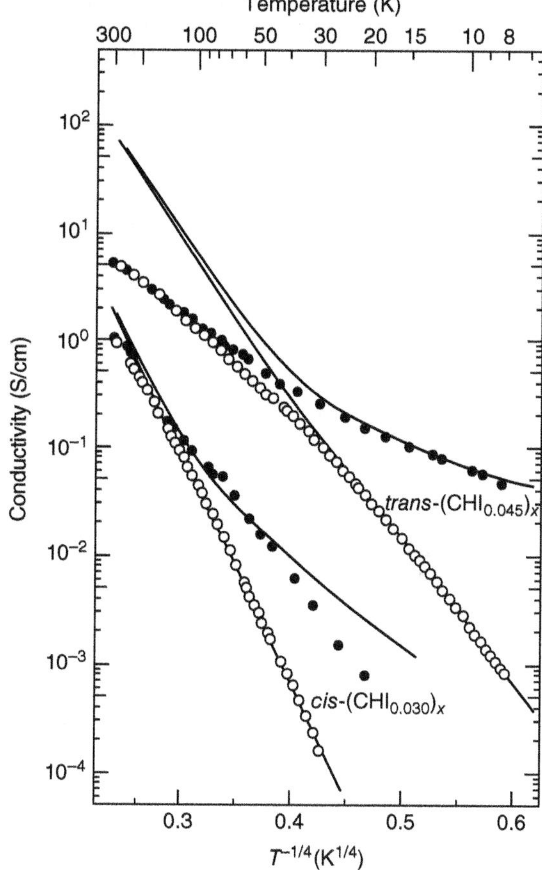

Figure 9.64 DC and microwave conductivity of iodine-doped polyacetylene and fit of extended-pair approximation [45]: ○, 30 Hz; •, 9.9 GHz.

dependency is found in the microwave region (10 GHz). In Figure 9.64 the temperature dependencies of DC (in this case 30 Hz is close enough to DC) and AC conductivity at 10 GHz are shown for two differently doped polyacetylene samples [47]. The abscissa scale is in $T^{-1/4}$, so that a straight line is obtained when Mott's law is fulfilled. It can be seen that, as expected, the AC data are situated above the DC data. The model parameters for the extended pair approximation can be obtained from a fit to the DC data, allowing the calculation of the microwave conductivity. Consequently the model predicts the experimental results reasonably well (with the exception of the values at high temperatures), where the approximation in the model overestimates both DC and AC conductivity (Mott's $T^{-1/4}$ law is a *low*-temperature approximation to VRH).

The model of VRH assumes a random distribution of localized states. If the distribution is not random and the defects tend to cluster, the model of *fluctuation-induced tunneling* (FIT) is more appropriate [60]. This model assumes metallic islands in an insulating matrix. Particles of carbon black or aluminum flakes in a thermoplastic polymer are typical examples. Other examples are the coexistence of doped and undoped regions in a conjugated polymer, crystalline and amorphous domains, or disordered fibers. Tunneling between large metallic particles is temperature independent. The tunneling probability is just a function of the barrier width and the barrier height. The number of electrons on a particle will fluctuate thermally, however, and the electrostatic potential will also fluctuate when the particles are small. Therefore tunneling between small particles is a superposition of phonon-assisted and temperature-independent processes. Hence, the VRH equation has to be replaced by Eq. (9.27), which contains T_1 as an additional parameter:

$$\sigma = \sigma_0 \exp\left[-\frac{T_0}{T_1 + T}\right] \tag{9.27}$$

It leads to constant conductivity for $T > T_1$ and to activated behavior for $T < T_1$, with the parameters σ_0, T_0, and T_1 depending on the geometry of the tunnel barrier and the size of the conducting particles. Experimental data on heavily doped polyacetylene (Type S) and a fit of Eq. (9.27) are shown in Figure 9.65.

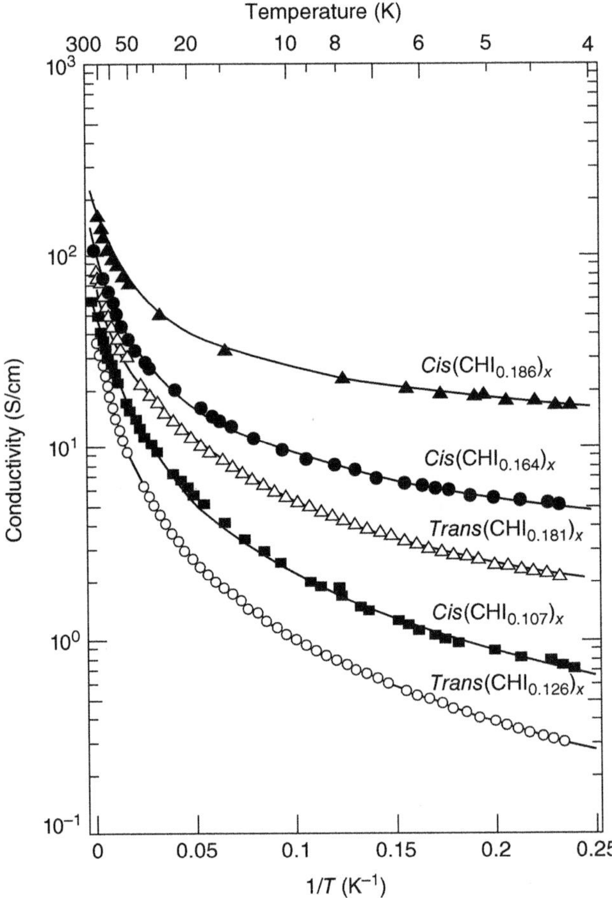

Figure 9.65 Temperature dependency of the DC conductivity of heavily (iodine) doped polyacetylene and fit of the model of fluctuation-induced tunneling. Solid curve: theoretical fit.

9.12 Highly Conducting Polymers

We mentioned that there are two types of conducting polymers at least in polyacetylene: Type S, in which charge transport occurs by hopping, and Type N, where there is band conductivity. Here we examine and discuss Type N polyacetylene. It should be pointed, however, out that there do exist other heavily doped, metallic-like polymers [61] (in the sense that the conductivity stays finite as $T \rightarrow 0$). But it isn't clear if these polymers represent the N-type nature of polyacetylene. Generally though, our discussion should apply to any polymer that may exhibit band-like conducting behavior.

At the ICSM meeting in 1986 in Kyoto, Naarmann announced that he modified the Shirakawa synthesis [62] and prepared a highly stretchable version of polyacetylene. After iodine doping conductivity values of about 100 000 S/cm could be obtained. In the following years he would even further increase the conductivity [63]. Naarmann sent several samples to the University of Bayreuth, and while none of these did exhibit conductivity values above 100 000 S/cm, $\sigma > 80\,000$ S/cm was confirmed [3].

Naarmann's idea was to slow down the polymerization process by aging the catalyst and by dissolving it in a viscous solvent such as silicon oil. Slower polymerization should lead to polymer films of higher perfection. Following this route, Tsukamoto et al. [49d, e] further improved the synthesis by using decaline as solvent. Subsequently many research groups tried to repeat Naarmann's or Tsukamoto's syntheses, but never succeeded to reach conductivity values of 100 000 S/cm or above. Today it is generally accepted that 20 000 or 30 000 S/cm can be obtained in a reproducible way, but to go beyond requires lucky circumstances that are not yet accessible to quantitative analysis. To obtain 30 000 S/cm polyacetylene, two further routes exist: either the catalyst is dissolved in cummene followed by removing most of the solvent before polymerization ("solvent-free route"), or the solvent is a liquid crystal, and the polyacetylene film grows epitactically (or quasi-epitactically) on its surface [64].

Some years ago the aim of a European joint project [65] was to synthesize polyacetylene with conductivities greater than 100 000 S/cm following one of the above routes, but 40 000 S/cm was the highest achieved. A further goal was to correlate the conductivity of highly conjugated polyacetylene with other measurable material parameters. But no correlation was found. Apparently, the conductivity is by far more sensitive to impurities and imperfections than any other parameter.

For an ideal single crystal of polyacetylene, we can predict its conductivity by theoretical means. The ideal crystal consists of "sufficiently long" well-ordered polyacetylene chains. It is doped to a high level, the dopant ions are between the chains, and their electrostatic potential is screened so that the electrons moving along the chains do not notice it. Because of the high doping and some interchain coupling, the Peierls transition is suppressed. This way a very anisotropic metal is obtained. The electrons mainly move along the chains and hop only very seldom to a neighboring chain. In case electrons are scattered by phonons, it is predominantly umklapp scattering, in which the electron momentum is reversed. Because of energy and momentum conservation, in highly anisotropic polyacetylene there are only very few phonons that can participate in umklapp processes, namely,

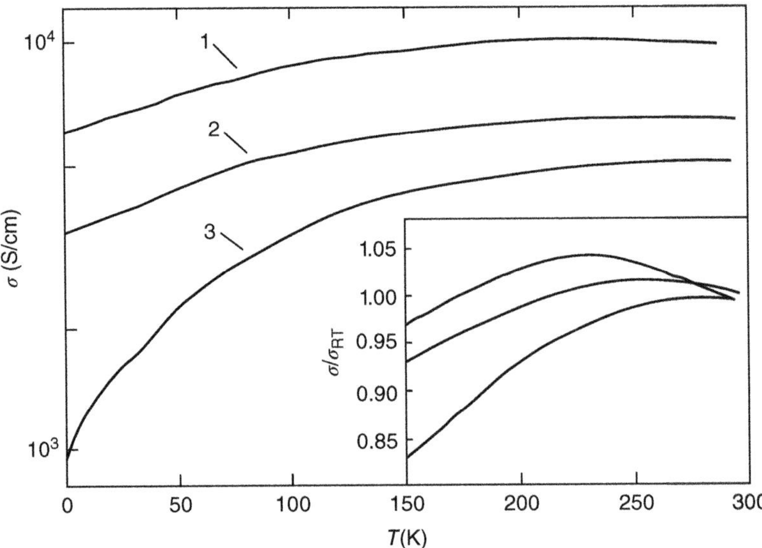

Figure 9.66 Temperature dependency of the conductivity of highly conducting polyacetylene samples. The curves refer to different iodine concentrations.

those phonons whose wavevector is twice the Fermi wavevector of the electrons $\boldsymbol{k}_F$. Because of the lack of appropriate phonons, the "theoretical" conductivity of polyacetylene is very high. Estimates yield as much as 2×10^7 S/cm at room temperature, which is about 30 times the conductivity of copper [66]. Moreover, $2\boldsymbol{k}_F$ phonons have a fairly large energy (large, compared with the sample temperature), so they can easily be frozen out; the conductivity of ideal polyacetylene will rise exponentially, approaching infinity at absolute zero.

Of course, the theoretical limit is not observed in experiments. But, at least for some samples, the temperature coefficient of the conductivity is negative [67]. Figure 9.66 shows a plot of the temperature dependence of the conductivity for several iodine-doped Naarmann polyacetylene samples. In the inset the maximum around 250 K is clearly seen.

There are several models assuming the coexistence of highly and less conducting domains in polyacetylene. The model of FIT has already been mentioned. Kaiser [68] proposed other models for heterogeneous polyacetylene samples. These models evaluate the thermopower as well.

The high conductivity of doped conjugated polymers has stimulated many ideas for application. Many of these lie in flexible and printable electronic and electro-optic devices. However, to realize these applications, a deeper understanding of charge transport is a matter of first priority.

9.13 Magnetoresistance

In a magnetic field the Lorentz force bends the paths of the charge carriers as discussed in Chapter 8. This leads to a reduction of the effective mean free path and thus normally to a decrease of the mobility and to an increase of the resistivity. The change of the resistance in a magnetic field is called *magnetoresistance*. In most metals this effect is rather small, just a few percent at room temperature. It increases at lower temperatures and reaches some limit.

The magnetoresistance depends on the band structure, the details of the scattering mechanisms, the morphology and homogeneity of the sample, and even on the shape of the sample. Evidently, dimensionality will play an important role. Interestingly, there are several conditions where *negative magnetoresistance* can occur. This means the resistance decreases when a magnetic field is applied.

Giant magnetoresistance (in the order of 100%) and *colossal magnetoresistance* are found in certain heterogeneous (layered) systems and are used to read data in hard disk drives and to store data in *magnetoresistive memories*. (The 2007 Nobel Prize in physics was awarded to Albert Fert and Peter Grünberg for the discovery of *giant magnetoresistance*.)

Let's look at some examples of magnetoresistive behavior in conducting polymers and carbon nanomaterials. In Figure 9.67, the resistance change of iodine-doped polyacetylene films in magnetic fields up to 7 T is plotted [69]. Depending on the doping level, we see both positive and negative magnetoresistances in the range of a few percent. As stated above, a positive magnetoresistance would be expected from bending the path of charge carriers by the Lorentz force. But also in hopping conductivity, we would expect a positive magnetoresistance, because the extension of the localized states will shrink and the electrons will have to hop farther. To explain a negative magnetoresistance, we have to evoke such effects as reduced scattering on the surface of polymer fibers (bundles) or week localization and anti-localization [70].

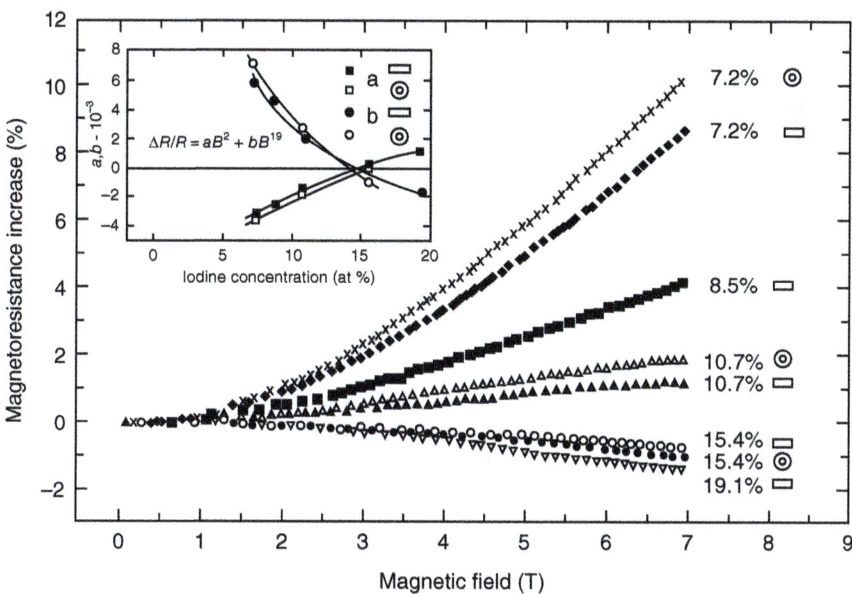

Figure 9.67 Positive and negative magnetoresistance in iodine-doped polyacetylene at 4.2 K [69].

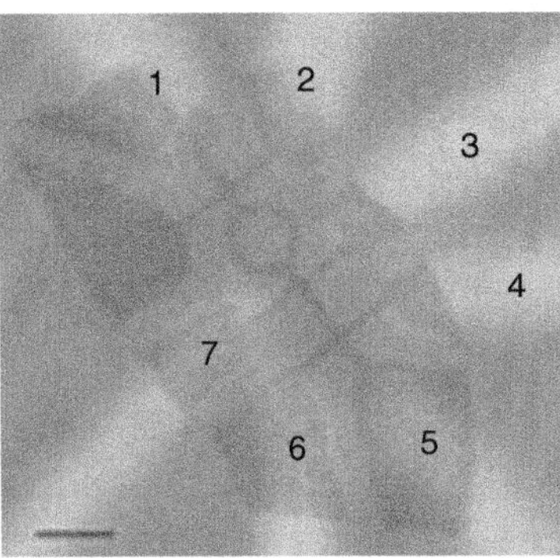

Figure 9.68 Graphene ring and contacts to study the Aharonov–Bohm effect (scale bar = 1 μm) [71].

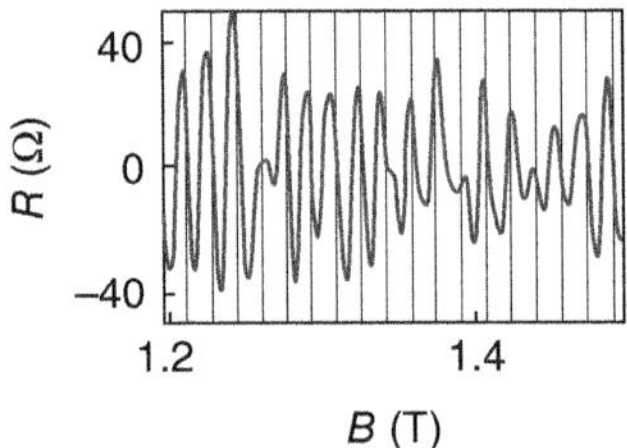

Figure 9.69 Aharonov–Bohm oscillations in graphene [72].

A simple argument for negative magnetoresistance is associated with interference of Bloch waves. Think for a moment of the heterogeneous spaghetti morphology we have shown for polyacetylene. This morphology can be seen as tiny loops throughout the volume. Of course, carriers moving through these loop-like conduits will interfere with each other from junction to junction creating an effect not unlike our modified, low-dimensional Drude metal of Chapter 8. A change in the overall interference pattern will be seen in a magnetic field because the different branches of a conducting loop in a magnetic field will pick up different phase shifts. This is the famous *Aharonov–Bohm effect*, applied over and over again throughout the heterogeneous structure. Figure 9.68 shows a micro-loop of graphene, patterned into graphene by electron beam lithography and plasma etching [71]. A very nice example of Aharonov–Bohm oscillations in graphene is shown in Figure 9.69 [72]: due to the periodic change between constructive and destructive interference where the two branches of the loop merge, the magnetoresistance oscillates, and depending on which side of the peaks we are, the magnetoresistance is either negative or positive. Of course our disordered network will average this effect over many, many loops and junctions. But if the fibers are "sufficiently crystalline" and the Bloch waves sufficiently coherent, negative magnetoresistance is a reasonable expectation. Similar arguments hold for disordered metals and semiconductors, where there are conductive loops around defect clusters.

A different kind of oscillation is seen in context with the quantum Hall effect. In that case there are steps in the Hall voltage and oscillations in the magnetoresistance (*Shubnikov–de Haas Oscillations*) whenever the Lorentz force bends the electron paths to closed circles and the circles hit the Fermi surface [73].

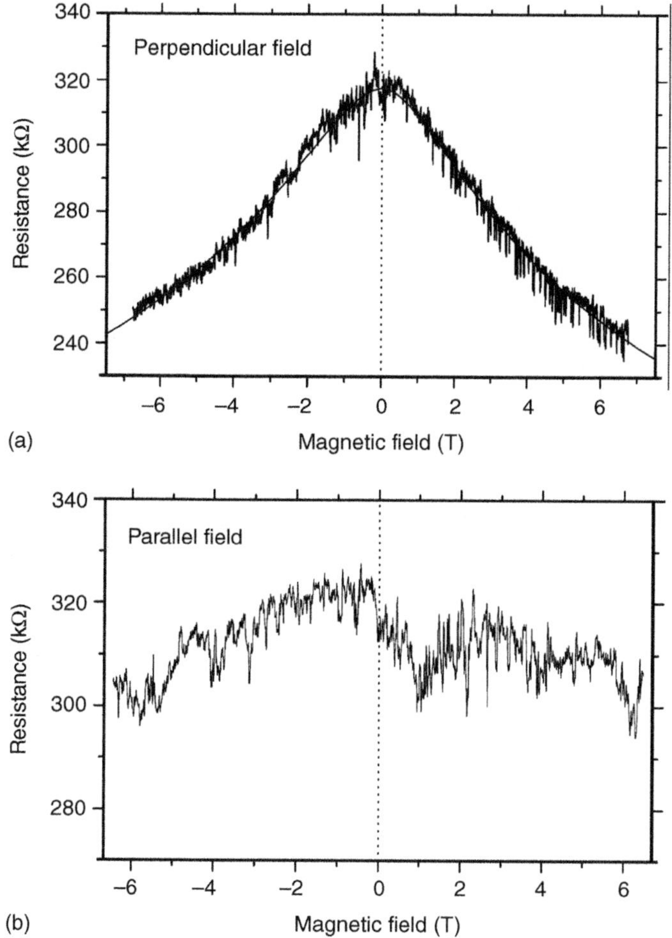

Figure 9.70 (a) Perpendicular and (b) parallel magnetoresistance of a SWNT rope, dependent on the orientation of the magnetic field. The measurements were carried out at liquid helium temperature [74].

The magnetoresistance of a rope (bundle) of single-walled carbon nanotubes is shown in Figure 9.70 [74]. If the magnetic field is perpendicular to the bundle, a pronounced negative magnetoresistance is observed, some 25%, perhaps again due to weak localization. In parallel field orientation the effect is smaller, and some oscillation is superposed, perhaps an indication of the Aharonov–Bohm effect (electron paths clockwise and counterclockwise around the tubes?) or universal conductance fluctuations? (Figure 9.70).

A very special experiment can be carried out on an individual single-wall carbon nanotube: *magnetochiral anisotropy* [75]. When a current is passing along carbon nanotubes, depending on how the nanotubes are rolled ("chiral angle"), there is a helical component of the current, and the nanotube acts like a tiny solenoid (Figure 9.71). This solenoid produces a magnetic field, and depending on the direction of the field with respect to the current flow, the solenoid field

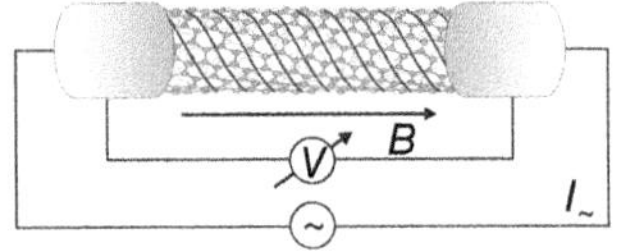

Figure 9.71 A very special example of magnetoresistance: *magnetochiral anisotropy* [75].

is added or subtracted from the applied magnetic field. The carbon atoms on the tube experience the total field, and the magnetoresistance depends on the direction of the current flow. The effect is very small, but it can be seen in the experiment, and the distribution of added or subtracted fields corresponds to the distribution of the nanotube chiralities.

Another magnetotransport measurement on individual nanotubes is the investigation of spin valves. The device is schematically shown in Figure 9.72a [77]: here the carbon nanotube is put in field-effect transistor configuration. But now the metal contacts forming source and drain are *ferromagnetic*, e.g. cobalt. From the magnetized source contact polarized electrons flow into the nanotube, i.e. all the spins of the conduction electrons point in the same direction. If the electrons don't change their spin (depolarize) while moving along the tube, they can only continue to the drain contact, if this is magnetized in the same way: the device is in the low resistance state. If source and drain are oppositely magnetized, polarized electrons cannot flow, and the device is in the high resistance state. If we put the device into a magnetic field and sweep the field strength, source and drain will flip their magnetization at different field strengths, because of different coercive forces in source and drain (due to slight geometrical differences in the contacts). The magnetoresistance behavior we expect to see is shown in Figure 9.72b, and what we actually see is depicted in Figure 9.72c.

To observe the signal in Figure 9.72c, the nanotube had to be cleaned, i.e. a fairly high electric current had to be passed through the nanotube so that it would warm up by Joule heating ("current-induced annealing") and adsorbed gas molecules would leave. Adsorbed gas molecules act as scattering centers, and in these scattering events the electrons can change their spin (spin-flip scattering).

Perhaps you have noted that in this section we have been talking about two different interactions of electrons with magnetic fields. In the first part the Lorentz force was important. This acts on the motion of the electrons: it bends the path of the electrons in an "orbital" effect. In the spin valve the field acts on the spin of the electrons. In general, a magnetic field will influence both spin and motion of the electrons.

In a strictly one-dimensional system, the path of the electrons cannot bend. So there should be spin effects only. And if conduction is by spinless solitons, there should be no magnetoresistance! Young Woo Park from Seoul National University and his team [78] compared the magnetoresistance of very thin and very well-ordered polyacetylene fibers with that of similar fibers of polyaniline and of polythiophene. Under certain conditions they found zero magnetoresistance up to 35 T in polyacetylene and 20–40% magnetoresistance in the other polymers.

In Park's experiments [73], electrical and magnetic fields and strain fields in the sample act together on the charge carriers typical for conjugated polymers: solitons, polarons, bipolarons, singlet excitons, and triplet excitons. A similar situation should exist in *field-induced organic light emitters* discussed below.

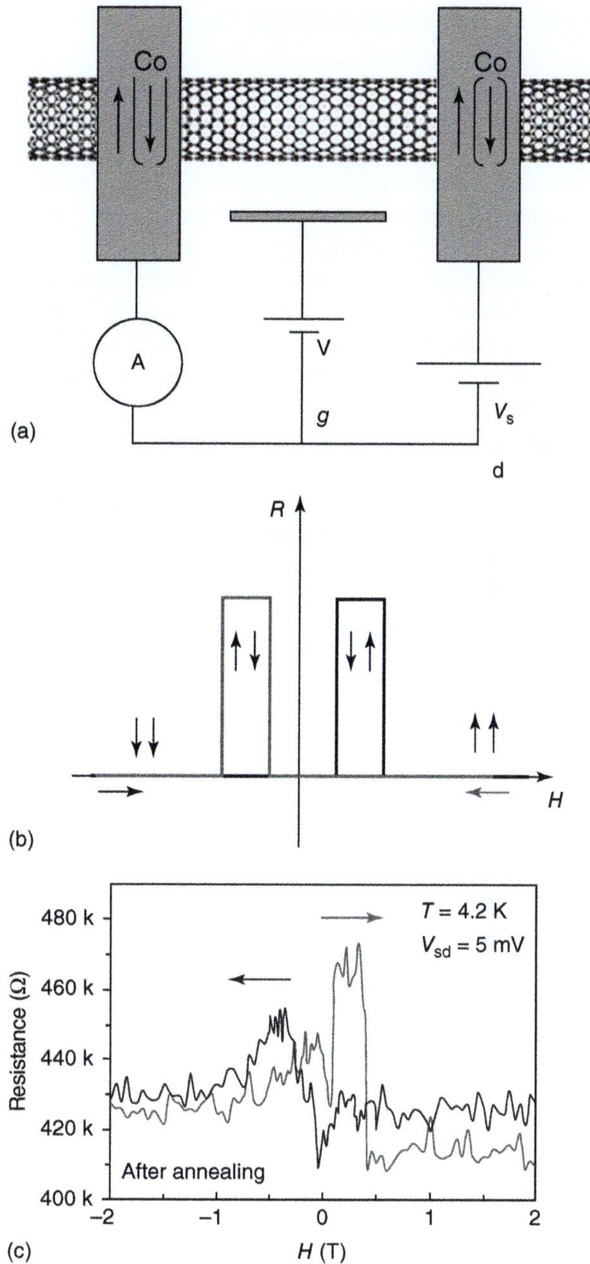

Figure 9.72 (a) A carbon nanotube spin valve: the device is like a nanotube field-effect transistor, but source and drain are ferromagnetic metals [75]. (b) Signal expected from a perfect spin valve [75]. (c) Signal of a real carbon nanotube spin valve [76].

They are driven by high frequency (some 50 kHz) AC electric fields, which couple to magnetic fields (as AC fields do according to Maxwell's equations). Recently a negative magnetoresistance has been suggested in these systems, and the negative magnetoresistance might be related to singlet–triplet conversion and to the high light output of the organic emitters [79].

9.14 Organic Molecular Devices[3]

There is more than one use of the term *molecular electronics or organic electronics*. In this text we tend to mean the use of synthetic metals and organic semiconductors to create micron-scale electronics. This is intended to distinguish from *molecular-scale electronics*, which refers to the use of individual molecules as active electronic elements. Molecular materials for electronics are not new of course. Even the etymological roots of electronics lead to a molecular material: to the natural resin amber, which in Greek is "electron" and where electrical phenomena were first observed (charging by friction). Commonly molecular materials have served as insulators. Other well-known molecular materials for electronics are photoresists used in the photolithographic process for the production of microelectronic devices. Of course such materials are most interesting when they represent the *active* part of a device or circuit; that is, they must be *electroactive (conjugated)* or *photoactive*. But given what we have just learned above, how do we implement such materials in electronics? Does it work in an analogous way to semiconductors? Exactly what are these organic devices?

9.14.1 Molecular Switches

There are an enormous number of *switching molecules*. In fact, every molecule has several excited states and transitions between these states can be regarded as "switching." Some molecules are bistable and are switched back and forth by external triggers; others need a trigger to switch only one way; they switch back spontaneously. The latter case is rather trivial. An example is the absorption of light and the subsequent decay of the excited state. But even this trivial behavior can be used for data processing (like the polyacetylene holographic computer). In this section we present two switching molecules: one because of its relationship to conjugated polymers and the other because of its complex conformational changes during switching – yielding interesting physics.

In Figure 9.73 the chemical structure of 7-piperonyl-7′, 7′-diapocarotene-7′-nitile (PCNC) is shown; a D–π–A molecule with a piperonyl group as donor and a cyano group as acceptor. As indicated before the excitation of this molecule corresponds to pushing an electron from the donor toward the acceptor. An additional electron on a polymer chain behaves like a soliton, and quantum chemical calculations show indeed that a phase-slip center in the charge distribution is created [80]. Solitons lead to states in the gap, which can be seen in optical absorption spectra.

Figure 9.74 shows the results of an investigation of photoinduced absorption in PCNC [81]. The sample is irradiated by light with quantum energy larger than the π–π^* gap so that many molecules are pumped into the excited state. The absorption difference between pump-on and pump-off is recorded.

---

3 Strictly speaking this section is not about transport at all. It is included to give some insight into how the application of polaronic materials might compare with those that utilize a more "bare charge" conduction mechanism.

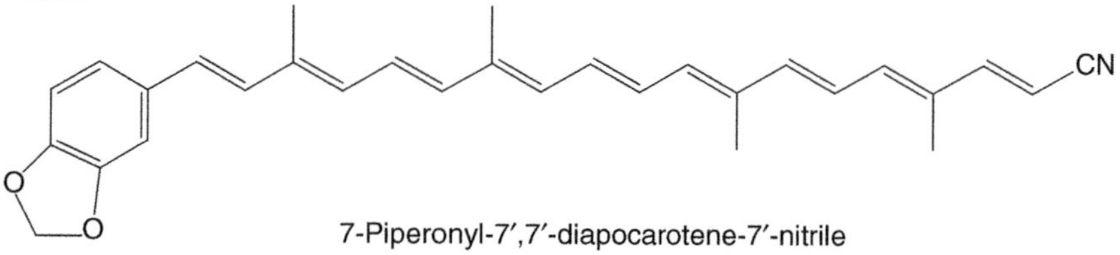

PCNC

CN

O

O

7-Piperonyl-7′,7′-diapocarotene-7′-nitrile

Figure 9.73 Donor–acceptor polyene, a D–π–A molecule for photoinduced absorption.

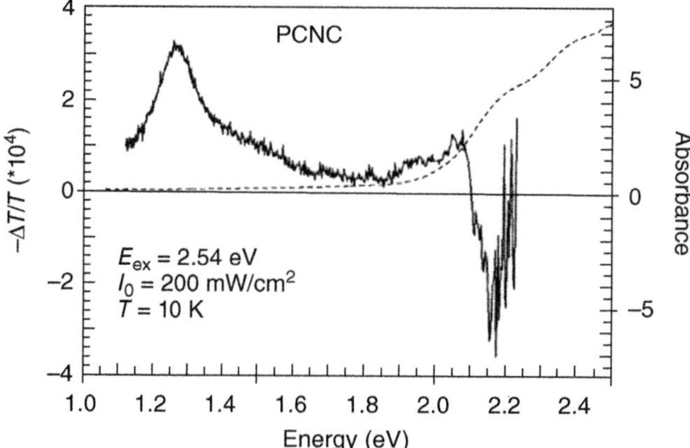

Figure 9.74 Photoinduced absorption of donor–acceptor polyene. The dashed line corresponds to normal absorption. Important features are the bleaching dip at 2.2 eV and the photoinduced peak at 1.2 eV, which corresponds to states within the π–π* gap [2].

The figure shows a photoinduced peak at about 1.2 eV and a bleaching dip, which is due to the reduction of the ground-state population by constant pumping. The peak results from states in the gap that do not exist in the ground state. It is reasonable to assign this peak to "solitons." We "switch" the molecules by irradiating with light, and we "read" the state of the molecule by absorption spectroscopy. The purpose of the experiment is not, however, to demonstrate the possibility of molecular switching. We want to investigate the excited state. From the position of the photoinduced peak, information can be obtained on whether the soliton concept can be applied reasonably well to short-chain polyenes, whether it will survive the attachment of donor and acceptor groups at the chain ends, etc. Short polyenes are usually treated by *ab initio* calculations [82] that do not explicitly make use of solitons or polarons. However, the short-chain approach must continuously fit to the long-chain approach.

The bianthrone molecule is shown above. The ground state is planar and puckered, and the excited state twisted and unpuckered. Puckering and twisting can be followed from the temporal evolution of the absorption spectrum. For this purpose the sample is irradiated by a short and intensive light pulse (1 ps, 1 mJ) that triggers the switching, and then wideband absorption spectra are taken in intervals of several picoseconds or nanoseconds. The experiment requires a fast optical pump-and-probe setup and is about as sophisticated as the soliton

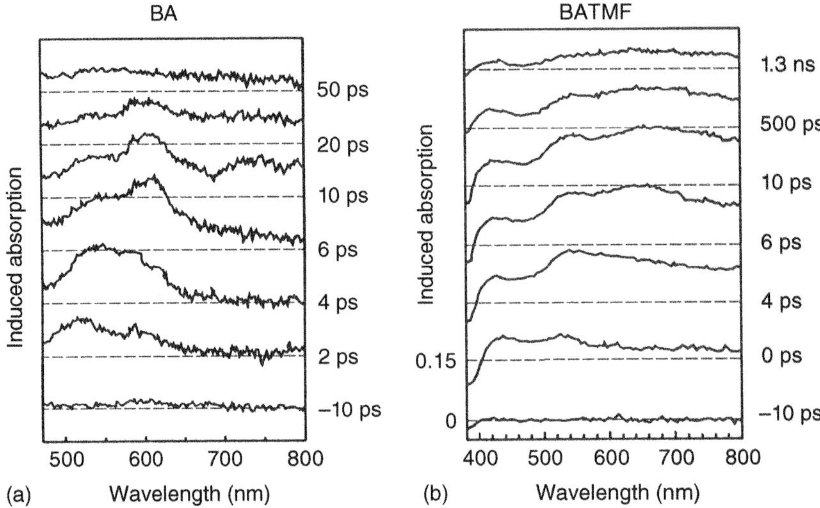

Figure 9.75 Temporal evolution of the absorption spectra of unsubstituted, BA (a) and substituted, BATMF (b) bianthrone. The time indicated is the delay of the white light pulse with respect to the switching pulse (pump pulse) [83].

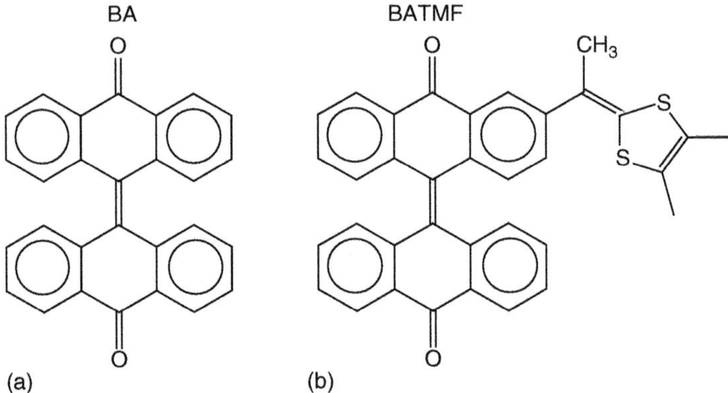

Figure 9.76 Chemical formula of (a) unsubstituted bianthrone (BA) and (b) bianthrone substituted with tetramethylfulvalene (BATMF).

lifetime investigation described in Section 5.9 (short and intensive laser pulses, picoseconds and nanoseconds delay lines, conversion into picoseconds white light pulses, diode array as spectrometer). Figure 9.75 shows the comparison of the time evolution of the absorption spectra of substituted and unsubstituted bianthrone (BATMF and BA, respectively).

The chemical structures are shown in Figure 9.76. The features in the absorption spectra can be assigned to twist and pucker changes [83]. As a striking result, we see that BATMF substitution stabilizes the excited state as shown in Figure 9.77. Systematic studies of this type should finally help to design molecules with time constants optimized to the intended application.

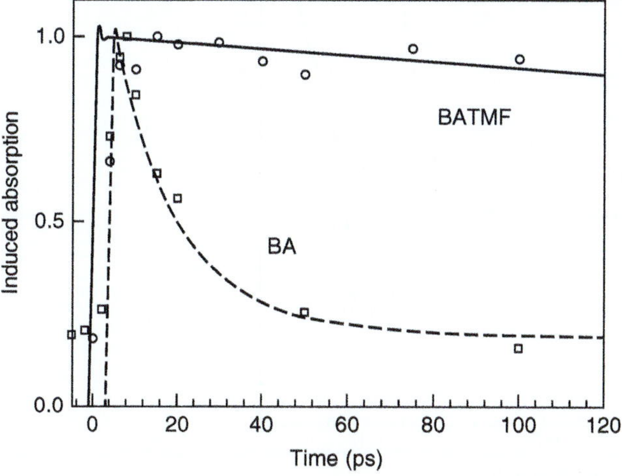

Figure 9.77 Decay of the excited state in substituted bianthrone (BATMF) compared with that of the unsubstituted bianthrone (BA) as an example of molecular engineering for optimizing decay time constants [83].

9.14.2 LB Diodes

An elegant use of heterojunctions based on Langmuir–Blodgett (LB) films has been demonstrated by Fischer et al. [84] wherein gold microelectrodes are covered with a few LB layers of a palladium phthalocyanine derivative, followed by layers of a perylene derivative, and finally followed by evaporated gold top electrodes. This last step must be done without destroying the delicate organic structure underneath (Figure 9.78). The device structure is small, so several of the diodes are placed on the silicon chip substrate to ensure pinhole-free can

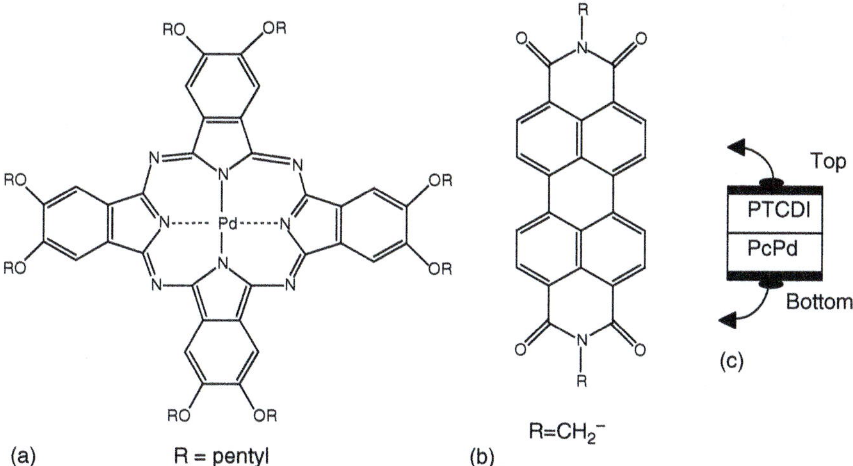

Figure 9.78 Rectifying organic heterolayers. (a) Palladium phthalocyanine. (b) Perylene derivative. (c) Arrangement of sandwiched layer structure. The substituent groups R in (a) and (b) have been attached to facilitate film formation in the LB technique.

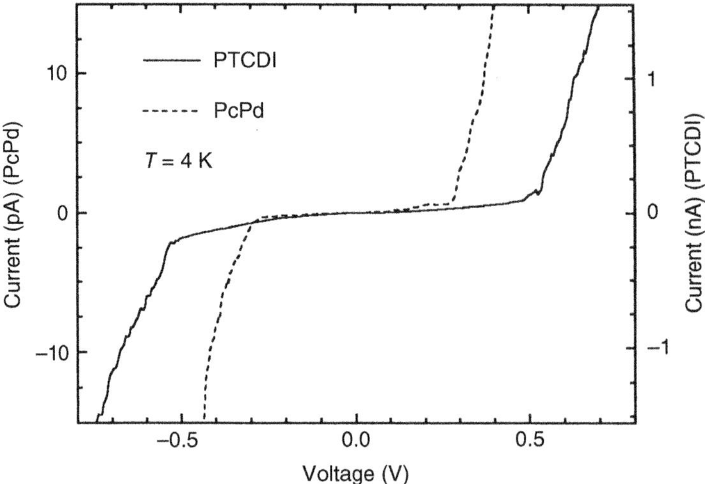

Figure 9.79 Symmetric current–voltage characteristic of Au/PcPd/Au and Au/PTCDI/Au sandwiches. The structures block up up to a threshold voltage and conduct above.

be achieved and current–voltage characteristics can be measured for multiple devices.

Figure 9.79 shows the current–voltage characteristics of Au/PcPd/Au (dotted lines) and of Au/PTCDI/Au (solid line) devices. These characteristics are symmetric, and current is blocking up to a well-pronounced threshold. The threshold is interpreted as the coincidence of the Au Fermi level with the π or π^* band edge in the organic layer.

Figure 9.79 shows the asymmetric characteristics of an Au/PTCDI/PcPd/Au device, with a positive threshold of 0.9 eV and a negative threshold of −0.5 eV. At a "working point" of 0.6 eV, a rectification ratio of several orders of magnitude is obtained. Additional interesting features are the steps in the I–V characteristic shown in Figure 9.80. These are interpreted as single-electron effects.

Tentatively the π electron systems of the phthalocyanine and perylene are summed as a system of quantum dots charged by single electrons and blocking further electrodes from following (unless the voltage is raised high enough to surpass the Coulomb barrier, hence the steps). Here a caveat has to be added: if the evaporated top electrode is not perfect, small gold particles might migrate into the LB film, and the features of Figures 9.79 and 9.80 can also be interpreted as tunneling between the electrodes and the gold nanoparticles.

9.14.3 Organic Light-Emitting Diodes

There are, in fact, several classes of light-emitting devices based on organic materials. The first type, an important one, is the OLED, which is the simplest of the active light-emitting devices discussed below and now commercially available in cell phones, displays, and some lighting applications. The second type of device is the AC-driven, field-activated organic emitting device. These devices are a much more recent development and have some unique features of their own. Finally,

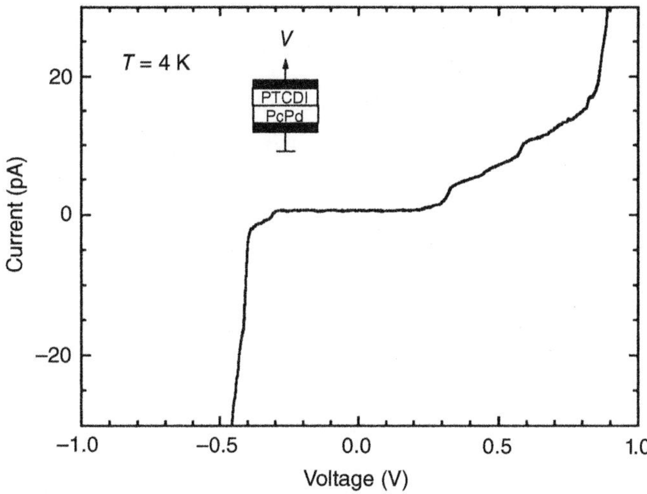

Figure 9.80 Asymmetric current–voltage characteristic of a Au/PTCDI/PcPd/Au heterostructure. Rectifying properties are observed when the device is operated between 0.5 and 0.9 V. The steps in the characteristic are interpreted as single-electron charging of the organic macromolecules.

Figure 9.81 The candle is among the first "organic light-emitting devices." A standard candle burning at about 0.1 g/min delivers roughly 80 W of heat and ~13 lm of light. This yields a luminous efficacy of roughly 0.16 lumens per watt (LPW). Today some still use the candle as a standard of measure – candela and candle power.

the third is organic laser. Electrically stimulated organic lasers have yet to be achieved, and there are some daunting barriers to overcome for this. However, optically stimulated organic lasers are widely studied. Notice that here we leave out chemiluminescent emitters since they are based on entirely different principles. Likewise, an obvious omission is the candle, which certainly qualifies as an organic light emitter (Figure 9.81).

Each of the classes cited above has a common underlying mechanism we are interested in that ties them with the dimensionality of the organic material, that is, the creation of excitations, polarons and excitons, and their subsequent decay with the emission of light. Organics, or if you prefer synthetic metals, present a slightly more complex set of circumstances than what occurs in inorganic solid-state light-emitting diodes or LEDs. This comes from excitations that can occur as singlet spin states or triplet spin states, and these have different lifetimes and recombination dynamics. There are other, more subtle differences as well. In this section we will examine the basic mechanisms one is faced with in organic emitters. But for a more detailed examination, the reader is encouraged to read the many treatises on organic emitters now available [85].

9.14.3.1 Fundamentals of OLEDs

OLEDs are the first molecular electronic light-emitting systems to reach the consumer market (organic lasers have been used in research labs for longer). OLED technology is now established in cell phone displays, flat panel displays, car dashboards, handheld electronics, and other applications. However, this doesn't mean that basic research has stopped. Indeed, since their introduction, growth in the research field has continued. Specifically improvements to lifetime, processing (for cost), flexibility, and robustness against laminate failure, as well as efficiency and color control (white), are being pursued. Figure 9.82 shows a schematic view of a standard research device.

The overall design of an OLED is conceptually quite simple. A thin organic film is sandwiched between two electrodes, one of them semitransparent and one reflective. A voltage is applied: one electrode injects electrons and the other injects holes into the film. The polymer lattice relaxes, and electrons and holes form solitons or polarons or whatever electron–lattice coupling is required. These charged carriers flow through the volume of the film as allowed by the applied potential, until a positive and a negative meet. Then the charge carriers form a weakly bound, short-lived pair (exciton) that will ultimately recombine, and light is emitted (luminescence). Generally speaking one must be careful to ensure that the positive charge carriers (holes) and negative charge carriers (electrons) meet in the middle of the emitting materials. This is known as the *recombination zone*

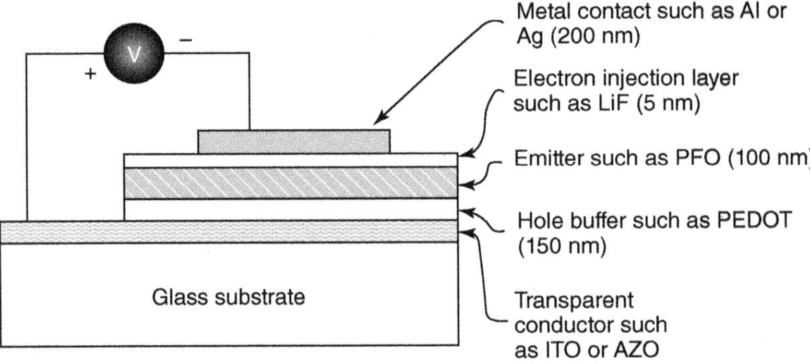

Figure 9.82 Schematic setup of an organic light-emitting diode (OLED).

of the device. However, electrons and holes (or their respective excitations) might move at very different rates in the materials. That is, they usually have different mobilities under the applied field. Holes are typically faster, so additional buffer layers are used to balance this charge mobility problem and to provide "energetic confinement" within the emitting layer. In Figure 9.82, we show the use of LiF to match the work functions of the injected electrons from metal to the emitter. This helps to lower the operational voltage and block holes from leaving without recombining. The use of polyethylene dioxythiophene (PEDOT) against the ITO does the same thing for the holes plus it gives the holes a long path to get into the emitter, allowing time for the slower electrons to make it into the middle of the emitter before they arrive. As mentioned earlier, recombination requires that electrons and holes be proximal, so if they happen to miss each other on their flight through the emitter, the buffer layers help to confine them by providing a "step-up" in potential energy. So the basic design of the OLED structure is really a balance of these three layers. If other colors are desired, then the emitter can be coupled with multiple emitters, and the structure becomes more complicated.

The injection process is outlined diagrammatically in Figure 9.83. There are two recombination channels: a radiative and a non-radiative one. For efficient OLEDs the radiative channel must be more pronounced. Non-radiative recombination generates heat and limits the lifetime of the device. Furthermore, carriers should not get lost by falling into traps or (as discussed in Section 9.14.3.1) by transferring to the opposite electrode without meeting a partner and recombining. Finally the polymer and other parts of the device must be sufficiently transparent to the generated light so that it does not get reabsorbed before leaving the OLED. Here we do note one fine point in our nomenclature. In semiconductor physics LED means light-emitting *diode*, and the diode is a p-/n-junction or Schottky junction. In this book we are discussing organics more broadly and so we use the more general expression: light-emitting *device*. With this interpretation of the letter "D," we want to indicate that the organic-emitting film might be a p-type

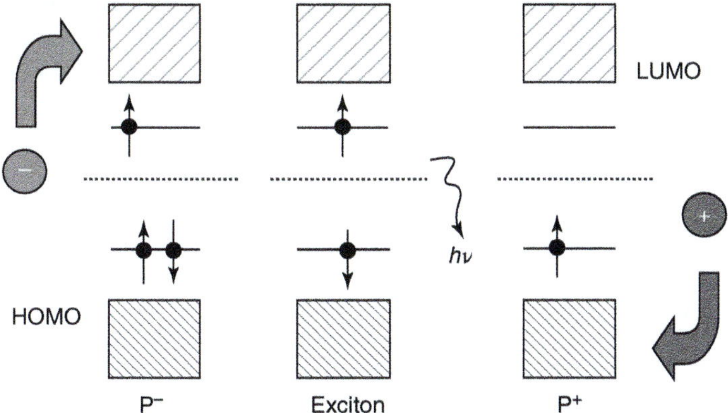

Figure 9.83 The injection of positive and negative charge results in charged polarons. The polaronic states sit within the HOMO–LUMO gap of the active material as discussed in Chapter 7. Radiative recombination yields light.

or n-type semiconductor, but it also could be an intrinsic semiconductor with an equal concentration of electrons and holes. In particular, both types of carriers will be injected for the OLED to work. Then the device would not be rectifying unless electrodes with different work functions are used.

In OLEDs we must consider two different types of organic emitters, those based on conjugated polymers and multilayer dye, or small molecule, OLEDs [86]. Reliable OLEDs were first reported by a group at Eastman Kodak: C.W. Tang and S. VanSlyke in 1987, using small molecules (Alq3) [87]. Burroughs et al. [31c] and several other groups [88] followed quickly with polymer-based devices. The behaviors and characteristics of these two types of devices are essentially the same although transport and recombination details do differ slightly.

OLEDs have several advantages over their inorganic counterparts. Among them is the ability to use simple processing steps – such as solution processing – to fabricate OLEDs, lowering overall costs. Another important advantage is the color tunability. Since the range of synthesis of the emitter molecule is practically unlimited, a very wide range of colors can be created in the emission spectrum. Aiding in this color tunability is the fact that many emitters can be created with relatively narrow or broad color emission, allowing for color overlap and color mixing when multiple emitter molecules are used. Recall that it takes more than a mixture of pure RED, GREEN, and BLUE, to create ALL the colors to which the human eye is sensitive.

The use of different color components actually comes in handy for other reasons. As noted previously, the excitons can form in a singlet or triplet spin state. Singlet states are relatively short-lived; they decay rapidly to yield visible light. However triplets are long-lived and decay through non-radiative processes. Generally, there are three triplets formed for every singlet, so this dramatically limits the potential efficiency of the "singlet only" device. In the late 1990s an elegant solution was found to address this problem [89]. Phosphorescent dyes containing a heavy metal complex was added to the emitting layer. These dyes were, for example, $Ir(ppy)_3$,[4] FirPic,[5] and others containing Ir. In these systems, the singlet excitons would decay quickly giving off light. But the triplet excitons would resonantly transfer their energy through *Dexter processes*[6] to the dye. Then, through spin–orbit coupling associated with the metal ion, the phosphorescent dye would emit light as well. This "secondary emission" can be very efficient, and it provides another complementary color to the singlet emission. Since the dyes are small and introduced in small amounts, there is typically no phase separation within the layer during operation.

In addition to the use of dyes, several other emitter additives have been widely studied. Again, as in the case of dyes, only small quantities of such impurities are blended into the emitter host. Among these, the use of nanoparticles such as quantum dots and carbon nanotubes are quite relevant to this discussion. Such additives can have multiple effects within the electroactive matrix such as

---

4 Tris[2-phenylpyridinato-C2,N]iridium(III).
5 Bis[2-(4,6-difluorophenyl)pyridinato-C2,N](picolinato)iridium(III).
6 See Exploring Concepts.

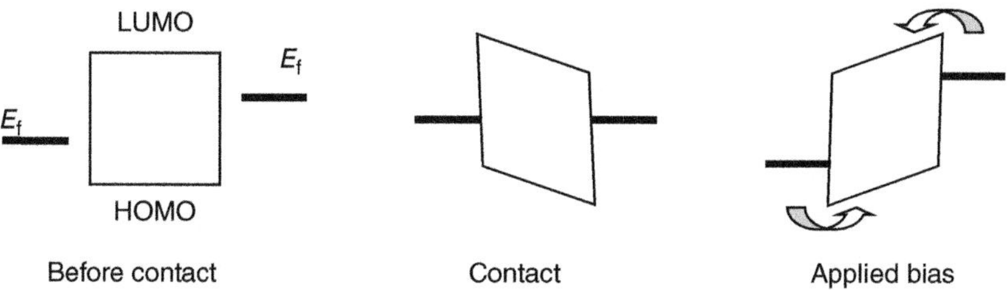

Figure 9.84 Band diagrams for applying contacts and subsequent injection of charge into organic layers.

providing color components, modifying energy exchange mechanisms through antennae effects, and modifying transport properties.

To more fully understand the operation of the OLED configuration, the problem is typically broken down into three parts: charge injection, charge transport, and charge recombination. Then the principles of *detailed balance* are used to account for each charge and each photon, giving the overall expected performance of the device. Doing this for double carrier injection has been a difficulty even in solid-state systems with long-range order. However, short-range order and hopping mechanisms make the problem considerably more difficult with organics. We explore these three parts schematically here.

Beginning with charge injection, we first imagine what local potential barriers to charge injection occur at the internal interfaces between the layers of the device. We mentioned the electronic efficiency factor above and this is related. Each interface may have trapping states and potential barriers that cost some of the applied voltage placed across the stack. Generally, the external contacts are present among the largest barriers. Shown in Figure 9.84 is an energy level diagram of a simple, single-layer OLED device, where the organic is depicted as a *fully depleted* semiconductor (i.e. no free charge). Notice that the bands are shown to be flat – no bending due to itinerant carriers.

Qualitatively, before the contacts are added, the Fermi levels of the cathode and anode are shown as E_f. When brought into contact with the semiconductor the Fermi levels of the contacts must align across the device. This results in a built-in potential, V_{bi}, across the organic layer. For an applied potential V_{app}, of less than V_{bi}, the electric field inside the organic layer opposes charge injection and forwards drift currents. When $V_{app} > V_{bi}$, current is injected over the barrier. Current density and irradiance in the device then increases rapidly. Notice that the level of the cathode to the polymer band edge might be quite different from the level of the anode to the band edge; thus to control the necessary voltage for injection, layers such as LiF are used to change the Fermi level of the contact material.

The exact *mechanism* of injection, that is, how the bare charge of the cathode and anode becomes the more exotic excitation of the organic system, is still not unambiguously understood. However, significant progress has been made in modeling the process since its first discovery. In fact, instead of the simple band as shown in our diagram, it is a little more precise to imagine the organic electronics as a distribution of isolated and localized states [90]. This absence of long-range order does have profound effects on all aspects of fundamental device operation: injection, transport, and recombination. Yet many concepts of the crystalline system can be transferred. For instance, organic/metal interfaces can be Ohmic or injection limited [91]. With very few exceptions, most attempts to model this injection have relied on Richardson–Dushman thermionic emission [92] or Fowler–Nordheim tunneling [93]. In more advanced treatments, the disorder of the polymeric materials, or the organic material, suggests that significant localization will occur near the injection electrode resulting in increased reflection probabilities, reduced injection rates due to dipole layers, and a steeper field dependence of the injected current. It is important however to note that not all simulations and models yield the same results. In fact, if injection can be modeled as tunneling for lower currents but thermionic at higher currents, then injection probabilities will vary widely depending on the regime of conductivity [94].

The second major step in understanding OLEDs is how the charge is transported in the organic layer. In fact, this general statement of transport in such materials has relevance well beyond connections to electronics applications [95], and we have discussed these mechanisms in this text. Recall that charge transport is typically viewed as a hopping process, where hopping takes place between sites that are statistically different in their surroundings. Formally, theories such as Bässler's (where disorder is distributed using a Gaussian function) have been quite successful in describing the overall features of transport [96]. The primary feature of such models is that the mobility of the carriers will be dependent on the applied electric field and is usually written as

$$\mu = \mu_0 \exp[(\beta q/(k_B T))\sqrt{F}] \tag{9.28}$$

$$\beta = \text{Poole} - \text{Frenkel factor}, F = \text{applied field} \tag{9.29}$$

Reflecting on our previous models for injection, it is quite easy to see that such a system should result in field-dependent mobilities. States introduced into the gaps will have trapping and detrapping times dependent on the applied field. If the mobility of the carrier is dominated by these trapping–detrapping events, then the overall effect will be to build in a nonlinear field dependence to the time of flight through some part of the material.

Finally, in terms of recombination, there is little doubt that Langevin-type rate equations dominate in the organic systems [97]. Yet, again, recombination mechanisms on a microscopic scale are still lacking. We do know that, in analogy to crystalline systems, microcavity effects are observed in these systems. Modification spectrally and spatially of spontaneous emission rates, using microcavities, suggests further correlations with solid-state systems [98]. Naturally, spin states must be considered as well as de-excitation routes that include phosphorescence (following on with our discussion above).

9.14.3.2 Materials for OLEDs

Electroluminescence in poly(*para*-phenylene vinylene) (PPV) was discovered by workers in Cambridge in 1990 [31c]. Since that time, a vast number of polymers have been demonstrated with high efficiencies, multiple colors, and other features. In fact, emissive polymers can be thought of as falling into classes: the phenylene vinylenes, the thiophenes, the pyridines, the polyfluorenes, and more. Each of these is based on a modification of a more simple semiconducting polymer. Side groups and alterations to the chain are used to alter the overall performance, coloration, and solubility of the polymers. There are now complete catalogues available from manufacturers of their electroluminescent products. These will reflect different color options, different balances between hole and electron mobilities at specified applied fields, etc.

Likewise, the options for resonantly matched, metal-containing dyes have expanded almost exponentially since they were first introduced into OLEDs. Many of these are still based on Ir. So for the device builder, one now can look up a wide range of options for designing the emitter layer to nearly any color and internal efficacy one wants.

More recently, attempts at making pure white emitters from a single polymer or a small number of components have been the focus of many chemists. An interesting approach to this is the copolymerization of different color centers along the same polymer strand. In such an approach the charge is injected at the highest energy to enter the blue component bands. The carriers quickly thermalize to fill the lower-lying energies where they decay to give the colors of each of these levels. When balanced appropriately, the outcome is a white emission [99].

9.14.3.3 Designs for OLEDs

As we stated above, the process of designing an efficacious emitter comes with lots of options. But it should be noted that having a really efficient emissive polymer does not always lead to a good OLED. As with any solid-state electronic device, the engineering of the OLED must take into account contact barriers, lifetime, diffusion lengths, and more. In our discussion we have already hinted at this. Here we can be a little more precise as to the design features necessary for OLEDs.

Generally speaking, the work function of the electrodes must be matched to that of the polymer of choice. This will dramatically effect the voltage at which the device operates. UPS (or some equivalent technique) measurements of the work function are necessary. Since the work function difference between commonly used metals such as Al and that of the polymer can be large, strategies are usually employed to modify the metal. This is usually easily done with an addition of a small amount of alkali metal at the metal contact to manipulate the metal work function value to match a needed value. Caution must be used with this approach though since some alkali metals will react with the polymer yielding unintended consequences. On ITO, the transparent hole injector, it is a little more difficult. The addition of a PEDOT layer over the ITO can help with this only a little. Additionally, the PSS usually used to dope the PEDOT will interact with any water that leaks into the system. This forms an acid that will attack polymers [100].

Furthermore, hole and electron mobilities are not usually the same. So it is necessary to use layer thicknesses arranged such that the electrons and hole arrive in the center of the luminescent layer at the same time. This is not as easy as it sounds, and careful time-of-flight measurements are necessary to optimize the layer thickness. Really thick luminescent layers will lead to non-radiative recombination and a reduction in performance.

Finally, not all electrons and holes that make it to the center of the active layer will recombine (either radiatively or not), but rather they miss. So what happens then? This can lead to what is known as "leakage currents." These are the carriers that pass straight through the device or that do not recombine in some way and make it to the opposite electrode. Thus, most designers will choose buffer materials that do not allow for the opposite charge to be injected into them. For example, ITO is an excellent hole conductor and quite good at injecting holes. However, it is a poor electron conductor and does not readily allow for electrons to "leak" through this contact. Further blocking layers can be added to the device to help prevent this parasitic process. Some choices are shown in Figure 9.81. By blocking the leakage currents, the carriers are confined to the emissive layer giving them more time to interact.

9.14.3.4 Performance of OLEDs

With continuing research and the advent of phosphorescent OLEDs together with new polymer or small molecule emitters, the performance metrics of OLEDs are a moving target. What are these metrics? We usually think of power efficiency and brightness. But this is an incomplete picture of how such devices might actually compare with other lighting sources. First of all, let's leave cost out of the equation for a moment. Then certainly we would characterize the device by:

1. Internal quantum efficiency (IQE), which is how well each injected electron is converted into a photon.
2. Luminous efficacy of radiation (LER), which is power in and power out of the emitter.
3. External power efficiency, which measures power in and power out of the device.
4. Color coordinates for given operational parameters (using some recognized scale such as the CIE).
5. Brightness.

For modern applications in lighting, usually the power efficiency is among the most important starting points. Note that here we make the subtle distinction between efficacy and power efficiency. The efficacy is only the light emerging from the emitter and the power entering the emitter, whereas the external efficiency includes how much the physical device itself blocks the light from leaving the emitter and the resistance of the device to the power entering it (the electrical efficiency factor). To determine the power efficiency of a device, we measure how much power is put into the device ($P = VI$) and then the amount of radiant power emitted that is within the response of the human eye (also known as the photopic response). This measure obviously includes the effects of the physical structure. If all the power that went into the device comes out in exactly the photopic spectrum, then one would achieve the maximum possible power efficiency of 683 lumens per watt (LPW). So any lighting device operating at 100 LPW is producing only a fraction of the light it could produce at maximum efficiency. For white-emitting OLEDs, a power efficiency of 80 LPW operating at 1000 Cd/m^2

in brightness has been reported. Such devices have internal quantum efficiencies of ~85%. Recall from above that for pure fluorescent emission this IQE can be no higher than 25%, so obviously these devices utilize phosphorescence. These devices can be quite bright as well with thousands of Cd/m^2 observed.

Importantly, the lifetime of OLED devices must be considered. Operating in air, only a few hours of emission can be expected. This is because of oxidation of the polymers, delamination of the components, and exposure to atmospheric water. However, when encapsulated, many thousands of hours can be obtained from an OLED, even under high brightness conditions. Indeed, lifetimes, efficiencies, and brightness have improved to the level that such devices are now used commercially. But this has been primarily in pixel formats or backlighting (low light levels). The use of OLEDs in overhead efficient panel lighting is still progressing but poses a deeper challenge.

9.14.4 Field-Induced Organic Emitters

There does exist an alternative to OLED. We use the phrase "field-activated" or "field-induced" electroluminescence to imply that the carriers used to recombine originate *from within the device*, and not from the external contacts. These devices were first fully investigated by Lee et al. in 2005 [101] with a number of variants on the structure coming quickly afterward [102]. Depending on the exact structure and on the authors of the papers, there have been several acronyms proposed for such devices including field-induced polymer electroluminescent lamp (FIPEL), field-activated organic electroluminescent lamp (FA-OEL), and AC-OEL. Some even refer to them as AC-OLEDs though this is confusing because one can operate a regular OLED in AC mode. To be as general as possible, in this text we will use the latter: AC-OEL.

A basic schematic of the AC-OEL is shown in Figure 9.85a. The semiconducting polymer emitter (or small molecule organic emitter) is placed between two insulating layers: above and below. The device is a capacitor of sorts with the emitting material in the middle so that charge cannot be injected into it from the electrodes. The device is driven by a time-dependent voltage or AC. Thus such devices are high impedance with the power coupling capacitively to the emitter. This makes *driving* field-activated devices more complicated than in the case of OLEDs. Specifically, the power received by the device is the time average of the Poynting vector, and this depends on the total capacitance of the system.

To understand how this "field activation" of the polymer creates light, consider the above simple form (Figure 9.85a). When an AC field is placed across the emitting layer, a polarization current will be established, $J_p \sim \mathrm{d}E/\mathrm{d}t$. One can think of this in terms of bound charge density ($\sigma_b \sim -\boldsymbol{n} \cdot \boldsymbol{P}$) at the interfaces in the device. When the field becomes sufficiently strong, the surface dipoles of this polarization current can dissociate (by a tunneling process) and lead to "free" excitonic carriers. These carriers can recombine and decay in the usual fashion. However, the field strength for this to occur is typically dangerously close to the breakdown field of the materials. So, one might at first guess that such devices always "teeter" on the brink of destruction. But, in all the references cited for this section, the field-activated devices work quite well without breakdown! That is,

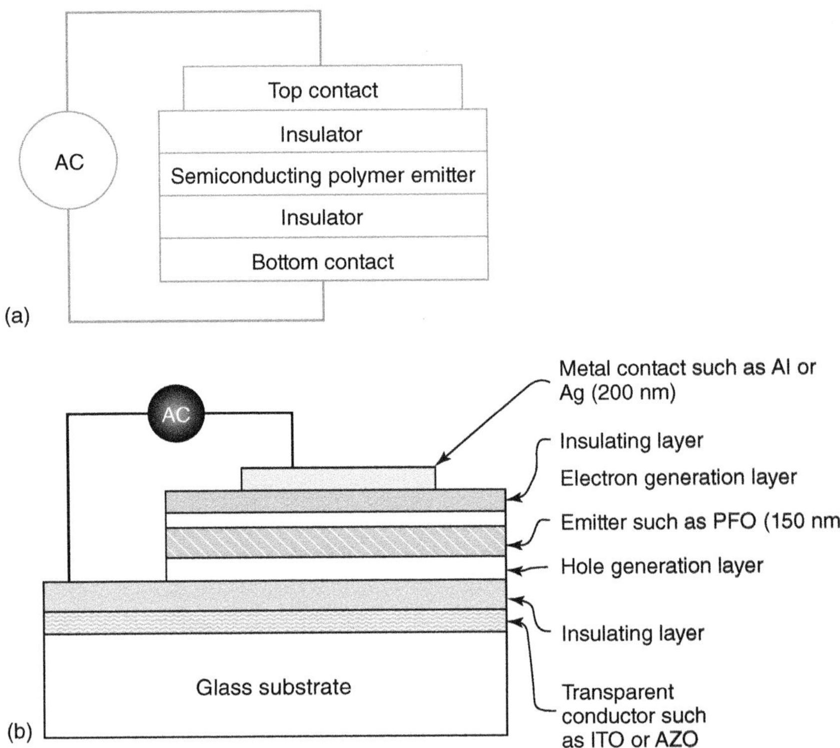

Figure 9.85 The field-activated organic light-emitting device. (a) A simple diagrammatic representation. (b) A research device typically used in literature.

at field densities lower than the breakdown fields of the insulators, they produce a stable light of reasonable brightness. Why? The answer comes from a modification of σ_b by interface states, together with defects within the polymer emitter. Defects and interfaces introduce localized electronic states within the gap of the emitter as we have already seen. Under the applied AC field, these become an additional component of the polarization current but are dissociated at lower fields (you should be able to determine why they "activate" at lower fields). These are the origin of the carriers, and the more of them you have, the brighter the device. Moreover, defects and interfaces provide for momentum conservation. We show these states in the band diagram of Figure 9.86. So for a semiconducting polymer placed between the two insulating layers and stimulated by a field density approaching 10^6 V/m, it will produce a low level, stable light at roughly the bandgap of the polymer. It is important to note that light production is typically asymmetric with respect to the input power cycle. This arises from the specific positions of the defect states and interface states, as well as the polymer chosen.

In the schematic of Figure 9.85b, we have also included charge generation layers (CGLs) as is typically seen in the literature [103]. These layers are used to enhance light emission and generally are not necessary to achieve a low light level from

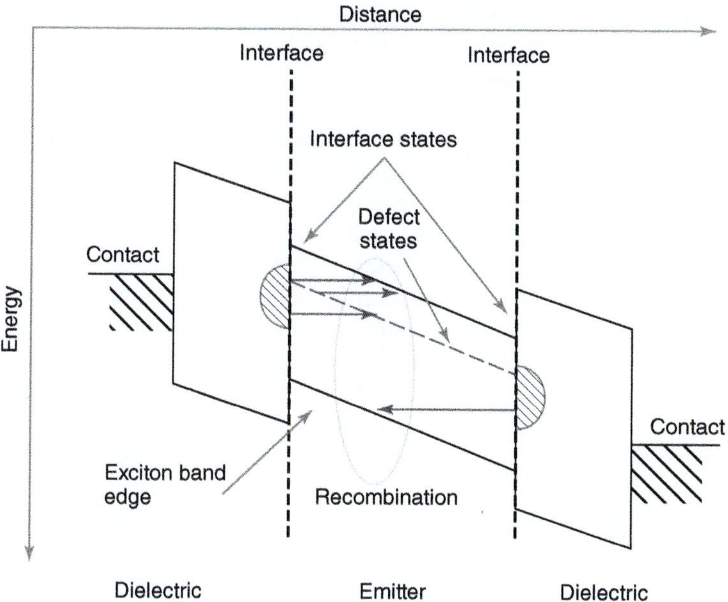

Figure 9.86 The proposed injection process for AC-OELs. This device uses a double insulating layer and an intrinsic semiconducting polymer. Such "non-doped" devices can produce up to 100 Cd/m^2 in any color that can be provided by an OLED.

such a device. However, they increase the number of carriers available to tunnel into the semiconductor by adding many more filled states at the interface.

Figure 9.86 shows a simple proposed band diagram for the "injection" process in an AC-OEL. This device is "neat," with no doping used to create "additional" defect states (as in OLED). It is important here to state that the process of recombination and light emission is similar to that of the OLED once the carriers are created. So the use of Ir complex dyes is also possible and has been demonstrated [104]. Curiously, however, whereas in the OLED the use of such dyes was strictly limited to <1% of the composition due to quenching, no quenching is observed in the AC-OEL until one reaches dye doping levels of around 30%! Several models have been suggested for this difference [105].

Of course, the light output of a neat AC-OEL is too low to be of interest to most applications. The enhancement of carrier concentration has been explored in several ways. To do this a high density of bandgap states must be added without dramatically decreasing carrier mobility. The approaches to this are varied. By adding nanoparticles to the emitter such as Au or carbon nanotubes (in a blend), one can concentrate the field locally while also adding a source of carriers [106]. Alternatively, the simple addition of highly electronegative and electropositive molecules blended into the emitter matrix can also introduce states. As in the case of nanoparticles, care must be taken to balance the injection of positive and negative carriers. Finally, the emitter layer can be externally "doped" by adding electronegative and electropositive layers such as shown in Figure 9.85 (CGL). Again, these must be chosen appropriately to lie within the bandgap of the polymer emitter and allow for facile charge injection into the emitting layer. This approach is a little more complicated, and models based on Zener tunneling from the CGL have been proposed [107]. This makes a lot of sense because injection can only occur under one polarization at a given interface – and the opposite polarization at the other interface. So injection of electrons occurs at the Zener breakdown field of the electronegative CGL/emitter junction and holes at the Zener breakdown field of the electropositive CGL/emitter junction. This happens at two different places in the power cycle.

The determination of efficacy, quantum efficiency, and other internal operating parameters is a little harder for the AC-OEL than in the case of OLEDs. However, brightness and external power efficiency is straightforward. The input power is simply the voltage times the current times the phase angle between the two. Output light is measured the same way as in OLEDs. To date AC-OELs have been demonstrated with an external power efficiency of 29.3 LPW (110.7 Cd/A) operating at a brightness of 20 500 Cd/m^2 and a white output [103]. This highlights an unusual feature of the AC-OEL as compared with the OLED; the power efficiencies are relatively high for high brightness. This is because the mechanism for the creation of carriers is more efficient at higher field densities and higher frequencies. However, loss in the dielectrics can limit the total power efficiency.

9.14.5 Organic Lasers and Organic Light-Emitting Transistors

The last class of light emitters we will discuss is that of organic lasers. As we will see in this section, we have included light-emitting transistors in the discussion because of their potential relevance to electrically stimulated organic lasers. However, there are excellent and detailed reviews of this topic and so our discussion is meant only as a brief introduction to concepts. For a deeper level review, we refer the reader to the references [108].

We begin with the organic lasers that are already well known and used widely: the optically pumped organic laser. Conceptually, the electroactive organic material is used as a gain medium in which to establish a population inversion of excited states. This must be placed into a resonator structure of some sort such as between mirrors, a distributed Bragg reflector, or on a fiber to form whispering gallery modes as shown in Figure 9.87 [109]. For such systems, an optical pump system must be used to excite the electroactive organic.

The basic photophysics of molecular materials, or organic semiconductors to be more precise, lends itself well to use as a laser medium. First of all, they tend to be strongly absorbing (light). As we will see in the photovoltaics section below, this means thin films (~100 nm) can absorb 90% or more of the light at the absorption band maximum. Since stimulated emission is directly related to absorption, this is a very important feature. The fluorescence spectra of organics can also be quite broad and very tunable using chemistry. This too can be quite advantageous when lasing at different lines is desired.

There are also drawbacks. For one, when used in a condensed (solid state) form, molecular materials can interact strongly among themselves. That is, they can form aggregates or excimers where the intermolecular interactions lead to quenching. This reduction in photoluminescence efficiency (photoluminescence

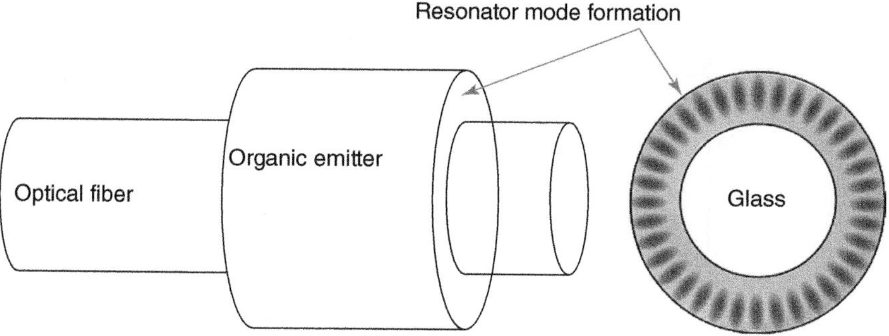

Figure 9.87 An "active" organic coating placed onto a fiber optic to form a whispering gallery mode resonator system.

quantum yield, PLQY) is detrimental to lasing. So as long as laser dyes are used in dilute solutions, where the active absorbing molecules are isolated from each other, there is no problem. But once they are placed in a thin film, for instance, strategies must be used to reduce intermolecular interactions. With small molecule dyes, blending into a transparent, noninteracting host medium works well. With polymers, typically large side groups must be added to isolate the strands. So, as discussed in the very beginning of this text, for laser applications one desires the "one-dimensional" behavior of the polymer to be preserved.

As is well known, lasing is based on the stimulated emission from an excited state in the system. This excited state can be prepared by first pumping it by a photon or exciting using injected charge. In organics, only the first way of preparing excited states has been demonstrated to yield lasing. The key element for gain in a medium to be achieved is that the emitted photon from this stimulated de-excitation has the same phase, frequency, and direction as the stimulating photon. Einstein showed that the cross sections for stimulated emission and absorption between two states are identical. This means finally to get gain from a medium, there needs to be more excited states for stimulated de-excitation than there are absorbing states. This is also known as a "population inversion." Given the highly disordered state of polymers in a film, one might be tempted to think it wouldn't be possible to achieve these conditions in such a material. However, not only can they be achieved, but also it turns out that such materials require relatively low amounts of pumping energy to do so (a low threshold to lasing).

To understand why, we must remember the many closely spaced excited states that are associated with organic molecules such as polymers. Remember, that as a practical matter, a population inversion cannot be achieved in a two-state system. However, with three or more states, it is possible to establish a population inversion. In particular if we consider the case of a four-level system as shown on Figure 9.86, absorption (or excitation) can occur between states A and B, whereas stimulated de-excitation occurs between C and D. If the transitions from B to C and D to A are properly balanced, then a population inversion can exist in the C to D transition even when most of the molecules are in the ground state. This means lasing can occur even for a relatively low rate of excitation (a low threshold, Figure 9.88).

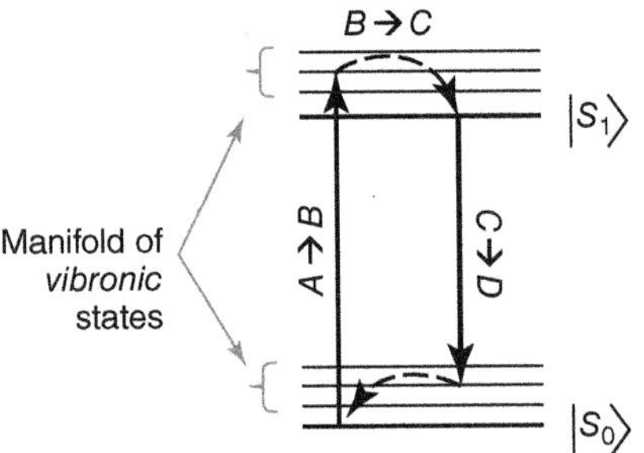

Figure 9.88 The energy diagram of a hypothetical four-level system built from the singlet and *vibronic* energy levels of a molecule.

In molecular materials (organics), this is exactly the situation we are faced with. If we associate the lines at A and C with the first excited singlet state of the system, the sublevels above A and C can be associated with *vibronic* states of the system. We have met vibronic states before, but we didn't call them by this name. They arise when a vibrational excitation of the molecule occurs that is coupled with the electronic energy of the electrons in the molecule. The term *vibronic* is a combination of the term *vibrational* and the term *electronic*. So if the molecule has a vibrational state into which it can be excited, that consequently raises the energy of the excited electronic state; the vibronic state will sit just above the excited singlet state. The relaxation (or vibronic cooling) of this excited vibrational state into the lowest vibrational state is rapid and also lowers the electronic state energy. De-excitation of the electron from the first excited singlet state occurs into a vibronic state just above ground-state energy. From there, it relaxes into the ground state. Typically such vibronic states are separated from the excited state above and the ground state below by about 0.2 eV, well above thermal energies. So there is little thermal excitation into the D state.

So the process is that a photon excites the organic from ground state into some vibrational level of the first excited singlet state manifold (A to B). A rapid transition follows from this vibrational state to the lowest energy of this singlet manifold (B to C). This is followed by a slower transition between the singlet excitation and the ground-state vibrational states (C to D). Finally, the vibrational state at D transits back to the ground state at A. Lasing can take place in the C to D transition, so population inversions are sensitive to the balance of transition times between B to C and D to A. In this way organic semiconductors can give rise to a four-level lasing system in which the threshold for lasing is low and the emission wavelength of the laser is longer than the absorption wavelength of the excitation photon.

There are, of course, many subtleties that we have not mentioned here, and the reader will notice that we have combined our conversation to include ALL organics: dyes, small molecules, polymers, etc. These general principles are the basis for lasing from such system when photons are used to prepare the excited states: "optical pumping." Indeed, there really isn't that much difference between solution-based dye lasers that we are familiar with in the lab and optically pumped solid-state organic lasers. An important aspect of the process is the molecule acting "isolated." For a more complete discussion of these processes, there are a number of excellent reviews [108].

However, this approach to lasing in solid-state organic systems is quite different from the laser diode (based on a standard inorganic LED). Specifically, electrically driven lasing in organics hasn't been fully realized and developed. As we have noted above, there have been great advances made in OLEDs. However, there are challenges with turning an OLED into an organic laser. Specifically there are three main areas in which researchers are focusing: current densities, losses due to contacts, and losses due to polaron and triplet formation.

9.14.5.1 Current Densities

A typical LED laser operates at $\sim 10^3$ A/cm^2 and an OLED operates at a lowly 10^{-1}–10^{-2} A/cm^2. This is simply not enough pumping to build an effective population inversion. Can we increase the current density? The very low mobility of organic conductors suggests this might be hard. There have been reports of very high current densities in organics from pulsed injection [108]. This may be one way around this limit.

9.14.5.2 Contacts

Another concern is the loss associated with contacts. The resonator structure of a thin film organic device typically means the laser output is oriented along the plane of the film. This not only gives a long interaction length with the gain material but also means that there is an equally long interaction length with the metal contact of the device. In each time the light interacts with the contact, some is absorbed, leading to loss from the resonator. Such losses can add up to become large and quench the lasing. So resonator design is strictly limited.

9.14.5.3 Polarons and Triplets

While the first two drawbacks may have some "work-arounds" the last one is far more serious. Optical excitation generally yields singlet excitons. These excitations, as we discussed above, are what are needed for lasing. Electrical injection takes the form of polarons that then recombine into excitons that can be either singlets or triplets. Triplets are a forbidden optical transition and so gain from such excitations will be orders of magnitude smaller than from singlets. Both polarons and triplets have associated absorptions that are quite strong. So filling the resonant cavity with these excitations introduces losses that increase as one approaches lasing due to the increasing opacity as current density is raised. For a material with mobility of 10^{-4} cm^2/V s, Tessler [110] estimates there are as many as 1000× more polarons than singlets and of course 3× the number of triplets. Importantly, the absorption of these excitations in most organic systems is rather broad and therefore overlaps with the lasing line.

All three of these issues are related directly or indirectly to the issue of low mobilities usually found in organic systems. So there is always a possibility of discovering new organic laser materials, but as we have already seen, these would have some fundamental differences to what we know of organics. However, there are other approaches that have been suggested.

One such "alternate route" is that of the "organic light-emitting transistor" (OLET) shown in Figure 9.89. Such devices have been touted for their potential uses in a wide variety of applications including electrically driven organic

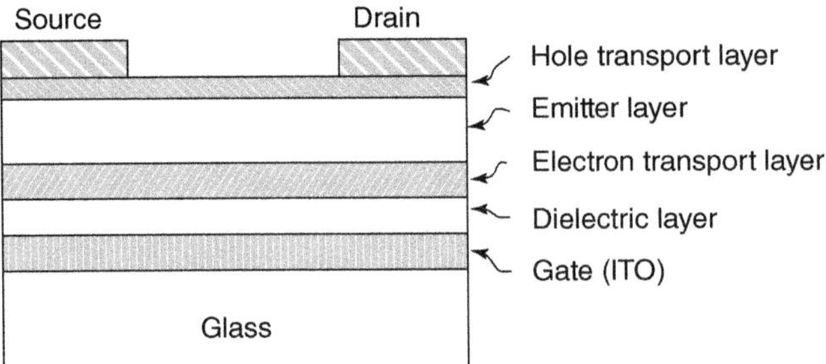

Figure 9.89 The structure of an organic light-emitting transistor is shown. These have been proposed as a potential route to electrically driven organic lasers.

lasers [111]. The OLET is a light-emitting form of transistor wherein holes are injected from the source electrode and electrons from the drain electrode, into a semiconducting polymer or organic material. A back gate electrode with a dielectric insulating layer separating it from the semiconducting channel above is added, as shown in the diagram. This allows one to switch the light emission on and off, without additional circuitry as required with the OLED, yielding a general utility in pixelated display technologies. However, the gate voltage also allows for the recombination region with its polarons and triplet states to be confined to a region near the dielectric. This means the light emission can be used to stimulate a population inversion elsewhere within the volume of the conduction channel. Schols et al. have hypothesized that this feature might be used in electrically driven lasers [112]. Technically, the volume that is luminescing due to recombination of injected charge (and where the strongly absorbing triplets and polarons are located) would not be lasing.

9.14.6 Organic Solar Cells

Related to the organic light emitting diode is the organic photovoltaic (OPV) cell and much of the discussion above, also applies to the construction of such cells. Highly doped conjugated polymers are metals, but undoped or lightly doped, they are semiconductors as pointed out above. Generally, in crystalline system like silicon, a p–n junction is formed between two thin layers of doped material. Light that enters creates an electron–hole pair that migrates randomly in the film until they come across the p–n junction. At this point (at the p–n interface), the electron–hole pair is separated, and we retrieve this as usable current. It is not easy to make p–n junctions in conjugated polymers, however, because the dopants, being interstitial, are very mobile. So they migrate and compensate. But Schottky barriers can be formed, just by evaporating a thin layer of a metal with proper work function onto the polymer. A Schottky barrier on polyacetylene or another conjugated polymer acts as a crude, and rather inefficient, solar cell. Such a device is shown schematically in Figure 9.90. A Schottky barrier works as a solar cell because at the polymer–metal interface there is a "built-in" electrical field created by electron exchange between metal and polymer. If the incoming light generates electron–hole pairs (an exciton), these carriers can be separated

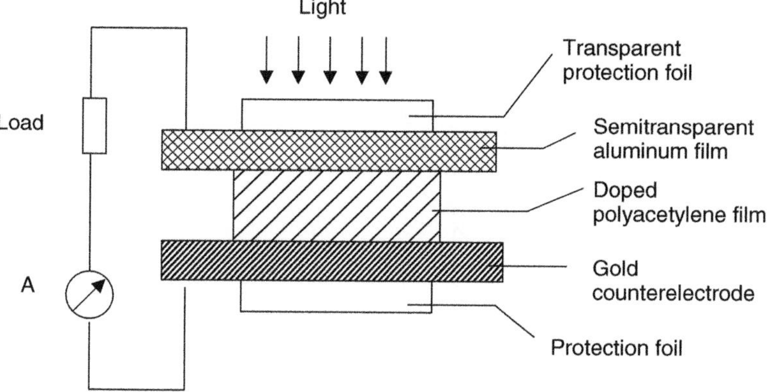

Figure 9.90 Schematic view of a polyacetylene – metal solar cell.

in the field if they make it to a separation junction before they recombine (giving off light). When the electrodes are connected to an external load the photocurrent can flow, and the field will be maintained. Power conversion efficiencies of ~1% (depending on design) can be achieved using such an approach.

There are challenges to this approach however. It is well known, for instance, that organics can provide exceptionally good chromophores. Indeed, the efficiency of a polymer to convert light into an electron–hole pair [113] is excellent, exceeding all but the best-known solid-state crystalline systems. However, it is usually observed that it is nearly impossible to remove substantial amounts of these excitations from the polymer as current, meaning that organics couple well to photons but make poor photovoltaic conversion materials. But why is this so? The difficulty comes in fundamental material properties that derive both from the polymer electronics and its structure in solid form. Migration lengths for the donor–acceptor excitations [114] are much shorter than the absorption lengths required to create the excitations. Thus the films thickness must be >100 nm to create an excitation, but the distance to an effective separation barrier must be <10 nm to for the electrons and holes to be removed as useful current. Otherwise they will recombine to give luminescence. Some creative ways have been devised to get around this difficulty, but to see how they work, we first need to go a little deeper into the terminology and metrics used in the field.

In the field of OPV devices, there are several basic terms that recur. The first is the IQE of the device. This is the number of electron–hole pairs actually separated per incoming photon, and it is usually presented as a function of wavelength. Second, and most often cited, is the external efficiency or the "power conversion efficiency," η_e,

$$\eta_e = V_{oc}[\mathrm{V}] \times I_{sc}(\mathrm{A/cm^2})\ \mathrm{FF}/P_{in}(\mathrm{W/cm^2}) \tag{9.30}$$

where, V_{oc}, I_{sc}, FF, and P_{in} are the open circuit voltage, the short circuit current, the filling factor (FF), and the incident power, respectively. The filling factor is determined by calculating the maximum power rectangular area under the I/V

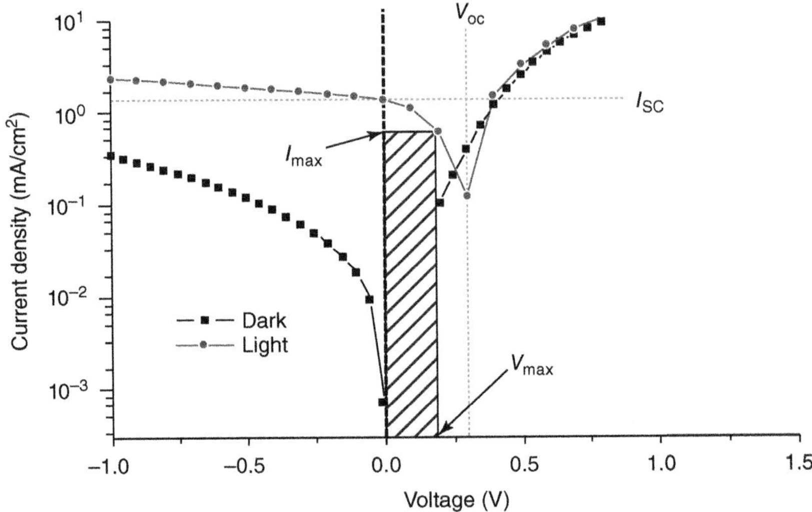

Figure 9.91 An I/V curve of a thin film photovoltaic cell under illuminated and dark conditions. The maximum power rectangle is drawn in as described in the text.

curve. Simply, it is given by

$$\text{FF} = V_{\text{p}} \times I_{\text{p}} / V_{\text{oc}} \times I_{\text{sc}} \tag{9.31}$$

where V_{p} and I_{p} are the intersections of the I/V curve with the maximum power rectangle. You can think of the filling factor as a quality factor of how well the device can drive a load.

Figure 9.91 shows a typical I/V curve of an illuminated and dark organic photocell. To get such a curve, you ramp the voltage placed on the device and collect the current under dark and illuminated conditions. The illuminated I/V will shift as seen above. The voltage the device would produce under infinite load (open circuit) is shown as V_{oc}. The current the device would produce when illuminated with no resistance in the load is shown as I_{sc} (for short circuit). But when the illuminated device is connected to a finite load, it produces less power than $V_{\text{oc}} \times I_{\text{sc}}$ (P_{max}). This is because the device is not an ideal current or voltage source. To compute how much power the device will produce under load, we first draw the rectangle against the 0 V line such that it intersects with the illumination curve as shown. There are an infinite number of such rectangles possible, and we choose the one with a maximum area possible, typically determined computationally. This is referred to as the "maximum power rectangle." The power that the device will produce is actually $V_{\text{max}} \times I_{\text{max}}$ as shown. However, this is generally given as the ratio of the areas of the maximum power rectangle to that of the "ideal case," which is known as the filling factor (FF) times P_{max}. In the case shown we have $0.2\,\text{V} \times 8.0 \times 10^{-4}\,\text{A} = 1.6 \times 10^{-5}\,\text{W}$. This gives FF = 0.23 when divided into the V_{oc} and I_{sc}, roughly. The filling factor is typically used as an indication of the quality of the photovoltaic and can range from 0.9's (for Si) to 0.7's (for many organics).

Importantly, our example here happens to be an organic thin film device. It is illuminated with a *solar standard* of 1.5 air masses (am 1.5 g). Such a standard simulates the average power of the solar spectrum at each wavelength as it occurs at the surface of the Earth (traveling through 1.5 air masses of our atmosphere). Such a path for the light to reach the Earth's surface corresponds roughly to a position on the globe equivalent to Michigan in the United States. The 1.5 irradiance curve used in this test corresponds to the ASTM G173 testing standard agreed to by international commissions on testing. This particular example has an external power efficiency of about 2.3%.

The offset (or open circuit voltage) in the system is roughly the difference in work functions of the contact and thin active film. The origin of the V_{oc} has been studied extensively by Friend and cowokers [115], among others. In such I/V studies, comparisons can be made between devices of the structure, ITO/PPV/Mg and ITO/PPV/Ca, for instance. The offset voltages in the I/V curve, that is, the open circuit voltages, were 1.2 and 1.7 V, respectively, for these two examples. These numbers are roughly the same as the difference in work function between Mg and ITO and Ca and ITO. However, in an ITO/PPV/Al device, one also obtains 1.2 V for the open circuit voltage. Oddly, this is significantly higher than the work function difference between ITO and Al. In this case a sizable Schottky barrier has occurred, effectively raising the injection energy cost. It is surprising that this is not so strongly seen in the case of the alkali metals for this system.

As mentioned above, an essential difficulty, and thus the focus of intense research, is the mismatch of length scales in these systems generally. The most successful approach to addressing this shortcoming is the use of a "bulk heterojunction." A BHJ device is formed with the introduction of a highly conducting "nanophase" such as fullerenes [116]. In this scheme, fullerene molecules (C_{60} or C_{70}) or conjugates such as PCBM are blended into the active absorber to aid in the separation of charge. The idea is to form a percolating network of the highly conductive material out of the layer. By using fullerene, the electrons are transferred to the nanophase and are transported out along its pathways. The holes exit through the polymer. This approach has been tried with conducting polymer phases as well as carbon nanotubes [117], but none have worked as well as fullerenes because of their strong acceptor quality.

The key to getting this interpenetrating network approach to work well is the morphology of the nanophase. The fullerene phase must be spread throughout the host without aggregation but with enough interconnectivity to allow for electron transport. Moreover, it must not disturb too much the π stacking of the polymer host phase or the mobility of the holes will suffer. While there have been numerous advances in the polymers used since OPVs were first introduced, the control over morphology for each system remains a point of "art." Specifically, once a new candidate polymer is developed, a significant amount of time and effort must go into creating the right morphology for it to work optimally. Multiple solvents, high-temperature annealing, and additional side groups of the polymer are all employed to optimize the polymer for use with an acceptor phase.

There are today quite literally hundreds of polymers under study for use in OPVs. The power conversion efficiencies have now been raised to over 11% and are still climbing. This is approaching the performance of amorphous Si cells. So will OPVs ever become a viable competitor? A solar cell made of single-crystal silicon has to run for several years until it has generated the amount of energy that has been used for the production of the cell. Counting the cost of the cell and its support apparatus, power generated by Si is just at $1 per Watt. There are some cells that are cheaper than this! Since conjugated polymers are synthesized by room temperature catalytic processes, they are energetically cheap to manufacture. But this is not the only economically relevant figure. Land costs and maintenance costs are equally important. Lifetime and replacement costs are also important. So even with higher efficiencies being achieved and the need for flexible and mobile power sources growing, it is becoming clear that organic solar cells will have to provide service lifetimes approaching that of Si and take up less space (or be integrated into building materials) to be able to compete effectively. This is simply a restatement of an old truth: "power production tends toward its most dense form in the marketplace."

9.14.7 Organic Field-Effect Transistors

In each of the examples above, we have seen how the disordered nature, of our one-dimensional systems, has modified standard notions about electronic devices. This applies to transistors made from organic semiconducting systems as well. More specifically, we might well imagine that the field-dependent mobility would express itself rather strongly in an all organic field-effect transistor (OFET). We met OLETs above. OFETs [118] have historically been prepared as a tool for investigating material properties of the organic films, but with improving material quality they are quickly approaching practical utility. Prepared in the form of thin films, these semiconductors are ideal for the creation of flexible, inexpensive field-effect transistors.

OFETs fall generally into two categories: small molecule and polymer. In both cases aspects of dimensionality are important. For small molecules laid down in a thin film, the molecules arrange themselves in such a way as to provide conducting one- or two-dimensional pathways through the film (depending on the molecule). We met such arrangements in Chapter 2. Polymers inherently conduct along the strand or by hopping from strand to strand. So both approaches share common themes. Figure 9.92 shows an all-polymer field-effect transistor [45]. It consists of a polymer film between a source and a drain contact and a gate on the backside. To avoid direct contact from the source or drain and the gate, the gate is protected by a thin insulating layer of silicon dioxide.

As the back gate of the device is biased, the total amount of charge allowed through the channel is changed. An example of drain current vs. drain voltage characteristics is shown in Figure 9.93 [118]. The characteristics demonstrate that all-polymer field-effect transistors "work." More important is presently the use of such devices as experimental tools, for example, to determine the mobility of charge carriers in thin organic layers.

The fundamental operation of the device follows in the same manner as the crystalline system [119]. As mentioned above, however, the mobility of charge is dominated by its hopping nature. Further, there is a relatively low number of charge carriers, and it is therefore rather difficult to build a significant depletion

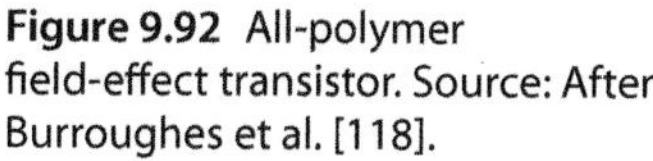

Figure 9.92 All-polymer field-effect transistor. Source: After Burroughes et al. [118].

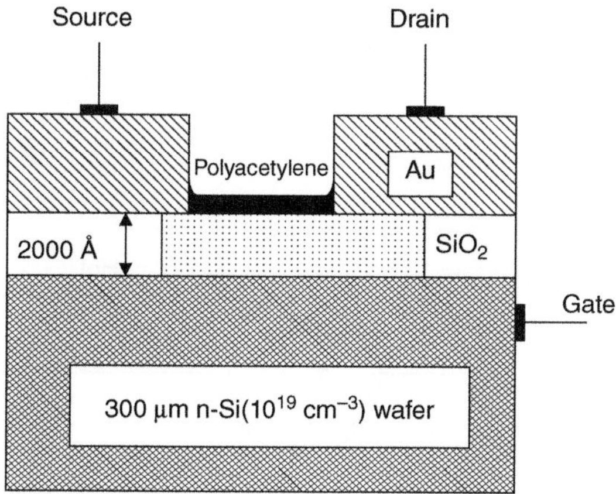

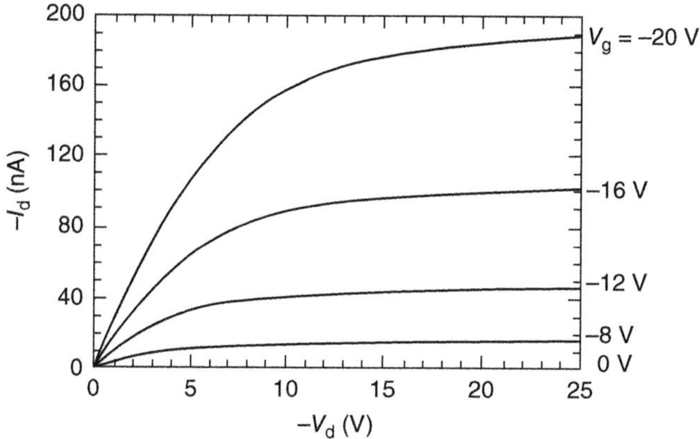

Figure 9.93 Drain current vs. drain voltage characteristics of an all-organic (sexithiophene) field-effect transistor. Source: After Garnier et al. [118].

region. These factors introduce limitations into the current applicability of the OFET structures. However, advances are being made rapidly.

9.14.8 Organic Thermoelectrics

Organics can provide a reasonable conductivity of electrical charge, but we would expect them to be poor thermal conductors. This is because generally in solid-state systems, the electronic contributions to thermal conduction are not as large as the phonon contribution and polymers have small phonon contributions. And this is indeed the case. Typically thermal conductivities in even highly electrically conducting polymers are vanishingly small.

This is true even in polymers filled with nanotubes. Nanotubes can have extremely high thermal conductivity on their own. But when put into a polymer matrix, the phonon energy must hop from nanotube to nanotube, introducing scattering into the flow of heat. So for such composites where electrical conductivities can approach 10^5 S/cm, thermal conductivities are still $<1\ \mathrm{W/m^2\ K}$.

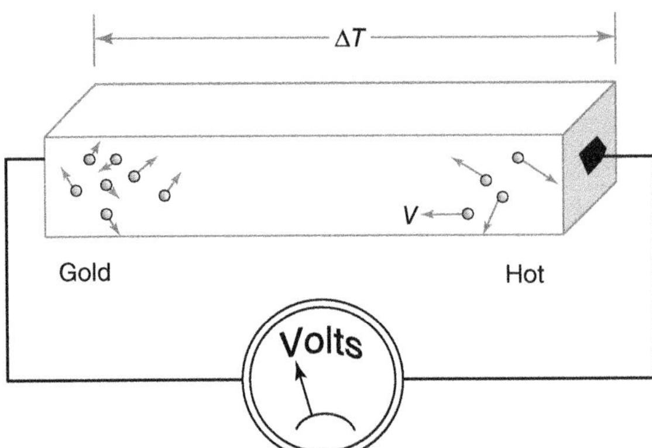

Figure 9.94 A schematic diagram of the Seebeck effect. The carriers behave as a gas shown here with velocity vectors ***v***.

Interestingly, this "scattering" is not very sensitive to nanotube loading within the matrix (above a certain minimum). For loadings around 50% by weight up to 100% nanotubes (a mat), the thermal conductivity of the matrix doesn't change very much [120].

Materials with high electrical conductivity and low thermal conductivity are typically excellent candidates for thermoelectrics. The *Seebeck effect*, one part of the *thermoelectric effect* in materials, is a thermodynamically reversible process that turns a temperature gradient into a voltage. A simple picture of this process might look like Figure 9.94.

In this simple model the carriers (electrons or holes depending on if the material is n type or p type) act as a gas of particles in the "box" of the material sample. When one end is heated relative to the other, the velocities of these particles are increased, allowing the carriers from the hot side to diffuse to the cold side more rapidly than vice versa. This results in a buildup of charge on one side of the sample and a *thermovoltage*. The current–voltage relation is tied together through this temperature gradient as

$$J = -\sigma\alpha\Delta T \tag{9.32}$$

where J is the current density, σ is the conductivity, and α is the Seebeck coefficient (otherwise known as the thermopower). Typical values for α are -100 to $+1000\,\mu\text{V/K}$, and these are an intrinsic property of the material. It is relatively straightforward to show that the power that can be delivered by a thermoelectric material is given as

$$P \sim \Delta T^2 \alpha^2 A_{\text{TE}}/\rho L \tag{9.33}$$

where ρ, resistivity; L, length of material; and A_{TE}, cross-sectional area of the material. Furthermore the maximum efficiency for converting heat power into electrical power by a thermoelectric material is limited by the Carnot efficiency. If we divide power out by power in, we can derive an expression for the efficiency based on a combination of material properties that we combine into a single expression ZT:

$$ZT = \alpha^2 T/\rho K \tag{9.34}$$

Table 9.1 Values comparing several different types of thermoelectric materials (BiTe, PEDOT, CNT composites).

	Bi_2Te_3 [121]	PEDOT [122]	CNT composite [123]
α (μV/K)	150–200	100–800	10–60
σ (S/m)	10^5	10^4–10^{-2}	10^6–10^{-1}
κ (W/m/K)	3	0.3	0.1–0.3
$\alpha^2\sigma$ (μW/m/K)	7800	100	10–100
ZT	1	0.2	0.02–0.1

where T is the temperature of the heat bath and K is the thermal conductivity. Z is the so-called figure of merit. But often Z is multiplied with T and in compilations like Table 9.1 the "ZT value" is used.

Of course this extremely simple picture overlooks many nuances, but clearly if the phonons generated in the hot part of the volume are also transported efficiently to the cold side, ΔT will be hard to maintain. Commercially the inorganic Bi_2Te_3 is among the best performers, but there are many inorganic material systems under study, including skutterudites and clathrates [124]. These materials exhibit strong phonon scattering while still transporting current. Of course as pointed out above, polymers can expect the same properties. While the inorganic systems such as Bi_2Te_3 above have ZT's around 1, highly conducting polymer systems such as PEDOT are quickly catching up ($ZT \sim 0.2$). A few literature values are shown in Table 9.1.

As an example of a thermoelectric power generator, Figure 9.95 shows the photograph of a "power felt." This is a stack of polymer films loaded with carbon nanotubes, electrically wired in series, and thermally in parallel. At a temperature difference of 100 °C (one end in ice water, the other in boiling water), a voltage of 50 mV can be measured [120].

9.15 Summary

At an MRS meeting back in the 1980s or so, it was once bragged that "given time, Si could match the properties of any of its technology rivals: GaAs, GaN, InP, etc."[7] This does seem to be coming true with the development of ultrapure Si, porous Si, and so on. From this same perspective, this chapter has introduced another "supermaterial" in electronics: conjugated polymers (plastics). And with the rapid progress made in the field, it is tempting to say that, given time, we will be able to build any type of device from polymers!

But how can this be? How can a low-dimensional system with such strange and wonderful coupling to its carriers become technologically relevant? As we saw, polymers can be thought of as an archetype for electron–phonon coupling

7 He was obviously in the wrong session!

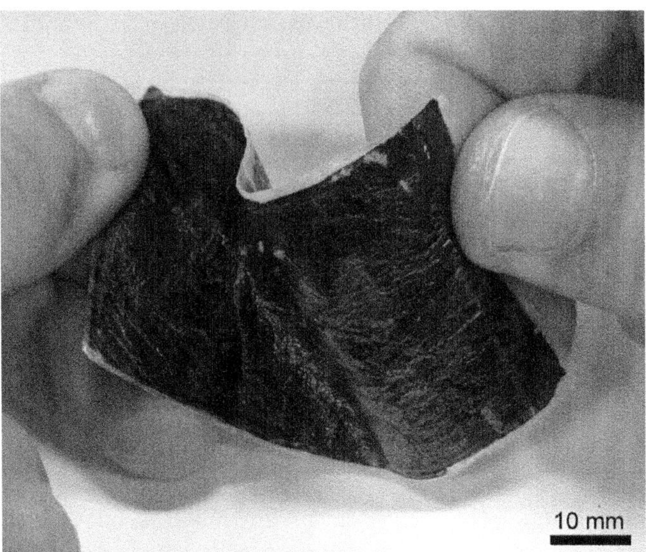

Figure 9.95 *Power felt*, stack of nanocomposite films containing carbon nanotube networks, delivering a thermopower of 90 mV for a temperature difference of 100 °C [120].

with the coupled states referred to as *solitons*. Such solitons have a number of properties not available to pure charges moving in a solid. These include different spin projections, spin–charge separation, and more. However, they can be seen as "just another charged particle" in the system like electrons or holes. To this end, we demonstrate that electronics and photonics based upon these strange conductors are, in reality, extremely similar to standard semiconductor electronics with the exception that they carry with them the mechanical properties of a plastic. In some cases, however, the additional properties of the carriers can yield surprisingly better devices than the crystalline counterparts.

Exploring Concepts

1 *Derivations*: Work through the derivation of the E_{gap} and the solution to the master equation above. Show explicitly where the *linear* δ term comes in and over what range.

2 *The Peierls equation*: Research *physical models* for the A term in the Peierls equation. How does it arise and what are reasonable values for it? For more reading, try Ref. [125].

3 *Spatial extent of domain walls*: What would limit the spatial extent of the domain wall in a conjugated polymer?

4 *Position of polaronic states*: Using the molecular orbital approach, explain why polaronic states will occur in the bandgap?

5 *VRH or FIT*: We have confined our discussion to polyacetylene on purpose. It must be obvious by now that many other systems exist to which VRH or FIT models can be applied. Consider bare mats of nanotubes. The morphology seems to be quite similar to that of the polymer discussed here.
(a) Look at the temperature-dependent conductivity data formats of multi-walled carbon nanotubes (in literature). Determine which model is most appropriate?
(b) Now compare this to mats of single-walled carbon nanotubes, which are mixtures of semiconductor and metallic nanotubes. Is there a change in your answer? What effect does doping have?

6 *Nanotube mats*: What is the highest conductivity achieved in nanotube mats?

7 *Two-state systems*: Why does a two-state system not allow for a population inversion?

8 *Optically vs. electrically pumped lasers*: Using $\sim 1\ \mu J/cm^2$ as the threshold for *optical* pump density and assuming 25% of the injected excitations are the desired singlets with a lifetime of 5 ns, can you determine the current density necessary to achieve lasing for an organic *electrically* pumped system? (Compute the photon density first and work from there.)

9 *Singlets*: Why is the relatively short (∼ns) excited state lifetime of singlets a problem for the current densities we are able to achieve in organics?

10 *Vibronics*: How fast are *vibronic* transitions?

11 *Organic resonators*: Investigate several resonator designs for organics and draw them out. Explain how they work.

12 *Gain in triplets*: Why do the triplet transition selection rules lead to lower gain from these excitations?

13 *Thermoelectric efficiencies*: Using ZT, come up with an expression for the maximum efficiency of a thermoelectric material. The expression should involve only intrinsic variables.

14 *Energy transfer*: So how does an exciton in a host polymer transfer its energy to a guest dye molecule? There are three ways: radiative energy transfer, Förster transfer, and Dexter transfer.
Radiative transfer occurs through the radiative deactivation of the donor molecule and subsequent reabsorption of the energy by the guest acceptor. The probability of the process goes as

$$P \sim [A]xJ$$

where $[A]$ is the concentration of the acceptor molecules, x is the specimen thickness, and J is the spectral overlap integral.

Förster transfer is very fast $<10^{-9}$ s resonant dipole–dipole coupling. Because it is dipole mediated, it can be relatively long range ~10 nm, and *resonant* means the transitions must be allowed by selection rules. So typically this applies to singlet–singlet transfer. The transfer rate constant from donor to acceptor is given as

$$K_{\mathrm{FT}}(r) \sim (1/\tau_{\mathrm{D}})(r_0/r)^6$$

where r_0 is the Förster radius given by

$$r_0 \sim \left[\frac{K^2 J Q_0}{n^4}\right]^{1/6}$$

K is an orientation factor between the dipoles, J is the spectral overlap integral, n is the refractive index, Q_0 is the quantum efficiency without energy transfer, τ_{D} above is the radiative lifetime of the donor, and r is the distance between the donor (D) and acceptor (A). Again, only "dipole-allowed transitions" are possible.

Dexter transfer requires the overlap of the wavefunctions of the donor and acceptor and is the dominant energy exchange mechanism for triplet–triplet exchange. It is typically associated with the exchange of an electron. Only singlet–singlet and triplet–triplet exchange is allowed by the mechanism and then only within an interaction radius of ~1 nm. The transfer rate constant is given as

$$K_{\mathrm{DT}} \sim \hbar P^2 J e^{-2r/L}$$

where J is the overlap integral, P and L are constants, and r is the distance between donor and acceptor. In Dexter transfer spin is conserved.

It is useful for the reader to review these mechanisms in the context of host–guest interactions and derive the formulae given above. Notice the differences in interaction length scales will place specific limits on how much dye can be used to achieve the maximum brightness in an OLED device.

References

1 Skotheim, T.A. (1986). *Handbook of Conducting Polymers*, vol. 1–2. New York: Marcel Dekker.

2 Baeriswyl, D. and Kiess, H. (1992). *Conjugated Conducting Polymers*, Springer Series in Solid-State Sciences, vol. 102. Heidelberg: Springer Verlag.

3 Przyluski, J. (1991). *Conducting Polymers – Electrochemistry*. Vaduz: Sci-Tech Publications.

4 Chien, J.C.W. (1984). *Polyacetylene – Chemistry, Physics, and Material Science*. Orlando: Academic Press.

5 Mair, H.J. and Roth, S. (eds.) (1989). *Elektrisch leitende Kunststoffe*, 2e. München-Wien: Carl Hanser Verlag.

6 (a) Kuzmany, H., Mehring, M., and Roth, S. (1985). *Electronic Properties of Polymers and Related Compounds*, Springer Series in Solid-State Sciences, vol. 63. Berlin, Heidelberg: Springer Verlag. (b) Kuzmany, H., Mehring, M., and Roth, S. (1987). *Electronic Properties of Conjugated Polymers*, Springer Series in Solid-State Sciences, vol. 76. Berlin, Heidelberg: Springer Verlag. (c) Kuzmany, H., Mehring, M., and Roth, S. (1989). *Electronic Properties of Conjugated Polymers III – Basic Model and Applications*, Springer Series in Solid-State Sciences, vol. 91. Berlin, Heidelberg: Springer Verlag; (d) Kuzmany, H., Mehring, M., and Roth, S. (1992). *Electronic Properties of Polymers – Orientation and Dimensionality of Conjugated Systems*, Springer Series in Solid-State Sciences, vol. 107. Berlin, Heidelberg: Springer Verlag.

7 (a) Przyluski, J. and Roth, S. (1987). *Electrochemistry of Conducting Polymers*, Materials Science Forum 21. Aedermannsdorf: Trans Tech Publications. (b) Plocharski, J. and Roth, S. (1989). *Electrochemistry of Conductive Polymers*, Materials Science Forum 42. Aedermannsdorf: Trans Tech Publications. (c) Przyluski, J. and Roth, S. (1993). *Conducting Polymers – Transport Phenomena*, Materials Science Forum 122. Aedermannsdorf: Trans Tech Publications; (d) Przyluski, J. and Roth, S. (1994). *New Conducting Materials with Conjugated Double Bonds: Polymers, Fullerenes, Graphite*, Materials Science Forum. Aedermannsdorf: Trans Tech Publications.

8 Aldissi, M. (1993). *Intrinsically Conducting Polymers: An Emerging Technology*. Dordrecht: Kluwer Academic Publishers.

9 Salaneck, W.R., Clark, D.T., and Samuelsen, E.J. (1991). *Science and Applications of Conducting Polymers*. Bristol: Adam Higler.

10 Roth, S. and Bleier, H. (1987). Solitons in Polyacetylene. *Adv. Phys.* 36: 385.

11 Heeger, A.J., Kivelson, S., Schrieffer, J.R., and Su, W.P. (1988). Solitons in conducting polymers. *Rev. Mod. Phys.* 60: 781.

12 Lu, Y. (1988). *Solitons and Polarons in Conducting Polymers*. Singapore: World Scientific.

13 (a) Schaumburg, K. et al. (1992). *Nanostructures Based on Molecular Materials* (ed. W. Göpel and C. Ziegler). Weinheim: Wiley-VCH. (b) Lara-Avila, S. et al. (2010). Bianthrone in a single-molecule junction: conductance switching with a bistable molecule facilitated by image charge effects. *J. Phys. Chem. C* 114: 20686.

14 Peierls, R.E. (1955). *Quantum Theory of Solids*. Oxford: Clarendon Press.

15 Friend, R.H. and Jérome, D. (1979). Periodic lattice distortions and charge density waves in one- and two-dimensional metals. *J. Phys. C Solid State Phys.* 12: 1441.

16 Chao, K.A. (1984). On a Seminar Tour through Europe.

17 Landau, L.D. and Lifshitz, E.M. (1980). *Statistical Physics I*. New York: Pergamon Press.

18 (a) Kuhn, H. (1948). Free-electron model of absorption spectra of organic dyes. *J. Chem. Phys.* 16: 840. (b) Kuhn, H. (1949). A quantum-mechanical theory of light absorption of organic dyes and similar compounds. *J. Chem. Phys.* 17: 1198.

19 (a) Longuett-Higgins, H.C. and Salem, L. (1959). Alternation of bond lengths in long conjugated chain molecules. *Proc. Roy. Soc. London* A251: 172.

(b) Salem, L. (1966). *The Molecular Orbital Theory of Conjugated Systems*. London: Benjamin.

20 (a) Goldberg, I.B., Crowe, H.R., Newman, P.R. et al. (1979). Electron spin resonance of polyacetylene and AsF_5-doped polyacetylene. *J. Chem. Phys.* 70: 1132; (b) Nechtschein, M., Devreux, F., Greene, R.L. et al. (1980). One-dimensional spin diffusion in polyacetylene, $(CH)_x$. *Phys. Rev. Lett.* 44: 356; (c) Weinberger, B.R., Ehrenfreund, E., Pron, A. et al. (1980). Electron spin resonance studies of magnetic soliton defects in polyacetylene. *J. Chem. Phys.* 72: 4749.

21 Pople, J.A. and Walmsley, S.H. (1962). Bond alternation defects in long polyene molecules. *Mol. Phys.* 5: 15.

22 (a) Su, W.P., Schrieffer, J.R., and Heeger, A.J. (1979). Solitons in polyacetylene. *Phys. Rev. Lett.* 42: 1698. (b) Su, W.P., Schrieffer, J.R., and Heeger, A.J. (1980). Soliton excitations in polyacetylene. *Phys. Rev. B* 22: 2099. (c) Rice, M.J. (1979). Charged π-phase kinks in lightly doped polyacetylene. *Phys. Lett. A* 71: 152.

23 Chiang, C.K. et al. (1977). Electrical conductivity in doped polyacetylene. *Phys. Rev. Lett.* 39: 1098.

24 (a) Brazovskii, S.A. (1978). Electronic excitations in the Peierls-Fröhlich state. *JETP Lett.* 28: 606. (b) Brazovskii, S.A. (1980). Self-localized excitations in the Peierls-Fröhlich state. *Sov. Phys. JETP* 51: 342.

25 (a) Baeriswyl, D. (1983). Conducting polymers: solitons or not? *Helv. Phys. Acta* 56: 639. (b) Roth, S., Ehinger, K., Menke, K. et al. (1983). Polyacetylene: la chasse aux solitons. *J. Phys.* C3: 69.

26 Takayama, H., Lin-Liu, Y.R., and Maki, K. (1980). Continuum model for solitons in polyacetylene. *Phys. Rev. B* 21: 2388.

27 Bredas, J.L. and Street, G.B. (1985). Polarons, bipolarons, and solitons in conducting polymers. *Acc. Chem. Res.* 18: 309.

28 Stafström, S. and Chao, K.A. (1984). Soliton states in polyacetylene. *Phys. Rev. B* 29: 2255.

29 Grupp, A., Höfer, P., Käss, H. et al. (1987). *Electronic Properties of Conjugated Polymers*, Springer Series of Solid-State Sciences, vol. 76 (ed. H. Kuzmany, M. Mehring and S. Roth). Heidelberg: Springer Verlag.

30 Su, W.P. and Schrieffer, J.R. (1980). Soliton dynamics in polyacetylene. *Proc. Natl. Acad. Sci.* 77: 5626.

31 (a) Friend, R.H. and Burroughes, J.H. (1989). Charge injection in conjugated polymers in semiconductor device structures. *Faraday Discuss. Chem. Soc.* 88: 213. (b) Kanicki, J. (1986). Polymeric semiconductor contacts and photovoltaic applications. In: *Handbook of Conducting Polymers* (ed. T. Skotheim). New York: Marcel Dekker. (c) Burroughs, J.H. et al. (1990). Light emitting diodes based on conjugated polymers. *Nature* 347: 539. (d) Pilkuhn, M.H. and Schairer, W. (1993). Light emitting diodes. In: *Handbook of Semiconductors*, 4e (ed. T.S. Moss and C. Hilsum). Amsterdam, North Holland: Elsevier. (e) Leising, G., Tausch, S., and Graupner, W. (1997). Fundamentals of electroluminescence in paraphenylene-type conjugated polymers and oligomers. In: *Handbook of Conducting Polymers* (ed. T. Skotheim, R. Elsenbaumer and J. Reynolds). New York: Marcel Dekker.

32 Jackino, R. and Schrieffer, J.R. (1981). Solitons with fermion number ½ in condensed matter and relativistic field theories. *Nucl. Phys. B* 190: 253.

33 (a) Schrieffer, J.R. (1987). The quantum hall effect: relation to quasi one dimensional conductors. *Synth. Met.* 17: 1. (b) Kivelson, S., Kallin, C., Arovas, D.P., and Schrieffer, J.R. (1986). Cooperative ring exchange theory of the fractional quantum hall effect. *Phys. Rev. Lett.* 65: 873.

34 Bak, P. and Jensen, M.H. (1980). Theory of helical magnetic structures and phase transitions in MnSi and FeGe. *J. Phys. C Solid State Phys.* http://iopscience.iop.org.

35 (a) Moessner, R. (2014). Magnetische Monopole in Spineis. *Phys. J.* 14: 41. (b) Castelnovo, C., Moessner, R., and Sondhi, S.L. (2008). Magnetic monopoles in spin ice. *Nature* 451: 42. (c) Castelnovo, C., Moessner, R., and Sondhi, S.L. (2012). Spin ice, fractionalization, and topological order. *Annu. Rev. Condens. Matter Phys.* 3: 35. (d) Harris, M.J., Bramwell, S.T., McMorrow, D.F. et al. (1997). Geometrical frustration in the ferromagnetic pyrochlore $Ho_2Ti_2O_7$. *Phys. Rev. Lett.* 79: 2554. (e) Morris, D.J. et al. (2009). Dirac strings and magnetic monopoles in the spin ice $Dy_2Ti_2O_7$. *Science* 326: 411.

36 Reichenbach, J., Kaiser, M., Anders, J. et al. (1992). Picosecond photoconductivity in $(CH)_x$ measured by cross-correlation. *Europhys. Lett.* 18: 251.

37 Jeyadev, S. and Cronwell, E.M. (1987). Polaron mobility in trans-polyacetylene. *Phys. Rev. B* 35: 6253.

38 Foote, D.P. (1992). *ME-News (13)* (ed. M. Petty). Durham: Center of Molecular Electronics.

39 Jeyadev, S. and Conwell, E.M. (1987). Soliton mobility in *trans*-polyacetylene. *Phys. Rev. B* 36: 3284.

40 Bardeen, J. and Schockley, W. (1950). Deformation potentials and mobilities in in non-polar crystals. *Phys. Rev.* 80: 72.

41 Reichenbach, J., Kaiser, M., and Roth, S. (1993). Transient picosecond photoconductivity in polyacetylene. *Phys. Rev. B* 48: 14104.

42 (a) Kivelson, S. and Heeger, A.J. (1981). Electron hopping conduction in the soliton model of polyacetylene. *Phys. Rev. Lett.* 46: 1344. (b) Kivelson, S. and Heeger, A.J. (1982). Electron hopping in a soliton band: conduction in highly doped $(CH)_x$. *Phys. Rev. B* 25: 3798. (c) Kaiser, A.B. and Park, Y.W. (2002). Conduction mechanisms in polyacetylene nanofibers. *Curr. Appl. Phys.* 2: 33. (d) Bhattacharyya, S., Saha, S.K., Chakravorty, M. et al. (2001). Frequency-dependent conductivity of interpenetrating polymer network composites of polypyrrole-poly(vinyl acetate). *J. Polym. Sci. B* 39: 1935. (e) Bhattacharyya, S., Saha, S.K., Mandal, T.K. et al. (2001). Multiple hopping conduction in interpenetrating polymer network composites of polypyrrole and poly(styrene-co-butyl acrylate). *J. Appl. Phys.* 89: 5547.

43 (a) Davidov, D., Roth, S., Neumann, W., and Sixl, H. (1983). ESR and conductivity studies of doped polyacetylene. *Z. Phys. B* 51: 145. (b) Peo, M. (1982). Anomales magnetisches Verhalten von metallisch leitendem Polyacetylen. PhD thesis. Konstanz: University of Konstanz. (c) Ikoma, T. et al. (2002). Spin soliton in a π-conjugated ladder polydiacetylene. *Phys. Rev. B* 66: 014423.

44 Nechstein, M., Devreux, F., Genoud, F. et al. (1986). Polarons, bipolarons and charge interaction in polypyrrole: physical and electrochemical approaches. *Synth. Met.* 15: 59.

45 (a) Ikehata, S. et al. (1980). Solitons in polyacetylene: magnetic susceptibility. *Phys. Rev. Lett.* 45: 1123. (b) Epstein, A.J., Rommelmann, H., Druy, M.A. et al. (1981). Magnetic susceptibility of iodine doped polyacetylene: the effect of nonuniform doping. *Solid State Commun.* 38: 179. (c) Matsui, T., Ishiguru, T., and Tsukamoto, J. (1999). Spin Susceptibility and its relationship to structure in perchlorate doped polyacetylene in the intermediate dopant-concentration region. *Synth. Met.* 104: 179.

46 (a) Chung, T.C., Moreas, F., Flood, J.D., and Heeger, A.J. (1984). Solitons at high density in trans-$(CH)_x$: collective transport by mobile, spinless charged solitons. *Phys. Rev. B* 29: 2341. (b) Matsui, T. and Ishiguro, T. (2001). Spin gap behavior and electronic phase separation in doped polyacetylene. *Synth. Met.* 117: 15.

47 Kivelson, S. and Heeger, A.H. (1985). First-order transition to a metallic state in polyacetylene: a strong-coupling polaronic metal. *Phys. Rev. Lett.* 55: 308.

48 Zhu, Q., Fischer, J.E., Zuzok, R., and Roth, S. (1992). Crystal structure of polyacetylene revisited: an X-ray study. *Solid State Commun.* 83: 179.

49 (a) Naarmann, H. (1987). Synthesis of new conductive polymers. *Synth. Met.* 17: 223. (b) Marikhin, V.A., Guk, E.G., and Myasnikova, L.P. (1997). New approach to achieving the potentially high conductivity polydiacetylene. *Phys. Solid State* 39: 696. (c) Liu, P.Y. and Chen, J.F. (1999). Preparation of polyacetylene with a high conductivity. *J. Mater. Sci. Technol.* 15: 387. (d) Tsukamoto, J., Takahashi, A., and Kawasaki, K. (1990). Structure and electrical properties of polyacetylene yielding a conductivity of 105 S/cm. *Jpn. J. Appl. Phys.* 29: 125. (e) Akagi, K., Suezaki, M., Shirakawa, H. et al. (1989). Synthesis of polyacetylene films with high density and high mechanical strength. *Synth. Met.* 28: 1.

50 Ehinger, K. (1984). Elektrische Transporteigenschaften von Polyazetylen. PhD thesis. Konstanz: University of Konstanz.

51 Epstein, A.J., Rommelmann, H., Abkowitz, M., and Gibson, H.W. (1981). Temperature dependent conductivity of polyacetylene. *Mol. Cryst. Liq. Cryst.* 77: 81.

52 Roth, S., Pukacki, W., Schäfer-Siebert, D., and Zuzok, R. (1992). *Science and Technology of Conducting Polymers* (ed. M. Aldissi). CRC Press.

53 (a) Schäfer-Siebert, D. (1988). Metallisch leitende Polymere: Der begrenzende Einfluss von konjugationsdefekten auf die elektrische Leitfähigkeit dargestellt am Polyacetylen. PhD thesis. Karlsruhe. (b) Schäfer-Siebert, D., Budrowski, C., Kuzmany, H., and Roth, S. (1987). Influence of the conjugation length of polyacetylene chains on the DC conductivity. *Synth. Met.* 21: 285.

54 Miller, A. and Abrahams, E.A. (1960). Impurity conduction at low concentration. *Phys. Rev. B* 120: 745.

55 Mott, N.F. and Davis, E.A. (1979). *Electronic Processes in Non-Crystalline Materials*, 2e. Oxford: Clarendon Press.

56 Roth, S. (1991). *Hopping Transport in Solids* (ed. M. Pollak and B.I. Shklovskii). Amsterdam: Elsevier Science Publishers.

57 Schäfer-Siebert, D., Budrowski, C., Kuzmany, H., and Roth, S. (1987). *Electronic Properties of Conjugated Polymers*, Springer Series in Solid-State Sciences, vol. 76 (ed. H. Kuzmany, M. Mehring and S. Roth). Heidelberg: Springer Verlag.

58 Ehinger, K., Summerfield, S., Bauhofer, W., and Roth, S. (1984). DC and microwave conductivity of iodine-doped polyacetylene. *J. Phys. C Solid State Phys.* 17: 3753.

59 (a) Summerfield, S. and Butcher, P.N. (1982). A unified equivalent-circuit approach to the theory of AC and DC hopping conductivity in disordered systems. *J. Phys. C Solid State Phys.* 15: 7003. (b) Summerfield, S. and Butcher, P.N. (1983). AC and DC hopping conductivity in impurity bands and amorphous semiconductors. *J. Phys. C Solid State Phys.* 16: 295.

60 Sheng, P. (1980). Fluctuation-induced tunneling conduction in disordered materials. *Phys. Rev. B* 21: 2180.

61 (a) Yoon, C.O., Reghu, M., Moses, D., and Heeger, A.J. (1994). Transport near the metal-insulator transition: polypyrrole doped with PF5. *Phys. Rev. B* 49: 10851. (b) Kim, G.T. et al. (1999). Conductivity and magnetoresistance of polyacetylene fiber network. *Synth. Met.* 105: 207. (c) Heeger, A.J. (2002). The critical regime of the metal-insulator transition in conducting polymers: experimental studies. *Phys. Scr.* 102: 30. (d) Bhattacharyya, S. and Saha, S.K. (2002). Room-temperature metallic behavior in nanophase polyacetylene. *Appl. Phys. Lett.* 80: 4612. (e) Martens, H.C.F., Brom, H.B., and Menon, R. (2001). Metal-insulator transition in PF6 doped polypyrrole: failure of disorder-only models. *Phys. Rev. B* 64: 201102. (f) Ahlskog, M., Mukherjee, A.K., and Menon, R. (2001). Low temperature conductivity of metallic conducting polymers. *Synth. Met.* 119: 457. (g) Martens, H.C.F., Reedijk, J.A., Brom, H.B. et al. (2001). Metallic state in disordered quasi-one-dimensional conductors. *Phys. Rev. B* 63: 073203. (h) Chapman, B., Buckley, R.G., Kemp, N.T. et al. (1999). Low-energy conductivity of PF6-doped polypyrrole. *Phys. Rev. B* 60: 13479. (i) Clark, W.G. et al. (1999). High field NMR studies of conduction electron dynamics in metallic polypyrrole-SF6. *Synth. Met.* 101: 343. (j) Ahlskog, M. and Menon, R. (1998). The localization-interaction model applied to direct-current conductivity of metallic conducting polymers. *J. Phys. Condens. Matter* 10: 7171. (k) Fukuhara, T., Masubuchi, S., and Kazama, S. (1998). Pressure-induced metallic resistivity of PF6-doped poly(3-methylthiophene). *Synth. Met.* 92: 229.

62 Shirakawa, H. and Ikeda, S. (1971). Infrared spectra of polyacetylene. *Polym. J.* 2: 231.

63 (a) Naarmann, H. and Theophilou, N. (1987). New process for the production of metal-like, stable polyacetylene. *Synth. Met.* 22: 1. (b) Naarmann, H. (1990). The development of electrically conducting polymers. *Adv. Mater.* 2: 345. (c) Naarmann, H. (1987). *Electronic Properties of Conjugated Polymers*, Springer Series of Solid-State Sciences, vol. 76 (ed. H. Kuzmany, M. Mehring and S. Roth). Heidelberg: Springer Verlag.

64 Akagi, K., Katayama, S., Ito, M. et al. (1989). Direct synthesis of aligned cis-rich polyacetylene films using low-temperature nematic liquid crystals. *Synth. Met.* 28: D51.

65 BRITE/EURAM (1990–1995). Action HICOPOL (Highly Oriented Highly Conducting Polymers), Graz, Karlsruhe, Montpellier, Nantes, Strasbourg, Stuttgart.

66 (a) Pietronero, L. (1983). Ideal conductivity of carbon π polymers and intercalation compounds. *Synth. Met.* 8: 285. (b) Kivelson, S. and Heeger, A.J. (1988). Intrinsic conductivity of conducting polymers. *Synth. Met.* 22: 371. (c) Kryszewski, M. and Jeszka, J. (2003). Charge carrier transport in heterogeneous conducting polymer materials. *Macromol. Symp.* 194: 75. (d) Kohlman, R.S. et al. (1997). Limits of metallic conductivity in conducting polymers. *Phys. Rev. Lett.* 78: 3915. (e) Heeger, A.J. (1989). Charge transfer in conducting polymers, striving toward intrinsic properties. *Faraday Discuss. Chem. Soc.* 88: 203.

67 (a) Park, Y.W., Yoon, C.O., and Lee, C.H. (1989). Conductivity and thermoelectric power of newly processed polyacetylene. *Synth. Met.* 28: D27. (b) Plocharski, J., Pukacki, W., and Roth, S. (1994). Conductivity study of stretch oriented new polyacetylene. *J. Polym. Sci. B* 32: 447. (c) Pukacki, W., Zuzok, R., Roth, S., and Göpel, W. (1992). *Electronic Properties of Polymers – Orientation and Dimensionality of Conjugated Systems* (Kirchberg IV), Springer Series in Solid State Sciences, vol. 107 (ed. H. Kuzmany, M. Mehring and S. Roth). Heidelberg: Springer Verlag.

68 (a) Kaiser, A.B. (1992). *Electronic Properties of Polymers – Orientation and Dimensionality of Conjugated Systems* (Kirchberg IV), Springer Series in Solid States Sciences, vol. 107 (ed. H. Kuzmany, M. Mehring and S. Roth). Heidelberg: Springer Verlag. (b) Kaiser, A.B. (1993). Conductivity and thermopower of doped polyacetylene and polyaniline. *Mater. Sci. Forum* 122: 13.

69 (b) Röss, W., Philipp, A., Seeger, K. et al. (1983). Magnetoresistance and Corbino resistance of iodine-doped polyacetylene. *Solid State Commun.* 45: 933.

70 (a) Abrahams, E., Anderson, P.W., Licciardello, D.C., and Ramakrishnan, T.V. (1979). Scaling theory of localization: absence of quantum diffusion in two dimension. *Phys. Rev. Lett.* 42: 673. (b) Dynes, R.C. (1982). Localization and correlation effects in metals and semiconductors-experiment. *Physica* 109 and 110B: 1857. (c) Altshuler, B.L., Aronov, A.G., and Lee, P.A. (1980). Interaction effects in disordered Fermi systems in two dimensions. *Phys. Rev. Lett.* 44: 128. (d) Altshuler, B.L., Khmelnit'zkii, D., Larkin, A.I., and Lee, P.A. (1980). Magnetoresistance and Hall effect in a disordered two-dimensional electron gas. *Phys. Rev. B* 22: 5142.

71 Yoo, J.S., Park, J.W., Skakalova, V., and Roth, S. (2010). Shubnikov-de Haas and Aharonov-Bohm effects in graphene nanoring structure. *Appl. Phys. Lett.* 96: 143112.

72 Huefner, M. et al. (2009). Investigation of the Aharonov-Bohm effect in a gated graphene ring. *Phys. Status Solidi B* 246: 2756.

73 Novoselov, K.S. et al. (2005). Two-dimensional gas of massless Dirac fermions in graphene. *Nature* 438: 197.

74 McIntosch, G.C. et al. (2002). Orientation dependence of magneto-resistance behavior in a carbon nanotube rope. *Thin Solid Films* 417: 67.

75 (a) Rikken, G.L.J.A., Raupach, E., Krstic, V., and Roth, S. (2002). Magnetochiral anisotropy. *Mol. Phys.* 100: 1155. (b) Krstic, V. et al. (2003). The electrical magnetochiral effect. In: *Molecular Nanostructures* (ed. H. Kuzmany, J. Fink, M. Mehring and S. Roth). New York: American Institute of Physics. (c) Krstic, V., Weis, J., and Roth, S. (2003). Magnetotransport through single-walled carbon nanotubes. *Synth. Met.* 135: 799.

76 Liang, C.W., Sahakalkan, S., and Roth, S. (2008). Electrical characterisation of the mutual influence between gas molecules and single-walled carbon nanotubes. *Small* 4: 432.

77 Sahakalkan, S. (2014). Spin transport in carbon nanotubes. PhD thesis. Stuttgart and Tübingen: Max Plank Institute.

78 (a) Choi, A. et al. (2010). Suppression of the magneto resistance in high electric fields of polyacetylene nanofibers. *Synth. Met.* 160: 1349. (b) Park, Y.W. (2010). Magneto resistance of polyacetylene nanofibers. *Chem. Soc. Rev.* 39: 2428. (c) Choi, A. et al. (2012). Probing spin-charge relation by magnetoconductance in one-dimensional polymer nanofibers. *Phys. Rev. B* 86: 155423.

79 This is hypothesized based on similar phenomena seen in organic light emitting diodes with externally applied fields. See: (a) Hu, B. and Wu, Y. (2007). Tuning magnetoresistance between positive and negative values in organic semiconductors. *Nat. Mater.* 6: 985–991. (b) Xu, J., Smith, G.M., Dun, C. et al. (2015). Tunable distribution of singlet and triplet excitons in a self-coupled magnetic field. Presented at the Kirchberg Winterschool (7 March 2015).

80 Shen, Y.Q., Göhring, W., Hagen, S. et al. (1990). NMR and Pariser-Parr-Pople investigations of donor-acceptor substituted polyenes. *J. Mol. Electron.* 6: 31.

81 (a) Filzmoser, M. (1991). Optische Untersuchungen an Donator-Akzeptor modifizierten Polyenen. PhD thesis. Graz: Universität Graz. (b) Filzmoser, M. and Roth, S. (1991). Photoinduces absorption investigations on donor-acceptor polyenes. *Synth. Met.* 41: 1263.

82 (a) Hudson, B.S., Kohler, B.E., and Schulten, K. (1982). Linear polyene electronic structure and potential surfaces. *Excited States* 6: 1. (b) Kohler, B.E. (1995). *New Conducting Materials with Conjugated Double Conds: Polymers, Fullerenes, Graphite* (ed. J. Przyluski and S. Roth). Aedermannsdorf: Materials Science Forum, Trans Tech Publikations.

83 Anders, J. et al. (1993). Excited state transient spectroscopy of anthracene based photochromic systems. *Synth. Met.* 55–57: 4820.

84 Fischer, C.M., Burghard, M., Roth, S., and von Klitzing, K. (1994). Organic quantum wells: molecular rectification and single electron tunneling. *Europhys. Lett.* 28: 129.

85 Buckley, A. (ed.) (2013). *Organic Light-Emitting Diodes (OLEDs): Materials, Devices and Applications*, Electronic and Optical Materials. Woodhead Publishing.

86 Tsutsui, T. and Saito, S. (1993). *Intrinsically Conducting Polymers – An Emerging Technology* (ed. M. Aldissi). Dordrecht: Kluwer Academic Publishers.

87 Tang, C.W. and VanSlyke, S.A. (1987). Organic electroluminescent diodes. *Appl. Phys. Lett.* 51 (12): 912.

88 (a) Braun, D. and Heeger, A.J. (1991). Visible-light emission from semiconducting polymer diodes. *Appl. Phys. Lett.* 58: 1982. (b) Ohmori, Y., Uchida, M., Muro, K., and Yoshino, K. (1991). Visible-light electroluminescent diodes utilizing poly(3-alkylthiophene). *Jpn. J. Appl. Phys.* 30: L1938. (c) Ohmori, Y., Uchida, M., Muro, K., and Yoshino, K. (1991). Blue electroluminescent diodes utilizing poly(alkylfluorene). *Jpn. J. Appl. Phys.* 30: L1941. (d) Grem, G., Leditzky, G., Ullrich, B., and Leising, G. (1992). Realization of a blue-light emitting device using polyphenylene. *Adv. Mater.* 4: 36. (e) Burn, P.L., Holmes, A.B., Kraft, A. et al. (1992). Synthesis if a segmented conjugated polymer chain giving a blue-shifted electroluminescence and improved efficiency. *J. Chem. Soc. Chem. Commun.* (1): 32. (f) Burn, P.L. et al. (1992). Chemical tuning of electroluminescent co-polymers to improve emission efficiencies and allow patterning. *Nature* 356: 47. (g) Gustafsson, G., Cao, Y., Klavetter, F. et al. (1992). Flexible light-emitting diodes made from soluble conducting polymers. *Nature* 357: 477. (h) Brown, A.R. et al. (1993). *Intrinsically Conducting Polymers – An Emerging Technology* (ed. M. Aldissi). Dordrecht: Kluwer Academic Publishers. (i) Karg, S., Reiss, W., Dyakanov, V., and Schwoerer, M. (1993). Electrical and optical characterization of poly(phenylene-vinylene) light emitting diodes. *Synth. Met.* 54 (437).

89 (a) Yang, X., Neher, D., Hertel, D., and Daubler, T. (2004). Highly efficient single-layer polymer electrophosphorescent devices. *Adv. Mater.* 16 (2): 161. (b) Baldo, M.A. et al. (1998). Highly efficient phosphorescent emission from organic electroluminescent devices. *Nature* 395: 151. (c) Baldo, M.A., Lamansky, S., Burrows, P.E. et al. (1999). Very high-efficiency green organic light-emitting devices based on electrophosphorescence. *Appl. Phys. Lett.* 75: 4. (d) Adachi, C., Baldo, M.A., Thompson, M.E., and Forrest, S.R. (2001). Nearly 100% internal phosphorescence efficiency in an organic light-emitting device. *J. Appl. Phys.* 90: 5048.

90 Bässler, H. (1993). Charge transport in disordered organic photoconductors (a Monte Carlo simulation study). *Phys. Status Solidi B* 175: 15.

91 Abkowitz, M., Facci, J.S., and Rehm, J. (1998). Dimensional scaling of 1/f noise in the base current of self-aligned polysilicon emitter bipolar junction transistors. *J. Appl. Phys.* 82: 2670.

92 Ashcroft, N.W. and Mermin, N.D. (1976). *Solid State Physics*, 363. New York: Holt, Rinehart, Winston.

93 Fowler, R.H. and Nordheim, L. (1928). Electron emission in intense electric fields. *Proc. Royal Soc. A* 119: 173.

94 (a) Garstein, Y.N. and Conwell, E.M. (1996). Field-dependent thermal injection into a disordered molecular insulator. *Chem. Phys. Lett.* 255: 93.

(b) Arkhipov, V.I., Emelianova, E.V., Tak, Y.H., and Bässler, H. (1998). Charge injection into light-emitting diodes: theory and experiment. *J. Appl. Phys.* 84: 848. (c) Bussac, M.N., Michoud, D., and Zuppiroli, L. (1998). Electron injection into conjugated polymers. *Phys. Rev. Lett.* 81: 1678. (d) Emtage, P.R. and O'Dwyer, J.J. (1966). Richardson-Schottky effect in insulators. *Phys. Rev. Lett.* 16: 356.

95 Borsenberger, P.M. and Weiss, D.S. (1998). *Organic Photoreceptors for Electrophotography*. New York: Marcel Dekker.

96 (a) Hertel, D., Scherf, U., and Bässler, H. (1998). Charge carrier mobility in a ladder-type conjugated polymer. *Adv. Mater.* 10: 1119. (b) Redecker, M., Bradley, D.D.C., Inbasekaran, M., and Woo, E.P. (1998). Nondispersive hole transport in electroluminescent polyfluorene. *Appl. Phys. Lett.* 73: 1565.

97 Langevin, P. (1903). *Ann. Chem. Phys.* 3: 289.

98 (a) Takada, N., Tsutsu, T., and Saito, S. (1993). Control of emission characteristics in organic thin-film electroluminescent diodes using optical-microwave structure. *Appl. Phys. Lett.* 63: 2032. (b) Wittman, H.F., Gruner, J., Friend, R.H. et al. (1995). Microwave effect in a single-layer polymer light-emitting diode. *Adv. Mater.* 7: 541. (c) Tokito, S., Noda, K., and Taga, Y. (1996). Strongly directed single mode emission from organic electroluminescent diode with microcavity. *Appl. Phys. Lett.* 68: 2633.

99 Aiamsen, P., Carroll, D.L., and Phanichphant, S. (2011). Synthesis and optical properties of light-emitting polyfluorene derivatives. *Mol. Cryst. Liq. Cryst.* 539 (88).

100 Czerw, R., Carroll, D.L., Woo, H.S. et al. (2004). Nanoscale observation of failures in organic light-emitting diodes. *J. Appl. Phys.* 96: 641.

101 Lee, S.B., Fujita, K., and Tsutsui, T. (2005). Emission mechanism of double-insulating organic electroluminescence device driven at AC voltage. *Jpn. J. Appl. Phys.* 44: 6607.

102 (a) Perumal, A. et al. (2011). Novel approach for alternating current (AC)-driven organic light-emitting devices. *Adv. Funct. Mater.* 22: 210. (b) Sung, J., Choi, Y.S., Kang, S.J. et al. (2011). AC field-induced polymer electroluminescence with single wall carbon nanotubes. *Nano Lett.* 11: 966. (c) Chen, Y., Xia, Y., Smith, G.M. et al. (2012). Emission characteristics in solution-processed asymmetric white alternating current field-induced polymer electroluminescent devices. *Appl. Phys. Lett.* 102: 013307.

103 Chen, Y. et al. (2014). Solution-processable hole-generation layer and electron-transporting layer: towards high-performance, alternating-current-driven, field-induced polymer electroluminescent devices. *Adv. Funct. Mater.* 24 (18): 2677.

104 Chen, Y. et al. (2014). High-color-quality white emission in AC-driven field-induced polymer electroluminescent devices. *Org. Electron.* 15: 182.

105 Chen, Y. et al. (2014). Solution-processed highly efficient alternating current-driven field-induced polymer electroluminescent devices employing high-k relaxor ferroelectric polymer dielectric. *Adv. Funct. Mater.* 24 (11): 1501.

106 Chen, Y., Smith, G.M., Loughman, E. et al. (2013). Effect of multi-walled carbon nanotubes on electron injection and charge generation in AC field-induced polymer electroluminescence. *Org. Electron.* 14: 8.

107 Fröbel, M., Hofmann, S., Leo, K., and Gather, M.C. (2014). Optimizing the internal electric field distribution of alternating current driven organic light-emitting devices for a reduced operating voltage. *Appl. Phys. Lett.* 104: 071105.

108 Samuel, I.D.W. and Turnbull, G.A. (2007). Organic semiconductor lasers. *Chem. Rev.* 107: 1272.

109 Svelto, O. (1998). *Principles of Lasers*, 4e. New York: Plenum Press.

110 Tessler, N. (1999). Lasers based on semiconducting organic materials. *Adv. Mater.* 11: 363.

111 Muccini, M. (2006). A bright future for organic field-effect transistors. *Nat. Mater.* 5: 605.

112 Schols, S., Verlaak, S., and Heremans, P. (2006). A novel organic light-emitting device for use in electrically pumped lasers. *Proc. SPIE* 6333: 63330U. https://doi.org/10.1117/12.680343.

113 Silinsh, E.A. (1970). On the physical nature of traps in molecular crystals. *Phys. Status Solidi* 3: 817.

114 (a) Sariciftci, N.S., Smilowitz, L., Heeger, A.J., and Wudl, F. (1992). Photoinduced electron transfer from a conducting polymer to buckminsterfullerene. *Science* 258: 1474. (b) Yu, G., Gao, J., Hummelen, J.C. et al. (1995). Polymer photovoltaic cells: enhanced efficiencies via a network of internal donor-acceptor heterojunctions. *Science* 270: 1789.

115 Marks, R.N., Halls, J.J.M., Bradley, D.D.C. et al. (1994). The photovoltaic response in poly(p-phenylene vinylene) thin-film devices. *J. Phys. Condens. Matter* 6: 1379.

116 For a review in this field see, for example:Hadziioannou, G. and van Hutten, P.F. (eds.) (2000). *Semiconducting Polymers*. Weinheim: Wiley-VCH.

117 Kymakis, E. and Amaratunga, G.A.J. (2002). Single-wall carbon nanotube/conjugated polymer photovoltaic devices. *Appl. Phys. Lett.* 80: 112.

118 (a) Horowitz, G. (1990). Organic superconductors for new electronic devices. *Adv. Mater.* 2: 287. (b) Garnier, F., Horowitz, G., and Fichou, D. (1989). Conjugated polymers and oligomers as active materials for electronic devices. *Synth. Met.* 28: C705. (c) Burroughes, J.H., Friend, R.H., and Allen, P.C. (1989). Field-enhanced conductivity in polyacetylene-construction of a field-effect transistor. *J. Phys. D. Appl. Phys.* 22: 956. (d) Burroughes, J.H., Jones, C.A., Lawrence, R.A., and Friend, R.H. (1990). *Conjugated Polymeric Materials, Opportunities in Electronics, Optoelectronics, and Molecular Electronics*, NATO ASI Series E: Applied Science, vol. 182 (ed. J.L. Bredas and R.R. Chance). Dordrecht: Kluwer Academic Publishers. (e) Paloheimo, J., Kuivalainen, P., Stubb, H. et al. (1990). Molecular field-effect transistor using conducting polymer Langmuir-Blodgett films. *Appl. Phys. Lett.* 56: 1157. (f) Paloheimo, J., Punkka, E., Kuivalainen, P. et al. (1989). New dimensions for semiconductor devices: polymeric field-effect transistors. *Acta Polytech. Scandinavica*, Electrical Engineering Series 64: 178. (g) Garnier, F., Deloffre,

F., Yassar, A. et al. (1993). *Intrinsically Conducting Polymers – An Emerging Technology* (ed. M. Aldissi), 107. Dordrecht: Kluwer Academic Publishers.

119 See for example:Sze, S.M. (1981). *Physics of Semiconductor Devices*, 2e. New York: Wiley Interscience Publishers.

120 (a) Hewitt, C.A., Kaiser, A.B., Roth, S. et al. (2011). Varying the concentration of single walled carbon nanotubes in thin film polymer composites, and its effect on thermoelectric power. *Appl. Phys. Lett.* 98: 183110.
(b) Hewitt, C.A., Kaiser, A.B., Roth, S. et al. (2012). Multilayered carbon nanotube/polymer composite based thermoelectric fabrics. *Nano Lett.* 12 (3): 1307.

121 Goldsmid, H.J., Sheard, A.R., and Wright, D.A. (1958). The performance of bismuth telluride thermojunctions. *Br. J. Appl. Phys.* 9: 365.

122 Massonnet, N. et al. (2014). Improvement of the Seebeck coefficient of PEDOT:PSS by chemical reduction combined with a novel method for its transfer using free-standing thin films. *J. Mater. Chem. C* 2: 1278.

123 Yu, C., Choi, K., Yin, L., and Grunlan, J.C. (2011). Light-weight flexible carbon nanotube based organic composites with large thermoelectric power factors. *ACS Nano* 5: 7885.

124 Snyder, G.J. and Toberer, E.S. (2008). Complex thermoelectric materials. *Nat. Mater.* 7: 105.

125 Kennedy, T. and Lieb, E.H. (1987). Proof of the Peierls instability in one dimension. *Phys. Rev. Lett.* 59: 1309.

10

Correlation and Coupling

In conducting polymers, we saw how the *coupling* between a distortion on a one-dimensional (1D) lattice and a charge associated with the bonds at that distortion can move together and act as a single *composite* particle. However, this particle, because of its partner distortion, can take on properties that a bare charge could not. These phenomena are, quite generally, the nature of *correlation* and *coupling* in solid-state systems. Now, the reader will notice that we have used these two words together, but they really describe slightly different aspects of an *interaction*. Electrons that act in concert with each other are said to be *correlated*. However the exact mechanism of *coupling* their behaviors together may or may not be known completely. So, as you can see, the use of these terms can be a little nuanced.

In this chapter we go a little further into the ways in which a material's electronic and/or vibrational properties are linked. Our approach here is surely quite heuristic: to examine a property of a system and find a physical picture that explains how this property comes about. This traditional approach gives a remarkably good "seat-of-the-pants"[1] understanding of correlation in materials. But it is a *hodge-podge*. The properties we are looking at are all a natural result of the many-particle nature of the complete Hamiltonian for the system and its many-particle wavefunction. Whatever we find in our "seat-of-the-pants tour" should converge to a description given by the full many-particle wavefunction description, if we were only clever enough to solve for it.

10.1 The Metal-to-Insulator Transition and the Mott Insulator

It should be obvious that electrons moving about in a solid don't pass through each other, and they don't bounce about like small hard-core marbles. They do interact through long-range Coulomb forces, however, sometimes rather

1 A great old phrase borrowed from a dear friend, Worth Seagondollar, who is no longer with us. It was his contention that if you couldn't explain things on the back of a napkin or an envelope, then you simply couldn't explain it. Worth was a professor of physics at NCSU, having previously worked on the Manhattan Project. I, [Carroll] have always been confronted by the simple, brutal reality of this statement.

Foundations of Solid State Physics: Dimensionality and Symmetry, First Edition.
Siegmar Roth and David Carroll.

strongly. These situations can be termed *systems with strong electron–electron correlation*. The classic example of this might be the case of materials that should be metals but are in fact insulators.

As we have seen in our discussions of band structure, electrons that move through the solid originate with the atoms, which become ionized in a way. In the case of transition metal monoxides (such as CoO, NiO, and MnO), the conduction band of the solid is frequently "composed of" d orbitals[2] that do not extend far from the lattice site. Thus, there should be a significant level of localization to these electrons – you can think of the modulus of the Bloch wave as having a large electron density near the atomic sites generally. If one calculates the band structure of a material like NiO or similar, we find that the Fermi level falls in a partially filled band, so there should be no trouble with calling this a *metal*. Perhaps the bands are rather flat, and there is this degree of "localization," but that just leads to a large m^* and shouldn't change the conduction properties otherwise. Unfortunately stoichiometric NiO is an insulator. So what went wrong?

In 1949 Mott introduced a model for interactions that might solve this puzzle [1]. Remember that if there are *no interactions* between the electrons at all, then the relevant length scale between ionic lattice positions is simply the lattice constant/s "d_{ij}." This length scale will normalize everything in the physics to the distances it takes the electron to travel from lattice site to lattice site under whatever transport model we might choose (except Drude of course, which doesn't recognize the lattice). However, an electron localized even momentarily at a specific lattice site *does* repel the other electrons at other lattice sites unless the Coulomb interaction is effectively screened by the polarization of a surrounding environment. Thus, there is a second length scale introduced into the problem that has to do with the extent of this screened Coulomb potential. You might allow, as a guess, this to be the Bohr radius: "a_0." In such cases the "free electrons" tend to delocalize over the lattice to lower their state energy, but the electron–electron repulsive interactions work against this delocalization, forcing localization on lattice sites. This reduces the density of states (DOS) at the Fermi level and a *Mott gap* opens up. This is the essence of the Mott argument:

$$d_{ij} \gg a_0 \text{ is an insulator} \tag{10.1a}$$

$$d_{ij} \ll a_0 \text{ is a metal} \tag{10.1b}$$

There should actually be some critical value of d_{ij} at which a metal-to-insulator transition (MIT) occurs. To see this a bit more clearly, we follow Mott and introduce a *pseudo-order parameter*:

$$\varepsilon = I - E \tag{10.2}$$

where I is the ionization energy of the solid and E is the electron affinity. So if there are no electron–electron interactions in the solid, $I = E$ and $\varepsilon = 0$. However, if there are such interactions, clearly $I > E$ and $\varepsilon > 0$. Note as d decreases, $d_{ij} \to$ o, ε increases, and the system will undergo a transition from metallic to insulating. We might think this could happen in two ways: path a (continuous) and path b (discontinuous) as seen in Figure 10.1.

2 In the sense of LCAO.

So our picture is one in which the electrons are held in their places so to speak. This means there is an energy gap from the "frozen state" to a state in which transport can take place. The DOS splits into two bands, known as a *lower Hubbard band* and an *upper Hubbard band*. The energy gap that occurs is given by the deceptively simple term

$$E_{\text{Mott gap}} \sim U - 2zt \tag{10.3}$$

where U is the Coulomb energy, z is the number of nearest neighbor atoms, and t is the so-called *transfer integral* usually computed using tight binding models of band structure. The gap occurs when U is large enough.

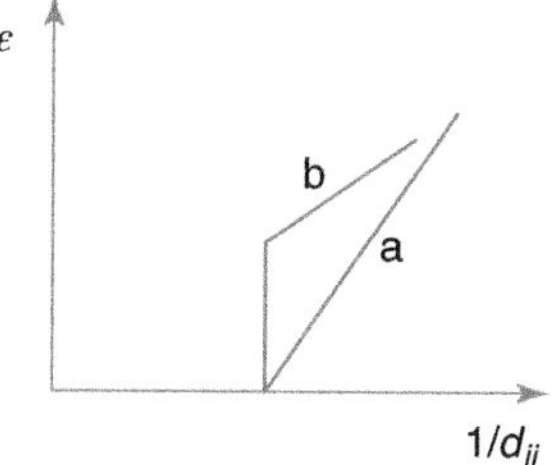

Figure 10.1 Two possible paths that a system can proceed toward a metal-to-insulator transition as the lattice constant is changed. Path a represents a continuous phase transition; however the discontinuous path b represents a first-order phase transition in the system. Mott argued for the second scenario.

So, we have used the terms *Hubbard this* and *Hubbard that*. This refers to the mathematical model that Hubbard came up with to describe Mott's idea: the *Hubbard model*. This model is seen as a static lattice occupied by electrons with spin-up or spin-down hopping between sites. Figure 10.2 shows this schematically. A site can be occupied by zero, one, or two electrons – but, in the case of two electrons, the spins must be antiparallel to satisfy Pauli. In a configuration that has two antiparallel spin electrons, there is an increased energy due to the Coulomb repulsion. This additional energy is given as U (the same U as above). In the model, the electrons are free to hop from site to site, and when they do their spin projection doesn't change. The amplitude for this hopping comes from the overlap integral between the two wavefunctions, and this is denoted as $-t$. Finally, a chemical potential is usually added to the model so that the energy increases as the number of electrons in the configuration increases.

The full formulation of this problem is really better done using *second quantization*, with creation and annihilation operators as opposed to analytical wavefunctions. While this is not particularly difficult, it does sort of miss the point of our present discussion: "seat-of-the-pants pictures of mechanisms."[3] So, we will set the problem here, but to see it in mathematical detail, you must go to one of the many specialized texts on the topic. Here we take a moment to say that this model has been applied now to a wide range of problems, not just Mott insulators and their analogues. The impact of Hubbard-like models on optical lattices to gauge theories should not be underestimated. This apparently again points to a

3 At this point in the text, many problems are approached with advanced and nuanced mathematical techniques that can sometimes obscure the underlying physics we are discussing. This is an issue of philosophical perspective. We, the authors, wish to provide a background that allows the reader to picture how some process is taking place. The point of this is to gain the kind of understanding that allows for more detailed discussions with specialists in the subtopics or to launch ones own enterprise in learning about those subtopics. So we see the mathematical apparatus as artifice and the statement of explanation about the physics as the underlying truth approaching some reality of the universe. This perspective is not universally held: but hey, "our book, our rules…"

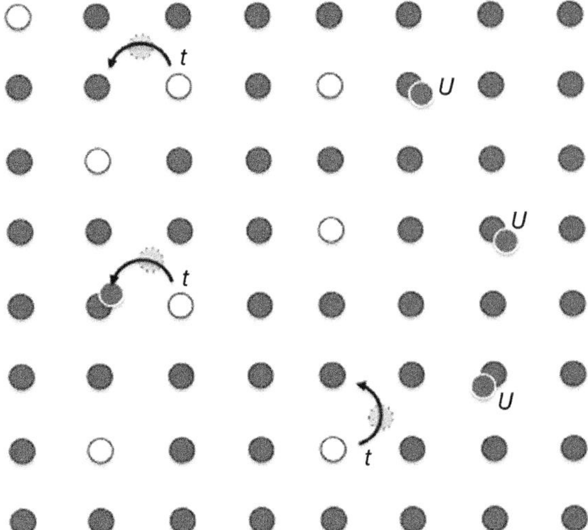

Figure 10.2 A picture of the Hubbard model. The lattice is shown as the empty circles, and the colored circles are the occupying electron population. Notice that when two electrons end up on a site, the configuration energy goes up by U. Of course, this is the 2D analogue of the idea, and one can easily imagine how this could be thought of in 3D or in 1D. In the example shown three sites are doubly occupied, two sites have an electron hopping onto an unfilled site to singly occupy it, and a single site has an electron hopping onto a sit that is already occupied by an electron, thereby making it become doubly occupied as well. One might also image a large chemical potential in this system since there are nearly enough electrons to occupy all sites. For fewer electrons, such a chemical potential would decrease as would the probability of double occupancy.

universality in the descriptive language of physics with many of our mathematical models being appropriate for more than one problem.

10.1.1 The Hamiltonian

For our discussion of Mott insulators, we start with a Hamiltonian of the system: the Hubbard Hamiltonian (HH). Of course we wish to avoid the full many-particle description, since it is generally not easily solved, and we make approximations that will describe the essence of our hopping picture (Figure 10.2). A reasonable first guess, Hubbard suggested, might look something like this:

$$H = H_{\text{KE}}(-t) + H_{\text{Int}}(U) - H(\mu) \tag{10.4}$$

The first term is a "kinetic energy," and its sum preserves spin projection (it does not allow for the same spin on a single site). This is obviously a function of the hopping integral ($-t$). In other words all KE, like a good beer, is wrapped up in the "hops." The second term is the "potential energy" of electrostatic repulsion. This term is a sum that goes through every lattice site and adds the constant U for every site that has two electrons, arranged with antiparallel spins of course. The last term is the chemical potential and accounts for the filling of the bands with electrons. The case where there is one electron per site is the "half-filling" case, because it has half the total number of electrons per site as the maximum case, which would be two electrons per site. In fact, this is the most interesting case, and many Hubbard model studies begin here since it is the case that tends to show the interesting physics such as the Mott gap.

10.1.2 The Lattice and Antiferromagnetic Ordering

Because the t integral is sensitive to spin and the U is added to each site with more than one electron, the Hubbard model allows for the system to relax into spin-ordered phases as shown in Figure 10.3. The *bipartite* lattice configuration can be divided into two sublattices: A and B, where the nearest neighbor of each A lattice site is a B lattice site and *vice versa*. Thus, bipartite lattices can be given to antiferromagnetic ordering as seen in Figure 10.3. We note that square and hexagonal lattices are both of the bipartite type, whereas the triangular lattice is not. Physically there is just no way the triangular lattice sites can be arranged such that sites of different spins are always nearest neighbors. In the triangular lattice there will always be two sites with similar spin next to each other. Thus we refer to antiferromagnetic ordering on the triangular lattice as *frustrated*.

10.1.3 Other Considerations: The Particle-Hole Symmetry (PHS)

This HH has an interesting *particle-hole symmetry* (PHS) in bipartite lattices. Notice that the overlap integral for an electron hopping in one direction must be equivalent to a hole hopping in the opposite direction. Thus, we must conclude that this part of the Hamiltonian makes no distinction. The PHS allows us to relate properties of the HH to different values of parameters in the system. It also forms the basis of constructing mappings between attractive and repulsive HHs. You will find it used frequently in *quantum Monte Carlo simulations*.

This (PHS) symmetry has a rather important impact on the behavior of bipartite systems. It appears when we consider the exchange of electrons for holes in

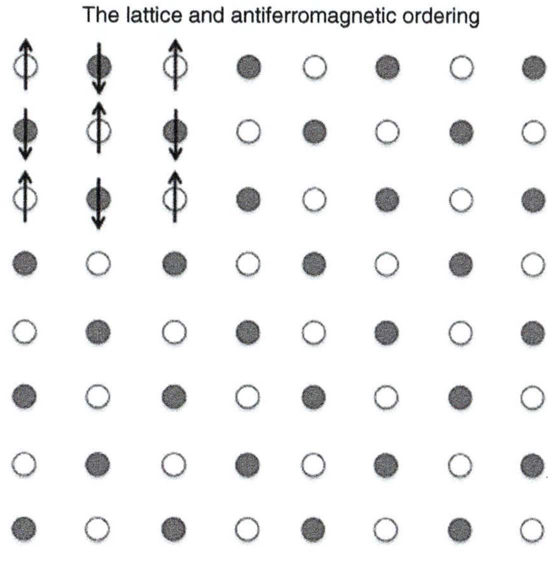

The lattice and antiferromagnetic ordering

Figure 10.3 The *bipartite* ordered lattice of the *half-filled* Hubbard model. The spin projections of the electrons are shown in the upper left, and each of the sites has an electron.

the Hubbard model. Formally, this is done through a *particle-hole transformation* (PHT) in which creation and annihilation operators are switched. However, it is clear that, for the bipartite lattice, the effect of switching electrons and holes is twofold. First, lattice sites of sublattice A are transformed to sublattice B. This is because t relates the two nearest neighbor sites, and so all hopping transitions take place between these two sublattices. If we run this backward, a hole goes from B to A as an electron goes from A to B. Secondly, eigenstate occupation numbers are switched: $n = 1$ occupied states become unoccupied states, $n = 0$. Thus, in our HH, the kinetic energy term must look exactly the same under an exchange of electrons to holes. If we further take the step to shift the chemical potential and internal energy by a trivial constant, we can redefine U such that $+\frac{1}{2}U$ and $-\frac{1}{2}U$ is the energy associated with the double occupancy and single occupancy of a given sublattice site. This leaves the HH for holes equivalent to that of the electrons except for the chemical potential.

So, under a PHT, the density operator ρ transforms to $1 - \rho$, and the HH transforms to the same HH with a negative sign in front of the μ term. As a result at $\mu = 0$ we have half-filling in the system, and $\rho = 1$ for any value of t, T, and U. This implies that the phase diagram for the bipartite system is symmetric around half-filling. This fact can be useful since the square lattice HH is sometimes applied to cuprate superconductors (the so-called type II superconductors). Often in such models a t' integral is included that connects sites across the diagonal of the square lattice. This means that sits on the same sublattice are connected, thereby breaking the PHS. So, the properties of the HH are not the same for $\mu < 0$ and $\mu > 0$. This, in fact, correctly captures the observation that n-doped and p-doped cuprates have strikingly different properties.

10.1.4 The Hubbard Model in Lower Dimensions

Despite the relative simplicity of the Hubbard model, physicists have not been able to fully (exactly) solve for its properties in the *thermodynamic limit*, for systems of two or three dimensions. Here, the *thermodynamic limit* means systems with large numbers of sites and electrons. Indeed, this remains a kind of "holy grail" for condensed matter physicists in general. Of course, this doesn't mean we can't solve the easier problems, and they can be quite instructive. To see this let's look at the *two-site Hubbard problem*.

The *two-site Hubbard model* is the simplest nontrivial example of the properties of the Hubbard. It is useful for understanding the binding of some binary molecules and, through extension, some low-dimensional structures. In our case (the two-site model), there are 16 possible configurations of electrons on two sites. They are 1 with no electrons, 4 with one electron (an up or down electron on each of the two sites), 6 with two electrons (1 with two up electrons on different sites, 1 with two down electrons, and 4 with an up electron and a down electron), 4 with three electrons, and 1 with four electrons. Notice that we have included possibilities with more than two electrons on a site since there are four available, two from each of the sites.

Electrons are conserved of course, and they don't change their spin when they hop in our model. So we can classify configurations into different groupings wherein each element of a group is somehow equivalent. We will solve for the relative energy states of each separately. Now to start, the configuration with no electrons is not like any other configuration: it has an energy of zero. The two configurations with an up electron are equivalent to each other, as are the two configurations with a single down electron.

We next have to solve the Schrödinger equation for the single up electron (and single down electron) case. We'll use a notation for the wavefunction that looks like this:

$$|\psi\rangle = a|\uparrow, 0\rangle + b|0, \uparrow\rangle \tag{10.5}$$

where a and b are complex numbers that satisfy a normalization condition that the sum of their squares equals one, $|\uparrow, 0\rangle$ is the configuration with the up electron on the first site and with the second site open, and $|0, \uparrow\rangle$ is the other configuration with the up electron on the second site.

The Schrödinger equation, $H|\psi\rangle = E|\psi\rangle$, uses our Hubbard Hamiltonian H, and E are the energies of the stationary states. In this case, H is a two-by-two matrix, and we need to solve the eigenvalue equation

$$\begin{bmatrix} 0 & -t \\ -t & 0 \end{bmatrix} \begin{bmatrix} a \\ b \end{bmatrix} = E \begin{bmatrix} a \\ b \end{bmatrix} \tag{10.6}$$

and the solutions look like

$$a = b = 1\sqrt{2} \tag{10.7}$$

for $E = -t$ and

$$a = 1/\sqrt{2};\ \ b = -1/\sqrt{2} \tag{10.8}$$

for $E = t$.

We already have an interesting finding: in the delocalized solutions, the spin is not at one site or another. Instead the spin is delocalized between the two sites. There is a *symmetric state* with an energy of $-t$ and an *antisymmetric state* with energy t. So, if the system is in either one of these states, the probability of finding a spin at a given site is simply 1/2.

Our next group of states to consider is the slightly more complicated situation of configurations with two electrons. If the spins are on different sites and both are pointing up or (equivalently) pointing down, we have another energy 0 configuration. Remember that to be on the same site, they must be antiparallel. So, there are four configurations with one up spin and one down spin. These configurations are equivalent to each other. In our notation, the wavefunction looks like

$$|\psi\rangle = a|\uparrow\downarrow, 0\rangle + b|\uparrow, \downarrow\rangle + c|\downarrow, \uparrow\rangle + d|0, \uparrow\downarrow\rangle \tag{10.9}$$

The Schrödinger equation gives

$$\begin{bmatrix} U & -t & -t & 0 \\ -t & 0 & 0 & -t \\ -t & 0 & 0 & -t \\ 0 & -t & -t & U \end{bmatrix} \begin{bmatrix} a \\ b \\ c \\ d \end{bmatrix} = E \begin{bmatrix} a \\ b \\ c \\ d \end{bmatrix} \tag{10.10}$$

All the algebra of this eigenvalue problem is left for the reader, but the result is that the lowest energy state with one up and one down electron has a lower energy than the states obtained with two electrons of the same spin projection. This causes an *antiferromagnetic* tendency in the system; spins on neighboring sites tend to want to point in opposite directions. Of course, physically, this occurs because the electrons have the ability to hop back and forth. So, as we guessed above in the bipartite lattice, the two-site Hubbard model with one electron per site should be a molecular *anti*ferromagnet at low temperatures.

Now we skip ahead a bit; the states obtained from configurations with three electrons have energies of $U - t$ or $U + t$, and the state with four electrons has an energy of $2U$. That should be all of them. Once one has all the energies of the possible system configurations, then the probability of each configuration is weighted against the Boltzmann distribution to get the *statistical mechanical* behavior of the system. The two sources of randomness in the system are *quantum mechanical* and *statistical* (thermal fluctuations). This is true as we scale the system up as well. Here we have considered only the two-site model, but what if we include more sites? Then two things must be true: (i) there must be some adjustment to the chemical potential as the number of electrons is increased (we ignored this in our simple two-site model), and (ii) the matrix eigenvalue equation becomes exponentially large, leading to computational problems.

10.1.5 Real One-Dimensional Mott Systems

Can such an insulator really occur in one dimension? What systems would become a Mott insulator in 1D? In our previous Gedanken experiment, we can immerse our monatomic chain into a screening medium, thereby mitigating such effects. This allowed us to discuss polyene chains and conductive polymers as if the *electron–lattice coupling* were the dominant interaction, and the simplification introduced many relevant concepts in a didactic way: *ad usum Delphini.*[4] But, in many real 1D systems, this Mott-like *electron–electron interaction* is as large as the *electron–lattice coupling*, and the expectation of a Peierls gap is not valid, even as a rough approximation. This is particularly true for some conjugated polymers [2]. The study of 1D Mott insulators has become important in trying to understand *fractionalized quasiparticles*. These are excitations within the solid that appear to "split" charge and spin into separately acting objects. Of particular interest is the argument that spin–orbit splitting is a 1D signature; see, for example, the Sr_2CuO_3 system [3].

At this point there are so many excellent reviews and extensions of the Hubbard model to problems tangential to the Mott insulator that we are better off leaving the reader here to their own discoveries. But Mott–Hubbard models are very useful and should be a part of every working condensed matter physicist's language.

---

4 Dauphin was the title of the French crown prince. Obscene parts of classical books were censored to protect the child's soul. Such books were marked "ad usum Delphini."

10.2 The Superconductor

The *Hubbard model* has set us up for a discussion on systems to which it has been widely applied: *superconductors*. One of the most stunning observations in solid-state electronic transport of the past century must be that of *superconductivity*. Historically, the field of *1D metals* was motivated by the search for high-temperature superconductors as we have seen. And just as in the case of Mott, the effect arises when electrons *correlate*.

10.2.1 The Basic Phenomena

Only a decade after Thomson's discovery of the electron (1887) and Drude's model of the free electron gas (1900), Heike Kamerlingh Onnes discovered that the resistivity of mercury "suddenly vanishes" when the sample is cooled slightly below the boiling point of liquid helium [4]. In Figure 10.4 we see the data that Onnes interpreted as a *phase transition* and where he first coined the term *superconductivity*. So far, >40 elements have been found to become superconducting at low temperatures, for some only at very low temperatures (for example, Rh at 3.2×10^{-4} K), and others only at very high pressures in addition to the cooling (e.g. Si: $T_c = 5.4$ K at $p > 110$ kbar; Se: $T_c = 6.9$ K at $p > 130$ kbar). There are also an astonishing 1000 or more superconducting alloys and compounds known [5]. Two types of superconductors have been identified, rather conveniently, namely, type I and type II. You can tell the difference by the slightly different behaviors in magnetic fields of the two types.

> Who would a superconductor wish to be; must answer me these three questions ... with apologies to Monty Python

There are *three* important parameters that *define* superconductivity – to have a superconductor you must have the following:

1. A *critical temperature* at which the material becomes superconducting. This is the huge drop in resistivity as seen in Figure 10.4. The critical temperature is usually denoted $T_{c.}$

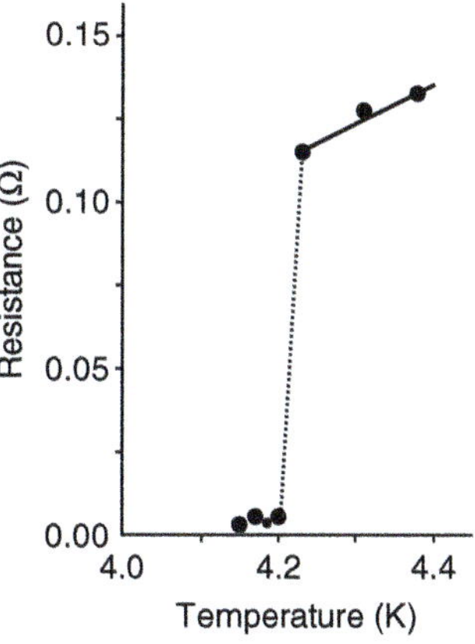

Figure 10.4 Heike Kamerlingh Onnes' discovery of superconductivity in 1911. At 4.15 K the resistance of a mercury sample dropped from some tenth of an ohm to less than 10^{-5} Ω [4].

2. A *critical magnetic field* above which superconductivity breaks down. This is accompanied by the *Meissner effect* or the complete repulsion of the magnetic field by the superconductor. By now, you have all seen the "permanent magnet floating above the superconductor" pictures. But these demonstrations are an important part of determining if you truly have a superconductor. Moreover, type II superconductors have an interesting caveat to this rule: they allow some flux lines to penetrate the superconductor, but only in bunches, and these result in current vortices.
3. A *critical current density* that is the maximum current density a superconductor can support, and beyond which it reverts to a normal conductor.[5]

Let's look at these requirements a little more closely. The first is a *critical temperature.* The structure of the conductivity vs. temperature is discussed in more detail in specialty texts on superconductivity [6] as well as numerous review articles [5, 7]. For us however, in presenting Onnes' discovery, we begin by noting that he used words to the effect of the resistance "suddenly vanishes." The disappearance of the resistance and resulting *persistent current* is very exciting. From our general knowledge of metallic conductivity, we may not be surprised to find zero resistivity in a perfect, defect-free three-dimensional (3D) metal at absolute zero, when all phonons are frozen out (*ballistic* transport). However, zero resistance at a finite temperature, say, at 4.15 K as in Hg, or at 135 K as in the system Hg–Ba–Ca–Cr–O, or even at room temperature, would contradict our intuition. Perhaps this is because our intuition is built upon single particle state energetics *without correlation.* But as we have just seen, correlation is a natural and real part of materials systems. So maybe we shouldn't be so surprised after all.

The sharp "turn on" of these effects can be a bit confusing however. The *superconducting phase transition* is remarkably sharp. The German word *Sprungtemperatur* expresses this sharpness better than the English term *critical temperature* (sprung = jump). It is difficult to imagine that all possible electron scatterers suddenly vanish or move out of the way and much easier to assume that the electrons change in such a way that they do not scatter anymore. This *phase transition* in the electron gas is essential for the presently accepted theory of superconductivity, which was formulated in 1957 by Bardeen, Cooper, and Schrieffer (BCS) [8]. Note the time: almost half a century passed between H.K. Onnes' discovery and the *BCS theory.*

Next we have the *Meissner effect* or the total repulsion of magnetic field from the volume of a superconductor. In 1933, Meissner and Ochsenfeld [9] found that a superconductor *expels* an applied magnetic field so that within the superconductor the magnetic field is zero. More specifically, below T_c the flux density within the superconductor is $\boldsymbol{B} = 0$. Recall

$$\boldsymbol{B} = \boldsymbol{H} + 4\pi\boldsymbol{M} \tag{10.11}$$

In the perfect conductor:

$$m\ddot{r} = -e\boldsymbol{E}$$

5 Some scientists will take exception here. Many do not consider the critical current a fundament sign of superconductivity. However, if we were in a lab looking for proof of such a state, we would feel happier if we saw this.

and current density is given as:

$$J = -en_s\dot{r}$$

where n_s is the number of superconducting charges. So:

$$\dot{J} = \frac{n_s e^2}{m}E.$$

From Faraday's Law:

$$\nabla \times E = -\frac{1}{c}\frac{\partial B}{\partial t}$$

and:

$$\nabla \times \frac{\partial J}{\partial t} = -\frac{n_s e^2}{cm}\frac{\partial B}{\partial t}.$$

And Amper's law:

$$\nabla \times B = \frac{4\pi}{c}J$$

which then gives:

$$\nabla \times \nabla \times \frac{\partial B}{\partial t} = -\frac{4\pi n_s e^2}{mc^2}\frac{\partial B}{\partial t}$$

Applying a vector identity and another of Maxwell's eEquations:

$$\nabla \times \nabla \times C = \nabla(\nabla \cdot C) - \nabla^2 C$$
$$\nabla \cdot B = 0$$

we get:

$$\nabla^2\left(\frac{\partial B}{\partial t}\right) = \lambda^{-2}\left(\frac{\partial B}{\partial t}\right)$$
$$\lambda = \sqrt{\frac{mc^2}{4\pi n_s e^2}}.$$

This λ is known as the *penetration length.* Let's solve this differential equation in the 1D case where the dimension x moves from a half space of superconductor, to a half space of normal conductor. We get:

$$\frac{\partial B}{\partial t} = \left(\frac{\partial B}{\partial t}\right)_{x=0} e^{-x/\lambda}.$$

And $B = 0$ for the superconductor, so the susceptibility $\chi = \partial M/\partial H$ yields $\chi = -\frac{1}{4\pi}$. If one increases the field applied to a superconductor, it eventually destroys the superconducting state. The system is driven back into the normal conducting state. In *type I* superconductors, there is no intermediate state separating the superconducting and normal states when the field is increased. However, in the *type II* superconductors, a mixed state occurs. This state typically appears before the transition from superconducting into normal state. In this *mixed state,* the magnetic field does partially penetrate the material through the formation of an array of *flux tubes.* These flux tubes are simply tubes of normal state that enclose the flux of the field as it passes through the superconductor. However, it is curious that the magnetic flux carried by these tubules is quantized: some multiple of the magnetic flux quantum

$$\Phi_0 = \frac{hc}{2|e|} \tag{10.12}$$

From a theoretical point of view, the perfect conductivity of a superconductor comes from electron–electron mediated by phonons. We will discuss more of this in a moment. However, we are starting our conversation with the presumption that perfect conductivity is a natural outgrowth of the *many-body problem* for these materials. Thus, we might conclude that the perfect diamagnetism of the Meissner effect is an outgrowth of some Maxwellian boundary conditions associated with the perfect conductivity. Can we justify this view?

No! As we can see from the derivation panel, the magnetic field inside a *perfect conductor* is constant over time, not exactly the Meissner effect (that says it must be *zero* everywhere). What's the difference? Well if we applied a constant field $\boldsymbol{B}_o$ above T_c and then cooled the material to below T_c, the Meissner effect suggests the field would be expelled from the superconductor, $\boldsymbol{B} = 0$, whereas in this *perfect conductor* the field would simply remain $\boldsymbol{B}_0$. This little Gedankenexperiment tells us something rather important: the superconductor is NOT just a perfect conductor. The perfect diamagnetism comes from something more; it is fundamental to the superconducting state.

To capture this phenomenology, the London brothers introduced a model that arbitrarily eliminates the time derivatives from above to yield

$$\nabla^2 \boldsymbol{B} = \lambda^{-2} \boldsymbol{B} \tag{10.13}$$

In fact, this equation correctly reflects the Meissner effect and perfect diamagnetism. Coupled with Ampere's law,

$$\nabla \times \boldsymbol{J} = -\frac{n_s e^2}{mc} \boldsymbol{B} \tag{10.14}$$

It easily relates $\boldsymbol{B}$ and $\boldsymbol{J}$. And since

$$\boldsymbol{B} = \nabla \times \boldsymbol{A} \tag{10.15}$$

we get the so-called *London equation* in the Coulomb gauge:

$$\boldsymbol{J} = -\frac{n_s e^2}{mc} \boldsymbol{A} \tag{10.16}$$

Recall the Coulomb gauge $\nabla \cdot \boldsymbol{A} = 0$ ensures $\boldsymbol{\nabla} \cdot \boldsymbol{J} = 0$ from the continuity equation.

This London equation implies something else about our system's ground state: it is *rigid*. Why? Well, in the ground state the total momentum of the system, according to Bloch, is zero:

$$\langle \psi \mid \boldsymbol{p} \mid \psi \rangle = 0 \tag{10.17}$$

A *rigid* state is one that seems to *spatially translate* together. So lets imagine that this is the case. An applied field ($\boldsymbol{A}$ for its vector potential) results then in

$$\boldsymbol{p} = m\boldsymbol{v} - e\boldsymbol{A}/c \tag{10.18}$$

and the expectation value

$$\langle v \rangle = eA/mc \tag{10.19}$$

This just reduces to the London equation

$$J = -en_s\langle v \rangle \tag{10.20}$$

And finally, we have the existence of a *critical current*. Again, we may be tempted to assign this to the current value in which the magnetic field created by the flowing current becomes so large that the superconductor might no longer be able to expel it. Thus the superconductivity would be destroyed. But as in the case of the Meissner effect, *such an assumption would be misleading*. In fact, as we will see later in our BCS discussion, it is actually associated with the natural energy scales within the system at which the electrons can no longer form the superconducting state.

10.2.1.1 In What Compounds Has Superconductivity Been Observed?

In Table 10.1 we show the wide range of materials classified as superconducting. These include elements, alloys and intermetallic compounds, organic charge-transfer salts, fullerenes, and oxides (also referred to as *ceramic superconductors* or *cuprates* since many of these involve copper oxide compounds). For each class we selected a material among the highest T_c values known to date. Notice that there is a rather big jump between the ceramics and all other classes of superconductors. Indeed, the ceramic type II superconductivity is thought to originate differently from that of other type I superconductors.

10.2.2 A Basic Model

There are actually two main components that comprise the microscopic theory of superconductivity introduced by Bardeen, Cooper, and Schrieffer, otherwise known as the BCS theory. The first component is the proposal that within the lattice there can exist an effective attractive potential between two electrons with opposite momenta and opposite spin.[6] This potential will lead to the formation

Table 10.1 "Classes" of superconducting materials.

Class	Material with highest T_c	T_c (K)	References
Elements	Nb	9.2	[5a]
Alloys and intermetallic compounds	Nb_3Ge	23.2	[5b]
Polymers	$(SN)_x$	0.26	[5c]
Organic charge-transfer salts	κ-$(ET)_2Cu[N(CN)_2Cl]$	12.8	[5d]
Fullerenes	Cs_2RbC_{60}	33	[5e]
Oxides ("ceramics")	Hg–Ba–Ca–Cr–O at 150 kbar	135–160	[5f]

6 A BCS-like approach can be formulated using entirely repulsive forces as well.

of bound states that we call *Cooper pairs*. The second component of this theory is that the system can form a single, coherent ground state with a condensation of these Cooper pairs, known as a *superconducting condensate*, which is responsible for all the above observed phenomena.

10.2.2.1 How Does an Attractive Potential Show Up Between Two Negatively Charged Particles?

Let's imagine for a moment an electron moving happily through a lattice as though it were located on a path exactly between the atoms. Momentarily, the electron should illicit a dielectric response from the atoms at the lattice sites wherein the charge upon the site becomes polarized, forming a slight dipole. But the electron moves on its way, and this appears as a very subtle wake of displaced charge behind the electron. Now let's say that another electron followed the first, and "seeing" the atomic displacement ripple left by that first electron, it is attracted to the momentary positive charge it "sees" in front of it. The result is that each electron undergoes some retardation and a dispersion of the positive charge ripple unless their speeds match. However, if the electrons are heading in opposite directions so that momentum in the center of mass frame of reference is zero, then there can be a short period of time in which the charged displacement is not dispersed and the two electrons can exist in a resonance (a short-lived bound state) with the charged distortion between them (Figure 10.5).

Now, we usually use the terminology that *electron binding into Cooper pairs is mediated by phonon exchange*. After all, lattice distortions are merely phonons.

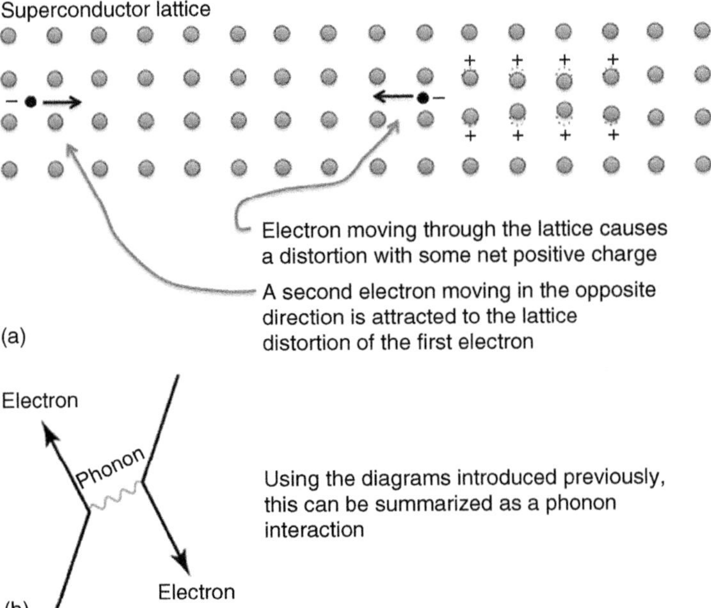

Figure 10.5 The BCS ansatz in picture form. (a) The lattice deformation and resulting attractive potential between electrons; (b) the Feynman diagram for the phonon exchange.

But Cooper pair formation is fleetingly transient. After a very short time, each electron in a given pair goes on to form new Cooper pairs with other electrons. The distortion of the lattice is short lived and is sometimes called a *virtual phonon* because its lifetime is too short to propagate through the lattice like a normal phonon. This process continues with the newly formed Cooper pairs. The end result is that each electron in the solid is attracted to every other electron in the solid, thereby forming a collective network held together by interactions – but more on that in a moment.

Cooper pairs are unusual *composite* particles. They do not stick together like two protons forming a hydrogen molecule, but rather they are paired in a more general way: they are *correlated*. This means that their spins and momenta are *coupled* so that the electrons belonging to one pair move in opposite directions, thus pairing momentum $\overline{p}$ with momentum $-\overline{p}$. The *pair correlation* is effective over a characteristic length, called the *coherence length* ξ. Usually ξ is between 1000 Å and 1 μm, whereas the distance between two electrons in a solid is on the order of 1 Å. Consequently, the Cooper pairs interpenetrate largely, and it is perhaps not too difficult to understand that an ensemble of interpenetrating Cooper pairs behaves very differently from a gas of noninteracting electrons.

Let's not be too intimidated by these new particles. In conjugated systems we saw an example of electron–phonon coupling: the Peierls distortion. Superconductivity is another example. Cooper pairs are abstract entities: *momentum-correlated electrons*. But they are no more exotic than Peierls "pairs": electrons and distortions paired in direct space. Besides, the pairing of particles that are the opposite (in momentum and spin) is little different from pair bonding among many human couples.

10.2.2.2 Cooper Pair Binding

Much of the essence of the Cooper pair formation can be captured in a simple two-body quantum mechanical problem. To see this we consider two electrons interacting through $V(\boldsymbol{r}_1 - \boldsymbol{r}_2)$. The eigenvalue problem is

$$\left[-\frac{\hbar^2}{2m}(\nabla^2_{r_1} + \nabla^2_{r_2}) + V(\boldsymbol{r}_1 - \boldsymbol{r}_2)\right]\psi(\boldsymbol{r}_1, \boldsymbol{r}_2) = E\psi(\boldsymbol{r}_1, \boldsymbol{r}_2) \tag{10.21}$$

Using relative displacement, $\boldsymbol{r} = \boldsymbol{r}_1 - \boldsymbol{r}_2$, and center of mass coordinates, $\boldsymbol{R} = ½(\boldsymbol{r}_1 + \boldsymbol{r}_2)$,

$$\left[-\frac{\hbar^2\nabla^2_R}{2m^*} - \frac{\hbar^2\nabla^2_r}{2\mu} + V(\boldsymbol{r})\right]\psi(\boldsymbol{r}, \boldsymbol{R}) = E\psi(\boldsymbol{r}, \boldsymbol{R}) \tag{10.22}$$

with an effective mass m^* and reduced mass μ. The potential does not depend on $\boldsymbol{R}$, so solutions have the form

$$\psi(\boldsymbol{r}, \boldsymbol{R}) = \psi(\boldsymbol{r})e^{i\boldsymbol{K}\cdot\boldsymbol{R}} \tag{10.23}$$

reducing the eigenvalue problem to

$$\left[-\frac{\hbar^2\nabla^2_r}{2\mu} + V(\boldsymbol{r})\right]\psi(\boldsymbol{r}) = \tilde{E}\psi(\boldsymbol{r}) \tag{10.24}$$

and

$$\tilde{E} = E - \frac{\hbar^2 K^2}{2m^*} \tag{10.25}$$

The lowest energy that can be obtained is one in which $\boldsymbol{K} = 0$ or all center of mass momentum is zero. In this case the Cooper pair electrons will have exactly opposite momenta: $p_r = 0$. Now the spatial part of the wavefunction ($\psi(\boldsymbol{r})$) can have one of two symmetries either even, $\psi(\boldsymbol{r}) = \psi(-\boldsymbol{r})$, or odd, $\psi(\boldsymbol{r}) = -\psi(-\boldsymbol{r})$. However, the *total wavefunction* must be *antisymmetric* because the electrons transpose their momenta ($\boldsymbol{K} \to -\boldsymbol{K}$). Therefore they must form a spin singlet for a spatially even wavefunction or a spin triplet for a spatially odd wavefunction.

To work our problem further, we will introduce the Fourier transform of the eigenvalue problem

$$\psi(\boldsymbol{k}) = \int d^3r\, \psi(\boldsymbol{r})e^{-i\boldsymbol{k}\cdot\boldsymbol{r}} \tag{10.26}$$

which then gives

$$\frac{\hbar^2 k^2}{2\mu}\psi(\boldsymbol{k}) + \int d^3r\, V(\boldsymbol{r})\psi(\boldsymbol{r})e^{-i\boldsymbol{k}\cdot\boldsymbol{r}} = E\psi(\boldsymbol{k}) \tag{10.27}$$

$$\int \frac{d^3q}{(2\pi)^3} V(\boldsymbol{q}) \int d^3r\, \psi(\boldsymbol{r})e^{-i(\boldsymbol{k}-\boldsymbol{q})\cdot\boldsymbol{r}} = \left(E - \frac{\hbar^2 k^2}{m}\right)\psi(\boldsymbol{k}) \tag{10.28}$$

$$\int \frac{d^3k'}{(2\pi)^3} V(\boldsymbol{k} - \boldsymbol{k}')\,\psi(\boldsymbol{k}') = (E - 2\varepsilon_k)\psi(\boldsymbol{k}) \tag{10.29}$$

Here we have changed $\boldsymbol{q} = \boldsymbol{k} - \boldsymbol{k}'$ and defined the free electron energy as $\varepsilon_k = \frac{\hbar^2 k^2}{2m}$.

Now we note that we can have a *bound state* between the two electrons when $E < 2\varepsilon_k$:

$$\Delta(\boldsymbol{k}) = (E - 2\varepsilon_k)\psi(\boldsymbol{k}) \tag{10.30}$$

where

$$\Delta(\boldsymbol{k}) = -\int \frac{\mathrm{d}^3k'}{(2\pi)^3} \frac{V(\boldsymbol{k}-\boldsymbol{k}')}{2\varepsilon_k - E} \Delta(\boldsymbol{k}') \tag{10.31}$$

This is just the Schrödinger equation written in a slightly bizarre way. Now looking back at our pictorial model for the lattice interaction with the electrons, we imagine a slightly attractive potential mediated by this lattice distortion (i.e. the phonon):

$$V(\boldsymbol{k}-\boldsymbol{k}') = -V_0 \text{ for } \varepsilon_{k'}, \varepsilon_k < \hbar\omega_\mathrm{D} \tag{10.32}$$

The potential is zero otherwise. Recall that ω_D is the Debye frequency. We seek a solution with a constant Δ, so $\Delta(k) = \Delta$. This implies an even spatial wavefunction and thus a spin singlet. In this two-electron system, we can define the *DOS per spin* as

$$\rho(\varepsilon) = \frac{m^{3/2}}{\sqrt{2}\hbar^3\pi^2}\sqrt{\varepsilon} \tag{10.33}$$

This yields

$$\Delta = \frac{V_0 \Delta m^{3/2}}{\sqrt{2}\hbar^3\pi^2} \int_0^{\omega_\mathrm{D}} \frac{\mathrm{d}\varepsilon\sqrt{\varepsilon}}{2\varepsilon - E} \tag{10.34}$$

$$1 = \frac{V_0 m^{3/2}}{\sqrt{2}\hbar^3\pi^2}\left[\sqrt{\omega_\mathrm{D}} - \sqrt{-\frac{E}{2}}\arctan\left(\sqrt{\frac{2\omega_\mathrm{D}}{-E}}\right)\right] \tag{10.35}$$

which will now determine the value of the bound state energy $E < 0$ as a function of V_0. This suggests the existence of a minimum value of V_0 to have a bound state generally, as we would expect. So we take the limit as $E \to 0-$ to get this minimum:

$$V_{0,\min} = \frac{\sqrt{2}\hbar^3\pi^2}{m^{3/2}\sqrt{\omega_\mathrm{D}}} \tag{10.36}$$

Now we know how strong this attraction must be to create such a bound pair. But we have overlooked an important caveat. The only electrons that will be affected by this attractive potential will sit at the Fermi level of the band structure. We can account for this by setting up an attractive potential for the unfilled states just above the Fermi level. Following the same steps as before,

$$V(\boldsymbol{k}-\boldsymbol{k}') = -V_0 \tag{10.37}$$

$$\varepsilon_{k'} - \varepsilon_F;\ \varepsilon_F < \hbar\omega_\mathrm{D} \tag{10.38}$$

$$\hbar\omega_\mathrm{D} \ll \varepsilon_F \tag{10.39}$$

$$\Delta(\boldsymbol{k}) = \Delta \tag{10.40}$$

$$\Delta = V_0\rho(\varepsilon_F)\Delta \int_{\varepsilon_F}^{\varepsilon_F+\omega_D} \frac{d\varepsilon}{2\varepsilon - E} \tag{10.41}$$

$$\frac{2}{V_0\rho(\varepsilon_F)} = \ln\left(\frac{2\varepsilon_F - E + 2\omega_D}{2\varepsilon_F - E}\right) \tag{10.42}$$

In the limit $V_0\rho(\varepsilon_F) \ll 1$, E is close to $2\varepsilon_F$ and

$$2\varepsilon_F - E + 2\omega_D \approx 2\omega_D \tag{10.43}$$

Defining

$$E_b \equiv 2\varepsilon_F - E \tag{10.44}$$

we get

$$E_b = 2\omega_D e^{-2/V_0\rho(\varepsilon_F)} \tag{10.45}$$

This seems to indicate that in the case where there is a well-defined Fermi surface separating the occupied and unoccupied states, a bound state can exist regardless of the strength of V_0. This is of course a very different situation to our expectations from free electron gasses. It is this bound state we refer to as the *Cooper pair.*

Recall also that our final total energy, E, will include the center of mass motion:

$$E = E_{K=0} + \frac{\hbar^2 K^2}{4m} \tag{10.46}$$

$$E = 2\varepsilon_F - E_b + \frac{\hbar^2 K^2}{4m} \tag{10.47}$$

In the limit $E \to 2\varepsilon_F$, we can still observe a bound state:

$$K = \frac{2}{\hbar}\sqrt{mE_b} \tag{10.48}$$

But notice that such a state has a limit. The current density cannot exceed

$$J = n_s e\frac{\hbar K}{m} = 2n_s e\sqrt{\frac{E_b}{m}} \tag{10.49}$$

This *critical current density* was mentioned above in our tests for superconductivity.

10.2.2.3 The BCS Ground State

Now let's talk about the *many-particle superconducting ground state* itself. The creation of Cooper pairs is clearly, energetically favorable: the mechanism proposed does form a bound state. But can we say what the preferred state of the *system* is if it is permitted to create such Cooper pairs? We might want to simply assume all the electrons that can pair up will pair up since we gain energy advantage this way. But we immediately run into a problem with this; Cooper pairs are stabilized by the *Fermi surface.* If we remove all the electrons, the Fermi surface will collapse, and pairing cannot occur. So we want a system state that maximizes the Cooper pair numbers, thereby yielding the lowest ground state energy, and yet still allows for all of these Cooper pairs to be stable (a well-formed Fermi surface). Notice that here we are talking about a sort of ground state of the system.

This challenge is exactly what BCS takes up. It will construct a many-particle wavefunction that is essentially a multiplicative combination of the possible "paired particle" states. But the theory proposes a highly restrictive set of criterion with which to choose the paired electron states to be used. The pair states that will be used in our many-particle wavefunction will have electrons like this: $|k\uparrow\rangle, |k\downarrow\rangle, |-k\downarrow\rangle, |-k\uparrow\rangle$. So, the many-particle wavefunction will be made up of pairs where the electrons have opposite $\boldsymbol{k}$ values and opposite spins, just like we postulated earlier. This means there are 16 possible single particle combinations (recall the Hubbard model). BCS further restricts which one of these combinations we may choose for our construction of the many-particle wavefunction: singlets ONLY! So in the end we would write down the many-particle wavefunction as

$$\psi_0^s = \prod_k \left[u_k|0_k 0_{-k}\rangle + v_k|\psi_k\psi_{-k}\rangle\right] \tag{10.50}$$

And we have used a notation here such that:

$|\psi_k\psi_{-k}\rangle$ is the presence of a pair with opposite momentum and opposite spins.
$|0_k 0_{-k}\rangle$ is the absence of such a pair (or a hole).
$u_k u_k^* + v_k v_k^* = 1$ is the normalization condition for the coefficients.
$|v_k|^2$ is the probability that a Cooper pair of momentum $\boldsymbol{k}$ is in the ground state.
$|u_k|^2$ is the probability that it is not.

Notice that our linear combination to form the many-particle wavefunction is definitely a little strange: it involves the wavefunction of the *electrons in the Cooper pairs + their mirror images in the Fermi sea* once they have been removed or their holes. Thus, we think of this many-particle state as some phase-related combination of the electrons and holes of the system. But the electrons and holes are sort of grouped together. Each pair of $\pm\boldsymbol{k}$ electrons has the antisymmetric nature of two fermions because we chose to use only singlets. But the wavefunction has a total momentum of zero and a total spin of zero because we have included the hole below the Fermi level. The $\pm\boldsymbol{k}$ grouping is simply a composite particle (or quasiparticle) with a bosonic character. The state is a coherent collection of all of these bosons (again with zero total momentum). This is somewhat analogous to a Bose–Einstein condensate of bosons, but of course here we are using composite particles. However, the statistics seem to be concerned only with the bosonic or fermionic character, not whether the particles are composite or not.

In terms of our defined coefficients above, we can say the following:

$$v_{-k} = v_k \text{ and } u_{-k} = u_k \tag{10.51}$$

The normal (nonsuperconducting) state is described as

$$|\boldsymbol{k}| < k_F \rightarrow \begin{cases} |v_k| = 1 \\ |u_k| = 0 \end{cases} \tag{10.52}$$

$$|\boldsymbol{k}| > k_F \rightarrow \begin{cases} |v_k| = 0 \\ |u_k| = 1 \end{cases} \tag{10.53}$$

at $T = 0$ all states below k_F are filled.

Following BCS, we have made a guess as to the nature of the many-particle wavefunction. Now we need a Hamiltonian to apply it to. This should look like

$$H = H_0 + H_{\text{phonon}} \tag{10.54}$$

The first term here comes from the single particle picture interactions with a frozen lattice just like we have already covered. The second term describes the electron interactions mediated by phonons (the "ionic motion" picture we used above) and is where the new physics comes from.

So

$$H_0 = \sum h(\boldsymbol{r}_i),\ H_{\text{phonon}} - \frac{1}{2}\sum\sum V_{\text{phonon}}(\boldsymbol{r}_i - \boldsymbol{r}_j) \tag{10.55}$$

And the single particle wavefunctions, $|\psi_k\rangle$, are eigenvectors of H_0 with eigenvalues ε_k. These form a complete orthonormal set of eigenvectors spanning the variable space of the many-particle system. ε_k is usually referred to as the "kinetic energy" though that is really not exactly what it is, since it has the energies of all the single particle electrons in it. The convention is to measure ε_k against the Fermi level. The matrix elements of the first part of the Hamiltonian using these single particle wavefunctions to build Cooper singlets look like

$$\begin{aligned} &u_k u_{k'}^* \langle 0_{k'} 0_{-k'} \mid H_0 \mid 0_k 0_{-k}\rangle\ u_k v_{k'}^* \langle \psi_{k'} \psi_{-k'} \mid H_0 \mid 0_k 0_{-k}\rangle \\ &v_k u_{k'}^* \langle 0_{k'} 0_{-k'} \mid H_0 \mid \psi_k \psi_{-k}\rangle\ v_k v_{k'}^* \langle \psi_{k'} \psi_{-k'} \mid H_0 \mid \psi_k \psi_{-k}\rangle \end{aligned} \tag{10.56}$$

The primes are used to distinguish between two different $\boldsymbol{k}$ values and thus different indices in the matrix. Notice that only the terms that do NOT have a $\mid 0_k 0_{-k}\rangle$ give a nonzero value for the matrix element. So the nonzero terms look like this:

$$v_k v_{k'}^* \langle \psi_{k'} \psi_{-k'} \mid H_0 \mid \psi_k \psi_{-k}\rangle = |v_k|^2 \delta(\boldsymbol{k} - \boldsymbol{k}')\varepsilon_k + |v_k|^2 \delta(\boldsymbol{k} + \boldsymbol{k}')\varepsilon_k \tag{10.57}$$

and our first term in the Hamiltonian gives us

$$\langle \psi_0^s \mid H_0 \mid \psi_0^s \rangle = \sum_{\boldsymbol{k}} 2\varepsilon_k |v_k|^2 \tag{10.58}$$

What about the second term? For these interactions it is clear that a process like the one described by our Feynman diagram in Figure 10.6 is taking place. Consider a $\boldsymbol{k}/-\boldsymbol{k}$ electron pair in the ground state. The interaction of the pair with a phonon scatters it out of the ground state to be replaced by a $\boldsymbol{k}'/-\boldsymbol{k}'$ pair that was not originally a part of the ground state.

So for this second term in the Hamiltonian, the only nonzero matrix elements are of the form

$$[v_{k'}^* \langle \psi_{k'} \psi_{-k'} | u_k^* \langle 0_k 0_{-k} |] \boldsymbol{H}_{\text{phonon}} [v_k | \psi_k \psi_{-k}\rangle u_{k'} | 0_{k'} 0_{-k'}\rangle] \tag{10.59}$$

Following through with this,

$$V_{kk'} = \langle \psi_{k'} \psi_{-k'} | \boldsymbol{V}_{\text{phonon}} | \psi_k \psi_{-k} \rangle \tag{10.60}$$

$$E_0^s = \sum_{\boldsymbol{k}} 2\varepsilon_k |v_k|^2 + \sum_{k}\sum_{k'} V_{kk'} u_k^* v_{k'}^* u_{k'} v_k \tag{10.61}$$

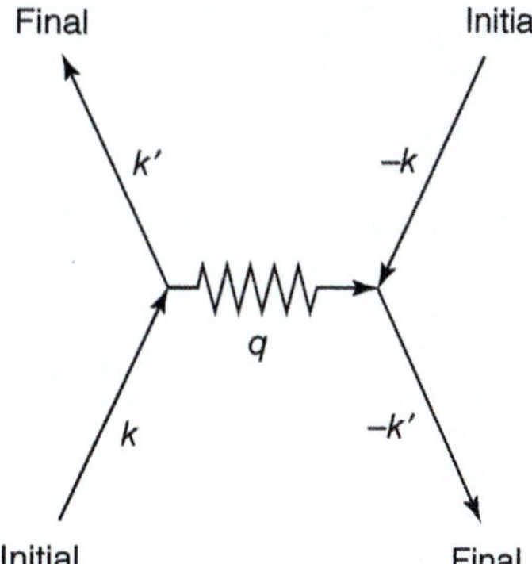

Figure 10.6 A $k/{-k}$ electron pair interacts with a phonon to scatter into a $k'/{-k'}$ pair.

This is our BCS ground state energy, and we must still get some idea of what the coefficients u_{k} and v_{k} might be. Remember, physically they stand for the amount of each single particle state that is mixed into our pair states (linear combination) that we are using to build the ground state (multiplicative combination) of the system. There is a pretty standard way of doing this: variational treatment of $E_0^{(s)}$ with one of the states. We will require $E_0^{(s)}$ to be a minimum with respect to variations in v_{k}^*.

The ground state energy — Stationary requirement

$$\frac{\partial E_0^{(s)}}{\partial v_{\mathrm{k}}^*} = 2\epsilon_{\mathrm{k}} v_{\mathrm{k}} + \sum_{\mathrm{k}'} V_{\mathrm{k'k}} u_{\mathrm{k'}}^* u_{\mathrm{k}} v_{\mathrm{k'}} + \sum_{\mathrm{k}'} V_{\mathrm{kk'}} \frac{\partial u_{\mathrm{k}}^*}{\partial v_{\mathrm{k}}^*} v_{\mathrm{k'}}^* u_{\mathrm{k'}} v_{\mathrm{k}} = 0$$

Which reduces to $\dfrac{\partial u_{\mathrm{k}}^*}{\partial v_{\mathrm{k}}^*} = -\dfrac{v_{\mathrm{k}}}{u_{\mathrm{k}}}$ (using normalization and assumptions above)

Defining $\Delta_{\mathrm{k}} = \sum_{\mathrm{k}'} V_{\mathrm{k'k}} u_{\mathrm{k'}}^* v_{\mathrm{k'}}$, $\Delta_{\mathrm{k}}^* = \sum_{\mathrm{k}'} V_{\mathrm{kk'}} u_{\mathrm{k'}} v_{\mathrm{k'}}^*$

Yields $2\epsilon_{\mathrm{k}} v_{\mathrm{k}} u_{\mathrm{k}} + \Delta_{\mathrm{k}} u_{\mathrm{k}}^2 - \Delta_{\mathrm{k}}^* v_{\mathrm{k}}^2 = 0$

The variational treatment in the derivation box to the left gives a characteristic equation that relates the kinetic energy term to the coefficients and the Δ that contains the potential of interaction. Notice this is true for each $\boldsymbol{k}$. And, following the derivation presented by Kaxira [10], we now choose specific forms for the coefficients. (Parenthetical note: if you are following along with your "second quantized approach to BCS handbook," you will notice a striking parallel between the choices of form for these coefficients and the creation and annihilation operators). Anyway, let's say

$$u_k = \cos\left[\frac{\theta_k}{2} e^{i\omega_k/2}\right] \quad v_k = \sin\left[\frac{\theta_k}{2} e^{-i\omega_k/2}\right] \tag{10.62}$$

These satisfy the normalization conditions, and ω_k is the phase difference that might exist between the two coefficients.

$2\varepsilon_k \sin\frac{\theta_k}{2}\cos\frac{\theta_k}{2} = \Delta_k^*\sin^2\frac{\theta_k}{2}e^{-iw_k} - \Delta_k\cos^2\frac{\theta_k}{2}e^{iw_k}$	We start off by substituting our expressions for the u's/v's into the characteristic equation
$= Re\left[\Delta_k^*\sin^2\frac{\theta_k}{2}e^{-iw_k} - \Delta_k\cos^2\frac{\theta_k}{2}e^{iw_k}\right]$	Since the left side is real
$2\varepsilon_k \sin\frac{\theta_k}{2}\cos\frac{\theta_k}{2} = \left(\mid \Delta_k \mid \sin^2\frac{\theta_k}{2} + \mid \Delta_k \mid \cos^2\frac{\theta_k}{2}\right)\cos(w_k)$	
$\varepsilon_k \sin\theta_k + \mid \Delta_k \mid \cos\theta_k \cos(w_k) = 0$	Get out your half-angle trig tables and you can reduce the above to this. Notice the smallest value you can get for Δ_k occurs when $\cos(w_k)$ goes to ± 1. So use this value to get two roots…
$\sin\theta_k = -\frac{\lvert\Delta_k\rvert}{\zeta_k},\quad \cos\theta_k = \frac{\varepsilon_k}{\zeta_k}$ $\sin\theta_k = \frac{\lvert\Delta_k\rvert}{\zeta_k},\quad \cos\theta_k = -\frac{\varepsilon_k}{\zeta_k}$	Where we have defined: $\zeta_k \equiv \sqrt{\varepsilon_k^2 + \lvert\Delta_k\rvert^2}$
$\lvert u_k\rvert^2 = \frac{1}{2}\left(1+\frac{\varepsilon_k}{\zeta_k}\right),\quad \lvert v_k\rvert^2 = \frac{1}{2}\left(1-\frac{\varepsilon_k}{\zeta_k}\right)$	Using the top root (the lowest energy so it is the ground state whereas the second is the first excited states) we derive the following relations
$\lvert u_k\rvert\lvert v_k\rvert = \frac{1}{2}\left(1-\frac{\varepsilon_k^2}{\zeta_k^2}\right)^{1/2} = \frac{1}{2}\frac{\lvert\Delta_k\rvert}{\zeta_k}$	

To simplify our picture and get some insight into its meaning, let's consider the very simple case wherein we drop the $\boldsymbol{k}$ dependence in the u and v variables:

$$|u|^2 = \frac{1}{2}\left(1+\frac{\varepsilon}{\sqrt{\varepsilon^2+|\Delta|^2}}\right), |v|^2 = \frac{1}{2}\left(1-\frac{\varepsilon}{\sqrt{\varepsilon^2+|\Delta|^2}}\right) \tag{10.63}$$

Recall that the coefficients correspond roughly to the occupancy of Cooper pair states. $\varepsilon = 0$ represents the system state in which both coefficients have the value of ½. In other words, if these are behaving as a distribution function (as we are claiming they are), then the $\varepsilon = 0$ point is equivalent to the Fermi level of the system. Notice that the spread in the magnitude of the coefficient from 0 to 1 takes place over an energy range of roughly Δ centered on $\varepsilon = 0$. So the occupation of Cooper pairs is significant only at energies near the Fermi level.

Next we examine what happens to the DOS in the system. We first recognize that the number of states as we go from normal to superconducting must be conserved:

$$g(\zeta)d\zeta = g(\varepsilon)d\varepsilon \rightarrow g(\zeta) = g(\zeta)\frac{d\varepsilon}{d\zeta} \tag{10.64}$$

If we approximate $g(\varepsilon)$ as g_F (DOS at the Fermi level), we get

$$g(\zeta) = \frac{g_F|\zeta|}{\sqrt{\zeta^2-|\Delta|^2}}, \text{for } |\zeta| > |\Delta| \tag{10.65}$$

$$(\zeta) = 0, \text{for } |\zeta| < |\Delta| \tag{10.66}$$

So what have we learned about this fascinating state of interaction and correlation? Firstly, in the many-body ground state from BCS, the total linear momentum is zero since the linear momentum of all Cooper pairs must be zero. This means that all paired electrons in this collective state travel in opposite directions. A *current* flowing in the superconductor shifts the total moment from zero rigidly, so on average, one electron in a Cooper pair has a slightly larger momentum than its pair. They still travel in opposite directions however.

Secondly, a small amount of energy is needed to destroy the superconducting state and make it normal. This energy is called the energy gap. Though we didn't derive it in our discussions here, at absolute zero the superconducting gap is

$$2\Delta(0) = \frac{7}{2} k_B T_c \tag{10.67}$$

Thirdly, causing just one of these electrons to collide and scatter from atoms in the lattice means the whole network of electrons must be made to collide into the lattice. This is energetically very costly. So the collective behavior of all the correlated electrons in the solid prevents collisions with the lattice, since nature prefers energy minimization. Here, the minimum energy state is to have no collisions with the lattice.

10.2.2.4 Supplementary Thoughts

We should also reflect more carefully on a few points before completely leaving this topic. The first of these is that when doing our "arguments" above, we chose to allow only singlet states to participate in the Cooper pairs, so-called *s-wave superconductivity*. But of course this is not strictly necessary. We can bend the rules a bit so that triplets and other spin states can be used in the construction of the basis states for the Cooper pairs. For instance, *d-wave* superconductivity involves quite different selections of spin states, but again they condense into a bosonic ground state at low temperatures through the phonon coupling mechanism.

The next point we should really emphasize is in this so-called electron–phonon coupling; it really is a phonon we are talking about here. This fact is reflected in the BCS expression for the critical temperature:

$$T_c = 1.13 \frac{\eta\omega_D}{k_B} \exp\left[-\frac{1}{N(E_F)V^*}\right] \tag{10.68}$$

where ω_D is the Debye frequency (a characteristic phonon frequency), $h = 2\pi\hbar$ is Planck's and k_B is Boltzmann's constant, $N(E_F)$ is the electronic DOS at the Fermi energy, and V^* is a constant characterizing the electron–phonon interaction.

This very simple expression can be used to explain the high critical temperatures of classes of superconductors. For example, in the A15 superconductors (mentioned in earlier chapters), the Fermi level is assumed to be within a narrow band dominated by the d orbitals of the transition elements (for example, Nb in Nb_3Ge). Hence, the DOS at the Fermi energy is large, and T_c is large. Similarly, the increase of T_c in alkali-doped fullerene crystals when going from Na to K to Cs can be interpreted as follows: the fullerene lattice "swells" by alkali intercalation. The larger the ionic radius of the dopant, the greater the swelling. This leads to a smaller overlap of the molecular π orbitals and thus to narrower bands and, according to our Eq. (10.68), to higher T_c. So this little equation can be useful in setting trends among superconductors.

But as we said, it is the phonon in BCS theory that we want to emphasize for a moment. The relevance of phonons can be tested by looking for an *isotope effect*. In our critical temperature Eq. (10.68), T_c is proportional to the Debye frequency ω_D. This frequency should vary with the square root of the atomic mass M. Consequently, we expect

$$T_c \sim M^{-1/2} \tag{10.69}$$

M can be varied by exchanging it with another isotope. The bonding will look exactly the same, so force constants do not change. While it is a bit expensive to do this, the test has been carried out for many different superconductors, and in general T_c does vary as $M^{-1/2}$. So far, any occasional deviation from this rule has been explained through a known isotopic dependency of V^*.

Naturally, as we have come to expect, in 1D and organic superconductors, we find ourselves having to differentiate between other interactions between the electrons and the phonon exchange. In Chapter 2 we mentioned briefly that the occurrence of the BCS superconductivity gap stops the martensitic distortion. Competition between various low-temperature instabilities is a salient feature of 1D metals. Arguments for or against any such hypotheses are typically rooted in studies of this isotope effect. Specifically, in organic superconductors, the carbon isotope C-12 (natural abundance 99%) is replaced by C-13 (natural abundance 1%), or protons (natural abundance 99.985%) can be replaced by deuterons (natural abundance 1.5×10^{-4}%).

But there is more to this "superconducting gap." We have already seen that in the Peierls case, electron–phonon coupling leads to a gap at the Fermi level in the electronic DOS. And, similarly, in superconductivity there is also a gap at the Fermi level. The formation of Cooper pairs removes states within $\varepsilon_F \pm \Delta$, and thus a gap is created. But in Peierls systems we also encountered new particles: the solitons. Their energy states are in the Peierls gap and, in the ideal case, exactly at the midgap. The question might be posed, where in Figure 10.7 the Cooper pairs ought to be drawn. But, unlike solitons, Cooper pairs cannot be depicted in a diagram of the single-electron DOS!

The concepts and analysis above forms the basis from which superconductivity derives. There are, however, further subtleties that make the parallel with Peierls

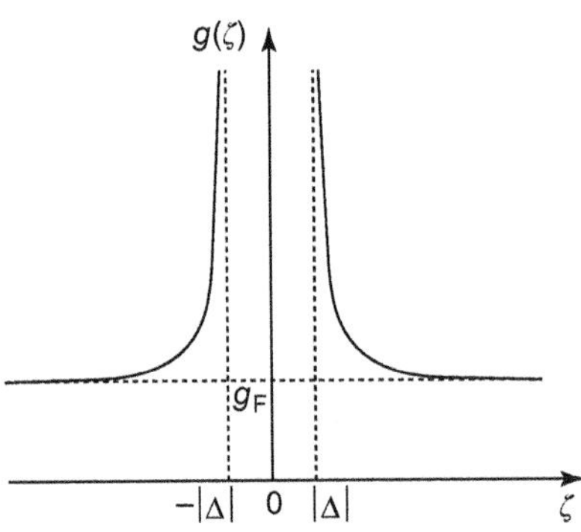

Figure 10.7 Graph of the density of states in the BCS model near the Fermi level. Notice the onset of the gap with magnitude 2|Δ|.

systems even more compelling. Most importantly, as already hinted at above, the division of classes of superconductors suggests that there are different phenomena (other than simply higher T_c) observed in the compound oxides as for the elemental superconductors. In fact, this is quite right; the so-called "high T_c" materials are quite different. Though the fundamental mechanisms of pairing leading to a superconducting gap are still valid, other curious characteristics arise in these materials. As an example, we restate the situation of the critical magnetic field. As stated above, the elemental superconductor is sensitive to applied magnetic fields with a certain critical field, for a given temperature, destroying the superconducting state. Below that critical field strength $\boldsymbol{B}_c$, the type I (elemental) superconductor is a perfect diamagnet. In the "type II" or compound superconductors, there are two such distinctive field values: $\boldsymbol{B}_{c1}$ and $\boldsymbol{B}_{c2}$. For a magnetic field $\boldsymbol{B}$ less than $\boldsymbol{B}_{c1}$, the superconductor exhibits a type I response to the field. For a magnetic field $\boldsymbol{B}$ greater than $\boldsymbol{B}_{c2}$, the superconducting state is destroyed as in the case of type I again. However, something unique occurs when the magnetic field is between $\boldsymbol{B}_{c1}$ and $\boldsymbol{B}_{c2}$. In this case, the superconductor has zero resistance but allows partial flux penetration! This is referred to as a vortex state and is pictured as cores of normal material in which the magnetic field penetrates, surrounded by material in the superconducting state responding to that field. Naturally, as the applied magnetic field increases, the number of normal cores will also increase until, eventually, the material can no longer sustain anymore. At this point $\boldsymbol{B}_{c2}$ has been reached and the superconducting state collapses.

But why should such an intermediate state exist in one type of superconductor and not the other? The answer comes from a more careful examination of the structure of the type II superconductors. This is where the subtle parallel to Peierls systems occurs and where dimensionality begins to play an important role. In most of the type II high T_c materials, the ceramic's crystal structure is composed of **two-dimensional (2D) planes** of conducting oxide. For these examples of high T_c materials, the conducting set of metal oxide planes is surrounded by a set of rare earth elements – yttrium, lanthanum, neodymium, gadolinium, and erbium. These surrounding species play important roles to modify the current-carrying planes. For instance, in the compound $YBa_2Cu_3O_{7-x}$ (a black orthorhombic material, where x is the variable for oxygen content), CuO planes act as the 2D planes of conductivity. And, in fact, the critical temperature of the superconducting state depends intimately on the overall stoichiometry of the CuO sheets with the T_c rising as the oxygen content rises. But this is not to say that a layer of CuO_x will be a superconductor on its own – if you could separate out such a single sheet. It was recognized early on that a charge reservoir was necessary for the establishment of electron pairing. This is the role of the surrounding crystal structure. Moreover, the heavy ions can also play some part in modification of specific phonon states. These are roles that are, in fact, quite similar in nature to that of the dopant ions in conducting polymer systems – just add one more dimension. Although the $YBa_2Cu_3O_{7-x}$ compound (generally abbreviated YBCO) is one of the most widely studied, this behavior seems to hold true for the rest of the high-temperature superconductors as well [11]. This parallel is not accidental. Superconducting behavior should be expected in some organic systems based upon similar mechanisms.

We are not quite finished with our discussion of inorganics. Clearly the example of the ceramic superconducting systems has shown that superconductivity can occur under conditions not previously anticipated, and even under conditions we would have thought prevented it from occurring (after all gadolinium is magnetic!). This happens again in the case of magnesium bromide, MgB_2 [12]. This common compound becomes superconducting at 39 K, an amazingly high T_c for a metallic material. The puzzling little compound, which practically every chemist has on his/her shelf, also presents us with a real mystery – more than one superconducting gap. But again, much of the mystery can be resolved in arguments of dimensionality. Like the type II cuprate superconductors above, MgB_2 is a layered compound. Only in the cuprate case, the conducting planes are insulators when undoped at ordinary temperatures. However, MgB_2 is always a metal. Structurally, the boron of this compound forms the simple honeycomb pattern found in graphite, whereas the Mg sits between planes centered on these hexagons. At first glance, one might think that a straightforward application of type II superconductor theories is all that is needed to explain the phenomena related to MgB_2, but as it happens, this doesn't do it. MgB_2 simply doesn't act like a type II superconductor altogether. In the end, it was our old friend BCS theory that did it.

In traditional BCS theory, pairing can be seen as a coupling to the lattice where a single electron emits and reabsorbs a phonon. This gives rise to the enhanced electron mass observed. But in MgB_2 these two values were apparently different. It was quickly realized that more than one type of electron might be involved in the pairing [13]. As in graphite, the honeycomb lattice is held together with σ bonds in plane. The π bonds above and below the graphene lattice allow for the movement of charge, and the same is true of boron with the π bonds forming weak pairs. But boron has fewer electrons than carbon, and not all the σ bond states are filled. Thus, a lattice vibration in the boron plane has a much more dramatic effect on electron pairing. Surprisingly not all the electrons are needed to form strong pairs confined to the plane. Since two different populations of electrons are acting to create the pairing effect, clearly there are multiple energies that can break the pairing – thus multiple superconducting gaps. Stated another way, electrons on different parts of the Fermi surface form pairs with different binding energies. The two types of pairs are coupled, and at 39 K the superconducting gaps converge, destroying superconductivity altogether.

10.2.3 Superconductivity Measurements Are Tricky

Have you synthesized a new superconductor? Because superconductivity is a sharp phase transition, the effect is theoretically quite stark when resistance or magnetic susceptibility is followed as a function of temperature. But complications arise because real samples are frequently small, oddly shaped, brittle, and hydroscopic or pyrophorous. Moreover it is often that low temperatures, high pressures, and magnetic fields have to be applied simultaneously, making the experiments tough to do. And of course, there is the fact that bulk materials can have multiple phases and impurities. Many times has a sample turned out to be nonsuperconducting, but there was a small superconducting contamination, which manifested itself as a little kink in the resistance vs. temperature curve, and then – following this hint – a new superconductor was found.

Four lead methods are typically employed in resistance measurements, but since we are mainly interested in resistance jumps and not in absolute values, it is not necessary to apply four point contacts or the van der Pauw technique. Often measurements can be carried out on pressed or "sufficiently well-squeezed" pellets because the contact resistance between the grains is bridged by the *Josephson effect* (superconducting tunneling through barriers; see [6]).

In some systems, such as organic superconductors and fullerenes, the coherence length might be very small, and the overall resistance of a granular film would not change although the core of the granuli could be superconducting. A superconductivity test independent of external and internal contact resistances is the measurement of the magnetic susceptibility. This method makes use of the Meissner effect, which expels the applied magnetic field. The superconducting quantum interference device (SQUID) magnetometer is the typical tool for this test, and susceptibility vs. temperature curves is straightforward. An alternative is the mutual induction measurement using small concentric coils, as indicated in Figure 10.8. The whole device can be made so small that it will fit through the outlet of a liquid helium container. The inner coil is excited by a reference signal of a lock-in amplifier, and the induction signal picked up at the outer coil is fed to an amplifier input. The sample is placed inside the coils, and, when it becomes superconducting, the induction suddenly changes (the inductance of a coil depends on the susceptibility of the material inside the coil). To calibrate the device, a small piece of a known superconductor is placed into the coil together with the sample. Since the inductance change is proportional to the volume becoming superconducting, this method allows one to estimate the superconductive volume fraction of the sample.

As a typical example of the use of the induction technique, consider the work done by Roth et al. of this book to measure superconductivity in powder samples

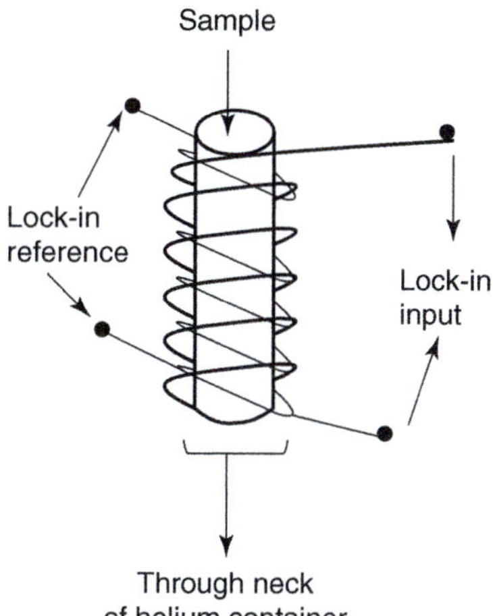

Figure 10.8 Simple homemade mutual inductance device to measure superconducting transitions through the neck of a standard helium container.

of potassium-intercalated graphite. This material is very pyrophorous. Therefore the potassium graphite powder was mixed with vacuum grease under argon atmosphere. This was sufficient to keep an oxygen-free environment for further sample handling. The paste was squeezed into the coils together with a short piece of iridium wire stuck in as a calibrator (T_c of iridium is 0.14 K). The device was put into a dilution refrigerator, and both the iridium and the potassium graphite transitions were easily observed. Unfortunately no transitions were observed when several potassium-doped conjugated polymers were treated the same way, but the investigations were not sufficiently systematic to completely exclude the possibility of superconductivity in potassium-doped polymers in the dilution refrigerator temperature range.

Zero resistance and ideal diamagnetism are convincing evidences of superconductivity. But the experimental methods discussed above do not yield absolute values; they reveal changes in the conductivity and changes in the susceptibility. Therefore precautions must be taken before relating the observed changes to the occurrence of superconductivity. A trivial piece of advice is to measure controls, the empty sample holder, or nonsuperconducting dummy samples. The inductance method is especially sensitive to the presence of solder (this contains lead) or other superconductive additives in the coil vicinity. Another precaution is checking the superconducting state by the application of magnetic fields. Superconductivity is destroyed by magnetic fields. High critical temperature implies high critical field with the details depending on the type of the superconductor. If T_c is very low, in the mK range, the earth magnetic field might be strong enough to destroy superconductivity. On the other hand, the critical fields of high T_c superconductors can be as high as 100 T or above. For "ET" salts with $T_c \sim 10$ K, the critical field is about 20 T. When resistance or susceptibility anomaly is observed, which is independent of the applied magnetic field, we should hesitate with assigning it to superconductivity.

10.2.4 Superconductivity and Dimensionality

As we have seen, the superconducting state is a delicate interplay of lattice instability and electron correlation. So it is a reasonable guess that dimensionality could play some role in the correlation lengths of a system. Clearly lattice stability can be more easily tuned in low-dimensional systems. Indeed, the quest for high-temperature superconductors has been one of the driving forces in the field of 1D metals. For a long time, intermetallic compounds with A15 structure held the world record in high T_c. Because of the van Hove singularities in their 1D band structure, there are sharp peaks in the electronic DOS. If the Fermi level happens to fall into such a peak (which can happen when materials are sufficiently doped), $N(E_F)$ will be large and we get a high T_c. The same singularities lead to lattice instabilities, to soft values for V^*, and again to high T_c. Of course, there is always an upper limit for T_c: at some point the lattice will become too unstable and fall apart.

In Chapter 2 we presented Little's proposal [14] of a conjugated chain with appropriate side groups that should superconduct. Electrons moving along the chain would be spin and momentum correlated by excitations in the side groups. These excitations take the role of the phonons in a traditional BCS superconductor. According to Little's estimates, this mechanism could lead to T_c values as high as room temperature. Synthesis efforts following Little's course have not lead to new superconducting materials. In fact, $(SN)_x$, a polymer composed of sulfur–nitrogen chains (Chapter 2), is the only polymer that has become superconducting so far. Its transition temperature is very low (T_c = 0.26 K). In addition, the chains are close; they sufficiently interact to suppress one-dimensionality. However, superconductors, where 1D aspects are most pronounced, are the Bechgaard salts (Chapter 2). Here again, the competition with the other instabilities typical for 1D metals limits T_c to a few kelvin. This same competition exists to some degree in the oxide superconductors [15], as discussed above, but, as we now know, some aspects of one-dimensionality or low dimensionality also exist in the cuprates.

10.2.5 More on Organic Superconductors

Organic superconductors can be subdivided into 1D, 2D, and 3D solids provided that fullerenes are classed with the organics; in Table 10.1 fullerenes are listed as a separate class [16]. This subdivision is based on the topology of the Fermi surface. As a reminder, 1D, 2D, and 3D Fermi surfaces are shown in Figure 10.9. In low-dimensional solids the Fermi surfaces are open, and they consist of two parallel planes for 1D solids and cylinders for 2D solids. Weak interactions in higher dimensionality (interchain and interplane) lead to warping and barrel-like distortions until the Fermi surface finally becomes closed; in this case we speak

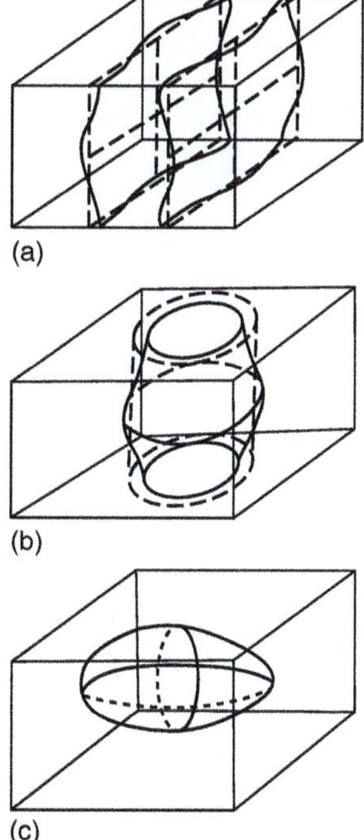

Figure 10.9 Different shapes of Fermi surfaces: quasi-one-dimensional (a), quasi-two-dimensional (b), and anisotropic three-dimensional (c). Source: After Jérome [7].

of a 3D solid. The study of the Fermi surface has been nicknamed "fermiology" and is widely used to understand the properties of 3D metals.

Among the most important experimental tools in *fermiology* are oscillations in the magnetoresistance known as *Shubnikov–de Haas* (SdH) *oscillations* and oscillations in the magnetic susceptibility known as *de Haas–van Alphén* (dHvA) *oscillations*. These oscillations occur when the resistance or the susceptibility is measured as a function of the magnetic field. The magnetic field exerts a lateral force on moving electrons (Lorentz force), as in the Hall effect. At sufficiently low temperatures and with the field strong enough, the electrons move in circles (Landau orbits) that are further split by the Zeeman energy associated with the spin of the electron in the field. The radius (and thus energy) of a Landau orbit depends on the magnetic field, and the resulting energy spectrum is made up of these Landau levels – each separated by the cyclotron energy. In each Landau level, the cyclotron and Zeeman energies as well as the number of electron states all increase linearly with applied $\boldsymbol{B}$-field strength. Thus, as the $\boldsymbol{B}$-field strength increases, the Landau level moves to higher energies. Eventually, with increasing $\boldsymbol{B}$-field strength, the energy level of the electrons in the Landau levels reaches the Fermi energy, and the level becomes depopulated as the electrons are now free to move as a current. So, the passage of the Landau orbits through the Fermi surface leads to oscillations in resistivity and other properties dependent on the electron density at the Fermi level.

In previous decades high-purity crystals of silver, gold, copper, etc. were studied extensively using this method; a recommended monograph is Shoenberg's book [17]. Some of the organic charge-transfer salts form very perfect single crystals, and these materials have led to a renaissance in *fermiology* [7]. As an example we show the SdH oscillations of $\beta_H(ET)_2I_3$ in Figure 10.10 [9, 18]. The dimensionality can be deduced from the angular dependency of the oscillations when the sample is rotated in the magnetic field. (In 1D metals with totally open Fermi surfaces – Figure 10.9a – oscillations are not observed because the electrons cannot move in circles!)

10.2.5.1 One-Dimensional Organic Superconductors

The Bechgaard salt $(TMTSF)_2PF_6$ (TMTSF, tetramethyltetraselenafulvalene)is the organic charge-transfer salt in which superconductivity was observed for the

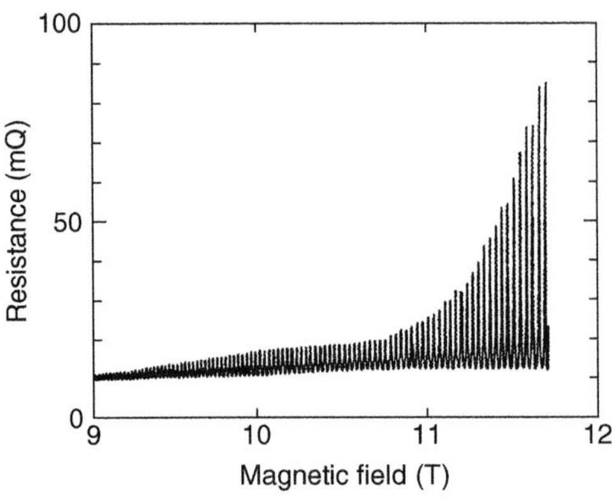

Figure 10.10 Shubnikov–de Haas oscillations in $\beta_H(ET)_2I_3$. Source: Kang et al. 1989 [18]. Reproduced with permission of American Physical Society

Table 10.2 Critical temperatures of TMTSF superconductors.

Compound	T_c (K)	Pressure (kbar)
$(TMTSF)_2PF_6$	1.1	6.5
$(TMTSF)_2AsF_6$	1.1	9.5
$(TMTSF)_2SbF_6$	0.36	10.5
$(TMTSF)_2TaF_6$	1.35	11
$(TMTSF)_2ReO_4$	1.2	9.5
$(TMTSF)_2FSO_3$	2.1	6.5
$(TMTSF)_2ClO_4$	1.4	Ambient

Source: After Williams et al. [7].

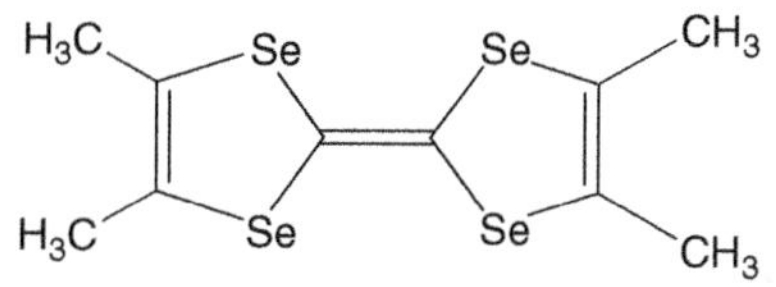

Figure 10.11 TMTSF, tetramethyltetraselenafulvalene.

first time [19]. A pressure of 6.5 kbar had to be applied, and superconductivity occurred at $T_c \sim 1.1$ K. The Fabre–Bechgaard salt $(TMTTF)_2ClO_4$ (TMTTF, tetramethyltetrathiafulvalene) is the only member of the 1D family that becomes superconducting at ambient pressure ($T_c = 1.4$ K [20]). The transition temperatures of TMTSF superconductors are listed in Table 10.2.

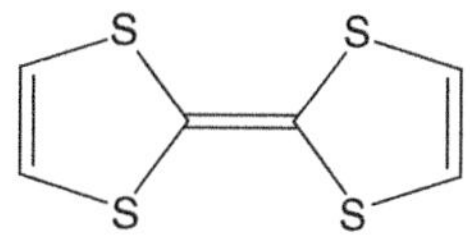

Figure 10.12 TTF, tetrathiafulvalene.

The two compounds TMTSF and TMTTF are similar with the S atoms of TMTTF being replaced by selenium atoms in TMTSF as seen in Figure 10.11. Both derive from tetrathiafulvalene (TTF) with methyl groups added (Figure 10.12).

The anions PF_6, AsF_6, SbF_6, TaF_6, ReO_4, and ClO_4 act as electron acceptors, but beyond that they do not participate in metallic conductivity or superconductivity. Conductivity is exclusively due to the overlapping of the π orbitals of the cations.

This overlap is different from the p orbital overlap forming the π bands in conjugated polymers. The different types of overlaps are schematically shown in Figure 10.13. Overlapping in conjugated polymers occurs sidewise, along the polymers axis, and leads to very wide bands, $W \sim 10$ eV, while overlapping in charge-transfer salts is top to bottom, along the stacking axis, and leads to rather narrow bands, $W \sim 1$ eV. Conjugated polymers are *intramolecular* 1D conductors, while charge-transfer salts are *intermolecular* conductors. (There is also an intermolecular, for example, interchain overlapping in conjugated polymers with $W \leq 1$ eV and an interstack overlap in TMTSF charge-transfer salts with $W \ll 1$ eV.)

Perhaps selenium/organic superconductors are not very spectacular, because selenium itself is a superconductor, albeit at very high pressures ($T_c = 6.9$ K for $p > 130$ kbar). The ET salts are sulfur-based organic superconductors and

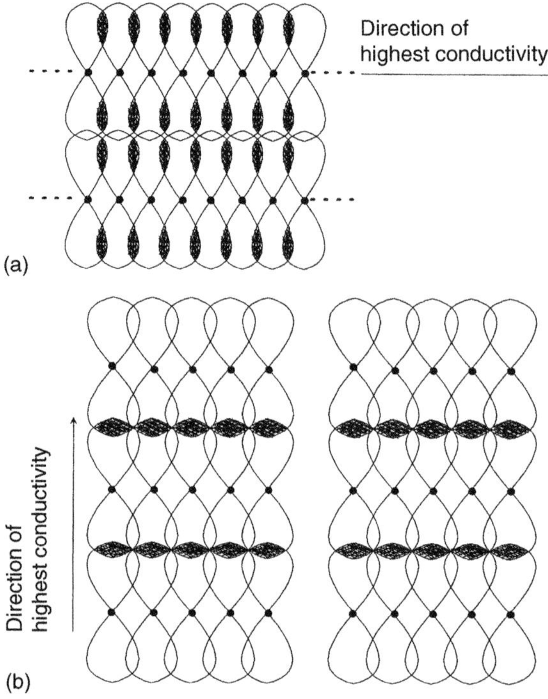

Figure 10.13 Different overlaps of the electronic wavefunctions forming the conduction bands in conjugated polymers and in molecular crystals. (a) Overlap of atomic p orbitals forming the π band of conjugated polymers and the molecular π orbitals in aromatic systems (benzene, fulvalene, etc.). Two chains are shown. (b) Overlap of the atomic p orbitals (or the molecular π orbitals) forming the conduction band of molecular crystals (e.g. of charge-transfer salts). Two stacks are shown.

from that point of view perhaps more exciting. The importance of the TMTSF compounds lies in their one-dimensionality and in the possibility they offer to study the competition between superconductivity and other 1D instabilities. $(TMTTF)_2Br_2$ becomes superconducting at 26 kbar; $T_c = 0.8$ K. It is the first sulfur-based superconductor observed in the $(TM)_2X$ series [21].

Figure 10.14 shows a "generalized phase diagram" of the $(TM)_2X$ series, where TM stands for tetramethylene derivatives of thio- or selenofulvalene and X denotes inorganic anions (PF_6, Br, ClO_4). We move along the abscissa

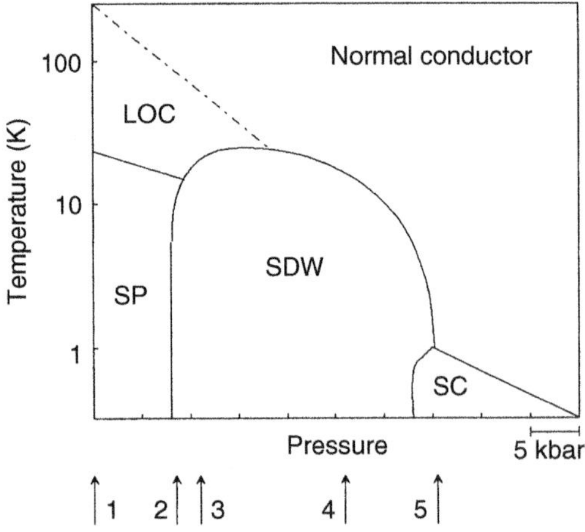

Figure 10.14 Generalized phase diagram of the $(TM)_2X$ series of one-dimensional solids. Source: After Jérome [18]. The arrows indicate the position of the respective substances at ambient pressure: (1) $(TMTTF)_2PF_6$; (2) $(TMDTDSF)_2PF_6$ (tetramethylene dithiodiselenofulvalene); (3) $(TMTTF)_2Br$; (4) $(TMTSF)_2PF_6$; (5) $(TMTSF)_2ClO_4$; SC, superconductivity; SDW, spin density wave; SP, spin Peierls; LOC, localized states.

by applying external pressure or by changing the chemical composition and thus exerting "internal pressure." The arrows indicate where the respective compounds are located at ambient external pressure. The region in which superconductivity occurs is denoted SC. An important feature of this diagram is the vicinity of superconductivity (SC) and spin density waves (SDWs). At ambient pressure, only $(TMTTF)_2ClO_4$ shows a normal-to-superconducting transition upon cooling. All other compounds go into an insulating state, which can be suppressed by about 12 kbar for $(TMTSF)_2PF_6$. The TMTSF compounds have been extensively reviewed in the references [22]. Further 1D superconducting systems can be found in Refs. [5, 7, 23].

10.2.5.2 Two-Dimensional Organic Superconductors

In 1D superconductors, the competition with the other instabilities seems to limit the critical temperature. Therefore stronger interstack coupling appears to be desirable. The ET salts exhibit fairly strong lateral coupling between the cations so that they are 2D rather than linear. As evidenced from *fermiology* the Fermi surface is cylindrical, and the conductivity is isotropic in the plane of the donor molecules (the in-plane to out-of-plane anisotropy is about 10^3).

Figure 10.15 shows the chemical structure of bis(ethylenedithio)tetrathiafulvalene (BEDT-TTF), otherwise known as ET. It is a sulfur-based relative of TTF and contains eight instead of four chalcogen atoms. $(ET)_2ReO_4$ was the first sulfur-based organic superconductor ($T_c = 2$ K at $p > 4.5$ kbar [24]), κ-$(ET)_2Cu[N(CN)_2]Br$ is the superconductor of the group with the highest ambient pressure T_c (11.6 K) [25], and κ-$(ET)_2Cu[N(CN)_2]Cl$ with $T_c = 12.8$ K at a pressure of 0.3 kbar [5] holds the absolute T_c record in this competition. Table 10.3 summarizes the ET superconductors. Note that the ET donor can be combined with a large variety of acceptors and the system exhibits a rich diversity in structures. There are groupings, like the β and the κ family, but for the unskilled observer the structure is not easy to comprehend. We note that ET layers exist in both families (in the β family ET molecules form a honeycomb-like sulfur network; in the κ family it is more complicated). These donor layers are separated by layers of acceptor molecules. For further details we refer to Refs. [7e, f].

The layered structure of ET salts advances these substances into the vicinity of the high-temperature ceramic copper oxide superconductors, which also consist of "metallic" layers (in this case CuO_2 layers) separated by "inert" spacers and counterions. In both classes of superconductors, the coherence length ξ is very anisotropic. In κ-$(ET)_2Cu[N(CN)_2]Br$, $\xi_{\text{perpendicular}} = 37$ Å within the plane of the donor layer and $\xi_{\text{parallel}} = 4$ Å perpendicular to the plane. The latter value is quite remarkable because it is much smaller than the distance (~15 Å) between the organic layers [7], thus stimulating discussions on unconventional interplane coupling mechanisms.

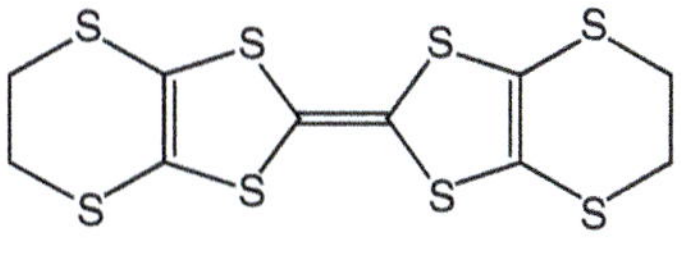

Figure 10.15 BEDT-TTF or ET – bis(ethylenedithio)tetrathiafulvalene.

Table 10.3 Critical temperatures of ET superconductors.

Compound	T_c (K)	Pressure (kbar)
$(ET)_2ReO_4$	2.0	4.5
β-$(ET)_2I_3$	1.4	Ambient
β*-$(ET)_2I_3$	8.0	0.5
γ-$(ET)_3(I_3)_{2.5}$	2.5	Ambient
θ-$(ET)_2I_3$	3.6	Ambient
κ-$(ET)_2I_3$	3.6	Ambient
α-$(ET)_2I_3$	7–8	Ambient
(α/β)-$(ET)_2I_3$	2.5–6.9	Ambient
β-$(ET)_{1.96}(MET)_{0.04}I_3$	4.6	Ambient
β-$(ET)_2IBr_2$	2.8	Ambient
β-$(ET)_2AuI_2$	4.98	Ambient
κ-$(ET)_4Hg_{3.8}Cl_8$	1.8 5.3	12 29
κ-$(ET)_4Hg_{2.89}Br_8$	4.3 6.7	Ambient 3.5
$(ET)_2Hg_{1.41}Br_4$	2.0	Ambient
α-$(ET)_2[(NH_4)Hg(SCN)_4]$	1.15	Ambient
$(ET)_3Cl_2 \cdot 2H_2O$	2.0	16
κ-$(ET)_2Cu(NCS)_2$	10.4	Ambient
κ-$(ET)_2Ag(CN)_2 \cdot H_2O$	5.0	Ambient
κ-$(ET)_2Cu[N(CN)_2]Br$	11.6	Ambient
κ-$(ET)_2Cu[N(CN)_2]Cl$	12.8	0.3

Source: After Williams [7e].

10.2.5.3 Three-Dimensional Organic Superconductors

In Chapter 1 we introduced fullerene as a zero-dimensional modification of carbon: in a C_{60} ball, 60 π electrons are confined in a sphere about 10 Å in diameter. So a fullerene molecule can be regarded as a quantum dot. But 60 is already a fairly large number, so the C_{60} molecule is really a "mini-solid." And, if we think of it as a rolled-up graphene plate or as the coiled-up chain of a conjugated polymer, then we might expect to find both polyacetylene and graphene aspects in its behavior. But these mini-solid C_{60} molecules can be used to build a regular array in *three dimensions: fullerite* crystal.[7]

As with conjugated polymers and graphite, fullerite can be intercalated with various dopants. In conjugated polymers and graphite, both acceptors and donors can be intercalated, but in fullerite donor doping is preferred (so alkali metals are pretty commonly used). Figure 10.16 shows a plane of the doped fullerite crystal. The hatched circles mark interstitial sites (tetragonal and octahedral) that can be

7 Some scientists prefer the term Fullerene crystal, but we find that confusing since fullerene itself is a crystal of sorts. Of course our structure for the 3D array may not be chemically consistent with the fullerite name, but it does some the mini/macro solid ambiguity.

Figure 10.16 Crystal structure of fullerene. The hatched spheres present the interstitial sites that can be occupied by dopant ions. Source: After Lüders [26].

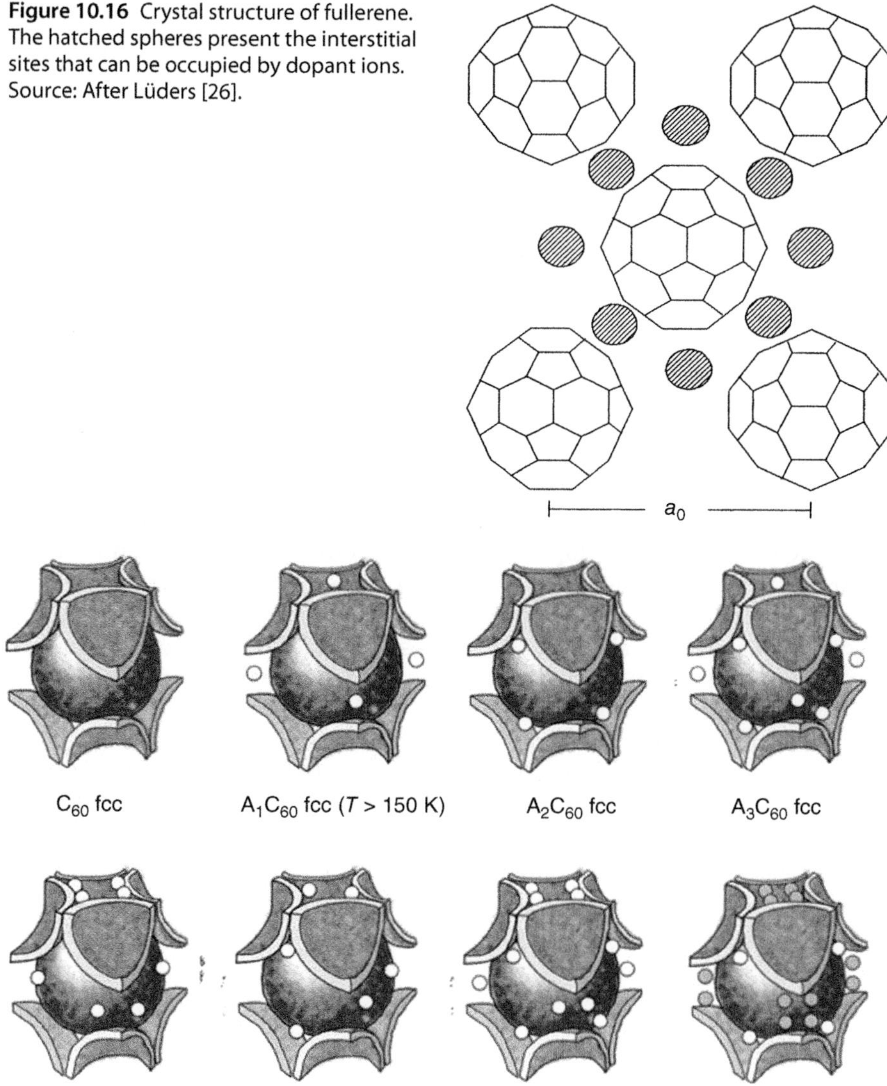

Figure 10.17 Summary of structure types of alkali-intercalated C_{60}. The A15 structure has only been observed in the alkaline earth phase, Ba_3C_{60}. Source: After Fleming et al. [27].

occupied by dopant ions. In Figure 10.17 we present a more sophisticated view of the spatial arrangement of C_{60} balls and dopant ions for a variety of doping stoichiometries.

Alkali graphite intercalation compounds become superconducting, usually with a critical temperature below 1 K. In the Cs–Bi graphite system, T_c values up to almost 5 K have been observed. But alkali-doped fullerites are also superconductors, and surprisingly the critical temperature of these compounds is as high as 33 K. Thus, they beat the ET charge-transfer salts by a factor of 3!

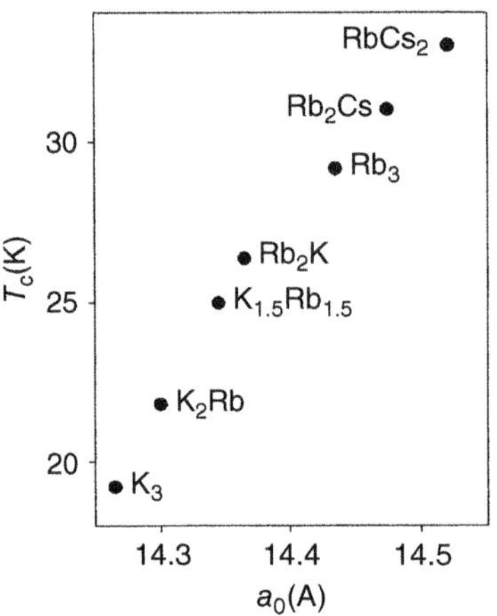

Figure 10.18 Variation of T_c with the lattice parameter a_0 for various compositions of A_3C_{60}. Source: After Lüders [26].

For a comparison of graphite and fullerite superconductors, see Refs. [26, 28]. We have already mentioned that the C_{60} balls in a fullerite crystal move further apart when the material is doped. This increases the DOS at the Fermi surface and thus T_c increases. The behavior is nicely shown in Figure 10.18, where the critical temperatures of A_3C_{60} are plotted vs. the lattice parameter a_0 (in A_3C_{60} the symbol A stands for alkali metal). An intensive study of fullerene compounds has been underway for some time now, and scientists have moved to working on chemically modifying the carbon balls using cycloaddition schemes. It might well turn out that with the presently highest critical temperature of 33 K [29], the maximum T_c has not yet been reached for this materials class [30].

It should be recognized, however, that the developments of smaller fullerene derivatives such as C_{20} [31] and C_{36} have suggested modifications to our view of three dimensionally based organic superconductors. Specifically, these smaller cage molecules can have extraordinarily high DOS [32] at the Fermi level. Further, they exhibit significantly stronger electron–phonon coupling constants than their larger cage counterparts (C_{60}, C_{70}, C_{80}) [33].

10.2.6 Trends

In research and development it is not possible to predict the future by extrapolating the past. Discoveries are unforeseen; otherwise they would not be called discoveries. In retrospect, most evolutions begin exponentially, continue linearly, and then saturate. But there is no way of foretelling the saturation level from the exponential start. Nevertheless it is fascinating to look at the historic development in a certain field of science and to speculate on the future. In this respect it does not matter whether the selection of historic events is biased or not, because we use the data only as a ladder for our fantasy.

A selection of historic events in organic (and high T_c) superconductivity is given in Table 10.4. If the highest T_c values are plotted vs. the year of the respective observation and if the various compounds are summarized into teams, races can be arranged. Figure 10.19 shows the competition of metallic, organic, oxide [34], and fullerite superconductors. We have the impression that the oxides will

Table 10.4 Chronology of organic superconductors.

Year	Organic superconductors
1962	Synthesis of TCNQ
1970	Synthesis of TTF
1973	TTF–TCNQ ("first organic metal")
1978	Synthesis of ET
1979	Bechgaard salts $(TMTSF)_2X$, superconducting under pressure, $T_c \sim 1\,K$
1981	$(TMTSF)_2ClO_4$, first organic superconductor at ambient pressure, $T_c \sim 1.4\,K$
1983	β-$(ET)_2ReO_4$, first sulfur-based organic superconductor under pressure, $T_c \sim 1\,K$
1985	Synthesis of C_{60}
1986	Superconductivity in La–Ba–Cu–O
1987	T_c above liquid nitrogen temperature
1987	κ-$(ET)_2Cu(SCN)_2$, $T_c \sim 10.4\,K$
1990	κ-$(ET)_2Cu[N(CN)_2]Br$, $T_c \sim 11.6\,K$ κ-$(ET)_2Cu[N(CN)_2]Cl$, $T_c \sim 12.8\,K$ at 0.3 kbar
1990	C_{60} in large quantities available
1991	Superconductivity in alkali-doped C_{60}
1993	$T_c = 133\,K$ in Hg–Ba–Ca–Cu–O
1994	$T_c = 160\,K$ under pressure in Hg–Ba–Ca–Cu–O

Source: Partially after Williams et al. [7f].

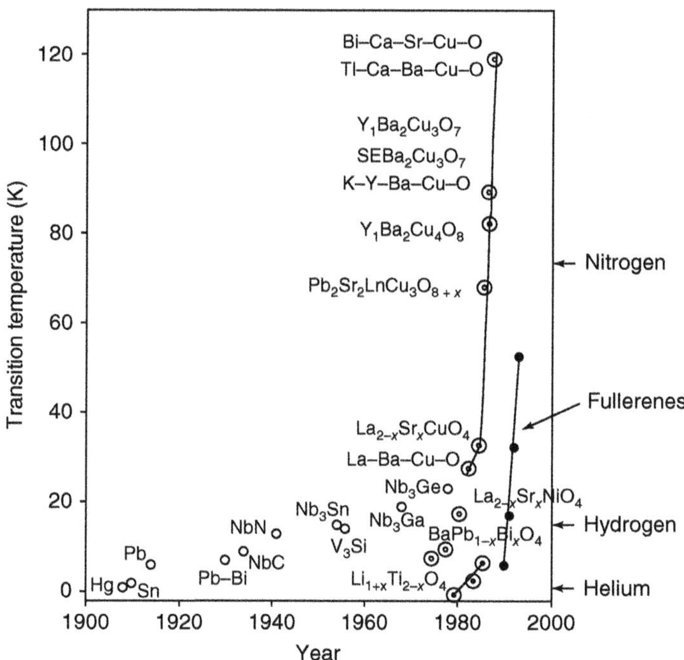

Figure 10.19 History of superconductivity; transition temperatures are plotted vs. the year of discovery (O, metallic, 1979, organic [not shown], O, oxidic, •, fullerenic). On the right-hand scale, the boiling points of helium, hydrogen, and nitrogen are marked.

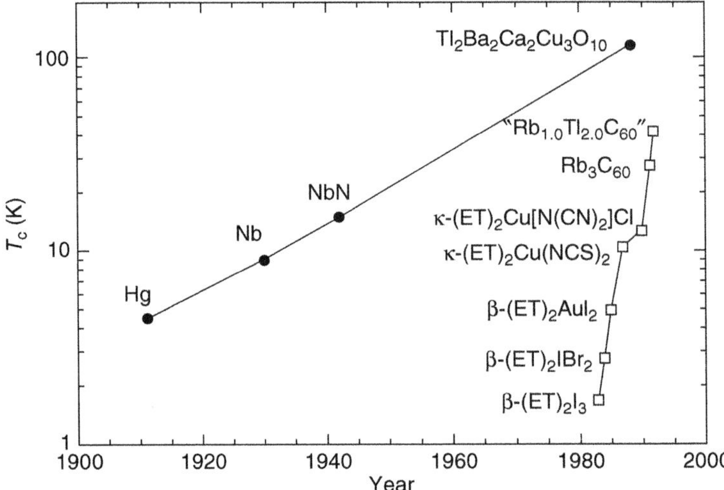

Figure 10.20 Superconductivity transition temperature T_c vs. year of discovery illustrating the race between inorganic and organic superconductors. Source: After Williams et al. [7].

win, if the goal were to reach room temperature superconductivity. In the organic vs. inorganic race shown in Figure 10.20, the organics seem to overtake soon.

However, no matter how one looks at it, the introduction of dimensionality to the study of superconducting systems has led to significant advances. We are quickly approaching the ability to design the materials properties based on these "dimensionality" principles.

10.3 The Charge Density Wave

When metals are cooled to very low temperatures, we have seen that they may undergo transitions to new phases that order through collective interactions. Superconductivity is one example but it surely isn't the only one. There are, for example, interactions within our many-particle state that lead to excitations and ordering of the charge density or spin density. At times these processes can compete with superconductivity, and at times superconductivity can exist simultaneously with these excitations. So our next illustration is the *charge density wave* (CDW).

10.3.1 The Charge Density Wave and Peierls

Recall that in our discussions of the Peierls transition, we treated it as a structural phase transition wherein the atoms rearrange and simultaneously the electronic structure of the system changes. The atomic rearrangement leads to new reflections in X-ray and neutron diffraction patterns. Recall also that a precursor of the *Peierls transition* is the *Kohn anomaly,* which is phonon softening at the wavevector $\boldsymbol{q}$, corresponding to two times the Fermi wavevector $\boldsymbol{k}_F$. The change of the electronic structure causes an MIT. The Peierls transition is typified in this 1D behavior and occurs only if the Fermi surface is planar or at least contains parallel sections (known as *nesting*). Finally, remember we considered a (hypothetical) monatomic alkali chain and then applied this Peierls concept to *conjugated polymers* by replacing the alkali atoms with CH radicals. Thus bond alternation was explained as a Peierls transition. The π electron density along the chain appears as a sinusoidal wave.

10.3.1.1 Modulation of the Electron and Mass Densities

The *CDW* is a collective ordering of the many-particle state with spatial modulation of the conduction electron density and an associated modulation of the lattice atomic positions. As in the case of superconductivity, CDWs come about due to an instability in the metallic Fermi surface that involves electron–phonon coupling. And this leads to charge density fluctuations that look like

$$\rho = \rho_0 \cos(\boldsymbol{q} \cdot \boldsymbol{r} + \phi) \tag{10.70}$$

Another way to phrase this is that the CDW state is, again, electrons coupling with the lattice. This coupling produces a charge density fluctuation quite similar in functional form to that seen in the conjugated systems that we explained with the Peierls transition.

Electron–phonon coupling is particularly favorable when the phonon modes soften, such as at the Kohn anomaly; thus CDWs are frequently found in low-dimensional systems. This coupling between the charge density and mass density is *the* essential feature of the CDW, but, unfortunately, the term "charge density wave" does not sufficiently stress this coupling. However, the term "charge density – mass density wave" is simply too clumsy.

10.3.1.2 Starting with Polymers

In the case of conjugate polymers, the modulation of the π electron density is rather trivial because there must be a higher electron density at the double bonds. Conjugated polymers, however, are not ideal Peierls systems. In Peierls systems, electron–phonon interactions are dominant, and in conjugated polymers, electron–electron interactions are equally as important. To account for this more complicated situation, the bond alternation in polymers is called a *bond order wave* (BOW) rather than *CDW*.

Alkali metal chains and conjugated polymers are 1D systems with half-filled bands: there is one electron per lattice site, and in a completely filled band, there would be two electrons, one with spin-up and the other with spin-down. In a half-filled band, the elementary cell doubles at the phase transition, and the Brillouin zone is reduced by a factor of 2. The Kohn anomaly occurs at the phonon vector $\boldsymbol{q} = 2k_F$ that is the end of the first Brillouin zone in the undistorted lattice. The Peierls concept, however, is not restricted to half-filled bands. In the hypothetical "polyfractiolene" we encountered band filling of 1/3, changing the structure by tripling the unit cell. More complicated is the situation in KCP, where the band filling is 5/6. Consequently, $2\boldsymbol{k}_F = 10/6\boldsymbol{a}^*$, which can be transferred into the first Brillouin zone by subtracting a reciprocal lattice vector $\boldsymbol{a}^*$ so that we get the result $\boldsymbol{q} = 2/3\boldsymbol{a}^*$. This fits quite well to the position where the Kohn anomaly is observed by inelastic neutron scattering in this system.

10.3.1.3 A Gap Is Introduced

With the lattice distortion going as

$$u_n = u_0 \cos(qz + \phi) \tag{10.71}$$

and $q = 2k_F$ or $\lambda_c = \pi/k_F$, n is the atom index, and then an energy gap will open up at $\pm k_F$ as we saw in the case of phonons generally. However these distortions are coupled to the electrons, and in the 1D case, the elastic energy cost to modulate the atomic positions is less than the energy gain in the conduction electrons. Thus the CDW becomes the preferred ground state of the system at low temperatures. At higher temperatures the energy gain of the conduction electrons is reduced by thermal excitations across the gap, making the metallic state stable. A second-order phase transition with temperature exists between the CDW state and the metallic state in such systems. This transition is exactly the *Peierls transition*.

10.3.1.4 The Order Parameter

The CDW state is often described in terms of a complex order parameter given as

$$\psi = \Delta e^{i\phi} \tag{10.72}$$

where Δ determines the size of the energy gap as well as the displacement of the atomic positions u_0 and ϕ is the phase lag between the CDW and the underlying lattice itself. We discussed order parameters earlier in Chapter 8 on lattice order as representing the probability of finding some observable at some position within the system. In this case variations in ϕ and Δ can occur due to collective excitations of the many-particle system known as *phasons* and *amplitudons*, respectively. Clearly such excitations are quantized into their own quasiparticles.

10.3.1.5 Phase Dynamics, Pinning, Commensurability, and Solitons

One should be a little careful about reading too much into the single particle state diagrams for dispersion. We didn't give one here for CDWs, but we do for superconductors above, and these can be misleading. There is an energy gap in both cases, but neither case is a semiconductor as we might suggest using the logic of our single particle arguments. This is because both have rigid, collective charge transport modes. In the superconductors we used this as an argument for reduced scattering generally. In CDWs something similar occurs. For CDW systems, when a field is applied, the CDW can slide rigidly with respect to the underlying lattice. Oscillation of the atoms locally produces a traveling potential landscape. The electrons are coupled to and move with this traveling potential, producing a current from the CDWs. This *phase slip* is the primary mechanism for transport within the state.

Sliding without resistance of a CDW is only possible if the following two conditions are fulfilled:

1. The crystal is perfect.
2. The CDW is incommensurate or the crystal is made of jellium.

Jellium assumes a continuous distribution of the positive charges instead of discrete atoms. *Incommensurability* means that the ratio between the wavevector and the reciprocal lattice vectors is an irrational number. In both cases the CDW cannot *register* with the crystal lattice, and hence the energy is independent of the phase of the CDW.[8] A mechanical analogue of "registering" is the cogwheel and the bicycle chain. In many physical systems registering also occurs when higher harmonics match, but it does not when there is no rational ratio between the periodicities. An example of incommensurability is the rotation of the earth around its axis and the revolution around the sun. To compensate this incommensurability, we have to add intercalary days in leap years.

The CDW in polyacetylene registers strongly. To move it, a very high barrier must be overcome. This barrier corresponds to the cleavage of all double bonds. If the wavelength of the CDW is three times the lattice constant or perhaps 3/5, registering will be much less pronounced. If CDW and lattice are incommensurate, no registering will be possible. Of course, since the rational numbers are infinitely dense, the exact distinction between commensurate and incommensurate is rather artificial, but for practical purposes a CDW is incommensurate when the attraction to the crystal lattice is smaller than the thermal energy or other energy fluctuations.

Impurities, crystal boundaries, and crystal imperfections will *pin* the CDW. For example, a CDW cannot slide over chain ends. Any other irregularity will also prevent the CDW from sliding, because impurities prefer to remain either at the crest or in the valleys of the CDW. The interactions between individual impurities with the CDW sum up to a pinning force, and the CDW will slide only in case the force exerted by an applied electric field is larger than the pinning force. This leads to a *threshold field* for CDW motion. Below threshold the CDW does not slide; above threshold it will, however not without resistance. Moving the CDW in the presence of impurities leads to finite resistivity. Figure 10.21 shows the temperature dependency of the conductivity of TTF–TCNQ [36]. The conductivity increases upon cooling as expected for a metal, but below 80 K the increase grows abnormally large and probably is a precursor of sliding CDW conductivity.

At 53 K a transition to a pinned CDW occurs and the sample becomes insulating. In some samples an even greater conductivity increase just above 53 K has been observed [37], and reports of this "giant conductivity" have had an enormous impact on the scientific community.

To demonstrate the existence of a threshold field, we can examine TaS_3 as an example [38]. In TaS_3 the Peierls transition to a pinned CDW occurs at 220 K. In Figure 10.22 the electric field dependency of the conductivity at 130 K is shown. The conductivity is normalized to the room temperature value.

Up to the threshold field of $E_T \sim 0.3$ V/cm, the conductivity is negligibly small. At E_T the conductivity rises abruptly and saturates at high field values that correspond to the normal state conductivity. Saturation at normal state values is plausible from the equations relevant for CDW conductivity, which are identical to the conductivity in the Drude model of free electrons when the sliding

8 Here the meaning of the word "register" is taken from color printing, where the pads have to register to ensure correct superposition. Incommensurability was originally a theological term implying that there is no common measure for the grace of God and human merits meaning that one cannot be offset against the other.

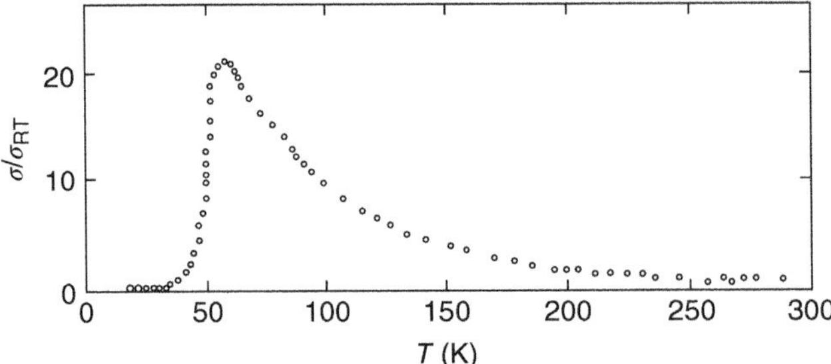

Figure 10.21 Temperature dependency of the conductivity of TTF–TCNQ. Source: After Cohen et al. [35]. The ordinate is the conductivity normalized to the room temperature value. Just above the metal-to-insulator transition at 53 K, an unusual increase of the conductivity is observed, which in some samples took the form of "giant conductivity."

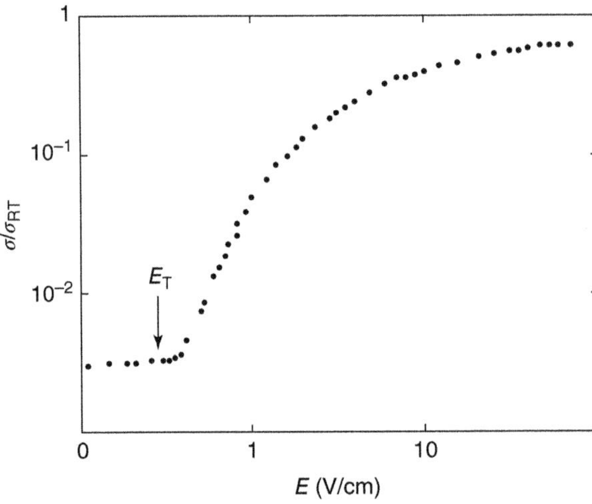

Figure 10.22 CDW conductivity of TaS_3 as a function of the applied electric. Below E_T the conductivity is negligibly small; above E_T sliding sets in, but the conductivity is not free of resistance. It remains finite and saturates at values expected for normal state conductivity ($\sim\sigma_{RT}$). Source: After Grüner and Zettl [39].

velocity corresponds to the drift velocity. Another consequence of pinning is the frequency dependency of the conductivity. Pinning makes it hard for a CDW to slide, but still the CDW can oscillate fairly easily. This leads to low DC and high AC conductivity.

Rice et al. [40] have predicted nonlinear charged excitations in pinned CDWs. They obtained the following equation for the phase ϕ of the CDW:

$$\frac{\partial^2\phi}{\partial t^2} - c_0^2\frac{\partial^2\phi}{\partial x^2} + \omega_0^2\frac{\mathrm{d}V}{\mathrm{d}\phi} = 0 \tag{10.73}$$

where ω_0 is the oscillation frequency of the CDW (determined by far-infrared measurements) and c_0 is the phason velocity as derived in the Lee, Rice, and Anderson [41] analysis of the A_ mode in Figure 10.22. The quantity V in the last term of Eq. (10.73) is the potential energy due to the registering of the CDW to the lattice. Consequently V is a periodic function, and Eq. (10.73) is the sine-Gordon equation mentioned in the general discussion on solitons. We know that the sine-Gordon equation has solitonic solutions. A soliton in a CDW is a phase-slip center, as depicted in Figure 10.23.

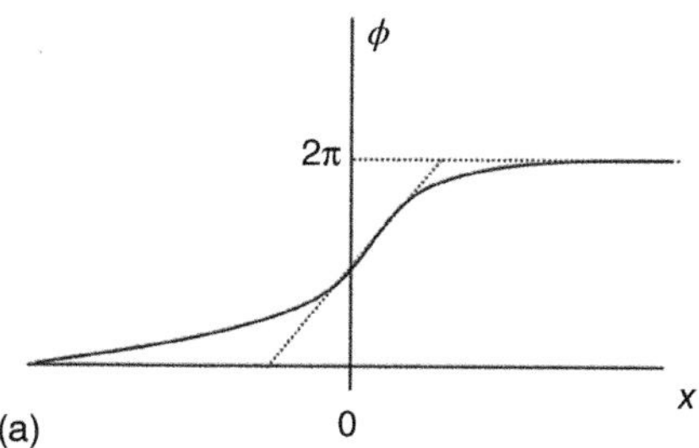

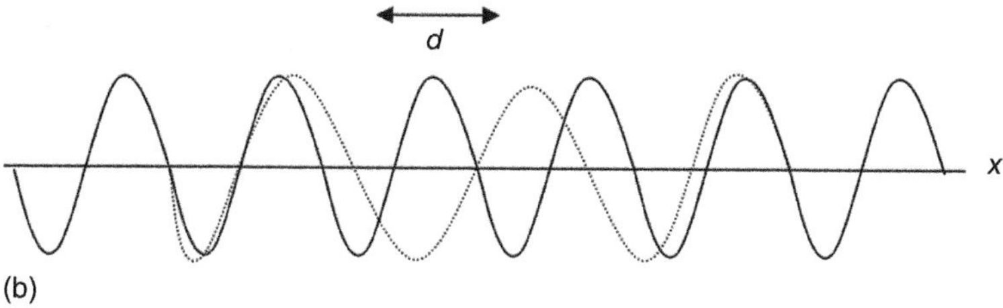

Figure 10.23 Phase slip center (soliton) in a pinned CDW. (a) Spatial change in the phase ϕ. Note the similarity to the bond alternation parameter in polyacetylene in Figure 5.23. (b) Charge density wave with local phase f.

It can be demonstrated that its charge amounts to $\pm 2e$, equivalent to that of two electrons. (The integral over a harmonic wave is zero, of course, but at the distortion the negative and positive parts of the wave differ in width, and integrating then leads to $\pm 2e$.) The conductivity of a CDW system below threshold is not exactly zero. It is believed that the residual conductivity is at least partially due to solitons. Soliton conductivity in pinned CDWs is similar to creep phenomena in metallurgy. Long before the shear forces are strong enough to allow for sliding of crystal slabs along lattice planes, the crystal deforms by the motion of dislocations.

If the charge density slides (above threshold), a very peculiar noise is generated and superimposed to the DC current. A spectral analysis reveals that this noise consists of a very narrow peak and its higher harmonics (*narrow band noise*). As an example, see the noise spectrum of TaS_3 in Figure 10.24. Again there is a close analogy to shear deformation of metals. When some metals (for example, tin or cadmium) are bent, they emit a characteristic acoustic noise ("Zinngeschrei," tin cry).

A classical single particle model [41] for the interpretation of narrow band noise is the so-called "washboard model" (Figure 10.25). The CDW slides as a rigid entity over rigidly fixed impurities. It corresponds to a ball rolling down a wiggled slope (washboard). This simple model explains both the threshold field and the dependence of the noise frequency on the CDW current as in Figure 10.26.

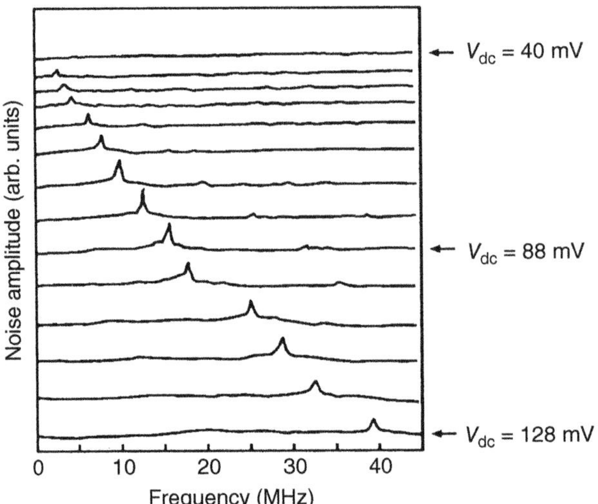

Figure 10.24 Noise spectrum for sliding CDW conduction in TaS_3. The narrow band noise is seen as a pronounced peak that shifts to higher frequencies as the sliding velocity of the CDW increases (along with increasing bias voltage). Source: After Grüner and Zettl [39].

Figure 10.25 Classical washboard model to demonstrate that noise frequency increases with the sliding velocity.

10.3.2 Peierls and Coulomb Interactions: Spin Interactions

The CDW with wavevector $\boldsymbol{q} = 2\boldsymbol{k}_\mathrm{F}$ is just one representative of a large variety of electron-driven lattice distortions in 1D solids. In the phase diagram of the Bechgaard salts, SDW are presented, and in some systems there is a CDW with $\boldsymbol{q} = 4\boldsymbol{k}_\mathrm{F}$ instead of $\boldsymbol{q} = 2\boldsymbol{k}_\mathrm{F}$. These interactions are predicted when the mechanisms of coupling become more complicated and *include more details.*

10.3.2.1 $4k_\mathrm{F}$ Charge Density Waves

A $4\boldsymbol{k}_\mathrm{F}$ *CDW* can occur when the Coulomb interaction between the electrons is large. To understand why, we turn to our friend from above: Hubbard [42].

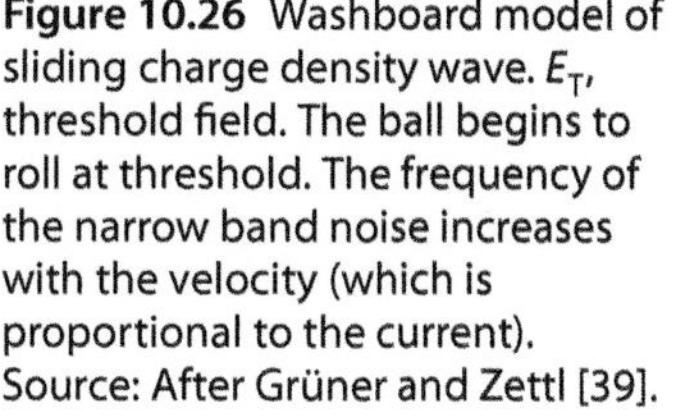

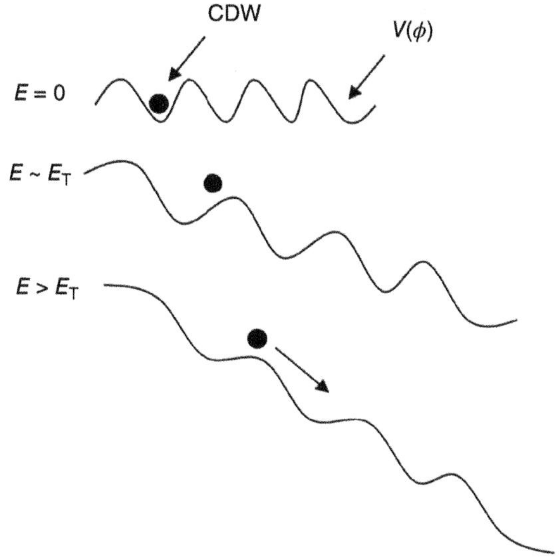

Figure 10.26 Washboard model of sliding charge density wave. E_T, threshold field. The ball begins to roll at threshold. The frequency of the narrow band noise increases with the velocity (which is proportional to the current). Source: After Grüner and Zettl [39].

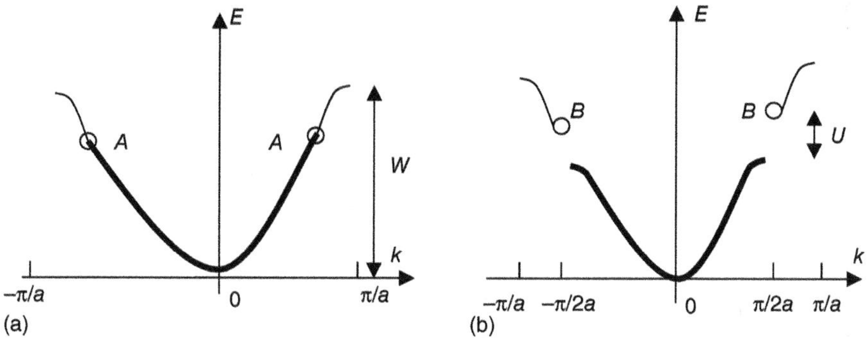

Figure 10.27 Creation of a Coulomb gap in a half-filled band due to electron–electron interaction. (a) Without Coulomb interaction. (b) Coulomb interaction switched on. Source: After Kagoshima et al. [43].

This model is one of the most simple ways to see how the interactions between electrons can give rise to collective phenomena in solids such as metal–insulator transitions, magnetic ordering, and in some cases superconductivity.

In the Hubbard model, if U is large, say, on the order of the bandwidth, all electrons are localized. If the band is half-filled, there will be one electron on each lattice site. An additional electron will require the energy U, because it will have to be placed where there exists another electron already, and hence it will be repelled. The repulsion creates a Coulomb gap of width U at the Fermi surface; the system is not a metal but a Mott–Hubbard insulator. The Coulomb gap is illustrated in Figure 10.27. The electronic band structure $E = E(k)$ is plotted with the filled part of the band indicated by a thick line. The Fermi energy is at point A. An additional electron will not be found at A but at B, which is higher by the Coulomb repulsion energy U.

Now to discuss the *$4k_F$ CDW*, we take a look at a linear chain with one electron per three atoms. The electrons will tend to be equally spaced. A lattice distortion

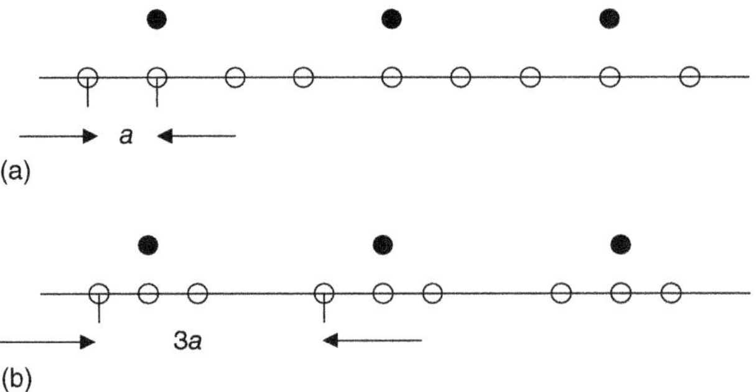

Figure 10.28 $4k_F$ charge density wave in a chain with one electron per three atoms and strong Coulomb interaction. (a) Undistorted chain. (b) $4k_F$ charge density wave. Source: After Kagoshima et al. [43].

in which the atoms are grouped by three and brought closer to the electrons will minimize the energy. In contrast to the Peierls distortion with $\boldsymbol{q} = 2\boldsymbol{k}_F$, the distortion shown in Figure 10.28 occurs at $\boldsymbol{q} = 4\boldsymbol{k}_F$. This can be easily explained: the wavelength of the CDW is $\lambda = 3\boldsymbol{a}$, and the wavevector is $\boldsymbol{q} = 2\pi/3\boldsymbol{a}$. The band filling results in $\alpha = 1/6$ because a completely filled band would have two electrons per site (spin-up and spin-down). In our case, however, there is only one electron per three sites. In a linear chain the Fermi vector is proportional to the filling factor $\boldsymbol{k}_F = \alpha\pi/\boldsymbol{a}$; hence in our case, $\boldsymbol{k}_F = (1/6)(\pi/\boldsymbol{a})$, and consequently $\boldsymbol{q} = 4\boldsymbol{k}_F$. A more general way of achieving this result is to say that strong Coulomb interaction lifts the spin degeneracy of the electrons. The electron system decouples into two subsystems with the states of one of the subsystems being energetically so far away that they can be disregarded. A subsystem band with one electron per atom is completely filled, not half-filled. All $\boldsymbol{k}$ vectors of the electronic states have to be multiplied by two, and the CDW occurs at $\boldsymbol{q} = 4\boldsymbol{k}_F$.

10.3.2.2 Spin Peierls Waves

The linear array of equidistant electrons in Figure 10.28b still has spin degrees of freedom, and the $4\boldsymbol{k}_F$ CDW is not the ground state. The spins can order ferromagnetically (all parallel) or antiferromagnetically (up and down alternating). The magnetically ordered spin chain is known as the *Heisenberg spin chain*. It has collective excitations, the *magnons*. If the equidistant electron arrangement is changed into a pair arrangement as indicated in Figure 10.29, the *magnon* energy can be lowered in a way similar to the lowering of the electron energy in the Peierls transition. The lattice adjusts to the paired electrons, the elementary cell is doubled, and the lattice distortion is again found at $\boldsymbol{q} = 2\boldsymbol{k}_F$. Because of the analogy the transition within the $4\boldsymbol{k}_F$ CDW is called a *spin Peierls* (SP) transition.

10.3.2.3 Spin Density Waves

The SP state is different from the SDW, which is accompanied by *no* lattice distortion [44]. An SDW actually is a split CDW: it consists of a wave for spin-up electrons and another one for spin-down electrons with a phase shift of 180° between the spin-up and the spin-down waves. The splitting is brought about by

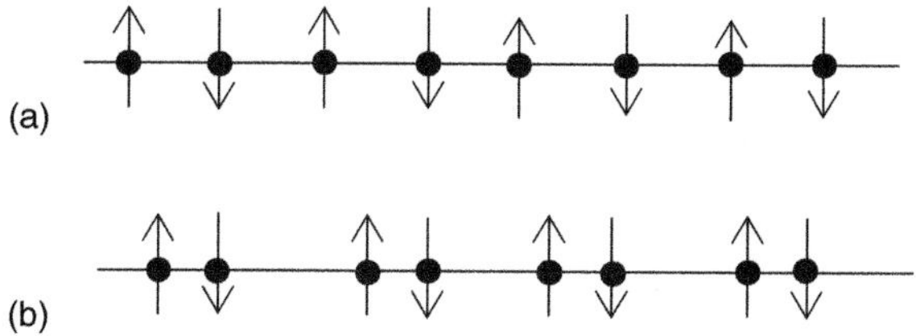

Figure 10.29 Spin Peierls transition in a $4k_F$ charge density wave. The $4k_F$ CDW leads to a chain of equidistant electrons (Heisenberg spin chain) that can order magnetically. The magnetic energy can then be lowered by the approach of the electrons in pairs.

Table 10.5 Various instabilities in the CDW family.

Acronym	$2k_F$ CDW	$4k_F$ CDW	SDW	SP
Full name	$2k_F$ charge density wave	$4k_F$ charge density wave	Spin density wave	Spin Peierls state
Wavevectors	$2k_F$	$4k_F$	$2k_F$	$2k_F$
Lattice distortion	Yes	Yes	No	Yes

Source: Partially after Alcacer [45].

electron–electron interactions. Because of the phase shift, the corresponding lattice distortions interfere destructively, and there is no net distortion. Therefore, an SDW cannot be detected by structural investigations (X-ray or neutron scattering); however, local susceptibility measurements turn out to be very powerful such as electron spin resonance (ESR) and nuclear magnetic resonance (NMR), the latter because of the hyperfine interaction through which the electronic spins relax the nuclear spins [45].

A summary of the various instabilities in the CDW family is given in Table 10.5. At low temperatures these instabilities compete with each other and with superconductivity. Sometimes this competition is expressed in terms of "g-ology," using the electron scattering parameters g_1 and g_2 as axes of a coordinate system and assigning the most divergent instabilities to the respective fields in the g_1–g_2 plane. The parameter g_1 is the amplitude for backward and g_2 for forward scattering. In a more complete diagram, g_3 is used for Umklapp scattering. In this diagram two types of superconductivity occur: the singlet superconductivity (SS) and the triplet superconductivity (TS). SS is the well-known "conventional" BCS superconductivity, where the spins of the electrons in a Cooper pair are opposite so that the net spin is zero. In TS the spins are parallel, summing up to a total spin $S = 1$, which can take three orientations relative to an external field (+1, 0, and −1, hence triplet). In the phase diagram of the Bechgaard salt superconductors, we noted that superconductivity is adjacent to SDW, and in Figure 10.30 we find the superconducting neighbor of SDW to be TS rather than SS. This observation has led to speculations that the superconductivity in Bechgaard salts may be of the triplet type rather than singlet [23, 46].

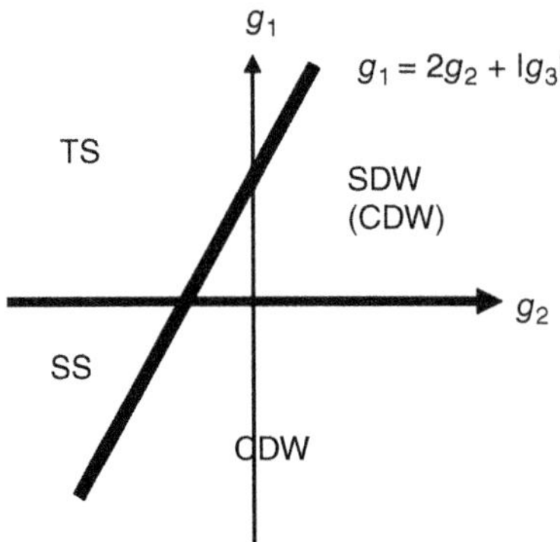

Figure 10.30 "g-Ology" of one-dimensional instabilities. g_1 is the amplitude for backward, g_2 for forward, and g_3 for Umklapp scattering. TS, triplet superconductivity; SS, singlet superconductivity; SDW, spin density wave; and CDW, charge density wave.

10.3.3 Phonon Dispersion: Phase and Amplitude in CDWs

Figure 10.31 shows the phonon dispersion relation of a system susceptible to the Peierls transition. In Figure 10.31a we see the dispersion relation far above the Peierls transition. In Figure 10.31b the Kohn anomaly develops at $\boldsymbol{q} = 2\boldsymbol{k}_{\mathrm{F}}$, as demonstrated in the example of KCP. In Figure 10.31c the phonons at the Kohn anomaly become completely soft, and the lattice rearranges. In this rearrangement $2\boldsymbol{k}_{\mathrm{F}}$ turned into a new reciprocal lattice point, and the first Brillouin zone is now limited by $\pm\boldsymbol{k}_{\mathrm{F}}$. The parts outside the first Brillouin zone are indicated by dotted lines in Figure 10.31d. These outside parts can be transferred into the first Brillouin zone by subtraction of the reciprocal lattice vector ($\pm\boldsymbol{k}_{\mathrm{F}}$). The reduced zone scheme is shown in Figure 10.31e.

Here there are two phonons with energy $\omega = 0$ at wavevector $\boldsymbol{k} = 0$. This contradicts one of the fundamental theorems of lattice dynamics. Since "*nicht sein kann, was nicht sein darf*,"[9] the lattice lifts this degeneracy and pushes one phonon branch (A_+) upward, which in the terminology of lattice dynamics is called an optical phonon branch (there are many optical phonon branches allowed), whereas A_- remains the acoustic phonon branch. There is only one acoustic phonon branch permitted (see Chapter 5 on phonons). From the point of view of lattice dynamics, this is not unusual. Rearrangements of the phonon branches occur at many structural phase transitions. But in the world of CDWs, we call A_- the *phase mode* or *phasons*, and A_+ the *amplitude mode* or *amplitudons* [41]. Given our discussion of these two names and their origins in

9 From a favorite poem *Die unmögliche Tatsache* by Christian Morgenstern (1871–1914):

Palmström, etwas schon an Jahren,
wird an einer Straßenbeuge
von einem Kraftfahrzeuge
Überfahren.
… (Fahrverbot) …
Und er kommt zu dem Ergebnis:
"Nur ein Traum war das Erlebnis.
Weil," so schließt er messerscharf,
"nicht sein kann, was nicht sein darf."

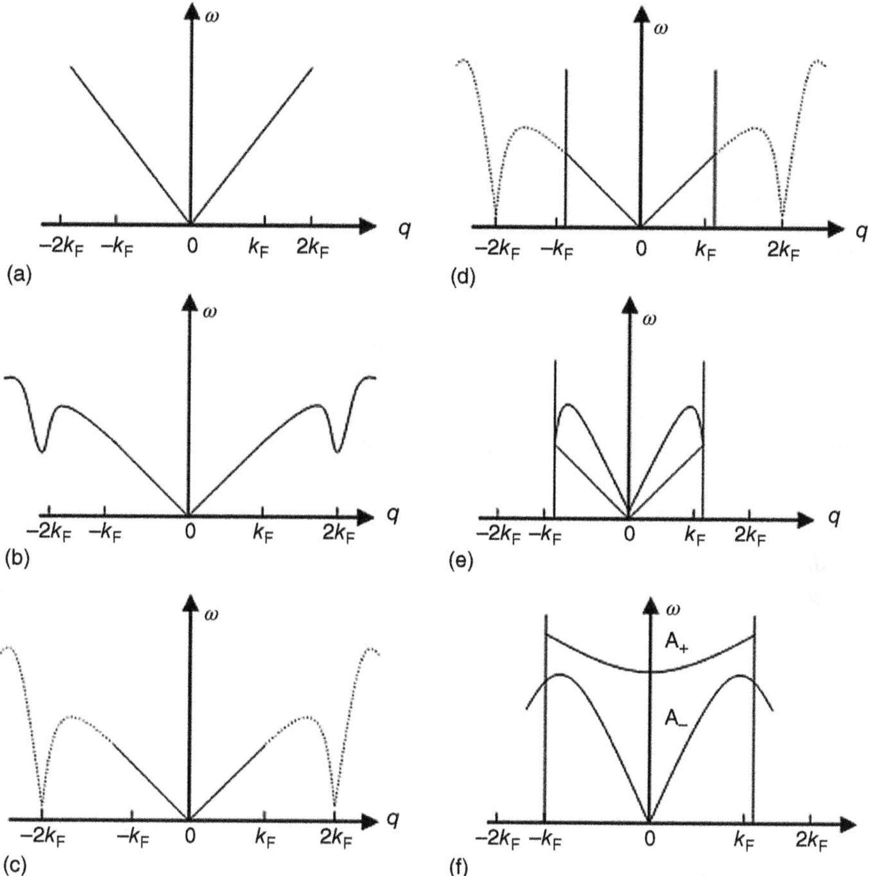

Figure 10.31 Phonon dispersion relation in a crystal with Peierls transition. (a) Undistorted phonon dispersion relation at high temperature. (b) Development of Kohn anomaly. (c) Phonons at Kohn anomaly become completely soft. (d) Lattice rearrangement. Dotted parts of the dispersion relation lie outside the first Brillouin zone of the new lattice. (e) Phonon dispersion relation in the reduced zone scheme. (f) Lifting of the $q = 0$ degeneracy and development of an amplitude mode (A_-) and a phase mode (A_+) in CDW terminology.

the order parameter above, it shouldn't be hard to link the two concepts together. An inspection of the CDW reveals that the *amplitude mode* corresponds to a modulation of the CDW amplitude and the *phase mode* to a phase modulation.

Of course, now, with our more detailed look into these excitations, the idea of a phason with $\omega = 0$ at $\boldsymbol{q} = 0$ turns out to be quite important. The term $\boldsymbol{q} = 0$ corresponds to that sliding of the CDW as a whole we were claiming above. Moreover, this motion needs no energy! This is the renowned *Peierls–Fröhlich mechanism* of superconductivity. In a real crystal the CDW is always associated with lattice imperfections, but a CDW in a perfect crystal would be a *Peierls–Fröhlich superconductor*.

10.3.4 More on Peierls–Fröhlich Mechanisms

Figure 10.31b shows the electronic band of a Peierls system. For comparison the band structure of metal with half-filled band is shown in Figure 10.31a. In the Peierls system there is a bandgap of width Δ at the Fermi wavevector $\boldsymbol{k}_{\mathrm{F}}$. According to the Peierls–Fröhlich mechanism, the CDW moves through the crystal lattice. From Section 10.3.3 we know that in a perfect crystal this sliding motion does not require energy (for the phason mode $\omega = 0$ at $\boldsymbol{q} = 0$). Strictly speaking, this is only true in the jellium approximation, where the positive ions do not form a discrete lattice but are continuously spread over the solid – or for an incommensurate CDW where $\boldsymbol{k}_{\mathrm{F}}$ is not a rational fraction of a reciprocal lattice vector. The flow without resistance stems from the fact that the energy is independent from the position (phase) of the CDW as well as from the absence of inelastic scattering due to the Peierls gap. CDWs in real crystals, however, associate with impurities. In Figure 10.31b the band structure of the system with a sliding CDW is shown. The CDW moves with the sliding velocity v_{s}, leading to a displacement of the planes of the Fermi surface by $\boldsymbol{q}$:

$$q = m^* v_{\mathrm{s}} / \hbar \tag{10.74}$$

where m^* is the effective mass of the electrons and the motion of the CDW corresponds to the current:

$$I_{\mathrm{CDW}} = n e v_{\mathrm{s}} \tag{10.75}$$

where n is the number of electrons in the sample. The sliding velocity v_{s} corresponds to the drift velocity in the usual derivation of the electric conductivity. The existence of a gap does not prevent conductivity, provided the electric field can force the gap to become asymmetric (at different positions in $+\boldsymbol{k}$ and $-\boldsymbol{k}$ directions). In this case the gap excludes loss due to inelastic scattering processes, and hence the current flows without resistance. In an insulator the electric field cannot push the gap into asymmetric positions, because the gap originates from structural features unable to slide along the crystal (Figure 10.32).

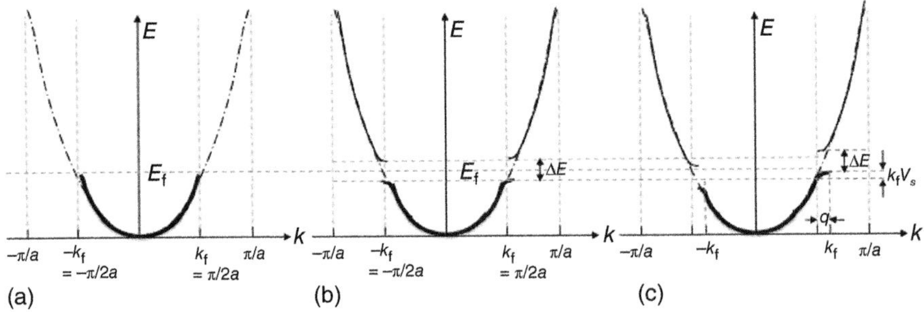

Figure 10.32 (a) Band structure of metal, half filled, no gap. (b) Electron band with Peierls gap Δ a Fermi wavevector $\boldsymbol{k}_{\mathrm{F}}$. (c) Displacement of the electron system with uniform velocity v_{s} and shift of Fermi vector by $\boldsymbol{q}$, leading to asymmetric gap positions and a positive sum of electron $\boldsymbol{k}$ vectors.

10.3.5 Spin Density Waves and the Quantized Hall Effect

The symmetry of TM_2X Bechgaard salts is triclinic, but most physical features can be better explained by assuming orthorhombic symmetry with axes *a*, *b*, and *c*. The direction of highest conductivity is along the stacking axis that we assign to the *a* direction. The crystal is not only anisotropic with respect to the *a* direction and *bc* plane; the *bc* plane is also anisotropic itself. For $(TMTSF)_2X$ compounds the ratio of the bandwidths is $t_a{:}t_b{:}t_c = 100 : 10 : 0.3$; $t_a \sim 0.2\,\text{eV}$ (the "*t*'s" are the transfer integrals, and the bandwidth is $W = 4t$). Because of the finite values of t_b and t_c, the Fermi surface is not planar but warped. There is still sufficient nesting so that $(TMTSF)_2PF_6$ is in the CDW state at ambient pressure and low temperatures. The application of the modest pressure of 12 kbar will increase t_b and destroy the nesting condition. Consequently, the CDW is suppressed, and the sample becomes superconducting.

Applying a sufficiently high magnetic field (above the critical field) will suppress superconductivity, and the sample should behave like a normal metal. But if the field is applied in *c* direction (the direction of lowest conductivity), the CDW will be reactivated [47]. This means the magnetic field restores the nesting condition! The effect is called a *field-induced spin density wave* (FISDW). An amazing property of the FISDW state is a stepwise field dependency of the Hall voltage [35], very similar to what is observed in the *von Klitzing effect* (quantum Hall effect, QHE) [47]. Examples for $(TMTSF)_2ClO_4$ and $(TMTSF)_2PF_6$ are shown in Figure 10.33.

In ordinary SDW, nesting occurs for the wavevector $\boldsymbol{q} = 2\boldsymbol{k}_F$. In high magnetic fields the electronic DOS is modified by the Landau quantization. For example,

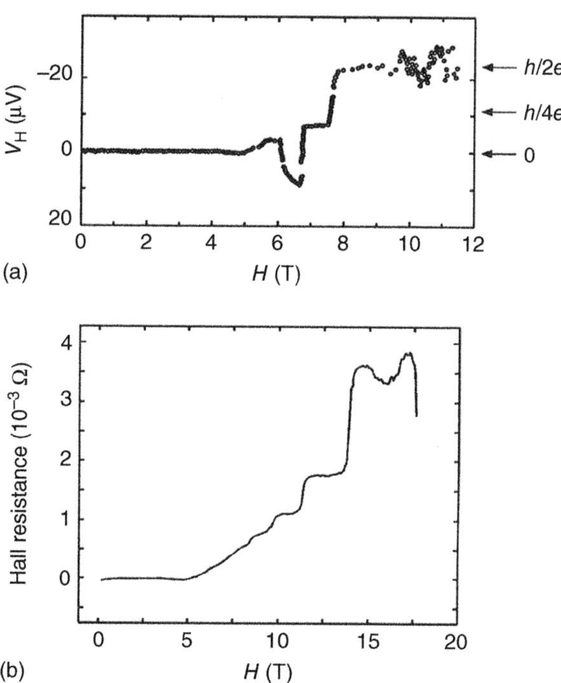

Figure 10.33 Quantized Hall effect in the field-induced spin density wave systems. (a) $(TMTSF)_2ClO_4$: $T = 0.5$ K, ambient pressure. (b) $(TMTSF)_2PF_6$: $T = 0.1$ K, $p \sim 8$ kbar [35, 46].

electrons move in closed orbitals of quantized energy, and the DOS between the Landau levels is zero. Interference between Landau levels and the Fermi surface leads to the dHvA and SdH oscillations discussed in the context of "fermiology." Because of the Landau quantization, the wavevector $\boldsymbol{q}$ of the SDW has to be modified to

$$q_{\mathrm{FISDW}} = 2k_{\mathrm{F}} + NeHb/h \tag{10.76}$$

where N is an integer, H is the applied field, and b is the lattice spacing in b direction. For q_{FISDW} there are new nesting conditions that depend on the magnetic field. A direct consequence of the field-dependent $\boldsymbol{q}$ vector is the quantization of the Hall resistance at $\rho_{\mathrm{H}} = h/2Ne^2$. Because nesting is a fairly crude geometrical overlap, the steps in the FICDW-QHE are by far less sharp than in the von Klitzing effect.

10.4 Plasmons

Plasmons are typically discussed in sections on the optical properties of solids. However, they too are a collective oscillation state of the conduction electrons and are therefore interesting as interaction physics. Indeed, there is some evidence that they may play an important role in the theory of type II superconductors: ceramics that are composed of 2D sheets of conductor [48].

10.4.1 The Drude Model and the Dielectric Function

Plasmons are quantized, as are all correlated phenomena. However, it is rather straightforward to see how volume plasmons might arise using semiclassical approaches and the Drude model we have already introduced. We start with free electrons interpenetrating a set of static ion cores in a lattice: a unique sort of plasma. Notice that the cores are static. As in the Drude model, they are ignored except as a smeared-out restoring force and perhaps the source of damping through scattering. Therefore coupling to phonons will play no role in constructing our *plasmon*.

Following the Drude model as previously defined, there is a characteristic time for scattering, τ, and a characteristic frequency of collision, $\gamma = 1/\tau$. (For systems that approximate a free electron gas, these numbers are roughly 10^{-14} and 100 THz, respectively.) The equation of motion of a given electron in this gas of electrons, when it is exposed to an electric field, is classically

$$m\ddot{\boldsymbol{x}} + m\gamma\dot{\boldsymbol{x}} = -e\boldsymbol{E} \tag{10.77}$$

So here we notice a subtlety, the $\boldsymbol{E}$ field that is applied, is *seen* directly by the electron. But the electron is embedded in a sea of other electrons along with an ionic background. As usual Drude-like models ignore this as an averaging of effects generally. But we will return to it in Chapter 11. For now lets move forward like this and assume that the applied field has a harmonic character: $\boldsymbol{E}(t) = \boldsymbol{E}_0\mathrm{e}^{-i\omega t}$. This is handy because such an assumption can be used to generate any form of driving force through Fourier's theorem. If you remember your basic mechanics, this yields solutions that look like $\boldsymbol{x}(t) = \boldsymbol{x}_0 e^{-i\omega t}$ where x_0 is complex. Applying boundary conditions, substituting, and doing a little algebra,

$$\boldsymbol{x}(t) = \frac{e}{m(\omega^2 + i\gamma\omega)}\boldsymbol{E}(t) \tag{10.78}$$

This is the familiar-driven harmonic oscillator with damping. Now if we set up our coordinates such that $\boldsymbol{P} = -ne\boldsymbol{x}$, we get

$$\boldsymbol{P} = \frac{-ne^2}{m(\omega^2 + i\gamma\omega)}\boldsymbol{E} \tag{10.79}$$

And, if you remember your first-year electrodynamics,

$$\boldsymbol{D} = \varepsilon_0\left(1 - \frac{\omega_p^2}{\omega^2 + i\gamma\omega}\right)\boldsymbol{E} \tag{10.80}$$

But an important definition has slipped in there:

$$\omega_{\mathrm{p}}^2 = \frac{ne^2}{\varepsilon_0 m} \tag{10.81}$$

This is the *plasma frequency* of the free electron gas. Curiously this allows us to write a dielectric function for the free electron gas as well:

$$\varepsilon(\omega) = 1 - \frac{\omega_{\mathrm{p}}^2}{\omega^2 + i\gamma\omega} \tag{10.82}$$

Now you can see why the topic is typically discussed in *optical properties* instead of *collective behavior*.

10.4.2 The Significance of the Plasma Frequency

What is so significant about the plasma frequency? To understand that, recall that from Maxwell's equations for the propagation of a wave in the media, we get

$$\nabla \times \nabla \times \boldsymbol{E} = -\mu_0 \frac{\partial^2 \boldsymbol{D}}{\partial t^2} \tag{10.83}$$

This is with no driving force applied. And this transforms to

$$\boldsymbol{K}(\boldsymbol{K}\cdot\boldsymbol{E}) - \boldsymbol{K}^2\boldsymbol{E} = -\varepsilon(\boldsymbol{K},\omega)\frac{\omega^2}{c^2}\boldsymbol{E} \tag{10.84}$$

in the Fourier domain (you may have to remind yourself with Griffiths [39]). We notice here that $\varepsilon(\boldsymbol{K}, \omega)$ is presented as a Fourier component; it is not necessarily local. In the case of transverse waves, $\boldsymbol{K}\cdot\boldsymbol{E} = 0$, and

$$\boldsymbol{K}^2 = \varepsilon(\boldsymbol{K},\omega)\frac{\omega^2}{c^2} \tag{10.85}$$

In the case of longitudinal waves, this implies $\varepsilon(K, \omega) = 0$. Thus, longitudinal collective oscillations can occur only at the poles of ε. Indeed we might ask what happens at ω_{p}? Well, according to our equation for the dielectric function of the gas, in the small damping limit, $\varepsilon(\omega_{\mathrm{p}}) = 0$ at $\boldsymbol{K} = 0$. The excitation must correspond directly to a collective, longitudinal oscillation mode. And $\boldsymbol{D} = \varepsilon_0\boldsymbol{E} + \boldsymbol{P} = 0$ and $\boldsymbol{E} = -\boldsymbol{P}/\varepsilon_0$ is a pure depolarization field.

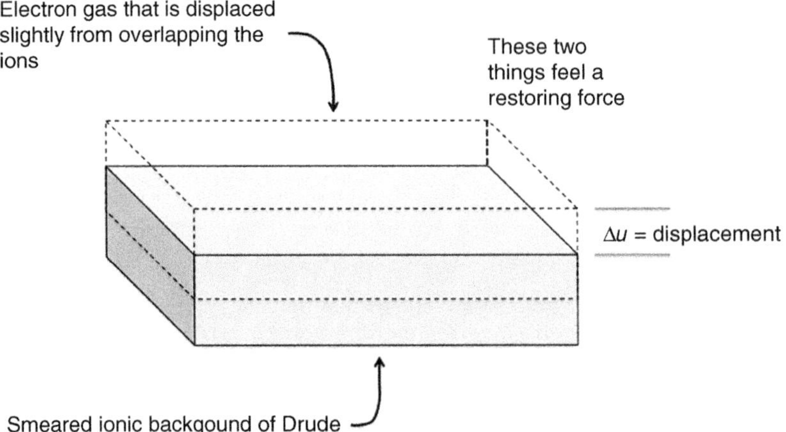

Figure 10.34 A classical interpretation of the origins of plasmon oscillations.

To better interpret, or picture, what these oscillations that can propagate through the solid with a frequency of ω_p are, we turn to our simple diagram in Figure 10.34. Here we simply image that the positive and negative charges in the Drude model fill the volume of the solid uniformly and the positive (ionic cores) can be separated uniformly from the electrons. We ask what is the restoring force and what are the dynamics of this electron fluid as it goes crashing back into the positive volume? This is actually a widely assigned mechanics problem and not too hard to solve. If we displace the electrons by u, then the charge on the top and bottom slabs will be $\sigma = \pm neu$. This gives us a field of $E = neu/\varepsilon_0$. The equation of motion looks like

$$nm\ddot{u} = -ne\boldsymbol{E} \tag{10.86}$$

$$nm\ddot{u} = -n^2e^2u/\varepsilon_0 \tag{10.87}$$

$$\ddot{u} + \omega_p^2 u = 0 \tag{10.88}$$

So ω_p is the natural free oscillation frequency of such a system. We note that we have treated all electrons as though they are moving in phase with each other. This is a $\boldsymbol{K} = 0$ long wavelength limit of motion. The collective oscillations are known as *plasmons* (*volume plasmons* to be precise).

We will return to plasmons in the optical properties section of our Chapter 13. However, it is interesting to note that the longitudinal properties of the *volume plasmon* prevent it from coupling directly to transverse electromagnetic waves. It can only be stimulated by impact with charged particles and indeed is widely used in electron energy loss spectroscopy as a characterization tool. Further, the decay of the volume plasmon occurs only through energy transfer to individual electrons. This is known as *Landau damping*.

10.5 Composite Particles and Quasiparticles: A Summary

We began our study of electronic properties using band pictures that enumerated single particle states of electrons in the solid. As we began to look more closely, interactions between electrons themselves, electrons and lattice vibrations, and (as we will see) electrons and the full electromagnetic field were necessary to explain the phenomena observed in solid-state systems. The problem is do we explain these as modified single particle states? They really aren't this; they are collective states where we have used some illustrations to help us understand the dominant physics. It was really Landau who formulated a rescue to this conundrum. Landau's basic idea was as follows: for a complicated system of strongly interacting particles like electrons, we could reduce its description into a picture of weakly interacting composite particles. These particles have their origins in the collective behavior or correlation of a background field. This idea is widely accepted by physicists today, and we have an "-on" (magnon, plasmon, polaron, etc.) for practically everything. In actual fact the approach is not really that different from modern quantum field theory in elementary particle physics.

We have loosely called these "-ons" *quasiparticles* because they represent a hybrid of sorts. They are not an individual particle as we think of one. They are not elementary or even that well localized. However they do carry momentum and have properties like spin and obey particle statistics. The examples given in this section have tried to help the reader understand how such composite structures come about and how we typically treat them in theories of the solid state. But this is by no means exhaustive, and the astute reader will by now have realized that delving into the specifics of any particular collective behavior becomes very specialized very quickly.

Exploring Concepts

1 *The Hubbard Model*: There are many useful refinements to the Hubbard model, and the math is really rather simple. By looking through the literature:
   (a) Determine how the model would change when a magnetic field is introduced.
   (b) Determine how the model would change with a chemical potential added.
   (c) What systems did the Hubbard model originally address and why?

2 *MgB_2*: The MgB_2 honeycomb lattice is constructed from partially filled σ bonds.
   (a) From our argument above, how many superconducting gaps would you expect in MgB_2? Read through the references and see if your guess is correct.

(b) Draw out this structure as carefully as possible. (It can also be found in the references.) In the simple spring picture we presented in Chapter 3, what does this partial filling do to the effective spring constants? How would this change the phonon dispersion? Why and how might this change the electron–phonon coupling?

3 *Phase Diagrams*: Our phase diagram above isn't quite complete. Look through the literature at the different $(TM)_2X$ compounds, and see if you can place them on the diagram. Now consider a compound using an anion that isn't in your list. Can you make a guess as to where it might appear?

4 *Nanotubes and Superconductivity*: Our argument regarding A15 compounds relies on the existence of van Hove structures in the bands together with doping such that the Fermi level approaches one such peak. Another material that would certainly present the same opportunity is the carbon nanotube. But these materials also present a very different set of symmetries and potential electron–phonon coupling mechanisms.
(a) Would you expect for superconductivity to be found in single-walled carbon nanotubes? If so, describe the dominant model for the behavior.
(b) What about multiwalled nanotubes? Would the presence of a second wall disturb the cooperative state? Why?
(c) Find sources in literature that back up these notions. Notice that the claims of superconductivity at higher temperatures are for very small diameter nanotubes. Why would this be?

5 *Peierls–Fröhlich Mechanism*: There are now several exactly solved models of the Peierls–Fröhlich mechanism in one dimension. A quick literature search will discover that a central feature is to understand how such a system will behave with disorder. Consider now a single defect within a 1D conjugated system. What effect will this have on the formation of domain walls? Describe the effects as the number of these defects increase.

References

1 Mott, N.F. (1949). The basis of the electron theory of metals, with special reference to the transition metals. *Proc. Phys. Soc. London, Sect. A* 62: 416.
2 (a) Baeriswyl, D. (1992). *Conjugated Conducting Polymers*, Springer Series in Solid-State Sciences, vol. 102 (ed. H. Kiess). Heidelberg: Springer Verlag. (b) Przyluski, J. and Roth, S. (eds.), Materials Science Forum*New Conducting Materials with Conjugated Double Bonds: Polymers, Fullerenes, Graphite.* Aedermannsdorf: Trans Tech Publications.
3 Schlappa, J. et al. (2012). Spin-orbital separation in the quasi-one-dimensional Mott insulator Sr_2CuO_3. *Nature* 485: 82.
4 Onnes, H.K. (1911). Communications – Leiden, 120b.

5 (a) Buckel, W. Table on Superconducting Elements, p. 20 op.cit. [2]. (b) Testardi, L.R., Wernick, J.H., and Royer, W.A. (1974). Superconductivity with onset above 12 K in Nb–Ge sputtered films. *Solid State Commun.* 15: 1. (c) Greene, R.L., Street, G.B., and Suter, L.J. (1975). Superconductivity in polysulfur nitride $(SN)_x$. *Phys. Rev. Lett.* 34: 577. (d) Williams, J.M. et al. (1990). From semiconductor–semiconductor transition (42 K) to highest-T_C organic superconductor, κ-$(ET)_2Cu[N(CN)_2]Cl$ (T_C = 12.5 K). *Inorg. Chem.* 29: 3272. (e) Tanigake, K. et al. (1991). Superconductivity at 33 K in $Cs_xRb_yC_{60}$. *Nature* 352: 222. (f) Schilling, A., Cantoni, M., Guo, J.D., and Ott, H.R. (1993). Superconductivity above 130 K in the Hg–Ba–Ca–Cu–O system. *Nature* 363: 56.

6 (a) Buckel, W. (1984). *Supraleitung – Grundlagen und Anwendungen*, 3rde. Weinheim: Physik-Verlag. (b) Buckel, W. and Kleiner, R. (2004). *Supraleitung*, 6the. Wiley-VCH.

7 (a) Kresin, V.Z. and Little, W.A. (1990). *Organic Superconductivity*. New York: Plenum Press. (b) Saito, G. and Kagoshima, S. (1989). *The Physics and Chemistry of Organic Superconductors*. Berlin: Springer-Verlag. (c) Ishiguro, T. and Yamaji, K. (1990). *Organic Superconductors*. Berlin: Springer-Verlag. (d) Khurana, A. (1990). Highest T_C continues to increase among organic superconductors. *Phys. Today* 43: 17. (e) Williams, J.M. et al. (1991). Organic superconductors – new benchmarks. *Science* 252: 1501. (f) Williams, J.M. et al. (1992). *Organic Superconductors (including Fullerenes): Synthesis, Structure, Properties, and Theory*. Englewood Cliffs: Prentice-Hall, Inc. (g) Williams, J.M. et al. (1984). Ambient-pressure superconductivity at 2.7 K and higher temperatures in derivatives of beta$(ET)_2IBr_2$: synthesis, structure, and detection of superconductivity. *Inorg. Chem.* 23: 3839. (h) Kini, A.M. et al. (1990). A new ambient-pressure organic superconductor, kappa $(ET)_2Cu[N(CN)_2]Br$, with the highest transition temperature yet observed (Inductive Onset T_c = 11.6 K, Resistive Onset=12.5 K). *Inorg. Chem.* 29: 2555. (i) Schlueter, J.A. et al. (1994). The first organic cation-radical salt superconductor (T_c = 4 K) with an organometallic anion: superconductivity, synthesis and structure of kappa $(ET)_2M(CF_3)4(C2H_3X_3)$. *J. Chem. Soc. Chem. Commun.* 1599. (j) Geiser, U. et al. (1996). Superconductivity at 5.2 K in an electron donor radical salt of bis(ethylenedithio)tetrathiafulvalene (BEDT-TTF) with the novel polyfluorinated organic anion beta$(ET)_2SF_5CH_2CF_2SO_3$. *J. Am. Chem. Soc.* 118: 9996. (k) Williams, J.M. et al. (1991). New high-T_c benchmarks for organic and fullerene superconductors. *Physica C* 355: 185. (l) Schlueter, J.A. et al. (1994). Superconductivity up to 11.1 K in three solvated salts composed of $[Ag(CF_3)_4]$ – and the organic electron-donor molecule bis(ethylenedithio)tetrathiafulvalene (ET). *Physica C* 233: 379. (m) Jérome, D. (1991). The physics of organic semiconductors. *Science* 252: 1509. (n) Brooks, J.S. (1993). Organic conductors and superconductors: new directions in the solid state. *MRS Bull.* 18 (8): 29. (o) Jérome, D. (1994). *Organic Conductors* (ed. J.P. Farges). New York: Marcel Dekker. (p) Jérome, D. (1999). *Advances in*

Synthetic Metals: Twenty years of Progress in Science and Technology (ed. P. Bernier, S. Lefrant and G. Bidan). Amsterdam: Elsevier Science.

8 Bardeen, J., Cooper, L.N., and Schrieffer, J.R. (1957). Theory of Superconductivity. *Phys. Rev.* 108: 1175.

9 Meissner, W. and Ochsenfeld, R. (1933). Ein neuer Effekt beim Eintritt der Supraleitung. *Naturwissenschaften* 21: 787.

10 Kaxiras, E. (2003). *Atomic and Electronic Structure of Solids.* New York: Cambridge University Press.

11 Ford, J.P. and Saunders, G.A. (1997). High temperature superconductivity: ten years on. *Contemp. Phys.* 38: 63.

12 Nagamatsu, J., Nakagawa, N., Muranaka, T. et al. (2001). Superconductivity at 39 K in magnesium diboride. *Nature* 410: 63.

13 Choi, H.J., Roundy, D., Sun, H. et al. (2002). The origin of the anomalous superconducting properties of MgB_2. *Nature* 418: 758.

14 Little, W.A. (1964). Possibility of synthesizing an organic superconductor. *Phys. Rev. A* 134: 1416.

15 (a) Bednorz, J.G. and Müller, K.A. (1986). Possible high TC superconductivity in the Ba–La–Cu–O system. *Z. Phys. B* 64: 189. (b) Wu, M.K. et al. (1987). Superconductivity at 93 K in a new mixed-phase Yb–Ba–Cu–O compound system at ambient pressure. *Phys. Rev. Lett.* 58: 908. (c) Schilling, A., Cantoni, M., Gud, J.D., and Ott, H.R. (1993). Superconductivity above 130 K in the Hg–Ba–Ca–Cu–O system. *Nature* 363: 56.

16 Takehiko, I., Yamaji, K., and Saito, G. (1998). *Organic Superconductors,* Springer Series in Solid-State Sciences, 2nde, 88.

17 Shoenberg, D. (1984). *Magnetic Oscillations in Metals.* Cambridge University Press.

18 Kang, W., Montambaux, G., Cooper, J.R. et al. (1989). Observation of giant magnetoresistance oscillations in the high-TC phase of the two-dimensional organic conductor β-$(BEDT\text{-}TTF)_2I_3$. *Phys. Rev. Lett.* 62: 2559.

19 Jérome, D., Mazaud, A., Ribault, M., and Bechgaard, K. (1980). Superconductivity in a synthetic organic conductor $(TMTSF)_2PF_6$. *J. Phys. Lett.* 41: L95.

20 Bechgaard, K., Carneiro, K., Olsen, M. et al. (1981). Zero-pressure organic superconductor: di-(tetramethyltetraselenafulvalenium)-perchlorate [$(TMTSF)_2ClO_4$]. *Phys. Rev. Lett.* 46: 852.

21 Balicas, L. et al. (1994). $(TMTTF)_2Br$: the first organic superconductor. *Adv. Mater.* 6: 762.

22 (a) Bulaevskii, L.N. (1988). Organic layered superconductors. *Adv. Phys.* 37: 443. (b) Conwell, E.M. (1988). *Highly Conducting Quasi-One-Dimensional Organic Crystals.* San Diego, CA: Academic Press. (c) Saito, G. and Kagoshima, S. (1989). *The Physics and Chemistry of Organic Superconductors.* Berlin: Springer-Verlag.

23 (a) Kini, A.M. et al. (1990). A new ambient-pressure organic superconductor, kappa $(ET)_2Cu[N(CN)_2]Br$, with the highest transition temperature yet observed (inductive onset T_c = 11.6 K, resistive onset = 12.5 K). *Inorg. Chem.*

29: 2555. (b) Schlueter, J.A. (1994). The first organic cation-radical salt superconductor (T_c=4 K) with an organometallic anion: superconductivity, synthesis and structure of kappa $(ET)_2M(CF_3)_4(C_2H_3X_3)$. *J. Chem. Soc. Chem. Commun.* 1599. (c) Geiser, U. et al. (1996). Superconductivity at 5.2 K in an electron donor radical salt of bis(ethylenedithio) tetrathiafulvalene (BEDT-TTF) with the novel polyfluorinated organic anion beta $(ET)_2SF_5CH_2CF_2SO_3$. *J. Am. Chem. Soc.* 118: 9996. (d) Schlueter, J.A. et al. (1994). Superconductivity up to 11.1 K in three solvated salts composed of [Ag(CF3)4]- and the organic electron-donor molecule bis(ethylenedithio)tetrathiafulvalene (ET). *Physica C* 233: 379. (e) Jérome, D. (1994). *Organic Conductors* (ed. J.P. Farges). New York: Marcel Dekker. (f) Bernier, P., Lefrant, S., and Bidan, G. (eds.) (1999). *Advances in Synthetic Metals Twenty Years of Progress in Science and Technology*. Amsterdam: Elsevier Science.

24 Parkin, S.S.P. et al. (1983). Superconductivity in a new family of organic conductors. *Phys. Rev. Lett.* 50: 270.

25 Kini, A.M. et al. (1990). A new ambient-pressure organic superconductor κ-$(ET)_2Cu[N(CN)_2]Br$, with the highest transition temperature yet observed (inductive onset T_C = 11.6 K, resistive onset = 12.5 K). *Inorg. Chem.* 29: 2555.

26 Lüders, K. (1993). *Chemical Physics of Intercalation*, NATO ASI Series (ed. P. Bernier, J.E. Fischer, S. Roth and S.A. Solin). New York: Plenum Press.

27 Fleming, R.M., Zhou, O., and Siegrist, D.W.M.T. (1994). *Progress in Fullerene Research* (ed. H. Kuzmany, J. Fink, M. Mehring and S. Roth). Singapore: World Scientific.

28 Przyluski, J. and Roth, S. (1995). New materials: conjugated double bond systems. *Proceedings of the 5th International Autumn School*, Warsaw, Poland (26–30 September 1994). Trans Tech Publications.

29 Tanigake, K. et al. (1991). Superconductivity at 33 K in $Cs_xRb_yC_{60}$. *Nature* 352: 222.

30 (a) Taliani, C., Ruani, G., and Zamboni, R. (1993). *Fullerenes: Status and Perspectives*. Singapore: World Scientific. (b) Kroto, H.W., Fischer, J.E., and Cox, D.E. (1993). *The Fullerenes*. Pergammon Press. (c) Koruga, D., Hameroff, S., Wither, J. et al. (1993). *Fullerene C_{60}: Physics, Nanobiology, Nanotechnology*. Amsterdam: Elsevier. (d) Billups, W.E. and Ciufolini, M.A. (1993). *Buckminsterfullerenes*. Weinheim: Wiley-VCH. (e) Gärtner, S. (1992). Festkörperprobleme 32. In: *Advances in Solid State Physics*, vol. 32, 295. (f) Kuzmany, H., Fink, J., Mehring, M., and Roth, S. (1993). *Electronic Properties of Fullerenes*, Springer Series in Solid-State Sciences, vol. 117. Heidelberg: Springer Verlag. (g) Kuzmany, H., Fink, J., Mehring, M., and Roth, S. (1994). *Progress in Fullerene Research*. Singapore: World Scientific. (h) Dreselhaus, M.S., Dresselhaus, G., and Ecklund, P.C. (1995). *Science of Fullerenes*. New York: Academic Press.

31 Prinzbach, H. et al. (2000). Gas-phase production and photoelectron spectroscopy of the smallest fullerene, C_{20}. *Nature* 407: 60.

32 Grossman, J.C., Louie, S.G., and Cohen, M.L. (1999). Solid C_{36}: crystal structure, formation, and effect of doping. *Phys. Rev. B* 60: R6941.

33 Devos, A. and Lannoo, M. (1998). Electron-phonon coupling for aromatic molecular crystals: possible consequences for their superconductivity. *Phys. Rev. B* 58: 8236.

34 For reviews on oxide superconductors see for example: (a) Aranda, M.A.G. (1994). Crystal Structure of copper-based high-T_C superconductors. *Adv. Mater.* 6: 905. (b) Narlikar, A.V. (1989–1994). *Studies of High Temperature Superconductors*, vol. I–XI. New York: Nova Science Publishers. (c) Ginsberg, D.M. (1990–1992). *Physical Properties of High Temperature Superconductors*, vol. I–III. Singapore: World Scientific.

35 (a) Ribault, M., Jérome, D., Tuchendler, J. et al. (1983). Low-field and anomalous high-field Hall effect in $(TMTSF)_2ClO_4$. *J. Phys. Lett.* 44: 953. (b) Oshima, K., Suzuki, M., Kikuchi, K. et al. (1984). High field magneto-conductivity in $(TMTSF)_2ClO_4$. *J. Phys. Soc. Jap.* 53: 3295. (c) Kagoshima, S. (1996). Low-dimensional organic conductors under high maghetic fields. *J. Phys. I* 6: 1787. (d) Chaiken, P.M. et al. (1983). Tetramethyltetraselena-fulvalenium perchlorate, $(TMTSF)_2ClO_4$, in high magnetic fields. *Phys. Rev. Lett.* 51: 2333. (e) Kornilov, A.V. et al. (2002). Novel phases in the field-induced spin-density-wave state in $(TMTSF)_2PF_6$. *Phys. Rev. B* 65: 060404. (f) Lebed, A.G. (2000). Does the "quantized nesting model" properly describe the magnetic-field-induced spin-density-wave transitions? *JETP Lett.* 73: 141. (g) Kang, W., Aoki, H., and Terashima, T. (1999). Evidence of new split FISDW transitions – transport properties. *Synth. Met.* 103: 2119. (h) Valfells, S. et al. (1996). Quantum Hall transitions in $(TMTSF)_2PF_6$. *Phys. Rev. B* 54: 16413. (i) Kagoshima, S. (1996). Low-dimensional organic conductors under high magnetic fields. *J. Phys.* 6: 1787. (j) Cooper, J.R., Kang, W., Auban, P. et al. (1989). Quantized Hall effect and a new field-induced phase transition in the organic superconductor $(TMTSF)_2PF_6$. *Phys. Rev. Lett.* 63: 1984. (k) Vuletic, T. et al. (2001). Influence of quantum Hall effect on linear and nonlinear conductivity in the FISDW states of the organic conductor $(TMTSF)_2PF_6$. *Eur. Phys. J. B.* 21: 53. (l) Eom, J., Cho, H., and Kang, W. (1999). Quantum Hall effect and anomalous transport in $(TMTSF)_2PF_6$. *J. Phys. IV* 9 (P10): 191. (m) Hannahs, S.T., Brooks, J.S., Kang, W. et al. (1989). Quantum Hall effect in a bulk crystal. *Phys. Rev. Lett.* 63: 1988.

36 Cohen, M.J., Coleman, L.B., Garito, A.F., and Heeger, A.J. (1974). Electrical conductivity of tetrathiofulvalinium tetracyanoquinodimethane (TTF) (TCNQ). *Phys. Rev. B* 10: 1298.

37 Coleman, L.B., Cohen, M.J., Sandman, D.J. et al. (1973). Superconducting fluctuations and the peierls instability in an organic solid. *Solid State Commun.* 12: 1125.

38 Grüner, G. and Zettl, A. (1985). Charge density wave conduction: a novel collective transport phenomenon in solids. *Phys. Rep.* 119 (3): 117.

39 Griffith, D.J. (2014). *Introduction to Electromagnetism*, 4the. Harlow Essex: Pearson Education Limited.

40 Rice, M.J., Bishop, A.R., Krumhansl, J.A., and Trullinger, S.E. (1976). Weakly pinned Fröhlich charge-density-wave condensates: a new, nonlinear, current-carrying elementary excitation. *Phys. Rev. Lett.* 36: 432.

41 Lee, P.A., Rice, T.M., and Anderson, P.W. (1974). Conductivity from charge or spin density waves. *Solid State Commun.* 14: 703.

42 (a) Hubbard, J. (1963). Electron correlations in narrow energy bands. *Proc. R. Soc. London, Ser. A* 276: 238. (b) Hubbard, J. (1964). Electron correlation in narrow energy bands III. An improved solution. *Proc. R. Soc. London, Ser. A* 281: 401.

43 Kagoshima, S., Nagasawa, H., and Sambongi, T. (1982). *One-Dimensional Conductors*, Springer Series in Solid-State Sciences, vol. 72. Heidelberg: Springer Verlag.

44 For a basic review on spin density wave physics see: (a) Grüner, G. (1994). The dynamics of spin-density waves. *Rev. Mod. Phys.* 66: 1. More modern developments in spin waves and their relationship to low dimensional systems can be found in: (b) Konno, R. (2003). The self-consistent renormalization theory of spin fluctuations for itinerant antiferromagnetism in quasi-one-dimensional metals. *Physica B* 329: 1288. (c) Gabovich, A.M., Voitenko, A.I., and Ausloos, M. (2002). Charge- and spin-density waves in existing superconductors: competition between Cooper pairing and Peierls or excitonic instabilities. *Phys. Rep.* 367 (6): 583. (d) Tomio, Y., Dupuis, N., and Suzumura, Y. (2001). Effect of nearest- and next-nearest neighbor interactions on the spin-wave velocity of one-dimensional quarter-filled spin-density-wave conductors. *Phys. Rev. B* 64: 125123. (e) Mazumdar, S., Clay, R.T., and Campbell, D.K. (2001). The nature of the insulating state in organic superconductors. *Synth. Met.* 120: 679. (f) Sengupta, K. and Dupuis, N. (2000). Effective action and collective modes in quasi-one-dimensional spin-density-wave systems. *Phys. Rev. B* 61: 13493. (g) Tsuchiizu, M. and Suzumura, Y. (1999). Competition between SDW and SC states in two-Hubbard chains. *Synth. Met.* 103: 2189.

45 Alcacer, L. (1994). *Organic Conductors* (ed. J.P. Farges). New York: Marcel Dekker.

46 (a) Caron, L.G. (1994). *Organic Conductors* (ed. J.P. Farges). New York: Marcel Dekker. (b) Solyom, J. (1979). The Fermi gas model of one-dimensional conductors. *Adv. Phys.* 28: 201. (c) Fuseya, Y., Kohno, H., and Miyake, K. (2003). New varieties of order parameter symmetry in quasi-one-dimensional superconductors. *Physica B* 329: 1455. (d) Fuseya, Y., Onishi, Y., Kohno, H., and Miyake, K. (2002). Novel type of pairing in quasi-one-dimensional superconductivity. *Physica B* 312: 36. (e) Lee, I.J. et al. (2002). Triplet superconductivity in an organic superconductor probed by NMR knight shift. *Phys. Rev. Lett.* 88: 017004. (f) Emery, V.J. (1979). *Highly Conducting One-Dimensional Solids* (ed. J.T. Devreese, R.P. Evrard and V.E. van Doren). New York: Plenum Press. (g) Bourbonnais, C. and Caron, L.G. (1991). Renormalization group approach to quasi-one-dimensional conductors. *Int. J. Mod. Phys. B* 5: 1033. (h) Haddad, S., Charfi-Kaddour, S., Nickel, C. et al. (2003). Renormalization of the hopping parameters in quasi-one-dimensional conductors in the presence of a magnetic field. *Eur. Phys. J. B* 34: 33. (i) Kishine, J.I. and Yonemitsu, K. (2002). Dimensional crossovers and phase transitions in strongly correlated low-dimensional electron systems: renormalization-group study. *Int. J. Mod. Phys. B* 16 (5): 711. (j) Zanchi, D. and Schulz, H.J. (2000). Weakly

correlated electrons on a square lattice: renormalization-group theory. *Phys. Rev. B* 61: 13609. (k) Vescoli, V., Degiorgi, L., Henderson, W. et al. (1998). Dimensionality-driven insulator-to-metal transition in the Bechgaard salts. *Science* 281: 1181.

47 Kwak, J.J., Schriber, J.E., Greene, R.L., and Engler, E.M. (1981). Magnetic quantum oscillations in tetramethyltetraselenafulvalenium hexafluorophosphate [$(TMTSF)_2PF_6$]. *Phys. Rev. Lett.* 46: 1296.

48 Bill, A., Morawitz, H., and Kresin, V.Z. (2003). Electronic collective modes and superconductivity in layered conductors. *Phys. Rev. B* 68: 144519.

Intermission

The **solid** is composed of a lattice of atoms with their electrons interacting through the electromagnetic force. We have reduced the problem of explaining phenomena in solids into those aspects that are related to the ***motion of the lattice*** and those related to the ***motion of the independent electrons***. From this we can explain thermal transport to band structure and some simple electronic transport, as you have seen. But, when ***correlation*** and ***coupling*** is added to our picture, more exotic behaviors such as superconductivity emerge. Thus, we have a perspective for our *essentials*: (lattice + electrons + correlation). This approximation to the actual physics of the solid's many-body wavefunction is pretty good and very useful. But it can look a little like a *hodgepodge* of ideas thrown together.

Nevertheless this *adiabatic approximate* approach to the study of solids is important for building simple expectations from very complex phenomena. Sometimes the local/simple explanation is needed to really "see" the movie in your head, unlike our professor below. The answer to the mirror question is quite simple, in fact, but only presents itself as so when you actually play with the mirror.

While trying to explain why a mirror inverts left and right but not top and bottom, the professor went into reciprocal space. Solid-state physicists differ from other people by their familiarity with reciprocal space.

Foundations of Solid State Physics: Dimensionality and Symmetry, First Edition.
Siegmar Roth and David Carroll.

This approach of *playing* with the combined, independent aspects of (lattice + electrons + correlation) can be carried a bit further to include the solid's interaction with *electromagnetic* fields. We end our journey with an introduction to ***optical*** and ***magnetic*** response in the solid state. Using the mathematical and conceptual tools developed so far, we can build some useful models of spins and magnets, electric dipoles and dielectrics, and curious correlations and coupling in our systems. We are looking only for the basics here: the *essentials*. Indeed, these topics merit whole texts to themselves. Finally our theme of *dimensionality* will continue to play a strong role in phenomena, even here.

Now on with the show...

11

Magnetic Interactions

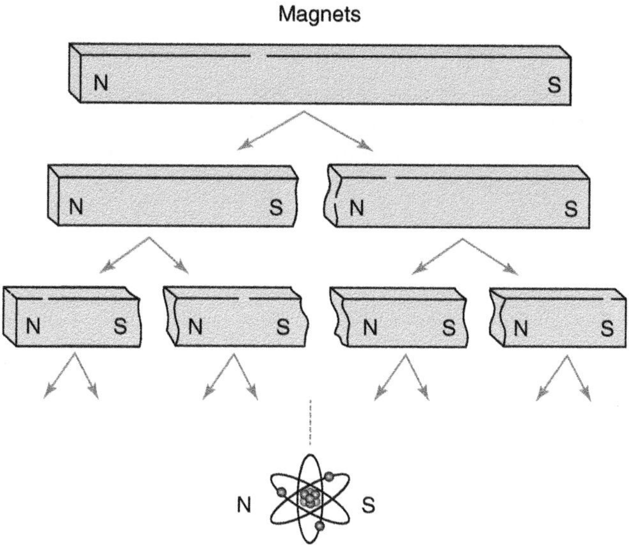

Split a magnet and you get another magnet. Touch some metals with a magnet and they too become magnets. Who didn't play with magnets as a child? But why do they act the way they do? What is the mechanism? Understanding this will require that we accept an astonishing coordination of magnetic moments of the atomic scale. Thus, magnetism is inherently a *quantum mechanical* feature of materials. Yes, that is right. Classically magnets do not exist! It is a good thing nature is unaware of this limitation.

The Basics

1. Atoms have their own internal magnetic fields. They behave like little dynamos with the field sourced from "tiny loops of current." The atomic-scale fields look mostly like dipoles, and so we can assign each atomic unit a dipole moment $\mathbf{m}_{LS}$.
2. Atoms or molecules with these magnetic dipole moments condense into solids, and their moments can be aligned collectively. Locally $\mathbf{m}_{LS} \rightarrow \mathbf{m}(\mathbf{r})$ and

Foundations of Solid State Physics: Dimensionality and Symmetry, First Edition.
Siegmar Roth and David Carroll.

globally $\mathbf{m}(\mathbf{r}) \rightarrow \mathbf{M}$. Thus, the solid takes on a "north–south"-poled magnetic field of a bar magnet.

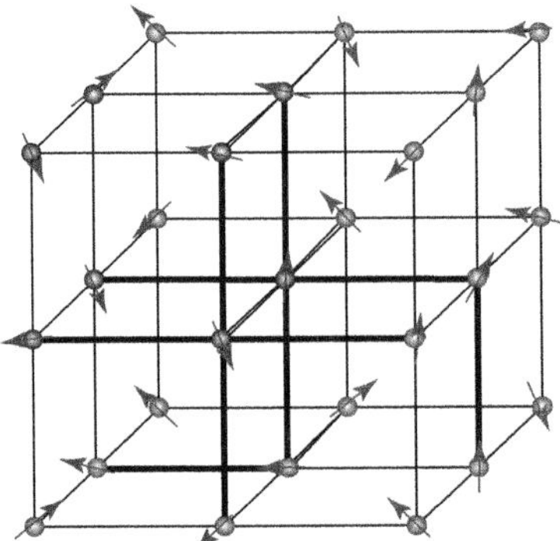

3. Alignment can happen when a **B** field is applied, and at times this induced state can be *frozen in* even when the field is removed. Other times the solid relaxes back to a nonmagnetic state. *Frozen-in* magnetism means atoms are not isolated magnetically.

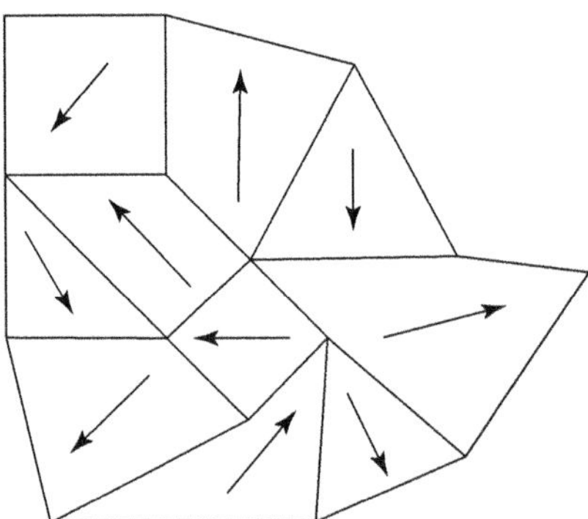

4. Magnetic moment interaction is not static. The lattice of moments can interact with other lattice denizens, such as phonons. This allows for phenomena such as spin waves (*magnons*) to form.

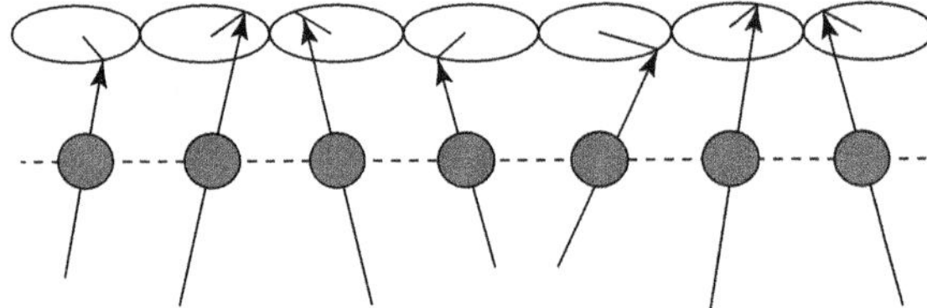

5. Finally, the specific bonding and fields of the surrounding crystal can mediate magnetic moment interactions. This gives rise to more exotic mechanisms of magnetic domain formation and supports large frozen-in fields. Superexchange and Ruderman–Kittel–Kasuya–Yosida (RKKY) magnets fall into this category.

11.1 Magnetism of the Atom

The magnetic moment of an atom *appears* to arise from the "current-like" components of angular momentum (**L**) and spin (**S**) and combined (**J**).[1] As shown in Figure 11.1, these components are responsible for all *paramagnetic* response of a material to an applied **B** field. A third "current-like" component that comes from induced changes in the **L** due to an applied **B** field gives rise to *diamagnetic* response in the magnetic moment:

$$\mathbf{m}_{\text{spin}} = -(g_e\mu_B/\hbar)\mathbf{s} \tag{11.1a}$$

$$\mathbf{m}_{\text{orbital}} = -(\mu_B/\hbar)\mathbf{l} \tag{11.1b}$$

g_e (g) is the *Landé g factor* and μ_B the familiar *Bohr magneton.*

Atoms can have multiple electrons with different orbital angular momentum, depending on the energy state of the orbital. Spins can project in different directions as can the angular momentum vectors. For this we apply *Hund's rules* [1]:

1. The *total* spin **S** of a system is given by the vector sum that yields the largest possible value of the magnitude of spin angular momentum ($\hbar S$) that is consistent with the Pauli principle: $S = \sum_i s_i$. Another way to say this is atoms in their ground states tend to have as many unpaired electrons as is possible. This is seen in the simple example of Figure 11.2.
2. The *total* angular momentum **L** is also given by the vector sum that maximizes the orbital angular momentum ($\hbar L$) that is consistent with Pauli and with rule #1: $L = \sum_i \mathrm{l}_i$. Again, an easier way to say this might be that for a given *multiplicity* or spin configuration among orbitals, the term with the largest value of L is the lowest in energy.

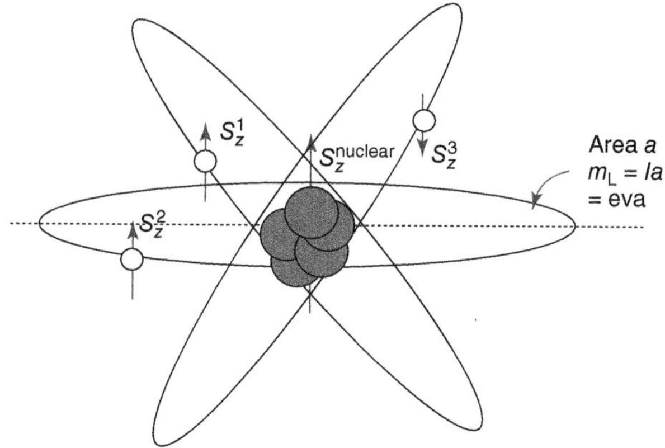

Figure 11.1 For a free atom or ion, dipole-like magnetic fields are generated by the spins of electrons, the spin of the nucleus, and the orbital angular momentum of the electrons. Each part has its own specific magnetic dipole moment.

1 *Caution*: The electron is visualized as spinning on an axis, but the quantum of spin is an internal degree of freedom that cannot really be described this way. Why? Well choose any axis against which to measure the electron spin. You will find values of up/down spin no matter what. The only time the electron seems to choose a specific axis direction is when something else, like a measurement, a magnetic field, or the m of the system, is placed in the background. But "seeing" spinning balls is pretty useful in playing the movie in our head.

This one is also quite easy to understand. Essentially it claims that if the electrons are all orbiting in the same direction (which means they all add up to give the atom a large total angular momentum), then they will meet each other less often than if they orbited in opposite directions to each other. In this way their total coulomb repulsion is less (on average) when L is large, and so the same direction state has a lower energy. It is therefore preferred by the atom. This is known as an *orbit–orbit* interaction. An example is given in Figure 11.3.

3. Finally, the angular momentum and spin momentum $\mathbf{L}$ and $\mathbf{S}$, which result from the applications of rules #1 and #2, combine to give the total angular momentum $\mathbf{J}$ of the atomic system. $\mathbf{J} = \mathbf{L} + \mathbf{S}$ for shells less than half-filled. $\mathbf{J} = \mathbf{L} + \mathbf{S}$ half-filled or more. This is called *spin–orbit* coupling. Again this isn't really so hard to understand:

$$\mathbf{m} = \mathbf{m}_{\text{spin}} + \mathbf{m}_{\text{orbit}} = -(g_e \mu_B \mathbf{J})/\hbar = -\mu_B(2\mathbf{S} + \mathbf{L})/\hbar \tag{11.2}$$

in terms of the magnetic moments. Note that these are vector sums: up vs. down matters here. Also we have written this in terms of the Landé factor:

$$g_e = [1 + \mathbf{J}(\mathbf{J}+1) + \mathbf{S}(\mathbf{S}+1)]/2\mathbf{J}(\mathbf{J}+1) \tag{11.3}$$

This factor comes about due to electron–electron shielding in specific orbital geometries.

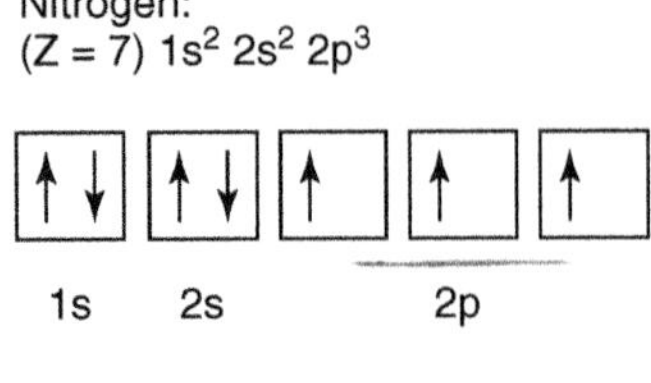

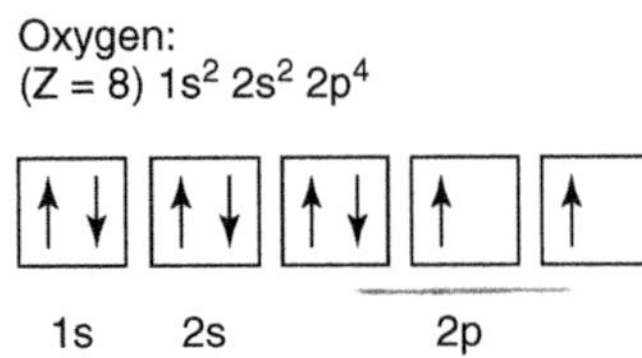

Figure 11.2 All of the orbital sublevels of the 2p level are filled with unpaired electrons as in nitrogen before they are allowed to pair up as in oxygen. This occurs simply because the paired state allows the electrons to approach each other more closely and thus it has a higher energy. This is a *spin–spin* interaction.

An atom with 4f^{10} outer shell

This means 10 electrons in the n = 4 l = 3 subshell (4f)

m_L −3 −2 −1 0 1 2 3

Backward

Figure 11.3 The electrons will fill the subshell so as to make the L largest. Here we use the example of an atom with an outer shell of 4f$^{10.}$ Using spectroscopic notation this means 10 electrons are in the outer 4f shell. Recall that we number the shells as s = 0, p = 1, d = 2, f = 3: so this is the l = 3 orbitals. m_l is the magnetic moment.

4. We should add a fourth *sub*rule: avoid details that might obscure the larger points while simultaneously making you look foolish. For example, $\mathbf{m}_{\text{nucleus}} = g_I \mu_N \mathbf{I}/\hbar$ ($\mathbf{I}$ = nuclear angular momentum, μ_N = nuclear magneton, and g_I = nuclear g factor). Nuclear spin ordering tends to be rather irrelevant for temperatures above 1 K or so; thus we ignore it. It is surprising how very useful this fourth rule can be.
   m changes when atoms become solids.
   In isolated atoms it is clear that the partially filled outer shells (both their spin and their orbit) contribute to the magnetic moment **m**. Filled shells do not contribute to the magnetic moment of the atom, although they certainly will contribute to the diamagnetic response through Larmor diamagnetism as we shall see. However, when the isolated atoms (with a finite **m**) are condensed into a solid, the proximity of other atoms can effect the spin and angular momentum states of their outer shell electrons. Specifically, the local electric and magnetic fields within the crystal (the *crystal field*) will readjust the potential energy environment around lattice points, and thus the energy states of the outer shell electrons, in such a way as to modify orbits and spins. In some cases, such as in bonding, the **m** of the isolated atom is entirely quenched.[2] But in other cases, only the orbital part is quenched, and the spin part survives intact or only partially modified.

The simple rules for this modification of atomic **m** ($\mathbf{m}_{LS}$) are as follows:

1. For free atoms or ions with partially filled s or p shells: **m** comes solely from the electrons in these unfilled shells. In the solid, these magnetic moments are all completely quenched. There are a few interesting exceptions, but they are rare, and generally, the orbital and spin parts of **m** in the free atom are gone.
2. For free atoms and ions with partially filled d-shell electrons, f-shell electrons, etc. and for the transition metals, rare earths, and so on, the story can be much more complicated. For a particular atom, we can say generally that the d and f electrons will frequently find themselves closer to the nucleus or ionic core and are rather well shielded by orbitals with more spatial extent. Thus the degree of their modification is small and can be expressed by a "quality factor" known as the *effective Bohr magneton* "p":

$$p = g[J(J+1)]^{1/2} \tag{11.4}$$

   J here is the total angular momentum quantum number of the free atom, and g is the Landé factor.

Experimentally, a material placed within a magnetic field will respond to the field, rearranging atomic spins and orbits, such that its *collective response field* adds or subtracts from the original field. In linear response regimes we might write something like

$$\mathbf{M} = \chi \mathbf{B} \tag{11.5a}$$

$$\mathbf{M} = \chi' H \tag{11.5b}$$

2 It is probably more correct to say that the **J**'s that make up **m** are no longer good quantum numbers for this *molecular* system.

M here is the magnetic moment of the whole solid in question, and it is this M that goes into the familiar E&M expressions such as $\boldsymbol{H} = (1/\mu_0)(\mathbf{B} - \mathbf{M})$. **B** is the total field (or flux density as some call it) and $\boldsymbol{H}$ is the auxiliary field. The χ's are the susceptibilities, and they are where the materials physics is to be found, including *crystal field* effects.

11.2 The Crystal Field

The features of what has become known as "crystal field theory" can be illustrated with an example. Here we start with an ionic species that has a partially filled outer 3d shell. So this is a transition metal ion in a crystal. Regardless of the crystal structure we might try to fit this ion into, we have an interesting problem: it has lobes in its wavefunction that have specific directions and symmetries. The crystal that we might try to place it in also has a specific symmetry depending on where the other bonding atoms or ions are located. These two symmetries may not be so well coordinated.

An example of the above might be Cr^{3+}, which can form crystals with both octahedral and tetrahedral symmetries (depending on the compound). Figure 11.4 shows the 3d orbitals laid out in the octahedral lattice unit cell.

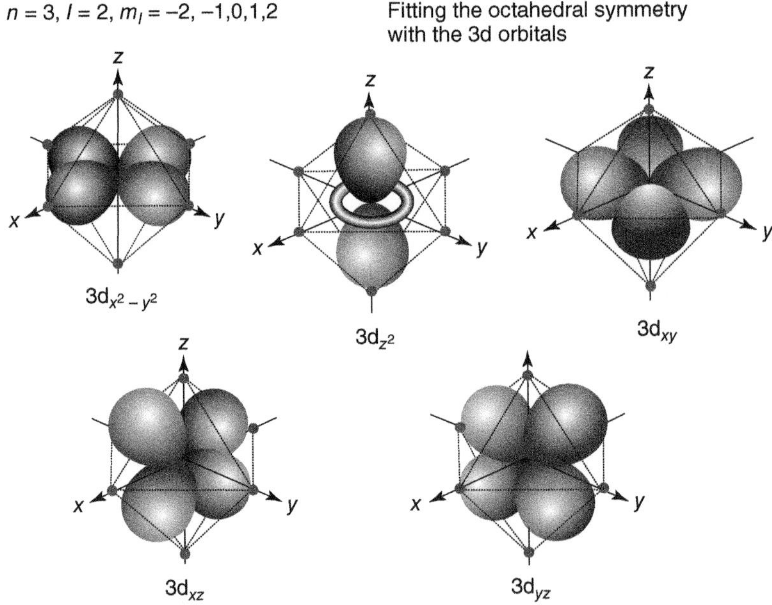

Figure 11.4 There are five possible d orbitals of a transition metal 3d ion as seen here. However, these orbitals can only be fit inside the octahedra of the lattice unit cell in specific ways. The system is ionic, so when the lobes of the orbitals align with the ions situated on the vertices of the octahedral, the energy of that orbital is shifted with respect to unaligned orbitals.

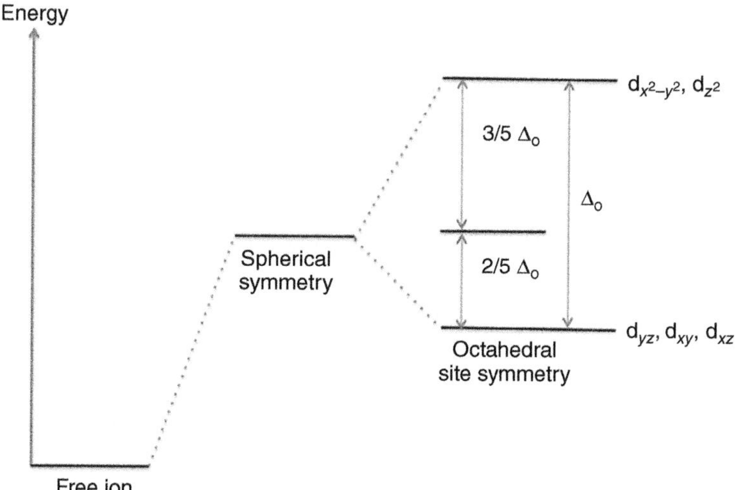

Figure 11.5 Energy level splitting (crystal field splitting) due to mismatches in the coordination between orbital symmetry and crystal symmetry. This is shown for a 3d shell in an octahedral site, but the reader should try to work out the tetrahedral case.

For an octahedrally coordinated site, there are six nearest neighbor anions as seen in Figure 11.4. Notice that the orbitals of the 3d transition metal project out into space but are well defined with respect to each other. That is, $d_x2 - d_y2$ and d_z2 (also known as the e_g orbitals where the *e* stands for doubly degenerate) will point in the direction of anions when the t_{2g} states (meaning the triple degenerate orbitals) – d_{xz}, d_{xy}, d_{yz} – will point into the voids between anions. Notice that the energy levels of the two degenerate e_g orbitals are increased due to interactions with the anions. So the two e_g orbitals will still be degenerated but higher in energy than the t_{2g} orbitals. The t_{2g} orbitals also remain degenerate. They too are shifted in energy from the free ion case, just not as much. This situation is seen in Figure 11.5.

In Figure 11.5 the "aligned" and "anti-aligned" orbitals are placed into equal energy groupings. For the fivefold degenerate 3d orbitals, we have a splitting of Δ_0 between the two groups. These new levels are referred to as *Stark levels.*

The completely degenerate spherically symmetric level would exist if we simply shifted the ion's energy by some mean field. Relative to this level, the e_g's increase in energy by $3/5\Delta_0$. However, the t_{2g}'s drop in energy by $2/5\Delta_0$. There can also be further splittings due to next nearest neighbors, and obviously different crystal structures, trigonal, tetragonal, etc. give different values of Δ and different symmetries.

But why this configuration? Couldn't we simply choose a configuration in which the orbitals are slightly misaligned with all anions? The answer is that this will likely not occur in nature. The ion seeks to configure itself, orbital occupation and direction, such that its energy in that particular site is the lowest. This configurational energy is referred to as the *crystal field stabilization energy* (CFSE), and it depends strongly on the occupation of the split states and the size of the splitting.

11.3 Magnetism in Condensed Systems

From classical E&M, the fields from differential elements of volume carrying a specific **m** are added together as vectors (the vector integral's value at point P). The approach is to treat the sum of little **m**'s as a big **M**, a dipole field. This is justified by looking at a *multipole expansion* of the field of a "magnetized" object. Higher-order magnetic poles decay rapidly with distance, and there are no monopole terms. So everything can be thought of in terms of dipoles (Figure 11.6).

The nature of that **m** or **M** can be quite different from material to material. That is, the response of materials to an applied field can range from opposing that field to adding to it. Figure 11.7 gives an example of two possible responses, each of which is frequently seen in the lab.

So how do materials vary so much in the magnetic response? The answer lies with the exact mechanism of spin and orbital alignment that is responsible for the magnetic response to start with. The subtleties discussed above actually make more of a difference in the final field than you might think. Let's see some examples.

11.3.1 Paramagnetism

Imagine a box full of spins, but there aren't very many (see Figure 11.8). Now if this box were full enough or it was small enough, such that the spins were really close, then there would be a torque between the spins that went like $\boldsymbol{\tau} = \mathbf{m} \times \mathbf{B}$ where **B** is the field of the nearest other spins at the site of **m**. But in this instance, our box is large, and the density of spins is small: the distance between any two spins is really rather large.

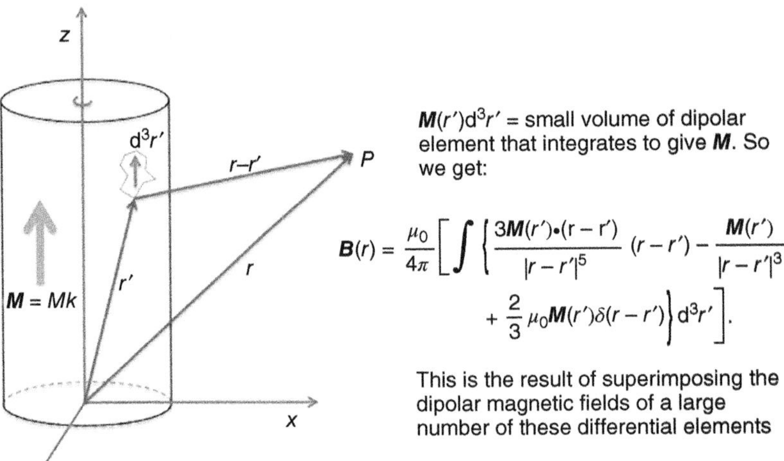

Figure 11.6 A schematic layout of how to compute the **B** field when the source is a bunch of little dipolar fields in a volume of d^3r'. The integral sums all of these contributions vectorially at P.

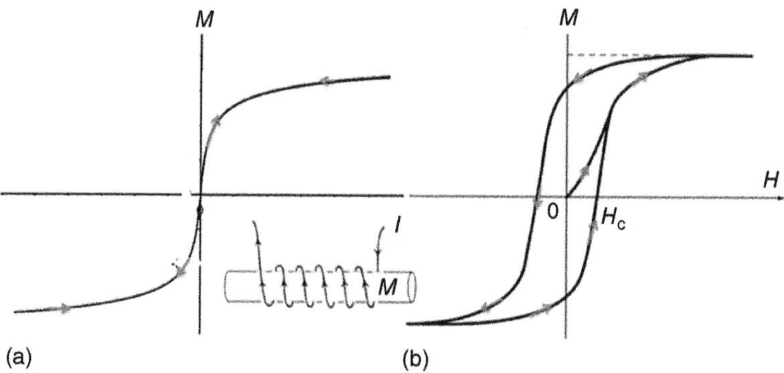

Figure 11.7 The response of a material (**M**) to an applied field (**H**) can have a memory or not as seen here. The small arrows are marking the direction of magnetic field increase or decrease. (a) No residual field and (b) residual field.

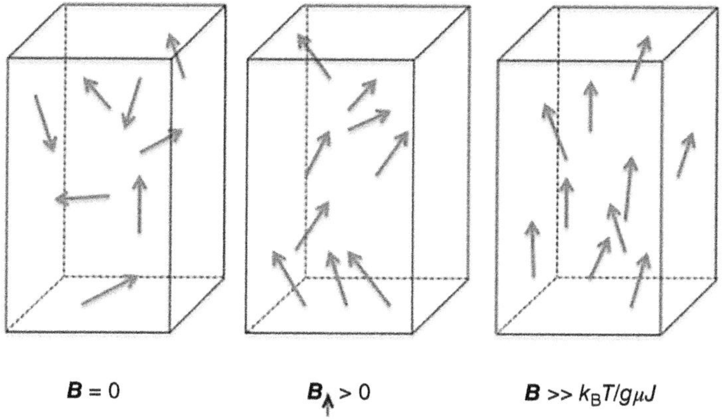

Figure 11.8 The classic paramagnetic model: the spins cannot interact between themselves, only with the externally applied field and with the heat bath. Notice here we have allowed spin to take on more than just an up or down spin projection.

If an external magnetic field is applied across our box, it is clear that the spins within the box will align *with* that field. The internal energy of the system will look like $U = -\mathbf{m} \cdot \mathbf{B}$ for each of the $\mathbf{m}$'s. Remember, here the $\mathbf{B}$ is from the outside. Of course we must keep in mind that this whole thing will sit in some heat bath, and so thermal agitations act upon the spins to misalign them. This energy is $\sim k_B T$, and for the spins to align, it must be less than the alignment energy from above, $g\mu JB$.

This is the basic model for a *paramagnetic* response. That means a magnetic response in materials where the localized magnetic moments are present but there is no macroscopic magnetization ($\mathbf{M}$) in zero field: $\mathbf{H} = 0$. Notice that when we increase the applied field, $\mathbf{M}$ increases, but when we decrease the field, $\mathbf{M}$ goes away completely due to the thermal energy as shown in Figure 11.7a.

The role thermal energy plays in the total magnetization of a paramagnet is illustrated in Figure 11.8. Spins do not tend to align completely unless applied fields are very large. When the "aligning field" is removed, thermal agitation to the individual spins will eventually destroy any residual magnetism in the system. In other words, the field generated by the spins themselves is not enough to keep them all aligned together.

But surely this must all depend on the temperature of the spin system. Yes. Below a specific temperature, paramagnets can exhibit permanence if interactions are large enough. This temperature is referred to as the *Curie temperature*, and the paramagnet that allows for such a limited inter-spin interaction is known as a *Curie paramagnet* (a paramagnet with a *Curie point*). It is typically associated with a thermodynamic phase transitions for spin systems and is sensitive to the density of spins, order and dimension of the system, and the magnitude of the spins, $\mathbf{m}$. So it follows the thermodynamic laws of phase transitions, scaling, and critical phenomena as expected. Indeed, placing spins on a lattice and tweaking interaction has led to numerous useful models of spin waves, spin transport, etc.

11.3.1.1 Curie Paramagnets

Curie paramagnetism is based upon an ensemble of *single*, localized electron spins in a magnetic field applied along some axis "z." Thus the $|\uparrow\rangle$ and $|\downarrow\rangle$ states are split in energy by $\approx 2\mu_B B$.

Boltzmann statistics determines the population of the $|\uparrow\rangle$ state relative to the $|\downarrow\rangle$ state within the ensemble, for a given temperature and field strength. So, if we have $n = (n^\uparrow + n^\downarrow)$ electrons per unit volume, the *induced magnetization* along $\boldsymbol{z}$ is $(n^\uparrow - n^\downarrow)\mu_B$. $n^{\uparrow,\downarrow}$ are the Boltzmann populations of the spin-up and spin-down energy levels. They are proportional to $\exp(\pm\mu_B B/k_B T)$. So

$$M = c\mu_B(e^{\mu_B B/k_B T} - e^{-\mu_B B/k_B T}) \tag{11.6}$$

and

$$n = c(e^{\mu_B B/k_B T} + e^{-\mu_B B/k_B T}) \tag{11.7}$$

The average z-component of the moment per atom defined as

$$\langle m_z\rangle = \mu_B(n^\uparrow - n^\downarrow)/(n^\uparrow + n^\downarrow) \tag{11.8}$$

can be written then as

$$\langle m_z\rangle = \mu_B(e^x - e^{-x})/(e^x + e^{-x}) \tag{11.9}$$

$$x = \mu_B B/k_B T \tag{11.10}$$

Notice we can write

$$M = n\mu_B \tanh x \tag{11.11}$$

And at room temperature,

$$\mu_B B \ll k_B T \tag{11.12}$$

x is small and so $\tanh x \approx x$. This approximation gives the now famous *Curie law* expression for the susceptibility. Using the definition $\chi = \mu_0 M/B$, we get

$$\chi = n\mu_0\mu_B^2/k_B T \tag{11.13}$$

This is sometimes written as

$$\chi = C/T \tag{11.14a}$$

$$C = \frac{n\mu_0\mu_B^2}{k_B} \tag{11.14b}$$

is the *Curie constant*.

So, according to the Curie law, the susceptibility diverges as $T \to 0$.

11.3.1.2 The Weiss Correction

The Curie law works well for paramagnetic response at higher temperatures, but it fails at low temperatures. This failure comes from the very considerations we have given above. Detailed derivations of the Curie–Weiss law, which are a little more accurate, can be found in the literature. But here we show that the "fix" is not so complicated.

First we start with the Curie constant. The Curie law above was derived for single spin states, but in fact we can extend this to consider the J states of the atoms, or domains, involved. We do this by replacing spin with the total angular momentum of the atom:

$$C = (n\mu_0\mu_B^2/3k_B)(g^2J(J+1)) \tag{11.15}$$

where we have now included the Landé g factor and the $J(J+1)$ angular momentum quantum number.

Next, we must consider the crystal field effects. We have done this somewhat in "C," but we must also recognize that the total magnetic field felt by an atom in the field is $B + \lambda M$ where λ is the *Weiss molecular field constant*. Thus,

$$\chi = \frac{M\mu_0}{B} \to \frac{M\mu_0}{B+\lambda M} = \frac{C}{T} \tag{11.16}$$

giving

$$\chi = \frac{C}{T - C\lambda/\mu_0} \tag{11.17}$$

This is the *Curie–Weiss law* and the $C\lambda/\mu_0$ term is the *Curie temperature*, T_c. So, far from T_c, approaching from the $T > T_c$ side, we can write

$$\chi = \frac{C}{T - T_c} \tag{11.18}$$

The Curie temperature represents a phase transition from the paramagnetic phase where an applied field is required to align the "spins" to a system with enough self-energy to stay aligned at temperatures below T_c. This means as we approach the phase transition T_c in T, the system becomes "critical," and a more complete theory is needed. From statistical mechanics we can already guess what that outcome will be:

$$\chi \sim \frac{1}{(T - T_c)^\gamma} \tag{11.19}$$

where γ is a critical exponent. We note that the assumptions that went into getting T_c change a little from right at the transition to far from it:

$$T_{c,\text{critical}} \neq T_{c,\text{far from critical}} \tag{11.20}$$

After all, we did use a rather simple reasoning to get this value. So physicists will usually assign the symbol Θ for T_c at higher temperatures, recognizing that it is a little different from the value right at criticality. This does make a lot of sense when you consider that T_c has the λ, molecular field, term in it. Clearly as the correlation lengths begin to diverge in the phase transition, this crystal field approximation must be changing significantly.

11.3.1.3 Free-Electron Magnets

So far we have considered only localized electrons, but what if they are delocalized? They still have spin, but do they act as paramagnets?

A reasonable approximation or model for metallic systems can be made by first considering the magnetic moments of free atoms and then extrapolating to a free-electron model. Clearly, the electron moments of atoms across the "metallic" part of the periodic table varies with atomic number Z, with some outer shells being fully paired such as those with even Z (alkaline earths) and others not. Atoms such as dysprosium or holmium can have a magnetic moment as high as $\sim 10\mu_B$. Iron, which has an electronic configuration of (Ar)$3d^6 4s^2$, has only four of the 3d electrons unpaired, so its spin magnetic moment is $\sim 4\mu_B$.

Now let's consider what happens when we bring a set of iron atoms together as an example. This is done in Figure 11.9. As the atoms condense to an ideal distance

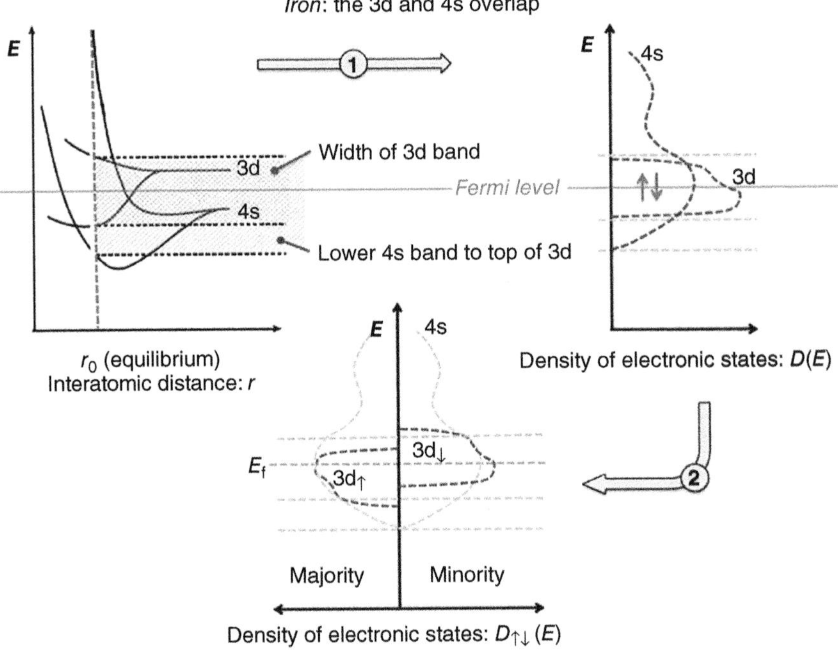

Figure 11.9 The formation of Fe bands from the 3d and 4s electrons. With no internal interactions the bands fill with up and down spin electrons evenly. But, as the internal spin–spin interaction is "turned on," the 3d band splits into two subbands representing spin majority and spin minority carriers.

apart (labeled r_0) to form a crystal, bands are formed just as we have seen before. If we think in terms of the LCAO methods already introduced, then it is easy to see that different parts of the band structure will be derived from different orbitals, and these sometimes overlap. This is illustrated in Figure 11.9 by the shading. Now for iron, the 3d and 4s outer shells form band states that surround the Fermi level, which means they are the ones taking part in conduction properties. And there is no gap, so these electrons are free and delocalized, like a Drude metal.

The states below this Fermi level are filled, but in our model, the electrons filling are allowed to "remember" their spin states as the band is formed. So, what does this mean? The complete overlap of the narrow 3d band with the broader 4s will result in some charge transfer between the bands 4s → 3d. Thus, the atomic iron configuration becomes something closer to (Ar) $3d^{7.4}$ $4s^{0.6}$. If there are *no* internal interactions (spin–spin interactions), then the density of states $D(E)$ and state filling diagram would look like that at the end of the arrow marked "1" in Figure 11.9. This is simply the integral of all states filled to the Fermi level, and at the Fermi level the unpaired electrons can respond to applied fields as any paramagnet would. However, if we follow the arrow marked "2," an internal field is introduced in a way similar to that used when we derived the Curie–Weiss law. Here there is a very strong internal interaction between spins. In this case it is clear that the band will spontaneously split into two subbands, distinguished by their spin projections. We have used a standard graphical artifice to illustrate this in Figure 11.9. Since the inner shells are all perfectly paired electrons and the 4s are mostly paired, contributions to these subbands come from the 3d band.

The case of large internal interactions and subband formation is referred to as *ferromagnetism*, of which Fe is an excellent example. The energy shift and spin selection comes about due to the interaction of spins with the internal alignment field. The offset and filling of the subbands results in a permanent magnetic moment, and so, Fe can be a permanent magnet. Indeed, in the subbands, Fe can have a 3d configuration of $3d^{4.8\uparrow}$ $3d^{2.6\downarrow}$, a total in unpaired electrons of 2.2 [2].

These simple models do not represent a comprehensive approach to paramagnetic behavior and can even fail to adequately address such simple cases as dilute composites of magnetic nanoparticles or semiconductors with dilute magnetic impurities. However, they can be recast into a number of interesting sub-models, any one of which is designed to address specific applications or observations in the lab. Most recasts involve subtle shifts or semiclassical treatments of the basic paramagnet idea above. They include *Brillouin paramagnets, Langevin paramagnets,* and *Van Vleck paramagnets,* among others. The reader is invited to follow the references given to learn more about their specific applications [2]. But no matter which pathway you go down, you will start with the basic models above for paramagnetism.

11.3.2 Diamagnetism

Atoms get their permanent magnetic fields by their electronic spin and unpaired electrons in the outer energy shells. Keep in mind that this last source is just the loop of an electron running around its orbit to make a current. Above we have assumed that as an external field is applied, the magnetic moments of these sources align *with* the field, *adding* to its strength: the *paramagnet*. But of course the field might also modify the orbit in some way as to perturb the current and thus the moment associated with the outer shell electrons. The difference in the current before and after the modification of the orbit might be thought of as an additional current added to the system. Lenz's law suggests that the current, thus induced, will always be associated with field generation that opposes the applied field. This is the principle of *diamagnetism*. Diamagnetism *opposes* the establishment of a field from the outside.

To do this, the system must allow for some smooth modification of the "angular momentum current." Different mechanisms have been proposed for different materials systems from metals with free electrons to diffuse isolated systems in insulator crystals. The first and most simple of these models is known as *Langevin diamagnetism*, and it is rather powerful in its simplicity.

As illustrated in Figure 11.10, imagine that we have an "isolated" atomic spin system with **J** as total angular momentum and **L** and **S** as orbital and electronic spin components, respectively. If we allow for the orbitals in this atom's outer shells to "tilt" a bit, then the easiest and most obvious way to "modify" its current is to allow for the precession of the magnetic moment (the current loop) around the axis of the applied field. This is known as the Larmor precession, and you have probably already seen it in basic classical mechanics. In Figure 11.10, the gray arrow is the applied field **B**, defining the axis about which a torque will occur. That torque, $\boldsymbol{\tau} = \mathbf{m} \times \mathbf{B}$, is a twist of the magnetic moment of the angular momentum current (i_c), about the **B** axis. **m** sweeps out the black circle at the top.

Imagine now that the charge e^- is spread out over the circle of the angular moment orbit's path. This means the precession sweeps out a cone of smeared charge above and below the zero axis on the graph, giving rise to another current.

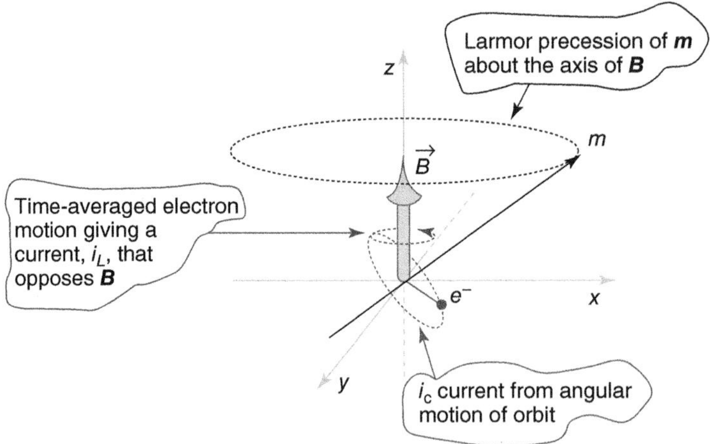

Figure 11.10 The Langevin model of diamagnetic response. An isolated atom with orbitals free to precess will allow the establishment of a current that generates an internal (or atomic-scale) dipole moment that opposes the applied field **B**. The mechanism is much like that of a gyroscope in a gravitational field. And the resulting current is centered about the nucleus of the atom.

This induced precessional current opposes the field **B** as best it can. Of course, the precession can go as fast as it would like to oppose B, and this happens to be the *Larmor frequency*:

$$\omega = eB/2m_e \tag{11.21}$$

Naturally we have to assume that the Larmor frequency is much smaller than that of the orbital motion. Also, in metals with free electrons, we must replace this idea with the cyclotron frequency, which is, in fact, twice that of the Larmor frequency. But the idea remains; the new current opposes the establishment of the field.

Naturally, the inner orbitals can take part in this as well. So if we have an atom as described with an atomic number Z, then

$$I_{\text{induced}} = (-Ze)[eB/4\pi m_e] \tag{11.22}$$

This current loop will have a magnetic moment:

$$\boldsymbol{\mu} = -[Ze^2\mathbf{B}/4m_e]\langle \rho^2 \rangle \tag{11.23}$$

The $\langle \rho^2 \rangle$ term is perpendicular distance of the electron orbit from the field axis. To convert this over to simply the average orbital radius $\langle r^2 \rangle$ for a spherically symmetric set of shells, $\langle \rho^2 \rangle = 2/3\langle r^2 \rangle$. This leaves us with the interesting semiclassical Langevin result:

$$\chi = N\mu/B \tag{11.24}$$

$$\chi = -[\mu_0 NZe^2/6m_e]\langle r^2 \rangle \tag{11.25}$$

For solids with few free electrons, this result is reasonable. Now we just have to find $\langle r^2 \rangle$. (For free-electron metals things are a little more complicated.) Such values can be calculated using quantum mechanics, and they are related to the spatial extent of specific orbitals.

11.4 Dia- and Para-Foundations of Other Magnets

The semiclassical presentations and results above actually agree rather well with the more formal quantum mechanical treatment for these properties. Notice that these models are referred to as semiclassical because they must invoke some quantum property, like spin or orbitals that do not decay, to make the models with classical forces work. Our simple visualizations have identified and estimated three primary sources of magnetic moment in any atom. Further, the atom condensed into a solid can either act to add to any applied field or oppose it. It now seems clear how to predict just what it will do and with what strength. We may also now suppose that there exists some special thermal energy balanced against the energy of ordering that allows for permanent fields from solids such as in *ferromagnetism*.

To do all this that came at a cost, we had to accept some rather significant simplifications – beyond just the use of classical forces and mathematics. As we saw

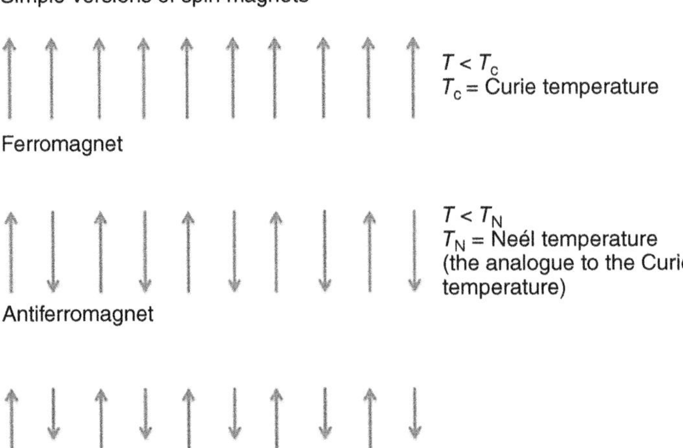

Figure 11.11 A schematic version of different *spin magnets* that shows the idealized ordering of spins in the crystal. Notice here we introduce the *Neél temperature,* which is the exact analogue of the Curie temperature for the *antiferromagnet*. It separates the ordered spin system from the paramagnetic phase.

with ferromagnetism, the biggest simplification is that we ignored interactions between the atomic-scale **m** sources (so-called *spin–spin interactions*).[3]

Interactions between the sources of the field internally can yield a wide range of magnetic behavior. Most of the time we can think of such magnetic systems as paramagnetic or diamagnetic with an interaction between the **m**'s added in. This interaction can, for instance, be different for **L** and **S**, change with temperature or other variables, have different strengths from material to material, or have different strengths between nonequivalent lattice sites in the same material. To a large extent, the specifics of these interactions can be used to explain a range of magnet types such as seen in Figure 11.11.

11.5 Mechanisms of Interaction: Spin Models

How exactly do the moments of adjacent atoms interact? What is the mechanism and interaction strength? What is the interaction distance? What new phenomena do we get when we add this to the picture? These questions are important to truly understand the wide range of magnetic phenomena we encounter in the lab. They have also yielded highly detailed spin models that have found use not only in magnetism but also in certain aspects of field theory, information theory, and brain science, as well as quantum computing. Indeed, there are many physicists

3 We are speaking loosely here. *Spin–spin interaction* here means the interaction between the two neighboring (or next nearest neighboring) *magnetic moments.* Many of the models we will develop will focus on systems of spins with only an up or down orientation and no possibility for precessional response. Here the dynamics of collective behavior is particularly sensitive to Spin–spin interactions where the spin is the more pure concept of internal spin quantum number = $\pm \frac{1}{2}\hbar$.

that have made whole careers from spin models without ever even touching a magnet (except those on their home refrigerators).

11.5.1 The Mean Field Model

An informal attempt to address spin–spin interactions has already been made in this text using a *mean field* approach. Recall that we wrote $\mathbf{B}_\mathrm{m} = \lambda \mathbf{M}$ and added it to the applied field, giving a total field $\mathbf{B}$. This represents the interactions of spins with the mean magnetization caused by all other spins. The approach gave rise to the *Curie temperature,* T_c. Below T_c, spontaneous magnetization can occur, induced alignment can be frozen in to the system, hysteresis occurs in the magnetization curve, etc. This temperature is the boundary between an ordered *ferromagnetic* phase and a disordered paramagnetic phase. With a little algebraic manipulation, we can show

$$\lambda = \frac{T_\mathrm{c}}{C} = \frac{3k_\mathrm{B}T_\mathrm{c}}{Ng^2S(S+1)\mu_\mathrm{B}^2} \tag{11.26}$$

and

$$\mu_0 M = \chi(B_\text{applied} + B_m) \tag{11.27}$$

11.5.2 Ising, Heisenberg, XY, and Hopfield[4]

There is a whole class of spin models that weights nearest neighbor interactions (spin–spin interactions) the most strongly. As in lattice sums, this removes the necessity of infinite series and accounts for the fact that dipoles do not interact over long distances. Generally speaking, the details of some specific interaction chosen for study are associated with some sort of configurational energy. This is analogous to a thermodynamic internal energy from which thermodynamic properties of the model can then be predicted. This energy must be partitioned appropriately as the system approaches equilibrium with a heat bath or a phase transition between different configurations typically. So, from this point of view, spin models become very interesting when trying to understand the thermodynamics and kinetics of aligned magnetic moment systems. In other words, they may help us capture the essential properties of magnets as they go through their *phase evolution.*

11.5.2.1 Ising Models

Ising models are an extremely well-studied set of spin models proposed by Lenz in 1920 and solved by Ernst Ising in one dimension (1925) [3] and in the two-dimensional (2D) square lattice by Lars Onsager (1944). So, they have been around for a while!

To see how they work, we start off with some set of lattice sites Λ. It doesn't really matter how you arrange them, and they can be 1, 2, 3, or higher in dimension. (The higher ones are helpful for quantum field theory). So, the lattice is d-dimensional. At each lattice site $k \in \Lambda$, we place a discrete (up or down)

4 Please note that in different models we use different variables to stand for spin: σ and S. This is because they frequently stand for slightly different things, depending on how the model is being used.

spin: $\sigma_k \in \{+1, -1\}$. Once a spin is assigned to every lattice site, we call this *a spin configuration*. Obviously, if there are many sites, then there are many unique spin configurations possible.

Between all adjacent or nearest neighbor sites (i, j), there is an interaction: J_{ij}. Further, at each site j there is an external magnetic field h_j (we had to use lower case h here because of the H for the Hamiltonian). The energy of this *spin configuration*, σ, is given by the Hamiltonian

$$H(\sigma) = -\sum_{\langle i\, j\rangle} J_{ij}\sigma_i\sigma_j - \mu\sum_j h_j\sigma_j \tag{11.28}$$

The first sum is over pairs of adjacent spins, counting each once. Note that the sign in the second term should be positive because the electron's magnetic moment is antiparallel to its spin, but the negative term is conventional [4]. The probability of this spin configuration occurring is given by the Boltzmann distribution

$$P_\beta(\sigma) = \frac{\mathrm{e}^{-\beta H(\sigma)}}{Z_\beta} \tag{11.29}$$

$$Z_\beta = \sum_\sigma \mathrm{e}^{-\beta H(\sigma)} \tag{11.30}$$

$$\beta = (k_\mathrm{B}T)^{-1} \tag{11.31}$$

To predict the expectation of some observable f, that is, its mean value for this lattice at a given temperature,

$$\langle f\rangle_\beta = \sum_\sigma f(\sigma)P_\beta(\sigma) \tag{11.32}$$

$P_\beta(\sigma)$ is the probability that, in equilibrium, the system is in the σ spin configurational state. Ising models are usually classified according to the sign of the interaction:

$$J_{ij} > 0, \text{ the interaction is called } \textit{ferromagnetic} \tag{11.33a}$$

$$J_{ij} < 0, \text{ the interaction is called } \textit{antiferromagnetic} \tag{11.33b}$$

$$J_{ij} = 0, \text{ the spins are } \textit{noninteracting} \tag{11.33c}$$

So, in the ferromagnetic Ising model, spins want to be aligned. That is to say, the spin configurations where most or all of the adjacent spins have the same sign (up or down) have higher a probability of occurring. In the antiferromagnetic case of the model, "adjacent spins with opposite sign" is the more probable spin configuration. This all depends on the J and how much it raises or lowers energy, depending on spin alignment. So, it defines the ordering phase of the magnet. And, if we ask about the external field, we get something similar:

$$h_j > 0, \text{ the spin site } j \text{ desires to line up in the positive direction} \tag{11.34a}$$

$$h_j < 0, \text{ the spin site } j \text{ desires to line up in the negative direction} \tag{11.34b}$$

$$h_j = 0, \text{ there is no external influence on the spin site} \tag{11.34c}$$

These models can be useful even without considering the external field. In this case the model is symmetric under switching the value of the spin at all the lattice sites. Or we can assume that the $J_{ij} = J$ as another simplification. This gives a translationally invariant ferromagnet on a d-dimensional lattice.

In one dimension, the Ising models do not allow for a phase transition generally [2]. However, in two and three dimensions, the models do allow for a phase transition between an ordered and a disordered phase as demonstrated by Peierls in 1936 [2]. In 1944, Onsager showed that the correlation functions and free energy of the 2D Ising model are determined by a *noninteracting lattice fermion*. Onsager in 1949 and Yang in 1952 demonstrated an expression for *spontaneous magnetization* in these systems [2]. So, clearly the study of these models has had a surprising longevity.

11.5.2.2 Heisenberg Models

Heisenberg models carry the idea of the Ising model, a little further. Instead of having spin-up or spin-down on each lattice site, we now allow for multiple spin projections. Actually, this can be formulated in two ways, with *semiclassical* or *quantum* spins. (These are not so different.) In either case we begin with a Hamiltonian that can look something like

$$H(\sigma) = -\sum_{\langle i\,j \rangle} J_{ij}\sigma_i \cdot \sigma_j - \mu \sum_j h_j \cdot \sigma_j \tag{11.35}$$

The *exchange integral*, J_{ij}, can be taken as a constant, J, for i and j nearest neighbors and zero otherwise, or it can have specified values.[5] Other terms are as in the Ising model except the σ_i, which is now a unit vector of the site spin. (As an aside, sometimes J is redefined as $2J$ so that the factor of 2 will cancel the factor of ½ necessary in front of the summation to keep from double counting the i, j pairs. So the above equation is correct as it stands with all combinations of i and j, but J is ½ the interaction energy.)

11.5.2.3 XY models

XY models are really rather simple: classical spins of unit length in a 2D lattice that can rotate in the plane of the lattice. Of course, the model can be cast such that these 2D spin vectors sit in a 3D lattice or higher, if those sorts of things interest you. However, typically, higher-order XY models leave the spin degrees of freedom at 2D.

For the standard version of the XY model, following the reasoning used for Heisenberg and Ising, we can write an interaction and *configuration energy* as

$$H(s) = -\sum_{i \neq j} J_{ij} s_i \cdot s_j - \mu \sum_j h_j \cdot s_j = -\sum_{i \neq j} J_{ij} \cos(\theta_i - \theta_j) - \sum_j h_j \cos \theta_j \tag{11.36}$$

This is, of course, doing nearest neighbor sums but could be extended if desired. The *configuration probability* of statistical observables becomes

$$P(s) = \mathrm{e}^{-\beta H(s)}/Z \tag{11.37}$$

$$Z = \int_{[-\pi,\pi]^\Lambda} \prod_{j \in \Lambda} \mathrm{d}\theta_j \mathrm{e}^{-\beta H(s)} \tag{11.38}$$

5 *Usually*: $h = gB\mu_0 H$

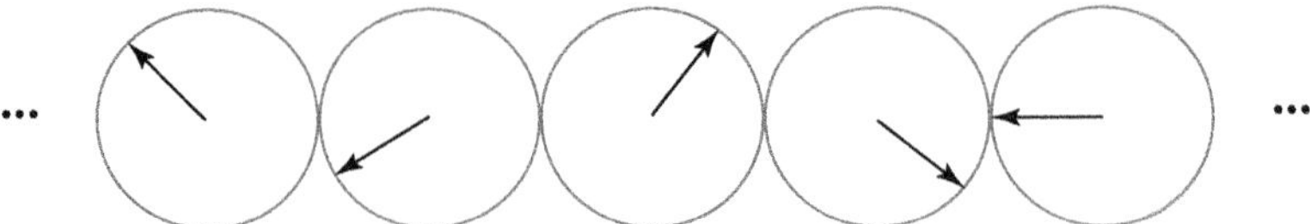

Figure 11.12 Visualization of a 1D XY model from the top.

where Λ is the set of all lattice points and Z represents the traditional partition function from statistical mechanics. Remember that $\beta = 1/k_B T$.

Just to see how this might work, let's look at the 1D case (Figure 11.12).

The partition function integral above becomes

$$\begin{aligned} Z &= \int_{-\pi}^{\pi} d\theta_1 \cdots d\theta_L e^{\beta J \cos(\theta_1 - \theta_2)} \cdots e^{\beta J \cos(\theta_{L-1} - \theta_L)} \\ &= 2\pi \prod_{j=2}^{L} \int_{-\pi}^{\pi} d\theta'_j e^{\beta J \cos\theta'_j} = 2\pi \left[\int_{-\pi}^{\pi} d\theta'_j e^{\beta J \cos\theta'_j} \right]^{L-1} \end{aligned} \tag{11.39}$$

where L is the last in the series of spins and θ'_j is simply $\theta_j - \theta_{j+1}$. This means all statistical observables have a form

$$Z = -\frac{1}{\beta} \ln \int_{-\pi}^{\pi} d\theta'_j e^{\beta J \cos\theta'_j} = -\frac{1}{\beta} \ln[2\pi I_0(\beta J)] \tag{11.40}$$

where I_0 is the modified Bessel function of the first kind.

XY models are particularly important for the study of critical phenomena and phase transitions and in our understanding of how dimension plays a role in criticality. Though this is not a text on statistical mechanics, you might recall that criticality is reflected in many fields of physics and exhibits what is known as traits of *universality*. This simply means the functional approaches to divergences in the theories of particle physics, solid-state physics, etc. are all the same. Now this could be a statement of something deeper in the universe. Or it may be a statement about our limited understanding of the mathematics that we use to describe our universe. Either way, XY models have been used to provide a little glimpse into this question.

Of course in terms of this text, they are also important. After all, we are trying to understand solids even as dimension is being restricted. The *Mermin–Wagner theorem* states that 2D spin systems such as XY and Heisenberg models cannot have phase transitions [5]. As it happens, 1D spin systems, as shown above, can also not show phase transitions. But "spin" can stand for many things in the universe, so these are quite interesting results generally. For the solid-state physicist, it suggests rather strongly that if we observe a 1D magnet (permanent magnetic moment), then we should be a little careful. There is likely something in a higher dimension stabilizing that structure in place.

But we aren't done quite yet. As it happens, there is a caveat. It is frequently the case that being too dogmatic about what a phase transition is or is not can lead us to miss something important. Let's consider the 2D XY system with nearest neighbor interactions only. At very low temperatures ($T < T_c$) T_c is a kind of transition temperature demarking the boundary between two different kinds of behaviors. For now we shall just say that it relates to spin–spin *correlation length.* We can show that the correlation function can be written, using the above equations, as

$$\langle \mathbf{S}_i \cdot \mathbf{S}_j \rangle \propto r^{-\eta} \tag{11.41}$$

and

$$\lim_{T \to T_c} \eta = 1/4 \tag{11.42}$$

Though we haven't derived it explicitly here, this result does come directly from the XY Hamiltonian for the case where the externally applied field is zero ($h_j = 0$).

At high temperatures ($T > T_c$), spins are strongly uncorrelated. So much so that even nearest neighbor spins have almost no correlation. In this case,

$$\langle \mathbf{S}_i \cdot \mathbf{S}_j \rangle \propto e^{r/\xi} \tag{11.43}$$

where ξ is the correlation length and it diverges at T_c.

So there do exist two phases:

1. *A low-temperature phase*: This has some "quasi" long-range order where most spins are aligned. The correlation function decays as a *power law.*
2. *A disordered high-temperature phase*: Here, the correlation function decays *exponentially.*

At the critical temperature, T_c, a type of transition occurs known as the *Kosterlitz–Thouless transition.* The Kosterlitz–Thouless transition is not a typical second-order phase transition where the first derivative of the internal energy is continuous. But it can be quite useful in understanding low-dimensional, finite-sized magnetic structures.

Using this model it is also possible to show, with a little diligence, that

$$C_V = \frac{1}{k_B T^2}(\langle E^2 \rangle - \langle E \rangle^2) \tag{11.44}$$

and

$$\chi = \frac{1}{k_B T}(\langle M^2 \rangle - \langle M \rangle^2) \tag{11.45}$$

11.5.2.4 Hopfield Models

As a curious aside, we could mention the Hopfield network model. Such a network is a form of interconnected spin system, very much like the Heisenberg model. It was first described by Little in 1974 and popularized by John Hopfield in 1982 [6] as a way to model content-addressable ("associative") memory systems. Surprisingly, this spin system uses binary threshold nodes (like Ising), but they capture some interesting features of human memory including their ability to make mistakes (convergence to false minima)!

Briefly, all spin pairs (i, j) in the model network are connected, and the connection is assigned an interaction weight ω_{ij}. These connection weightings have some rules: (i) $\omega_{ij} = 0$ for $i = j$, no self-connections, and (ii) $\omega_{ij} = \omega_{ji}$. Now instead of having nearest neighbor and thermal considerations to "flip" spins, the Hopfield model "updates" spin sites following rules based on numerous interactions. So, *updating* is performed if

$$s_i \leftarrow \begin{cases} +1 & \text{if } \sum_j w_{ij} s_j \geq \theta_i \\ -1 & \text{otherwise} \end{cases} \tag{11.46}$$

Here, s_i is the site value of the ith site. θ_i is the "threshold value" of the ith site. Here, clearly, this threshold depends on the collective influence of many spin sites.

Interestingly, the Hopfield model assigns a scalar number to each configuration taken by such a network. This is usually referred to as the configuration energy, and it is given as

$$E = -\frac{1}{2}\sum_{i,j} w_{ij} s_i s_j - \sum_i \theta_i s_i \tag{11.47}$$

The curiosity here is that this is so similar to the Heisenberg model we have introduced above. Indeed, the energy function in the Hopfield model belongs to the more general class of Ising spin models (as does the Heisenberg model energy). They are a specialized case of the mathematical creatures called *Markov chains*. Yet the approach has been indispensible in the modeling of memory processes in basic brain function.

11.5.3 Spin Wave and Magnons

An interesting outcome of spin models is that we can anticipate the excitations of spin systems and their dynamics. This leads naturally to *spin waves* and *magnons.*

11.5.3.1 Spin Waves

Spin waves are collective magnetic excitations of some system of spins. There are, in fact, many theoretical models that support the formation of spin waves, each sensitive to the exact nature of the exchange interaction used and each applicable to some specific observation in magnetics. There are really a couple of ways to picture the spin wave. The first is to use the classical *Heisenberg magnet,* and the second uses *itinerate electron magnets,* both discussed above. As with most basic texts, we will present the *Heisenberg magnet* approach here. It gives a more "intuitive" feel for the behavior of these excitations.

We begin by considering only a classical Heisenberg model where the spin is a classical vector. In this treatment, assume that **L** has been quenched by the lattice so that only the **S**'s are considered (yes, you can have antiferromagnetic spin wave solutions if this assumption is not made). We note that a quantum treatment will yield the same result. Each atom carries a magnetic moment: $\vec{\mu}_i = -g\mu_B \vec{S}_i$. The energy that this μ_i observes is $-\vec{\mu}_i \cdot \vec{B}$ when placed in a field B. And we know that from our *mean field model* above, the field that surrounding spins create is

$$\mathbf{B}_i^{\text{effect}} = -\frac{2}{g\mu_B}\sum_j J_j \mathbf{S}_j \tag{11.48}$$

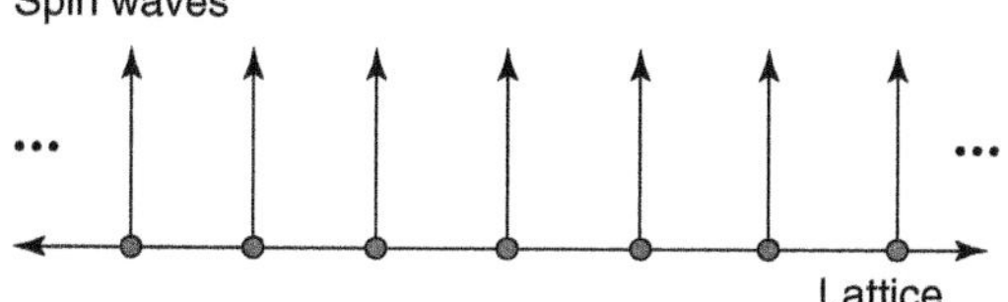

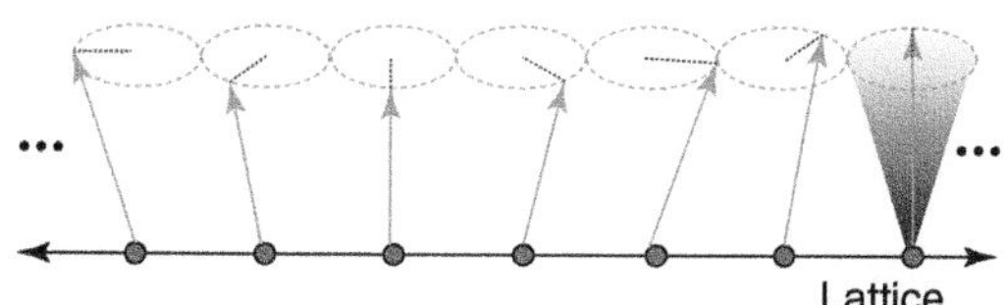

Figure 11.13 A simple visualization of the spin wave in a Heisenberg magnet.

Remember that this also gives a torque, $\tau_i = \mu_i \times \mathbf{B}_i^{\text{effect}}$, and from this we can construct equations of motion. Generally we can then give a spin–spin expression for torque:

$$\hbar\frac{\mathrm{d}\mathbf{S}_\mathrm{i}}{\mathrm{d}t} = 2\sum_j J_j(\mathbf{S}_\mathrm{i} \times \mathbf{S}_j) \tag{11.49}$$

From here let's simplify to just the ***1D*** system shown in Figure 11.13. This means j in the summation runs only along a string of spins, and we can approximate nearest neighbor interactions ($J_i = J$) and reduce the sum

$$\frac{\mathrm{d}S_n}{\mathrm{d}t} = J(\mathbf{S}_n \times \mathbf{S}_{n-1} + \mathbf{S}_\mathrm{n} \times \mathbf{S}_{n+1}) \tag{11.50}$$

or in matrix from

$$\begin{pmatrix} \frac{\mathrm{d}S_n^x}{\mathrm{d}t} \\ \frac{\mathrm{d}S_n^y}{\mathrm{d}t} \\ \frac{\mathrm{d}S_n^z}{\mathrm{d}t} \end{pmatrix} = J\left[\begin{vmatrix} \hat{e}_x & \hat{e}_y & \hat{e}_z \\ S_n^x & S_n^y & S_n^z \\ S_{n-1}^x & S_{n-1}^y & S_{n-1}^z \end{vmatrix} + \begin{vmatrix} \hat{e}_x & \hat{e}_y & \hat{e}_z \\ S_n^x & S_n^y & S_n^z \\ S_{n+1}^x & S_{n+1}^y & S_{n+1}^z \end{vmatrix}\right] \tag{11.51}$$

From here we consider just the x-component (since all the others will work the same except for the occasional sign change):

$$\begin{aligned}\frac{\mathrm{d}S_n^x}{\mathrm{d}t} &= J[S_n^y S_{n-1}^z - S_n^z S_{n-1}^y + S_n^y S_{n+1}^z - S_n^z S_{n+1}^y] \\ &= J[S_n^y(S_{n-1}^z + S_{n+1}^z) - S_n^z(S_{n-1}^y + S_{n+1}^y)]\end{aligned} \tag{11.52}$$

Small amplitude excitations means $S_n^z \approx S, S_n^{x,y} \ll S$. So

$$\frac{\mathrm{d}S_n^x}{\mathrm{d}t} = J[S_n^y 2S - S(S_{n-1}^y + S_{n+1}^y)] = JS[2S_n^y - S_{n-1}^y - S_{n+1}^y] \tag{11.53a}$$

$$\frac{\mathrm{d}S_n^y}{\mathrm{d}t} = -JS[2S_n^x - S_{n-1}^x - S_{n+1}^x] \tag{11.53b}$$

$$\frac{dS_n^z}{dt} = 0 \tag{11.53c}$$

For this set of equations, we introduce a *solution ansatz* that is plane wave in form:

$$S_n^x = uSe^{i(nka-\omega t)} \tag{11.54a}$$

$$S_n^y = vSe^{i(nka-\omega t)} \tag{11.54b}$$

The nka part of the exponent accounts for the fact that the spins are on a lattice with lattice parameter "a."

Copy the *solution ansatz* into the equations of motion and we get

$$-i\omega uSe^{i(nka-\omega t)} = vJS^2[2 - e^{-ika} - e^{+ika}]e^{i(nka-\omega t)} \tag{11.55}$$

$$-i\omega vSe^{i(nka-\omega t)} = uJS^2[2 - e^{-ika} - e^{+ika}]e^{i(nka-\omega t)} \tag{11.56}$$

$$-i\omega u = 2vJS(1 - \cos ka) \tag{11.57}$$

$$-i\omega v = 2uJS(1 - \cos ka) \tag{11.58}$$

$$\begin{pmatrix} i\omega & 2JS(1-\cos ka) \\ 2JS(1-\cos ka) & -i\omega \end{pmatrix} \begin{pmatrix} u \\ v \end{pmatrix} = 0 \tag{11.59}$$

This gives a nontrivial solution of

$$\begin{vmatrix} i\omega & 2JS(1-\cos ka) \\ 2JS(1-\cos ka) & -i\omega \end{vmatrix} = 0 \tag{11.60}$$

$$\omega = 2JS(1 - \cos ka) \tag{11.61}$$

which is known as the *magnon dispersion relation* and is the dispersion relation for Heisenberg spin waves in 1D. It describes the wave setup as in Figure 11.14.

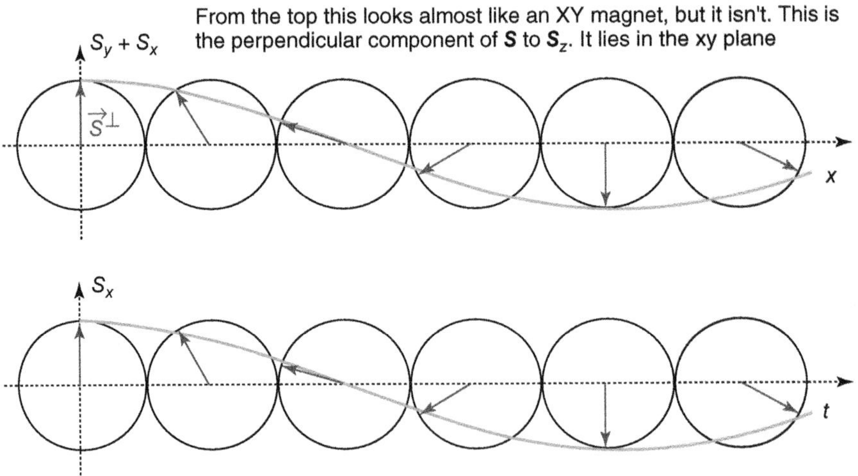

Figure 11.14 This can be drawn in many ways. But the precession of the x and y components of the spin around the circle at the top of the cones of Figure 11.13 forms a wave. The *ansatz solutions* describe this wave.

It is no accident that these solutions share a strong resemblance with the phonon modes and, indeed, they *quantize* in the same way. Imagine that boundary conditions are imposed such that the wave must fit within the lattice. Then k's become quantized as do the angles that the neighboring spins can make with each other: that is, the spin can only advance by a set phase angle from circle to circle. These quantized spin wave modes are referred to as *magnons*. They represent the low energy excitations of the Heisenberg spin magnet in any dimension, though we have only worked the 1D case here.

11.5.3.2 Thermodynamics

Since the solutions do have such a resemblance to the phonon modes, let's examine the thermodynamics of such a wave system. This will help illustrate the quantization as well as show that such waves carry energy and have an effect in the magnetization of the system. We start with internal energy:

$$E = \sum_k \hbar\omega_k \left(n_k + \frac{1}{2}\right) \tag{11.62}$$

$$U = \langle E \rangle = \sum_k \hbar\omega_k \left(\langle n_k \rangle + \frac{1}{2}\right) = E_0 + \sum_k \frac{\hbar\omega_k}{e^{\beta\hbar\omega_k} - 1} \tag{11.63}$$

In the limit $T \to 0$, only low energy magnons ($k \to 0$) are excited:

$$\omega = 2JS(1 - \cos ka) \approx JSa^2k^2 \tag{11.64}$$

with:

$$\sum_k \cdots = \frac{V}{(2\pi)^3} \int d^3k \cdots \tag{11.65}$$

$$U = E_0 + \frac{V}{(2\pi)^3} \int \frac{4\pi\hbar JSa^2k^4}{e^{\beta\hbar JSa^2k^2} - 1} dk \tag{11.66}$$

$$x = k\sqrt{\frac{hJSa^2}{k_B T}} = k\sqrt{\frac{D}{k_B T}} \tag{11.67}$$

$$dx = dk\sqrt{\frac{D}{k_B T}} \tag{11.68}$$

$$\begin{aligned} U &= E_0 + \frac{V}{(2\pi)^3} \int 4\pi x^4 \left(\frac{k_B T}{D}\right)^2 \frac{D}{e^{x^2} - 1} \sqrt{\frac{k_B T}{D}}\, dx \\ &= E_0 + \frac{V}{2\pi^2} D^{-3/2} (k_B T)^{5/2} \int x^4 \frac{dx}{e^{x^2} - 1} \end{aligned} \tag{11.69}$$

$$\int_0^\infty x^4 \frac{dx}{e^{x^2} - 1} = \frac{3\sqrt{\pi}}{8} \zeta(5/2) \approx (1.3419) \frac{3\sqrt{\pi}}{8} \tag{11.70}$$

$$\zeta(s) = \sum_{k=1}^{\infty} \frac{1}{k^s} \tag{11.71}$$

$$C_V = \left(\frac{\partial U}{\partial T}\right)_V \propto \left(\frac{k_B T}{D}\right)^{3/2} \tag{11.72}$$

But notice that the exponent is different from the phonon case. Well, we would expect this, right? The dispersion is a little different.

And for the magnetization

$$U = E_0 + \sum_k \hbar\omega(k)\langle n_k \rangle \tag{11.73}$$

keep in mind the k is a vector space (we have simplified the notation here):

$$U = E_0 + \sum_k \hbar\omega(k)\frac{1}{2S}\langle |S_k^x|^2 + |S_k^y|^2 \rangle \tag{11.74}$$

$$S_n = \sum_k S_k e^{ik\cdot r_n} \tag{11.75}$$

So

$$\frac{1}{2S}\langle |S_k^x|^2 + |S_k^y|^2 \rangle = \langle n_k \rangle \tag{11.76}$$

Classically this is the same as saying

$$E = \frac{1}{2}m\dot{x}^2 + \frac{1}{2}m\omega^2 x^2 \tag{11.77}$$

$$\langle E \rangle = \frac{1}{2}m\langle \dot{x}^2 \rangle + \frac{1}{2}m\omega^2\langle x^2 \rangle = m\omega^2\langle x^2 \rangle \leftrightarrow \hbar\omega\langle n \rangle \tag{11.78}$$

$$\langle x^2 \rangle \leftrightarrow \langle n \rangle \tag{11.79}$$

Or in words, the number of spin wave excitations in a given mode is equivalent to the average squared amplitude of the equivalent classical oscillator.

Continuing on with magnetization,

$$M(T) = \frac{g\mu_B}{V}\sum_n \langle S_n^z \rangle = \frac{g\mu_B}{V}\sum_n \langle \sqrt{S^2 - [(S_n^x)^2 + (S_n^y)^2]} \rangle \tag{11.80}$$

We look closer at

$$\sqrt{S^2 - [(S_n^x)^2 + (S_n^y)^2]} \tag{11.81}$$

for the case where

$$(S_n^x)^2 + (S_n^y)^2 \ll S^2 \tag{11.82}$$

$$T \to 0 \tag{11.83}$$

$$\begin{aligned}\sqrt{S^2 - [(S_n^x)^2 + (S_n^y)^2]} &= S\sqrt{1 - \frac{[(S_n^x)^2 + (S_n^y)^2]}{S^2}} \\ &\approx S\left\{1 - \frac{[(S_n^x)^2 + (S_n^y)^2]}{2S^2}\right\}\end{aligned} \tag{11.84}$$

$$\begin{aligned}M(T) &\approx \frac{g\mu_B}{V}\sum_n \left\langle S\left\{1 - \frac{[(S_n^x)^2 + (S_n^y)^2]}{2S^2}\right\}\right\rangle \\ &= \frac{g\mu_B}{V}\left(NS - \sum_n \left\langle \frac{[(S_n^x)^2 + (S_n^y)^2]}{2S} \right\rangle\right)\end{aligned} \tag{11.85}$$

$$M(T) = \frac{g\mu_B}{V}\left(NS - \sum_n \frac{1}{2S}\langle (S_n^x)^2 + (S_n^y)^2 \rangle\right) \tag{11.86}$$

$$M(T) = \frac{g\mu_B}{V}\left(NS - \sum_n \langle n_k \rangle\right) \tag{11.87}$$

So, the magnetization is reduced because the spin system does not align as fully when it is supporting waves. We substitute from above

$$M(T) = \frac{g\mu_B}{V}\left(NS - \frac{V}{(2\pi)^3}\int \frac{4\pi k^2 \mathrm{d}k}{e^{\beta\hbar J S a^2 k^2} - 1}\mathrm{d}k\right) \tag{11.88}$$

with

$$x = k\sqrt{\frac{hJSa^2}{k_B T}} = k\sqrt{\frac{D}{k_B T}} \tag{11.89}$$

$$\mathrm{d}x = \mathrm{d}k\sqrt{\frac{D}{k_B T}} \tag{11.90}$$

so

$$M(T) = \frac{g\mu_B}{V}\left(NS - \frac{V}{2\pi^2}\left(\frac{k_B T}{D}\right)^{3/2}\int \frac{x^2 \mathrm{d}x}{\mathrm{e}^{x^2} - 1}\right) \tag{11.91}$$

$$M(T) = M(T=0)\left(1 - \frac{V}{2\pi^2 NS}\left(\frac{k_B T}{D}\right)^{\frac{3}{2}} \frac{\sqrt{\pi}\zeta(3/2)}{4}\right) \tag{11.92}$$

The result is known as *Bloch's $T^{3/2}$ law*. This small exercise nicely illustrates a few of the more important features of the spin wave system. It should be obvious that extensions into higher dimension can be made for which dispersion branching (as seen for phonons) will also occur.

11.5.3.3 The Particle Nature of Magnons

There are a few more things you should know before we leave magnons:

1. Magnons carry an angular momentum of $1\hbar$.
2. Magnons carry a magnetic moment of $1g\mu_B$, which corresponds to a spin flip in a solid. Because of this integer spin, surprisingly, the Magnon is a boson!
3. The wavelengths for magnons are typically 0.6–2.0 nm.
4. In the very long wavelength limit $k \to 0$, $\omega \to 0$, $T \to 0$, a long-range order is stable. This is associated with spontaneous breaking of continuous symmetries in the system. (For example, the Heisenberg Hamiltonian has a continuous rotational symmetry that can be spontaneously broken depending on the dimension, *d*.)

   Such "ordered states" with $k \to 0$, $\omega \to 0$ are guaranteed in systems with broken continuous symmetries such as our classical Heisenberg spin model, and they are known as *Goldstone modes* or *Goldstone bosons*. Of course, physically, it isn't so hard to visualize what may be going on here. Imagine that the excitation has such a long wavelength that it can fit exactly ½ λ onto the lattice. At this point there is little difference between the alignment of this situation and that of the ground state we started with.

11.5.3.4 Stoner Excitations

We did mention at the beginning of the subsection that there were two ways to formulate the spin wave: one using discretely placed lattice spins and the other using the free-electron models of magnets we introduced above. Magnetic excitations in the itinerate electron systems are usually first described using *Stoner excitations*, which is an electron–hole excitation wherein the electron and the hole occupy two different, opposite spin subbands [7]. There are two configurations of such excitations: the majority hole with minority-electron state and minority hole with majority-electron state. Though we describe this as a single particle state excitation, these itinerate electron bands are correlated, and the *Stoner excitations* can form a continuum of states spread across the Brillouin zone. Herring and Kittel [8] have worked out a complete theory on how this is equivalent to a spin wave system and further yields the $T^{3/2}$ law as expected.

11.5.3.5 Coupling to the Electromagnetic Field: Magnon–Photon Coupling

The time-dependent magnetic field of the spin wave can couple with the time-dependent magnetic field of an incident electromagnetic wave. This is in direct analogy to the *polariton* in the next chapter. The interacting system can have its own dispersion curve, and it behaves "particle like" [9]. Of course, the coupling in the magnetic system is significantly weaker than that of the electric (polariton) system, and so the dynamics can have regimes of chaotic behavior. The temptation is to name such a beast a magneton to be consistent with polariton, but unfortunately we have already used this word for μ. Perhaps *magnoniton* would do? You will have to wait for our discussion of polaritons to see why this is so interesting from a general perspective.

11.6 More Complicated Situations

The models above present relatively simple and straightforward pictures of the interactions between lattice sites with an $\boldsymbol{L}$ and $\boldsymbol{S}$. However, magnetic phenomena in materials can get significantly more complicated than this. Here we will briefly mention some of those complications.

11.6.1 Double Exchange

Magnetic exchange (that J integral above) is typically simplified as much as possible to make the sums and integrals easy. But what about the case where neighboring lattice sites are not equivalent and electrons (spin carriers) are physically shared through bonding? This can certainly complicate the idea of a simple overlap of wavefunctions as the interaction integral implies.

The *double exchange mechanism* is just this type of magnetic exchange. In this mechanism, ions with different oxidation states share charge. And when they do, the spin from the charge of the "donor" ion effects the spins of the charges in the "acceptor" ion [10]. The ease or difficulty that an electron has in exchanging between the two ionic species, together with how much energy is gained or lost

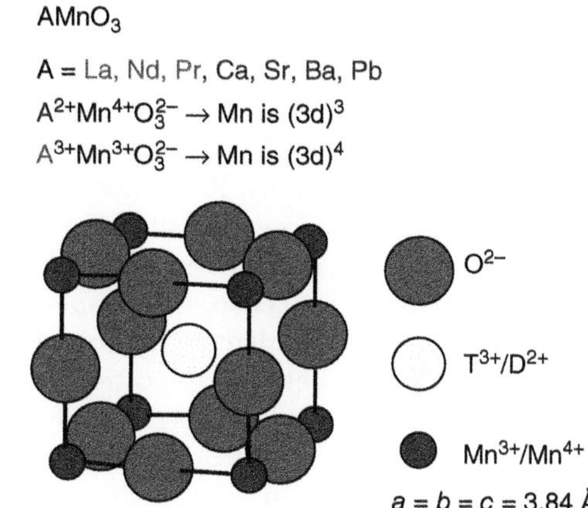

Figure 11.15 A common double exchange example $AMnO_3$ where A is used to create different valencies in the Mn.

in the new spin configuration, determines (for the most part) whether *double exchange* materials are ferromagnetic or antiferromagnetic.

A really commonly found example for double exchange is the manganates: $AMnO_3$. The typical structure and the Mn–O–Mn double exchange units are shown in Figure 11.15. To make this work, two different oxidation states, or outer valencies in a more general sense, must find themselves pairing up through the oxygen (in this case). So exactly what *A* is in the structure is quite important. Here in Figure 11.15, the manganate would need to place unit cells of the T^{3+}/D^{2+}, as shown, in proximity.

In the $AMnO_3$ example, the Mn e_g orbitals overlap with the O 2p orbitals, and one of the Mn ions has more electrons than the other. In the ground state, electrons on each Mn ion are aligned according to Hund's rules as seen in Figure 11.16. In the double exchange mechanism, the O gives a spin-up electron to Mn^{+4}, leaving behind a vacant orbital. This orbital is then filled by an electron from Mn^{+3}. So, the electron hops between neighboring metal ions, and, importantly, it retains its spin. Clearly, the electron movement between ions is facilitated if the electrons on the "accepting ion" do not have to change spin direction in order to conform with Hund's rules when the "visiting" electron arrives. In the case where the new spin configuration reduces the overall energy on the acceptor site, this energy savings can lead to ferromagnetic alignment of neighboring ions.

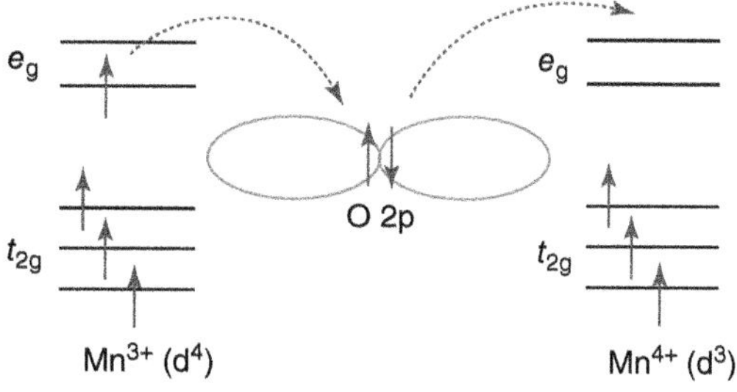

Figure 11.16 The double exchange mechanisms provide for the transfer of spin between the two ionic species. The hopping process back and forth, together with the energy gain from Hund's rules when a stable spin configuration is achieved, allows for the formation of ferromagnetic order.

11.6.2 Super Exchange

The *Kramers–Anderson superexchange* (or just *super exchange*) "looks" similar to double exchange in that it involves the sharing of electrons. But looks can be deceiving. Double exchange occurs when the ions have different numbers of valence electrons generally. Remember the electrons must hop back and forth between "donor" and "acceptor" sites. *Superexchange* occurs when nearest neighbor cations have the same valency, but they are linked together by a nonmagnetic intermediary anion (also, strictly speaking, the double exchange mechanism doesn't need an intermediary). The job of the intermediary is to maintain the spins that are shared between the cations and enforce spin selection rules across the superexchange "unit." This typically leads to a strong ferromagnetic or, more commonly, antiferromagnetic coupling and system alignment.

We show the common superexchange picture using the common example MnO in Figure 11.17. From this diagram, and the naïve model we have presented so far, one might guess that the correlation of spin projection in space and the overall symmetry of the bonding configuration are important for superexchange mechanisms. That is, if the spin is to be maintained through the nonmagnetic intermediary, this is clearly easier if the bonding is collinear as shown but may be harder if the bonds are at 90°, for instance. And indeed, this is the case. In fact, spin configurations in superexchange systems are quite sensitive to direct exchange mechanisms competing with superexchange in the system, variations in the cation–anion–cation bonding angles, and circumstances when spin–orbit coupling becomes large.

A set of semiempirical rules called the *Goodenough–Kanamori rules* [11] are typically used in providing insight into the qualitative magnetic properties for a wide range of materials. They use orbital symmetry and electron occupancy of the overlapping orbitals to predict spin selection and ordering. A full quantum superexchange theory generally assumes localized orbitals (as opposed to free-electron bands) and Hubbard-like transfer of electrons (with hopping integrals and Hubbard energies). Thus, the Pauli exclusion principle dominates spin–spin interactions.

11.6.3 RKKY

You might recall that early on we eliminated the nuclear spin from our considerations. However, the nuclei of atoms can certainly have spin and a magnetization. This is the basic consideration of the RKKY models. More specifically, the RKKY

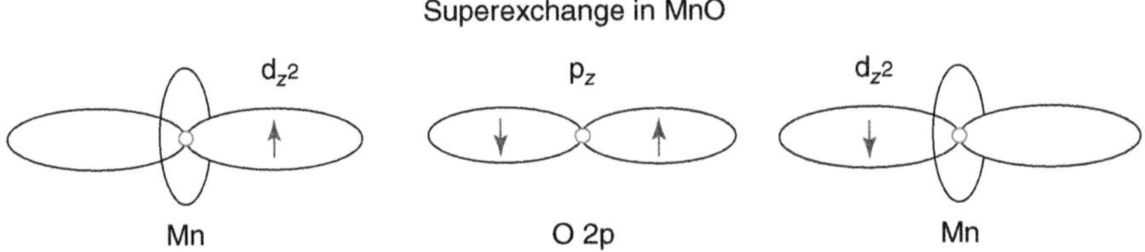

Figure 11.17 *Superexchange* in the MnO system.

interaction is the $J/t \gg 1$ limit of double exchange (t is that Hubbard hopping integral if you remember) and is suitable for nuclear spin interaction as well as the interaction of inner d- and f-shell electrons in a metal.

In the simplest terms the Hamiltonian of the system includes a spin–isospin term referred to hyperfine coupling:

$$H = H_{\mathrm{el}} + \sum_i A\mathbf{S}_i^{\mathrm{el}} \cdot \mathbf{I}_i^{\mathrm{nucl}} \tag{11.93}$$

The H_{el} term is the electron Hamiltonian for the system.

A particularly important success of RKKY theory has been to explain giant magnetoresistance (GMR). GMR is the coupling between thin layers of magnetic materials separated by a nonmagnetic spacer material. For such systems, magnetization will oscillate between ferromagnetic and antiferromagnetic as a function of the distance between the layers, as predicted by RKKY [12].

11.7 Time Reversal Symmetry

If you have been following along closely, you may have noticed an unusual problem with our descriptions of magnetic materials to this point. If we consider a spin vector (*J*), it has an associated current with it, as we have explained. Of course, the right-hand rule relates the current and spin directions. Now imagine the rules and equations for this system running backward in time (t goes to $-t$). Or in quantum terms, we apply the time reversal operator to the systems' wavefunction. We like for the basic rules of our universe (those not involved with entropy) to run the same backward and forward in time: think Newton's laws. The result here is to reverse the direction of the flow of the current. But this operation would require the reversal of the spin since it aligns or anti-aligns with that field. Spin is not a time-dependent part of the dynamics, and a reflection is required to make the dynamics of t and $-t$ look like time-reversed versions of each other. Said colloquially, $-t$ is not the backward running version of t for this system! In a solid, consider, as an example, ferromagnetic ordering. When the ferromagnetic phase sets in (let's say after a phase transition), that ferromagnetic crystal has a definite spin direction and not the reverse one. This selection of a spin projection itself breaks *time reversal symmetry*.

In this context of condensed matter physics, *time reversal symmetry breaking* (TRSB) has generally become associated with something in the system behaving like a magnetic field. Consider cases like Sr_2RuO_4 and UPt_3. Both have spin triplet, rather than spin singlet, Cooper pairs. The superconducting *order parameter* in Sr_2RuO_4 is a complex function given by

$$\Psi = |\Psi_0(\theta)| e^{-i\phi(\theta)} \tag{11.94}$$

$|\Psi_0|$ is the amplitude and ϕ is the phase of this order parameter. Both of these parameters can generally vary around the Fermi surface, which is parameterized by a θ that denotes where you are on the Fermi surface. The superconducting order parameter of Sr_2RuO_4 in Figure 11.18 is widely believed to have $p_x \pm ip_y$

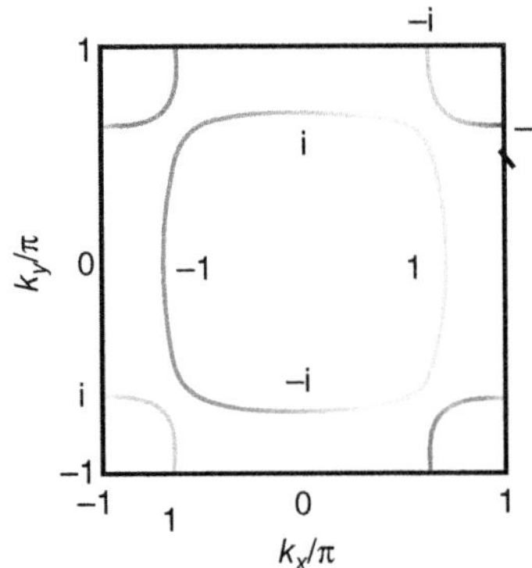

Figure 11.18 Two of the Fermi surfaces of strontium ruthenate [11], showing the continuous evolution of the phase around the Fermi surface, as $\theta = 0, \pi/2, \pi, 3\pi/2$.

symmetry. This corresponds to a phase that continuously winds around the Fermi surface.

Any gradient in the phase corresponds to a current for a superconductor. This is because the phase difference appears like a translation of the carriers. If this gradient is built into the order parameter itself, the material will naturally and in the absence of any external field feature a persistent and circulating current in the superconducting state. As we know, *current* → *internal magnetic field* → *TRSB*. So, when these materials enter the superconducting state, something like a *magnetization* appears, and this breaks time reversal symmetry for the spin pairs. These edge currents have yet to be directly observed for these systems, likely due to domain structure. But the magnetizations have been observed using the polar Kerr effect and muon spin relaxation, and the onset of the Kerr rotation corresponds directly to T_c.

So! You might say: what does it matter if the dynamics of these Cooper pairs is not time reversible? Perhaps the most interesting consequence of time reversal symmetry in such systems, that is, from ferromagnets to superconductors, is the appearance of a *Goldstone mode* that has zero energy at the zero momentum transfer in the spin wave spectrum.

11.8 Summary

This chapter has presented a "curio cabinet" of magnetic models and phenomena. We have not developed all of these concepts because some are quite specific and the physicist goes into the details when these specifics become relevant to a given application (there are many exceptions to any given approach). However, it is quite important to understand how they are related to each other and the deeper message that they bring to our conversation. And that message is that spin systems unlock the power of condensed matter systems to illustrate the deeper meanings of the fundamental symmetries in the universe and their relationship to dimension and scale.

Exploring Concepts

1 *A Little Experimental Fun:* Imagine you have two physically identical metal bars. One has a permanent magnetic moment, and one is a soft magnet that has a magnetic response but does not have a permanent moment. You work in a poorly funded lab, and while this lab is completely shielded from external magnetic fields (you cannot sense the Earth's magnetic field), you have access only to a single piece of string. Can you tell which is the permanent magnet and which is not? Explain. If you are having a little trouble with this one, we can help you out by giving you one more tool. What other simple tool that is not a magnetometer or the like might you be able to use if you could just run down to the supply room?

2 *Be Careful with Models of Spinning Balls of Charge:* Many of the models have taken great liberty in visualizing the magnetic moment of constituents as related to the swirling and spinning of electric charge at the atomic level. There is questionable wisdom in this. For instance, from this we might conclude that the magnetic responses of ^{12}C and ^{13}C are the same since they differ only by the addition of a charge-less neutron in the nucleus. But while we know that ^{12}C has no magnetic moment,^{13}C certainly does, making it useful in ^{13}C dating. In fact the uncharged neutron possesses a magnetic moment of −1.913 nuclear magnetons. Do a little digging and see why this is.

3 *Thermodynamics and Magnets:* We have seen so-called "two-state systems" before in this text. Here in magnetism, we have a very natural set of circumstances to give rise to such systems. Imagine naively that we have an atom with a large nuclear magnetic moment. The outer shell of this atom has a spin aligned with this nuclear field, and a spin anti-aligned, causing a subtle and small energy splitting – a two-state system. These *hyperfine interactions* between nuclear and electronic spins in magnetically ordered atoms and paramagnetic salts can have important effects on the thermodynamics of the system.

(a) Show that for such a two-state system, the heat capacity can be written as

$$C = \left(\frac{\partial U}{\partial T}\right)_\Delta = k_B \frac{(\Delta/T)^2 e^{\Delta/T}}{(1 + e^{\Delta/T})^2}$$

where Δ is related to the splitting energy:

$$k_B\Delta$$

(b) Examine this function for $T \gg \Delta$ and $T \ll \Delta$; explain the trends. Show for the former

$$C \cong k_B(\Delta/2T)^2 + \cdots$$

which is the experimental sign that a two-state system might be present. The peaks that occur in such plots are known as *Schottky anomalies.*

4 *Conduction Electrons:* Charles Kittel, in his now famous book, points out that the susceptibility of an electron gas in the conduction band of a system at absolute zero can be thought of in terms of its energy minimization. This clever approach goes something like this.
Let the spin-up N$^+$ and spin-down N$^-$ concentrations in this band at $T = 0$ be written as

$$N^+ = \frac{1}{2}N(1+\zeta);\;\; N^- = \frac{1}{2}N(1-\zeta)$$

where N is the total concentration of electrons and ζ is some slight deviation away from half and half.

(a) Show then that in an applied magnetic field B,

$$E^+ = E_0(1+\zeta)^{5/3} - \frac{1}{2}N\mu B(1+\zeta)$$

where

$$E_0 = \frac{3}{10}N\varepsilon_F$$

and similarly for E^-.

(b) Now here comes the tricky part. *Minimize* the total energy $E = \mathrm{E}^+ + \mathrm{E}^-$, with respect to ζ, to show that there is a nonzero equilibrium value in the applied field.

(c) Show that the magnetization is then

$$M = \frac{3}{2}\frac{N\mu^2 B}{\varepsilon_F}$$

(d) Now let's approximate electron–electron spin interactions like this. Assume that electrons with a parallel spin, in the population, interact with an energy $-V$ (V is a positive number). Further assume that electrons with antiparallel spin alignment do not interact at all. Show that the energy of the spin-up band (this population has a different and separate energy, so it has its own band now) is

$$E^+ = E_0(1+\zeta)^{5/3} - \frac{1}{8}VN^2(1+\zeta)^2 - \frac{1}{2}N\mu B(1+\zeta)$$

and similarly for E^-.

(e) From this result minimize the total energy again with respect to ζ, as you did before to show that

$$M = \frac{3N\mu^2 B}{2\varepsilon_F - \frac{3}{2}VN}$$

in the limit of $\zeta \ll 1$. This is an enhancement of the susceptibility by the interaction.

(f) Now allow $B \to 0$. Show that for

$$V > 4\varepsilon_F/3N$$

the total energy is unstable if $\zeta = 0$. As Kittel points out, this means that the ferromagnetic state $\zeta \neq 0$ has a lower system energy than the paramagnetic state, and the V value is known as the *Stoner condition*. But this is reasoned for the $T = 0$ condition. How would having a small but nonzero temperature change such a result? Naturally, you may want to break your answer into two regions: $k_B T < V$ and $k_B T > V$.

5 *Magnons in Different Dimension:* As we have suggested throughout this text, the dimensionality of a material – *the physical dimension of a material object* – will influence in that material's intrinsic properties. For properties that depend on highly directional internal interactions, such as spins in a solid, this is particularly true. Consider for a moment the magnetic excitations we have just been studying. They can occur in a lattice of any dimension d. If they have a dispersion relation of the form $\omega_k = ak^n$, where n is any number and "a" is a simple prefactor, then show that the specific heat of the solid can be written as $\sim T^{d/n}$. Explain why this is.

6 *Rough Estimates:* Consider a single component ferromagnet. Let's say that the crystal structure is FCC and it has a lattice constant of 0.35 nm. Further we will suppose that each atom contributes an orbital moment of $2\mu_B$.
(a) Using any method above you choose, estimate the magnetization of this system?
(b) Relate this magnetization to a surface current density.

7 *Frames of Reference:* This problem will require you to remember a little of your special relativity. Recall that we have previously introduced the Lorentz force in the Hall effect. However, let's now consider an electron with a spin moving under the influence of an applied electric field **E**. Consider the situation in which the electro travels ballistically and the spin is initially aligned with the direction of motion.
(a) Show that the magnetic field seen by the electron in the frame in which the electron is not moving is given by

$$\mathbf{B}^* = -(1/c^2)v \times \mathbf{E}$$

(b) Show that this leads to a nutation of the electron spin about the **E** axis. This is known as the *Rashba effect*.
(c) While this is true for an electron in a solid or in an electron beam, it does have implications for spin orientation of a flowing current in a solid. What are these implications?

References

1 For examples Gersten, J.I. and Smith, F.W. (2001). *The Physics and Chemistry of Materials*, 1e. Wiley-Interscience.
2 Coey, J.M.D. (2009). *Magnetism and Magnetic Materials*. Cambridge: Cambridge University Press.
3 Ising, E. (1924). Beitrag zur Theorie des Ferro- und Paramagnetismus. Dissertation. Hamburg.

4 See Baierlein, R. (1983). *Newtonian Dynamics*. McGraw-Hill College.
5 Mermin, N.D. and Wagner, H. (1966). Absence of ferromagnetism or antiferromagnetism in one- or two-dimensional isotropic Heisenberg models. *Phys. Rev. Lett.* 17: 1133.
6 A fascinating read of this work is presented by MacKay, D.J.C. (2003). *Information Theory, Inference and Learning Algorithms*. Cambridge University Press. ISBN: 0521642981.
7 Stoner, E.C. (1938). Collective electron ferromagnetism. *Proc. R. Soc. London, Ser. A* 165: 372.
8 Herring, C. and Kittel, C. (1951). On the theory of spin waves in ferromagnetic media. *Phys. Rev.* 81: 869.
9 Kusminskiy, S.V., Tang, H.X., and Marquardt, F. (2016). Coupled spin-light dynamics in cavity optomagnonics. *Phys. Rev. A* 94: 033821.
10 Zener, C. (1951). Interaction between the d-shells in the transition metals. II. Ferromagnetic compounds of manganese with perovskite structure. *Phys. Rev.* 82: 403.
11 (a) Goodenough, J.B. (1955). Theory of the role of covalence in the perovskite-type manganites [La, M(II)]MnO_3. *Phys. Rev.* 100: 564. (b) Goodenough, J.B. (1958). An interpretation of the magnetic properties of the perovskite-type mixed crystals $La_{1-x}Sr_xCoO_{3-\lambda}$. *J. Phys. Chem. Solids* 6: 287. (c) Kanamori, J. (1959). Superexchange interaction and symmetry properties of electron orbitals. *J. Phys. Chem. Solids* 10: 87.
12 (a) Parkin, S.S.P. and Mauri, D. (1991). Spin engineering: direct determination of the Ruderman-Kittel-Kasuya-Yosida far-field range function in ruthenium. *Phys. Rev. B* 44 (13): 7131. (b) Yafet, Y. (1987). Ruderman-Kittel-Kasuya-Yosida range function of a one-dimensional free-electron gas. *Phys. Rev. B* 36 (7): 3948.

12 Polarization of Materials

When we presented *electronic transport in solids*, discussion centered on degenerately doped semiconductors and metals. "Well of course" the astute reader may cry. "After all, it is pointless to study fields placed across insulators where nothing will happen!" But in actual fact something generally does happen: *polarization* of the crystal. And this happens not only in fully insulating, covalently, or ionically bonded solids but also in some semiconductors. Indeed, in any solid for which a local electric field, $\mathbf{E}_{\text{loc}}$, at the atomic/molecular scale can be established, some level of polarization can result.

Clearly, where there are many itinerate electrons free to move around, establishing a local field is difficult. Most fields internal to the solid will be canceled by the motion and accumulation of macroscopic charge distributions. Therefore not much polarizes in things like metals. But in insulators and semiconductors with a low carrier density, polarization can be large and very useful.

Polarization is literally the rearrangement of local bound charge to form a *multipole* field in response to the applied local field. It is typically characterized by a response field: $\mathbf{p}$ microscopically and $\mathbf{P}$ macroscopically. Because opposite charges attract, the response field that is established typically opposes the local applied field, thereby reducing the net total electric field in the solid by some amount.

12.1 Simple Atomic Models

So how do we get $\mathbf{P}$ from $\mathbf{p}$, and how do we get $\mathbf{p}$ to begin with? In the case of an insulator, an electric field placed across the volume will establish a force that rearranges charge distributions at the atomic or molecular level in a solid. This can happen in many ways:

1. Electrons are perturbed from their orbits around ionic cores to make a asymmetric charge distribution.
2. Polar molecular components of the lattice rotate their orientation to anti-align with the applied field.
3. In ionic lattices (like alkali halides or salts), the ions can displace to form dipole-like fields that anti-align to the applied field.
4. Charge can accumulate at interfaces in materials: *interface polarization* $\mathbf{P}_{\text{int}}$.

Foundations of Solid State Physics: Dimensionality and Symmetry, First Edition.
Siegmar Roth and David Carroll.

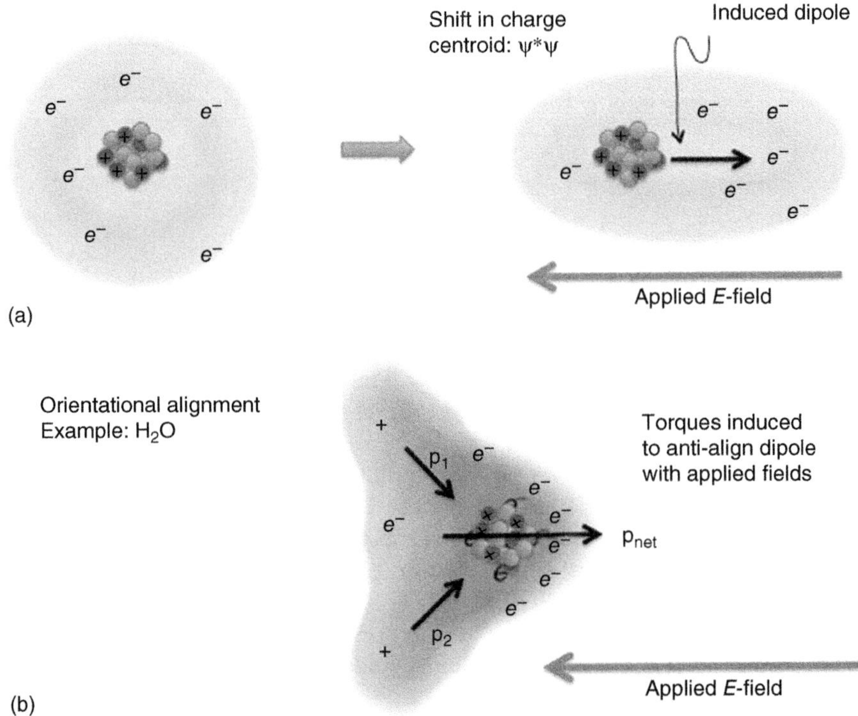

Figure 12.1 Schematic representations of the induced dipole (a) and the permanent dipole that is reoriented in the applied field (b).

Actually, (2) and (3) are similar ideas. So we could break "mechanisms" into *induced polarization* at the atomic/molecular level and ionic *displacement mechanisms* within the lattice. Indeed, polarization is likely to be a combination of all of these effects (or some variants). The two we will consider as dominant are shown in Figure 12.1.

12.1.1 Linearity in the Response

A typical simplification is a *dipole-like* nature local response field to the local applied field. The actual structure of the field near the atoms or molecules is obviously more complicated [1]. However, in a multipole expansion of local response fields, charge neutrality insists the monopole term is zero, so the dipole term tends to dominate field structure.

As the dipolar field structure is a simplification, so too is the strength of the response to an applied field. We can expand this in a power series:

$$|p| \sim \sum_{i=0}^{\infty} a_i |E|^i \tag{12.1}$$

The proportionality shown is written with the total resultant macro-field **E**, not the field that established **p**. Recall that **D** is only *like* this establishing field (displacement field) [2]. $\mathbf{E}_{\text{loc}}$ can also be used in the expression, but this atomically local field is a composite of applied fields and resulting charge shifts. It is not easy to know. However, as vector fields, they are related additively. For small "fields" (**D**) in many materials, **p** is nearly linear with the applied local fields ($\mathbf{E}_{\text{loc}}$) and thus the total resultant field (**E**). We have cleverly named such a medium a *linear* medium.

Thus, we can write **p** in a linear response framework: $\mathbf{p} = \alpha\mathbf{E}$. α is the *atomic polarizability*, and it is exactly the propensity of each of the atomic building blocks to create a local dipole **p** in the field **E**. Note that **E** in the equation represents the total resultant field, which we will claim is related to the local field $\mathbf{E}_{\text{loc}}$. $\mathbf{E}_{\text{loc}}$ is shown in Figure 12.1:

$$\mathbf{P} = \sum_{i=1}^{N} \mathbf{p}_i = \sum_{i=1}^{N} \alpha_i \mathbf{E}_{\text{loc},i}$$

$\mathbf{E}_{\text{loc},\,i}$ is the local field at the ith lattice position.
$\mathbf{E}_{\text{loc}}$ is related to E (the total resultant macro-field):

$$\mathbf{E}_{\text{loc}} = \mathbf{E} + \frac{\mathbf{P}}{3\varepsilon_0}$$

is the Lorentz relation (from crystals of cubic symmetry)

Thus,

$$\frac{\chi}{\chi + 3} = \frac{\varepsilon - 1}{\varepsilon + 2} = \frac{1}{3\varepsilon_0} \sum_{i=1}^{N} \alpha_i$$

which are the now famous Clausius–Mossotti relations.

$$\chi_{\text{e}} = \frac{N\alpha/\varepsilon_0}{1 - N\alpha/3\varepsilon_0}$$

$$\alpha = \frac{3\varepsilon_0}{N}\left(\frac{\varepsilon_{\text{r}} - 1}{\varepsilon_{\text{r}} + 2}\right)$$

Or these can be written a little differently so as to highlight the proportionality constants. (We have now worked the sum for identical lattice sites, changed to ε_{r} from ε, and added the subscript e to χ to remind us that this is atomic.)

Naturally, we imagine that all the small dipole fields associated with each differential unit of volume (or unit cell) of the material add together under the anti-aligning influence of some applied field. This results in the macroscopic material's response to the said field as is seen in Figure 12.2. This simple polarization model suggests that if we have a dipole density of, say, N dipoles per unit volume with a dipole strength of p, then we would observe a polarization per unit volume of approximately $\mathbf{P} = N$, $\mathbf{p} = \alpha N\mathbf{E}$. So we might assume that α and χ_{e} are related as $\chi_{\text{e}} = N\alpha/\varepsilon_0$. For a very diffuse N, this is a pretty good approximation in fact. Unfortunately, we have a self-consistency problem as N becomes denser. The equation we used above to state the linearity of **p** with **E** fails to illuminate

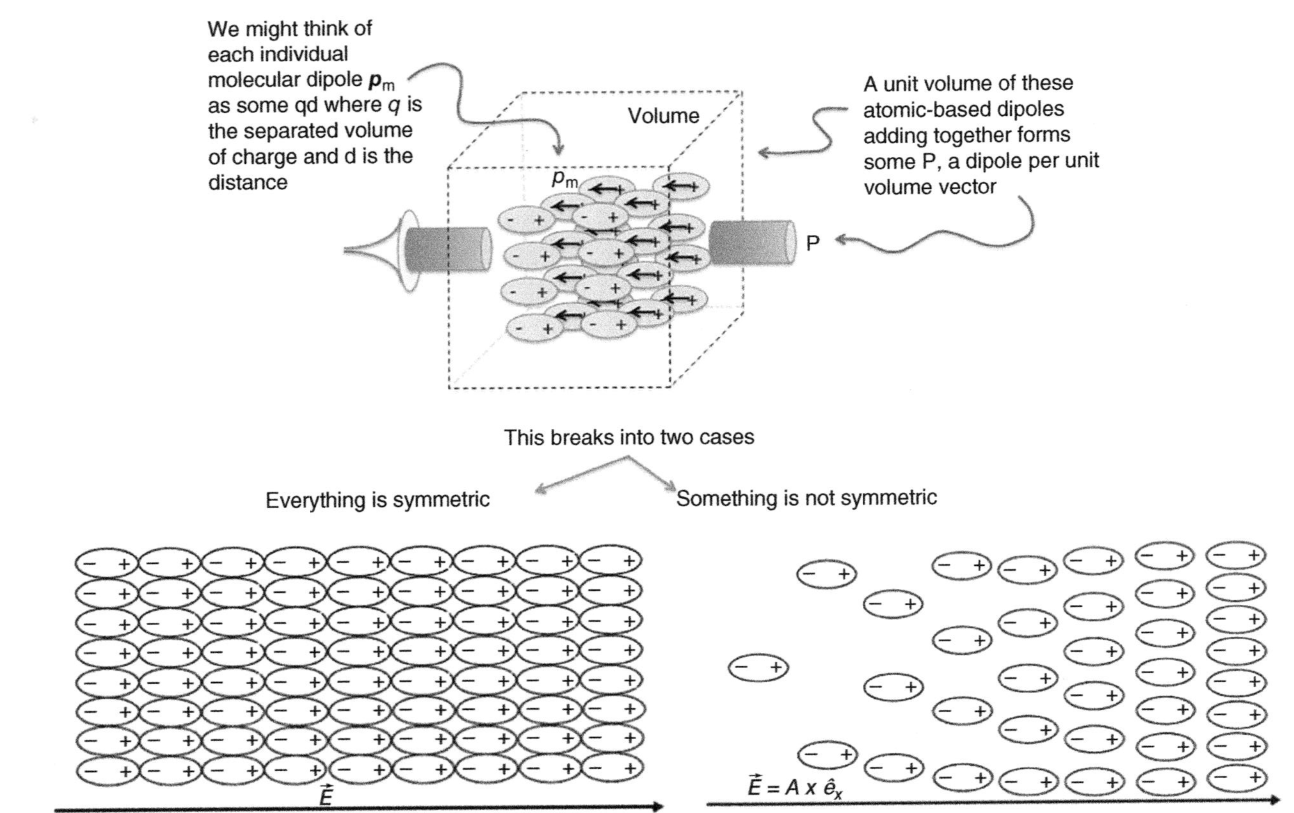

Figure 12.2 Regardless of how the dipoles come to be formed, when they align, they will be subsequently randomized by thermal energy from phonons. Further their alignment will either be symmetric wherein all the dipole density is constant and the applied field is a constant, or they will align with gradients in density or field, leading to the so-called *bound charges*.

the effects of the dipole that we are polarizing on that effective field. In other words, when we polarize some atomic element within the volume, the field from that element adds to the fields of the other elements. This changes the **E** that made our polarized element to begin with. So everything must be allowed to relax into some lowest energy state in a self-consistent way. This is actually derived in nearly every text on electromagnetism that exists. So we will simply state the result here. These are known as the *Clausius–Mossotti* equations.

This combined response quantity of all the **p**'s goes into the makeup of the macroscopic ε, as does in the case of some semimetals and semiconductors, any itinerate charge[1] or bound charge associated with any material's polarization. Let's just remember that bound charge distributions in polarized media are typically associated with surfaces wherein the dipoles terminate, leaving an unbalanced charge exposed, and with gradients in the local dipole density. Recall that

$$\sigma_b = \mathbf{P} \cdot n \text{ (surface bound charge density)} \tag{12.2}$$

$$\rho_b = -\nabla \cdot \mathbf{P} \text{ (volume bound charge density)} \tag{12.3}$$

Importantly, such bound charge densities can occur near buried interfaces, grain boundaries, and other inhomogeneities within the solid. Because of nonstoichiometric local environments at such grain boundaries and interfaces, induced polarization can take place there as well. These two additional mechanisms were mentioned above (number 4 in our list) but can produce spurious results when simple methods are used to determine ε for a given material.

12.1.2 Relating the Fields

It is common to see the macro-response field **P**, written in terms of the material-based bound macro-charge distributions and related to the resultant total field in the medium (see Figure 12.3). Using our *auxiliary field* (**D**), we write the standard expression from E&M relating **P** and **E**:

$$\mathbf{D} = \varepsilon_0 \mathbf{E} + \mathbf{P} \tag{12.4}$$

where **P** is generally derived from microscopic details of local fields within the solid as above. Again, this is general for dielectrics. We must have some knowledge of the type of material (semiconductor, metal, or insulator) to construct and solve Maxwell's equations for a specific problem. This idea is shown in Figure 12.3. Notice that we have included the bound and free charge densities $\sigma_{\text{bound/free}}$ and $\rho_{\text{bound/free}}$ as well as the polarization field **P** defined only up to the boundaries of the material in typical E&M style.

Classically, **P** is defined in a macroscopic way and represents a macro-response to an unspecified applied field. When we come to microscopic derivations of **P**, life is not so simple since the way in which charge in the microenvironment is paired can leave the precise definition of **P** slightly ambiguous [3]. That is why classical approaches tend to treat the solid's properties in terms of phenomenological parameters: ε, μ, and σ.

1 *Itinerate* charge is *mobile* charge within the unfilled band of a solid.

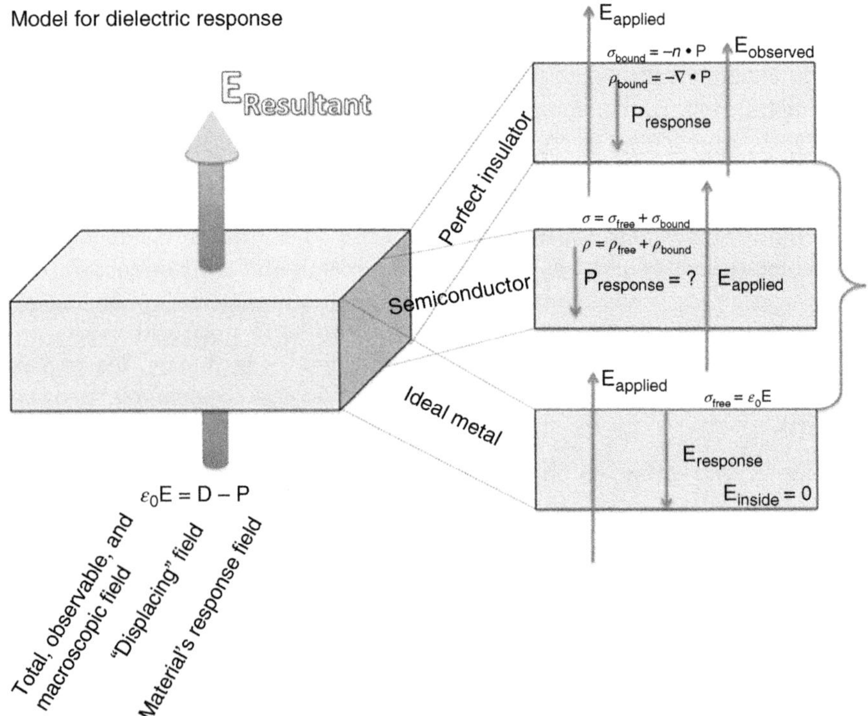

Figure 12.3 The basic breakdown of static dielectric response for different materials from the insulator to a metal is shown. Here we have used Maxwell's term "displacing field" for *D*, but this is slightly inaccurate since *D* doesn't represent a "real field." It is only *like* the generating field. Remember also that *P* is not really found in the metal (where the response is NOT from microscopic dipoles) and is only defined internally for the insulator.

Then, $\mathbf{P} = \varepsilon_0 \chi_e \mathbf{E}$ for linear dielectric response where χ_e is what is known as the *susceptibility*. Using the expression above,

$$\mathbf{D} = \varepsilon_0\mathbf{E} + \mathbf{P} = \varepsilon_0\mathbf{E} + \varepsilon_0\chi_e\mathbf{E} = \varepsilon_0(1 + \chi_e)\mathbf{E} \tag{12.5}$$

$$\varepsilon \equiv \varepsilon_0(1 + \chi_e) \tag{12.6}$$

$$\varepsilon_r \equiv (1 + \chi_e) = \varepsilon/\varepsilon_0 \tag{12.7}$$

$$\mathbf{D} = \varepsilon\mathbf{E} \tag{12.8}$$

ε_r (or sometimes κ) is known as the *relative permittivity*, and it can be used as a comparative standard.

It is plain to see that the density of dipoles and the dipole strength will play an important role in determining the macro-field associated with polarization response. Remember too that the $|\mathbf{p}| = qx$ where q is the displaced charge and x is the actual displacement. So when we run across a table like that of Table 12.1, it can be placed into context.

Curiously, note the extraordinarily large value found for the perovskite $BaTiO_3$. This is actually not that uncommon for perovskites (ABX_3) and is found

Table 12.1 The relative permittivities for several substances.

Material	k
Air	~1
Polystyrene	2
Polyester	4
Mica	5
Alumina	9
Polyvinylidene fluoride (PVDF)	10
Inorganic glasses	6–20
Distilled water	80
Rutile	80–170
Barium titanate	7000

in $CaTiO_3$, $PbTiO_3$, $PbZrO_3$, etc. Such materials are of technological importance not only for their dielectric response, but also many of them have *piezoelectric* and *pyroelectric* properties. Why should this be? If the reader refers back to the structure of this system, you will recall that the smallest cations (Ti^{4+}) occupy the octahedral interstices. The other cations (such as Ba^{2+}) occupy the dodecahedral interstices. This yields two situations in the crystal: (i) a high density of dipoles and (ii) the cations are able to move around in their interstices more freely than normal, so the dipoles can be larger.

12.2 Temperature Dependence

There are, of course, some caveats associated with this anti-alignment when the atomic or molecular sites are already polar (as in the second example above). This is due to temperature and randomization. Specifically the dipoles are not perfectly aligned but are distributed about the **E** axis as a Langevin function, as we might expect in a heat bath.

We can detail this point a little further as below. Note that it follows essentially the same route as it would if this were spins sitting in a magnetic field, so it is a good idea to get a good picture in your head as you will use it again. Quite simply the internal energy of an ensemble of these dipoles is largest when they are all aligned and smallest when they are all anti-aligned. For a single dipole this is expressed as the dot product shown. But each of the dipoles in the ensemble will "feel" different perturbations from the flow of phonons through the system, transferring heat. The thermal average will be the probability of finding a **p** pointing antiparallel. We thus use this probability to estimate the total energy of the system at temperature. This is not entirely accurate since we have not accounted for the effects of the dipole field itself. But it is a useful estimate.

Interaction energy: $U = -\mathbf{p} \cdot \mathbf{E}_{loc}$

The probability of finding a dipole $p(\theta)$ oriented at angle θ to the field is given by Boltzmann:

$$p(\theta) \sim e^{-U/k_B T} = e^{[(pE_{loc}\cos\theta)/k_B T]}$$

We have ignored nearest neighbor interactions in this equation. Technically, this is only for very diffuse dipoles like in a gas. But it will give us a little guidance anyway.

The average component of the dipole moment parallel to the field:

$$\mathbf{p}_{\parallel} = < p\cos\theta >$$

The thermal average:

$$= p\frac{\int_{-\pi/2}^{\pi/2}\cos\theta\exp[(pE_{loc}\cos\theta)/k_B T]\,2\pi\,\sin\theta\,d\theta}{\int_{-\pi/2}^{\pi/2}\exp[(pE_{loc}\cos\theta)/k_B T]2\pi\,\sin\theta\,d\theta}$$

$$= p(\coth x - 1/x) = pL(x)$$

$L(x)$ is the Langevin function.

12.3 Time Dependence: $\varepsilon(\omega)$

Our reasoning so far has been based on a static field approximation. Within the solid, the field is applied, and the dipoles respond adiabatically. However, it is surely more reasonable to suggest that it takes some time for the dipole to respond in the instant that a field is introduced. Likewise it is reasonable to assume that if the field is quickly removed, there exists some restoring force that allows the system to relax back into its nonpolarized form. This time-varying field is exactly the kind of situation we find ourselves in when an electromagnetic wave propagates through the solid or if we drive the material with a high frequency AC field. The electric field drives a polarization in one direction and then in the other, going through zero field as it completes a driving cycle. If the **p** is the result of a driven oscillator, then $\varepsilon = \varepsilon(\omega)$ since the response of a driven oscillator typically depends on the natural frequencies of the system and the driving frequency.

There is a famous and rather simple model for this polarization cycle, taken from classical mechanics.[2] It is based on the driven harmonic oscillator. The electric field of the E&M wave is the driver: $\sim qE_0 e^{i\omega t}$. The displaced charge has a restoring force linear with displacement magnitude (stretched bonds as springs, for instance). When the applied field goes through zero, the electrons will ring back and forth about the ionic core (it too may move depending on the situation). If we turn the field off altogether, we suspect that the polarization will return to a

2 Note that this model is written and described in terms of induced polarization. There are equivalent ways of treating the rotation of molecules that are already polar.

$\mathbf{P}_{net} = 0$ state eventually. To get our model to do this, we add in a damping term: $-m\gamma\ \mathrm{d}x/\mathrm{d}t$.

The *micro*-dipole goes as $|\mathbf{p}| = qx(t)$, and it is the $x(t)$ that we seek from a dynamics point of view. So we do a little Newtonian manipulation as seen in Figure 12.6. Physically the damping term can be ascribed to different mechanisms from radiation damping to the small effects of the magnetic field (Lorentz force). The charge and distance components are related to whichever mechanism we are discussing: induced polarization, ionic motion, etc.

Starting with a polarized driver,[3] we write the force and the solution to the damped harmonic oscillator equation (Figure 12.4) in complex notation

$$\tilde{x}(t) = \tilde{x}_0 \mathrm{e}^{-i\omega t} \tag{12.9}$$

The displacement is complex, and its real part expresses the real displacement of charge. If we substitute this solution into our classical, driven, simple harmonic oscillator differential equation,

$$\tilde{x}_0 = \frac{q/m}{\omega_0^2 - \omega^2 - i\gamma\omega} E_0 \tag{12.10}$$

$$\tilde{p}(t) = q\tilde{x}(t) = \frac{q^2/m}{\omega_0^2 - \omega^2 - i\gamma\omega} E_0 \mathrm{e}^{-i\omega t} \tag{12.11}$$

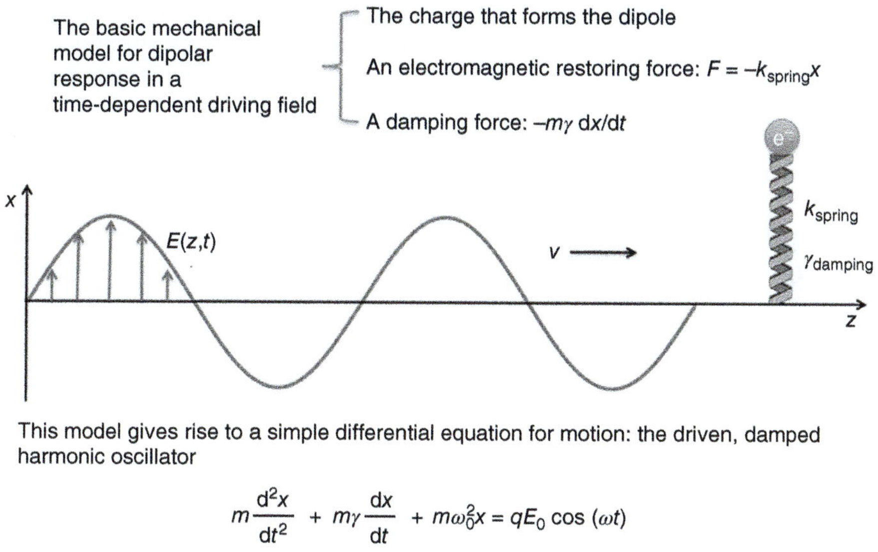

Figure 12.4 The mechanical model for a driven dipolar response in materials along with the resulting differential equation of motion it produces. This old and familiar differential equation relates the driving frequency ω to the response frequency (natural resonance) of the system ω_0.

3 Polarized in the sense of the direction of the electric field. We use the same word for this and the polarization of the dipoles unfortunately.

This introduces a natural response frequency into the system: ω_0. It is the frequency that **p** "rings" with when there is no driving force.[4] The driving force of the applied field has a frequency of ω. Since any material might be made up of more than one type of bond or more than one type of atomic site, we may safely assume that there will be some number of these resonances indexed here by j. f_j has been introduced as a weighting factor to account for the proportion of the total that is represented by each resonance:

$$\tilde{\mathbf{P}} = \frac{Nq^2}{m}\left(\sum_j \frac{f_j}{\omega_j^2 - \omega^2 - i\gamma_j\omega}\right)\tilde{\mathbf{E}} \tag{12.12}$$

$$\tilde{\mathbf{P}} = \varepsilon_0 \tilde{\chi}_e \tilde{\mathbf{E}} \tag{12.13}$$

Using the additive field equation, we would assign the complex ε the value

$$\tilde{\varepsilon}_r = 1 + \frac{Nq^2}{m\varepsilon_0}\sum_j \frac{f_j}{\omega_j^2 - \omega^2 - i\gamma_j\omega} \tag{12.14}$$

It would seem that our dielectric is a function of ω, and that function has "poles" around the natural resonances of the material. The damping constant can play a rather large role in how ε behaves, and it can be dominated by any number of mechanisms as we will see.

If the driver is an E&M wave, then the velocity of that wave in this medium depends on ω now, with $\varepsilon = \varepsilon_r\varepsilon_0$. For monochromatic plane wave solutions, the change in $\varepsilon(\omega)$ with frequency leads to a convolution of real and imaginary parts:

$$\tilde{\mathbf{E}}(z,t) = \tilde{\mathbf{E}}_0 e^{-i(\tilde{k}z - \omega t)} \tag{12.15}$$

$$\tilde{k} \equiv \sqrt{\tilde{\varepsilon}\mu_0}\omega \tag{12.16}$$

$$\tilde{k} = k + i\kappa \tag{12.17}$$

$$\tilde{\mathbf{E}}(z,t) = \tilde{\mathbf{E}}_0 e^{-\kappa z} e^{-i(kz - \omega t)} \tag{12.18}$$

$$\alpha \equiv 2\kappa; \quad n = ck/\omega \tag{12.19}$$

From this it is typical to define an *extinction or attenuation coefficient* (α) as well as the *index of refraction* (n). The α term describes the loss of field near the resonance, and the n describes the slowing of the wave as it moves forward.

However, it is plotting out the dielectric or its associated functions such as α and n that really shows what these poles on the ε_r equation do. As seen in Figure 12.5, we have chosen to show α and n because these are most commonly measured as a function of ω, but certainly one can see that the ε curve would be consistent.

4 We should be a little careful here; we are still talking about a quantum system, and the distortion of the electron orbit does not represent a true eigenstate of the unperturbed system. A classical system "rings." It isn't entirely clear what the quantum system does. But we are going to go with it since we know from experience that this works.

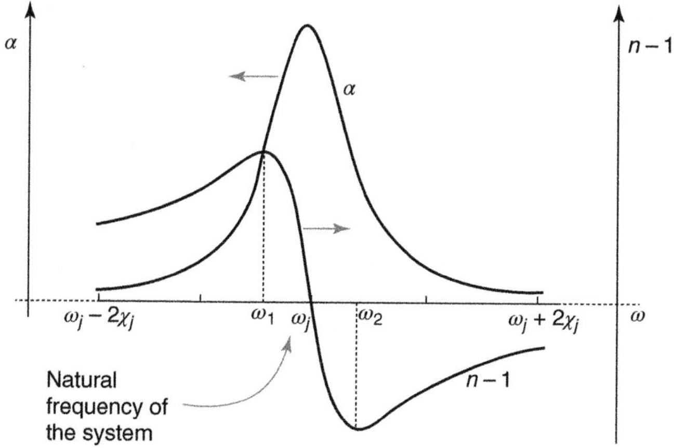

Figure 12.5 The response of the system as a function of ω.

12.4 A Familiar Equation in Optics

We can put this into a form that the reader may have already seen in an optics class. Starting with our complex wavenumber,

$$\tilde{k} = \frac{\omega}{c}\sqrt{\tilde{\varepsilon}_r} \cong \frac{\omega}{c}\left[1 + \frac{Nq^2}{2m\varepsilon_0}\sum_j \frac{f_j}{\omega_j^2 - \omega^2 - i\gamma_j\omega}\right] \tag{12.20}$$

and solving for n and α, we get

$$n = \frac{ck}{\omega} \cong 1 + \frac{Nq^2}{2m\varepsilon_0}\sum_j \frac{f_j(\omega_j^2 - \omega^2)}{(\omega_j^2 - \omega^2)^2 + \gamma_j^2\omega^2} \tag{12.21}$$

$$\alpha = 2\kappa \cong \frac{Nq^2\omega^2}{m\varepsilon_0 c}\sum_j \frac{f_j\gamma_j}{(\omega_j^2 - \omega^2)^2 + \gamma_j^2\omega^2} \tag{12.22}$$

Assume a relatively low density of dipoles and weak damping:

$$n = 1 + \frac{Nq^2}{2m\varepsilon_0}\sum_j \frac{f_j}{\omega_j^2 - \omega^2} \tag{12.23}$$

Taking

$$\frac{1}{\omega_j^2 - \omega^2} = \frac{1}{\omega_j^2}\left(1 - \frac{\omega^2}{\omega_j^2}\right)^{-1} \cong \frac{1}{\omega_j^2}\left(1 + \frac{\omega^2}{\omega_j^2}\right) \tag{12.24}$$

we get

$$n = 1 + \left(\frac{Nq^2}{2m\varepsilon_0}\sum_j \frac{f_j}{\omega_j^2}\right) + \omega^2\left(\frac{Nq^2}{2m\varepsilon_0}\sum_j \frac{f_j}{\omega_j^4}\right) \tag{12.25}$$

which reduces to the equation in optics describing dispersion:

$$n = 1 + A\left(1 + \frac{B}{\lambda^2}\right) \tag{12.26}$$

12.5 Understanding the Context

What does this all mean? Remember when we said we would return to that damping constant? In fact, *damping* and *polarization times* are our "looking glass" into physical mechanisms of polarization. We really should expect a few specific features of the polarizing systems (dielectric or semiconducting):

1. Accumulation of charge at interfaces is likely to be quite slow among our list of potential mechanisms. So we should see $\mathbf{P}_{\mathrm{int}}$ contributions to $\mathbf{P}_{\mathrm{net}}$ only at the lowest frequencies. Moreover, as the driving frequency increases beyond the time frames needed for full charge saturation at the interfaces, the dielectric response would simply drop as is seen in the first step of Figure 12.6 unlike that seen in Figure 12.5.
2. Similarly, the torsional rearrangement of permanent dipoles needed for the $\mathbf{P}_{\mathrm{d}}$ mechanism in the lower part of Figure 12.1 will be slow, and it too will have a saturation value above which we will observe a simple drop-off in dielectric response. Remember that such a mechanism would be dependent on temperature and saturation should represent full anti-alignment.
   By analogy, the two mechanisms above might be seen as critically damped systems mechanically.
3. However, ionic motion and electronic induced polarization are both far more local; they induce less local strain on the lattice, and they are far quicker than the first two mechanisms. The ionic motion is, of course, slower than the electronic polarization by virtue of the mass of the moving object. By analogy again, we might suggest these mechanisms to be under-damped mechanical systems, thereby yielding *resonance* phenomena at specific frequencies. The "spikes" in dielectric function, index, and attenuation as the resonance is approached, as in Figure 12.5, are typical of such a mechanical system.
   We might construct a graph (Figure 12.6) using this simple mechanical model, changing the masses, the γ's, and the ω's, to accommodate differences in the mechanisms we have been discussing.

One could argue that we are just applying a mechanically driven oscillator system to anything we might find a resonance for in the solid, and this is surely the case. But, in fact, it is rather useful.[5] Certainly, there are more sophisticated approaches to each of these different mechanisms for polarization that specialists in the field use. But their starting point is inevitably our mechanical oscillator.

12.6 The Dielectric Function and Metals

In the case of a metal, surely the idea of a dielectric response of a metal seems a little ridiculous – there is nothing to polarize in the sense of our picture presented

5 In actual fact, for any given field of physics, there are only a few problems scientists can work out in closed form. For kinetics the simple harmonic oscillator is one of those. The trick is to find a way to make every mechanics problem a simple harmonic oscillator problem.

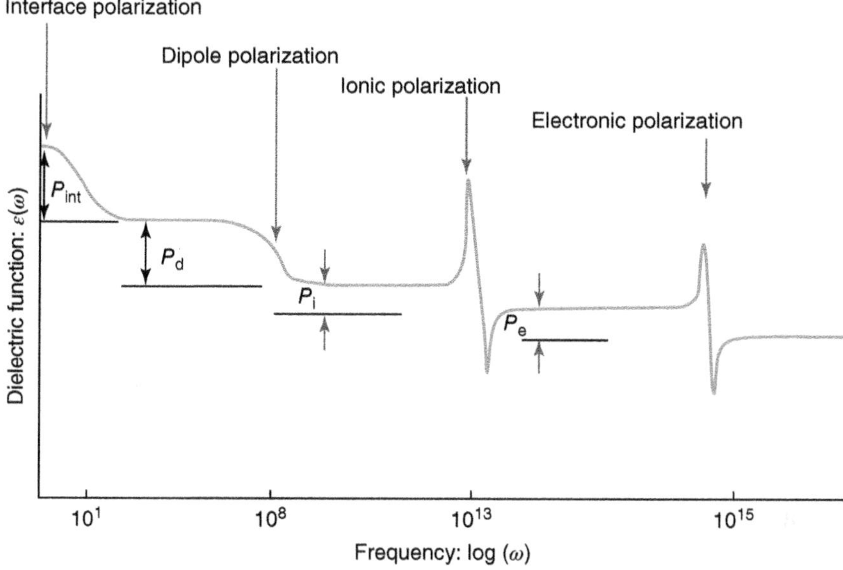

Figure 12.6 Different response times from different mechanisms result in *resonances* across the frequency spectrum. The first two steps, P_{int} (for interfaces) and P_d (for alignment of permanent dipoles), don't really "look" like resonances but rather like limiting cutoffs for response, whereas the other two do. In the analogues of a mechanical oscillator with damping, we might say the first two are critically damped (depending on the number of free charge and density of the dipoles) and the latter two are under-damped.

above. When the electrons are free, they go wherever they want as fast as they want, right? This presumably means that there are no internal local fields due to charge accumulation on the material's boundaries (canceling the external source of field altogether), and so nothing happens to the atoms.

This is certainly reasonable for an applied static **E** or **B** field, when a time-dependent AC field is applied, such as in the case of an E&M wave, an electron gas can certainly have a "polarizing" response. Why? It is simply because if the frequency is high enough and the time scales are short enough, full charge accumulation is never fully achieved. Moreover, we have addressed this already in Chapter 10 where we apply the very same mechanical oscillator model to come up with *plasmons*. That section could have just as easily been added here instead of there.

12.7 Piezoelectrics, Pyroelectrics, and More

Clearly, crystalline or semicrystalline solids with noncentrosymmetric unit cells can have permanent dipoles associated with those unit cells. And, when permanent dipoles exist within the solid, their collective behavior can be hidden by randomization, driven by the surrounding heat bath or accumulated surface and volume charges. As we have seen an applied electric field can align these dipoles, yielding a substantial dielectric response. But there can be other mechanisms that align the dipoles as well. Since such effects are widely used and studied intensively in solid-state physics labs, this might be a good place to mention them. Note, however, that there exist many specialized texts for these topics, and so we will introduce only the essentials here.

Piezoelectric systems have a curious aspect of their crystal symmetry in that if they undergo distortion in one or more of the principle directions of the lattice, the dipoles will align, giving rise to a polarization: **P**. This could of course happen for a number of different mechanical reasons. Shown in Figure 12.7 is a typical unit cell for the class of piezoelectrics known by its most famous member PZT. The Curie temperature, T_c, mentioned in the diagram is the temperature at which the electrostatics in the system allows it to fall into a minimum energy configuration that has a dipole moment. Above this temperature, the system is cubic. Notice that the crystal is constructed of a ridged set of bonds, cantilevered off in various directions. While the dipoles can be arranged up or down under the T_c, when pressure is applied along one crystal axis, the dipoles will align, giving a macroscopic P.

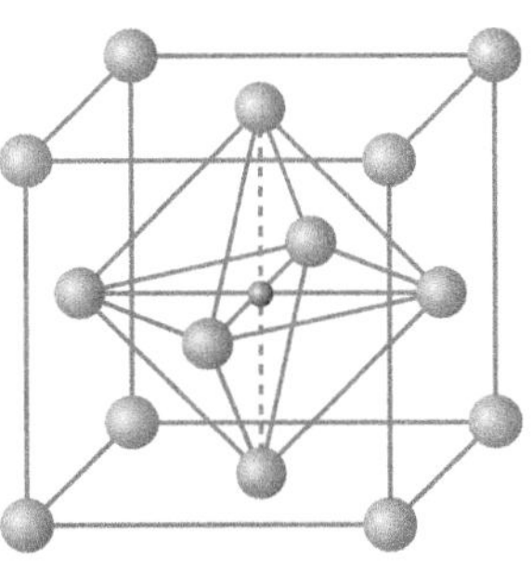

Above the Curie temperature T_c, many of these crystals have a cubic symmetry in the arrangement of the positive and negative charge

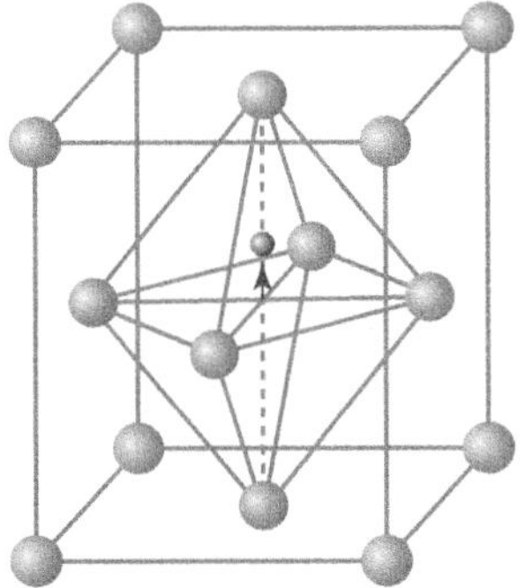

But, below T_c, the crystal is tetragonal (orthorhombic) and has an electric dipole

A^{2+} = Pb, Ba, etc. (large, divalent metal ion)

O^{2-} = oxygen

B^{4+} = Ti, Zr, etc. (smaller, tetravalent metal ion)

A typical crystal structure such as PZT, for piezo-active materials

Figure 12.7 The piezo-active crystal PZT (and its family) shown in its cubic and tetragonal forms.

The *direct piezo effect* is given by

$$P = \mathrm{d}\sigma \tag{12.27}$$

$$\underbrace{(\text{Electric polarization})}_{\substack{(\text{C/m}^2)\\ \text{First rank vector}}} = \underbrace{(\text{Piezoelectric coefficient})}_{\substack{(\text{C/N})\\ \text{Third rank tensor}}}\underbrace{(\text{Mechanical stress})}_{\substack{(\text{N/m}^2)\\ \text{Second rank tensor}}}$$

There is also an inverse effect wherein a field is applied that leads to internal strain:

$$\varepsilon = \mathrm{d}E \tag{12.28}$$

$$\underbrace{(\text{Mechanical strain})}_{\substack{(\text{Dimensionless})\\ \text{First rank vector}}} = \underbrace{(\text{Piezoelectric coefficient})}_{\substack{(\text{m/V})\\ \text{Third rank tensor}}}\underbrace{(\text{Electric field})}_{\substack{(\text{V/m})\\ \text{Second rank tensor}}}$$

As would be expected, the effect is strongly dependent on the direction the force is added and the crystal directions.

Written in a tensor, like indices are summed and they correspond to crystal directions. Hidden within the "d" coefficients are the mechanisms responsible for the alignment in a given crystal system and their magnitudes. Such coefficients usually require detailed atomistic models to unravel:

$$P_i = d_{ijk}\sigma_{jk} \tag{12.29}$$

$$\varepsilon_{jk} = d_{ijk}E_i \tag{12.30}$$

We can get a little more insight into the relations between polarization and the work of deformation from the first law of thermodynamics:

$$\mathrm{d}U = \sigma_{ij}\,\mathrm{d}\varepsilon_{ij} + E_k\mathrm{d}P_k + T\,\mathrm{d}S \tag{12.31}$$

The middle term is the electric work done internally to the crystal. The Gibbs free energy is given as

$$G = U - TS - E_kP_k + \sigma_{ij}\varepsilon_{ij} \tag{12.32}$$

$$\mathrm{d}G = -\varepsilon_{ij}\,\mathrm{d}\sigma_{ij} - P_k\,\mathrm{d}E_k - S\,\mathrm{d}T \tag{12.33}$$

The Gibbs energy is used as the state function because the system is defined completely by σ, E, and T, where

$$\mathrm{d}G = \left(\frac{\partial G}{\partial \sigma_{ij}}\right)_{E,T}\mathrm{d}\sigma_{ij} + \left(\frac{\partial G}{\partial E_k}\right)_{\sigma,T}\mathrm{d}E_k + \left(\frac{\partial G}{\partial T}\right)_{\sigma,E} S\,\mathrm{d}T \tag{12.34}$$

$$\frac{\partial G}{\partial \sigma_{ij}} = -\varepsilon_{ij};\ \frac{\partial G}{\partial E_k} = -P_k;\ \frac{\partial G}{\partial T} = -S \tag{12.35}$$

This allows us to enumerate several different effects that will be found:

$$\frac{\partial^2 G}{\partial \sigma_{ij}\partial E_k} = -\frac{\partial \varepsilon_{ij}}{\partial E_k}\quad \text{Converse piezoelectric effect} \tag{12.36a}$$

$$= \frac{\partial^2 G}{\partial E_k \partial \sigma_{ij}} = -\frac{\partial P_k}{\partial \sigma_{ij}}\quad \text{Direct piezoelectric effect} \tag{12.36b}$$

$$\frac{\partial^2 G}{\partial \sigma_{ij} \partial T} = -\frac{\partial \varepsilon_{ij}}{\partial T} \quad \text{Thermal expansion coefficient} \tag{12.37a}$$

$$= \frac{\partial^2 G}{\partial T \partial \sigma_{ij}} = -\frac{\partial S}{\partial \sigma_{ij}} \quad \text{Piezocaloric effect} \tag{12.37b}$$

$$\frac{\partial^2 G}{\partial E_k \partial T} = -\frac{\partial P_k}{\partial T} \quad \text{Pyroelectric coefficient} \tag{12.38a}$$

$$= \frac{\partial^2 G}{\partial T \partial E_k} = -\frac{\partial S}{\partial E_k} \quad \text{Electrocaloric effect} \tag{12.38b}$$

We have discussed the piezo-response and its converse above. Thermal expansion was discussed in the chapter on phonons. However, for such crystals, we have three new effects that can be observed: *piezocaloric*, which is like the inverse of thermal expansion for a piezo-crystal, and the coupled pair of *pyroelectric* and *electrocaloric* effects. Again the last two are like inverses of each other.

The *pyroelectric* response in piezo-materials is the change in the macroscopic polarization vector with a change in temperature. Usually this is written as

$$\Delta P = p_i \Delta T \tag{12.39}$$

The p_i is called the *pyroelectric coefficient* and it does depend on crystal orientation. ΔP is a vector while ΔT is the change in temperature. Its inverse, the *electrocaloric effect*, would suggest that changing the polarization by the application of a time-dependent electric field would change the temperature of the system. Indeed, the electrocaloric effect can be rather subtle and has been observed in piezo-active polymers such as polyvinylidene fluoride (PVDF) and its copolymers [4].

12.7.1 The h-BN Example

Of course our presentation here has dealt with three-dimensional, standard materials. What is the effect of dimensionality on such piezo/pyroelectric behavior? A good example of this is found in atomically thin sheets of hexagonal boride: h-BN. Shown in Figure 12.8 is such a sheet of h-BN, and you will notice that it is noncentrosymmetric in plane.

We might expect that such a crystal structure would yield a piezo-response. However, due to its threefold symmetry and the fact that it is completely flat, it shows no in-plane or out-of-plane dipoles to align. Recall that we discussed earlier that threefold symmetries can lead to frustrated ordering. Yet there is a simple way to alter this situation. If hydrogen or fluorine is introduced to the h-BN sheet, binding to the lattice sites of the crystal, the flatness is distorted, and an out-of-plane dipole is formed. This two-dimensional system becomes quite impressively piezoelectric [6].

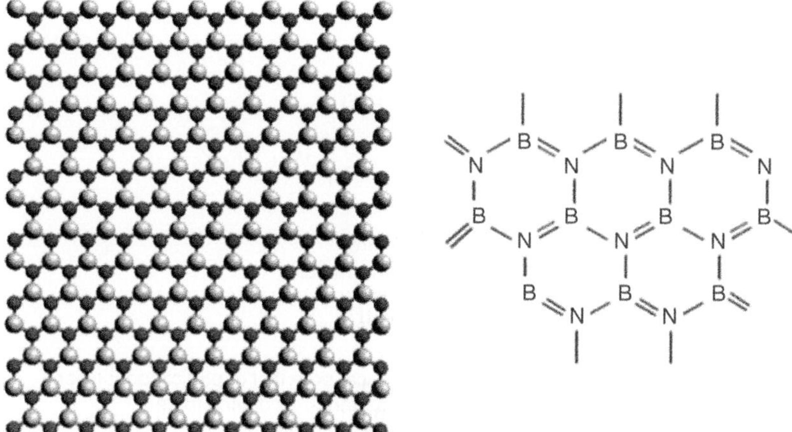

Figure 12.8 Arranged similarly to graphene, h-BN forms a hexagonal sheet of atoms. However, there is a broken symmetry since the A and B sublattices of the graphene structure are occupied by two different kinds of atoms. Notice also the alteration of double and single bonds among the B and N atoms. Sheets of h-BN as well as nanotubes of the material are now easily synthesized [5].

12.8 Summary

Electromagnetic fields can strongly couple to both insulators and (for AC fields) to metals through the polarization of the lattice. As we stated above, semiconductors will lie somewhere in between these two (with some nuances that are discussed in the next chapter). A simple phenomenological model of aligning dipoles and mechanical oscillators can be used as a good starting point for understanding the temperature and frequency dependencies of these polarization phenomena.

In materials that have a permanent dipole moment due to crystal structure, the Gibbs state function has provided some guidance as to how macroscopic polarization and stress–strain constraints are linked. This added level of complexity can make a detailed understanding of the effects of such internal local fields very difficult. Yet we are able to extract some predictive phenomenological statements in such systems through the use of simple atomistic models. Importantly, our understanding of such effects is not hampered by dimensionality. As we have demonstrated low-dimensional structures such as the polymers of PVDF and the h-BN sheets can both express symmetries that lead to piezo-response.

Exploring Concepts

1 *The Onsager Relations*: In Section 12.7 we introduce a thermodynamic approach to *inverse (or complementary) processes* for such noncentrosymmetric systems as piezoelectrics. This is why the section was included in

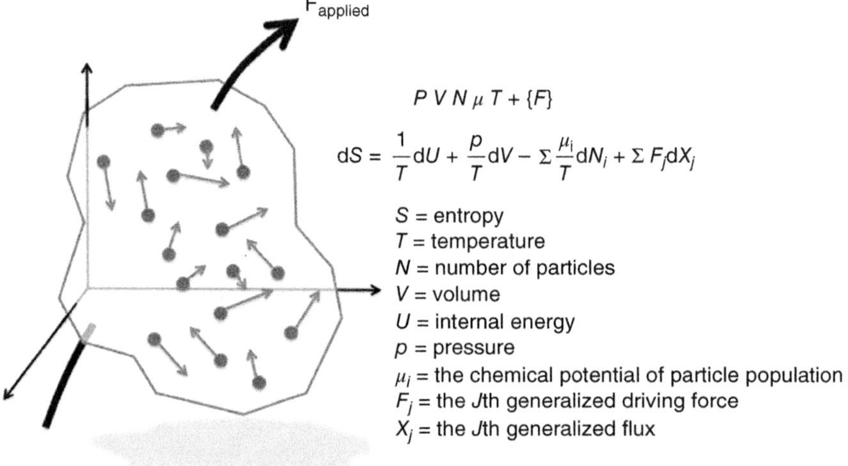

Figure EC12.1 The system is defined by thermodynamic entities as seen. But these vary across the volume and through time. They lead to a nonequilibrium-based flow of entropy in system to equilibrium. In transport, polarization, piezoelectrics, etc., our question is: what happens to this relaxation when a force (a field of some kind) is placed across the system?

this chapter. However, the *thermodynamic driving forces* and *current fluxes* we describe are far more general, so if we are a little more careful, we can make this approach (and this exercise) also of relevance to the chapter on **transport** (Chapter 8).

Consider a thermodynamic system, not necessarily in equilibrium. The idea is shown in Figure EC12.1.

To answer this, recall that the first law of thermodynamics can be stated in terms of the entropy state function:

$$T\frac{\partial S}{\partial t} = \frac{\partial E}{\partial t} - \mu\frac{\partial N}{\partial t}$$

And we define two specific kinds of currents in the system – particles and energies:

$$\mathbf{J}_N = N\nu \quad \text{(partial current)}$$
$$\mathbf{J}_E = E\mathbf{J}_N/N \quad \text{(energy current)}$$

yielding two conservation equations

$$\frac{\partial N}{\partial t} = \Delta \cdot \mathbf{J}_N$$

$$\frac{\partial E}{\partial t} = \Delta \cdot \mathbf{J}_E = F \cdot \mathbf{J}_N$$

F is this external set of forces added to the system as shown in Figure EC12.1. They can be an applied electric field to a system of charge, an applied magnetic field to a system of currents, etc.

If we combine all of this through that first law,

$$T\frac{\partial S}{\partial t} + \Delta \cdot \mathbf{J}_E - \mu\nabla \cdot \mathbf{J}_N = F \cdot \mathbf{J}_N$$

And this relationship is important because it says for such processes as we are discussing that the rate of *entropy* and *heat* generation in this relaxation is given as

$$\frac{\mathrm{d}S}{\mathrm{d}t} = \frac{\partial S}{\partial t} + \nabla \cdot \left[\frac{\mathbf{J}_E - \mu\mathbf{J}_N}{T}\right] = \frac{F \cdot \mathbf{J}_N}{T} - \nabla\left(\frac{\mu}{T}\right) \cdot \mathbf{J}_N + \nabla\left(\frac{1}{T}\right) \cdot \mathbf{J}_E$$

and

$$\frac{\mathrm{d}Q}{\mathrm{d}t} = \frac{T}{V}\frac{\mathrm{d}S}{\mathrm{d}t}$$

If our force is the Lorentz force, for example, $-e\mathbf{E} - e\mathbf{v}/c^2x\mathbf{B}$,

$$\left.\frac{\mathrm{d}Q}{\mathrm{d}t}\right|_{\text{Lorentz}} = \left[-eE - \nabla\mu - \frac{\nabla \cdot T}{T}(E-\mu)\right] \cdot \frac{\mathbf{J}_N}{V}$$

These serve as a sort of continuity equation for entropy and suggest an important aspect we didn't cover in transport or in dielectric response. When fields are applied to systems of electrons, phonons, magnons, or whatever, the response of the system is in keeping with the creation, flow, and conservation of entropy. This means that applied fields and flowing currents come along with changes in thermodynamic entities such as temperature, generally.

But there is more. In general we associate a flux (a movement or current of something) with each force applied:

$$X_j = \frac{\mathrm{d}}{\mathrm{d}F_j}[\partial Q/\partial t]$$

(a) Take the force to be an applied electric field. Show that its corresponding (or conjugate) current is the electrical current.

(b) Take the force to be a temperature gradient. Derive and discuss the meaning of this flux.

If there are numerous forces that are applied to the system, as is typical, and if they are reasonably weak (in the regime of linear response – this means the current responds to the force linearly), then some rather nonobvious results are accomplished. This is when temperature gradients lead to electrical field and other "cross effects" are observed. Consider the linear response form

$$X_i = \sum_j L_{ij}F_j$$

These coefficients actually correspond to the physical coefficients for the physical phenomena we have grown to know and love: resistivity in the case of E-forces, the piezo-constants in the case of applied pressures, and so on. But notice that these coefficients do not have just diagonal nonzero elements in a matrix and that L_{ij} relates the flux of j in response to the force i.

The *Onsager relations* relate these coefficients and in doing so relate specific conjugate behaviors physically. As a result of a system's microscopic reversibility in the thermodynamic sense, the relations explain why physical processes in the solid state, such as the Seebeck effect, have a conjugate process, the Peltier effect. They are stated as

$$L_{ij}(\mathbf{B}) = L_{ji}(-\mathbf{B})$$

Here we have stated it in terms of the B field, which enters into the force equations as a pseudo-vector, and thus in time-reversing systems with B fields, one must include a negative sign. So too must this be done with the linear response coefficients. More importantly, there are some conditions for which the *Onsager relations do not work*, but generally they are a statement about symmetry and conservation.

Exercises (a) and (b) deal with only one force and flux in the system. Now let's look at the two so that we can understand the cross-coupling:

(c) Following along our presentation of the thermoelectric effect from the transport section, translate this into the language of the Onsager expressions. Identify the two generalized thermodynamic forces and the subsequent current fluxes. Construct a matrix of $\{L\}$ with these two forces, and show that Onsager's expression holds true for thermoelectrics. You may need a little outside reading for this.

2 *Quartz*: As it happens, quartz, the crystalline form of SiO_2, is quite a technologically important material. It is strongly piezoelectric in nature.

(a) We make watches from the stuff, and it is used as a timing mechanism. Do a little background reading (Walter Caddy might be a good place to start), and describe how you would use a quartz crystal to keep time.

(b) Large quartz deposits have been used to explain ghostlike lights that appear over some mountains (like those of Brown Mountain in NC, USA). How do you think this might work? [Joe Nickell, The Brown Mountain Lights: Solved! (Again!), Skeptical Inquirer Volume 40.1, January/February 2016].

3 *Electrocalorics*: The electrocaloric effect is actually the inverse of the pyroelectric effect in the same way that the Seebeck effect is the inverse of the Peltier effect (see Exercise 12.1). Thus, whereas a time-dependent temperature gradient can give rise to an electric field, so too can an electric field give rise to a temperature gradient in time. Using the thermodynamic entities described above:

(a) Identify and describe the thermodynamic forces and flux currents in this conjugate pair of phenomena.

(b) Describe a cooling cycle as a heat engine based on this effect. You may need to go to the literature on this one for a little help.

4 *Nanometals*: We have discussed the dielectric response of isotropic and homogeneous materials and have used the charge separability of the atom or molecule's electrons and nuclei as the source of dipolar fields. Imagine now that we have a composite material. Instead of being made up of pone polymer or ceramic, it is instead made up of a highly insulating polymer host, and it has well dispersed throughout its volume gold nanoparticles of around ~10 nm in diameter. The thickness of our film is more than 1 μm, and using a good deal of microscopy, we have determined that none of the little metal spheres of Au touch each other.

(a) Using the model we have established above as a guide.
(b) Suggest a way of estimating the overall dielectric response of such a material.
(c) Compare the results of this estimate with the same host polymer but without the NPs.
(d) Finally describe the frequency response you might expect from this system. How might this change as the average diameter of the particles gets larger?

5 *Transistor Dielectrics*: We discussed the structure and function of organic thin-film transistors (OTFTs) in a previous chapter, so to do this exercise you will need to look back a bit. Often, such devices have relatively thick gate dielectrics with a capacitance per unit area of typically less than 100 nF/cm^2.

(a) This means that OTFTs often require relatively large gate-source voltages of about 10 V or more to work. Why? Write an expression that relates the on–off channel voltage in terms of the applied gate voltage and its dielectric constant and thickness.
(b) For many applications in mobile devices, where small batteries are used, significantly lower operating voltages are needed. This usually means going to thinner dielectrics with higher dielectric constants. There are two main ways to do this. The first is to make high dielectric polymers, and the second is to use composites. Using a literature search, describe both of these approaches with details as to the actual material compositions. (You might find the work of Dr. H. Klauk useful here.)
(c) What do you anticipate the effects of these approaches might be on the frequency response of the devices? Why?

References

1 There is an outstanding treatise written on the subject of macroscopic polarization that is a must read for seriously understanding this topic. Resta, R. (1994). Macroscopic polarization in crystalline solids: the geometric phase approach. *Rev. Mod. Phys.* 66: 899.
2 One should not get too carried away with seeing this D (displacement) field as the field that started p. In some cases it can be like it, but in others it can seem not very physical indeed. For an interesting discussion see Griffiths, D.J. (2017). *Introduction to Electrodynamics*, Chapter 4, 4e. Cambridge University Press.
3 See Maurder for a more full discussion of this point Marder, M.P. (2015). *Condensed Matter Physics*, 2e. Wiley.
4 Scott, J.F. (2011). Electrocaloric materials. *Annu. Rev. Mater. Res.* 41: 229–240. https://doi.org/10.1146/annurev-matsci-062910-100341.

5 Golberg, D., Bando, Y., Huang, Y. et al. (2010). Boron nitride nanotubes and nanosheets. *ACS Nano* 4 (6): 2979–2993.
6 Noor-A-Alam, M., Kim, H.J., and Shin, Y.H. (2014). Dipolar polarization and piezoelectricity of a hexagonal boron nitride sheet decorated with hydrogen and fluorine. *Phys. Chem. Chem. Phys.* 16: 6575.

13

Optical Interactions

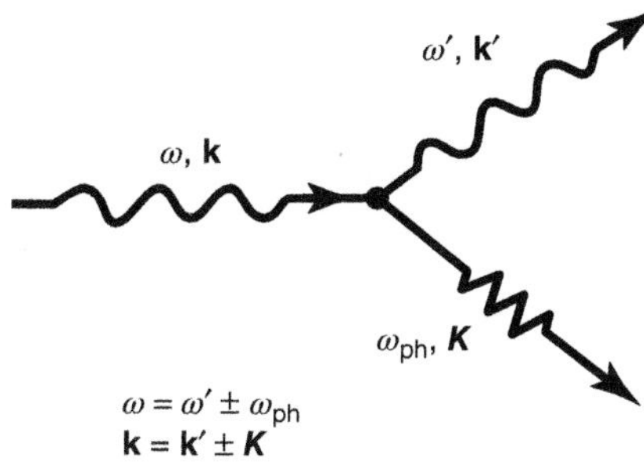

Photon scattering from phonons: *Brillouin* scattering with acoustic phonons or *polarition* scattering with optical phonons (Raman)

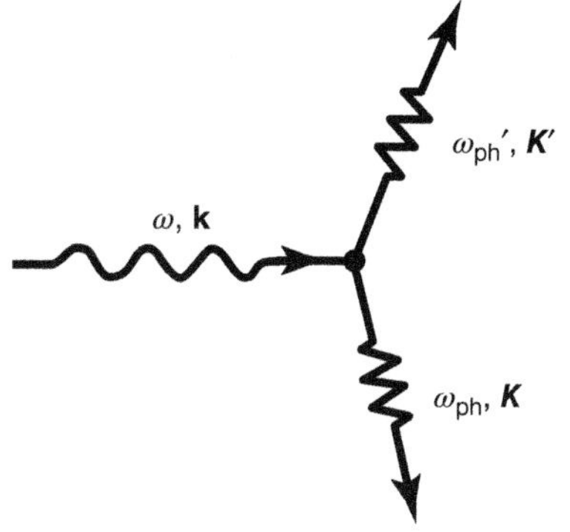

Two phonon infrared absorption and electron absorption leading to conduction electrons or electron emission

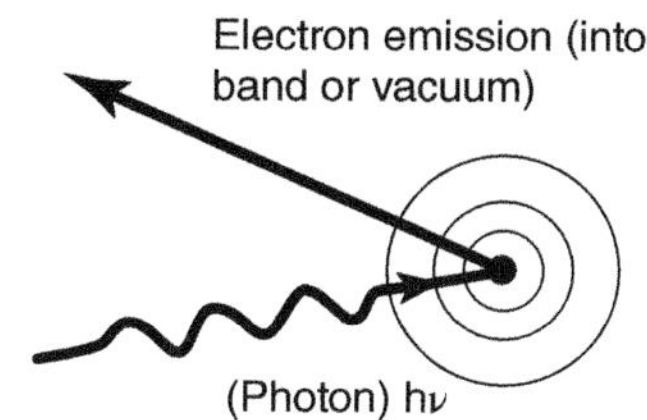

There are several ways to view light's interaction with solids. We can think of discrete photons interacting with a collection of quasiparticles in a way analogous to particle–particle scattering. The Feynman diagrams above encourage this view. Alternatively, a classical picture can emerge wherein the time-dependent response of the solid to an incident electromagnetic field is considered through Maxwell's equations. Then, many-body models for electrical polarization, electron–electron interactions, etc. are applied through the interaction volume in place of quasiparticles.

Foundations of Solid State Physics: Dimensionality and Symmetry, First Edition.
Siegmar Roth and David Carroll.

Naturally, according to the correspondence principle, these pictures are equivalent, so either will do. But, in actual practice, the physicist in the lab typically calls upon both perspectives to understand the electro-optic behavior of a solid. Our approach here is to present this amalgam of photon and field language, as typically heard in the physics community, in as useful a way as possible.

In Chapter 12, we began a discussion on the time dependence of dielectric response in a domain of typical E&M waves. The magnetic response and its correlative, *permeability*, weren't explored since it is clear that in "normal materials" the influence of the ***B*** field on *observable optics* is usually small compared with the electric field. So, we usually do think of material optics as *Optics = Dielectric Response to an E&M wave.*[1] And this really is too bad; such simplifications often obscure the more elegant physics. For example, metamaterials that combine components of negative permittivity and negative permeability often yield a negative indexed material (bending light backward from Snell's law) [1]. *Optics* today is a dynamic and evolving field due to the demonstrations of such materials systems – materials that no one thought should exist!

In a *solid-state physics* text though, our focus is different. If *optics* is how a material changes a light wave, then *optical interactions* (to our mind) must be how the light wave modifies a material. Our goal is to explore the behaviors of a solid's electrons and ions when exposed to an *optical field* (a photon). And what we failed to examine previously is the interplay between the E&M field and a material's *polarization* in the context of many-body physics and correlation.[2] Optical-field-*mediating* correlated behaviors include *polaritons, giant oscillator strength phenomena, plasmonics, and more.* So, if we want to know why structures from *azo*-linkages to carbon nanotubes couple strongly to the optical field, then we must know something of this type of correlation. You can think of Chapters 12 and 13 to be related in a way analogous to Chapters 8 and 10; there we began with single particle states to describe the current in a conductor and ended up with highly correlated behaviors that we called quasiparticles.

Unfortunately, the universe[3] places limits on space-time. So, the reader must already be familiar with electromagnetism[4] to get the most from this presentation. This means a familiarity with Maxwell's equations and their solutions, reflection, refraction, Snell's law, Beer's law, etc. A good intermediate undergraduate course will do nothing too taxing. But we will state as fact some results from classical electromagnetic theory (Figure 13.1).

We can expect that the strongest coupling between materials and E&M waves will occur at the concordance between natural resonances of the material at the microscopic level and the frequencies of the E&M wave. But what are those resonances, and to what part of the materials system are they attributable? Plasma oscillations? Elastic oscillations? Other? What *collectively* couples to the time-dependent fields $\mathbf{E}(\mathbf{r},t)$ and $\mathbf{B}(\mathbf{r},t)$, and how strongly?

---

1 We are leaving out magneto-optic phenomena like the Kerr rotation. But for simple a lens, prism, mirror, you can see what we mean.

2 Here we mean quantum correlation or the propensity for forming quasiparticles.

3 By "universe" here, we mean "publisher."

4 Here we make reference to the intermediate level course usually taken by US college juniors.

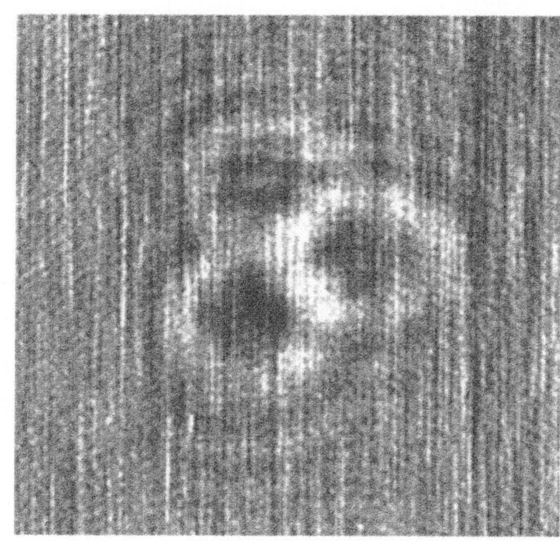

Figure 13.1 Near-field image of three silver nanoparticles (~10 nm in diameter each). Source: Courtesy Nanotech Center WFU.

13.1 Maxwell and the Solid (Review)

Briefly recall that Maxwell's equations lead to the basic solution in vacuum of a traveling E&M wave. This traveling electromagnetic wave (light/radiation/photon) is a delicate and self-perpetuating dance between the electric field and the magnetic field. The wavelike solution exists in materials through which the wave transverses as well. But the waves in a vacuum and waves in a solid are subtly different.

13.1.1 In a Vacuum

As a reminder, Maxwell's four first-order differential equations – ($\nabla\times$ **E**/**B** =), ($\nabla\cdot$ **E**/**B** =) – are usually combined into two second-order differential equations to show that the connections between **E** and **B** lead to a traveling wave solution:

$$\nabla^2\mathbf{E} = \mu_0\varepsilon_0\frac{\partial^2\mathbf{E}}{\partial t^2},\quad \nabla^2\mathbf{B} = \mu_0\varepsilon_0\frac{\partial^2\mathbf{B}}{\partial t^2} \tag{13.1}$$

There are many wave solutions for these equations. The exact solution for a specific problem is uniquely determined by the boundary conditions that might be applied. We can state *general* solutions by suggesting that any waveform that *could* be a solution to these equations would be subject to Fourier's theorem. This means it can be expanded in terms of a set of basis functions that adequately covers the space of interest. In Cartesian coordinates this usually means choosing sines and cosines:

$$B_0 = \frac{k}{\omega}E_0 = \frac{1}{c}E_0 \tag{13.2}$$

Shown in Figure 13.2 are the so-called *plane wave solutions*. The fact that we have identified c with $1/(\mu_0\varepsilon_0)^{1/2}$ indicates that this wave moves in a vacuum. The terms in the second-order wave equation also indicate that there are no sources

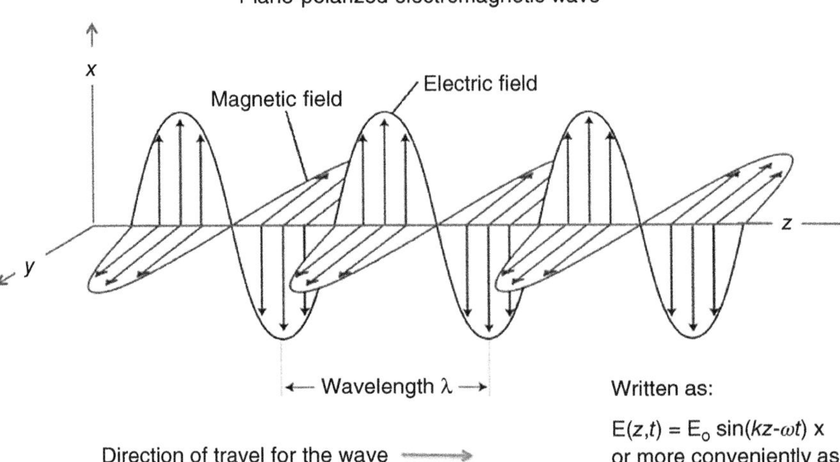

Figure 13.2 A plane polarized E&M wave *in vacuo*. The tilde over the E in the exponential expression indicates that the amplitudes are complex written this way. This wave is moving in **+z** and polarized along **x**. The magnetic field that goes along with it is in phase, perpendicular to **E**, with **k** pointing along **z** and **B** along **y**, and a magnitude.

of charge or current in the system. This is a freely propagating plane wave. Curiously, given time, this simple self-sustaining structure can make its way across the universe!

13.1.2 In a Material

$$\nabla \times \vec{H} - \frac{1}{c}\frac{\partial \vec{D}}{\partial t} = \frac{4\pi}{c}\vec{j}$$
$$\nabla \times \vec{E} + \frac{1}{c}\frac{\partial \vec{B}}{\partial t} = 0$$
$$\nabla \cdot \vec{D} = 0$$
$$\nabla \cdot \vec{B} = 0$$
$$\vec{D} = \epsilon \vec{E}$$
$$\vec{B} = \mu \vec{H}$$
$$\vec{j} = \sigma \vec{E}$$

In a material, Maxwell's equations change. This is due to polarization and magnetization effects. In simplest terms $\mu_0\varepsilon_0$ becomes $\mu\varepsilon$, and there can be bound charge and current distributions as well. The four expressions of Maxwell are shown at the lower left. Here we have taken the case where there are no free charges anywhere.

Notice that we have switched to the use of cgs units. Thus, unlike the equations in Chapter 12, we have the additional prefactors $1/c$ and $4\pi/c$. This really adds nothing to the physics, but it does make the units look a little nicer at the end. Also, many of the references we cite do it this way, so we are trying to make it a little easier to follow along in them.There are *constitutive equations* that go with Maxwell's equations in materials. These contain assumptions about the relation of **P** to **E** and **D**. This is the specific information that we saw in Chapter 12. Obviously here, these constitutive equations are stating that we are considering linear dielectrics:

$$\nabla^2\vec{E} = \frac{\epsilon\mu}{c^2}\frac{\partial^2\vec{E}}{\partial t^2} + \frac{4\pi\sigma\mu}{c^2}\frac{\partial\vec{E}}{\partial t}$$

$$\nabla^2\vec{H} = \frac{\epsilon\mu}{c^2}\frac{\partial^2\vec{H}}{\partial t^2} + \frac{4\pi\sigma\mu}{c^2}\frac{\partial\vec{H}}{\partial t}.$$

The equations are then combined as before to yield the second-order wave equations shown. σ is the conductivity of the medium, and a new linear term appears in the equations. For an *insulator* or undoped semiconductor at low temperatures, $\sigma = 0$, and these expressions describe the propagation of the E&M wave within the medium where the velocity is different from that of c shown in Figure 13.3. *Metals,* however, have an uncountably large number of freely moving electrons. This yields a large σ. The linear term in the equation then gives a real exponential decay, and propagation is damped out rather quickly as in Figure 13.3. *Semiconductors* are seemingly between metals and insulators. They have itinerate carriers and polarization.

Finally, there are *exceptional situations* where dimension, length scale, and topology also defines response (a materials plus object-level response). If a metal nanoparticle is so small that not many electrons are available for response, the field inside may no longer be zero! A *nonconnected* assembly of these nanoparticles might then have a finite dielectric response unlike most metals. Carbon nanotubes have extraordinarily long carrier phase coherence. Thus, their *antenna properties* (or oscillator strengths) are exceeding large. This makes them better antennae than one would expect from classical electromagnetism, by orders of magnitude. Such examples challenge our classical notions.

13.1.3 A General Solution in the Solid

Propagating solutions to Maxwell's equations are usually broken down into those for *dielectrics,* those for *metals,* those for *semiconductors,* and the *special cases* like nanoscale objects, random scattering media, phase coherent arrays, and metamaterials (you get the idea). Shown in Figure 13.3 are the solution cases for dielectrics and metals as a reminder.

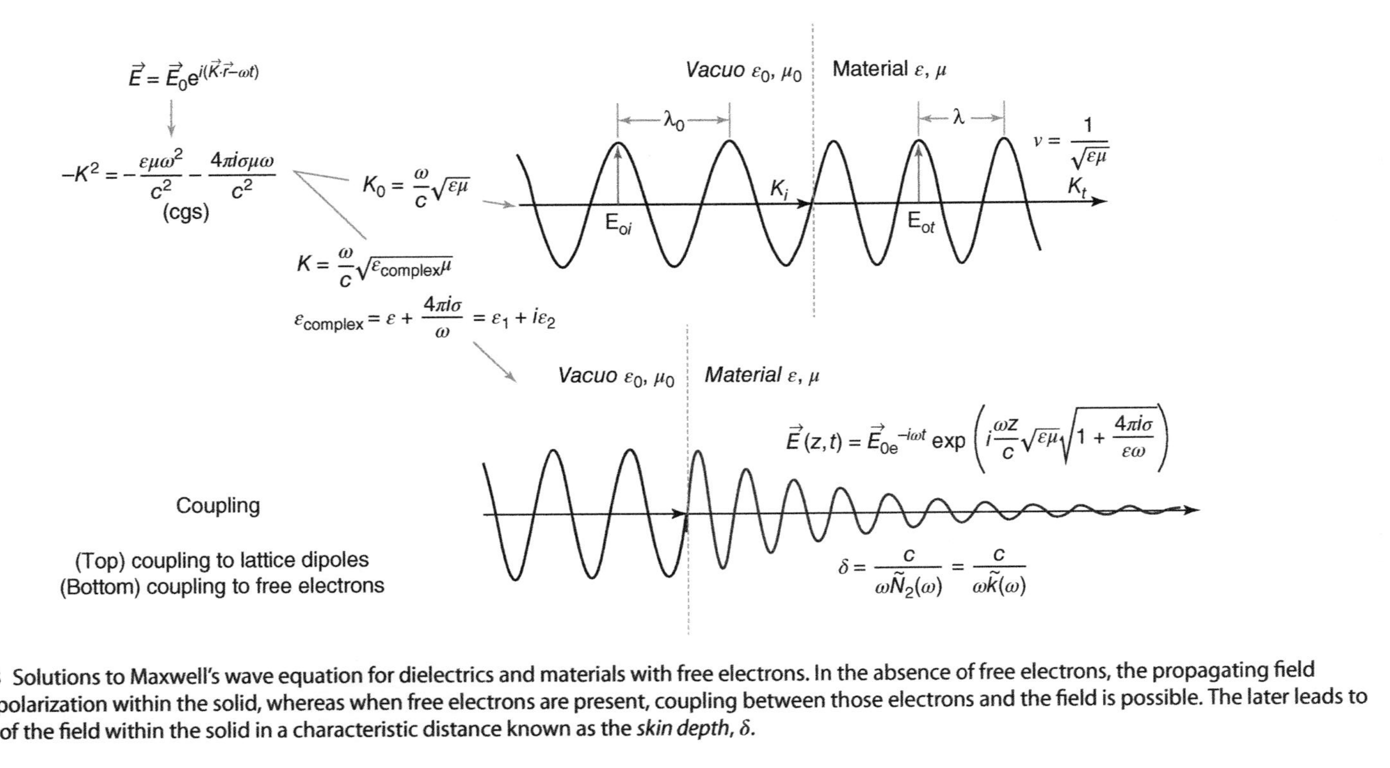

Figure 13.3 Solutions to Maxwell's wave equation for dielectrics and materials with free electrons. In the absence of free electrons, the propagating field couples to polarization within the solid, whereas when free electrons are present, coupling between those electrons and the field is possible. The later leads to a damping of the field within the solid in a characteristic distance known as the *skin depth*, δ.

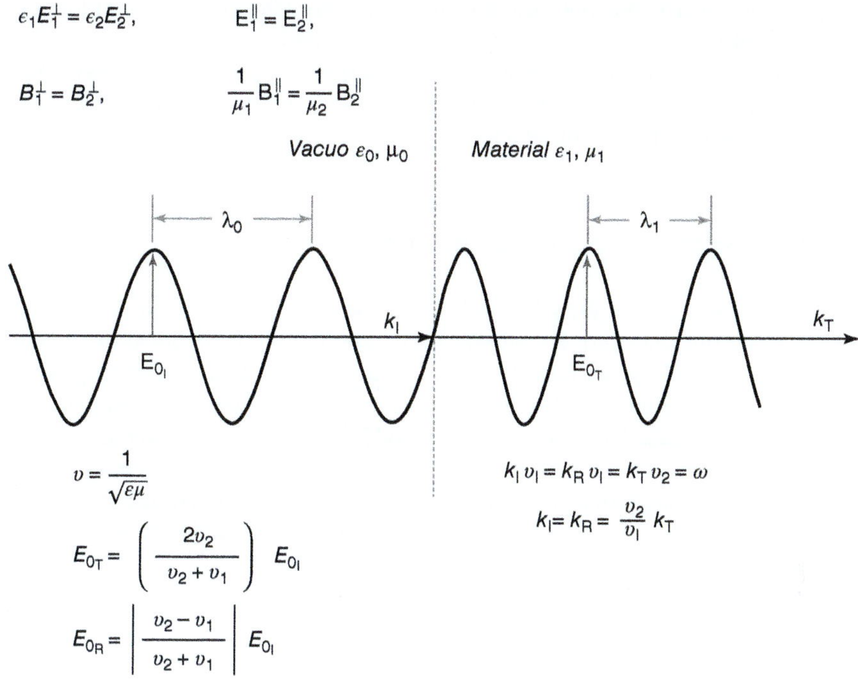

Figure 13.4 The classical picture of light/condensed matter interactions. (Top) The boundary conditions applied to the sine wave solutions of Maxwell's equation. (Bottom left) The velocity and the electric field strength for the reflected (R), incident (I), and transmitted (T) wave. (Bottom right) The frequency is constant set by the source, while the wavelength and wavenumber changes because the velocity is changing in the material.

In using these solutions, boundary conditions are applied, wherein the spatial polarization of the incoming electromagnetic wave, with respect to the surface normal, is the important factor. Shown in Figure 13.4 is the familiar geometrical construct for applying such boundary conditions at a dielectric surface. The problem is reduced to solving for components of the electric/magnetic field that lie in the plane of the surface and components that lie perpendicular. The standard notation to refer to the parallel (||) and perpendicular ($\perp$) components of the **E** and **B** fields is shown. We have not shown the reflected wave in Figure 13.4. But, of course, reflected and transmitted waves will add vectorially to give the incoming wave a sort of conservation principle. With this construction and geometrical arrangement, it is possible to derive a whole range of optical phenomena from reflection to Snell. This is done in many E&M textbooks such as Griffiths and is what is usually referred to as *geometrical optics* [2]. Such a simple approach is quite useful, but in reality the response of a material to the translating fields of a *photon* is far more nuanced and interesting than this. There is much, much more to the interaction between photons and materials than simple geometrical effects.

13.1.3.1 A Fun Notational Fact

When talking about spectroscopy, scientists will often refer to "p- and s-polarized light." This refers to the relative orientation of the **E**-field polarization with respect to the surface normal and the incoming ray (the plane of incidence). *p-Polarized* light is understood to have the electric field vector pointing parallel to the plane of incidence on a surface (p for parallel), and *s-polarized* light has the electric field vector perpendicular to this plane of incidence (s for *senkrecht*, the German word for perpendicular).

13.2 Polarization Coupling: Polaritons

Let's begin to expand our view by examining how the photon field couples to *polarization* within the solid. By *couples* here, we mean that the local fields will add vectorially to yield some new dynamics of the system, and we are supposing that spatially modulated and distributed dipole fields in materials will provide an opportunity for this. In gross terms, the velocity of an E&M wave in a solid changes to $v = (\varepsilon\mu)^{-1/2}$; the *frequency* remains fixed, with λ and k changing to compensate. This change in ε_0 to ε is a dielectric response (a polarization) to the optical field locally. But for systems where correlation is possible, things are not so simple.

13.2.1 Phonons with Electrical Polarization

Clearly (Chapter 12), there are many different ways in which atomic-scale charge separation can yield local electric fields (let's call them polarization fields). Naturally, this depends on the kind of atoms and their order and bonding within the solid. But, in a crystal, everything is bonded somehow to every other thing, and it is easy to imagine that atomic displacement will have significant impact on local polarization. This, in turn, has implications for the collective vibrational modes of a system.

How might that be? Consider the case of a two-atom basis one-dimensional lattice like the one in Figure 13.6. In Chapter 5 we identified this system's specific modes of vibration, analyzed the atom's relative motion to each other, and plotted the dispersion curves as shown in Figure 13.6 (lower left). But what if such a system had a natural dipole moment built in? Let's suppose an ionic bonding between the masses. So M_1 and M_2 are opposite ions with an electrostatic interaction in a kind of 1D rock salt. In this case the treatment of Chapter 5 is a little less accurate. There are now long-range forces that interact between distant sites, and our *nearest-neighbor assumption* is inaccurate. We mentioned this problem in ionic lattice sums previously.

The unique long-range interactions introduced in this geometry will lead to anharmonicity in the oscillations. But, we might, quite fairly, imagine that we should still expect a *near-usual* set of acoustic and optical mode branches in the dispersion characteristics. And there is more. Recall that we have longitudinal and transverse wave polarizations (spatial) for the branches. In our previous approximations, we treated these as *degenerate*. So, for example, the single *optical longitudinal* (*OL*) and the two *optical transverse* (*OT*) waves had the same ω and k for each energy in the dispersion relation. However, in the optical branch, the

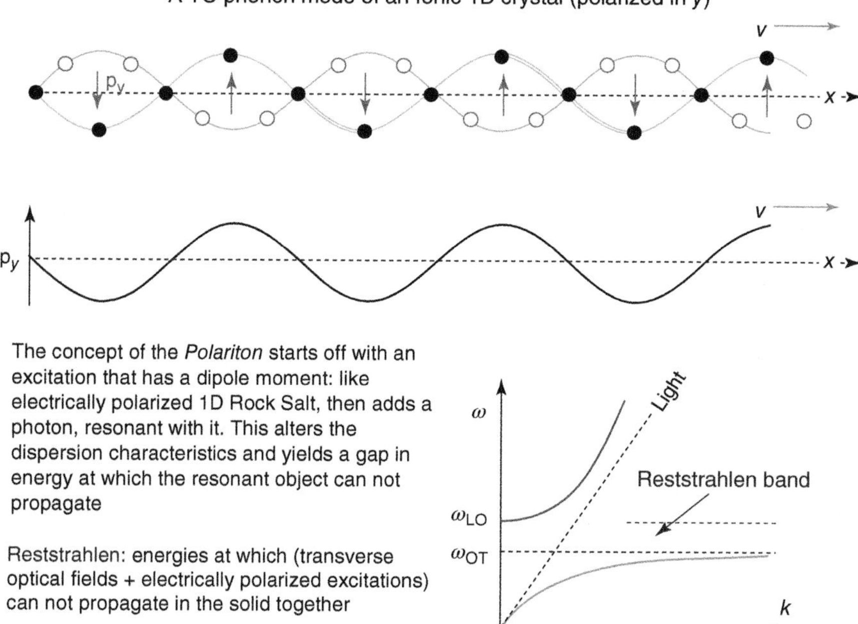

Figure 13.5 A transverse, optical phonon mode with positive and negative ions that make up two sublattices. In the optical phonon band, these move out of phase. Electrical polarization follows this motion, and **p** is now a function of position and time along the line of ions. The middle graph shows this polarization wave as it corresponds to the phonon wave above it. When this object is coupled or resonant with a photon in the space, the objects can move together supporting each other in their motion. The joint object's motion is described by *dispersion curves* just as we did in the case of electrons and phonons before.

atomic motion is such that the two ions (M_1 and M_2) move in antiphase with each other – an out-of-phase vibration of the sublattices. An example of this out-of-phase motion for a transverse wave is shown in Figure 13.5 (top). As a consequence, we note that the local polarization associated with atomic length scales (p_y) flips back and forth, forming a transverse wave of its own as seen in Figure 13.5 (middle). Thus, we can anticipate important consequences for the net polarization of the lattice, P, and the dielectric properties of the system. But it should be readily apparent that if the phonon wave were *longitudinal*, the polarization (p) wave formed would point along the direction of the strand since that is the direction of the atomic motion and thus the charge separation. And, since dipole–dipole interaction energies depend on the orientation of the interacting dipoles, it is further clear that the interaction energy of the dipoles in the transverse wave must be different from the interaction energy of the dipoles in the longitudinal wave. This lifts the degeneracy in the optical branch quite clearly. Indeed, the characteristic frequencies might now be labeled ω_L and ω_T, and they have quite different values for a given energy or k. Note that the effect itself is not very large and affects the self-energy of both longitudinal and transverse modes without any other applied fields around.

Since the optical phonon modes generally involve antiphase motion of the atomic sublattices, positive and negative ions in our example, the wave in p isn't surprising. It is simply one of the resonant conditions (normal modes) of the lattice, and notice the wavelength of p need not be the same as that of the phonon *carrying* it. Moreover, one might expect such systems of polarization waves to interact throughout such solids, forming resonances and quasiparticles. But all of this would happen without the presence of external fields and energies that depend on the specific system of interest.

13.2.2 Phonons Meet Photons

Now that we have introduced the idea of an electrically polarized phonon, let's examine how such an object might interact with an electromagnetic wave. So, in other words, we want a true *optical interaction* for such a solid. Of course, we really don't much care how the electrical polarization of the phonon takes place: if the electromagnetic wave induces it or if it is rock salt and is already present. We are mainly interested in how an electromagnetic wave (a photon) might interact with this phonon + polarization wave (a quasiparticle).

We begin by first noting that an electric field vector pointing longitudinally would only shift all the ions of one sign in one direction *not* forming a wave. So, it would appear that transverse electromagnetic waves *couple* (add) with OT phonon modes. In fact, longitudinal E&M waves do not couple with OT or LO phonons, and transverse E&M waves do not couple with LO phonons.

Secondly, true *coupling interactions* imply that there exists some sort of resonance between the photon and the phonon, i.e. a shared frequency or resonant frequency. As in Figure 13.5 (in the lower two graphs), the dashed lines are the dispersion of the light or photon, and these intersect with the optical branch phonons showing where a common frequency and common k might be. This perhaps makes it a little easier to understand why the names "acoustic" and "optical" were chosen for the branches of the phonon dispersion spectrum.

Finally, as we will show, there emerge natural conditions at which the OT phonon + polarization + the transverse electromagnetic wave will wish to propagate together. There are, however, gaps in these conditions for which no propagation of the combined or resonant object is allowed, known as the *reststrahlen band*. Further, the coupling of our two particles *does not* give a dispersion curve of the photon plus a dispersion curve of an electrically polarized phonon added together. Instead it gives a wildly different dispersion curve of a new quasiparticle that arises from this resonance, known as a *polariton*. This is actually shown in Figure 13.5 (lower right), but we return to it in more detail below.

Notice we have omitted the *longitudinal modes*. This is because photons do not couple with them, so they are not a part of our optical interactions study. But these modes together with any electrical polarization they may carry with them can be excited by or interact with itinerate charge injected into the solid (so $\nabla \cdot \mathbf{D} \neq 0$). The injected charge essentially "plows" its way through the lattice, creating its own polarization wave as it goes. Thus, there can be coupling between the itinerate charge and a polarization wave + phonon.[5] We refer to such an object as a *polaron*.[6]

13.2.3 The Phonon–Polariton

Here, we make a little bit of clarification. What is being described is not necessarily the stimulation of a polarization wave by a photon (view from causality), but rather a resonance between a photon and a transverse phonon polarization wave irrespective of the initial states. The *composite beast*, or *resonance*, is known more formally as a *phonon polariton*, and it should be seen as a quantum superposition of the photon and the *polar phonon*.[7] Notice that unlike some presentations, we use the phrase *phonon polariton* because there are different kinds of *polaritons*, and we wish to distinguish between them.

Now we have a very simple picture of the *phonon polariton* above [3]. It consists of a transverse electromagnetic wave and a phonon on a lattice that can support an electrical polarization (dipoles) at the unit cell scale. To understand the nature of coupling between the two as well as the dispersion characteristics of the coupled (or resonant) state, we need to write down a mechanical expression that correlates the undulation of the field with the motion of the ions. So we start with Newton's law for the ions of the rock salt chain (at general position $\boldsymbol{r}$) and include a sine wave electric field for the photon as a driving force (Eq. (13.3)). In this approach, the electric field of the photon is the only local electric field being considered. We ignore the relatively smaller contribution made by the polarized lattice locally (the little dipoles between every two sites). Of course if we wanted a more complete treatment, this would have to be included. But it isn't needed to get to the basic physics.

Beginning here:

$$m\ddot{\boldsymbol{r}} = -\kappa \boldsymbol{r} + e\mathbf{E} = -m\omega^2 \boldsymbol{r} \tag{13.3}$$

$$\mathbf{E} = \mathbf{E}_0 \mathrm{e}^{-i\omega t} \quad \text{(the optical field)} \tag{13.4}$$

We note that the fields associated with the motion of Eq. (13.3) must satisfy Maxwell:

$$\nabla \times \boldsymbol{H} = \frac{1}{c}\dot{\boldsymbol{D}} = \frac{1}{c}(\dot{\boldsymbol{E}} + 4\pi\dot{\boldsymbol{P}}) = -\frac{i\omega}{c}(\boldsymbol{E} + 4\pi\boldsymbol{P}) \tag{13.5}$$

$$\nabla \times \boldsymbol{E} = -\frac{1}{c}\dot{\boldsymbol{H}} = -\frac{i\omega}{c}\boldsymbol{H} \tag{13.6}$$

$$\boldsymbol{P} = N' e\boldsymbol{r} + n\alpha \boldsymbol{E} \tag{13.7}$$

5 Notice here that we are being careful to refer to phonon + polarization wave since the polarization wave is superimposed on the phonon with wavelengths related to that of the phonon. They are not separate things.

6 With careful reflection one can see a certain analogy in this "polaron" and in the polymer version we introduced a few chapters ago.

7 We have to be a little careful with language here. A polarized phonon can be one of the polarizations of the phonons motion (transverse vs. longitudinal), or it can be a phonon with a polarization wave. Let's call the second a *polar phonon* for now.

$$N' = \text{number of phonon modes}$$
$$n = \text{electron concentration}$$
$$\alpha = \text{polarizability} \tag{13.8}$$

Expressing in Cartesian coordinates with the wave moving along z and $\mathbf{E}$ pointing in x, we know we will have plane wave solutions "inside" this 1D solid (bringing the $\boldsymbol{K}\cdot\boldsymbol{r}$ out of $\mathbf{E}_0$ above as Kz), but we don't know the fields:

$$\begin{aligned} E_x &= E_x^0 e^{-i(\omega t - Kz)} \\ H_y &= H_y^0 e^{-i(\omega t - Kz)} \\ P_x &= P_x^0 e^{-i(\omega t - Kz)} \\ r_x &= r_x^0 e^{-i(\omega t - Kz)} \end{aligned} \tag{13.9}$$

To get to the fields (or rather the restrictions on $\omega(K)$ that makes them consistent with each other, we substitute these proposed solutions into Maxwell, and we get four equations with four unknowns to solve for

$$\begin{aligned} &iKH_y - \frac{i\omega}{c}E_x - \frac{4\pi i\omega}{c}P_x = 0 \\ &-iKE_x + \frac{i\omega}{c}H_y = 0 \\ &-\omega^2 r_x + \frac{\kappa}{m}r_x - \frac{e}{m}E_x = 0 \\ &P_x - N'er_x - n\alpha E_x = 0 \end{aligned} \tag{13.10}$$

This leaves the determinate

$$\begin{vmatrix} \omega/c & -K & 4\pi\omega/c & 0 \\ K & -\omega/c & 0 & 0 \\ e/m & 0 & 0 & \omega^2 - \kappa/m \\ -n\alpha & 0 & 1 & -N'e \end{vmatrix} = 0 \tag{13.11}$$

which yields a characteristic equation whose roots are the conditions for which the fields will be consistent:

$$\omega^4[1 + 4\pi n\alpha] - \omega^2\left[c^2K^2 + \frac{\kappa}{m} + \frac{4\pi N'e^2}{m} + \frac{4\pi n\alpha\kappa}{m}\right] + \frac{K^2c^2\kappa}{m} = 0 \tag{13.12}$$

This is the dispersion relation, $\omega(K)$, or alternatively $\varepsilon(\omega)$ once it is simplified. To simplify and examine limits, we define

$$\omega_{\mathrm{T}}^2 \equiv \frac{\kappa}{m} \tag{13.13}$$

And for $\omega \gg \omega_{\mathrm{T}}$,

$$\begin{aligned} P_\infty &= n\alpha E \\ \varepsilon_\infty &= 1 + 4\pi P_\infty / E \\ \varepsilon_\infty &= 1 + 4\pi n\alpha \end{aligned} \tag{13.14}$$

whereas for $\omega \ll \omega_{\mathrm{T}}$, $\varepsilon \to \varepsilon_0$. And for $\omega = 0$,

$$\boldsymbol{r} = e\mathbf{E}/\kappa \tag{13.15}$$

So,

$$\mathbf{P}_0 = \left[\frac{N'e^2}{\kappa} + n\alpha\right]\mathbf{E}$$

$$\varepsilon_0 = 1 + 4\pi\left[\frac{N'e^2}{\kappa} + n\alpha\right]$$

$$\varepsilon(\omega) = 1 + 4\pi\left[\frac{N'e^2}{\kappa - m\omega^2} + n\alpha\right] \quad (13.16)$$

So we get for our determinant of the matrix

$$\omega^4\varepsilon_\infty - \omega^2[c^2K^2 + \omega_T^2\varepsilon_0] + \omega_T^2c^2K^2 = 0 \quad (13.17)$$

which has two solutions:

$$\omega^2 = \frac{1}{2\varepsilon_\infty}(\omega_T^2\varepsilon_0 + c^2K^2) \pm \left(\frac{1}{4\varepsilon_\infty^2}(\omega_T^2\varepsilon_0 + c^2K^2)^2 - \frac{\omega_T^2K^2c^2}{\varepsilon_\infty}\right)^{1/2} \quad (13.18)$$

There is an important characteristic that stands out right away: there are two (OT) solutions for ω for every value of k (we just said that)! One set of values forms the *upper polariton band* (UPB), and the other a *lower polariton band* (LPB). So, for a given K there are two possible states – one of high energy and one relatively lower – but both of which represent a transverse polariton. Pictorially, one might think of this situation in terms of the relative phase relationships between the superimposed wavefunctions of the interacting entities. Graphically the dispersion looks like Figure 13.6.

Notice that for small wave vectors the positive solution of the above equation is

$$\omega^2 = \frac{1}{\varepsilon_\infty}(\omega_T^2\varepsilon_0 + c^2K^2) \quad (13.19)$$

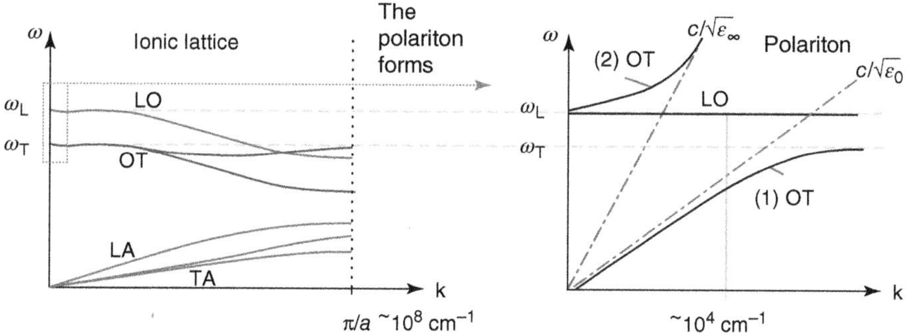

Figure 13.6 The coupled excitation of the optical transverse phonon **to** the electromagnetic radiation is what is referred to as a *phonon polariton*. Above is the transition of the dispersion curve from the phonons of the 1D ionic crystal to this *phonon polariton* where the photon–phonon has become resonant. It shows a splitting of the optical phonon modes and a band of energies where phonon polaritons cannot propagate but are rather absorbed. The LO phonon mode is also shown with the upper polariton branch. Notice that the diagram right is the section of the left diagram highlighted in the box along the ω-axis.

$$\omega_L^2 \equiv \frac{\omega_T^2 \varepsilon_0}{\varepsilon_\infty} \tag{13.20}$$

$\omega_{T/L} = cK/\sqrt{\varepsilon_{0/\infty}}$ are shown as a straight thin dashed lines on the graph in Figure 13.6, and in the long wavelength limit, the dispersion curve's upper and lower branches can be seen to approach these values. Notice that the lower branch of the polariton curve is pinned below the $\omega_T - cK/\sqrt{\varepsilon_0}$ axis lines and the upper branch is pinned above the $\omega_L - cK/\sqrt{\varepsilon_\infty}$. This gives us some insight into how we might physically picture the evolution of the states as a function of ω: the two lines define the assignment of "phonon-like" and "photon-like" behaviors. Clearly when the "resonance object" is approaching the ω_T/ω_L line asymptotically, it has a dispersion like an uncoupled phonon. If it approaches the photon curves, then it is acting more like a photon.

The ω_L value shown on Figure 13.5 at the intercept of the upper branch is defined by the zero of $\varepsilon(\omega)$. As $\varepsilon(\omega)$ goes negative there is no propagation because **K** becomes imaginary for a real ω. This situation occurs through the gap between ω_{OT} and ω_{LO}. Thus, this region gives strong absorption because the exponent in the wave solution becomes real. Curiously, ω_{LO} is also the frequency of the LO phonon modes of the lattice, so it has two meanings (see Exploring Conceptssection). Eq. (13.19) is actually known as the Lyddane–Sachs–Teller equation, and it was originally derived for a two-atom basis on a cubic lattice. Notice that as ω_T goes to 0 (soft phonon modes), $\varepsilon(0)$ becomes infinite. This is a characteristic of the interactions between **p**-waves we spoke of above, leading to *ferroelectricity*.

13.2.4 The Plasmon Polariton

As we have already described in Chapter 10, *plasmons* are the oscillations of the electrons in a plasma. These oscillations lead to traveling waves, and they have, each, a well-defined frequency and wave vector. Moreover, the *plasmon* plasma wave is a longitudinal wave. Thus the displacement of the electrons (relative to the positive ionic background) is parallel to the direction of propagation of the wave. This means that the transverse fields of the photon do not couple with the plasmon in any obvious way.

When we consider the electron sea moving relative to the ionic background in a density wave, it seems obvious that there will be alternating positions of relative positive and negative charges along the wave spatially. Therefore, we expect that the plasmon would have a local dipole moment as part of its excitation in a way analogous to that of the ionic phonon. So, we ask, "are there some set of conditions that would allow for a photon to couple to this excitation forming a polariton?" We have already suggested that the answer is "no" for most normal circumstances, but we all know that nature will frequently find a way around our objections.

To see nature's answer, we note that plasmons themselves are found in a variety of different configurations, both 2D (nanoplatelets) and 3D. Figure 13.7 gives a diagram of some of these. The dispersion characteristics of the plasma wave are quite different in each case, but it is the localization and restriction of electron flow across the object or within the material that provides the opportunity

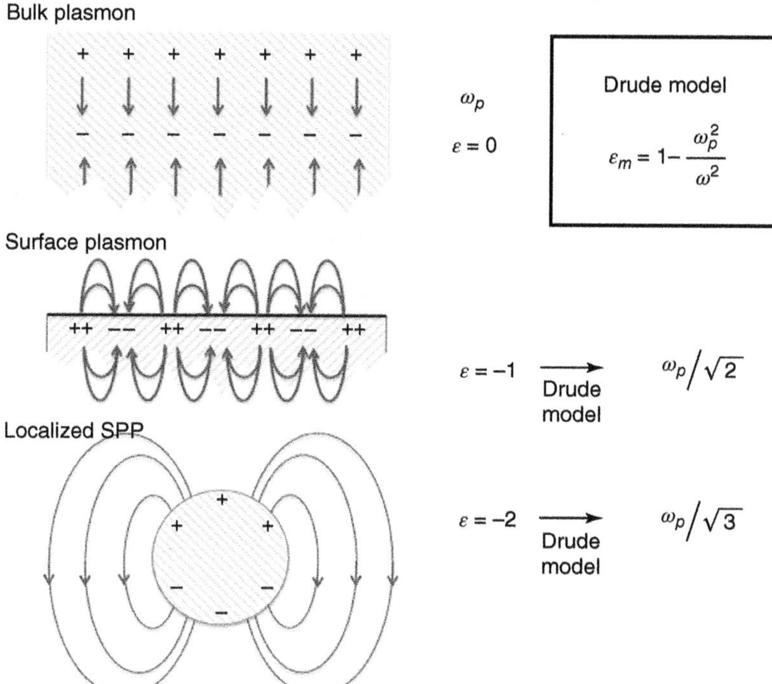

Figure 13.7 Models of the bulk plasmon, the surface plasmon, and the localized surface plasmon on a nanoparticle. On the right is given the results of applying Maxwell's equations to the Drude model for ε and the frequency range of each type of plasmon. These are, for example, the eigenfrequencies that the plasmonic system can take on in the case of surface plasmons: $0 < \omega < \omega_p/\sqrt{2}$.

for coupling this wave with light (the formation of the polariton). To be more specific:

1. Plasmons localized to the interface between two materials (*surface plasmons)* can form the *surface plasmon polariton (SPP).* This is because a photon can be incident on the interface at an angle. A *p-polarized* photon (polarized in the plane of incidence) can be decomposed into two components: one perpendicular to the interface and the other parallel to the interface. The parallel component can excite a surface plasmon because this component now forms a longitudinal wave. The photon and surface plasmonic state (sometimes called the "surface plasmon resonance" (SPR)) must have the same frequency and wave vector. So, the component of the E&M wave vector parallel to the interface has to be equal to that of the SPR. To observe an SPP experimentally, the absorption of an incident E&M wave (of constant frequency) is measured as a function of the angle of incidence on the interface. By changing the angle of incidence, the component of the photon wave vector parallel to the interface is "dialed in." The presence of a peak in the absorption vs. angle curve will indicate the excitation of an SPP. By varying the incident photon frequency and determining the angle of incidence at which the SPP is observed (and hence

the wave vector of the SPP), the entire dispersion curve of the SPP can be mapped.

2. Plasmons with some sort of bulk confinement such as found in a metamaterial (or a material with hyperbolic bands) can form bulk-like plasmon polaritons. But there must be edges to current flow in one way or another. Here, we do *not* mean a collection of internal SPPs. Instead the coupling of photon–plasmon states in electron-confined spaces allows for unique self-interactions and gives rise to an entirely new part of the dispersion curve for plasmon polaritons. Such polaritons are far less common but are quite important for understanding losses in metalenses and some types of meta-antennae. Moreover, they also provide a route to the transport of photons through a metallic-like system as a polariton.

There are similarities between the concepts of the *plasmon polariton* and the *phonon polariton*. In Chapter 10 the dielectric response of the Drude metal was expressed with ω_p = plasma frequency and γ = system damping:

$$\varepsilon_m = 1 - \frac{\omega_p^2}{\omega^2}; \ \omega_p = \sqrt{\frac{ne^2}{\varepsilon_0 m}} \tag{13.21}$$

This is the simplest description of the plasma oscillation. It is shown schematically in Figure 13.7 and graphically in Figure 13.8.

Using the dielectric expression above, we return to Maxwell and solve for the equations of the dispersion curve. Of course, if we follow the steps of the phonon polariton dispersion exactly, then we would be deriving a bulk-like *plasmon polariton* and a bulk-like $\varepsilon(\omega)$ where we have not yet given any boundary conditions (Figure 13.9):

$$\omega^2 = \omega_p^2 + c^2k^2 \tag{13.22}$$

But this dispersion implies that the bulk plasmon polariton frequency is always greater than the plasma frequency ω_p. This means that no transverse electromagnetic wave with a frequency smaller than ω_p can be transported through the material.

So we have a dispersion curve of a bulk plasmon polariton, where such objects are allowed to exist. Next, we have the plasmon polariton associated with surface waves. The picture one should have is as in Figure 13.10.

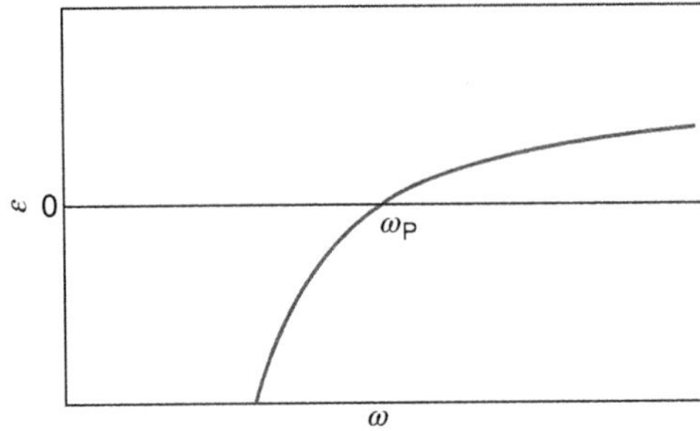

Figure 13.8 An approximation to the dielectric function of the electron gas. Notice that the function goes negative for frequencies below ω_p.

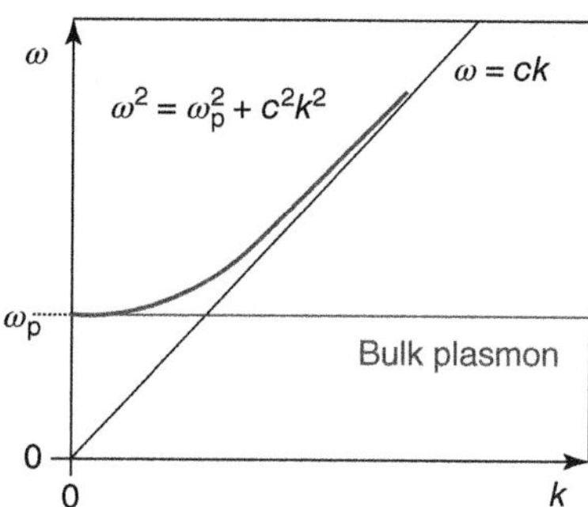

Figure 13.9 The bulk dispersion curve derived from applying Maxwell (not shown). The dispersion equation for the bulk properties of the bulk plasmon polariton is in Eq. (13.22).

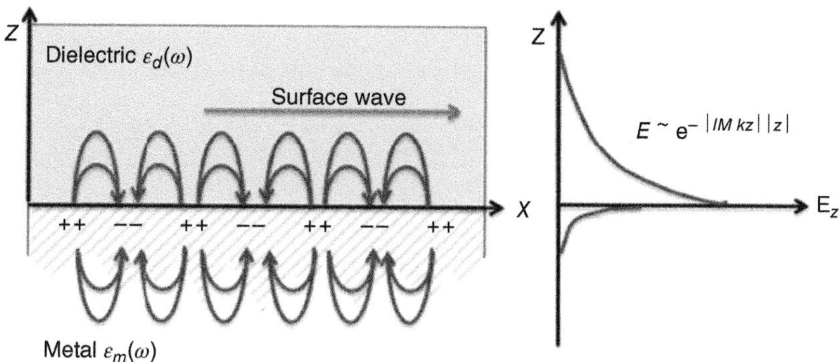

Figure 13.10 The surface plasmon polariton is a coupling between the photon and the surface plasmon at the interface between a metal and a dielectric. However, the electric field does not extend far beyond this interface.

In the dispersion relations for a wave in the geometry of Figure 13.10 **configuration**, we apply $\nabla \times \mathbf{H} = \varepsilon c^{-1}\ \partial\mathbf{E}/\partial t$ assuming a wavelike solution for the fields. Keeping in mind that the k's and the ε's are different on different sides of the interface, we then have to use boundary matching for **E** and **H** components. This gives us the following components:

$$E_y = H_x = H_z = 0 \tag{13.23}$$

$$H_d = (0, H_{yd}, 0)\mathrm{e}^{i(k_{xd}x + k_{zd}z - \omega t)}$$

$$E_d = (E_{xd}, 0, E_{zd})\mathrm{e}^{i(k_{xd}x + k_{zd}z - \omega t)}$$

$$H_m = (0, H_{ym}, 0)\mathrm{e}^{i(k_{xm}x + k_{zm}z - \omega t)}$$

$$H_m = (E_{xm}, 0, E_{zm})\mathrm{e}^{i(k_{xm}x + k_{zm}z - \omega t)} \tag{13.24}$$

as we go from metal to dielectric. And our boundary conditions at $z = 0$ are

$$\varepsilon_m E_{zm} = \varepsilon_d E_{zd} \tag{13.25}$$

$$E_{xm} = E_{xd} \tag{13.26}$$

$$H_{ym} = H_{yd} \tag{13.27}$$

which yield

$$k_{xm} = k_{xd} \tag{13.28}$$

so

$$k_{zm}H_{ym} = -\varepsilon_m \frac{\omega}{c} E_{xm} \tag{13.29}$$

$$k_{zd}H_{yd} = \varepsilon_d \frac{\omega}{c} E_{xd} \tag{13.30}$$

$$k_{xm}H_{ym} = -\varepsilon_m \frac{\omega}{c} E_{zm} \tag{13.31}$$

$$k_{xd}H_{yd} = -\varepsilon_d \frac{\omega}{c} E_{zd} \tag{13.32}$$

From these we get

$$\frac{k_{zm}H_{ym}}{k_{zd}H_{yd}} = -\frac{\varepsilon_m E_{xm}}{\varepsilon_d E_{xd}} \tag{13.33}$$

$$\frac{k_{zm}}{k_{zd}} = -\frac{\varepsilon_m}{\varepsilon_d} \tag{13.34}$$

$$\frac{k_{zd}}{\varepsilon_d} + \frac{k_{zm}}{\varepsilon_m} = 0 \tag{13.35}$$

and

$$k_x^2 = \left(\frac{\omega}{c}\right)^2 \frac{\varepsilon_m \varepsilon_d}{\varepsilon_m + \varepsilon_d} \tag{13.36}$$

$$k_{zd}^2 = \left(\frac{\omega}{c}\right)^2 \frac{\varepsilon_d^2}{\varepsilon_m + \varepsilon_d} \tag{13.37}$$

$$k_{zm}^2 = \left(\frac{\omega}{c}\right)^2 \frac{\varepsilon_m^2}{\varepsilon_m + \varepsilon_d} \tag{13.38}$$

where the k_x is real and the k_z's are imaginary. Thus the dispersion curve looks like the one in Figure 13.11.

However, we note that propagation is going to be along the interface. And the more astute student who is adept at E&M will notice that this doesn't look so different from other waveguide problems. However, there are a few things that should be recognized:

1. Only *radiative* surface plasmons are coupled with propagating E&M waves. *Non-radiative* surface plasmons do not couple with propagating E&M waves.
2. Perfectly flat surfaces allow only SPPs that are always non-radiative! Thus, rough surfaces will be needed to see emission from plasmon polaritons.
3. In contrast to conventional waveguides, the electric field on either side of the interface is *evanescent*.

It has probably become clear that we could make a far more general statement about polaritons. The *polariton* can be generally defined as the mixing (in the

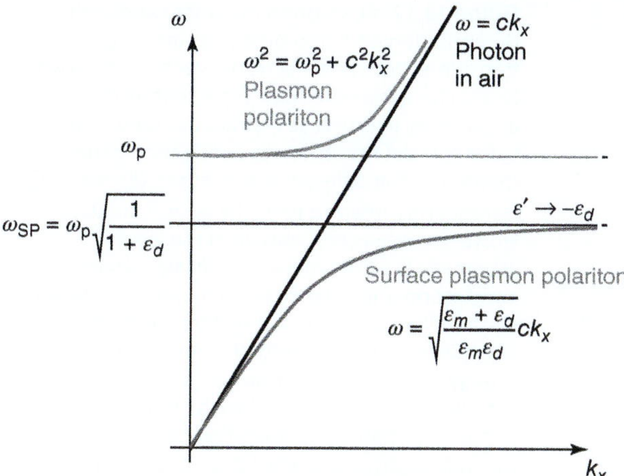

Figure 13.11 The dispersion curve of plasmon polaritons. Notice that the ω_p is the volume or bulk plasmon frequency. The ω_{SP} is the collective nonpropagating surface oscillations.

sense of a quantum superposition of states) of a photon with any excitation of the solid that has a polar component. We have seen examples in *phonon polaritons* (for ionic crystals) and *plasmon polaritons*, but there are many others. We will examine one final example of this: *exciton polaritons*. First, however, we need to introduce some new ideas.

13.3 Optical Transitions, Excitons, and Exciton Polaritons

Things can get complicated when direct transitions between electronic states are allowed: bandgap absorption of a photon. It is even worse when optically active defects (defects that absorb a photon and release a charge into a band) are involved. But this is exactly what happens in semiconductors and small bandgap insulators. In fact, the absorption and emission of light from solids is an entire field of technology today. So, we need a few good models to help guide us.

13.3.1 Transitions

Recalling from our previous discussions on semiconductor band structure, we discussed transitions between bands. The electron absorbs a photon or some thermal energy and is promoted from a lower band into a higher energy state (presumably another band unless there are defect states around). Shown in Figure 13.12 are two versions of the process that we are discussing: the direct

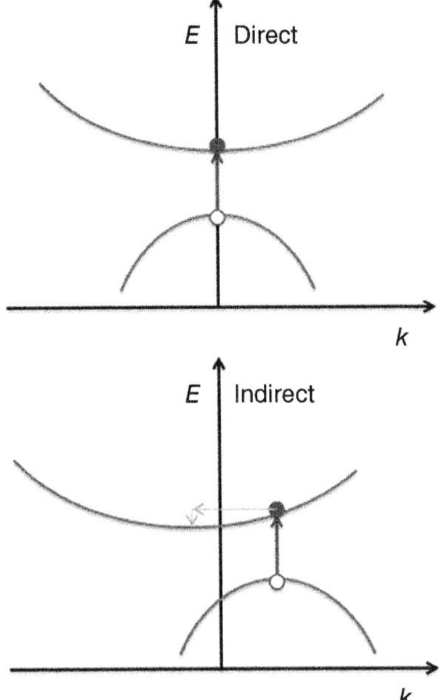

Figure 13.12 Absorption processes and their de-excitation emission processes are simply transitions between electronic states in the solid. Shown here is a schematic representation of absorption, in one case a direct excitation from valence band maximum to conduction band minimum. The difference in energies ($E_g = E_c - E_v$) is frequently referred to as the *optical bandgap*. However, the process shown in the lower schematic is a little more complicated. Here conduction and valence band minima and maxima do not align. So, either a direct transition might take place if the incoming photon has enough energy, in which case a phonon is generated and the electron is thermalized to the bottom of the conduction band, or a transition between energy minima might take place if momenta can be balanced with an existing phonon in the system. So, for the indirect transition the semiconductor can absorb or emit a phonon.

and indirect bandgap transitions. You may also recall that these two different processes are the result of momentum conservation in the electronic transition and that the direct one leads to luminescence while the other indirect transition typically does not:

$$\hbar\omega = E_{\mathrm{f}} - E_{\mathrm{i}} \pm \hbar\omega_q \tag{13.39}$$

where the f and i subscripts are for final and initial states. Either way, the probability for this process is different from that of the direct process. Thus, the absorption coefficient in the two cases will be different, as will the photon emission (de-excitation or *recombination*) characteristics.

Notice that an electron and hole are produced by such transitions, but they may have very different effective masses. The energy difference between the valence band maximum and conduction band minimum is called the energy gap E_{g} for both cases, but the top is a *direct gap* and the bottom an *indirect gap*.

Of course, we are all now familiar with Beers law: $I \sim I_0 \mathrm{e}^{-\alpha z}$. This is a simple rule in which the physics "model making" is locked up in α, the absorption coefficient. Seen schematically in Figure 13.13, α can help us understand which processes are dominate in photon–solid interactions.

Indeed, α can be a bit like the $\varepsilon(\omega)$ described above: it provides a specific way of measuring the effects of the models we propose. Unfortunately, direct first-principles calculations of this quantity (for most processes) can be pretty detailed, so we tend to rely on approximations using *Fermi's golden rule* from basic quantum mechanics.

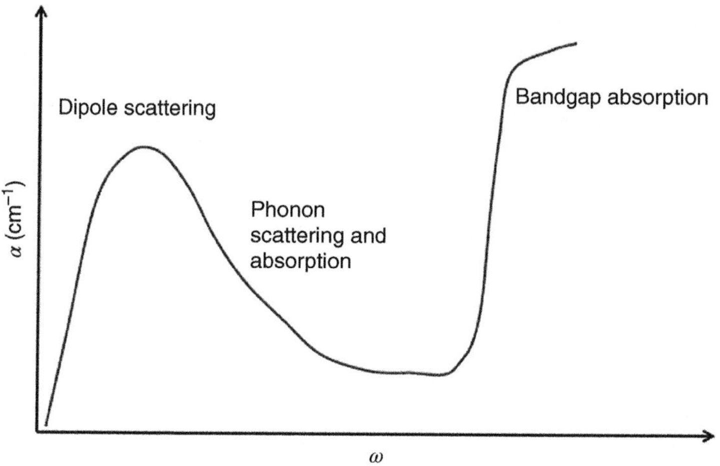

Figure 13.13 The absorption coefficient varies with frequency of light being used, and it can incorporate many different absorption and scattering mechanisms. Shown here is an idealized curve of α. There is the simple scattering from defects and inclusions in the crystal, and, of course, we have already seen how phonons can effect incoming photons. Plasmons would also be in here if such an entity existed in the solid along with any other kind of quasiparticle. However at the bandgap energy, electronic transitions between bands begin to dominate.

The many faces of α.

Free Carrier Absorption
absorption of an excited state carrier into another excited state

- Semiconductor $\alpha_{\text{abs}}(\omega) \sim \omega^{-2}$
- Metal at low frequency $\alpha_{\text{abs}}(\omega) \sim \omega^{\frac{1}{2}}$

Direct Interband Transition
absorption of valence band electron into conduction band excited state

- Allowed transitions $\alpha_{\text{abs}}(\omega) \sim \dfrac{(\hbar\omega - E_g)^{\frac{1}{2}}}{\hbar\omega}$
- Forbidden transitions $\alpha_{\text{abs}}(\omega) \sim \dfrac{(\hbar\omega - E_g)^{\frac{3}{2}}}{\hbar\omega}$

Indirect Interband Transition
phonon or other quasiparticle assisted absorption processes

- General Coeff. $\alpha_{\text{abs}}(\omega) \sim (\hbar\omega - E_g \pm \hbar\omega_q)^2$

The α dependence on ω for different processes, as the result of a basic transition rate analysis, is listed above. Full derivations of these results can be found in a number of texts devoted to optical properties of solids [4]. There are, however, two things of note that should be emphasized. The first is that reflection and transmission spectroscopy *does* obviously provide a powerful tool in identifying dominant interaction mechanisms for materials. The second point is that these results are worked out for bulk, 3D, infinite crystals. They change drastically when considering systems of finite size in 1D or 2D!

13.3.2 Carbon Nanotubes: An Example

This last point is of particular interest to this text. The absorption characteristics as measured in the absorption coefficient can be strongly influenced by the dimensionality of the materials under consideration. Nowhere is this better demonstrated than with carbon nanotubes.

We note first that absorption in a thin film of aligned carbon nanotubes is strongly polarization dependent. When the electric field of the incoming photon is aligned with the axis of the nanotube, absorption is quite strong, whereas when the electric field is aligned perpendicular to the nanotube axis, the absorption is considerably weaker. Thus, they act like a polarization film, and an absorption matrix must be calculated to account for this orientational dependence. As we already know carbon nanotubes can be metallic or semiconducting depending on diameter and chirality, so for the metallic nanotubes, this makes absolute sense in terms of a simple Drude model. In terms of the semiconducting variety, it is a little harder to understand but becomes more apparent when we think in terms of the available k-states into which an electron may be promoted.

We may then ask, "in the case of optimal polarization for absorption, do carbon nanotubes behave as we might expect carbon, or any other semiconductor or metal, to behave in terms of α?" Again, the answer is not quite. A detailed calculation and analysis of the absorption coefficients of whole families of nanotubes has been performed by Malić et al. [5]. This study found that the absorption coefficient for an arbitrary zigzag tube goes as ω^{-2} across the frequency spectrum even though such tubes do have a finite bandgap. This looks more like free carrier absorption in 3D systems. However the study underestimates the strength of the absorption (when compared to experiment). This could be contributed to the extremely large dephasing length in nanotubes that allow them to behave as antennae generally. In such systems, the whole object contributes to the *oscillator strength* of the transition [6].

13.3.3 Color Centers and Dopants

Besides direct and indirect band transitions, another important mechanism to consider in absorption and recombination phenomena is that of defect and dopant atoms in the lattice. We have previously discussed defect states and dopant states. They lead to states within the bandgap of semiconductors and insulators. But now we must consider the potential of such states to be optically active. Recall that the energy difference between the band edge and the gap state is typically less than $E_c - E_v$. This means that sub-bandgap light can excite such transitions and that very specific emission profiles will be associated with them.

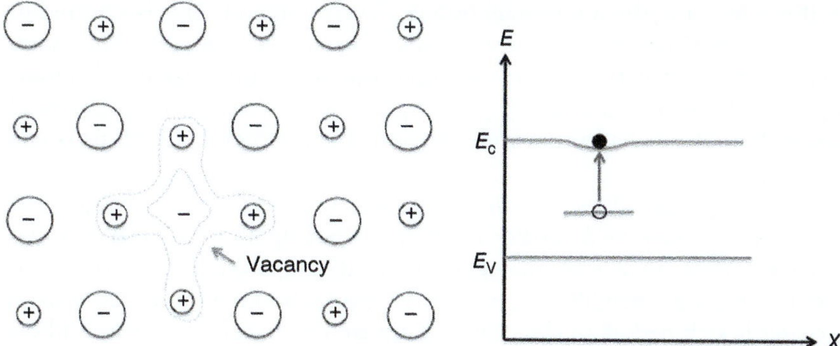

Figure 13.14 A negative ion vacancy yields a gap state. Relaxation around the vacancy gives a complex potential landscape (gray) that can trap an electron in the vicinity of the vacancy. The Earnshaw theorem says this electrostatic trap cannot be stable. So there is a finite lifetime.

In ionic crystals (i.e. alkali halides), electrons can be trapped at negative ion vacancies that have an effective positive charge. We can think of this in terms of Figure 13.14, a simple model of electronic transition from a filled defect state into the conduction band with the local Coulomb field and Pauli exclusion[8] localizing the electron near that defect.

The electron trapped in such a localized state will typically have an *s-like* or *p-like* wavefunction, so it is rather hydrogenic in nature. The different excited states of this "hydrogenic atom" are linked by optical transitions. Absorption and de-excitation due to these hydrogenic transitions will usually impart some color to the crystal, so they are known as *Farbe centers* or *F-centers* (Farbe is the German word for color). A variation on this theme is particularly advantageous in many gemstones.

There is a curiosity in the transitions of *F*-centers. That is, there is a pronounced shift in wavelength between the absorbed photon and the emitted photon. This is due to the nature of the trapping potentials involved. When a photon is absorbed, the vacancy-trapped electron is in a state that is defined by the local potential well. This is strongly influenced by the relaxation of other ions in the location of the defect. However when the electron is excited, the local potentials change. The excitation process is fast. The change in the relaxation of the surrounding ions is slow. Thus a *Franck–Condon effect* takes place; the transition happens, and then the local ions respond, changing the energy of the state the electron moved to. Thus, the difference in the absorption and emission energies is directly related to the energy of configuration change in the lattice locally. This also means that F-centers can be quite sensitive to temperature.

A vacancy is energetically the favored defect to occur. But there are many ways in which this can happen and, thus, many different types of optically active defect centers. First, defects, much like their atomic counterparts, like to cluster in the solid. Defects typically form in such a way as to be charge neutral. So when an ion is removed from a site and is displaced into an interstitial location of the

8 We include Pauli exclusion here to include the idea that the electron in the conduction band may move through a Hubbard-like hopping mechanism.

lattice, the defect and the ion remain bound electrostatically. This is referred to as a *Frenkel pair*. However, when nearby vacancies, one positive (where a cation should be) and one negative (where an anion should be), link together Coulombically, this is known as a *Schottky pair*. Again, because of the local potential landscape, the optical absorption and emission characteristics of such pairs can be unique.

This clustering also gives rise to the proximal location of F-centers. For instance, the F_2*-center* or *M-center* is two neighboring anionic vacancies that bind two electrons in a way analogous to a helium atom. An F_3*-center* or *R-center* is three such neighboring anionic vacancies binding three electrons. If an F_2*-center* is stripped of an electron, this appears as an ionized system: a hole bound by two negative charges. This is referred to as a V_k*-center*. Curiously, the positive analogue of the *F-center* does not exist. That is, there isn't such a case where a cation vacancy traps a hole. Quite simply this is because the concept of the hole is a many-body-derived concept. It only makes sense over several lattice positions but not over the length scales of a single vacancy.

13.3.4 Excitons

In a simple single particle picture, absorption leads to the creation of an electron in the conduction band with a hole left behind in the valence band. Both are free to move in the crystal. But we note that the electron and hole are charged particles. Further, at temperatures that are low relative to the bandgap energy, any semiconductor or insulator would have a relatively small number of free carriers in the conduction band. So it wouldn't be too surprising to suggest that the electron and hole might lower their combined energy by binding together. The lack of free carrier density would prevent effective screening (for $k_B T \ll E_g$). This means the Coulomb field could certainly produce such binding. In 1931 Yakov Frenkel argued that the electron–hole pair created in this way would behave as a single charge neutral entity that he called an *exciton* [7]. Indeed, the nanotube absorption example that was given above is dominated by such excitons.

In general, the exciton must be considered in two different limits of the band structure. The first is the case of very weak binding. Such binding can occur when the conduction and valence bands are relatively flat and curved slightly differently, yielding slightly different dispersion characteristics of the hole and electron. Such excitons are referred to as *Mott–Wannier excitons*, and the electron and hole orbit each other at large distances. Alternatively, the second case is when the electron and hole are very tightly bound, so as to occupy practically the same lattice position. In this case, typically, critical points exist between bands, and the dispersion of the electron and hole is equal. This is called a *Frenkel exciton*.

In a quantum mechanical treatment of such bound states, the Schrödinger equation is usually split into a hydrogenic part and a center of mass part for the motion of the bound pair. However, just as in the case of the F-centers above, this can be a little misleading since such a treatment relies on a single particle picture of electronic band states when in fact we are dealing with a many-body problem in general. Thus, as expected, there are a number of caveats in exciton behavior that arise. While these go a bit beyond our present discussion, effects such as giant oscillator strength in self-trapped excitons are now widely studied [8].

13.3.5 Exciton Polaritons

One of these "unusual properties" found in excitons is their ability to interact with a photon to form an *exciton polariton*. As we know, the polariton concept requires a polar excitation in the crystal with which the photon may interact. The moving exciton provides exactly this. Clearly the displacement of charge results in a local dipole moment. If orbital frequencies are some multiple of the frequency of some incident photon, then a superposition of fields can occur, yielding a polariton. This is an exceedingly simple mechanical picture of course, but it does suggest a dispersion curve that divides up into regions of photon-like and exciton-like behaviors just as we saw above for plasmons and phonons.

13.4 Kramers–Kronig[9]

To finish our "tour" of the optical properties in solids, we should examine the models and data above from the experimental point of view. We have already mentioned processes such as Raman and Brillouin scatterings, and these are really quite useful. However, a particularly simple but important tool is simply absorption and reflection. We have written many of our models in terms of the dielectric response, but how is the connection between absorption and reflection made? The connection is, if fact, not so hard [9].

We begin with the relationship between **E** and **D** for some particular frequency in a solid:

$$\mathbf{D}(\boldsymbol{x}, \omega) = \varepsilon(\omega)\mathbf{E}(\boldsymbol{x}, \omega) \tag{13.40}$$

giving the Fourier transforms

$$D(\boldsymbol{x}, t) = \frac{1}{\sqrt{2\pi}} \int_{-\infty}^{\infty} D(\boldsymbol{x}, \omega) \mathrm{e}^{-i\omega t} \mathrm{d}\omega \tag{13.41}$$

$$D(\boldsymbol{x}, \omega) = \frac{1}{\sqrt{2\pi}} \int_{-\infty}^{\infty} D(\boldsymbol{x}, t') \mathrm{e}^{i\omega t'} \mathrm{d}t' \tag{13.42}$$

$$E(\boldsymbol{x}, t) = \frac{1}{\sqrt{2\pi}} \int_{-\infty}^{\infty} E(\boldsymbol{x}, \omega) \mathrm{e}^{-i\omega t} \mathrm{d}t \tag{13.43}$$

$$E(\boldsymbol{x}, \omega) = \frac{1}{\sqrt{2\pi}} \int_{-\infty}^{\infty} E(\boldsymbol{x}, t') \mathrm{e}^{i\omega t'} \mathrm{d}t' \tag{13.44}$$

So

$$D(\boldsymbol{x}, t) = \frac{1}{\sqrt{2\pi}} \int_{-\infty}^{\infty} \varepsilon(\omega) E(\boldsymbol{x}, \boldsymbol{\omega}) \mathrm{e}^{-i\omega t} \mathrm{d}\omega \tag{13.45}$$

9 This requires a good knowledge of complex integrals.

$$D(x,t) = \frac{1}{2\pi}\int_{-\infty}^{\infty} \varepsilon(\omega)e^{-i\omega t}d\omega \int_{-\infty}^{\infty} E(x,t')e^{i\omega t'}dt' \tag{13.46}$$

$$D(x,t) = \varepsilon_0\left[E(x,t) + \int_{-\infty}^{\infty} G(\tau)E(x,t-\tau)d\tau\right] \tag{13.47}$$

where

$$G(\tau) = \frac{1}{2\pi}\int_{-\infty}^{\infty}\left[\frac{\varepsilon(\omega)}{\varepsilon_0} - 1\right]e^{-i\omega\tau}d\omega = \frac{1}{2\pi}\int_{-\infty}^{\infty}\chi_e(\omega)e^{-i\omega\tau}d\omega \tag{13.48}$$

is called the "susceptibility kernel."

Since

$$\varepsilon(\omega) = \varepsilon_0[1 + \chi_e(\omega)] \tag{13.49}$$

and

$$\chi_e = \frac{\varepsilon}{\varepsilon_0} - 1 = \frac{\omega_p^2}{\omega_0^2 - \omega^2 - i\gamma_0\omega} \tag{13.50}$$

from our section of dielectrics, then,

$$G(\tau) = \frac{\omega_p^2}{2\pi}\int_{-\infty}^{\infty}\frac{e^{-i\omega\tau}}{\omega_0^2 - \omega^2 - i\gamma_0\omega}d\omega \tag{13.51}$$

This integral can be worked using contour methods. First, we find the roots of the denominator:

$$\omega_{1,2} = \frac{1}{2}\left[-i\gamma \pm \sqrt{-\gamma^2 + 4\omega_0^2}\right] \tag{13.52}$$

or

$$\omega_{1,2} = -\frac{i\gamma}{2} \pm \omega_0\sqrt{1 - \frac{\gamma^2}{4\omega_0^2}} = -\frac{i\gamma}{2} \pm \nu_0 \tag{13.53}$$

where $\nu_0 \approx \omega_0$ when $\omega_0 \gg \gamma/2$, and these poles are in the lower half plane since γ is dissipative (negative). This then gives

$$G(\tau) = (2\pi i)\frac{\omega_p^2}{2\pi}\oint_C \frac{e^{-i\omega\tau}}{(\omega - \omega_1)(\omega - \omega_2)}d\omega \tag{13.54}$$

We close the integral in the upper half plane and restrict $\tau < 0$. This is to ensure that the integrand goes to zero at the boundary at infinity where ω has a positive imaginary part. Since the integral encloses no poles in the upper part, $G(\tau) < 0$ vanishes. This is essentially ensuring causality.

Now closing the integral on the lower half plane, $\tau > 0$, we get

$$G(\tau) = \omega_p^2 e^{-\frac{\gamma\tau}{2}}\frac{\sin\nu_0}{\nu_0}\Theta(\tau) \tag{13.55}$$

where Θ is the Heaviside function.

Finally we use Cauchy's theorem again. **E** and G are real. So integrating by parts,

$$\frac{\varepsilon(\omega)}{\varepsilon_0} - 1 = i\frac{G(0)}{\omega} - \frac{G'(0)}{\omega^2} + \cdots \tag{13.56}$$

and so we conclude

$$\varepsilon(-\omega) = \varepsilon^*(\omega^*) \tag{13.57}$$

Therefore in the upper half plane, we have

$$\frac{\varepsilon(z)}{\varepsilon_0} - 1 = \frac{1}{2\pi i}\oint_C \frac{\frac{\varepsilon(\omega')}{\varepsilon_0} - 1}{\omega' - z} d\omega' \tag{13.58}$$

Let

$$z = \omega + i\delta \tag{13.59}$$

$$\delta \to 0_+ \tag{13.60}$$

that is, deform the contour along the real axis just below the singular point. Then from the Plemelj relation:

$$\frac{1}{\omega' - \omega - i\delta} = \frac{P}{\omega' - \omega} + i\pi\delta(\omega' - \omega) \tag{13.61}$$

Thus through substitution,

$$\frac{\varepsilon(\omega)}{\varepsilon_0} = 1 + \frac{P}{i\pi}\int_{-\infty}^{\infty} \frac{\frac{\varepsilon(\omega')}{\varepsilon_0} - 1}{\omega' - \omega} d\omega' \tag{13.62}$$

This then leaves us with the famous Kramers–Kronig relations

$$\mathrm{Re}\left(\frac{\varepsilon(\omega)}{\varepsilon_0}\right) = 1 + \frac{P}{\pi}\int_{-\infty}^{\infty} \frac{\mathrm{Im}\left(\frac{\varepsilon(\omega')}{\varepsilon_0}\right)}{\omega' - \omega} d\omega' \tag{13.63a}$$

$$\mathrm{Im}\left(\frac{\varepsilon(\omega)}{\varepsilon_0}\right) = -\frac{P}{\pi}\int_{-\infty}^{\infty} \frac{\mathrm{Re}\left(\frac{\varepsilon(\omega')}{\varepsilon_0}\right) - 1}{\omega' - \omega} d\omega' \tag{13.63b}$$

This tells us that if we know the whole absorptive spectrum, then we can derive the reflective spectrum.

13.5 Summary

So, as we already know, there are many quasiparticles that occur when electrons and holes interact or when the lattice vibrates. These are correlated many-body effects, and they have their own dispersion characteristics and dynamics. But it would also appear that there exist a number of equally exotic creatures that occur when the electromagnetic field is in superposition with a solids' electronic or vibrational excitations. We have seen excitons and polaritons of different types. These can be viewed as the interaction between a photon and a quasiparticle of nonoptical origin. Alternatively they can be pictured as the superposition of a crystal excitation and an electromagnetic wave. They arise specifically because of the ability of the E&M field to produce propagating solutions in the solid. To be sure, we have not exhausted all the possibilities here. But we have produced the reasoning for the occurrence of such optical interactions generally.

What have we left out? We have not presented the massive amount of work that revolves around spin–orbit effects and the use of heavy ions to create optical effects.

Exploring Concepts

1 *Phonon–Polaritons*: In our discussion of the frequencies of the LO phonon modes and the dispersion characteristics of the phonon polariton in Figure 13.6, we made the statement that the ω_{LO} has two meanings in this context.

(a) What is meant by this statement? What are the two meanings we have hinted at?

(b) Using the dispersion curves in Figure 13.6, show how these two meanings or interpretations of ω_{LO} converge to each other – in other words they are part of the say phenomenon.

2 *Plasmon–Polaritons*: In the development of organic photovoltaics and organic light-emitting devices, it is well known that reflection and absorption from the interfaces that couple the light into or out of the device is one of the most serious sources of loss in device performance. A quick literature search will reveal that organic light-emitting devices (OLEDs) and organic photovoltaics (OPVs) both suffer from losses arising from SPPs. But, why these? Why not bulk plasmons? Do a detailed study of the literature to explain the reason that these excitations are so important. Then demonstrate using the dispersion curves above exactly why this might impact OPVs and OLEDs so much.

3 *Polarization Current*: Now that we have some models for how polarization occurs from the last chapter, we can make some sense of what the *polarization current* is physically. Recall this is defined as

$$J_{\text{pol}} = \partial P / \partial t$$

and as light enters a solid, such a time-varying P is established. But can the polarization current lead to energy absorption? Why or why not? Prove your assertion.

4 *Plasma Oscillations*: Have you ever noticed that physicists typically refer to excitations in the solid-state electron gas, as *plasma oscillations*, not *plasma waves*? Why is this?

To see why, show that the group velocity, $v_g = \partial\omega/\partial\,(2\pi/\lambda)$, is given by

$$v_g = c\left[1 - \left(\frac{\omega_p}{\omega}\right)^2\right]^{1/2} = nc \text{ at the plasma frequency}$$

This means the group velocity is zero (there is no propagation of energy) at the plasma frequency. To get a nonzero group velocity, the q dependence of the dielectric function must be taken into account.

5 *Interband Transitions*: Consider a system with numerous interband transitions (bound electrons). For such a system show that the conditions for plasmon resonance is given as

$$\varepsilon(\omega) = [1 + \delta\varepsilon^{\mathrm{b}}(\omega)]\{1 - [(\omega_p^*)^2/\omega^2\,]\} = 0$$

where

$$(\omega_p^*)^2 = \frac{4\pi N e^2}{m(1 + \delta\varepsilon^{\mathrm{b}})}$$

6 Franz–Keldysh
Consider a simple semiconducting system with bandgap E_g. Then the absorption spectrum of such a material should look like Figure EC13.1.
For direct bandgap semiconductors this is already quite clear from what we have learned so far. But what happens when a strong electric field is placed across the material? This question was addressed independently by W. Franz and L.V. Keldysh in 1965. Two effects were found to occur.
The absorption coefficient below E_g is now nonzero. In fact it is given as

$$\alpha(\hbar\omega) \propto \exp\left(-\frac{4\sqrt{2m_{\mathrm{e}}^*}}{3|e|\hbar\varepsilon}(E_{\mathrm{g}} - \hbar\omega)^{\frac{3}{2}}\right)$$

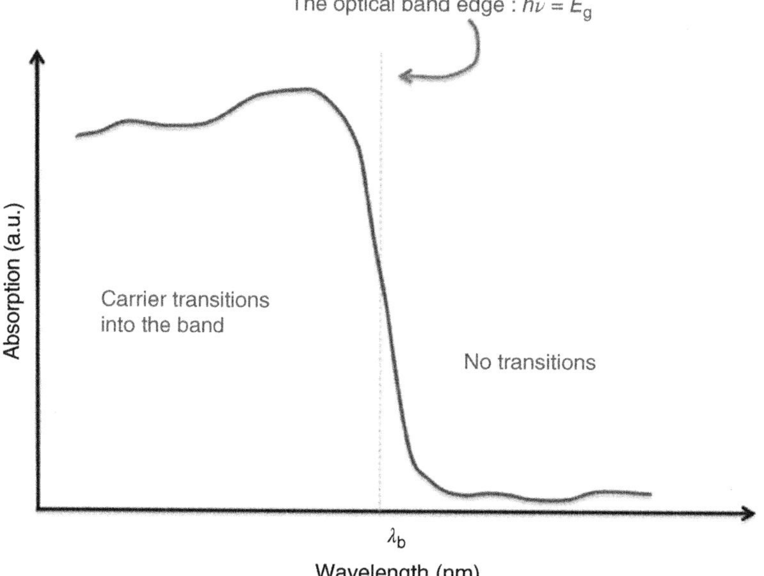

Figure EC13.1 The typical rendition of the optical band edge. This edge corresponds with the energy at which sufficient absorption of photons can be detected due to interband transitions. Notice that the "edge" is never really very sharp. There are defects and other effects that allow for "slightly sub-bandgap" light to be absorbed in quantity.

decreasing exponentially with

$$(E_g - \hbar\omega)$$

So the optical band edge shifts to a lower energy as the applied field in increased. Secondly, the absorption coefficient for

$$\hbar\omega > E_g$$

is modulated by an oscillatory function: *Franz–Keldysh oscillations*. Taken together these are known as the *Franz–Keldysh effect*:

(a) Using the Kramers–Kronig relations, reason that the applied field and increasing absorption will increase the refractive index of the material (this modulation of the optical constants by an electric field is known as the *electro-optic effect*).
(b) Estimate the electric field strength it takes to redshift the absorption band (position of the absorption band edge) of GaAs by 0.01 eV. The effective mass in this system is $0.067m_e$.
(c) Do a little more background reading, and give a good model for the physics of the Franz–Keldysh effect. What is happening?

7 Magnetic Franz–Keldysh
For direct transition semiconductors, as we described in Exercise 13.6, can we expect a magnetic analogue? The answer is yes. We know that an applied magnetic field flux $\boldsymbol{B}$ will result in Landau orbits. Let's say we put the B field in the z-direction. In the conduction band the electron energies will be quantized as Landau levels in x–y but free in z:

$$E^n(k_z) = \left(n + \frac{1}{2}\right)\frac{e\hbar B}{m^*} + \frac{\hbar^2 k_z^2}{2m^*}$$

Taking $E = 0$ as the top of the valence band, the energy levels look like

$$E_n^e(k_z) = E_g + \left(n + \frac{1}{2}\right)\frac{e\hbar B}{m_e^*} + \frac{\hbar^2 k_z^2}{2m_e^*}$$

$$E_n^h(k_z) = -\left(n + \frac{1}{2}\right)\frac{e\hbar B}{m_h^*} - \frac{\hbar^2 k_z^2}{2m_h^*}$$

for electrons and holes, respectively. When illuminated and an interband transition takes place, the electron is promoted into the conduction band leaving a hole in its place, both falling into Landau levels. The k_z values in the transition will remain unchanged since the photon carries little momentum with it. But surprisingly the Landau level index, n, must also stay the same. So the transition energy has to look like

$$\hbar\omega = E_n^e(k_z) - E_n^h(k_z) = E_g + \left(n + \frac{1}{2}\right)\frac{e\hbar B}{\mu} + \frac{\hbar^2 k_z^2}{2\mu}$$

with μ being the reduced mass. This will have an oscillatory effect on the absorption coefficient. Specifically a high absorption coefficient is expected for any transition that satisfies the above equation when $k_z = 0$. This we get a series of peaks at the Landau energies:

$$\hbar\omega = E_g + \left(n + \frac{1}{2}\right)\frac{e\hbar B}{\mu};\ n = 0, 1, 2, \dots$$

Moreover, we expect the absorption band edge to shift to higher energy by the amount:

$$\hbar eB/2\mu$$

(a) Using the fact that the B field classically creates circular orbits of angular frequency eB/m shows that the transition selection rule is given as $\Delta n = 0$.
(b) Estimate the magnetic band edge shift from a 3.6 T field in GaAs (you have to look up the hole mass).
(c) Draw the expected absorption spectrum for a 1D direct gap semiconductor, and compare this to the magneto-FK effect. Why is the magnetic system a good model for 1D semiconductors?

References

1 Veselago, V.G. (1968). The electrodynamics of substances with simultaneously negative values of ε and μ. *Sov. Phys. Usp.* 10 (4): 509–514. https://doi.org/10.1070/PU1968v010n04ABEH003699.
2 Griffiths, D. (2017). *Introduction to Electrodynamics*, 4e. Cambridge University Press.
3 Huang, K. (1951). Lattice vibrations and optical waves in ionic crystals. *Nature* 16: 7779.
4 Dresselhaus, M., Dresselhaus, G., Cronin, S.B., and Gomes Souza Filho, A. (2018). *Solid State Properties: From Bulk to Nano*. Springer.
5 Malić, E., Hirtschulz, M., Milde, F. et al. (2006). Analytical approach to optical absorption in carbon nanotubes. *Phys. Rev. B* 74: 195431.
6 Choi, S., Deslippe, J., Capaz, R.B., and Louie, S.G. (2013). An explicit formula for optical oscillator strength of excitons in semiconducting single-walled carbon nanotubes: family behavior. *Nano Lett.* 13 (1): 54.
7 Frenkel, J. (1931). On the transformation of light into heat in solids. I. *Phys. Rev.* 37: 17.
8 Song, K.S. and Williams, R.T. (1996). *Self-Trapped Excitons*. Springer-Verlag.
9 Arfken, G.B., Weber, H.J., and Harris, F.E. (2013). *Mathematical Methods for Physicists: A Comprehensive Guide*, 7e. Waltham, MA: Academic Press (Elsevier).

14
The End and the Beginning

So there you have it. What more can we say? Well quite a bit really. We have actually left out a whole field of solid-state physics known as **kinetics** or the motion of atoms and ions through the lattice acting under a variety of driving forces. So if you wanted to know how a battery works, exaggerated grain growth, or dislocation pinning, this book won't tell you. You need to look for a text on kinetics. There is also the field of fluids and fluid dynamics, superfluids, and plasmas – all forms of matter that have condensed. But what we have given is the foundation of electronics, electro-optics/optoelectronics, and magneto-electronics. And at the time of the writing of this text, all of these fields are undergoing a revolution based in the dimensionality and length scale of electronic solids.

However, it seems a requirement to end a text with some words of *wisdom* regarding its use. So, we will add the following.

Technological revolutions do exist. They are triggered by major innovations. The steam engine, railroad, electricity, and artificial fertilizers are examples. Economists speak of Kondratiev cycles [1]. It is not clear that one-dimensional metals, conducting polymers, or organic superconductors will initiate a new Kondratiev wave, at least not in the way it was originally conceived. Not even a spray-on room temperature superconductor would do that. Of course, material scientists are excited when giant conductivity in TTF–TCNQ, solitons in polyacetylene, superconductivity in cuprates, or mass production of carbon nanotubes are reported. They will have open-ended discussion meetings till early in the morning, they will file hundreds of research proposals, and they will write thousands of publications. But neither the scientists nor the administrators of funding agencies should be disappointed if five, ten, or even twenty years later, electrical cables are still made of copper and computer chips and are still made of silicon. The more a new technology differs from the established methods, the higher the innovative step is required for its acceptance. A new material that is just a little bit better will have a hard time replacing conventional materials, because it is not only competing in regular costs but also has to compensate for the depreciation of the investments for the old product. Consequently, for quick success, a new material must have at least an "order-of-magnitude advantage" somewhere.

For our *traditional* uses of electronic materials, it doesn't seem as though one-dimensional metals and exotic materials represent such a "quantum step"

Foundations of Solid State Physics: Dimensionality and Symmetry, First Edition.
Siegmar Roth and David Carroll.

in performance.[1] A possible exception is light-emitting devices for large-scale conformal displays. Therefore it is likely that these materials will slowly spread into economic niches rather than replace existing materials for the application areas we have discussed.

But what about the real "unknown?" What about the application areas we have not discussed? As the twenty-first century marches on, it is becoming more clear that we cannot always predict the direction of technological development. It is not always just an improvement to old technology – but sometimes it is an avenue that was never before considered. Some of these avenues may be only accessible by exotic, topological, or low-dimensional materials. Consider, for instance, that the overall chemistry and structure of an organic one-dimensional metal is not that different from many biological molecules. Moving forward, we might expect this line to blur even more – between biological and technological. As Smalley once said, "Silicon has rigid bonding requirements and this leads to technology. Carbon has flexible bonding requirements and this leads to life." It would seem then that the very difference between biology and technology lies at the center of what this book has been about. At the printing of this text, scientists are introducing stretchable electronics, electronics embedded in tattoos, and electronic systems that can be integrated into the body's organs and derive their power from biobatteries. From the management of disease to a brain–electronic interface, paradigms are being challenged as to exactly what the next step is in our technological evolution. Perhaps the most intriguing of all is quantum information processing and artificial intelligence.

Will these exotic and complex materials systems play a role in this technological evolution? We don't know yet. But one thing is for sure – the explosion of interest in such systems is still in its infancy, and their potential to change our lives remains largely unexplored.

Reference

1 Kondratiev, N.D. (1925). *The Major Economic Cycles*, 1. Economic Bulletin of the Conjuncture Institute.

1 Quantum step, quantum leap, or quantum jump – in German "Quantensprung," Sprung for a sudden change like Sprungtemperatur in superconductivity. There are some people who criticize the misuse of this metaphor, arguing that a quantum leap is very small. The typical energy of a light quantum is in the order of 1 eV, and 1 eV is nearly nothing compared with the energy in the fuel tank of a car. (The chemical energy of 50 l of gasoline is about 1 GJ (gigajoule, 10^9 J), and 1 J is about 6×10^{12} eV.) How can we say that a quantum step is a big change? But from the point of view of an individual molecule, 1 eV is huge, and it is about the energy we need to create a soliton–antisoliton pair in polyacetylene. In the context of this book, a quantum step is the difference between "true" and "false!"

Index

Foundations of Solid State Physics: Dimensionality and Symmetry, First Edition.
Siegmar Roth and David Carroll.

q

t

Printed and bound by CPI Group (UK) Ltd, Croydon, CR0 4YY